MOLECULAR EVOLUTION

an
annotated
reader

E.A. Terzaghi
A.S. Wilkins
D. Penny

MASSEY UNIVERSITY
PALMERSTON NORTH
NEW ZEALAND

JONES AND BARTLETT PUBLISHERS, INC.
Boston Portola Valley

A series of Books in Biology
Consulting Editor: *Cedric I. Davern*

Editorial offices: Jones and Bartlett Publishers, Inc., 30 Granada Court, Portola Valley, CA 94025
Sales and customer service offices: Jones and Bartlett Publishers, Inc., 20 Park Plaza, Boston, MA 02116

Library of Congress Cataloging in Publication Data

Main entry under title:
Molecular Evolution.
Bibliography
Includes index
1. Chemical evolution. I. Terzaghi, Eric A.
II. Wilkins, Adam S. III. Penny, David
QH371.M75 1984 577 83-17541
ISBN 0-86720-021-9

ISBN 0-86720-021-9

Printed in the United States of America
10 9 8 7 6 5 4 3 2 1

CONTENTS

PREFACE

The study of evolution at the molecular level, as revealed in DNA, RNA and protein sequences, measures change without prejudice as to its significance, and provides a basis for its unambiguous quantification; it is thus a calibration of the mode and tempo of evolution.

This book is the outgrowth of a university course that attempts such an approach to the study of evolution. During the four years that we have taught this course we have come to appreciate anew the special contribution that molecular biology is making to evolutionary theory through both its central ideas and its analytical techniques. Because the field of evolution has now reached that stage of maturity in which many or most of the major questions have been posed, and where the form of the answers can begin to be perceived, it seems appropriate to take stock of where we stand. An additional motivation for bringing out this anthology stems from the current controversy over the scientific status of the ideas of Darwinian evolution.

Since the early 1960s there has been a symbiosis between molecular biology and classical evolutionary studies, wherein the calibration of genetic distance separating contemporary species in molecular terms complements the historical record provided by paleontological studies. This book has been designed principally for the student of evolution who has at least a nodding acquaintance with contemporary molecular biology. We hope that it can serve a valuable double function as an introduction to an important area of evolutionary studies, and as an instrument in the development of those critical abilities essential for the intelligent reading of the scientific literature. The book is divided into nine chapters, each containing key or representative papers on a common theme; this organization is designed to help the student connect the fundamental ideas and the techniques that together comprise the excitement of contemporary molecular biology.

In order to provide a context for the reprinted papers, each chapter is introduced by a summary of the development of the ideas in that particular area supplemented with a selected list of references to guide the reader into a deeper exploration of material that may be of interest. Key references for further reading are included where appropriate.

The selection of papers for reprinting reflects several practical and pedagogical compromises. Primary among the practical concerns is length. We feel that papers should be reprinted in their entirety, and, accordingly, conciseness has weighed heavily in the selection process. Our guiding concern has been to provide students with the context within which current literature may be understood and appreciated. Thus, papers have been selected on the grounds of either giving due recognition to those who have been responsible for initiating and developing ideas, or to those who have presented a particularly clear picture of the current state of understanding in each area covered. A set of problems is given at the end of each chapter. These problems are of three types--those on vocabulary items (both conceptual and methodological) that appear in the chapter; those based on the reprinted papers and covering points of logic, methodology, or data interpretation (these questions are designed to encourage critical reading of the material); and those based on recent research papers that were not included in the anthology. The problems are an adjunct to the reading material, but are highly recommended for the student eager to consolidate an understanding of the material.

There are several very different yet genuine debts of gratitude that we wish to acknowledge: to our students, Philip Calder, Duncan Graham and Anna Riddiford, who, by their enthusiastic response to both the subject matter and our approach, encouraged us to proceed with the project; to our respective spouses for very substantial assistance in the form of critical reading, indexing, typing, and general moral support; to Ric Davern and David Freifelder, for support and helpful suggestions along the way; to Thomas Jukes for a critical reading of the final manuscript; and to the authors of reprinted papers who readily gave their approval for reprinting their papers and who frequently gave helpful suggestions and comments. Finally, E. T. wishes to acknowledge with thanks the Biochemistry Department of the University of Washington and especially the Laboratoire de Biologie Moléculaire/INRA (Toulouse), France, whose warm hospitality and assistance contributed so much to his portion of this volume.

E. A. Terzaghi
A. S. Wilkins
D. Penny

October, 1983

INTRODUCTION

In the century following the publication in 1859 of *The Origin of Species* by Charles Darwin, the theory of evolution was augmented and modified by contributions from many disciplines--paleontology, geology, ecology, taxonomy, biosystematics, and, most importantly, genetics. The contemporary theory of evolution--Neo-Darwinism or the "modern synthesis"--has retained the essentials of Darwinism but is built on the genetic foundation of Mendelian laws that Fisher, Haldane, and Wright applied to changes in gene frequencies in populations. Without doubt, this general development would have been neither surprising nor displeasing to the originator of the theory.

In the last twenty years another discipline, molecular biology, has begun to contribute to the study of evolution. The findings from this area, designated simply as molecular evolution, have enriched our understanding of evolution and are leading to better comprehension of some of the foundations of the modern synthesis.

The methods of molecular biology are largely concerned with the analysis of genes, gene products, and genome structure, and these analyses have been used to test hypotheses about the way the genetic material originally arose and how it has changed in the course of evolution. This type of analysis was not possible with older methods of classical genetics and, as a result, many of the inferences about genetic change embodied in Neo-Darwinism were untested assumptions. To appreciate the usefulness of the approach of the molecular biologists to evolution, it is helpful to review the history of evolutionary theory briefly.

Many popular histories of the Darwinian revolution emphasize the controversy that immediately swirled up around the theory, particularly in the context of questions concerning human origins. It is often overlooked that acceptance of evolution by biologists as a fact was, though not unanimous, remarkably swift and widespread among both the older and the younger members of the scientific community (Hull *et al.*, 1978). The evidence that Darwin marshalled for the occurrence of evolution was considered compelling, and at least the broad outline of his proposed mechanism--that of directional change through natural selection of favoured variants within populations--was deemed highly plausible. However, some aspects of the theory remained controversial among those who acclaimed it. Darwin and some of his adherents (the "gradualists") believed that all evolutionary change, including the large-scale phylogenetic divergences, were produced by the cumulative effects of small genetic "variants". On the other side, in a camp led by Thomas Huxley and John Galton, were those people who felt that the major or "transspecific" evolutionary changes were "saltational"--that is, produced by genetic changes of large phenotypic effect ("macromutations").

This debate would have been placed in clearer focus, though not resolved, if the essentials of Mendelian genetics had become known immediately in 1866 when Mendel published his findings. When Mendelian genetics was brought to light in 1900, the debate between gradualists and saltationists became part of a new controversy. This second dispute between the Mendelians on one side (Galton, Bateson, de Vries) and the biometricians (Pearson and Weldon) on the other, was over the question of whether Mendelian genetics underlay all heredity. The resolution of this issue in the affirmative required another twenty years (Provine, 1971). The acceptance of Mendelian genetics as the basis of heredity, the elaboration of the chromosome theory of heredity, the discoveries of linkage and recombination, and the incorporation of the results from biosystematic and population studies, paved the way for the establishment of contemporary population genetics. The work of Fisher, Haldane, Wright, Dobzhansky and others in the 1920s and 1930s finally led to the modern synthesis. A detailed analysis of the development of the modern synthesis (or neo-Darwinism) is found in Mayr and Provine (1980). The resulting theory was the conversion of a vague and nonquantitative theory of hereditary change to a quantitative and mechanistically precise (though stochastic) theory of progressive genetic change in populations.

The fundamental tenets of Darwinism can be summarized as follows:

Hypothesis I. The great variety of forms of life on earth have originated from very few forms, perhaps even from a single organism; there is an evolutionary tree or an evolutionary history. This has been called "the theory of descent" by Darwin, or "the fact of evolution" by Julian Huxley.

Hypothesis II. There are three observable characteristics of living systems that explain this evolutionary history and two deductions from these facts (Huxley, 1942):

Fact 1. There is a tendency for all organisms to increase in a geometric ratio.

Fact 2. In spite of this potential, the numbers of a given species remain more or less constant.

Deduction 1. From Facts 1 and 2 there must be a struggle for existence, or competition for survival and reproduction.

Fact 3. Organisms within populations vary appreciably and randomly. Much of the variation is inherited.

Deduction 2. Some inherited variations will be advantageous in the struggle for survival, others unfavourable. Individuals with these variations increase or decrease in frequency, and this is cumulative from generation to generation. This is natural selection and will act constantly to maintain and improve organisms in their environment, and by implication, when combined with isolating mechanisms, is sufficient to explain speciation.

Enriched by contributions from other fields, this theory has proved to be extremely useful in explaining evolutionary phenomena and in prompting a wealth of experimental investigations. (Whether the theory explains too much is a point to which we shall return.) Field and laboratory studies have amply confirmed the importance of genetic variation and natural selection in the dynamic changes of populations. However, every major advance in science itself prompts new questions, and the modern synthetic or neo-Darwinian theory of evolution has been no exception.

The first set of questions about evolution concerns history. While a basic postulate of Darwinism and neo-Darwinism (Hypothesis I) is that all living forms that comprise an extraordinarily diverse set, are derived from just one or a few ancestral forms, the theory itself did not provided a direct method for deducing phylogenetic relationships. For this reconstruction of past events evolutionists were forced to fall back on the older methods of fossil analysis (where there was a good fossil record, as for the vertebrates), and of systematics (for groups that lacked an extensive fossil record--for example, soft-bodied invertebrates, many insects, plants, and bacteria). The results obtained by these methods have often been controversial, and for some groups of organisms few people agreed about the phylogenetic schemes that were proposed.

A second set of questions arose directly from neo-Darwinism because it stressed the importance of genetic variation in populations as a source of evolutionary change. These questions concerned the nature of this variation. For example, how much and what kinds of variation, both in terms of gene structure and phenotypic effects exist in populations? These questions then raised other questions about the way natural selection acts in populations. If most newly-occurring mutations are deleterious, then one would expect them to be promptly eliminated. (This mode of selection has been termed "stabilizing" or "normalizing" selection.) On the other hand, if most variants are either neutral or favorable in the appropriate genetic background, a great deal of this variation should accumulate in a population and be subject to drift or to positive forms of selection ("directional" or "diversifying" selection). Since the gene itself was a completely mysterious entity in the 1930s when population genetics was launched as a separate discipline, it is not surprising that questions about the nature of genetic variation could not be addressed directly.

A third type of question concerned an implicit assumption of Darwinism: that evolutionary change was gradualistic. This view was certainly the consensus but was not unanimous. Goldschmidt (1940) argued strongly that a major change in an evolutionary grade involves something like the saltational genetic changes of Thomas Huxley. The neo-Darwinians could not accept the idea that sudden major genotypic-and-phenotypic changes between generations played an important part in evolution. However, the assertion that major evolutionary change (macroevolution) utilizes genetic changes of a different kind from those that underlie slow phenotypic transformations of populations (microevolution), could not be disproved. Aside from certain obvious cases, such as polyploidy, the genetic basis of speciation and macroevolution still remains an unanswered question.

The advent of molecular biology in the 1950s and 1960s opened the way for an analysis of some of the implications of these genetic and evolutionary questions. The advances were both conceptual and technical (Lewontin, 1974). The conceptual advance came from the Watson-Crick model of DNA structure (Watson and Crick, 1953a, b). The gene was no longer a point on a chromosome, as it had been viewed in classical genetics, but was a form of chemical information--a specific string of nucleotides that prescribes a specific sequence of amino acids. Furthermore, the fact that during replication occasional errors will certainly occur implied that molecular evolution is inevitable. Finally, in principle the informational content of genes could be determined either directly by base sequence analysis or indirectly by amino acid sequence analysis (though the proof of colinearity of nucleic acid and protein sequences would take another decade). More recently, the importance of "noncoding" nucleotide sequences in signalling and controlling gene expression has been recognized and this provides another avenue for analysis of the genetic basis of phenotypic variation.

Several technical advances accompanied the conceptual breakthroughs. Foremost among these was the method of protein sequencing, developed by Sanger (1952). Other less laborious methods such as gel electrophoresis (Hunter and Markert, 1957; Prager and Wilson, 1971) have been valuable; these methods give less information than direct sequencing but have enabled large amounts of genetic variation to be detected. In addition, DNA hybridization (Hoyer *et al.*, 1964; Britten and Kohne, 1968, Chapter 1) provided an early means for comparing nucleic acids from different organisms. These methods yielded information about genome structure and organization, but have now been largely replaced by restriction fragment analysis (Southern, 1975), coupled with direct sequencing of the DNA fragments (Sanger and Coulson, 1975; Maxam and Gilbert, 1977). Sequencing has been of the greatest value in reconstructing evolutionary history.

A cornerstone of contemporary studies of phylogenetic history is the realization that in a sequence of nucleotides each gene carries a record of its own history. In the words of Zuckerkandl and Pauling (1965), "living matter preserves inscribed in its organization its own past history." While the accuracy of the replication process ensures a high probability that daughter copies of a gene will be identical to the parental sequence, one would predict from the processes of mutation, drift, and selection operating over spans of hundreds of millions of years that sequences would be slowly modified. At least some of this change can be accommodated with little or no meaningful biological consequence because of coding redundancy and the functional equivalence of some amino acids at less critical sites. Thus, homologous genes in two contemporary organisms derived from a common ancestor should have similar, but not identical, sequences and the differences should increase in proportion to the time elapsed since divergence. By comparing the "texts" (in the form of proteins) of such homologous genes, it becomes possible to deduce the phylogenetic histories of the genes themselves and to estimate relative rates of change (Dickerson, 1971). Furthermore, if the time of divergence is known from the fossil record, the rate of nucleotide substitution--that is, the rate of ticking of the "molecular clock"--can be estimated. This data can, in principle, be used to estimate times of divergence for lines less clearly defined in the fossil record.

The new methods have been equally helpful in assessing the mechanisms of evolutionary change and, in particular, for obtaining estimates of the amount of genetic variation existing within populations, a crucial piece of information required for testing Hypothesis II. Indeed, one of the first problems in evolutionary or population genetics in which the new techniques were employed was the study of genetic variation in populations. By applying starch-gel electrophoresis for the detection of electrophoretic variants of certain enzymes, a lower limit for the amount of variation of the genes encoding these enzymes could be evaluated. (It is a lower limit because the method does not detect nucleotide changes that do not alter amino acid sequences, and the early methods only detected amino acid changes that resulted in a changed electrical charge on the protein). Starch-gel electrophoresis was first applied to *Drosophila* populations (Hubby and Lewontin, 1966), and to humans (Harris, 1968, Chapter 1) and revealed greater populational genetic variation than expected on the basis of classical neo-Darwinism. This work spawned a tremendous number of similar studies, practically a field of genetics in itself, and sparked a controversy over the relative importance of positive selection and of neutral mutation and subsequent genetic drift as the principal driving force in gene evolution.

The discovery of the large amount of genetic variation in populations was one of the lines of evidence that Kimura (1968, Chapter 1) used when first suggesting that many, if not most, of the observed amino acid substitutions conferred no selective advantage or disadvantage on the organism. Such "neutral" substitutions have been called non-Darwinian (King and Jukes, 1969), because although Darwin considered the case of inherited variations of no selective advantage becoming fixed, subsequent versions of his theory focussed on the predominance of positive selection. The neutralist-selectionist debate continues, but as more data accumulates, it is increasingly clear that both drift of neutral mutations and positive selection play roles in the evolution of coding sequences, with the relative strengths of their contributions depending to a large extent on the functional restraints on the gene sequence being examined (Dickerson, 1971). Furthermore, if one accepts the suggestion that much of the change in morphological form and function during evolution may result from changes in the rate, time, and location of the expression of these structural genes (King and Wilson, 1975, Chapter 6), then much of the biological significance of the controversy may disappear.

The new techniques have also contributed to an understanding of the sources of variation that allow genome change. The significance of point mutations and genetic recombination as primary sources of variation has been supported, but the roles of other processes, such as the duplication and transposition of DNA sequences, have been revealed. The possible importance of gene duplication as a source of new sequences that can be altered was first pointed out in 1918 (Bridges, 1935) and later confirmed by analysis of the various hemoglobin proteins (Ingram, 1961, Chapter 1) and DNA sequences. The work on the origin of new protein functions (Chapter 5) gives details of molecular mechanisms that could lead to an increase in complexity during evolution; this work has been a major contribution of molecular evolution.

The importance of natural selection, the constructive force in evolution, has been affirmed in several ways by molecular analysis. Natural selection is a stabilizing force in the maintenance of coding functions but selection has a more creative role; an example is the selection of cell lines carrying a duplicated gene when a particular enzyme is present in limiting supply (Schimke *et al.*, 1978) or when a cell is presented with a novel substrate that a preexisting enzyme can utilize only very inefficiently (Hartley *et al.*, 1976, Chapter 5). In addition, Gilbert (1978, Chapter 2) has suggested that at least some modern polypeptides of complex function have arisen by the recombination of functionally simpler ancestral gene products. In addition to these examples of selection acting on populations of organisms to promote evolution of the genome, it has been possible to demonstrate the consequences of selection operating directly on replicating genomes *in vitro* (Mills *et al.*, 1967; Orgel, 1979, Chapter 9). Further study of the latter system may provide insights into the pathways involved in the initial evolution of nucleic acid based living systems.

An important remaining issue is whether the evolutionary processes involved in the origin of species (macroevolution) and the origin of varieties (microevolution) are distinct--an area of disagreement between Darwin and Thomas Huxley referred to previously. One aspect of this controversy concerns the rate of organismal evolution. Some modern authors have suggested that both Darwinian and neo-Darwinian theory require a slow continuous transformation of one species to another ("phyletic gradualism"). However, a careful reading of original texts shows that Darwin and Huxley recognized the possibility that on an evolutionary time scale, the morphological differentiation signaling speciation could occur rapidly. Therefore, we shall define "gradualism" as the idea that macroevolution depends only on the same kinds of genetic variation normally found in populations, which means that there are gradual, not abrupt, changes between generations.

The genetic basis of macroevolution is an important facet of the study of evolution; it is increasingly the major focus of controversy. It is known that the rate of DNA-sequence divergence occurs at about the same rate in groups that are morphologically conservative--for example, amphibia--and that are showing rapid evolutionary change--for example, mammals (King and Wilson, 1975; Wilson *et al*, 1974a, b). These findings focus the attention of both the developmental geneticist and the evolutionist on the nature of the genetic changes that produce the developmental changes. It has been suggested that the types of genetic change that cause major evolutionary change are a special class--regulatory changes in a broad sense--and that the molecular events responsible for these changes are eclipsed by the large background of neutral mutations that occur throughout the genome.

The importance of regulatory genes in development has been known for a long time, but it is clear that the term regulatory is being used in at least two ways in the research literature: (1) to describe gene regulation utilizing repressors, inducers and operators, as first described by Jacob and Monod (1961) and (2) to indicate genes that affect timing, rate, and extent of development of specific features of these organisms Huxley (1942). At present, little is known about the genetic basis of development in multicellular organisms and thus we are unable to determine whether such regulatory genes (in the Jacob-Monod sense) show intrapopulation variation similar to that of structural genes (Dickinson, 1980, Chapter 6). Because some regulatory genes may affect the activities of more than one gene, the behavior of regulatory genes in natural populations may possess special features, and this is now beginning to be recognized (Wallace, 1975; Hedrick and

McDonald, 1980).

The results of the molecular biological approach to evolution relate to certain philosophical questions about evolution, a point that is rarely brought out. Several authors have criticized aspects of evolutionary theory, but the criticisms that have been taken most seriously are those of Karl Popper, a leading philospher of science. He has claimed that evolutionary theory is not so much a fully scientific theory as it is a self-contained "metaphysical research program" (Popper, 1976). In other words, Popper suggests that the present theory of evolution has such vast explanatory powers and correspondingly weak predictive powers that no phenomenological observation or experimental procedure exists by which it might be proved false. The only real prediction of Neo-Darwinism, he says, is the gradualness of evolutionary change.

We suggest that this criticism can be answered in two ways. First, while modern evolutionary theory as a whole may not be falsifiable with a single simple test, it can be dissected into components that can individually be subjected to testing and potential falsification. In this respect, studies of evolution are not unlike other areas of investigation in science. This position has been taken by Dobzhansky *et al* (1977). Second, we maintain that the molecular approach can and does provide means of performing these tests.

One such testable area is that of evolution as history. If the phylogenetic trees constructed from fossil evidence and systematics are merely selfconsistent systems, then the phylogenies reconstructed from macro-molecular sequences would not be expected to bear any particular relationship to (let alone parallel) the morphologically derived trees. A failure to find such resemblances between trees derived from different data bases would undermine our confidence in being able to understand past evolutionary events. In fact, the parallels between the two types of phylogenetic reconstruction hold up well (Hoyer *et al*, 1964; Fitch and Margoliash, 1967) and confirm the idea of descent with modification. However, further quantitative investigation along these lines is needed (Penny *et al.*, 1982). In an important paper in 1978 Popper conceded that the existence of an evolutionary history is falsifiable. We consider the question of the occurrence of evolution to be of the same ilk as the existence of atoms.

Questions concerning the detailed mechanisms of evolution of genes and genomes are more difficult to test, but the role of natural selection, specifically, can be examined. Here too, molecular approaches may provide independent evidence that selection is a force in shaping the genome. The proofs are both analytical--based on comparisons of particular protein sequences (Perutz and Imai, 1980)--and experimental, in which the genomic changes following laboratory selection experiments are assayed. The strongest demonstrations are probably the *in vitro* RNA replication experiments (Orgel, 1979, Chapter 9) that give concrete evidence for the importance of variation and selection of favored variants in the evolutionary modeling of the genome.

In sum, the basic tenets of contemporary evolutionary theory have not only not been refuted by the molecular tests that have been applied, but, on the contrary, strongly confirmed. However, we consider it self-evident that molecular data alone are insufficient to solve the problem of speciation, which many consider to be the central problem of evolution.

In closing this Introduction, a few words on the structure of the book are in order. We have concentrated on proteins and nucleic acids, because they contain primary information about their own evolution. We will refer to the small noninformational molecules (such as cofactors) only where they relate to the evolution of nucleic acids and protein. We have begun with a chapter containing a selection of some of the foundation papers on molecular evolution. They have been significant either in focusing attention on some of the major questions, or in introducing some of the major techniques that have been used. These papers provide the background of the book. The main portion of the book following the classic papers is divided into eight chapters, each dealing with a specific topic. The order of the sections has been chosen so that one begins in the areas in which the questions and answers are clearest--the nature of the source of variation, eukaryotic genome structure, and evolution of coding sequences. The reader will then proceed to progressively more controversial and uncertain areas--evolution and selection of regulatory gene differences, origins of eukaryotic cells, origins and phylogenies of prokaryotic cells and finally to the most controversial subject, the origin of life. If, in following through this sequence, the reader closes the book with the feeling that molecular biology is making strong contributions to evolutionary theory, but that there is still much of interest to be studied, we will have accomplished our goal.

1 THE CLASSICS

early formulation of the key ideas

The decade of the 1960s was an extraordinarily fruitful period for investigation into the evolutionary implications of the concepts and techniques that had risen from molecular biology. The papers selected as "classics" are nearly all from this period and are significant in having presented either data within an explicit evolutionary context or important new data of particular relevance to studies of evolution. However, in going through these papers the reader must remember that none were created in a vacuum--they are all products of a developing climate of ideas, which is well described by Anfinsen (1959) and Jukes (1962).

The paper by Ingram (1961) was chosen because it serves as the point of departure for several of the principal lines of inquiry that we have developed as specific chapters in this book. In this paper, the fragmentary hemoglobin sequences available at that time explicitly pointed the way toward the use of sequences in the construction of gene and organism phylogenetic trees. This use of sequences and the implications of the resulting picture of structural gene evolution for understanding the dynamics of coding sequence alteration, are examined in more detail in Chapter 4. The application of sequence studies to organism phylogeny are taken up in Chapters 7 and 8. The discussion of the duplicated globin chains and their different functions leads one naturally to think about mechanisms of gene duplication and dispersion (considered further in Chapter 2) and about the role of these processes in the generation of complexity and specialization (examined in Chapter 5).

The paper by Harris (1966)was one of the first of many papers from numerous laboratories that showed that populations are genetically much more variable with respect to structural genes than had hitherto been believed. Kimura (1968), in his paper, offered a theoretical rationale for these observations; his model raised the possibility of functionally neutral amino acid substitutions giving rise to both a large amount of genetic variation and an apparently regular rate of evolution of proteins. The work led to a thorough re-examination of the relative role of drift and selection in the evolution of structural genes. The functional significance of sequence variation and the implications for selection and evolution constitute the central theme of Chapter 4.

The physical chemistry of the melting and reannealing of DNA provides the experimental and theoretical background for the paper by Britten and Kohne (1968), which marked the beginning of contemporary studies of the eukaryotic genome and its evolution. In this work, several classes of DNA sequences were distinguished by their copy number per genome. This discovery led to the establishment of criteria for sequence divergence and to investigations of the phylogenetic distribution of patterns of genomic interspersion. These points are presented in Chapter 3.

One consequence of the focus on sequence data measured in different ways was the recognition that there is often little correlation between the rate of amino acid substitution and the rate of morphological change. The latter, evolutionarily important (but experimentally less approachable) aspect of evolution presumably reflects a large and ill-defined class of regulatory changes, which are the topic of Chapter 6.

The final papers deal with some of the earliest evolutionary events. While the notion of the symbiotic origin of the eukaryotic cell is not new, the paper by Margulis (1968)was the first to make a clear and comprehensive statement of this idea in a form that has subsequently received substantial confirmation. More recent supportive evidence, particularly involving sequence information, is presented in Chapters 7 and 8.

A consideration of the origin and evolution of life, even while primarily focusing on macromolecules, must include some recognition of the fact that the entire process of evolution or of information expansion, is driven by the improvement of energy utilization. Reflecting the much earlier maturation of biochemistry, Horowitz (1945), in his paper, had considered the intriguing problem of the development of a biochemical pathway. This general topic is developed further in Chaper 5 and discussed briefly in Chapter 9.

Finally, we come to the very origin of life-- the transition from polymers, and aggregates of polymers to a system in which mutation and selection are intimately coupled. The paper by Crick (1968) contains an early discussion of the basic informational requirements of the primitive encoded and evolving systems. On the phenomenological side, the paper by Eck and Dayhoff (1966) presents an attempt to discern in the sequence of a contemporary polymer remnants of a much simpler primitive precursor, which may, in evolutionary terms, be close to the origin. These two complementary approaches are examined in greater depth in Chapter 9.

ON THE EVOLUTION OF BIOCHEMICAL SYNTHESES

By N. H. Horowitz

School of Biological Sciences, Stanford University, Calif.

Communicated April 23, 1945

Although it has been recognized for a long time that the biochemistry of the organism is conditioned by its genetic constitution, a more precise definition of this dependence has not been possible until recently. A considerable amount of evidence now exists for the view that there is a one-to-one correspondence between genes and biochemical reactions. This concept, foreshadowed in the work of Garrod[1] on human alcaptonuria, accounts in a satisfactory way for the inheritance of pigment formation in guinea pigs,[2] insects[3] and flowers,[4] and the synthesis of essential growth factors in *Neurospora*.[5] It appears from these studies that each synthesis is controlled by a set of non-allelic genes, each gene governing a different step in the synthesis. As to the nature of this control, it is probable that the primary action of the gene is concerned with enzyme production. That genes can direct the specificities of proteins has been shown in the case of many antigens,[6] while several mutations demonstrably affecting the production of enzymes have been reported.[6] Evidence on the postulated gene-enzyme relationship is in most cases, however, still circumstantial; this is partly because of technical difficulties involved in the study of synthetic, or free-energy consuming reactions *in vitro,* and partly because of the insufficiency of biochemical information on those reactions which happen to be susceptible of genetic analysis.

As a corollary of the above hypothesis, each biosynthesis depends on the direct participation of a number of genes equal to the number of different, enzymatically catalyzed steps in the reaction chain. In attempting to account for the evolutionary development of such a reaction chain one meets in a clear form the problem of explaining macroevolutionary changes in terms of microevolutionary steps. The individual reactions making up the chain are of value to the organism only when considered collectively and in view of the ultimate product. Regarded individually, intermediate substances cannot, in general, be assumed to have physiological significance, and the ability to produce them does not of itself confer a selective advantage. An example from *Neurospora* genetics will serve to illustrate this point. At the present time seven different genes are known to be concerned in the synthesis of arginine by the mold.[7] The inactivation of any one prevents the synthesis from taking place. On the basis of the above hypothesis, at least seven different catalyzed steps must occur in the synthesis. Several of the steps have been identified and controlling genes assigned to each. Two of the intermediates in the chain have been shown to be the amino acids ornithine and citrulline. Unlike arginine, neither of these substances is a general constituent of proteins. Aside from their function as precursors, they are apparently of no further use to the organism.

While the above example probably represents the general case, there are also well-known instances in which precursors serve independent functions. Thus, arginine, glycine and methionine are precursors of creatine in the rat,[8] but the synthesis goes through the non-functional intermediate, glycocyamine. On the other hand, acetylcholine may be synthesized from choline in one step.[9] In cases such as these, the problem is that of accounting for the synthesis of the precursors.

Since natural selection cannot preserve non-functional characters, the most obvious implication of the facts would seem to be that a stepwise evolution of biosyntheses, by the selection of a single gene mutation at a time, is impossible. It will be shown below that this is not a necessary conclusion, but that under special conditions the stepwise evolution of long-chain syntheses may occur. First, however, an alternative to stepwise evolution will be considered; that is, the origin of a new reaction chain through the chance combination of the necessary genes.

Although the probability of the origin of a useful character through the chance association of many genes may be small, it is never zero. Indeed, a consideration of the statistical consequences of the interaction of mutation, Mendelian inheritance, and natural selection has led Wright[10] to the conclusion that such chance associations may be of major importance in evolution. He has analyzed the evolutionary possibilities of various types of breeding structures and has shown that under certain conditions an extensive trial and error mechanism exists, whereby the species can test numerous combinations of non-adaptive genes. The breeding structure which most favors this type of evolution is that of a large population divided into many small, partially isolated groups. Within each group the cumulative effects of the accidents of sampling among the gametes are of major significance in determining gene frequencies, but the penalty of fixation of deleterious genes, ordinarily incurred under inbreeding, is avoided by exchange of migrants with other groups. The pressures of forward and reverse mutations, which between them determine an equilibrium frequency for non-adaptive genes in large, random-breeding populations, become of minor importance. As a consequence, a random drift of gene frequencies occurs. If, by chance, one group finds a particularly favorable combination of genes, a process of intergroup selection comes into play, whereby the favorable combination is spread to the population at large.

This model provides a means for the evolution of a new gene combination in spite of unfavorable mutation rates to active alleles and in the absence of selection of individual genes. It is thus favorable for the evolution of systems of individually non-adaptive, but collectively adaptive, genes. The effectiveness of the process would seem to be strongly dependent on the size of the gene combination required, however, decreasing approximately exponentially with increasing numbers of genes, other factors remaining constant. There would result a tendency toward the evolution of short reaction chains involving the recombination of molecular units already available. There is no doubt that a conservative tendency of this sort actually exists in nature. The wide variety of biologically important compounds built up on the pyrrole nucleus, to mention but one example, is a case in point.

The application of Wright's theory to the particular problem under consideration is limited by the fact that it operates only under biparental reproduction. It is probable that a large number of basic syntheses evolved prior to sexual reproduction. The universal distribution among living forms of certain classes of compounds—viz., the amino acids, nucleotides and probably the B vitamins—identifies them as essential ingredients of living matter. The synthesis of these substances must have evolved very early in geologic time, as a necessary condition for further progress, although loss of certain syntheses may have occurred in the later differentiation of some forms. It is therefore desirable to search for another solution of the problem applicable to compounds of this type, preferably one in which a minimum burden is placed on chance and a maximum one on directed evolutionary forces. It is thought that the following suggestion, while definitely a speculation, offers a possible solution along these lines.

In essence, the proposed hypothesis states that the evolution of the basic syntheses proceeded in a stepwise manner, involving one mutation at a time, but that the order of attainment of individual steps has been in the reverse direction from that in which the synthesis proceeds—i.e., the last step in the chain was the first to be acquired in the course of evolution, the penultimate step next, and so on. This process requires for its operation a special kind of chemical environment; namely, one in which end-products and potential intermediates are available. Postponing for the moment the question of how such an environment originated, consider the operation of the proposed mechanism. The species is at the outset assumed to be heterotrophic for an essential organic molecule, A. It obtains the substance from an environment which contains, in addition to A, the substances B and C, capable of reacting in the presence of a catalyst (enzyme) to give a molecule of A. As a result of biological activity, the amount of available A is depleted to a point where it limits the further growth of the species. At this point, a marked selective advantage will be enjoyed by mutants which are able to carry out the reaction $B + C = A$. As the external supplies of A are further reduced, the mutant strain will gain a still greater selective advantage, until it eventually displaces the parent strain from the population. In the A-free environment a back mutation to the original stock will be lethal, so we have at the same time a theory of lethal genes. The majority of biochemical mutations in *Neurospora* are lethals of this type.

In time, B may become limiting for the species, necessitating its synthesis from other substances, D and E; the population will then shift to one characterized by the genotype ($D + E = B$, $B + C = A$). Given a sufficiently complex environment and a proportionately variable germ plasm, long reaction chains can be built up in this way. In the event that B and C become limiting more or less simultaneously, another possibility is opened. Under these circumstances symbiotic associations of the type $(F + G \neq C, D + E = B)(F + G = C, D + E \neq B)$ will have adaptive value.

This model is thus seen to have potentialities for the rapid evolution of long chain syntheses in response to changes in the environment. As has been pointed out by Oparin[11] the hypothesis of a complex chemical environment is a necessary corollary of the concept of the origin of life through chemical means. The essential point of the argument is that it is inconceivable that a self-reproducing unit of the order of complexity of a nucleoprotein could have originated by the chance combination of inorganic molecules. Rather, a period of evolution of organic substances of ever-increasing degree of complexity must have intervened before such an event became a practical, as distinguished from a mathematical, probability. Or, put in another way, any random process which can have produced a nucleoprotein must at the same time have led to the production of a profusion of simpler structures. Oparin has considered in some detail the possible modes of origin of organic compounds from inorganic material and cites a number of known reactions of this type, together with evidences of their large-scale occurrence on the earth in past geologic ages. He concludes that in the absence of living organisms to destroy them highly complex organic systems can have developed. The first self-duplicating nucleoprotein originated as a step in this process of chemical evolution. The origin of living matter by physicochemical means thus presupposes the existence of a highly complex chemical environment.

To summarize, the hypothesis presented here suggests that the first living entity was a completely heterotropic unit, reproducing itself at the expense of prefabricated organic molecules in its environment. A depletion of the environment resulted until a point was reached where the supply of specific substrates limited further multiplication. By a process of mutation a means was eventually discovered for utilizing other available substances. With this event the evolution of biosyntheses began. The conditions necessary for the operation of the mechanism ceased to exist with the ultimate destruction of the organic environment. Further evolution was probably based on the chance combination of genes, resulting to a large extent in the development of short reaction chains utilizing substances whose synthesis had been previously acquired.

[1] Garrod, A. E., *Inborn Errors of Metabolism*, Oxford University Press (1923).
[2] Wright, S., *Biol. Symposia*, **6**, 337–355 (1942).
[3] Ephrussi, B., *Quart. Rev. Biol.*, **17**, 327–338 (1942).
[4] Lawrence, W. J. C., and Price, J. R., *Biol. Rev.*, **15**, 35–58 (1940).
[5] Horowitz, N. H., Bonner, David, Mitchell, H. K., Tatum, E. L., and Beadle, G. W., *Am. Nat.*, in press (1945).
[6] Summarized in Wright, S., *Physiol. Rev.*, **21**, 487–527 (1941).
[7] Srb, A., and Horowitz, N. H., *Jour. Biol. Chem.*, **154**, 129–139 (1944).
[8] Summarized in Schoenheimer, R., *The Dynamic State of Body Constituents*, Harvard University Press (1942).
[9] Lipmann, F., *Advances in Enzymology*, **1**, 99–162 (1941).
[10] Wright, S., *Bull. Am. Math. Soc.*, **48**, 223–246 (1942). Contains summary of earlier papers.
[11] Oparin, A. I., *The Origin of Life*, trans. by S. Morgulis, Macmillan, New York (1938).

GENE EVOLUTION AND THE HÆMOGLOBINS

By Prof. VERNON M. INGRAM

Division of Biochemistry, Department of Biology, Massachusetts Institute of Technology, Cambridge, Mass.

THE four types of polypeptide chain which go to make up the molecules of the three normal human hæmoglobins are believed to be controlled by four independent genes[1]. The following article is an attempt to discuss the chemical and genetic relationship between these chains from an evolutionary point of view; to a lesser extent, because much less is known, the hæmoglobins of other vertebrates will also be mentioned. It should be emphasized that the main purpose of this discussion, and of the evolutionary scheme to be proposed, is to provide a basis for the discussion of the evolution of genes in general and hæmoglobin genes in particular. Questions are raised which will, in part, be answered soon by the chemical work on vertebrate hæmoglobins which is now proceeding in various laboratories (see, for example, ref. 2).

The study of the chemistry of the human hæmoglobins has already provided evidence of the kind of phenotypic effects to be expected from gene mutations as, for example, in the known single amino-acid substitutions in the peptide chains of abnormal hæmoglobins[1]. It is on such changes in protein structure that the forces of natural selection are assumed to act. From the evolutionary and the practical point of view it is convenient to study a commonly occurring protein, such as hæmoglobin, which is found in all vertebrates. Hæmoglobin is not only accessible for chemical study but is also sure to have played an important part in vertebrate evolution, because of its vital physiological function as the carrier of oxygen.

Most of the discussion which follows will centre around the chemical findings in the human hæmoglobins, since they are by far the best studied of vertebrate hæmoglobins. However, the basic similarities in chemical structure between man's hæmoglobins and those of other mammals and lower vertebrates are very striking[3]. All known vertebrate hæmoglobins—except those of the lamprey and the hagfish—are built on the same molecular pattern: they consist of four polypeptide chains of roughly 17,000 molecular weight, with one hæm group each. The hæmoglobin of the lamprey is peculiar in consisting of a single polypeptide chain of molecular weight 17,000. The hagfish hæmoglobin appears to be similar, or possibly a dimer of 34,000 molecular weight[3].

The peptide chains in the other vertebrates are of two different types; for example, human adult hæmoglobin consists of two so-called α-chains and two β-chains. There is usually interaction between the four hæm groups in their reaction with molecular oxygen; in other words, a plot of the degree of oxygenation of the hæmoglobin molecules against the partial pressure of oxygen is sigmoid in shape[4]. This fact enhances enormously the physiological efficiency of the hæmoglobin molecule as our oxygen carrier[5], since it favours complete saturation with oxygen in the lungs and complete discharge in the tissues. This great similarity which characterizes most vertebrate hæmoglobins suggests that we are studying that aspect of the evolution of a particular protein molecule which is concerned with detailed development of an already well-defined molecule. It is true that the hæmoglobin quadruples its size during this evolution (if we include the lamprey hæmoglobin), but the change is due to the aggregation of four fairly similar protein sub-units (peptide chains), rather than to an actual lengthening of a molecule. In the evolutionary scheme to be discussed it is suggested that the increase in complexity, and in diversity, of the hæmoglobins is an illustration of a more general process of gene evolution which results in an increase of the number of genes.

Human Hæmoglobins

The following striking situation is found in the human hæmoglobins:

(1) The three normal human hæmoglobins all possess a common half-molecule which may be written $-\alpha_2^A$. This formulation indicates that this half-molecule unit is composed of two normal α-peptide chains as first described for normal adult hæmoglobin *A*. However, the other half of each of the three hæmoglobins consists of different types of peptide chain, β, γ or δ, one type for each hæmoglobin. All four chains are of roughly equal size with a chain molecular weight of about 17,000. These three hæmoglobins may be formulated as follows:

Adult	= hæmoglobin *A*	$= \alpha_2^A\beta_2^A$
Fœtal	= hæmoglobin *F*	$= \alpha_2^A\gamma_2^F$
Kunkel's minor component[6]	= hæmoglobin A_2	$= \alpha_2^A\delta_2^{A_2}$

This list may have to be extended as the study of other minor hæmoglobins continues[7].

I postulate[1] that four genes are involved in the manufacture of these peptide chains, so that the *genotype* of a normal individual may be written as:

$$\alpha^A/\alpha^A,\ \beta^A/\beta^A,\ \gamma^F/\gamma^F,\ \delta^{A_2}/\delta^{A_2}.$$

In this view the products of the α-genes are common to all the human hæmoglobins.

(2) In total amino-acid composition[8] the four types of chains compare as follows:

α^A and β^A	–	differ in perhaps 21 out of nearly 140 amino-acids
γ^F and β^A	–	differ in perhaps 23 out of nearly 140 amino-acids
δ^{A_2} and β^A	–	differ in less than 10 amino-acids.

In addition, γ^F is the only one of these chains to contain the amino-acid *iso*leucine—four residues per chain.

(3) The α^A-chain begins with the amino-acid sequence[7] Val-Leu- . . . ; the β^A-chain and the δ^{A_2}-chain begin with Val-His-Leu . . . . The γ^F-chain begins with the sequence Gly-His-Phe-, which although different from the β-chain shows a clear affinity in the type of amino-acid with the beginning of the β^A-chain. In both cases, the first amino-acid is neutral and aliphatic, the second is the basic amino-acid histidine, and the third has a non-polar side-chain.

The peptide chain beginning with the Val-Leu- sequence (α-chain) has a counterpart[3] not only in all known mammalian hæmoglobins but also in the hæmoglobins of all the lower vertebrates so far examined, except for the hæmoglobin of the lamprey. This is not to say that the Val-Leu- chains of other animals resemble the human chain precisely, but rather that there might be a strong 'family resemblance'. It is of interest to attempt to explain the repeated occurrence and strange similarity of the

Val-Leu- (the α) chain in the vertebrates.

The following postulates are used in developing a scheme for the evolution of the hæmoglobin genes.

Postulate 1. Mutations of a gene result in either single or multiple amino-acid substitutions in the peptide chain which that gene controls, or inversions of part of the amino-acid sequence, or a combination of these possibilities. The new hæmoglobin peptide chains produced by such mutations are then either favoured or discarded in the course of natural selection.

Postulate 2. At several points in the course of evolution a gene for a particular hæmoglobin peptide chain has undergone *duplication*, followed by, or simultaneous with, translocation. The two initially equivalent genes have then evolved independently, governed by the selective pressure of their environment on their protein products. Such mechanisms have been previously postulated; for example, ref. 9. The role of gene duplication[9,10] in evolution has been discussed by Stephens[11]. His conclusion was that the case for duplication as an important factor in evolution was so far neither proved nor disproved.

It is of course equally possible to postulate the occurrence of chromosome duplication, with subsequent independent evolution of the initially identical chromosomes. Such a scheme would fit into the proposed hypothesis equally well.

It seems likely that α-, β- and γ-genes are located on different chromosomes, since they segregate independently, whereas β- and δ-genes appear to be linked[12].

Postulate 3. Once the α-chain, or rather the α_2-dimer, is required to fit precisely with at least two other dimers—β_2 and γ_2—the α-chain is no longer as free to be varied by mutation. It has become 'conservative' and its rate of evolution will be less than that of the other chains, as is perhaps seen in the persistence of the Val-Leu- beginning of the α-chain of the vertebrates.

In addition, one can postulate that mutations of the α-gene produce mutant α-chains which are more severely selected against in the later stages of evolution than are mutants of other chains. This is likely to be so, because in the later stages of evolution the α-chains participate in the formation of fœtal hæmoglobin, a vulnerable point in the life-cycle of the animal. It is a fact that mutants of the human α-chain never reach a high frequency of distribution, in contrast to some β-chain mutants, although we know of as many different kinds of α-chain mutants as of β-chain mutants.

Scheme for the Evolution of the Hæmoglobin Genes

We might suppose that originally the hæmoglobin molecule was rather like the present-day myoglobin molecule, that is, that it had a single peptide chain with a single hæm group, and that therefore it could not show hæm–hæm interaction. The size of this molecule might vary, but eventually it might be stabilized at a value of around 17,000 molecular weight.

At this stage of evolution, presumably earlier than the teleost fishes, the hæm protein inside the muscle cells is assumed to be the same as that in the circulation. The muscle hæm protein became myoglobin in the course of evolution; it retained a molecular weight of 17,000 and a complexity of only one hæm group and one peptide chain per molecule. It was, of course, still subject to mutational changes, as can be seen from the fact that its present-day amino-acid composition and sequence in a given animal often differs considerably from that of any of the analogous hæmoglobin chains[13]. For example, human myoglobin contains *iso*leucine, but no cysteine, whereas the reverse is true of human adult hæmoglobin.

On the other hand, one can foresee limits to the kind of mutations which would be tolerated. The X-ray studies of Kendrew[14] and of Perutz[15] show that the three-dimensional arrangement of chains in myoglobin and the hæmoglobin sub-units (also of 17,000 molecular weight) are remarkably similar, though not identical. This statement applies also to the two kinds of sub-units found in hæmoglobin itself. Presumably, mutational alterations in the course of evolution which would drastically affect the three-dimensional structure were not tolerated. Such considerations imply that the configuration of the peptide chains in myoglobin and hæmoglobin became stabilized early in evolutionary history, at least in its most important features.

During the evolution of the primitive hæmoglobin chain—provisionally called the 'α'-chain—there occurred a gene duplication followed or accompanied by translocation. From now on the two duplicate 'α'-chain genes could evolve independently—one to become the modern myoglobin gene, the other to become the α-chain gene of present-day hæmoglobin. Eventually, according to the scheme, the α-chain gene would evolve in such a way that its product, the α-chain, had the property of dimerization in solution to form α_2 molecules. Such a property would be favoured strongly, if it entailed, in addition, the possibility of hæm–hæm interaction between the two hæm groups of the new dimer molecule and therefore the possibility of more efficient oxygenation and deoxygenation. Once produced, such a mutation is unlikely to be lost in the further evolution of hæmoglobin.

At this stage, the sequence of the 'α'-chains is still variable within the dictates of structural requirements, since there is nothing yet to put additional restrictions on it.

We might next postulate that the genes of the α_2-chains duplicated again. After this gene duplication two types of dimer—α_2 and γ_2—would evolve side by side. Sooner or later, these chains would have evolved sufficiently to be able to form tetramers with even greater selective advantage because of the increased hæm–hæm interaction likely to be found in such tetramers. The characteristics of the genes responsible for the ability of the chains to form tetramers would certainly be fixed. This stage of hæmoglobin evolution seems to have been reached already in some teleost fishes, because they already possess a four-chain hæmoglobin[3].

The third gene duplication and translocation is pictured as occurring with the γ-chain gene, giving rise to a new γ-gene destined to evolve into the β-chain gene. At this gene duplication the property of forming tetramers is already firmly established. The new gene can develop along its own line to provide a hæmoglobin tetramer—$\alpha_2\beta_2$—particularly adapted for the adult body. On the other hand, the old γ-chain continues to develop and to provide half the molecule of the fœtal hæmoglobin ($\alpha_2\gamma_2$). It is the γ-chain gene, rather than the α-chain gene, which is said to duplicate here, because the γ-chain dimers, γ_2, have already the necessary complementariness for forming tetramers with α_2. This complementariness will be automatically a property of the product of the new gene. In addition, we shall see later that β- and γ-chains are more closely related to one another than either is to the α-chain. Therefore we might consider them to have diverged at a later stage.

At this point of evolution, three independent genes—α, β, γ—are assumed to be present, each one capable of forming chains which dimerize and which aggregate to the tetramers $\alpha_2\beta_2$ or $\alpha_2\gamma_2$. Hæm–hæm interaction is strongly present in the tetramers. Such a situation has an important effect on the further

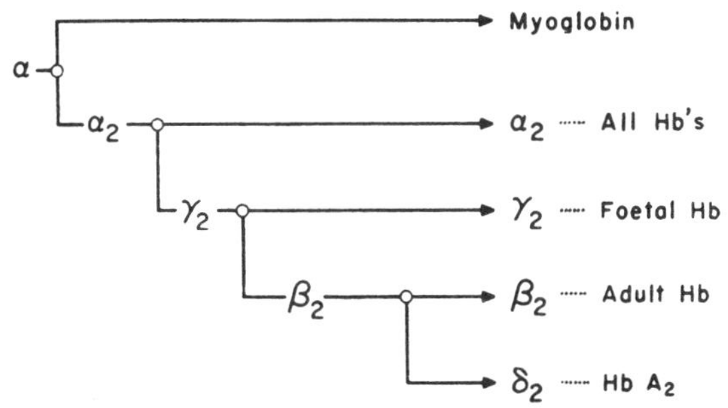

Fig. 1. Evolution of the hæmoglobin chains. The α-chain is the ancestral peptide chain. —○— indicates a point of gene duplication followed by translocation of the new gene

evolution of the α-chain. This chain, or rather its dimer, α_2, is required to fit with two different partners, β_2 and γ_2. As a result, less variation is allowed to the α-chain ; it has become 'conservative'. Perhaps such conservation is in part responsible for the apparently universal presence of α-chains beginning with Val-Leu- in the hæmoglobins of the higher vertebrates. There is an alternative explanation for the apparently greater stability of the α-genes ; this gene controls also the fœtal hæmoglobin and therefore may not undergo extensive mutational alterations, since the fœtus is a much more delicate organism. The very fact that any alteration in the α-chain gene, and therefore in the α-chain, seems to affect all types of hæmoglobin may be sufficient to explain the apparent 'conservatism'.

On the other hand, there is no *a priori* reason why different parts of a molecule as complex as a protein should develop at the same rate. The difference in apparent stability between the α-chain and the others could be no more than what would be expected as normal variation. It will be interesting to see just how similar the Val-Leu- chains of different vertebrates are.

At the fourth and last gene-duplication in the scheme we suppose that it is the β-gene which becomes duplicate, leading to the δ-chain genes controlling the δ-chains of hæmoglobin A_2. The origin of this δ-chain is placed near the end of the evolutionary scheme, because of its great chemical similarity to the β-chain[16]. Furthermore, the presence of a hæmoglobin A_2-like component seems to be confined to the higher primates[17]. In this view, hæmoglobin A_2 is a new hæmoglobin rather than an archaic one, as has often been supposed. It has been reported to have a higher affinity for oxygen[18] and perhaps it is a more efficient hæmoglobin, destined to replace eventually $\alpha_2\beta_2$ (ref. 19). Its proportion is certainly doubled in some thalassæmias[17], perhaps as a compensating mechanism. There is genetic evidence that the genes for δ- and for β-chains are linked[12], indicating perhaps that the process of translocation of the new δ-gene has not yet occurred.

Where does the lamprey fit into this scheme ? One form, *Petromyzon planeri*, has two fœtal and two (different) adult hæmoglobins[20]. The molecular weight of the adult mixture is given as 17,000 (ref. 3). Perhaps these lamprey hæmoglobins are the result of an independent evolution scheme similar to the one discussed in the present article, but which has never included the mutations which led to the formation of dimers and tetramers. We might regard the lampreys as having branched off before or just after the first gene duplication in the scheme of Fig. 1. Unfortunately, nothing seems to be known about the presence or absence of a separate myoglobin in the lampreys.

Discussion

The proposed scheme expresses a desire to explain the striking chemical similarity between α, β, γ and δ human hæmoglobin chains in terms of their evolution from a common precursor.

Braunitzer *et al.*[21] have published some preliminary findings of the amino-acid sequences in the first 30–40 positions of the human α- and β-chains (Table 1). Between the two sequences there is a strong 'family resemblance' of the degree expected for two gene products which have evolved independently, but from a common ancestor. Also Gratzer and Allison[8] have very recently discussed the possibility that the hæmoglobin chains might have evolved from a common ancestor.

It has been suggested to me that the four genes controlling the four chains might have evolved from four unrelated genes which originally controlled the synthesis of quite unrelated proteins ; thus there would have been no increase in the total number of genes. By a process of parallel evolution—successive and independent mutations—each of these four genes eventually changed so that it made one of the hæmoglobin peptide chains. The last stages of evolution, when pairs of genes were involved in producing parts of a complex molecule, must then be called on to produce similarity in chemical structure and in configuration. It is easy to conceive of a selective mechanism which would favour changes in various members of a molecular aggregate, such that the members might fit together better. However, it is hard to think of such a mechanism also selecting for chemical and structural similarity of the monomers. In addition, the ability to dimerize would have to arise *de novo* four times, an unlikely situation.

Although we cannot at the moment completely reject the idea of parallel evolution of the four genes from unrelated genes, it is not a very palatable one. The similarity between the human α-, β-, γ- and δ-chains is so great : the C-terminus of the chains ends with the sequence -tyrosyl-histidine or -tyrosyl-arginine[16,22] ; their common ability to incorporate an active hæm group ; their similar folding and chain length[14,15] ; absence of *iso*leucine in α, β and δ[8,16] ; and so forth. Altogether, an evolution from a common ancestor is more attractive.

Chemical Evolutionary Relationship Between the Chains

How many mutational steps occurred between the original 'α'-gene and the present-day human α-gene ?

Table 1. Comparison of the N-terminal Sequences of the α- and β-Peptide Chains of Human Hæmoglobin *A*.[21]

α-chain	Val-Leu-Ser-Pro-**Ala**-**Asp**-**Lys**-Thr-Asp-Val-**Lys**-Ala-Ala-Try-Gly-**Lys**-Val-Gly-Ala-His-Ala-Gly-Glu-Tyr-Gly-(Ala,Glu)-**Ala**-**Leu**-**Glu**-**Arg**-
	1 2 3 4 5 6 7 8 9 10 11 12 13 14 15 16 17 18 19 20 21 22 23 24 25 26 27 28 29 30 31
	Met-Phe-Leu-Ser-Phe-Thr-Pro-**Thr**-**Lys**-
	32 33 34 35 36 37 38 39 40
β-chain	Val-His-Leu-Thr-Pro-Glu-Glu-**Lys**-Ser-Ala-Val-Thr-Ala-Leu-Try-Gly-**Lys**-Val-Asp-Val-Asp-Glu-Val-Gly-Gly-Glu-Ala-Leu-Gly-**Arg**-Leu-
	1 2 3 4 5 6 7 8 9 10 11 12 13 14 15 16 17 18 19 20 21 22 23 24 25 26 27 28 29 30 31

Or between the original 'α'-gene and the human β-gene? It is more meaningful, perhaps, to ask how many mutational events might separate the present-day human α- and β-genes. Out of 140 amino-acids, we know from the composition that approximately 20 are different. This gives a minimum of 20 mutations of the kind effecting single amino-acid substitutions.

It is believed that the evolution of the vertebrates began some 5×10^8 years ago. Let us assume a mean generation life of 5 years among the vertebrates, which implies some 10^8 generations for the evolution of vertebrate hæmoglobins. A generally acceptable rate[24] per generation is 10^{-5}–10^{-6}, which leads to a figure of 100–1,000 mutations in the evolutionary history of a vertebrate hæmoglobin. This figure, which is itself a minimum one, is very considerably higher than the (minimum) number of mutations which appear to distinguish α-chains from β-chains and α-chains from γ-chains.

The only conclusion which can be drawn from such a calculation is that a more than sufficient number of mutational events have passed in the history of the vertebrates to account for present-day differences between the various parts of the hæmoglobin molecule in the human species.

Comparison of the Fingerprints of the Human α-, β-, γ- and δ-Chains

Rather than deduce the chemical relationship between the chains from their amino-acid composition, it is better to compare their 'fingerprints'[25]. The two-dimensional maps, or fingerprints, of the tryptic peptides derived from each of these four peptide chains give at least a crude idea of the similarity and dissimilarity in their amino-acid sequences and hence of the closeness of the chemical relationship between them.

Human α-Chains and β- or γ-Chains

With the exception of peptide number 21 (ref. 25), all the peptides (some 14 for each chain) occupy different positions on the fingerprints[23] of α-chains compared with β- or γ-chains, or they can be distinguished by such methods as extended paper ionophoresis. They are therefore different in their amino-acid sequences. Peptide 22 is free lysine. It does not follow that the differing peptides of the α- and of the β- or γ-chains are totally altered. They could contain short amino-acid sequences in common[21]. In any event, the indications are that the amino-acid differences between the α- and the β- or γ- chains are distributed throughout their length.

The general comment might be made that there is a familial likeness between the distribution of peptides on the fingerprints of α-, β- and γ-chains. Surely this will reflect a structural relationship (Table 1). We would suppose that portions of their amino-acid sequences are common to all three, as is the tertiary structure of the two types of chain of horse hæmoglobin, which correspond to the human α- and β-chains.

Human β-Chains and γ-Chains

Apart from peptides number 21 (a pentapeptide) and 22 (free lysine), the following peptides[25] occupy very similar positions[23] and would be closely related, if not identical: peptides 12, 14, 15β, 19, 20β, 26. The number is an impressive proportion of the whole chain. If it is correct that these peptides are indeed identical in the β- and γ-chains, it would mean that perhaps 9 peptides out of some 14 are shared by the chains. Some of these peptides are quite long. Their position along the chains is not known, except that peptide 26 is the third one from the N-terminus[21]. These conclusions underline the chemical similarity between β- and γ-chains, which can also be deduced from their amino-acid compositions.

Human β-Chains and δ-Chains

The recent work of Stretton[16] has shown that the β-chains and the δ-chains are very closely similar in their fingerprints. In fact, out of a total of some 14 tryptic peptides, only 4 peptides are different; they are numbers 5, 12, 25 and 26. The similarity between other pairs of peptides from the β- and δ-chains has been ascertained much more carefully by amino-acid analysis than has so far been done with the γ-chains. Moreover, the differences between peptides 12, 25 and 26 from β- and from δ-chains have been shown to involve only 4 single amino-acid differences, for example, serine-threonine (in two places), threonine-asparagine, glutamic acid-alanine. These four single amino-acid differences, together with an unconfirmed fifth one, are so far the only ones definitely known between this pair of chains, each of which is some 140 residues long—a very close chemical similarity indeed.

Apart from the tryptic peptides of the α-, β-, γ- and δ-chains, there is in each case the trypsin resistant 'core' to be considered which amounts to rather more than a quarter of the chain. Here the differences between the chains are far from clear and cannot yet be usefully discussed, except to say that the 'cores' all appear to be similar.

Fingerprints of hæmoglobins from other mammals (ref. 2, and Muller, C. J., unpublished) are recognizable as 'hæmoglobin'. It will be interesting to compare the degree of similarity between the mammalian hæmoglobins on one hand with, say, chicken or teleost hæmoglobin.

Recently, Zuckerkandl[2] has shown that fingerprints of hæmoglobin from the gorilla and chimpanzee are indistinguishable from those of human hæmoglobin. A greater number of differing peptide spots are observed in the less closely related orang-utan and rhesus monkey. These extraordinarily interesting findings give added proof of the validity of using a chemical study of vertebrate hæmoglobins to discover evolutionary relationships.

Conclusion

It appears that a sufficient number of generations has passed since the beginning of vertebrate evolution to allow for the known or suspected number of mutations which separate the α-, β- and γ-chains of human hæmoglobin.

The similarity in the fingerprints between β- and γ-chains of human hæmoglobin supports the suggestion that these chains diverged from one another at a later stage in evolution than either from the α-chain gene.

The close similarity between β- and δ-chains gives weight to the idea that the δ-chains of hæmoglobin A_2 are a recent evolutionary development from the β-chains.

The suggestion is made that a single primitive myoglobin-like hæm protein is the evolutionary forerunner of all four types of peptide chain in the present-day human hæmoglobins, and of the corresponding peptide chains in other vertebrate hæmoglobins. Such a scheme involves an increase in the number of hæmoglobin genes from one to five by repeated gene duplication and translocations; the scheme may thus illustrate a general phenomenon in gene evolution.

I acknowledge the many stimulating discussions with Drs. C. Levinthal, S. E. Luria, C. Baglioni, A. O. W. Stretton, J. V. Neel and P. S. Gerald. I am also grateful to Drs. W. B. Gratzer and A. C. Allison for allowing me to read their review paper[3] before publication.

Two books have been particularly stimulating: J. B. S. Haldane, *The Biochemistry of Genetics*; and C. B. Anfinsen, *The Molecular Basis of Evolution*.

No serious attempt has been made to survey critically the literature on gene evolution or duplication. This is partly in order to present a clearer and more provocative hypothesis and partly because I do not feel qualified to evaluate the numerous contributions to that field. Recent experimental work on the animal hæmoglobins has been admirably summarized in the Gratzer and Allison article[3].

This work was supported by grants from the National Science Foundation and the National Institutes of Health, U.S. Public Health Service.

Note added in proof. Further work by G. Braunitzer and his colleagues (*Z. physiol. Chem.*, in the press) shows that the similarity of the α- and β-chains continues throughout these chains.

1 Ingram, V. M., in *Genetics*, Macy Found. Symp., **141**, 147 (1959); *Hemoglobin and Its Abnormalities* (C. C. Thomas, in the press).
2 Zuckerkandl, E., Jones, R. T., and Pauling, L., *Proc. U.S. Nat. Acad. Sci.*, **46**, 1349 (1960).
3 Gratzer, W. B., and Allison, A. C., *Biol. Rev.*, **35**, 459 (1960).
4 Edsall, J. T., in *Hemoglobin* (National Research Council, Wash., 1958).
5 Riggs, A., *Nature*, **183**, 1037 (1959).
6 Kunkel, H. G., and Wallenius, G., *Science*, **122**, 288 (1955).
7 Schroeder, W. A., *Prog. Chem. Org. Nat. Prod.*, **17**, 322 (1959).
8 Stein, W. H., Kunkel, H. G., Cole, R. D., Spackman, D. H., and Moore, S., *Biochim. Biophys. Acta*, **24**, 640 (1957). Hill, R. J., and Craig, L. C., *J. Amer. Chem. Soc.*, **81**, 2272 (1959).
9 Lewis, E. B., *Cold Spring Harbor Symp. Quant. Biol.*, **16**, 159 (1951).
10 Bridges, C. B., *Science*, **83**, 210 (1936).
11 Stephens, S. G., *Adv. Genetics*, **4**, 247 (1951).
12 Cepellini, R., in *Biochemistry of Human Genetics*, 135 (CIBA Symp., 1959).
13 Rossi-Fanelli, A., Cavallini, D., and de Marco, C., *Biochim. Biophys. Acta*, **17**, 377 (1955).
14 Kendrew, J. C., Dickerson, R. E., Strandberg, B. E., Hart, R. G., Davies, D. R., Phillips, D. C., and Shore, V. C., *Nature*, **185**, 422 (1960).
15 Perutz, M. F., Rossmann, M. G., Cullis, A. F., Muirhead, H., Will, G., and North, A. C. T., *Nature*, **185**, 416 (1960).
16 Ingram, V. M., and Stretton, A. O. W. (submitted to *Nature* for publication).
17 Kunkel, H. G., Cepellini, R., Muller-Eberhard, U., and Wolf, J., *J. Clin. Invest.*, **36**, 1615 (1957).
18 Meyering, C. A., Israels, A. L. M., Sebens, T., and Huisman, T. H. J., *Clin. Chim. Acta*, **5**, 208 (1960).
19 Stretton, A. O. W., Ph.D. Thesis, Univ. Cambridge (1960).
20 Adinolfi, N., Chieffi, G., and Siniscalco, M., *Nature*, **184**, 1325 (1959).
21 Braunitzer, G., Liebold, B., Muller, R., and Rudloff, V., *Z. physiol. Chem.*, **320**, 170 (1960).
22 Guidotti, G., *Biochim. Biophys. Acta*, **42**, 177 (1960).
23 Hunt, J. A., *Nature*, **183**, 1373 (1959), and Ph.D. Thesis, Univ. Cambridge (1959); Hunt, J. A., and Lehmann, H., *Nature*, **184**, 872 (1959).
24 Haldane, J. B. S., *Biochemistry of Genetics*, 15 (Allen and Unwin, London, 1954).
25 Ingram, V. M., *Biochim. Biophys. Acta*, **28**, 539 (1958).

Reprinted from *Proc. Roy. Soc. Lond.* B **164**: 298–310.

C. GENETICS OF MAN

Enzyme polymorphisms in man

By H. Harris*

Department of Biochemistry and M.R.C. Human Biochemical Genetics Research Unit, King's College, Strand, London

There are a large number of different enzymes synthesized in the human organism, and many of these probably contain more than one structurally distinct polypeptide chain. If current theories about genes and proteins are correct we must suppose that the primary structure of each of these different polypeptides is determined by a separate gene locus, and that there are probably also other loci which are specifically concerned with regulating the rate of synthesis of particular polypeptides or groups of polypeptides. Furthermore, we may expect that genetical diversity in a human population will to a considerable extent be reflected in enzymic diversity. That is to say, in differences between individuals either in the qualitative characteristics of the enzymes they synthesize, or in differences in rates of synthesis.

The work I am going to discuss was largely aimed at trying to get some idea of the extent and character of such genetically determined enzyme diversity among what may be regarded as normal individuals. When my colleagues and I started on this line of work about three years ago the information available about this aspect of the subject was very limited. It had of course been recognized for quite a long time that there are many rare metabolic disorders, the so-called 'inborn errors of metabolism', which are due to genetically determined deficiencies of specific enzymes (Harris 1963). These conditions can in general be attributed to mutant genes which result either in the synthesis of an abnormal enzyme protein with defective catalytic properties, or in a gross reduction in rate of synthesis of a specific enzyme protein. By and large such genes appear to be relatively uncommon and have frequencies of between 0·01 and 0·001 in the general population. Heterozygotes often show a partial enzyme deficiency though they are usually in other respects quite healthy. A few cases are also known where a specific enzyme deficiency occurs quite commonly in certain populations. The most extensively studied example of this is glucose 6-phosphate dehydrogenase deficiency, and it seems likely that in this particular case the relatively high incidence in certain populations is attributable to a specific selective advantage which the deficiency may confer in situations where endemic malaria is an important selective agent (Motulsky 1964).

Virtually all these enzyme deficiencies have been identified in the first instance because of some more or less striking clinical or metabolic disturbance of which they were the cause. They therefore represent a highly selected group of mutants, and cannot be expected *per se* to provide us with any clear picture of how extensively genetically determined enzyme variation which does not result in overt pathological manifestations may occur in the general population. Nor can they provide us with any precise indication of what the general character of such 'concealed' variation might be. Whether it is, for example, mainly a matter of minor quantitative differences in rates of synthesis, attributable to genes at so-called 'regulator' or 'operator' loci, or whether qualitative differences involving enzyme structure are an important feature.

* Present address: The Galton Laboratory, University College London.

In attempting to tackle this rather general problem we have adopted a quite empirical and perhaps somewhat simple-minded approach. Our idea was to see whether, if we examined a series of arbitrarily chosen enzymes in normal individuals in sufficient detail, we would find genetically determined differences, and if so whether such differences were common or rare, and whether they were peculiar to one class of enzyme rather than another.

Because we would need to examine the selected enzymes in quite a large number of different people, and because we wished to carry out family studies on any enzyme differences that turned up, we were in the first instance largely forced to confine our attention to enzymes present in blood. We had, of course, also to make some decision about the kind of techniques we would utilize in looking for such differences. A wide variety of methods suitable for examining the many different properties of enzyme proteins are available, and it would have been impractical to attempt to utilize more than a few of these. In practice we have mainly relied in the first instance on the technique of starch gel electrophoresis. This is known to be capable, if one gets the conditions right, of detecting quite subtle differences in molecular charge and molecular size. It is however not designed to pick up other sorts of molecular differences, and it is also not very sensitive in the detection of small quantitative differences. Thus we could expect to detect at best only a proportion of all possible forms of enzyme variation.

Despite these limitations, we have found, during the course of examining in varying degrees of detail some ten arbitrarily chosen enzymes, three quite striking examples of genetically determined polymorphism.

Red cell acid phosphatase

The first enzyme we selected for study on this arbitrary basis was red cell acid phosphatase (Hopkinson, Spencer & Harris 1963, 1964). This enzyme is known to differ both in its pattern of substrate specificity and in its inhibition characteristics from the acid phosphatase present in other tissues, and it is thought to occur only in the erythrocyte. Its precise function, however, is not known.

A fairly simple method was developed for detecting the enzyme after electrophoresis in starch gel. The surface of the gel is incubated in a reaction mixture containing phenolphthalein diphosphate at pH 6·0, so that at any site of acid phosphatase activity free phenolphthalein is liberated. This can then be detected by making the surface of the gel alkaline, so that the sites of enzyme activity appear as bright red zones.

When haemolysates from a series of normal individuals were examined using this procedure, it was found that every sample showed more than one zone of enzyme activity. We were evidently dealing with what is now generally called a set of isoenzymes. Furthermore, there were clear-cut person-to-person differences in the number, the mobilities, and the relative activities of these isoenzyme components (figure 65). Five distinct phenotypes were soon identified. They are now referred to as A, BA, B, CA and CB, and in the British population occur with frequencies of about 0·13, 0·43, 0·36, 0·03 and 0·05 respectively. The initial family studies showed that these phenotypes are genetically determined and led to the hypothesis that three allelic genes (P^a, P^b and P^c) at an autosomal locus are involved (phenotypes A and B being produced by the homozygous genotypes P^aP^a and P^bP^b respectively, and phenotypes, BA, CA and CB by the heterozygous genotypes P^aP^b, P^aP^c and P^bP^c). The hypothesis predicted the occurrence of a sixth phenotype corresponding to the genotype P^cP^c. Gene frequency considerations indicated that this would be fairly uncommon (about 1 in 625 of the general

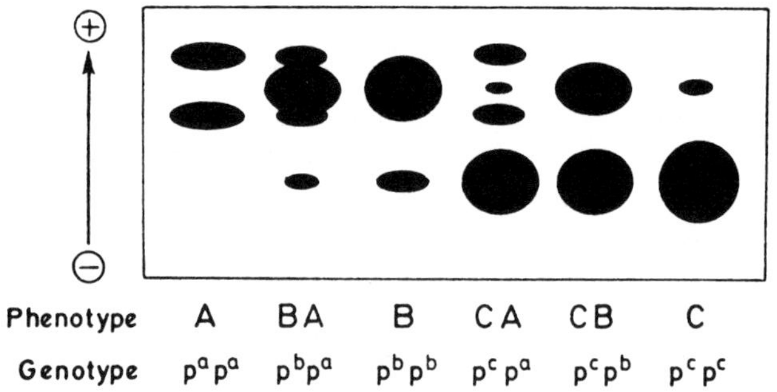

FIGURE 65. Diagram of isoenzyme components seen in the various red cell acid phosphatase phenotypes after electrophoresis at pH 6·0.

population), and indeed a few examples of what is probably this phenotype have now been observed (Lai, Nevo & Steinberg 1964).

With the five common phenotypes fifteen different mating types are possible and the segregation pattern in most of these has now been studied (table 7). These quite extensive family data have proved to be fully consistent with the hypothesis and the findings have also been confirmed by several other groups working with different populations. There seems little doubt therefore that these acid phosphatase variations reflect a polymorphism involving at least three alleles, and a variety of studies on the properties of the isoenzymes in the different phenotypes makes it appear reasonably certain that these alleles determine the synthesis of structurally different forms of the enzyme.

A particularly interesting feature of the polymorphism is that the qualitative differences between the phenotypes are reflected quantitatively by differences in the levels of enzyme activity (Spencer, Hopkinson & Harris 1964*a*). Levels of total acid phosphatase activity were determined by a standard method in a series of haemolysates from individuals of the different phenotypes, using *p*-nitrophenyl phosphate as substrate. Although there was considerable variation in activity between individuals of any one phenotype, nevertheless quite marked differences between the mean values for different phenotypes could be demonstrated (table 8). Using these values one may examine the question as to whether the quantitative

TABLE 7. SEGREGATION OF RED CELL ACID PHOSPHATASE PHENOTYPES IN 216 FAMILIES

parents	number of matings	children: A	BA	B	CA	CB	C	total
A × A	4	8	—	—	—	—	—	8
A × BA	25	31	25	—	—	—	—	56
A × B	11	—	20	—	—	—	—	20
A × CA	4	3	—	—	2	—	—	5
A × CB	5	—	2	—	5	—	—	7
BA × BA	51	24	52	21	—	—	—	97
BA × B	50	—	52	44	—	—	—	96
BA × CA	7	6	3	—	3	2	—	14
BA × CB	8	—	5	5	2	6	—	18
B × B	24	—	—	58	—	—	—	58
B × CA	9	—	7	—	—	16	—	23
B × CB	16	—	—	23	—	16	—	39
CA × CA	—	—	—	—	—	—	—	—
CA × CB	1	—	1	—	1	—	—	2
CB × CB	1	—	—	—	—	1	1	2
totals	216	72	167	151	13	41	1	445

TABLE 8. MEANS AND STANDARD DEVIATIONS OF RED CELL ACID PHOSPHATASE ACTIVITY IN INDIVIDUALS OF KNOWN PHENOTYPES

The activity is expressed as μM p-nitrophenol liberated in $\frac{1}{2}$ h at 37 °C/g haemoglobin.

phenotype	number of individuals	mean activity	standard deviation
A	33	122·4	16·8
BA	124	153·9	17·3
B	81	188·3	19·5
CA	11	183·8	19·8
CB	26	212·3	23·1

effects of the three postulated alleles are additive in a simple way or not. If they are additive one could expect the following relationships to be true:

$$(a)\quad \tfrac{1}{2}\overline{\text{A}}+\tfrac{1}{2}\overline{\text{B}} = \overline{\text{BA}},$$
$$(b)\quad \overline{\text{CA}}-\tfrac{1}{2}\overline{\text{A}} = \overline{\text{CB}}-\tfrac{1}{2}\overline{\text{B}},$$

where $\overline{\text{A}}$, $\overline{\text{BA}}$, $\overline{\text{B}}$, etc., are the mean values for the various phenotypes. It will be seen that the results support the idea of simple additivity rather well ($\tfrac{1}{2}\overline{\text{A}}+\tfrac{1}{2}\overline{\text{B}}$ = 155·35, $\overline{\text{BA}}$ = 153·9, $\overline{\text{CA}}-\tfrac{1}{2}\overline{\text{A}}$ = 122·6 and $\overline{\text{CB}}-\tfrac{1}{2}\overline{\text{B}}$ = 118·15). Estimates from these data of the average activity attributable to each allele are:

$$P^a \rightarrow \ 60{\cdot}7 \pm 1{\cdot}1 \text{ units},$$
$$P^b \rightarrow \ 93{\cdot}7 \pm 1{\cdot}0 \text{ units},$$
$$P^c \rightarrow 120{\cdot}3 \pm 3{\cdot}7 \text{ units}.$$

Somewhat unexpectedly the three values turn out to be very close to the simple ratio 2:3:4, and it is tempting to think that this may have some special significance in terms of enzyme structure.

It is of interest to note that if one determines red cell acid phosphatase activities in a series of randomly selected individuals one obtains a continuous unimodal distribution which, in fact, is not dissimilar in form to the distributions usually obtained when other enzymes in man are examined quantitatively in randomly selected populations. In particular the variance when related to the mean is of the same order of magnitude as is found with many other enzymes. In the present case however it is clear that the overall distribution represents a summation of a

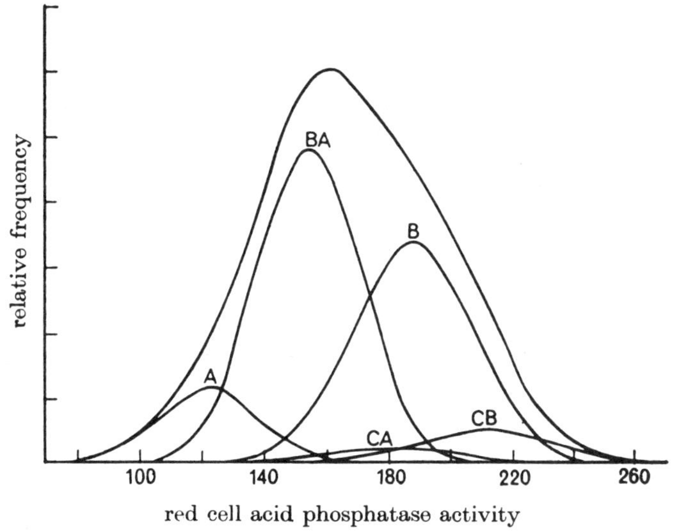

FIGURE 66. Distribution of red cell acid phosphatase activities in the general population (top line) and in the separate phenotypes. The curves are constructed from the values given in table 8 and from the relative frequencies of the phenotypes observed in a randomly selected population.

series of separate but overlapping distributions corresponding to each of the qualitatively different phenotypes (figure 66). Furthermore, the genetical component of the variance of the overall distribution can be largely if not entirely attributed simply to the effects of the three alleles. It is not unreasonable to suppose that the genetical component of other examples of continuous variation in enzyme levels may have a similar simple underlying basis.

Phosphoglucomutase

One of the main problems in studying enzymes by starch gel electrophoresis is the development of sensitive and specific methods for the detection of the zones of enzyme activity. The general approach has been to utilize substrates which will yield coloured products or products capable of reacting rapidly with some chemical included in the reaction mixture to give a coloured compound. Phenolphthalein diphosphate, for example, proved to be a useful substrate for red cell acid phosphatase, and naphthyl phosphates have been widely utilized for the study of other phosphatases. However, for the majority of enzymes this approach is not feasible because of their very restricted range of substrate specificity.

One way round this difficulty is to utilize other enzymes in the reaction mixture so as to build up a sequence of reactions culminating in the formation of some detectable substance. Such reaction mixtures are usually complex and often include six or more different interacting components whose relative concentrations require careful adjustment. However, the general method is proving extremely valuable and has opened up the possibility of examining many previously inaccessible enzymes. Our first successful application of this idea was with phosphoglucomutase, and it led to the discovery of another example of enzyme polymorphism (Spencer, Hopkinson & Harris 1964*b*).

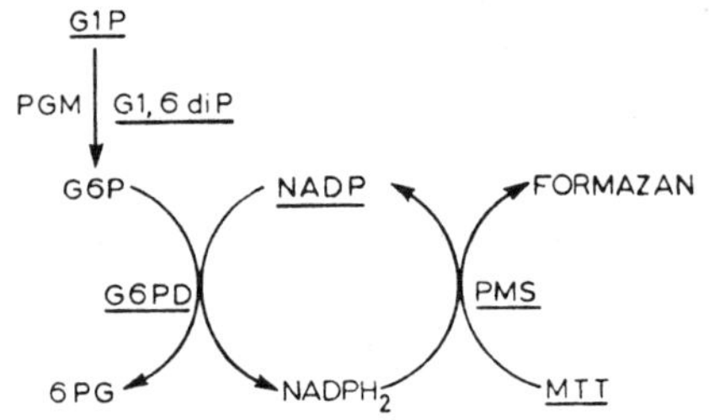

FIGURE 67. Sequence of reactions in the detection of phosphoglucomutase (PGM) after starch gel electrophoresis. The underlined components are contained in the reaction mixture. [*Key*: G1P, glucose 1-phosphate; G6P, glucose 6-phosphate; G1,6diP, glucose 1,6-diphosphate; 6PG, 6-phosphogluconate; G6PD, glucose 6-phosphate dehydrogenase; NADP and $NADPH_2$, oxidized and reduced nicotinamide adenine dinucleotide phosphate; PMS, phenazine methosulphate; MTT, tetrazolium salt.]

Phosphoglucomutase catalyses the reversible transfer of phosphate between glucose 1-phosphate and glucose 6-phosphate, and it has an important role in carbohydrate metabolism. Its detection following starch gel electrophoresis was accomplished via the sequence of reactions shown in figure 67. The underlined components are included in the reaction mixture and the sites of phosphoglucomutase activity are located by the deposition of a blue-coloured formazan formed by the reduction of the tetrazolium salt *MTT*.

When haemolysates from different individuals are subjected to starch gel electrophoresis and this reaction system is applied, one obtains the rather complex isoenzyme patterns shown in figure 68. At least seven different zones of activity (*a–g*) may be detected, and three quite distinct types of pattern can be identified in different individuals. These are referred to as PGM 1, PGM 2-1 and PGM 2. The phenotypes differ in the occurrence of components a, b, c and d; a and c being present in PGM 1 and PGM 2-1, but not PGM 2, while b and d are present in PGM 2-1 and PGM 2 but not PGM 1. Components e, f and g are present in all the three phenotypes.

In the British population the incidence of the three phenotypes has been found to be PGM 1 0·58, PGM 2-1 0·36 and PGM 2 0·06. Studies on the segregation of these phenotypes in more than 150 different families involving all the possible mating types make it clear that two autosomal alleles (*PGM*[1] and *PGM*[2]) determine these differences. Phenotypes PGM 1 and PGM 2 represent the homozygotes *PGM*[1] *PGM*[1] and *PGM*[2] *PGM*[2], and phenotypes PGM 2-1 the heterozygote *PGM*[1] *PGM*[2].

This suggests that the isoenzyme components a and c determined by *PGM*[1] may be molecular alternatives of components b and d determined by *PGM*[2]. Possibly these isoenzymes contain a common polypeptide chain, and the difference between the two homozygous phenotypes depends on a small structural difference in this, which involves perhaps a single amino acid substitution. If this is so then one would presume that this polypeptide chain is not present in the isoenzyme components e, f and g, as they appear to be uninfluenced by this gene substitution, and

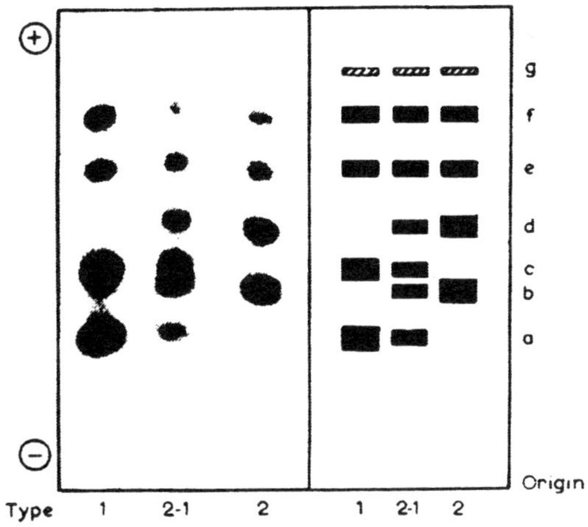

FIGURE 68. Photograph and diagram of phosphoglucomutase isoenzyme patterns obtained by starch gel electrophoresis at pH 7·4.

their structures are presumably therefore determined by other loci. Some support for this idea has recently been obtained by the discovery of an uncommon variant involving components e and f but not affecting a, b, c or d. This variant was found to segregate independently from phenotypes 1, 2-1 and 2, and the family study indicated that it was determined at a separate and not closely linked locus. It is also possible that other structurally distinct polypeptide chains may be contained in these isoenzyme components. These might, for instance, account for the mobility differences between isoenzymes a and c, or between e and f, and would if present imply the existence of further loci involved in the determination of this enzyme. No doubt structural studies on the isolated isoenzyme components will enable these questions to be resolved.

Unlike red cell acid phosphatase, phosphoglucomutase occurs in many different tissues, and it was therefore of some importance to see whether the isoenzyme components and the polymorphism found in erythrocytes also occurred elsewhere. It has in fact been possible to demonstrate that this is the case. The tissues studied included liver, kidney, muscle, brain, skin and placenta. The phosphoglucomutase isoenzymes have also been demonstrated in tissue culture cells grown *in vitro*. The tissue cultures were started from small skin biopsies from different individuals and were kept going for up to ten passages for more than three months. The cells were harvested at different times and the phosphoglucomutase examined. In each case the PGM phenotype found was the same as that originally observed in the red cells of the donor whose skin was used to start the culture.

Adenylate kinase

The third polymorphism discovered in the screening programme of arbitrarily selected enzymes involves the enzyme adenylate kinase (Fildes & Harris 1965). This catalyses the reversible reaction

$$2\,ADP \rightleftarrows ATP + AMP$$

and two procedures for detecting the enzyme after starch gel electrophoresis have been developed. Both of these require complex and different multienzyme reaction mixtures, but reveal the same pattern of zones of activity. It has been found that the enzyme as it occurs in erythrocytes, and also in skeletal muscle, includes several distinct isoenzyme components, and so far two discrete phenotypes have been recognized. One of these occurs in about 1 in 10 of the general population, and family studies indicate that the individuals showing it are heterozygous for two autosomal alleles. This work is still in its preliminary stages and it has not yet been possible to determine whether all or only some of the isoenzymes present are involved in the polymorphism.

Placental alkaline phosphatase

The various enzymes studied in this screening programme were selected because among other reasons they were known to occur in the erythrocyte, and this is obviously convenient if one wishes to carry out population and family investigations. However, the erythrocyte is a rather specialized cell type, and even though it is often possible to demonstrate that many red cell enzymes occur in essentially the same form in other tissues, it might be considered that a survey restricted to one cell type could present a somewhat biased picture of human enzyme variation in general. Analogous studies on enzymes localized to other tissues are for obvious reasons very much more difficult to pursue, and so far have not been carried out in any systematic way. However, it has been possible to investigate in some detail one particular example of a polymorphism involving what can be regarded as an organ-specific enzyme, and the results illustrate something of the possibilities and problems which such enzymes may present.

The enzyme is an alkaline phosphatase present in quite large amounts in the human placenta. It appears to be peculiar to this organ and to be different from the alkaline phosphatases present in other tissues such as liver, kidney and bone.

Following earlier work by Boyer (1961), it has now been possible to demonstrate that placentae may be classified into at least six distinct phenotypes according to the electrophoretic behaviour of the alkaline phosphatase they contain (Robson & Harris 1965). The electrophoretic patterns are illustrated in figure 69, and one may note that complete discrimination of the six phenotypes requires electrophoresis at two different pH's. The six phenotypes are referred to as S, F, I, SF, SI and FI. In types S, F and I most of the alkaline phosphatase activity is present in a single rapidly moving component, which however has a different mobility in each type. In the I phenotype, the characteristic component has a mobility very close to that of the F phenotype at pH 8·6, but is indistinguishable from that of the S phenotype at pH 6·0. In phenotypes SF, SI and FI, three such components are found, two of them in each case having mobilities similar to the component present in S, F or I,

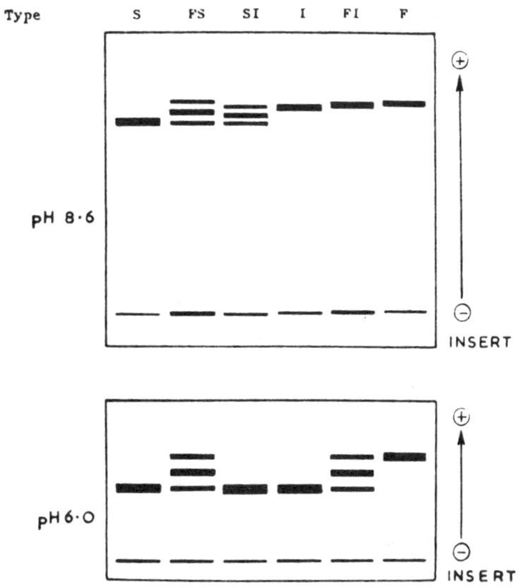

FIGURE 69. Placental alkaline phosphatase patterns obtained by electrophoresis at pH 8·6 and pH 6·0.

while the third has an intermediate mobility. There are reasons for thinking that this third component in these three phenotypes may represent a 'hybrid' enzyme containing polypeptide chains characteristic of the two other components present. In each of the six phenotypes at least one other component which migrates only very slowly may be seen. These slow components do in fact exhibit slight differences in mobility in the different phenotypes, and these can be shown to correlate with the more striking differences observed in the major and more rapidly moving components.

On the basis of the phenotypic patterns and their relative frequencies, it is possible to construct a simple genetical hypothesis which will account for these variations. This suggests that three autosomal allelic genes are concerned, phenotypes F, I and S representing the three homozygous genotypes, and phenotypes SF, SI and FI the corresponding heterozygotes. From the observed incidence of the six phenotypes in the British population, one may readily obtain values for the frequencies of the three postulated genes. They are 0·27, 0·09 and 0·64, and using these values one finds a very good agreement between the observed incidence of the phenotypes and those expected assuming a Hardy–Weinberg equilibrium (table 9)

It is of course obvious that family studies of the ordinary kind are impracticable in the case of a characteristic peculiar to the placenta. However it occurred to us

TABLE 9. OBSERVED AND EXPECTED NUMBERS OF PLACENTAL ALKALINE PHOSPHATASE TYPES IN A POPULATION SAMPLE ASSUMING A HARDY–WEINBERG EQUILIBRIUM

($p = 0{\cdot}27$, $q = 0{\cdot}09$ and $r = 0{\cdot}64$)

placental alkaline phosphatase type	expected incidence		expected numbers in population sample	observed numbers in population sample
S	r^2	0·410	135·9	141
SF	$2pr$	0·346	114·7	111
F	p^2	0·073	24·2	28
SI	$2qr$	0·115	38·2	32
FI	$2pq$	0·049	16·1	15
I	q^2	0·008	2·7	5
totals	$(p+q+r)^2$	1·001	331·8	332

TABLE 10. PLACENTAL ALKALINE PHOSPHATASE PHENOTYPES IN 130 DIZYGOTIC TWIN PAIRS

dizygotic twins		expected incidence assuming three alleles with frequencies p, q and r	expected incidence $p = 0{\cdot}27$ $q = 0{\cdot}09$ $r = 0{\cdot}64$		observed incidence	
like pairs						
S	S	$\frac{1}{4}r^2(1+r)^2$	35·80		39	
SF	SF	$\frac{1}{2}pr[pr+(1+p)(1+r)]$	25·34		27	
F	F	$\frac{1}{4}p^2(1+p)^2$	3·82		6	
SI	SI	$\frac{1}{2}qr[qr+(1+q)(1+r)]$	6·90		6	
I	I	$\frac{1}{4}q^2(1+q)^2$	0·31		1	
FI	FI	$\frac{1}{2}pq[pq+(1+p)(1+q)]$	2·24		3	
				74·41		82
unlike pairs						
S	SF	$pr^2(1+r)$	23·58		26	
S	F	$\frac{1}{2}p^2q^2$	1·95		0	
S	SI	$qr^2(1+r)$	7·85		9	
S	I	$\frac{1}{2}q^2r^2$	0·21		0	
S	FI	pqr^2	1·30		1	
SF	F	$p^2r(1+p)$	7·70		3	
SF	SI	$pqr(1+2r)$	4·60		4	
SF	I	pq^2r	0·18		0	
SF	FI	$pqr(1+2p)$	3·12		1	
F	SI	p^2qr	0·55		2	
F	I	$\frac{1}{2}p^2q^2$	0·03		0	
F	FI	$p^2q(1+p)$	1·09		0	
SI	I	$q^2r(1+q)$	0·73		1	
SI	FI	$pqr(1+2q)$	2·39		1	
I	FI	$pq^2(1+q)$	0·31		0	
				55·59		48

that we might be able to test the hypothesis by studying a series of pairs of placentae from dizygotic twins, because such twin pairs can be regarded as pairs of sibs. We were fortunate in being able to obtain such material from an extensive investigation of twin births which is being carried out in the Birmingham area under the general direction of Dr John Edwards. So far we have been able to examine the alkaline phosphatase types in 260 placentae from 130 dizygotic twin pairs. The findings are summarized in table 10. In 82 pairs the alkaline phosphatase phenotypes were the same, and in 48 pairs they were different. This result excludes the possibility that these placental phenotypes are determined by the maternal genotypes, because if this were so none of the pairs should have shown any differences.

If the phenotypes depend on the foetal genotype then one may test the hypothesis by calculating the expected incidence of the different sorts of sib pair using the gene frequencies previously obtained. This is shown in table 10, and it will be seen that there is quite good agreement between the numbers of the different sorts of twin pair observed, and those expected according to the hypothesis.

Thus the evidence strongly suggests that these placental alkaline phosphatase phenotypes are determined by the foetal genotype and that at least three autosomal alleles are concerned. The biological significance of the polymorphism is still quite obscure, but it might well be of importance in problems concerned with maternal–foetal interaction. It may also perhaps be worth considering other enzyme polymorphisms from this point of view. The phosphoglucomutase polymorphism, for example, can also be readily demonstrated in the placenta and here again it has been shown that the placental phenotype is determined by the foetal and not the maternal genotype.

Discussion

Although this work is still in its early stages, an interesting and perhaps in some ways an unexpected picture of enzyme variation in human populations is beginning to emerge.

In the course of examining some ten arbitrarily chosen enzymes, in none of which we had any particular reason to expect any degree of variation, and not all of which have been examined in great detail or by perhaps the most suitable methods, we have come across three quite striking examples of enzyme polymorphism. Although one can hardly draw firm numerical conclusions from such a small series, it seems likely, unless we have been excessively lucky in our choice of enzymes, that polymorphism to a similar degree may be a fairly common phenomenon among the very large number of enzymes that occur in the human organism.

Some idea of how extensive this diversity might be can be obtained by considering together the various enzymes which have been shown to exhibit some degree of polymorphism in our own population. Relevant data on seven such enzymes are given in table 11. For the present purpose only those variations where two or more allelic genes have been found to have frequencies greater than 0·01 have been included. In the case of one enzyme, serum cholinesterase, variation at two different loci fall into this category, so that eight loci are represented in all. Of these at least six can be regarded as 'structural' loci since the variation produced appears to involve qualitative differences. In the other two cases (serum cholinesterase E_2 and acetyl transferase) only quantitative differences in enzyme level have so far been identified, but it is possible that these may also reflect structural differences in the enzyme protein present.

Table 11. Enzyme polymorphism in the English population

enzyme	number of alleles with frequency greater than 0·01	frequency of commonest phenotype	probability of two randomly selected individuals being of the same phenotype	reference
red cell acid phosphatase	3	0·43	0·34	Hopkinson, Spencer & Harris (1963)
phosphoglucomutase	2	0·58	0·47	Spencer, Hopkinson & Harris (1964*b*)
placental alkaline phosphatase	3	0·41	0·31	Boyer (1961) Robson & Harris (1965)
acetyl transferase	2	0·50	0·50	Price Evans & White (1964)
adenylate kinase	2	0·90	0·82	Fildes & Harris (1965)
serum cholinesterase				
locus E_1	2	0·96	0·92	Kalow & Staron (1957)
locus E_2	2	0·90	0·82	Harris, Hopkinson, Robson & Whittaker (1963)
6-phosphogluconate dehydrogenase	2	0·96	0·92	Fildes & Parr (1963)
combined	—	0·037	0·014	—

Each of these variations occurs independently of the others so that quite a large number of different phenotypic combinations may be found in the general population. Indeed, the commonest of these will occur in less than 4% of people and the probability that two randomly selected individuals would be found to have the same combination of phenotypes is less than 1 in 70. Thus just taking into account this very limited series of examples, quite a high degree of individual differentiation in enzymic make-up is demonstrable. and it is of interest that most of this is probably attributable to variation in enzyme structure.

These different polymorphisms pose a variety of intriguing problems both in biochemistry and in genetics. One would like to know, for example, what is the precise nature of the structural differences between the variant forms of a given enzyme, and whether these are reflected in kinetic differences and in differences in functional activity. It is notable that several of these enzymes apparently occur in multiple molecular forms even in homozygous individuals. The recognition of these so-called isoenzyme systems is a fairly new development in enzymology and the further investigation of these particular examples, and of their variant forms, may well help to throw light on the general biological significance of this phenomenon.

One would also like to know why these different enzyme phenotypes occur with the particular frequencies that we observe, and why, as is the case, for example, with red cell acid phosphatase and placental alkaline phosphatase, the gene frequencies may vary quite widely from one population to another. Presumably selective differences are important here but at present we have virtually no idea what these might be. However, one may reasonably hope that, if the metabolic and functional differences which presumably derive from the various enzyme differences can be elucidated, this may provide us with some indication of what selective factors may be important.

References (Harris)

Boyer, S. H. 1961 Alkaline phosphatase in human sera and placentae. *Science*, **134**, 1002.

Fildes, R. A. & Harris, H. 1965 Genetically determined variation of adenylate kinase in man. (In preparation.)

Fildes, R. A. & Parr, C. W. 1963 Human red cell phosphogluconate dehydrogenases. *Nature, Lond.*, **200**, 890.

Harris, H. 1963 The 'inborn errors' today. In *Garrod's Inborn errors of metabolism*. Oxford University)Press.

Harris, H., Hopkinson, D. A., Robson, E. B. & Whittaker, M. 1963 Genetical studies on a new variant of serum cholinesterase detected by electrophoresis. *Ann. Hum. Genet., Lond.* **26**, 359.

Hopkinson, D. A., Spencer, N. & Harris, H. 1963 Red cell acid phosphatase variants: a new human polymorphism. *Nature, Lond.* **199**, 969.

Hopkinson, D. A., Spencer, N. & Harris, H. 1964 Genetical studies on human red cell acid phosphatase. *Amer. J. Hum. Genet.* **16**, 141.

Kalow, W. & Staron, N. 1957 On the distribution and inheritance of atypical forms of human serum cholinesterase as indicated by dibucaine numbers. *Canad. J. Biochem. Physiol.* **35**, 1305.

Lai, L., Nevo, S. & Steinberg, A. G. 1964 Acid phosphatases of human red cells: predicted phenotype conforms to a genetic hypothesis. *Science*, **145**, 1187.

Motulsky, A. 1964 Hereditary red cell traits and malaria. *Amer. J. Trop. Med. Hyg.* **13**, 147.

Price Evans, D. A. & White, T. A. 1964 Human acetylation polymorphism. *J. Lab. Clin. Med.* **63**, 394.

Robson, E. B. & Harris, H. 1965 Genetics of the alkaline phosphatase polymorphism in the human placenta. *Nature, Lond.* **207**, 1257.

Spencer, N., Hopkinson, D. A. & Harris, H. 1964*a* Quantitative differences and gene dosage in the human red cell acid phosphatase polymorphism. *Nature, Lond.* **201**, 299.

Spencer, N., Hopkinson, D. A. & Harris, H. 1964*b* Phosphoglucomutase polymorphism in man. *Nature, Lond.* **204**, 742.

Evolution of the Structure of Ferredoxin Based on Living Relics of Primitive Amino Acid Sequences

Abstract. *The structure of present-day ferredoxin, with its simple, inorganic active site and its functions basic to photon-energy utilization, suggests the incorporation of its prototype into metabolism very early during biochemical evolution, even before complex proteins and the complete modern genetic code existed. The information in the amino acid sequence of ferredoxin enables us to propose a detailed reconstruction of its evolutionary history. Ferredoxin has evolved by doubling a shorter protein, which may have contained only eight of the simplest amino acids. This shorter ancestor in turn developed from a repeating sequence of the amino acids alanine, aspartic acid or proline, serine, and glycine. We explain the persistence of living relics of this primordial structure by invoking a conservative principle in evolutionary biochemistry: The processes of natural selection severely inhibit any change in a well-adapted system on which several other essential components depend.*

Many of the principles of organic evolution have long been known and are productively used in the organization of biological concepts, but are seldom used to full advantage in biochemistry. In nature, biochemistry is included in biology. An organism is a functioning system composed of the structures, organs, tissues, and organelles of classical biology. These in turn are composed of metabolites, macromolecules, enzyme aggregates, and biochemical feedback systems. Biochemical details concerning these components have only recently become accessible. Potentially, a much greater amount of information relevant to evolution is available in biochemistry than in classical biology.

According to evolutionary theory, each structure or function of an organism is subject to occasional changes or mutations, but the infrequency of these mutations necessitates that they will almost always occur, and be selected for, one at a time. Each change or addition must be an improvement, or at least not too severe a disadvantage, in order that the processes of natural selection permit its survival. This limitation has a very conservative effect. If its ecological niche stays the same, a well-adapted organism strongly resists change. Thus we find familiar-looking fossil shells a third of a billion years old. If its niche changes, new functions evolve, but the most primitive structures tend to remain unchanged, since these older components have already come to be relied upon by several later additions. Any change in a very old component, even though it might be advantageous in some way, would coincidentally disturb so many other things that it would almost always be extremely disadvantageous to the organism. This conservatism is well illustrated in the amino acid sequences of proteins. For example, we can compare the amino acid sequences of cytochrome *c* from yeast (*1*) and from horse (*2*), position by position. In 64 of the 104 positions the amino acids in the two chains are identical. Between horse and human cytochrome *c* (*3*) there are only 12 amino acid differences.

When we consider evolution retrospectively, the constraints are even more severe. One basic evolutionary principle is that every living organism or structure or function had ancestors very similar to itself, but simpler. (This is true even if it had more complex *immediate* ancestors.) In a particular case there are generally only a few plausible slightly simpler ancestors. As we trace the changes in a structure or function back through time, we must bear in mind that *all* of the structures and functions of the cell may be simpler. We are then dealing with primitive components ancestral to those seen today.

The amino acid sequence of ferredoxin from *Clostridium pasteurianum*, a nonphotosynthetic anaerobic bacterium, has been reported (*4*). This protein seems to have arisen at an earlier times than many others which have been studied. We draw this inference from the following considerations.

1) Ferredoxin occurs in primitive anaerobic organisms, both photosynthetic and nonphotosynthetic (*5*). It must have been present in simpler organisms, the extinct common ancestors of these.

2) Ferredoxin contains iron and sulfur, bonded to the protein at its active site (*6*). Ferrous sulfide, FeS, is a widely dispersed mineral, a catalyst which would have been readily available to the most primitive organism.

3) The functions of ferredoxin are basic to cell chemistry. The reduction of ferredoxin is the key photochemical event in photosynthesis by chloroplasts (*5*). All the energy is channeled through this compound to other cellular energy-storage mechanisms. Ferredoxin is the most highly reducing stable compound so far found in the cell, having a reducing potential near that of molecular hydrogen (*5*). This suggests that its function may have evolved at a very early time when the earth's atmosphere was still strongly reducing. It reduces nicotinamide adenine dinucleotide (NAD) (*5*), a ubiquitous reducing agent in the cell. Therefore, it may be even more primitive than NAD. It catalyzes adenosine triphosphate (ATP) formation by radiation (*5*). This indicates possible relation to primitive energy transfer processes. It catalyzes the synthesis of pyruvate from carbon dioxide and acetylcoenzyme-A (*5*). This indicates its involvement with one of the simplest, most primitive synthetic processes in intermediary metabolism, the fixation of CO_2. It participates in nitrogen fixation (*7*) and hydrogenase-linked reactions (*5*).

4) Ferredoxin contains an unusually high proportion of the smaller, more thermodynamically stable (*8*) amino acids, such as glycine, alanine, cysteine, serine, and aspartic acid. Furthermore, their synthesis from inorganic substrates by autotrophic organisms requires only a small number of endothermic steps.

5) Ferredoxin is smaller than most other enzymes, having only 55 amino acid units. It appears to have some sort of repeating structure, so that it may once have been still smaller.

Let us now consider the amino acid sequence of ferredoxin. Applying the constraints imposed by the principles of evolution, can we find traces of its ancestry?

Figure 1 shows the reported sequence of ferredoxin (*4*) and some manipulations with it. For the purposes of our study, a single-letter notation (*9*) is much more suitable than the usual three-letter notation (row 1). If links 30 to 55 (row 3) are placed under links 1 to 29 (row 2), it is evident that the number of coincidences (row 4) far exceeds that which would be expected by chance ($P << .001$). It appears that this protein has evolved by doubling of the nucleic acid (gene) which determines it. Smithies, Connell, and Dixon showed how nonhomologous crossing over of chromosomes could produce this effect, and proposed it as the explanation of the apparent doubling which they detected in haptoglobin molecules (*10*). Because of the very high statistical improbability of the chance occurrence of the twelve coincidences, we consider for this study that the two halves of the ferredoxin sequence are in fact homologous, and attempt to decipher some details about their common ancestor. Presumably the ancestral sequence contained all those amino acid units which are common to both parts. Where the units are different, probably one of the two was present originally.

The ancestral sequence of 29 units must itself have had a simpler ancestor. An attempt was made to discover this by the same process, but no further regularities could be found by superimposing quarters of the chain.

Another kind of simplicity would be that the ancestor had fewer kinds of amino acids. During the evolution of the genetic code there must have been a time when the genetic mechanism could discriminate fewer amino acids. Perhaps those which coincide in the two halves of the ferredoxin sequence (glycine, cysteine, alanine, proline, valine, aspartic acid, and glutamine) may be survivors of this early stage. And perhaps when any of these occurs in either half, it is also likely to be a survivor of the earlier stage. Therefore, we record the positions and occurrence of these seven amino acids or serine (see below) in the paired ferredoxin sequences (row 5 of Fig. 1). In this arrangement, two regular patterns emerge. Cysteines occur in a cycle of three, and alanines occur in a cycle of four. A discontinuity in both patterns occurs at the midpoint, position 15. None of the other amino acids confirms a cycle of three. However, the cycle of four is confirmed by glycine, proline, aspartic acid, and serine. A repeating sequence of four amino acids (with a break at position 15) has been written in row 6. The combined halves agree with this cycle in 13 positions (row 7); they disagree in only four positions (row 8). Altogether 17 occurrences of these four simple amino acids in the living ferredoxin chain agree with this cycle, and five disagree with it. Figure 2 shows the reported sequence rewritten in groups of four for direct comparison with the repeating pattern.

The probabilities involved in a pattern of this kind are difficult to compute, because the pattern was discovered rather than predicted. Some sort of pattern can always be found if random data are examined in fine detail. In principle, one should modify the computed probability to reflect all the other patterns which would have been considered equally good if they had been discovered. This number is difficult to

```
1               5                   10                  15                  20                  25
Ala.Tyr.Lys.Ilu.Ala.Asp.Ser.Cys.Val.Ser.Cys.Gly.Ala.Cys.Ala.Ser.Glu.Cys.Pro.Val.Asn.Ala.Ilu.Ser.Gln.Gly.Asp.Ser.Ilu.

30                  35                  40                  45                  50                  55
Phe.Val.Ilu.Asp.Ala.Asp.Thr.Cys.Ilu.Asp.Cys.Gly.Asn.Cys.Ala.Asn.Val.Cys.Pro.Val.Gly.Ala.Pro.Val.Gln.Glu
```

```
    1       5         10        15        20        25        30        35        40        45        50        55
1.  A O K I A D S C V S C G A C A S E C P V N A I S Q G D S I F V I D A D T C I D C G N C A N V C P V G A P V Q E
                              1       5         10        15        20        25
2.                            A O K I A D S C V S C G A C A S E C P V N A I S Q G D S I

3.                            F V I D A D T C I D C G N C A N V C P V G A P V Q E

4.                                    A D   C     C G   C A     C P V   A     Q

5.                            A V   D A D S C V D C G A C A S V C P V G A P S Q G D S
                                              S                           V
6.                            A D S G A D S G A D S G A D D S G A D S G A D S G A D S G

7.                            A       A D S     D   G A     S         G A   S     D S

8.                                  D           S         A                     G
```

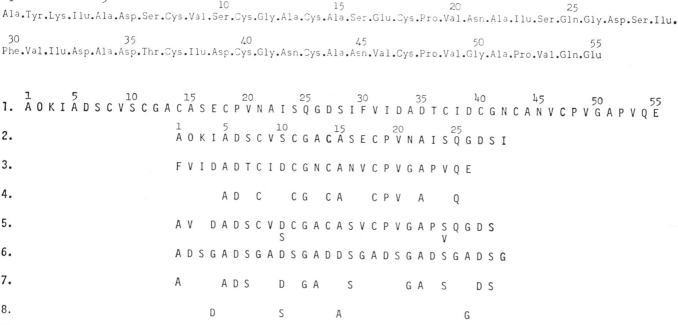

Fig. 1. Evidence of the primitive ancestry of ferredoxin. At the top the amino acid sequence of ferredoxin from *Clostridium pasteurianum* is shown in conventional notation (*4*). Row 1: The sequence is translated into a one-letter code (*9*), a notation more suitable for this type of study. Rows 2 and 3: The two halves of the chain compared by alignment. Row 4: Twelve amino acids which are identical in both halves. If the sequences were unrelated, one would expect only two or three such identities. The probability of finding as many coincidences as this by chance is negligible. Row 5: The same seven simple amino acids found in row 4, plus serine, whenever they occur in *either* chain. Row 6: A simple repeating sequence of four amino acids (with a discontinuity at position 15) from which row 5 appears to be derived. Row 7: Thirteen amino acids from row 5 which conform to row 6. Row 8: The four amino acids which do not conform to the cyclic pattern. The chance probability of as many coincidences as this would be very small if the pattern in row 6 had been independently given. Since it was derived from this study itself, the coincidence is less extreme, but still seems to be good evidence for the validity of the cyclic pattern.

Position				
1	A	*O*	*K*	*I*
30	*F*	*V*	*I*	D
5	A	D	S	*C*
34	A	D	*T*	*C*
9	*V*	S	*C*	G
38	*I*	D	*C*	G
13	A	*C*		
42	*N*	*C*		
15		A	S	*E*
44		A	*N*	*V*
18	*C*	P	*V*	*N*
47	*C*	P	*V*	G
22	A	*I*	S	*Q*
51	A	P	*V*	*Q*
26	G	D	S	*I*
55	*E*			
Cycle	A	D / P	S	G

Fig. 2. Evidence for a cycle of four in ferredoxin. The two half-chains, rows 2 and 3 in Fig. 1, are written in successive groups of four, with a break at positions 15 and 44. Alanine (A) occurs mainly in the first column, D and P in the second, S in the third, and G in the fourth. Exceptions are underlined. All other amino acids are shown in italics. This good fit appears unlikely to be due to chance, but a numerical evaluation of the probability would involve several arbitrary assumptions.

determine and must be somewhat arbitrary. However, in this case it does not seem to be very large, and the number of coincidences still seems beyond ordinary chance. We consider the observed number of coincidences to be good evidence for the pattern of a cycle of four.

Using a computer program, we have matched the sequence of ferredoxin against itself in all combinations, and against the various possible cycles. The result of this objective method is the same as found by inspection. Only cysteines occur in a cycle of three. Alanine, aspartic acid, serine, and glycine agree with a cycle of four; and proline in three places occupies the position of aspartic acid. This substitution is reasonable stereochemically, since the two side-chains have a very similar conformation, when aspartic acid folds into a hydrogen-bonded intramolecular ring. Aspartic acid is metabolically simpler to synthesize than proline, and therefore seems likely to have evolved earlier.

We will now use these patterns and the rules of evolution to reconstruct the history of the ferredoxin molecule. We first consider the prosthetic group, or inorganic part, in a prebiological environment. Then starting with an extremely primitive living system, we follow the development of the increasing intricacy of the protein.

At chemical equilibrium in an ideal gas mixture of a reducing nature at standard temperature and pressure, H_2S, CH_4, CO_2, H_2O, and N_2 predominate; ammonia and organic compounds occur in small amounts (*11*). At equilibrium, FeS is stable in this gas mixture. Possibly life may have organized about such a stable inorganic catalyst, one that could participate in capturing photons and in directing energy toward the reduction and fixation of CO_2 and toward the synthesis of pyrophosphate. In this photon-activated system, other, less stable catalysts may then have been synthesized, permitting development of more complex systems.

Regardless of how life originated, there was at one time a very primitive organism, far simpler than any known to be living today. It was capable of making and polymerizing some of the simplest amino acids and nucleotides. Perhaps the nucleic acids were made of only two nucleotides, simpler than the ones which occur in the present genetic mechanism. A variety of such polymers could be formed having useful structural or catalytic functions, for which natural selection preserved them.

One sequence of 12 nucleotides doubled and redoubled itself, making a longer, repetitive chain.

At about the same time, the primitive amino-acid-polymerizing mechanism of the cell began to utilize this nucleic acid chain as a template, and it coded for the amino acids alanine, aspartic acid, serine, and glycine. If these events occurred when the nucleotide chain was still only 12 units long, the resulting peptide would be A D S G, as in Fig. 3, row 1. After the nucleotide chain became longer, it coded for a simple repeating protein, A D S G A D S G . . . (row 2). This protein had some advantageous, perhaps structural, function in the cell, unrelated to the present energy-transfer function of ferredoxin. An aberration in the nucleotide sequence produced a break in the cycle (row 2, underlined).

The synthesizing abilities of the organism became more versatile and more efficient. The genetic code became more complex, so that the genetic mechanism was able to incorporate other amino acids. Mutations occurred which modified and complicated the particular amino acid sequence which we are following (row 3).

1. A D S G

2. A D S G A D S G A D S G A D D S G A D S G A D S G A D S G

3. A D S D A D S C V D C G A C A S V C P V G A P S Q G D S G

4. A D S D A D S C V D C G A C A S V C P V G A P S Q G D S G A D S D A D S C V D C G A C A S V C P V G A P S Q G **D S G**

5. A O K I A D S C V S C G A C A S E C P V N A I S Q G D S I F V I D A D T C I D C G N C A N V C P V G A P V Q E

Fig. 3. Proposed origin and evolution of ferredoxin (see text for fuller details). Row 1: Originally, in an extremely primitive organism, a short sequence of four of the simplest amino acids (alanine, aspartic acid, serine, and glycine) could be produced. Row 2: This sequence lengthened by doubling of the genetic material, and one discontinuity occurred (underlined). Row 3: The genetic code becoming more versatile, mutations (underlined) occurred, but only to relatively simple amino acids (the same four, plus cysteine, valine, proline, and glutamine). Iron sulfide was attached to the cysteines, which constituted the "active site" of the respiratory function of this primitive ferredoxin. This configuration still persists. Row 4: By "chromosome" aberration, the whole chain doubled. Row 5: The present more intricate genetic code having evolved, further mutations (underlined) to more complex amino acids occurred. The last three links were deleted. The result was the present sequence of ferredoxin from *C. pasteurianum* (*4*).

Cysteine was among these new amino acids added to our sequence. The sulfide bond of the cysteine unit became attached to iron sulfide. The protein thus cooperated in the photon-coupled catalytic function, which it still retains, and became protoferredoxin. On the principle that evolution proceeds one step at a time, we assume that the cell was already using iron sulfide as a catalyst, probably attached to cysteine alone, or to some peptide less suitable than protoferredoxin. This new attachment would merely have increased the efficiency of this function.

Eventually four cysteines were added by mutation, and two identical chains combined to make an intricate protein-iron-sulfide complex of greatly increased efficiency. It still retains essentially this structure.

The nucleic acid doubled in length by a process that was the prototype of a chromosome aberration, resulting in a protein of 58 units (row 4). In the three-dimensional structure, the effect of this change was to attach the two shorter chains end-to-end. They must already have been in a configuration which was only moderately disturbed by this new constraint. The attachment was an improvement but not a radical change. We predict that when the three-dimensional structure of ferredoxin is worked out, evidence will be found for the previous stage, with its two identical, cooperating, shorter chains. The three end units may have been lost at this time or later, to give the present total length of 55 units.

Many functions of the cell improved in efficiency and complexity, evolving new capabilities. The genetic mechanism also evolved and became capable of incorporating additional amino acids. Mutations occurred and were selected, each of which made a slight improvement in the overall function, until the present sequence was produced (row 5).

By this time there were many lines of descent in the phylogenetic tree, and different species must have produced different mutational variations on the earlier sequences. Comparison of amino acid sequences of many ferredoxins from diverse species should produce clear "living-fossil" evidence of the earliest stages of protein evolution.

At any of the stages, other aberrations may have occurred, resulting in additional duplication and separation of the genetic material, followed by mutations. This would create new genes, producing other proteins, which today may have varying degrees of similarity to ferredoxin.

Genes and their corresponding proteins have not only become more numerous but they have also become longer. Duplication of nucleic acids such as that inferred here in the ferredoxin gene may have been a major means of accomplishing this increase in length. If so, we may expect to see evidences of duplication in other protein sequences, when ways of recognizing distant homologous relationships become more precise than the mere counting of the few identical amino acids remaining. The diheme peptide of *Chromatium* may possibly be such a case (*12*).

Just as the salt composition of our tissue fluids is supposed to represent a stabilized sample of ancient sea water, so the simplest, metabolically most ancient components of cellular metabolism preserve some aspects of their original environment. In modern organisms, primitive reactions, such as those involving glutathione or coenzyme-A, operate under their primordial reducing conditions, isolated from the harsh outer environment by later adaptations. Such ancient systems are extremely conservative, because so many diverse later reactions have become intricately dependent on them that they are no longer "free" to evolve. A mutational change which might be beneficial in one way, in almost every case would be a strong disadvantage in many other ways. When such a mutation occurred, the process of natural selection would therefore reject it. This conservative principle enables us to comprehend why ferredoxin from a living organism could still retain detectable details of its ancient origin.

Thus, in organisms still living there may exist biochemical relics of the era encompassing the origin and evolution of the genetic mechanism. Determination of the sequences of proteins such as ferredoxin and of nucleic acids such as transfer RNA, whose prototypes must have functioned at this early time, should make possible a detailed reconstruction of the biochemical evolutionary events of this era.

RICHARD V. ECK
MARGARET O. DAYHOFF
National Biomedical Research Foundation, 8600 16th Street, Silver Spring, Maryland 20910

References and Notes

1. K. Narita, K. Titani, Y. Yaoi, H. Murakami, *Biochim. Biophys. Acta* **77**, 688 (1963).
2. E. Margoliash, E. Smith, G. Kreil, H. Tuppy, *Nature* **192**, 1121 (1961).
3. H. Matsubara and E. Smith. *J. Biol. Chem.* **237**, 3575 (1962).
4. M. Tanaka, T. Nakashima, A. Benson, H. F. Mower, K. T. Yasunobu, *Biochem. Biophys. Res. Commun.* **16**, 422 (1964).
5. D. I. Arnon, *Science* **149**, 1460 (1965).
6. W. J. Lovenberg, *Biol. Chem.* **238**, 3899 (1963).
7. L. E. Mortenson, *Proc. Nat. Acad. Sci. U.S.* **52**, 272 (1964).
8. All amino acids are highly unstable and decompose into CO_2, H_2O, CH_4, N_2, H_2 (and H_2S), in a reducing atmosphere, given a suitable catalyst. At thermodynamic equilibrium the smaller amino acids have a relatively higher concentration than the more complex ones. For example, in an ideal gas system containing C, H, O, N, and S in the proportions 20 : 50 : 30 : 400 : 1, glycine is present in 10^{-22} mole fraction. Others, in decreasing order of concentration are alanine, cysteine, serine, aspartic acid, and valine. The 14 other coded amino acids have concentrations of less than 10^{-30} mole fraction. These amounts are too low to be significant for the organization of living systems, 10^{-22} representing less than one molecule per droplet. The proportions, however, give some indication of the relative ease with which the amino acids might be made in a very simple, primitive system (M. O. Dayhoff, R. V. Eck, E. R. Lippincott, G. Nagarajan, in preparation).
9. A, alanine; C, cysteine; D, aspartic acid; E, glutamic acid; F, phenylalanine; G, glycine; H, histidine; I, isoleucine; K, lysine; L, leucine; M, methionine; N, asparagine; O, tyrosine; P, proline; Q, glutamine; R, arginine; S, serine; T, threonine; V, valine; W, tryptophan.
10. O. Smithies, G. E. Connell, G. H. Dixon, *Nature* **196**, 232 (1962).
11. M. O. Dayhoff, E. R. Lippincott, R. V. Eck, *Science* **146**, 1461 (1964).
12. K. Dus, R. G. Bartsch, M. D. Kamen, *J. Biol. Chem.* **237**, 3083 (1962); C. J. Epstein, and A. G. Motulsky, *Progr. Med. Genet.* **4**, 85 (1965).
13. This work was supported by NIH grants Nos. GM-08710 and GM-12168, and NASA contract 21-003-002.

21 February 1966

Evolutionary Rate at the Molecular Level

Calculating the rate of evolution in terms of nucleotide substitutions seems to give a value so high that many of the mutations involved must be almost neutral ones

by

MOTOO KIMURA

National Institute of Genetics, Mishima, Japan

COMPARATIVE studies of haemoglobin molecules among different groups of animals suggest that, during the evolutionary history of mammals, amino-acid substitution has taken place roughly at the rate of one amino-acid change in 10^7 yr for a chain consisting of some 140 amino-acids. For example, by comparing the α and β chains of man with those of horse, pig, cattle and rabbit, the figure of one amino-acid change in 7×10^6 yr was obtained[1]. This is roughly equivalent to the rate of one amino-acid substitution in 10^7 yr for a chain consisting of 100 amino-acids.

A comparable value has been derived from the study of the haemoglobin of primates[2]. The rate of amino-acid substitution calculated by comparing mammalian and avian cytochrome *c* (consisting of about 100 amino-acids) turned out to be one replacement in 45×10^6 yr (ref. 3). Also by comparing the amino-acid composition of human triosephosphate dehydrogenase with that of rabbit and cattle[4], a figure of at least one amino-acid substitution for every $2{\cdot}7 \times 10^6$ yr can be obtained for the chain consisting of about 1,110 amino-acids. This figure is roughly equivalent to the rate of one amino-acid substitution in 30×10^6 yr for a chain consisting of 100 amino-acids. Averaging those figures for haemoglobin, cytochrome *c* and triosephosphate dehydrogenase gives an evolutionary rate of approximately one substitution in 28×10^6 yr for a polypeptide chain consisting of 100 amino-acids.

I intend to show that this evolutionary rate, although appearing to be very low for each polypeptide chain of a size of cytochrome *c*, actually amounts to a very high rate for the entire genome.

First, the DNA content in each nucleus is roughly the same among different species of mammals such as man, cattle and rat (see, for example, ref. 5). Furthermore, we note that the G–C content of DNA is fairly uniform among mammals, lying roughly within the range of 40–44 per cent[6]. These two facts suggest that nucleotide substitution played a principal part in mammalian evolution.

In the following calculation, I shall assume that the haploid chromosome complement comprises about 4×10^9 nucleotide pairs, which is the number estimated by Muller[7] from the DNA content of human sperm. Each amino-acid is coded by a nucleotide triplet (codon), and so a polypeptide chain of 100 amino-acids corresponds to 300 nucleotide pairs in a genome. Also, amino-acid replacement is the result of nucleotide replacement within a codon. Because roughly 20 per cent of nucleotide replacement caused by mutation is estimated to be synonymous[8], that is, it codes for the same amino-acid, one amino-acid replacement may correspond to about 1·2 base pair replacements in the genome. The average time taken for one base pair replacement within a genome is therefore

$$28 \times 10^6 \text{ yr} \div \left(\frac{4 \times 10^9}{300}\right) \div 1{\cdot}2 \doteqdot 1{\cdot}8 \text{ yr}$$

This means that in the evolutionary history of mammals, nucleotide substitution has been so fast that, on average, one nucleotide pair has been substituted in the population roughly every 2 yr.

This figure is in sharp contrast to Haldane's well known estimate[9] that, in horotelic evolution (standard rate evolution), a new allele may be substituted in a population roughly every 300 generations. He arrived at this figure by assuming that the cost of natural selection per generation (the substitutional load in my terminology[10]) is roughly 0·1, while the total cost for one allelic substitution is about 30. Actually, the calculation of the cost based on Haldane's formula shows that if new alleles produced by nucleotide replacement are substituted in a population at the rate of one substitution every 2 yr, then the substitutional load becomes so large that no mammalian species could tolerate it.

Thus the very high rate of nucleotide substitution which I have calculated can only be reconciled with the limit set by the substitutional load by assuming that most mutations produced by nucleotide replacement are almost neutral in natural selection. It can be shown that in a population of effective size N_e, if the selective advantage of the new allele over the pre-existing alleles is s, then, assuming no dominance, the total load for one gene substitution is

$$L(p) = 2\left\{\frac{1}{u(p)} - 1\right\} \int_0^{4Sp} \frac{e^y - 1}{y}\,dy - 2e^{-4S}\int_{4Sp}^{4S} \frac{e^y}{y}\,dy + 2\log_e\left(\frac{1}{p}\right) \qquad (1)$$

where $S = N_e s$ and p is the frequency of the new allele at the start. The derivation of the foregoing formula will be published elsewhere. In the expression given here $u(p)$ is the probability of fixation given by[11]

$$u(p) = (1 - e^{-4Sp})/(1 - e^{-4S}) \qquad (2)$$

Now, in the special case of $|2N_e s| \ll 1$, formulae (1) and (2) reduce to

$$L(p) = 8N_e s \log_e(1/p) \qquad (1')$$

$$u(p) = p + 2N_e sp(1 - p) \qquad (2')$$

Formula (1′) shows that for a nearly neutral mutation the substitutional load can be very low and there will be no limit to the rate of gene substitution in evolution. Furthermore, for such a mutant gene, the probability of fixation (that is, the probability by which it will be established in the population) is roughly equal to its initial frequency as shown by equation (2′). This means that new alleles may be produced at the same rate per individual as they are substituted in the population in evolution.

This brings the rather surprising conclusion that in mammals neutral (or nearly neutral) mutations are occurring at the rate of roughly 0·5 per yr per gamete. Thus, if we take the average length of one generation in the history of mammalian evolution as 4 yr, the mutation rate per generation for neutral mutations amounts to roughly two per gamete and four per zygote (5×10^{-10} per nucleotide site per generation).

Such a high rate of neutral mutations is perhaps not surprising, for Mukai[12] has demonstrated that in *Drosophila* the total mutation rate for "viability polygenes" which on the average depress the fitness by about 2 per cent reaches at least some 35 per cent per gamete. This is a much higher rate than previously considered. The fact that neutral or nearly neutral mutations are occurring at a rather high rate is compatible with the high frequency of heterozygous loci that has been observed recently by studying protein polymorphism in human and *Drosophila* populations[13–15].

Lewontin and Hubby[15] estimated that in natural populations of *Drosophila pseudoobscura* an average of about 12 per cent of loci in each individual is heterozygous. The corresponding heterozygosity with respect to nucleotide sequence should be much higher. The chemical structure of enzymes used in this study does not seem to be known at present, but in the typical case of esterase-5 the molecular weight was estimated to be about 10^5 by Narise and Hubby[16]. In higher organisms, enzymes with molecular weight of this magnitude seem to be common and usually they are "multimers"[17]. So, if we assume that each of those enzymes comprises on the average some 1,000 amino-acids (corresponding to molecular weight of some 120,000), the mutation rate for the corresponding genetic site (consisting of about 3,000 nucleotide pairs) is

$$u = 3 \times 10^3 \times 5 \times 10^{-10} = 1{\cdot}5 \times 10^{-6}$$

per generation. The entire genome could produce more than a million of such enzymes.

In applying this value of u to *Drosophila* it must be noted that the mutation rate per nucleotide pair per generation can differ in man and *Drosophila*. There is some evidence that with respect to the definitely deleterious effects of gene mutation, the rate of mutation per nucleotide pair per generation is roughly ten times as high in *Drosophila* as in man[18,19]. This means that the corresponding mutation rate for *Drosophila* should be $u = 1{\cdot}5 \times 10^{-5}$ rather than $u = 1{\cdot}5 \times 10^{-6}$. Another consideration allows us to suppose that $u = 1{\cdot}5 \times 10^{-5}$ is probably appropriate for the neutral mutation rate of a cistron in *Drosophila*. If we assume that the frequency of occurrence of neutral mutations is about one per genome per generation (that is, they are roughly two to three times more frequent than the mutation of the viability polygenes), the mutation rate per nucleotide pair per generation is $1/(2 \times 10^8)$, because the DNA content per genome in *Drosophila* is about one-twentieth of that of man[20]. For a cistron consisting of 3,000 nucleotide pairs, this amounts to $u = 1{\cdot}5 \times 10^{-5}$.

Kimura and Crow[21] have shown that for neutral mutations the probability that an individual is homozygous is $1/(4N_e u + 1)$, where N_e is the effective population number, so that the probability that an individual is heterozygous is $H_e = 4N_e u/(4N_e u + 1)$. In order to attain at least $H_e = 0{\cdot}12$, it is necessary that at least $N_e = 2{,}300$. For a higher heterozygosity such as $H_e = 0{\cdot}35$, N_e has to be about 9,000. This might be a little too large for the effective number in *Drosophila*, but with migration between subgroups, heterozygosity of 35 per cent may be attained even if N_e is much smaller for each subgroup.

We return to the problem of total mutation rate. From a consideration of the average energy of hydrogen bonds and also from the information on mutation of *rIIA* gene in phage T_4, Watson[22] obtained $10^{-8} \sim 10^{-9}$ as the average probability of error in the insertion of a new nucleotide during DNA replication. Because in man the number of cell divisions along the germ line from the fertilized egg to a gamete is roughly 50, the rate of mutation resulting from base replacement according to these figures may be $50 \times 10^{-8} \sim 50 \times 10^{-9}$ per nucleotide pair per generation. Thus, with 4×10^9 nucleotide pairs, the total number of mutations resulting from base replacement may amount to $200 \sim 2{,}000$. This is 100–1,000 times larger than the estimate of 2 per generation and suggests that the mutation rate per nucleotide pair is reduced during evolution by natural selection[18,19].

Finally, if my chief conclusion is correct, and if the neutral or nearly neutral mutation is being produced in each generation at a much higher rate than has been considered before, then we must recognize the great importance of random genetic drift due to finite population number[23] in forming the genetic structure of biological populations. The significance of random genetic drift has been deprecated during the past decade. This attitude has been influenced by the opinion that almost no mutations are neutral, and also that the number of individuals forming a species is usually so large that random sampling of gametes should be negligible in determining the course of evolution, except possibly through the "founder principle"[24]. To emphasize the founder principle but deny the importance of random genetic drift due to finite population number is, in my opinion, rather similar to assuming a great flood to explain the formation of deep valleys but rejecting a gradual but long lasting process of erosion by water as insufficient to produce such a result.

Received December 18, 1967.

1 Zuckerkandl, E., and Pauling, L., in *Evolving Genes and Proteins* (edit. by Bryson, V., and Vogel, H. J.), 97 (Academic Press, New York, 1965).

2 Buettner-Janusch, J., and Hill, R. L., in *Evolving Genes and Proteins* (edit. by Bryson, V., and Vogel, H. J.), 167 (Academic Press, New York, 1965).

3 Margoliash, E., and Smith, E. L., in *Evolving Genes and Proteins* (edit. by Bryson, V., and Vogel, H. J.), 221 (Academic Press, New York, 1965).

4 Kaplan, N. O., in *Evolving Genes and Proteins* (edit. by Bryson, V., and Vogel, H. J.), 243 (Academic Press, New York, 1965).

5 Sager, R., and Ryan, F. J., *Cell Heredity* (John Wiley and Sons, New York, 1961).

6 Sueoka, N., *J. Mol. Biol.*, **3**, 31 (1961).

7 Muller, H. J., *Bull. Amer. Math. Soc.*, **64**, 137 (1958).

8 Kimura, M., *Genet. Res.* (in the press).

9 Haldane, J. B. S., *J. Genet.*, **55**, 511 (1957).

10 Kimura, M., *J. Genet.*, **57**, 21 (1960).

11 Kimura, M., *Ann. Math. Stat.*, **28**, 882 (1957).

12 Mukai, T., *Genetics*, **50**, 1 (1964).

13 Harris, H., *Proc. Roy. Soc.*, B, **164**, 298 (1966).

14 Hubby, J. L., and Lewontin, R. C., *Genetics*, **54**, 577 (1966).

15 Lewontin, R. C., and Hubby, J. L., *Genetics*, **54**, 595 (1966).

16 Narise, S., and Hubby, J. L., *Biochim. Biophys. Acta*, **122**, 281 (1966).

17 Fincham, J. R. S., *Genetic Complementation* (Benjamin, New York, 1966).

18 Muller, H. J., in *Heritage from Mendel* (edit. by Brink, R. A.), 419 (University of Wisconsin Press, Madison, 1967).

19 Kimura, M., *Genet. Res.*, **9**, 23 (1967).

20 *Report of the United Nations Scientific Committee on the Effects of Atomic Radiation* (New York, 1958).

21 Kimura, M., and Crow, J. F., *Genetics*, **49**, 725 (1964).

22 Watson, J. D., *Molecular Biology of the Gene* (Benjamin, New York, 1965).

23 Wright, S., *Genetics*, **16**, 97 (1931).

24 Mayr, E., *Animal Species and Evolution* (Harvard University Press, Cambridge, 1965).

Repeated Sequences in DNA

Hundreds of thousands of copies of DNA sequences have been incorporated into the genomes of higher organisms.

R. J. Britten and D. E. Kohne

The complementary structure of DNA plays a fundamental role in the cell. The complementary relations between nucleotide pairs are important not only in the duplication of DNA, but in the transcription and translation of genetic information. Matching of complementary nucleotide sequences is probably involved in genetic recombination as well as in other events of recognition and control within the cell.

It is a remarkable fact that separated complementary strands of *purified* DNA recognize each other. Under appropriate conditions they specifically reassociate (*1*). This phenomenon has supplied a useful tool for exploring the nature of molecular events within the cell and broader biological questions such as the relationships among species (*1–3*).

Simple complementary ribopolymer pairs were shown in 1957 to form a helical paired structure when mixed in solution (*4*). In 1960, DNA was dissociated into two strands, and the physical properties and biological activity of double-stranded DNA were then restored by incubation under appropriate conditions (*1, 5*). In 1961, virus-specific RNA, made by bacteria during viral infection, was shown to pair with the viral DNA (*6*). Techniques were developed for the immobilization of single-stranded DNA in cellulose (*7*), in agar (*8*), and on nitrocellulose filters (*9*). It then became possible to assay the reassociation of radioactively labeled single-stranded fragments of DNA or RNA with the immobilized DNA.

Reassociation of the DNA of vertebrates was observed in 1964 (*3*). The extent of reassociation between DNA strands derived from different species was shown to be a measure of the evolutionary relation between the species (*10*). However, measurements also showed that the nucleotide sequence pairing was imprecise even when DNA from a single species was reassociated (*11*).

Before these measurements were made it had been expected that it would be very difficult to observe the reassociation of the DNA of vertebrates and other higher organisms (*1*). The enormous dilution of individual nucleotide sequences in the large quantity of DNA in each cell was expected to make the reaction so slow that months would be required for its completion at practical concentrations with the DNA-agar method.

Investigation of this paradox was begun in our laboratory in early 1964, and shortly afterward the hypothesis was put forward (*12*) that some nucleotide sequences were frequently repeated in the DNA of vertebrates. This supposition was supported by the observation that 10 percent of the DNA of the mouse reassociated extremely rapidly. This fraction, identified as mouse-satellite DNA was shown by later measurements (*13*) to consist of a million copies of a short nucleotide sequence (*13*). Later work (*14, 15*) has shown that repeated nucleotide sequences are of very general occurrence.

In this article we describe selected measurements (*12–15*) that show most clearly the presence of repeated sequences and indicate some of their properties.

Conditions for Reassociation

The conditions for efficient reassociation of DNA were explored originally by Marmur *et al.* (*1*) and have since been studied in several laboratories (*15–17*). Briefly stated, the requirements are as follows. (i) There must be an adequate concentration of cations. Below 0.01*M* sodium ion, the reassociation reaction is effectively blocked. (ii) The temperature of incubation must be high enough to weaken intrastrand secondary structure. The optimum temperature for reassociation is about 25°C below the temperature required for dissociation of the resulting double strands. (iii) The incubation time and the DNA concentration must be sufficient to permit an adequate number of collisions so that the DNA can reassociate. (iv) The size of the DNA fragments also affects the rate of reassociation and is conveniently controlled if the DNA is "sheared" to small fragments (*18*). Thus, in order to achieve reproducible reassociation reactions the cation concentration, temperature of incubation, DNA concentration, and DNA fragment size must all be controlled (*19*).

The Measurement of Reassociation

Reassociation can be measured in a variety of ways, each depending on some easily detected physical difference between single-stranded (dissociated) DNA and double-stranded (reassociated) DNA (*1*). For example, dissociated DNA absorbs more ultraviolet light than reassociated DNA does. Double-strand DNA also has a greater

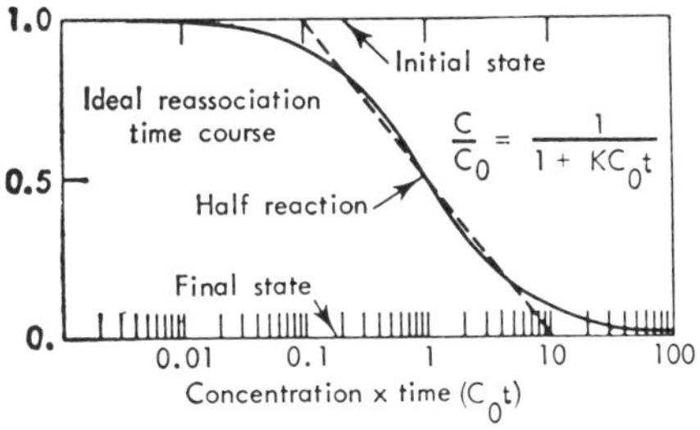

Fig. 1. Time course of an ideal, second-order reaction to illustrate the features of the log C_0t plot. The equation represents the fraction of DNA which remains single-stranded at any time after the initiation of the reaction. For this example, K is taken to be 1.0, and the fraction remaining single-stranded is plotted against the product of total concentration and time on a logarithmic scale.

The authors are staff members of the Department of Terrestrial Magnetism of the Carnegie Institution of Washington, Washington, D.C. 20015.

degree of optical activity than single-strand DNA.

In the DNA-agar method, reassociation is monitored by measuring the binding of labeled fragments of single-stranded DNA to long strands of DNA physically immobilized in a supporting substance. The immobilization prevents reassociation of the long DNA with itself. After incubation the unbound fragments are washed away, and the quantity of bound radioactive fragments is measured. It is now possible to measure the reassociation of DNA fragments with DNA immobilized on nitrocellulose filters (*20*). The rate of the reaction is markedly reduced compared to the rate in solution (*21*).

Another useful technique for measuring reassociation depends on the fact that double-stranded DNA can be separated from single-stranded DNA on a calcium phosphate (hydroxyapatite) column (*22*). Reassociation reactions can be followed by passing samples through hydroxyapatite and determining the amount of double-stranded DNA adhering to the column. This technique is particularly useful since DNA can be fractionated on a preparative scale on the basis of its ability to reassociate at a given C_0t, a parameter which may be explained as follows.

The Meaning of C_0t

Much of the evidence for repeated sequences depends on measurements of the rate of reassociation. In addition, the design of most reassociation experiments is strongly influenced by the time required to complete the process. The reassociation of a pair of complementary sequences results from their collision, and therefore the rate depends on their concentration. The product of the DNA concentration and the time of incubation is the controlling parameter for estimating the completion of a reaction. For convenience and simplification of language we have chosen to call this useful parameter C_0t, which is expressed in moles of nucleotides times seconds per liter (*23*).

Evidence is presented below that the DNA of each organism may be characterized by the value of C_0t at which the reassociation reaction is half completed under controlled conditions. The rates observed range over at least eight orders of magnitude. Therefore we have found it necessary to introduce a simple logarithmic method for the presentation of measurements of reactions over extended periods of time and wide ranges of concentration. For illustration, Fig. 1 shows the progress of an ideal second-order reaction plotted as a function of the product of the time of reaction and the DNA concentration on a logarithmic scale. On such a graph, reactions carried out at different concentrations may be compared, and the data may be combined to give a complete view of the time course of the reaction.

The symmetrical shape of an ideal second-order curve plotted in this way is pleasing and convenient. The central two-thirds of the curve follows closely a straight line, shown dashed. One useful indicator is the slope of this line which can be evaluated from the ratio of the values of C_0t at its two ends. This ratio is about 100 for an ideal reaction when estimated as shown on Fig. 1. If the ratio is much greater than 100, the reaction is surely heterogeneous; that is, species with widely different rates of reassociation are present.

Rate of Reassociation of DNA

Reassociation of a pair of complementary strands results from their collision. Therefore we expect the half-period for reassociation to be inversely proportional to the DNA concentration under fixed conditions for a particular DNA (*24*).

Further, one would expect for a given total DNA concentration that the half-period for reassociation would be proportional to the number of different types of fragments present and thus to the genome size (*25*). This expectation is exactly borne out in several cases. In Fig. 2, the time course of reassociation of a number of double-stranded nucleic acids is shown. Within the precision of the measurements, the reassociation of these various DNA's follows the time course of a single second-order reaction. In each case where it is applicable, the genome size (*25*) is marked with an arrow on the upper scale. Cairn's measurement (*26*) of the size of the *Escherichia coli* genome (4.5×10^6 nucleotide pairs) has been used to fix and locate this scale. The length of the T2 bacteriophage chromosome has also been carefully measured and found to

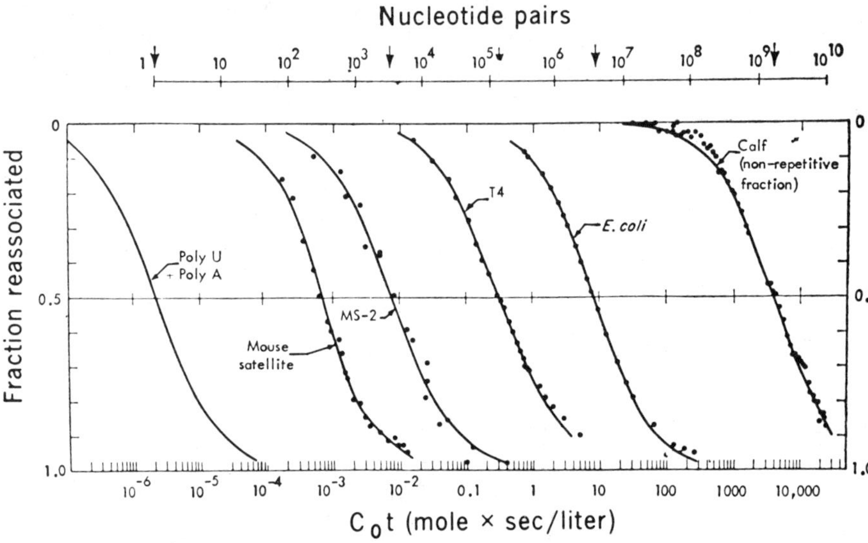

Fig. 2. Reassociation of double-stranded nucleic acids from various sources. The genome size (*25*) is indicated by the arrows near the upper nomographic scale. Over a factor of 10^9, this value is proportional to the C_0t required for half reaction. The DNA was sheared (*18*) and the other nucleic acids are reported to have approximately the same fragment size (about 400 nucleotides, single-stranded). Correction has been made (*19*) to give the rate that would be observed at 0.18*M* sodium-ion concentration. No correction for temperature has been applied as it was approximately optimum in all cases. Optical rotation was the measure of the reassociation of the calf thymus nonrepeated fraction (far right). The MS-2 RNA points were calculated from a series of measurements (*28*) of the increase in ribonuclease resistance. The curve (far left) for polyuridylic acid + polyadenylic acid was estimated from the data of Ross and Sturtevant (*29*). The remainder of the curves were measured by hypochromicity at 260 nm; a Zeiss spectrophotometer with a continuous recording attachment was used.

be 2 × 10^5 nucleotide pairs, and the size of T4 is similar (*27*). The total length of MS-2 viral RNA is 2.4 × 10^6 or 4000 nucleotide pairs in the double-stranded replicative form (*28*). The rate of reassociation (*29*) of the homopolymer pair [polyuridylic acid plus polyadenylic acid (polyU + polyA)] is consistent with the fact that these molecules are complementary in all possible registrations.

The proportionality between the C_0t (*23*) required for half-reassociation of the DNA and the genome size (*25*) is only true in the absence of repeated DNA sequences.

Figure 2 also shows the time course of reassociation for two fractions isolated from mammalian DNA. These fractions both follow the curve expected for a single second-order reaction, but one fraction reassociates more rapidly than the smallest virus, while the other reassociates 500 times more slowly than bacterial DNA. The former (mouse satellite DNA) represents 10 percent of the mouse DNA; its rate of reassociation indicates that the segment is roughly 300 nucleotide pairs and must be repeated about a million times (*13*) in a single cell. At the other extreme is a slowly reassociating fraction which includes about 60 percent of calf DNA. Its rate of reassociation is just that expected if it were made of unique (nonrepeating) sequences. The calf genome contains 3.2 × 10^9 nucleotide pairs (*30*).

Repeated Sequences in the DNA of Calf and Salmon

In order to obtain a fairly complete view of the repeated sequences in one organism, it is necessary to measure the degree of reassociation over a very wide range of C_0t. In Fig. 3 the reassociation of calf thymus DNA measured by the hydroxyapatite procedure is shown. The hydroxyapatite method is convenient for this purpose since the degree of reassociation can be directly determined by assay of the amount of DNA which is bound (*22, 15*). Samples were simply diluted into a convenient volume of 0.12*M* phosphate buffer (*31*) and passed over hydroxyapatite in a water-jacketed column at 60°C. A variety of tests (*15, 32*) have shown that under these conditions reassociated DNA is quantitatively bound, while not more than ½ percent of single-stranded DNA is adsorbed. The concentration of DNA present in the incubation mixture also can be varied over a wide range without interfering with the determination.

The hydroxyapatite binding measurements (Fig. 3) show that 40 percent of the calf DNA has reassociated before a C_0t of 2. Little if any reaction occurs in the next two decades of C_0t. Thus for calf DNA, there is a clear separation between DNA which reassociates very rapidly and that which reassociates very slowly.

The rapidly reassociating fraction in calf DNA requires a C_0t of 0.03 for half-reassociation, whereas the slowly reassociating fraction requires a C_0t of 3000. Thus the concentration of DNA sequences which reassociate rapidly is 100,000 times the concentration of those sequences which reassociate slowly. If the slow fraction is made up of unique sequences, each of which occurs only once in the calf genome, then the sequences of the rapid fraction must be repeated 100,000 times on the average.

The measurements shown on Fig. 3 were done in several series at different DNA concentrations. Nevertheless, the results are concordant. The points fall on a single curve with good accuracy. This establishes that the measured reassociation process results from a bimolecular collision. In turn, the rapidity of the early part of the reassociation reaction can result only from high concentrations of the reacting species. We

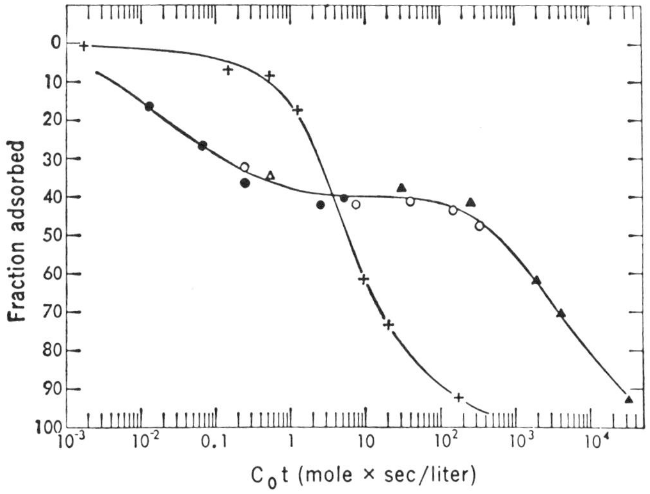

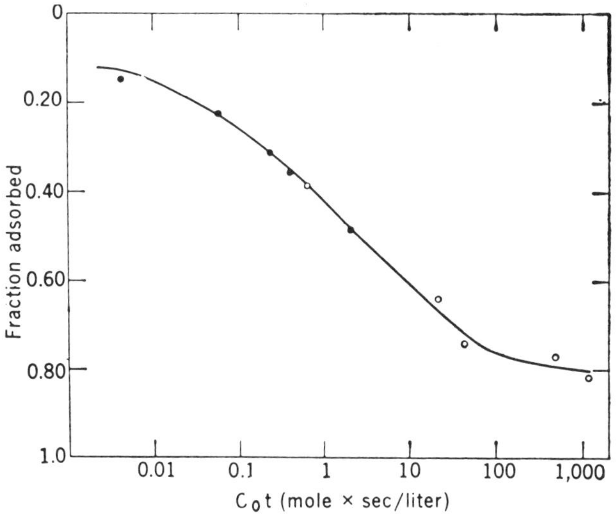

Fig. 3 (left). The kinetics of reassociation of calf-thymus DNA measured with hydroxyapatite. The DNA was sheared at 3.4 kilobars (*18*) and incubated at 60°C in 0.12*M* phosphate buffer (*31*). At various times, samples were diluted, if necessary (in 0.12*M* phosphate buffer at 60°C), and passed over a hydroxyapatite column at 60°C. The DNA concentrations during the reaction were (μg/ml): open triangles, 2; closed circles, 10; open circles, 600; closed triangles, 8600. Crosses are radioactively labeled *E. coli* DNA at 43 μg/ml present in the reaction containing calf thymus DNA at 8600 μg/ml. Fig. 4 (right). The kinetics of reassociation of salmon sperm DNA measured with hydroxyapatite. The DNA was sheared at 3.4 kilobars and incubated at 50°C in 0.14*M* phosphate buffer. The samples were diluted into 0.14*M* phosphate buffer at 50°C and passed over hydroxyapatite at 50°C. The DNA concentrations during the incubation were (μg/ml): closed circles, 8; open circles, 1600.

may conclude that about 40 percent of calf DNA consists of sequences which are repeated between 10,000 and a million times.

In one of the series of measurements shown on Fig. 3, in addition to the 8600 micrograms of sheared calf DNA per milliliter there was present 43 micrograms of P^{32}-labeled sheared *E. coli* DNA per milliliter, serving as an "internal standard." The simultaneous assay of the reassociation of the two DNA's that are present together controls a variety of possible experimental errors (*33*).

Figure 4 shows the reassociation of salmon sperm DNA measured with hydroxyapatite. Most of the salmon DNA appears to be made up of repeated sequences. The average degree of repetition is not as great as it is for the repeated fraction of calf DNA. The fact that the major part of the process extends over more than a factor of 10,000 in C_0t shows that many different degrees of repetition are present, varying from perhaps 100 copies to as many as 100,000 copies. The reassociation of the unique (single copy) DNA of this organism has not yet been observed. It would be expected to reassociate with a C_0t greater than 1000, and no measurements have yet been made in this region for salmon DNA.

The Occurrence of Repetitious DNA

With the observation of repeated DNA sequences in several vertebrate genomes the question arises: Are the DNA's of these creatures exceptional, or do repeated sequences occur generally among higher organisms? A limited survey was therefore carried out by the following procedure. (i) DNA was prepared and purified (*34*) from many organisms and then sheared to fragments consisting of 500 nucleotides (*18*). (ii) The DNA was dissociated in 0.12*M* phosphate buffer (*31*) and incubated at C_0t of 1 to 10 at 60°C. (iii) The solution was passed over a hydroxyapatite column equilibrated to 0.12*M* phosphate buffer and 60°C (*15, 32*). Under these conditions only the reassociated DNA becomes adsorbed to the column. (iv) The adsorbed DNA was eluted, and its reassociation kinetics were measured in the spectrophotometer.

The optical reassociation measurements for rapidly reassociating fractions prepared in this way from DNA of four organisms are shown in Fig. 5. All of these organisms contain repetitive DNA. The rate of reassociation of these fractions is very much faster than that calculated from the respective genome sizes. However, the reassociation pattern observed is quite different in each of the four cases. The curves for sea urchin DNA and calf DNA are probably representative of most of the repeated DNA in these organisms. However, the curve for mouse satellite DNA represents only the most repetitive fraction of the DNA from mouse cells (*13*), since the C_0t used before fractionation was very small. In the case of the onion DNA there is another repeating fraction which reacts more slowly than the fraction from onion shown in Fig. 5. This more slowly reacting fraction appears to have a repetition frequency between 100 and 1000.

Results obtained from DNA-agar experiments (*2, 3*) expand the list of higher organisms which contain repetitious DNA. The reassociation conditions of the DNA-agar technique yield a C_0t between 1 and 100. For DNA from higher organisms only repeated sequences will reassociate appreciably at these C_0t's. Therefore, the reassociation detected in DNA of higher organisms by the DNA-agar technique has been due to the reassociation of repetitious DNA. A list of organisms in which repeated DNA sequences have been found is shown in Table 1. Since so many types of organisms are represented it seems virtually certain that

Table 1. Occurrence of repititious DNA.

Protozoa
- Dinoflagellate (*Gyrodinium cohnii*)*
- *Euglena gracilis**

Porifera
- Sponge (*Microciona*)*

Coelenterates
- Sea anemone (*Metridium*) (tentacles)*

Echinoderms
- Sea urchin (*Strongylocentrotus*) (sperm)*†‡
- Sea urchin (*Arbacia*) (sperm)*†‡
- Starfish (*Asterias*) (gonads)*
- Sand dollar (*Echinarachnis*)‡

Arthropods
- Crab (*Cancer borealis*) (gonads)*
- Horseshoe crab (*Limulus*) (hepatopancreas)*

Mollusks
- Squid (*Loligo pealii*) (sperm)*

Elasmobranchs
- Dogfish shark (liver)*

Osteichthyes
- Salmon (sperm)*†‡
- Lungfish*†

Amphibians
- Amphiuma (liver, red blood cells, muscle)*
- Frog (*Rana pipiens*)†
- Frog (*Rana sylvatica*)‡
- Toad (*Xenopus laevis*) (heart, liver, red blood cells)
- Axolotl (*Ambystoma tigrinum*)‡
- Salamander (*Triturus viridescens*)‡

Birds
- Chicken (liver, blood)*†‡

Mammals
- Tree shrew‡
- Armadillo‡
- Hedge hog‡
- Guinea pig‡
- Rabbit‡
- Rat (liver)*†‡
- Mouse (liver, brain, thymus, spleen, kidney)*†‡
- Hamster‡
- Calf (thymus, liver, kidney)*†‡

Primates
- Tarsier‡
- Slow Loris‡
- Potto‡
- Capuchin‡
- Galago‡
- Vervet‡
- Owl monkey‡
- Green monkey‡
- Gibbon‡
- Rhesus†‡
- Baboon‡
- Chimpanzee*‡
- Human*†‡

Plants
- Rye (*Secale*)‡
- Tobacco (*Nicotiana glauca*)‡
- Bean (*Phaseolus vulgaris*)‡
- Vetch (*Vicia villosa*)‡
- Barley (*Hordeum vulgare*)*†
- Pea (*Pisum sativum* var. Alaska)*†
- Wheat (*Triticum aestivum*)*‡
- Onion (*Allium* sp.)*

* Rate of reassociation measured directly by hydroxyapatite fractionation or measurement of optical hypochromicity as a function of time or both. † Labeled, sheared fragments bind to DNA from the same species embedded in agar at a *Cot* so low that repetition must be present. ‡ Sheared nonradioactive fragments of DNA from the listed organism complete with the DNA-agar reaction (†) of a related species, reducing the amount of labeled DNA which binds to the embedded DNA.

repetitious DNA is universally present in higher organisms. In assembling this table we have made use of several sets of results (*3, 35*).

The species of bacteria (*E. coli, Clostridium perfringens, Proteus mirabilis*) that have been examined do not contain repetitious DNA detectable by our methods. In none of these cases was reassociated DNA of low thermal stability observed. In all cases the kinetic curve for reassociation apparently contained only the one major component. As a further check for repetitious DNA in *E. coli*, the first small fraction to reassociate ($C_0t = 0.5$) was isolated on hydroxyapatite and shown to reassociate at the same rate as most of the *E. coli* DNA. While the sensitivity of the test is high, the existence of a small amount of repetitious DNA cannot be ruled out. Optical measurement of the reassociation kinetics on unfractionated DNA from several viruses (simian virus 40 and bacteriophages T4 and lambda) has likewise given no evidence of repeated DNA sequences.

Only a very small repetitive fraction has been detected in DNA from *Saccharomyces cerevisiae*. Because of its small quantity and low native thermal stability it can tentatively be identified as mitochondrial DNA. At this moment, the relatively fragmentary evidence suggests that eukaryotes (except possibly yeast) contain repetitious DNA while prokaryotes do not. A great number of measurements will be necessary to ascertain the boundary between those life forms which do and those which do not have repetitious DNA.

Table 1 also describes interactions of DNA from a variety of tissues. We have seen no evidence for a variation in the pattern or amount of repeated sequences between different tissues of a given species or individuals of a species. In this work, sensitive tests for differences were not made; however, earlier experiments of McCarthy and Hoyer (*36*) with the DNA-agar method were specifically designed to detect variation of DNA from tissue to tissue. These may now be interpreted as showing that the repeated sequences in the mouse DNA occur to about the same extent in many tissues and in cultured cell lines.

The Precision of Repetition

When DNA strands which are not perfectly complementary reassociate, the resulting pairs have reduced stability. This effect supplies a method for measuring the degree of sequence difference among DNA strands. Measurements with artificial polymers indicate that, when 1 percent of the base pairs are not complementary, the temperature at which dissociation occurs (melting temperature, T_m) is about 1°C lower than that for perfectly complementary strands (*37, 12*). The data of Fig. 6 show that shearing bacterial DNA to small fragments, and its dissociation and reassociation, do not have a large effect on the melting temperature. This means that in the helically paired regions virtually all of the bases in re-

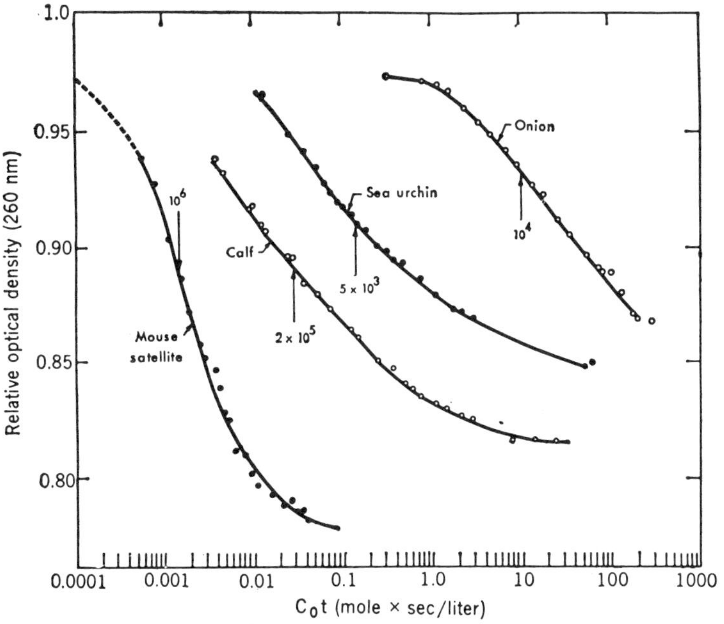

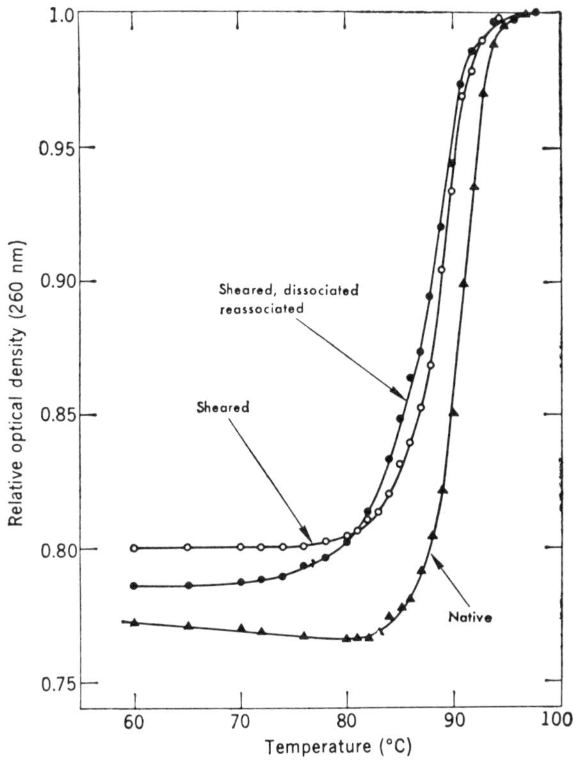

Fig. 5 (left). Optical reassociation curves of repetitious DNA fractions from various organisms. All of the fractions were purified on hydroxyapatite with only minor modifications in the procedure for each different DNA. The left scale gives the ratio of the optical density at 60°C to the initial value measured at 98°C. All of the fractions except that from onion were reassociated in 0.08*M* phosphate buffer at 60°C. The onion DNA was reassociated in 0.24*M* phosphate buffer at 60°C. The onion points were plotted a factor of 5 to the right to allow for the increased rate of reassociation, and give approximately the curve that would be observed in 0.08*M* phosphate buffer. The arrows permit estimation of the average degree of repetition in each case. They are located at the C_0t at which a fraction with the indicated degree of repetition would be half reassociated. The genome size and the amount of the rapidly reassociating fraction enter into the calculation in each case. **Fig. 6 (right).** Melting curves of *E. coli* DNA in 0.12*M* phosphate buffer. Open circles, native DNA sheared at 3.4 kilobars; closed circles, similarly sheared DNA dissociated (100°C, 5 minutes) and reassociated by incubation at 60°C in 0.12*M* phosphate buffer; triangles, native unsheared DNA. Shearing at 3.4 kilobars dissociates a part of the DNA, accounting for the somewhat greater hyperchromicity of the reassociated DNA.

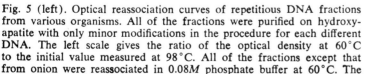

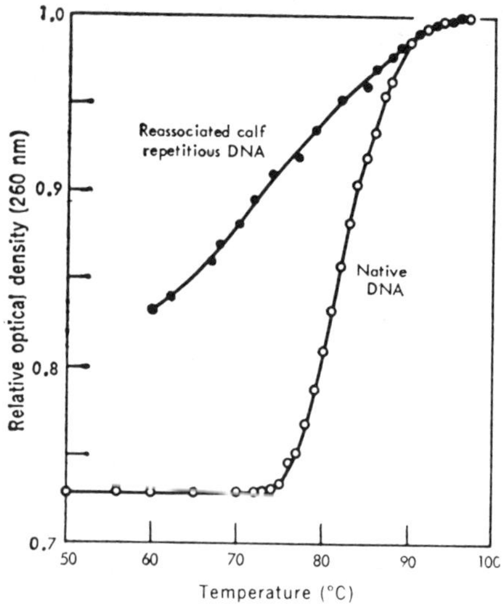

Fig. 7. DNA melting curves in 0.08*M* phosphate buffer. Open circles, unsheared native calf DNA; closed circles, reassociated calf repetitious DNA, sheared at 3.4 kilobars. In this solvent, single-stranded DNA gives an absorbancy change of only 3 percent from 60° to 95°C (see Fig. 9).

associated bacterial DNA are properly matched. Thus the strong reduction in thermal stability observed for calf DNA indicates actual dissimilarities in the paired sequences of the reassociated DNA.

Figure 7 shows the change in adsorbance with temperature of a repetitious fraction of calf thymus DNA (prepared by hydroxyapatite fractionation of DNA sheared at 3.4 kilobars). The optical density of this fraction changes over a wide range of temperature. Most of the thermal dissociation occurs below the temperature at which native DNA begins to dissociate. The change of absorbancy shows that base-paired structure has been formed in significant amount. The broad range of dissociation, in turn, indicates imprecise pairing. This observation confirms and extends earlier measurements with DNA-agar (*11*) in which more reassociation occurred at lower temperatures of incubation both for intraspecies and interspecies pairs. The temperature and salt concentration during a reassociation incubation establish a criterion of precision in that pairs form only if they are stable above the incubation temperature. Thus the incubation temperature determines which set of sequences will reassociate and controls the resulting melting temperature.

Raising the temperature of hydroxyapatite causes adsorbed double-stranded DNA to dissociate. The resulting single-stranded fragments may then be eluted. When dissociation is plotted against temperature, the profiles are very similar to those measured by change in ultraviolet absorbancy in free solution (*15, 22*).

Figure 8 shows reassociated repetitive salmon DNA fractionated with hydroxyapatite on the basis of its thermal stability. Sheared DNA from salmon was dissociated and then incubated at 50°C in 0.14*M* phosphate buffer (*31*) for a C_0t of 270, and the reassociated DNA was adsorbed on hydroxyapatite at 50°C, 0.14*M* phosphate buffer. The temperature of the column was raised in steps of 5°C, and at each temperature the dissociated DNA was washed from the column with 0.12*M* phosphate buffer. The resulting chromatogram (Fig. 8, dashed line) shows a broad range of thermal stability. In order to establish the specificity of the fractionation, samples eluted at 65° and 85°C were incubated again at 50°C. They were then readsorbed and reanalyzed as before.

The strand pairs formed during the second incubation are ordinarily not the ones that were originally eluted.

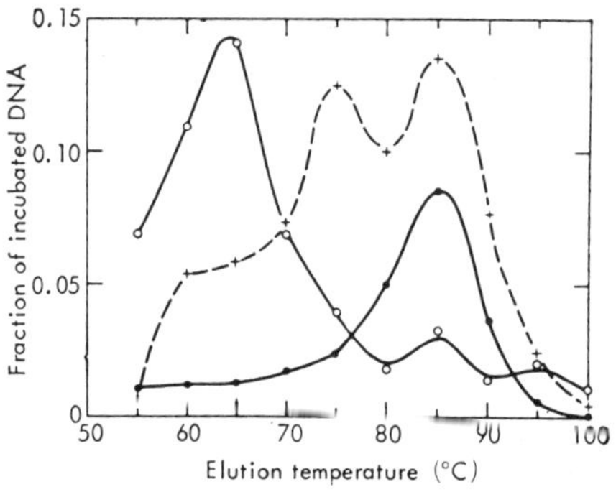

Fig. 8. Thermal fractionation on hydroxyapatite of reassociated salmon sperm DNA. DNA sheared at 3.4 kilobars was incubated at 50°C in 0.14*M* phosphate buffer (*Cot*. 370) and passed over hydroxyapatite at 50°C in 0.14*M* phosphate buffer. The adsorbed DNA was eluted by exhaustive washing (0.14*M* phosphate buffer) at intervals of 5°C (dashed line and crosses). To show specificity, four fractions (65°, 70°, 85°, 90°C) were again denatured (100°C, 5 minutes) and reincubated (50°C, 0.14*M* phosphate buffer, C_0t about 10) and readsorbed on hydroxyapatite at 50°C. Two of these were again thermally eluted from a column: open circles, 65°C fraction; closed circles, 85°C fraction. The other two were eluted with 0.4*M* phosphate buffer and melted in the spectrophotometer as shown in Fig. 9.

Instead, they are new duplexes formed by randomly assorted pairings among the selected set of strands. In each case, however, the same average degree of precision of relationship results. The portion eluting at 65°C shows a peak again at 65°C, and the 85°C portion peaks at 85°C. The degrees of sequence divergence are thus characteristic of these sets of fragments. Similar studies have been done with calf thymus DNA with entirely comparable results. In addition, experiments with labeled calf DNA fractions indicate that little sequence homology exists between precisely and imprecisely reassociating sets of repetitive DNA.

Length of Repeated Sequences

Are reassociated repeated sequences complementary only in short regions or are they complementary over most of their length? The thermal stability of a pair does not by itself answer this question since it appears that complementary sequences, 100 nucleotide pairs long, will have a thermal stability approaching that of very long complementary sequences (*38, 21*). However, ultraviolet hyperchromicity is a measure of the extent of sequence matching. Therefore, the hyperchromicity of a preparation of strand pairs gives a measure of the fraction of the total length which is complementary. Results for two such preparations are shown in Fig. 9. Native, completely complementary salmon DNA has a hyperchromicity of about 0.25. Single-stranded DNA has a hyperchromicity of 0.06 and melts mostly at lower temperatures, as shown by the top curve.

The 70° and 90°C fractions each have about half the hyperchromicity of native DNA. From this we may conclude in each case that the base-paired regions of the reassociated repetitious DNA include about half of the nucleotides of the fragments.

Several complicating factors interfere with a more firm conclusion. Reassociated fragments will, in general, have single-stranded ends, since two complementary fragments rarely terminate at the same points in the sequence. All degrees of overlap will occur, and for first pairing the expected hyperchromicity is between one-half and two-thirds that for native DNA. We do not

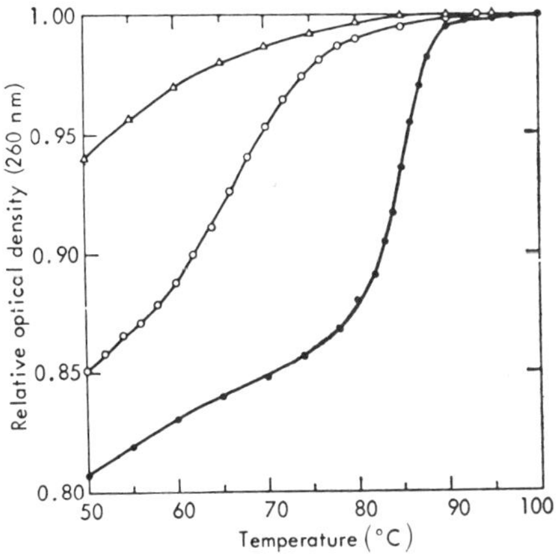

Fig. 9. Spectrophotometric melting curves in $0.14M$ phosphate buffer of fractions of salmon sperm DNA. Fractions were prepared as described in Fig. 8: closed circles, fraction eluted at 90°C; open circles, fraction eluted at 70°C. The upper curve (open triangles) is for the DNA which did not bind to hydroxyapatite (50°C, $0.14M$ phosphate buffer) in the first incubation and is therefore purely single stranded.

know the further extent of pair formation involving the single-stranded ends in these preparations. Finally, of course, repeated sequences have diverged from each other, and the unmatched nucleotides occurring within the paired sequences reduce the hyperchromicity.

These measurements are corroborated by the hyperchromicity we have observed for reassociated repetitive DNA fractions from many organisms. It usually falls between 0.10 and 0.20. Some examples are shown on Fig. 5. The few CsCl equilibrium centrifugation measurements we have made show a marked decrease in density upon reassociation which also implies a good extent of complementary pairing of repetitive DNA.

It appears that, on the average, repeated sequences are not extremely short (not less than 200 nucleotides) and may be much longer than our fragments, which average perhaps 400 nucleotides. In other words, wherever a region of sequence homology occurs between two fragments of the genome, it is likely to continue for at least several hundred nucleotides. It does not usually continue perfectly, however, since the reduced thermal stability observed implies that local interruptions of the homology must be scattered through the regions of homology.

Nonrepeated DNA of Higher Organisms

Somewhat more than half of the DNA from mouse or calf may be recovered in the single-stranded state after an extensive incubation (Fig. 3; $C_0t = 100$). The subsequent, very slow, reassociation of this fraction has been measured by three methods—hydroxyapatite adsorption (Fig. 3), optical rotation increase (370 nanometers) (Fig. 10), and hypochromicity (260 nanometers). The measurements establish that this fraction reassociates accurately and nearly completely. Experiments (*15*) indicate species dependence and therefore sequence specificity of the reassociation of the very slow fraction.

The curves shown on Fig. 3 give the results of a measurement of the rate of reassociation of the slow fraction of calf DNA in comparison with that of *E. coli* DNA. Labeled *E. coli* DNA was present with the calf DNA during the incubation as an "internal standard" (*33*). Thus these measurements yield a relatively precise measure of the concentration of complementary sequences in the *E. coli* DNA compared to those in the calf DNA.

The DNA content of the bull sperm is 3.2×10^9 nucleotide pairs per cell (*30*), and Cairn's measurement (*26*) gives 4.5×10^6 nucleotide pairs for the size of the *E. coli* genome. The ratio of these numbers is 710, and the ratio of the C_0t for half reaction of the slow part of the

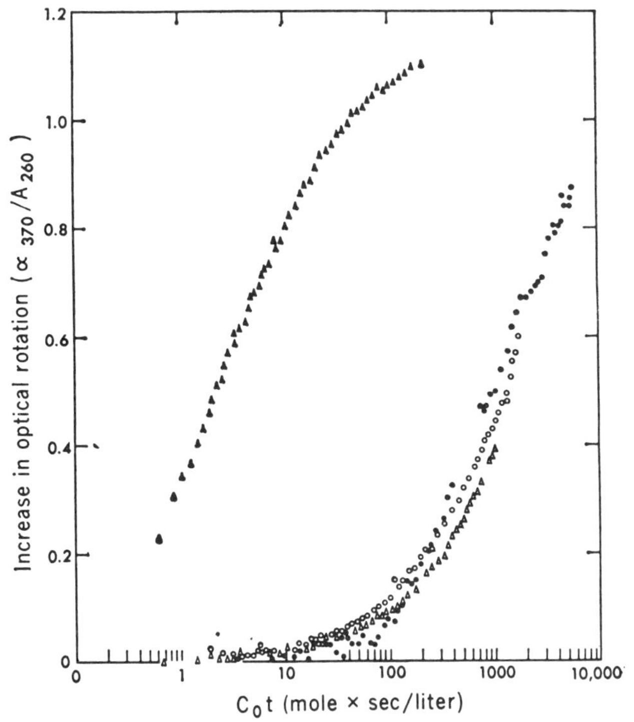

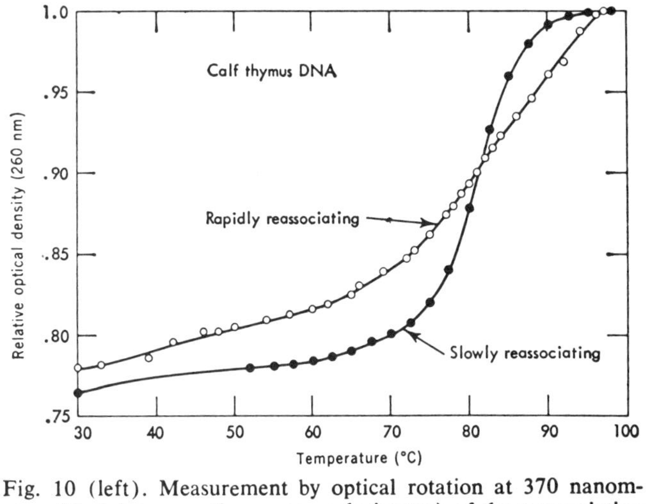

Fig. 10 (left). Measurement by optical rotation at 370 nanometers (Rudolph recording spectropolarimeter) of the reassociation of *E. coli* DNA and the slowly reassociating fractions of mouse and calf DNA ($0.24M$ phosphate buffer, 60°C). Closed triangles, *E. coli* DNA at 0.69 mg/ml in a 10-cm cell. Open triangles, mouse DNA fractionated to remove rapidly reassociating sequences. Open circles, calf thymus DNA fractionated to remove rapidly reassociating sequences. Closed circles, a second, similar, preparation of calf thymus DNA. The reaction rates shown here are nearly threefold greater than those shown in Fig. 2 because of the greater salt concentration. Fig. 11 (right). Hyperchromic melting curves for rapidly reassociating and slowly reassociating fractions of calf thymus DNA in $0.08M$ phosphate buffer. Fractions were prepared with hydroxyapatite and reassociated as described in the text.

calf DNA curve to that for *E. coli* is 690. This establishes with greater accuracy the conclusion drawn from Fig. 2 that under these conditions about 60 percent of the calf DNA does not exhibit repeated sequences. If the slowly reassociating sequences were nonrepeating (unique), they should also reassociate accurately since only the precisely complementary strand will be present. In other words, both the hyperchromicity and thermal stability of the reassociated pairs should approach that of native DNA of this fragment size.

Figure 11 shows the hyperchromic melting curves for very rapidly and very slowly reassociating fractions of calf thymus DNA. The slowly reassociating fraction was prepared by repeated incubation and passage through hydroxyapatite (C_0t = 100, 60°C, 0.12*M* phosphate buffer). It was then incubated extensively (C_0t = 1000, 60°C, 0.24*M* phosphate buffer), and the reassociated portion was isolated by binding to hydroxyapatite. The C_0t was sufficient only to reassociate about half of the DNA strands. Thus, little concatenation (*12*) had occurred, and the hyperchromicity had not achieved its final value. Nevertheless, the nonrepetitious fraction has almost 75 percent of the hyperchromicity of native high-molecular-weight DNA and a relatively sharp thermal transition. Much of the difference between this curve and that for native DNA (Fig. 7) is due to the fact that the DNA had been sheared before reassociation. Native high-molecular-weight and sheared *E. coli* DNA (Fig. 6) differ to about the same extent.

The rapidly reassociating fraction was eluted from the hydroxyapatite in a reassociated state and was diluted to the proper solvent and concentration for the measurement. A very brief incubation was used in this case, and the melting curve is strongly influenced by a particular fraction (*39*) which reassociates very rapidly, and apparently has a high guanine-cytosine content and melting temperature. A more typical melting curve of a repetitive DNA fraction prepared after a more extensive incubation is shown on Fig. 7.

Reassociation of unique sequences of calf DNA was originally observed in our laboratory by means of the spectropolarimeter. This instrument is particularly suitable since a high concentration of DNA is required both for the large C_0t and for an accurate measurement of the change in optical activity. The slowly reassociating DNA was prepared for this purpose from commercial calf thymus DNA by hydroxyapatite fractionation and lyophilization.

In the region of the spectrum above 300 nanometers there is a good contrast in specific rotation between the native and denaturated states, and little light is absorbed by the DNA. The reassociation of the purified slow fraction of both mouse and calf thymus DNA occurs at about 1/500 the rate of that of *E. coli* DNA under the same conditions (Fig. 10). Within the accuracy of the measurements this is the expected rate for a 60 percent fraction of mammalian DNA which has no repeated sequences. The reassociation of the unique sequences of mammalian DNA is therefore confirmed by measurement of the change in optical rotation.

Amphiuma is the organism with the largest known genome, having 8×10^{10} nucleotide pairs of DNA per haploid cell. The expected C_0t for half reassociation in the absence of repetitions would be 80,000 under the conditions of Fig. 2. *Amphiuma* DNA was fractionated on hydroxyapatite after shearing and incubation (C_0t = 80, 50°C, 0.14*M* phosphate buffer). Only 20 percent of the DNA was recovered in the slowly reassociating fraction. This fraction was incubated (60°C, 0.24*M* phosphate buffer, 5 milligrams per milliliter), and samples were analyzed on hydroxyapatite every few days. It exhibited the slowest reassociation reaction that has been observed, reaching half reassociation with a C_0t of 20,000. This apparent agreement with the expected rate cannot be considered definitive until a number of controls are done. Nevertheless, it appears likely that a significant fraction of the genome of even this creature is made up of unique sequences.

Patterns of Repetition

The rate of reassociation of the DNA of one organism can be evaluated over the whole course of the reaction (C_0t from 10^{-4} to 10^4). Individual measurements of reassociation at several concentrations are required, and fractionation of the DNA is useful. From these measurements the amount of DNA with various degrees of repetition may be calculated. The result is a repetition-frequency spectrogram for the DNA of the particular organism, such as the tentative one for mouse DNA shown in Fig. 12. This curve is correct in its broad aspects but has some indefiniteness in detail. The width of the peaks results in part from the difficulty of resolving reassociation rates that differ by less than a factor of 10.

The large peak at the left on Fig. 12 is due to the mouse satellite DNA (*13*). Such a class of DNA molecules which can reassociate with each other is called a family since the similarity in sequence implies a common origin. Correspondingly, other classes capable of reassociating are called families even though the precision of reassociation is less, presumably due to divergence of the members since the formation of the family.

Repetition-frequency spectra could also be derived for calf DNA and salmon DNA from Figs. 3 and 4. However, they would differ from that of Fig. 12. Neither would show the large isolated peak of 10^6 copies. The calf DNA would show a large broad peak in the region of 10^4 to 10^5 copies (40 percent of the DNA), little if any DNA with a small degree of repetition, and of course a large peak of unique DNA.

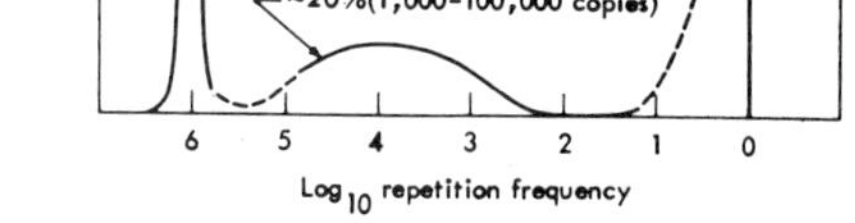

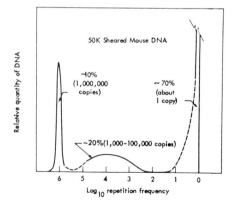

Fig. 12. Spectrogram of the frequency of repetition of nucleotide sequences in the DNA of the mouse. Relative quantity of DNA plotted against the logarithm of the repetition frequency. These data are derived from measurements of the quantity and rate of reassociation of fractions separated on hydroxyapatite. The dashed segments of the curve represent regions of considerable uncertainty.

What length of DNA sequence has been replicated to form families of repeated sequences? Forty percent of the calf DNA behaves as repeating DNA. The total quantity of repeated sequences is 1.3×10^9 nucleotide pairs per cell. Lengths of DNA totaling 13,000 nucleotide pairs copied 100,000 times would have about the same total quantity and average repetition frequency as the repetitious DNA of the calf. Such a homogeneous set of fragments would have the smallest possible information content that could be present in the repeated DNA of the calf. The situation is known to be more complex, however, and the potential information content of the repetitious DNA fraction is very much greater for the following reasons. (i) The nucleotide sequences of the members of a typical family are similar to each other but not identical. The differences may be of great genetic significance. (ii) A small amount of DNA probably occurs in families made up of long sequences repeated only a few times. (iii) The repeated sequences or fragments of them have been translocated into various parts of the genome, and their location and relationship to their neighbor sequences may be important.

In this regard an observation of Britten and Waring (*12*) with higher-organism DNA is significant. When 5 to 10 million dalton DNA strands are dissociated and incubated ($C_0t = 1$), large particles form creating a visible haze in solution, and most of the DNA may be sedimented to the bottom of a tube after centrifugation for 5 minutes at 10,000*g*. Apparently some regions of most of these DNA fragments are members of families of repeated sequences. Reassociation of the repeated sequences links the fragments into a large network. A number of such measurements with various fragment sizes indicate that repeated sequences are scattered throughout the genome. This extensive interspersion of members of families of repeated sequences may be related to their function. They could, for example, have regulatory or structural roles which would lead to such a distribution. This dispersion may, however, simply represent the degree of translocation of sequence fragments that has occurred during the evolution of higher organisms.

The Criteria for Repeated Sequences

The presence in a DNA solution of a fraction which reassociates rapidly indicates that certain nucleotide sequences are present at a higher concentration than the remainder. If the DNA's were derived from a single organism we can usually conclude that these sequences recur repeatedly in its DNA. The conclusion is essentially certain if the reassociation exhibits the variation of rate with concentration of a second-order collision-controlled reaction.

If such a concentration dependence is not shown other possibilities arise. For example, the rapidly reassociating DNA's could be in closed circular form, the two strands could be cross-linked, or a sequence might occur which contained its own complement and could renature by folding. An example of the latter is the satellite from crab DNA (*40*) which is principally an alternating sequence of adenine and thymine. In all of these cases, a bimolecular collision is not involved, and the reaction is extremely rapid under optimum conditions for reassociation. With the methods used in this work the rate would be so fast that it could not be observed. Thus, under our usual conditions the observation of a measurable rate of reassociation (faster than that expected from the genome size of the organism) is an almost certain indication of the presence of repeated sequences.

What are the limits of accuracy in the calculation of repetition frequency from a measured rate of DNA reassociation? If only a part of the length of fragments are complementary to each other the rate of reassociation will be reduced. The fraction of the nucleotides which are complementary in typical reassociated pairs of fragments containing repeated sequences is not well known but some limits can be set. The hyperchromicity observed for reassociated repetitive DNA ranges from just less than half up to nearly that characteristic for reassociated bacterial or viral DNA. This evidence implies that half or more of the nucleotides are complementary in typical reassociated repetitive DNA. We believe that under these conditions the reduction in the rate of reassociation is not large. However, the frequency of repetition may be somewhat underestimated in this work.

There is a possibility also of a potential overestimate in the quantity of repeated DNA sequences. When reassociation is assayed with hydroxyapatite, all strands of DNA which contain a sizable double-stranded region will bind. The minimum double-stranded region which will adsorb to hydroxyapatite under the conditions used is not known but is much smaller than the fragment size used. Thus, a certain fraction of the nonrepetitive DNA will be included in the measured repetitive fraction. If the repeated sequences occur in stretches which are long compared with the fragment size, this error will be small. Partly for this reason small fragments are used in this work.

There is evidence that homopolymer clusters occur in DNA (*41*). It has not yet been demonstrated that such clusters influence DNA reassociation. The quantity present in the DNA of higher organisms is not known. Nevertheless, in the following paragraphs we attempt an estimate of the maximum effect homopolymer clusters could have on the rate of DNA reassociation.

If the homopolymer clusters were long enough to form a large fraction of the length of a set of DNA fragments they would simply form an extreme example of repeated sequences. The reassociation rate constant would be that shown for the polyuridylic acid and polyadenylic acid pair on Fig. 2. The reassociation would appear to be instantaneous under our conditions, except with very low concentrations of DNA.

Very short homopolymer clusters might play two possible roles. They might cause fragments not complementary over the rest of their length to form stable structures which are paired only in the cluster region. These would form a class (not yet observed) of repetitive DNA with very low hypochromicity. Short homopolymer clusters might also increase the rate of nucleation of fragments which were fully complementary and thus increase their rate of reassociation.

The limit for the maximum possible factor of increase is less than twice the number of nucleotides in the homopolymer cluster. This can be seen by making the most favorable assumptions. (i) Nucleations in the homopolymer regions of otherwise noncomplementary fragments do not interfere (that is, they dissociate quickly); (ii) all nucleations occur independently and the rate of reassociation is proportional to the number of "in register" collisions possible; (iii) all possible "registrations" of the homopolymer cluster (twice the number of nucleotides in the cluster) lead to reassociation of major complementary regions if present.

Since condition (i) would not be met for homopolymer clusters longer than 20, the factor of increase in rate must be less than 40. Since condition (iii) is not likely to be met, the factor of increase is probably much less than that. This factor is not large enough to affect the conclusion about the general occurrence of repeated sequences.

Table 2. Characteristics of DNA reassociation.

Nonrepetitive DNA	Repetitive DNA
Source	
Bacteria Viruses	Vertebrates Invertebrates Higher plants Euglena Dinoflagellate
Rate of reassociation	
One rate, inversely proportional to DNA content per cell or particle	Many different rates. Slowest inversely proportional to DNA content per haploid cell. Fastest up to 10^6 times faster
Extent of reassociation	
Excellent, up to 90 percent reformed helices (no strong effect of fragment size)	Good if DNA cut into small fragments. Poor if DNA is of high molecular weight
Stability of reassociated DNA	
Temperature at which strands separate (T_m) almost equal to that of native DNA	Some with T_m near that native DNA and many lower degrees of stability
Particle size of reassociated DNA	
Several times the fragment size due to pairing of free single-stranded ends (concatenation)	Enormous, if DNA fragments are large, due to multiple interconnections (network formation)

Implications of Repeated Sequences

These studies have revealed new properties of the DNA of higher organisms which must be attributed to the repetition of nucleotide sequences. Some of these properties are summarized in Table 2. In general, more than one-third of the DNA of higher organisms is made up of sequences which recur anywhere from a thousand to a million times per cell. Thus the genetic material is not a collection of different and unrelated genes. A large part is made up of families of sequences in which the similarity must be attributed to common origin.

A minor degree of sequence repetition is to be expected from studies of protein sequences (*42*). The hemoglobin group shows similarities in sequences, and these similarities point to common origin of part or all of their structural genes. Trypsin and chymotrypsin also show similarities. There is evidence that, in some cases, different segments of the amino acid sequence of a given protein may have arisen by duplication and insertion of an earlier short segment. In addition to genetic evidence, banding patterns in polytene chromosomes show that gene duplication occurs (*43*). The genome sizes (*25*) of higher organisms range from 10^8 to 10^{11} nucleotide pairs (*30*). There is no doubt that a great increase in DNA content has occurred during the evolution of certain species.

These observations suggest that a degree of nucleotide sequence repetition might be observed in the DNA. It must be emphasized that they do not imply that DNA sequence repetition occurs on anything approaching the scale reported here. The very large number of members in the families of repeated sequences remains a most surprising feature for which an explanation must be sought. It may be reasonably predicted that large-scale new patterns of relationship among the proteins await discovery.

Certain minor classes of DNA probably consist of many copies of a short sequence. It appears likely that there are hundreds or thousands of similar ribosomal genes (*44*) and in certain cells, at least, thousands of similar, if not identical, copies of mitochondrial DNA (*45*). Taken together, such classes of DNA do not add up to more than a percent of the total DNA and, compared to the bulk of the repeated sequences, have a relatively low repetition frequency.

If many DNA sequences in a chromosome are similar to each other and adjacent, high rates of unequal crossing-over might occur. Although there is genetic evidence (*43*) that this occurs, it has not been considered common. Presumably higher organisms are protected from the lethal genetic events (*46*) that the families of repeated sequences might induce.

There is a certain amount of evidence that repeated sequences are genetically expressed. Pulse-labeled RNA (presumptive messenger) has been hybridized with the DNA of higher organisms (*47*). In most of these studies, hybrids were only observed with RNA that was complementary to families of repeated sequences. Due to the small C_0t, hybrids between RNA and nonrepeated DNA· sequences of higher organisms apparently did not occur.

The RNA populations made from repeated DNA sequences may have some role (perhaps regulatory) other than as messengers carrying structural information for protein synthesis. However, this is an unlikely (and certainly an unpopular) possibility. A good working hypothesis is that repeated sequences commonly occur in structural genes. In any case, transcription as complementary RNA is direct evidence for the genetic function of at least some of the repeated DNA sequences. In the course of embryonic development and during liver regeneration (*47*), changes occur in the pattern of types of hybridizable pulse-labeled RNA. These results suggest that during the course of differentiation different families of repeated sequences are expressed at different stages.

Origin and Age

The families of repeated sequences range from groups of almost identical copies (for example, mouse satellite DNA) to groups with sufficient diversity that, after reassociation, only structures of low stability are formed among the members. It seems likely that this situation has arisen from large-scale precise duplication of selected sequences, with subsequent divergence caused by mutation and the translocation of segments of certain member sequences. We cannot now describe the history of growth and divergence of any particular family of repeated sequences. However, the few measured properties of the repetitive DNA permit some inferences.

The extensive studies of Hoyer *et al.* (*3, 10, 11*) supply a measure of the repeated sequences held in common among different species. Because of the small C_0t used in their work only the reassociation of DNA sequences repeated in each organism was observed. These measurements were carried out at various temperatures (*11*), and the results were correlated with the period of time after divergence of the lines leading to the modern species (*10*).

These data show a low average melting temperature if strands of DNA from different species are reassociated. The longer the period after divergence of the species, the greater the reduction in thermal stability. This evidence indicates that the members of families of repeated sequences in the DNA of a species slowly change in nucleotide sequence.

It is an unlikely possibility that all the members of a family of repeated sequences in one organism undergo the same changes. This would involve either very severe selection on all the members or a complex event such as discarding all but one of the members of a family and then multiplying the remaining member 10,000 or 100,000 times. The much more appealing and simpler model is that the nucleotide sequences of the members of the families are not conserved by severe selection. The members may then change slowly and independently of each other leading, after a long period of time, to families with widely divergent members such as are observed.

In addition to the divergence of preexisting families, new families are produced in each species. Analysis of the present data on repetition frequency distribution (*14*) suggests that they result from relatively sudden events which we have called saltatory replications. Figure 13 symbolizes the resulting view of the history of families of repeated sequences. Along one axis is the time since formation of the family. Along the other axis is the temperature at which duplexes among the family members dissociate. The third axis represents the number of members. Thus the area of one of the peaks indicates the number of members of the family. The temperature at which the peak occurs is a measure of the extent of sequence difference among the members. The diagram is not intended to be quantitative, although we have used the estimates that are available for vertebrates. The frequency of events and rates of divergence are also probably very different for other phyla.

Even if saltatory replications are as rare as indicated on Fig. 13, certain stages of the process may be relatively common. We know nothing of the mechanism of the process, but the following steps seem necessary. (i) A sequence undergoes manyfold replication; (ii) the copies are integrated into the chromosome; (iii) they become associated with a favorable genetic element, and (iv) they are disseminated through the species by natural selection.

Each of the succeeding stages is likely to have a very low probability of occurrence, and thus the actual event of manyfold replication may occur fairly commonly and, in principle, be observable in individual organisms, in analogy to a somatic mutation. It does not seem impossible that, some time in the future, saltations may be artificially introduced into populations as mutations can already be.

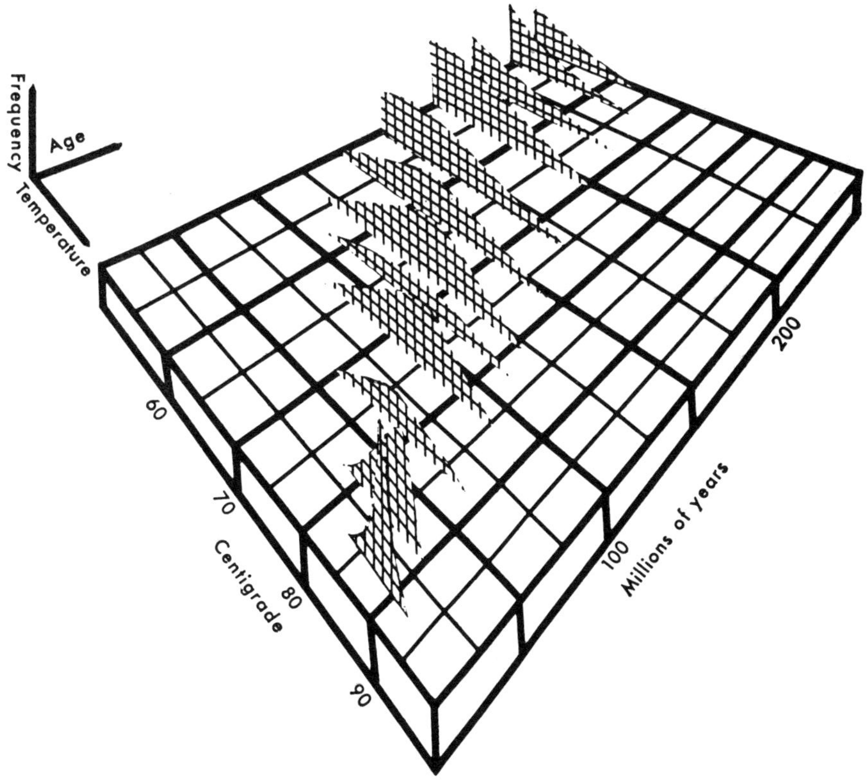

Fig. 13. A schematic diagram intended to suggest the history of the families of repeated sequences now present in the DNA of a modern creature. Each family is supposed to have originated in a sudden event (saltatory replication) at a time in the past shown on the right-hand scale. In the ensuing time, increasing divergence has occurred between the nucleotide sequences of the members of each family of repeated sequences. This divergence is represented on the left-hand scale by the thermal stability of reassociated pairs of DNA strands formed between members of each family. The height of the cross-hatched areas indicates the amount of DNA of a given thermal stability in a family of a particular age. Only a few of a potentially large number of families are indicated. The actual rate of divergence has not yet been well measured.

Speculation on Their Function

A concept that is repugnant to us is that about half of the DNA of higher organisms is trivial or permanently inert (on an evolutionary time scale). Furthermore, at least some of the members of DNA families find expression as RNA. We therefore believe that the organization of DNA into families of related sequences will ultimately be found important to the phenotype. However, at present we can only speculate on the actual role of the repeated sequences.

Multiple, nearly exact copies of a gene could provide higher rates of synthesis. This might be true for structural proteins required in large amounts and is very likely true for ribosomal RNA. Multiple similar copies could provide a class of similar protein chains as appear to occur in antibody proteins (*48*). However, their role could not be limited to the immune system since they occur in large quantities in the plants and other organisms in which antibodies have not been observed.

The DNA of each vertebrate that has been examined contains some families with 100,000 members or more. This very large number suggests a structural (*49*) or regulatory role. However, the significance of the very large number might be less direct. It might, for example, raise to a useful level the probability of some rare event such as the translocation of certain DNA sequence fragments into adjacent locations in the genome.

Saltatory replications of genes or gene fragments occurring at infrequent intervals during geologic history might have profound and perhaps delayed results on the course of evolution. In the following quotation Simpson (*50*) raises some relevant questions with regard to evolutionary history.

> The history of life is decidedly nonrandom. This is evident in many features of the record, including such points already discussed as the phenomena of relays and of major replacements at defined times. It is, however, still more striking in two other phenomena copiously documented by fossils. Both have to do with evolutionary trends: first, that the direction of morphological (hence also functional and behavioral) change in a given lineage often continues without significant deviation for long periods of time and, second, that similar or parallel trends often appear either simultaneously or successively in numerous different, usually related, lineages. These phenomena are far from universal; they are not "laws" of evolution; but they are so common and so thoroughly established by concrete evidence that they demand a definite, effective directional force among the evolutionary processes. They rule out any theory of purely random evolution such as the rather naive mutationism that had considerable support earlier in the twentieth century. What directional forces the data do demand, or permit, is one of the most important questions to be asked of the fossil record.

The appearance in a genome of many thousands of copies of a gene could have evolutionary significance. Perhaps not many copies would be actually expressed. Mutation, translocation, and recombination with other genes might yield new genetic potential. If the early effects were selectively advantageous, the repeated DNA sequences could be introduced into the population. The dynamics of selection for this set of genes would be fundamentally altered. Owing to the great multiplicity of copies, their selective elimination might be impossible.

Summary

The rate of reassociation of the complementary strands of DNA of viral and bacterial origin is inversely proportional to the (haploid) DNA content per cell. However, a large fraction of the DNA of higher organisms reassociates much more rapidly than would be predicted from the DNA content of each cell. Another fraction appears to reassociate at the expected rate. It is concluded that certain segments of the DNA are repeated hundreds of thousands of times. A survey of a number of species indicates that repeated sequences occur widely and probably universally in the DNA of higher organisms.

The repeated sequences have been separated from the remaining (unique sequence) DNA, and their physical properties have been studied. The range of frequency of repetition is very wide, and there are many degrees of precision of repetition in the DNA of individual organisms. During evolution the repeated DNA sequences apparently change slowly and thus diverge from each other. There appears to be some mechanism which, from time to time, extensively reduplicates certain segments of DNA, replenishing the redundancy.

References and Notes

1. J. Marmur, R. Rownd, C. L. Schildkraut, *Progr. Nucleic Acid Res.* **1**, 231 (1963).
2. B. J. McCarthy and E. T. Bolton, *Proc. Nat. Acad. Sci. U.S.* **50**, 156 (1963).
3. E. T. Bolton *et al.*, *Carnegie Inst. Wash. Year Book* **63**, 366 (1964); *ibid.* **62**, 303 (1963); B. H. Hoyer, B. J. McCarthy, E. T. Bolton, *Science* **144**, 959 (1964).
4. R. C. Warner, *J. Biol. Chem.* **229**, 711 (1957); A. Rich and D. Davies, *J. Amer. Chem. Soc.* **78**, 3548 (1956).
5. J. Marmur and D. Lane, *Proc. Nat. Acad. Sci. U.S.* **46**, 456 (1960).
6. B. D. Hall and S. Spiegelman, *ibid.* **47**, 137 (1961).
7. E. K. F. Bautz and B. D. Hall, *ibid.* **48**, 400 (1962).
8. E. T. Bolton and B. J. McCarthy, *ibid.*, p. 1390.
9. A. P. Nygaard and B. D. Hall, *J. Mol. Biol.* **9**, 125 (1964); D. Gillespie and S. Spiegelman, *ibid.* **12**, 829 (1965).
10. B. H. Hover, E. T. Bolton, B. J. McCarthy, R. B. Roberts, *Carnegie Inst. Wash. Year Book* **63**, 394 (1964); in *Evolving Genes and Proteins*, V. Bryson and H. Vogel, Eds. (Academic Press, New York, 1965), p. 581.
11. M. Martin and B. Hoyer, *Biochemistry* **5**, 2706 (1966); P. M. B. Walker and A. McLaren, *J. Mol. Biol.* **12**, 394 (1965).
12. R. J. Britten and M. Waring, *Carnegie Inst. Wash. Year Book* **64**, 316 (1965).
13. M. Waring and R. J. Britten, *Science* **154**, 791 (1966).
14. R. J. Britten and D. E. Kohne, *Carnegie Inst. Wash. Year Book* **66**, 73 (1967).
15. ———, *ibid.* **65**, 73 (1966).
16. J. A. Subirana and P. Doty, *Biopolymers* **4**, 171 (1966); J. A. Subirana, *ibid.*, p. 189; J. G. Wetmur and N. Davidson, *J. Mol. Biol.* **31**, 349 (1968); K. J. Thrower and A. R. Peacocke, *Biochim. Biophys. Acta* **119**, 652 (1966).
17. J. G. Wetmur, thesis, California Institute of Technology, Pasadena (1967).
18. The DNA used in this work has been sheared to a relatively uniform population of small fragments (about 400 or 500 nucleotides long) by passing it twice through a needle valve with a pressure drop of 3.4 kilobars. A specially built air-operated plunger pump was used to develop this high pressure. The DNA is denatured when sheared at 3.4 kilobars, unless a very high salt concentration is present to raise the temperature of melting (T_m). These small fragments give reproducible rates of reassociation and do not, under the usual conditions for reassociation, form large aggregates or networks.
19. For some purposes, correction can be made for different values of the parameters. There have been some quantitative measurements of the effect of temperature (*1, 15*) salt concentration (*15–17*) and fragment size (*16, 17*).
20. D. T. Denhardt, *Biochem. Biophys. Res. Commun.* **23**, 641 (1966).
21. B. J. McCarthy, *Bacteriol. Rev.* **31**, 215 (1967).
22. G. Bernardi, *Nature* **206**, 779 (1965); P. M. B. Walker and A. McLaren, *ibid.* **208**, 1175 (1965); Y. Miyazawa and C. A. Thomas, Jr., *J. Mol. Biol.* **11**, 223 (1965).
23. Here, C_0t is used as a noun and may be pronounced as the homonym of "cot." A C_0t of 1 mole × second/liter) results if DNA is incubated for 1 hour at a concentration of 83 μg/ml, which corresponds to an optical density of about 2.0 at 260 nanometers.
24. In our experience, the reassociation of purified sheared (*18*) DNA shows the concentration dependence expected for a second-order reaction. For DNA without repeated sequences, the time course also approximately follows second-order kinetics. While earlier measurements have suggested greater complexity, this is not supported by more recent work (*17*).

25. The word genome customarily means the genetic constitution of an organism. Here the genome size is taken to mean the haploid DNA content of a cell or virus particle. The number of different fragments will only be proportional to the genome size in the absence of repetition or unrecognized polyploidy.
26. J. Cairns, *Cold Spring Harbor Symp. Quant. Biol.* **28**, 43 (1963). Because of its apparent lack of repetition and measured genome size *E. coli* DNA is used as a reference for the comparison of reassociation rates of other DNA. The use of this reference in future work will permit comparison of rates where fragment size and other conditions affecting absolute rate may vary.
27. I. Rubenstein, C. A. Thomas, A. D. Hershey, *Proc. Nat. Acad. Sci. U.S.* **47**, 1113 (1961); J. Cairns, *J. Mol. Biol.* **3**, 756 (1961); E. Burgi and A. D. Hershey, *ibid.*, p. 458.
28. M. A. Billeter, C. Weissmann, R. C. Warner, *J. Mol. Biol.* **17**, 145 (1966).
29. P. D. Ross and J. M. Sturtevant, *J. Amer. Chem. Soc.* **84**, 4503 (1962).
30. B. J. McCarthy, in *Progr. Nucleic Acid Res. Mol. Biol.* **4**, 129 (1965); T. Mann, in *The Biochemistry of Semen and of the Male Reproduction Tract* (Wiley, New York, 1964), p. 147.
31. This phosphate buffer is composed of an equimolar mixture of Na_2HPO_4 and NaH_2PO_4. The indicated molarity is for the phosphate. The sodium-ion concentration is 1.5 times greater.
32. D. E. Kohne and R. Britten, in preparation.
33. The variables controlled in this way are salt concentration, temperature, and viscosity. In addition, any possible nonspecific interactions in the DNA at this high concentration will have similar effects on both the *E. coli* and calf DNA reassociation reactions. The half reaction C_0t for *E. coli* DNA in Fig. 3 is 6.0, whereas on Fig. 2 it is 8.0. Reactions usually appear twofold faster when assayed with the hydroxyapatite method as compared to the optical method since the fraction of fragments reassociated is measured in one case, while the fraction of total strand length reassociated is measured in the other case (*15*). On Fig. 3 there may be a 50 percent increase in the C_0t for half reaction for the data taken at 8600 μg/ml. This decrease in the rate of reassociation is due to the increased viscosity of the incubation solution.
34. The DNA was prepared by a combination of methods of J. Marmur, *J. Mol. Biol.* **3**, 208 (1961); K. I. Berns and C. A. Thomas, *Biophys. Soc. Abstr.* (1964); B. J. McCarthy and B. H. Hoyer, *Proc. Nat. Acad. Sci. U.S.* **52**, 914 (1964). Purity was tested in the spectrophotometer by melting in 0.12*M* phosphate buffer. We required that there be no measurable rise in optical density between 40° and 70°C, and that a normal melting curve was obtained with a hyperchromicity of at least 25 percent of the absorbancy at 98°C. Commercial calf and salmon DNA were utilized in some experiments, and no differences were observed with results obtained with DNA prepared from fresh tissue.
35. B. J. McCarthy and E. T. Bolton, *Proc. Nat. Acad. Sci. U.S.* **50**, 156 (1963); B. H. Hoyer, B. J. McCarthy, E. T. Bolton, *Science* **144**, 959 (1964); E. T. Bolton *et al.*, *Carnegie Inst. Wash. Year Book* **64**, 314 (1965); H. Denis, *ibid.*, p. 455.
36. B. J. McCarthy and B. H. Hoyer, *Proc. Nat. Acad. Sci. U.S.* **52**, 915 (1964).
37. E. K. F. Bautz and F. A. Bautz, *ibid.*, p. 1476; T. Kotaka and R. L. Baldwin, *J. Mol. Biol.* **9**, 323 (1964).
38. A. Rich, *Proc. Nat. Acad. Sci. U.S.* **46**, 1044 (1960); M. N. Lipsett, L. A. Heppel, D. F. Bradley, *J. Biol. Chem.* **236**, 857 (1961); M. N. Lipsett, *ibid.* **239**, 1256 (1964).
39. E. Polli, G. Corneo, E. Ginelli, P. Bianchi, *Biochim. Biophys. Acta* **103**, 672 (1965).
40. N. Sueoka, *J. Mol. Biol.* **3**, 31 (1961); *ibid.* **4**, 161 (1962).
41. H. Kubinski, Z. Opara-Kubinska, W. Szybalski, *ibid.* **20**, 313 (1966).
42. K. A. Walsh and H. Neurath, *Proc. Nat. Acad. Sci. U.S.* **52**, 884 (1964); K. Brew, T. C. Vanaman, R. L. Hill, *J. Biol. Chem.* **242**, 3747 (1967); E. Freese and A. Yoshida, in *Evolving Genes and Proteins*, V. Bryson and H. Vogel, Eds. (Academic Press, New York, 1965), p. 341.
43. C. B. Bridges, *J. Heredity* **26**, 60 (1935).
44. G. P. Attardi, P. C. Huang, S. Kabat, *Proc. Nat. Acad. Sci. U.S.* **53**, 1490 (1965); *ibid.* **54**, 185 (1965); F. Ritossa, K. Atwood, D. Lindsley, S. Spiegelman, *Nat. Cancer Inst. Monogr.* **23**, 449 (1966); S. A. Yankofsky and S. Spiegelman, *Proc. Nat. Acad. Sci. U.S.* **48**, 1466 (1962); H. Wallace and M. Birnstiel, *Biochim. Biophys. Acta* **114**, 296 (1966).
45. P. Borst, G. Ruttenberg, A. M. Kroon, *Biochim. Biophys. Acta* **149**, 140, 156 (1967); I. B. Dawid and D. R. Wolstenholme, *Biophys. J.*, in press; S. Granick and A. Gibor, in *Progr. Nucleic Acid Res. Mol. Biol.* **6**, 143 (1967).
46. C. A. Thomas, in *Progr. Nucleic Acid Res. Mol. Biol.* **5**, 315 (1966).
47. R. B. Church and B. J. McCarthy, *J. Mol. Biol.* **23**, 459, 477 (1967); D. D. Brown and J. B. Gurdon, *ibid.* **19**, 399 (1966); H. Wallace and M. Birnstiel, *Biochim. Biophys. Acta* **114**, 296 (1966); A. H. Whiteley, B. J. McCarthy, H. R. Whiteley, *Proc. Nat. Acad. Sci. U.S.* **55**, 519 (1966); H. Denis, *J. Mol. Biol.* **22**, 269, 285 (1966); V. R. Glisin, M. V. Glisin, P. Doty, *Proc. Nat. Acad. Sci. U.S.* **56**, 285 (1966); M. Nemer and A. A. Infante, *Science* **150**, 217 (1965); A. Spirin and M. Nemer, *ibid.*, p. 214; M. Crippa, E. Davidson, A. E. Mirsky, *Proc. Nat. Acad. Sci. U.S.* **57**, 885 (1967).
48. W. Gray, W. Dreyer, L. Hood, *Science* **155**, 465 (1967); S. Cohen and C. Milstein, *Nature* **214**, 449 (1967); G. M. Edelman and J. A. Gally, *Proc. Nat. Acad. Sci. U.S.* **57**, 356 (1967).
49. F. Crick (personal communication) has proposed that the repetitive segments of the DNA play their role structurally through interaction of these particular nucleotide sequences with the proteins of the chromosome.
50. G. G. Simpson, *This View of Life* (Harcourt, Brace and World, New York, 1964), p. 164.
51. We thank M. Chamberlin for excellent technical assistance; our colleagues R. B. Roberts, D. B. Cowie, E. T. Bolton, D. J. Brenner, and A. Rake for direct participation in the preparation of this article; B. H. Hoyer, D. Axelrod, and M. Martin for helpful discussion and for several DNA preparations; R. B. Roberts for proposing the schematic diagram (Fig. 13); F. R. Boyd and J. L. England for assistance in the design of the high pressure pump; and our colleagues and wives for patience and forbearance.

Reprinted from *Science* **161**: 1020–1022.

Evolutionary Criteria in Thallophytes: A Radical Alternative

Abstract. *The classical assumptions, upon which all previous phylogenies for the lower plants (Thallophytes) have been based, are claimed to be erroneous. An alternative view, that the eukaryotic cell arose in the late Precambrian from prokaryotic ancestors by a specific series of symbioses, is referred to here. Mutually consistent phylogenies, one for the prokaryotes, another for the lower eukaryotes, can be constructed on the basis of the symbiotic theory. The resulting prokaryote phylogeny is presented here; it is claimed to be more consistent with cytological data, measured DNA base ratios, and the fossil record than the several classical partial phylogenies for Thallophytes recently published.*

Klein and Cronquist have recently assembled data relevant to the possible phylogenetic relationships among the lower organisms (*1*). It is evident from their presentation of at least 14 different, and often mutually exclusive, "partial phylogenies" (*1*, figs. 20 and 22, a and b; scheme A, B, and C, p. 26, for example) that these new data do not clarify evolutionary relationships in the group as a whole. Taxonomic schemes should help us to make predictions. When we are told that a giraffe is a mammal, we infer that the female suckles her young. Without knowing anything else about *Acer pseudoplatanus* except that it is dicotyledonous, one can deduce that it photosynthesizes and that it has true leaves, roots and stems, flowers, and many other traits. These concepts, so obvious to the great evolutionists such as Simpson (*2*), have been often ignored by many new "biochemical evolutionists" [for example (*3*)] who tend to disregard whole organisms —the objects upon whose populations selection in the natural environment acts.

A new approach to phylogeny of the Thallophytes is obviously needed, and one such approach is suggested here. In Table 1 the principles upon which it is based are compared with the conventional ones of Klein and Cronquist (*1*). They have been discussed in much greater detail elsewhere (*4*). On the basis of these alternative principles a single, unified phylogenetic tree for almost all prokaryotic and eukaryotic organisms can be devised. The basic concept of the origin of eukaryotes from prokaryotes by a series of specific symbioses is outlined in Fig. 1. Details of possible derivations of various well-known and presumably natural prokaryotic groups are shown in Fig. 2. The scheme illustrating evolution of the various eukaryotic lines has already been published. For the details of the right side of Fig. 1, see fig. 1, p. 228 of (*4*). No attempt has been made here to use any but common names. Although, no doubt, there are errors in the details of the scheme, there is no datum known to the author that contradicts the idea. This is true of both the geological record (*5*) and modern biochemical data. [For example, see (*6*) for relationships between plastids and blue-green algae; see (*7*) for relationships between bacteria and blue-green algae, and (*8*) for a possible phylogenetic status of the

Table 1. Evolutionary criteria in Thallophytes.

Assumptions of Klein and Cronquist (*1*)	Alternative assumptions (*4*)
1. The basic dichotomy between organisms of the present-day world is between Animals and Plants.	The basic dichotomy between organisms of the present-day world is between Prokaryotes and Eukaryotes.
2. Photosynthetic eukaryotes (higher plants) evolved from photosynthetic prokaryotes (blue-green algae, "ur-algae").	Photosynthetic eukaryotes (higher algae, green plants) and nonphotosynthetic eukaryotes (animals, fungi, protozoans) evolved from a common nonphotosynthetic (amoebo-flagellate) ancestor. There is not now, nor was there ever, an "uralga."
3. The evolution of plants and their photosynthetic pathways occurred monophyletically on the ancient earth.	The evolution of photosynthesis occurred on the ancient earth in bacteria and blue-green algae; higher plants evolved abruptly from prokaryotes when the heterotrophic ancestor (2 above) acquired plastids by symbiosis.
4. Animals and fungi evolved from plants by loss of plastids.	Animals and most eukaryotic fungi evolved directly from protozoans.
5. Mitochondria differentiated in the primitive plant ancestor.	Mitchondria were present in the primitive eukaryote ancestor when plastids were first acquired by symbiosis.
6. The primitive plant differentiated the complex flagellum, the mitotic system, and all of the other eukaryote organelles.	Mitosis evolved in heterotrophic eukaryotic protozoans by differentiation of the complex flagellar system.
7. All organisms evolved from a primitive ancestor monophyletically by single steps.	All prokaryotes evolved from a primitive ancestor by single mutational steps; all eukaryotes evolved from a primitive eukaryote ancestor by single mutational steps. Eukaryotes evolved from prokaryotes by a specific series of symbioses.
8. Morphological, biochemical, and physiological characters are useful in classification of Thallophytes.	Only total gene-based biochemical pathways resulting in the production of some selectively advantageous markers are reliable "characters" in classification; morphology is useless in most prokaryotes (Fig. 2).
Result of Foregoing Assumptions	
Nothing predicted; no consistent phylogeny possible, many predicted organisms not found, for example "uralgae"; no correlation with fossil record possible; no presentation of phylogeny as a function of time elapsed is possible.	Major biochemical pathways predicted; consistent phylogeny constructed; biological discontinuity at Precambrian boundary predicted.

Table 2. The Four Kingdom Classification modified after Copeland (*16*).

Kingdom	Examples of organisms	Approximate time of evolution (millions of years ago	Major traits that environmental selection pressures acted on to produce	Most significant selective factor
Monera	All prokaryotes; bacteria, blue-green algae, actinomycetous fungi, and so forth	Early-Middle Precambrian (3000–1000)	Photosynthesis and aerobiosis	Solar radiation, increasing atmospheric oxygen concentrations
Protoctista	All "higher" (eukaryotic) algae: green, yellow-green, red and brown, and so forth; all protozoans, phycomycetous fungi, ascomycetes, basidiomycetes and so forth	Late Precambrian Early Paleozoic (1500–500)	Classical mitosis and meiosis: obligate recombination each generation; more efficient nutrition	Depletion of organic nutrients
Animalia	Metazoa: all animals developing from zygotes	Paleozoic (600 on)	Tissue development for heterotrophic specializations	Transitions from aquatic to terrestrial and aerial environments
Plantae	Metaphyta: all green plants (above green algae)	Paleozoic (600 on)	Tissue development for autotrophic trophic specializations	Transitions from aquatic to terrestrial environments

mitochondrion.] The fact that the DNA base ratios (*6, 9*) can be easily superimposed on the chart (Fig. 2) lends credence to the idea that the concept is correct.

If the genus *Cyanidium* (*1*, p. 219) had in fact neither mitochondria nor endoplasmic reticulum, it might have represented an inexplicable "uralgan" contradiction, for the theory (*4*) predict that no plastid-containing organisms without mitochondria ever evolved. However, recent electron micrographs (*10*) shows that *Cyanidium* is in these respects a typical eukaryote. That some eukaryotic algae may have lost their originally symbiotic plastids, and later reestablished new symbioses with somewhat different forms of blue-green algae, is indeed to be expected. Such anomolous symbioses could include not only *Cyanidium* but also *Cyanophora paradoxa* and *Glaucocystis nostochinearum* (*11*).

The amassed data fit the proposed phylogeny as well as they do any of the several schemes presented by Klein and Cronquist (*1*). Furthermore, the symbiotic theory enables one to make many predictions [for example, that the usual

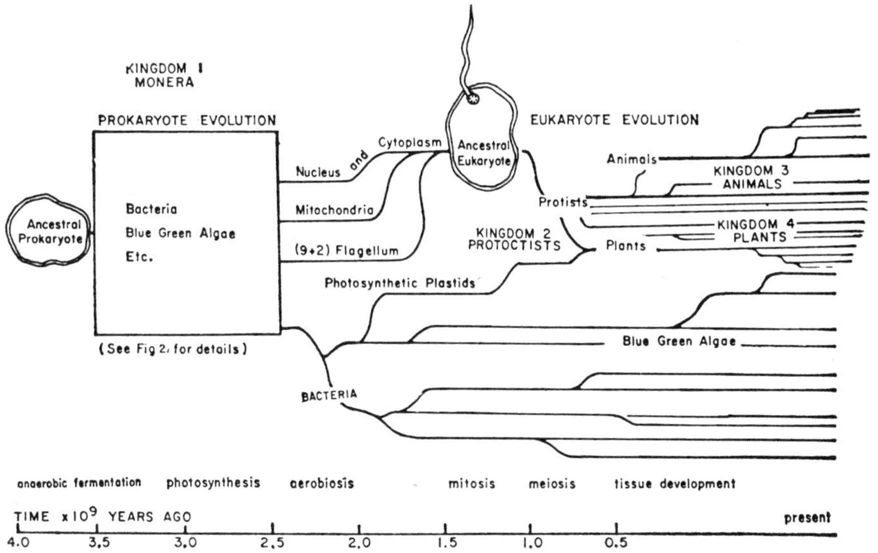

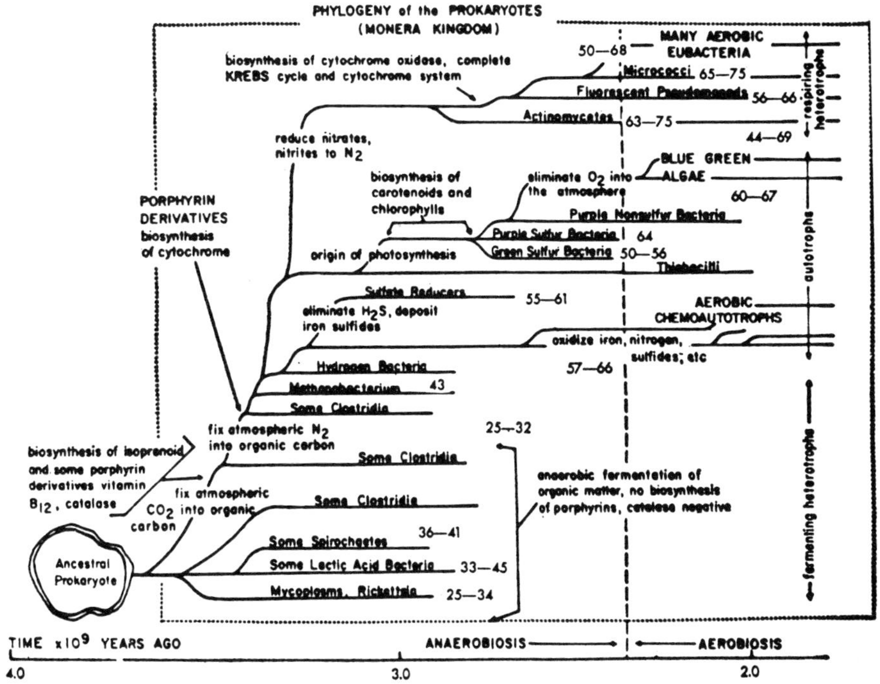

Fig. 1 (right, middle). Summary of the symbiotic theory of the origin of cells (*4*).

Fig. 2 (right, bottom). Summary diagram of the evolution in prokaryotes, based on principles discussed by Margulis (*4*). The numbers represent the ranges of DNA base ratios expressed in percentage of guanine plus cytodine.

pathway of the Krebs cycle and cytochrome electron transport oxidations will be found lacking in detail in prokaryotic chemo- and photoautotrophs, although present in all eukaryotic photoautotrophs; that all cells containing chloroplasts also have membrane-bound nuclei and can produce steroid derivatives; that all cells with the complex (9 + 2) flagellum of eukaryotic cells (*12*) must also contain cytochrome oxidase and other mitochondrial enzymes (*13*); that all cells with the "higher chromosomes" seen in classical mitosis (for example, red algae and ascomycetes) had a (9 + 2) flagellated ancestor and retain the relevant DNA of that "protoflagellum" (*4*) even if they lack visible (9 + 0) centrioles and basal bodies (*14*); that all eukaryotes potentially form the colchicine-sensitive protein of the microtubules (*15*); that all eukaryotic plant cells contain at least three different nonnuclear ("satellite") DNA's; and that steroid and flavonoid derivatives will be found only in relatively young sediments—much younger than those which first contain photosynthetically reduced carbon]. By challenging the students of the enormously diverse Thallophytes to find contradictions to the theory proposed here, perhaps some appropriately focused research will be stimulated. If it proves generally acceptable, the division of living organisms into four kingdoms proposed by Copeland (*16*) logically follows (Fig. 1 and Table 2).

LYNN MARGULIS

Department of Biology,
Boston University,
Boston, Massachusetts

References and Notes

1. R. Klein and A. Cronquist, *Quart. Rev. Biol.* **42**, 105 (1967).
2. G. G. Simpson, *Major Features of Evolution* (Columbia Univ. Press, New York, 1953).
3. S. Aaronson and S. H. Hutner, *Quart. Rev. Biol.* **41**, 13 (1966).
4. L. Sagan, *Theoret. Biol.* **14**, 225 (1967). This paper was written by the present author under a former name.
5. J. W. Schopf, in *McGraw-Hill Yearbook of Science and Technology* (McGraw-Hill, New York, 1967), p. 46.
6. M. Edelman, D. Swinton, J. Schiff, B. Zeldin H. Spstein, *Bact. Rev.* **31**, 315 (1967).
7. P. Echlin and I. Morris, *Biol. Rev.* **40**, 193 (1965).
8. P. Borst, A. M. Kroon, C. J. C. M. Ruttenberg, *Genetic Elements Properties and Function*, D. Shugar, Ed. (Academic Press, London, 1967), p. 81.
9. L. R. Hill, *J. Gen. Microbiol.* **44**, 419 (1966); J. Marmur, S. Falkow, M. Mandel, *Ann. Rev. Microbiol.* **17**, 329 (1963).
10. R. Troxler, personal communication.
11. W. T. Hall and G. Claus, *J. Phycol.* **3**, 37 (1967).
12. I. R. Gibbons, in *Formation and Fate of Cell Organelles*, K. B. Warren, Ed. (Academic Press, New York, 1967).
13. Subsequent dedifferentiation or secondary loss of mitochondria, for example, in yeast and trypanosomes may of course occur (*4*).
14. E. Stubblefield and B. Brinkley, in *Formation and Fate of Cell Organelles*, K. B. Warren, Ed. (Academic Press, New York, 1967); D. Mazia, *ibid.*; A. V. Grimstone, *ibid.*
15. G. Borisy and E. W. Taylor, *J. Cell. Biol.* **34**, 535 (1967).
16. H. F. Copeland, *Classification of the Lower Organisms* (Pacific Books, Palo Alto, Calif., 1956).
17. I thank Profs. R. Lewin and R. Estes for advice, and the Boston University Graduate School for support.

The Origin of the Genetic Code

F. H. C. Crick

Medical Research Council
Laboratory of Molecular Biology
Hills Road, Cambridge, England

(*Received 21 August 1968*)

The general features of the genetic code are described. It is considered that originally only a few amino acids were coded, but that most of the possible codons were fairly soon brought into use. In subsequent steps additional amino acids were substituted when they were able to confer a selective advantage, until eventually the code became frozen in its present form.

Introduction

The substance of this paper was originally presented at a meeting of the British Biophysical Society in London on 20 December 1966.

A very brief account appeared shortly after in a letter to *Nature* (Crick, 1967*a*). When this manuscript was in its first draft, Dr Leslie Orgel told me that he had already prepared a draft of a paper on a related theme. We therefore decided to publish our two papers together and have collated them to some extent to avoid overlap. We have not done this for all passages in the two papers which touch on the same topic, preferring on occasions to let our slightly different points of view be expressed as differences in treatment and emphasis. However, broadly speaking, each of us agrees with the opinion expressed by the other.

Since this paper was originally drafted a very full discussion has appeared in Woese's book *The Genetic Code,* which should be consulted for a fuller discussion of many of the points touched on here.

The Structure of the Present Genetic Code

The structure of the genetic code is now fairly well known. The code is a non-overlapping triplet code. Most, but not all, of the 64 triplets stand for one or another of the 20 amino acids and, in most cases, each amino acid is represented by more than one codon. The best present version of the code is shown in Table 1. This is taken from the 1966 Cold Spring Harbor Symposium on *The Genetic Code,* to which the reader is referred as a source of references for many of the topics discussed here.

Before starting on a detailed examination of this Table a few words of caution are necessary Although the code shown there has been mainly derived from studies on *Escherichia coli,* it must be very similar in such widely different organisms as tobacco plants and man. In what follows I shall assume, for convenience of exposition, that it is identical in all organisms, which is very far from being proved. In fact, it is probably untrue for the starting codons.

TABLE 1

The genetic code

1st ↓ 2nd →	U	C	A	G	↓ 3rd
U	PHE	SER	TYR	CYS	U
	PHE	SER	TYR	CYS	C
	LEU	SER	Ochre	?	A
	LEU	SER	Amber	TRP	G
C	LEU	PRO	HIS	ARG	U
	LEU	PRO	HIS	ARG	C
	LEU	PRO	GLN	ARG	A
	LEU	PRO	GLN	ARG	G
A	ILE	THR	ASN	SER	U
	ILE	THR	ASN	SER	C
	ILE	THR	LYS	ARG	A
	MET	THR	LYS	ARG	G
G	VAL	ALA	ASP	GLY	U
	VAL	ALA	ASP	GLY	C
	VAL	ALA	GLU	GLY	A
	VAL	ALA	GLU	GLY	G

This Table shows the "best allocations" of the 64 codons at the time of the Symposium. Some of these allocations are less certain than others. The two codons marked ochre and amber are believed to signal the termination of the polypeptide chain. The codons suspected of being concerned with chain initiation are not indicated here.

Again the function of the three presumed "nonsense" triplets is not known for certain. It is presumed that UAA (ochre) and UAG (amber) are signals for chain termination and probably UGA as well, at least in bacteria.

In *E. coli* there appears to be a special mechanism for initiating the polypeptide chain, involving formylmethionine and the codons AUG and GUG. The mechanism in higher organisms (if indeed a special one exists) is unknown.

Finally, it is uncertain whether there are ambiguous codons; that is, codons which represent more than one amino acid. Of course, it is known that mutations can produce errors in the translation mechanism and so make certain codons ambiguous, but it is not known whether ambiguity occurs "normally". Again in what follows I shall assume that this is not usually the case for present-day organisms.

The basic reason why one can ignore these complications and uncertainties for the moment is that the broad features of the genetic code are not likely to be greatly affected by them. What, then, are the properties of the code which require explanation?

There are some features which are of such a general type that they do not depend at all upon the details of the code.

They are:

(1) there are 4 distinct bases in the mRNA,

(2) each codon is a triplet of bases,

(3) only 20 of the numerous possible amino acids are used.

In examining Table 1, however, one is apt to take all these characteristics for granted. What, then, is special about the actual details of the genetic code?

The main features, which have frequently been commented on, are:

(4) The 20 amino acids are not distributed at random among the 64 triplets. In fact, several rules can easily be deduced from the Table. For example,

(a) XYU and XYC always code the same amino acid.

(b) XYA and XYG often code the same amino acid. The rare amino acids, methionine and tryptophan, which have only one codon each, appear to be exceptions to this rule.

(c) In half the cases (8 out of 16) XY· represents a single amino acid, where the · implies that all four bases are possible.

(d) In most cases the codons representing a single amino acid start with the same pair of bases. Thus the two codons for histidine both start with CA. There are three exceptions to this:

Leucine has CU· and UU^{A}_{G}.
Serine has UC· and AG^{U}_{C}.
Arginine has CG· and AG^{A}_{G}.

(e) If the first two bases consist only of G's and C's, then the four codons sharing the same initial doublet all code the same amino acid. That is, the meaning of these codons is independent of the third base. This is in fact true for all codons having C in the second position. More complicated rules along these lines can be produced for the remaining codons but they seem to me to be rather forced.

(5) Even allowing for the grouping of codons into sets, the amino acids do not seem to be allocated in a totally random way. For example, all codons with U in the second place code for hydrophobic amino acids. The basic and acidic amino acids are all grouped near together towards the bottom right-hand side of Table 1. Phenylalanine, tyrosine and tryptophan all have codons starting with U, and so on. It is very difficult not to imagine regularities in even a random grouping but nevertheless the general impression is that "related" amino acids have to some extent related codons (Epstein, 1966).

(6) The code is universal (the same in all organisms) or nearly so.

Why is the Code Universal?

Two extreme theories may be described to account for this, though, as we shall see, many intermediate theories are also possible.

The Stereochemical Theory

This theory states that the code is universal because it is necessarily the way it is for stereochemical reasons. Woese has been the main proponent of this point of view (see Woese, 1967). That is, it states that phenylalanine *has* to be represented by UU^{U}_{C}, and by no other triplets, because in some way phenylalanine is stereochemically "related" to these two codons. There are several versions of this theory. We shall examine these shortly when we come to consider the experimental evidence for them.

The Frozen Accident Theory

This theory states that the code is universal because at the present time *any change would be lethal*, or at least very strongly selected against. This is because in all organisms (with the possible exception of certain viruses) the code determines (by reading the mRNA) the amino acid sequences of so many highly evolved protein molecules that any change to these would be highly disadvantageous unless accompanied by many simultaneous mutations to correct the "mistakes" produced by altering the code.

This accounts for the fact that the code does not change. To account for it being the same in all organisms one must assume that all life evolved from a single organism (more strictly, from a single closely interbreeding population). In its extreme form, the theory implies that the allocation of codons to amino acids at this point was entirely a matter of "chance".

The Stereochemical Theory—Experimental Evidence

In its extreme form, the stereochemical theory states that the postulated stereochemical interactions are still taking place today. It should therefore be a simple matter to prove or disprove such theories.

Pelc and Welton (Pelc & Welton, 1966; Welton & Pelc, 1966) have suggested from a study of models that there is in many cases a specific stereochemical fit between the amino acid and the base sequence of its *codon* on the appropriate tRNA. Unfortunately, their models were all built backwards (Crick, 1967*b*) so their claims are without support. Such a theory implies that the expected codon sequence occurs somewhere on each tRNA. For example, no such sequence occurs in the tRNA for tyrosine either from yeast (Madison, Everett & King, 1966) or from *E. coli* (Goodman, Abelson, Landy, Brenner & Smith, 1968). In our opinion this idea has little chance of being correct.

A more reasonable idea is that the amino acid fits the *anticodon* on the tRNA. At least this has the advantage that it is always present. A model along these lines for proline has been briefly described by Dunnill (1966), but so far no detailed description has been published, nor has he extended his model-building to other amino acids.

The experimental evidence has already established that when the activating enzyme transfers the amino acid to the tRNA, the interaction is not solely with the anticodon and the common . . . CCA terminal sequence. This is shown by the fact that an activating enzyme from one species will not always recognize the appropriate tRNA from a different species although the anticodons must be very similar if not identical in different species (for a summary of the data, see Woese, 1967, p. 125). However, this does not preclude the idea that the interaction is partly with the anticodon and partly with some other part of the tRNA.

The best way to disprove the theory (if indeed it is false) would be to change the anticodon of some tRNA molecule and show that nevertheless it accepted the same amino acid from the activating enzyme. This has already been done for the minor tyrosine tRNA of *E. coli* whose anticodon has been changed (in an Su^+ strain) from GUA to CUA (Goodman *et al.*, 1968) although the experiments need to be done quantitatively. Further examples of such changes are likely to be reported in the near future. Until this is done we must reserve final judgement on the amino acid–anticodon interaction theory; but we consider it unlikely to be correct, except perhaps in a few special cases.

Even if it were established that the activating enzyme recognizes the anticodon, this would not by itself prove that the recognition is done by inserting the amino acid in a cage formed by the anticodon. Notice that the activating enzyme would have to release amino acid from its own recognition cavity and then insert it into the recognition site on the tRNA. Moreover, when the amino acid has been transferred to the tRNA and the activating enzyme has diffused elsewhere, the amino acid could not stay in the anticodon cage without blocking the interaction with the codon on the mRNA. None of this is impossible but it is certainly elaborate.

It is not easy to see at this stage what evidence would be needed to prove that the anticodon does indeed form a cage for the amino acid, though if the tRNA (or perhaps a fragment of it) could be crystallized it might be possible to see the amino acid sitting in such a position.

The present experimental evidence, then, makes it unlikely that every amino acid interacts stereochemically with either its codon or its anticodon. It by no means precludes the possibility that *some* amino acids interact in either of these ways, or that such interactions, even though now not used, may have been important in the past, at least for a few amino acids. We must now leave the system as it is today and turn to the examination of primitive systems.

The Primitive System

It is almost impossible to discuss the origin of the code without discussing the origin of the actual biochemical mechanisms of protein synthesis. This is very difficult to do, for two reasons: it is complex and many of its details are not yet understood. Nevertheless, we shall have to present a tentative scheme, otherwise no discussion is possible.

In looking at the present-day components of the mechanism of protein synthesis, one is struck by the considerable involvement of non-informational nucleic acid. The ribosomes are mainly made from RNA and the adaptor molecules (tRNA) are exclusively RNA, although modified to contain many unusual bases. Why is this? One plausible explanation, especially for rRNA, is that RNA is "cheaper" to make than protein. If a ribosome were made exclusively of protein the cell would need *more* ribosomes (to make the extra proteins, which would not be a negligible fraction of all the proteins in the cell) and thus could only replicate more slowly. Even though this may be true, we cannot help feeling that the more significant reason for rRNA and tRNA is that *they were part of the primitive machinery* for protein synthesis. Granted this, one could explain why their job was not taken over by protein, since

(i) for rRNA, it would be too expensive,

(ii) for tRNA, protein may not be able to do such a neat job in such a small space.

In fact, as has been remarked elsewhere, tRNA looks like Nature's attempt to make RNA do the job of a protein (Crick, 1966).

If indeed rRNA and tRNA were essential parts of the primitive machinery, one naturally asks how much protein, if any, was then needed. It is tempting to wonder if the primitive ribosome could have been made *entirely* of RNA. Some parts of the structure, for example the presumed polymerase, may now be protein, having been replaced because a protein could do the job with greater precision. Other parts may not have been necessary then, since primitive protein synthesis may have been rather inefficient and inaccurate. Without a more detailed knowledge of the structure of present-day ribosomes it is difficult to make an informed guess.

It is not too difficult to imagine that the early tRNA molecules had no modified bases (so that no modifying enzymes were needed), but it is much more difficult to decide whether activating enzymes were then essential. An attractive idea (suggested to us by Dr Oliver Smithies) is that the primitive tRNA was its own activating enzyme. That is, that its structure had a cavity in it which specifically held the side-chain of the appropriate amino acid in such a position that the carboxyl group could be easily joined on to the terminal ribose of the tRNA.

It is thus not impossible to imagine that the primitive machinery had no protein at all and consisted entirely of RNA. This is discussed at much greater length in the companion paper by Dr L. E. Orgel, where the importance of the ease of replication of nucleic acid is emphasized. We are faced with the question of the origin of all this RNA. Could the appropriate sequences have arisen by chance? We do not feel this is totally impossible, for three reasons.

(a) Some natural catalyst (such as a mineral) for random nucleotide polymerisation may exist. If this were so, RNA may have been made at very many places on the earth's surface over a very considerable period of time, so that altogether an enormous number of different sequences may have been synthesized. It is difficult to assess the value of this idea, since such a natural catalyst has not yet been discovered. Another possibility is that a crude template mechanism developed at an early stage. This is fully discussed in the companion paper.

(b) The mechanism of "random" synthesis may preferentially produce structures with multiple loops (this is also discussed in the companion paper) so that sequences of this sort (which are indeed found in tRNA and rRNA) may have been synthesized preferentially. Moreover, the actual base-pairs used in the base-paired regions may not be critical for their structures. In short, the synthesis of an acceptable rRNA and tRNA may not have been so unlikely as it seems at first sight.

(c) The base-sequences needed may have been repetitive. For example, the early tRNA molecules may have been very alike, only differing in the anticodon and in the region of the presumed cavity. For all we know, the structure of the large rRNA molecules may have been partly repetitive. These repetitions might have been produced rather easily if there were an RNA replicase available. Possibly the first "enzyme" was an RNA molecule with replicase properties. Thus a system based mainly on RNA is not impossible. Such a system could then start to synthesize protein and thus could evolve very rapidly by natural selection. We shall not discuss here the difficult problem of how the various components were kept together, that is, the origin of a cell.

The point of this sketch is to impress the reader with the great difficulty of the problem. It would certainly be easier if specific stereochemical interactions could occur between amino acids and triplets of bases, but even if these are possible the origin of the present ribosomal translation mechanism presents grave difficulties.

The Primitive Code

We must now tackle the nature of the primitive code and the manner in which it evolved into the present code.

It might be argued that the primitive code was not a triplet code but that originally the bases were read one at a time (giving 4 codons), then two at a time (giving 16 codons) and only later evolved to the present triplet code. This seems highly unlikely, since it violates the Principle of Continuity. A change in codon size necessarily makes nonsense of *all* previous messages and would almost certainly be lethal. This is quite different from the idea that the primitive code was a triplet code (in the sense that the reading mechanism moved along three bases at each step) but that only, say, the first two bases were read. This is not at all implausible.

The next general point about the primitive code is that it seems likely that only a few amino acids were involved. There are several reasons for this. It certainly seems unlikely that all the present amino acids were easily available at the time the code started. Certainly tryptophan and methionine look like later additions. Exactly which amino acids were then common is not yet clear, though most lists would include glycine, alanine, serine and aspartic acid. However, if stereochemical interaction played a part in the primitive code, this might select amino acids which were available but not particularly common. Again, it seems unlikely that the primitive code could code *specifically* for more than a few amino acids, since this would make the origin of the system terribly complicated. However, as Woese (1965) has pointed out, the primitive system might have used *classes* of amino acids. For example, only the middle base of the triplet may have been recognized, a U in that position standing for any of a number of hydrophobic amino acids, an A for an acidic one, etc.

Even though few amino acids (or groups of amino acids) were recognized, it seems likely that not too many nonsense codons existed, otherwise any message would have had too many gaps. There are various ways out of this dilemma. For example, as mentioned above, only one base of the triplet might have been recognized. Another possibility, however, is that the early message consisted not of the present four bases, but perhaps only two of them.

The Number of Bases in the Primitive Nucleic Acid

The only strong requirements for the primitive nucleic acid is that it should have been easy to replicate, and that it should have consisted of more than one base, otherwise it could not carry any information in its base sequence. One cannot even rule out the possibility that the base sequence of the two chains was complementary (as in the present DNA). Perhaps a structure is possible with only two bases in which the two chains run parallel (rather than anti-parallel) and pairing is like-with-like. It would certainly be of great interest if such a structure could be demonstrated experimentally.

Leaving this possibility on one side and restricting ourselves to complementary structures, we see that the number of bases must be even. If there were only two in the primitive DNA, the question arises as to which two. The obvious choices are either A with U (or T) or G with C. A less obvious possibility (suggested some time ago by Dr Leslie Orgel, personal communication) is A with I (where I stands for inosine, having the base hypoxanthine). It is not certain that a double helix can be formed having a random sequence of A's and I's on one chain and the complementary sequence (dictated by A–I or I–A pairs) on the other chain, but it is not improbable, especially as the RNA polymers poly A and poly I can form a double helix.

Several advantages could be claimed for this scheme. Adenine is likely to be the commonest base available in the primitive soup, and inosine could arise from it by deamination. Thus the supply of precursors might be easier than in the case of the other two alternatives, though how true this is remains to be established. Then again in a random (A, I) sequence I would presumably code in the same way as G does now, at any rate for the first two positions of the triplet. If we can use the present code as a guide (though we shall argue later that this may be misleading), it is noticeable that the triplets containing only A's or G's in their first two bases (the bottom right-hand corner of the Table) do indeed code for some of the more obviously primitive amino acids.

It is important to notice that a scheme of this sort (or even one with like-with-like pairing) does not violate the principle of continuity. To change over from an (A, I) double helix to one like the present one but having A, I, U and C, the only steps required are a change in the replicase to select smaller base-pairs, and a supply of the two new precursors. The message carried (by the "old" chain) is unaltered by this step. Gradually mutations would produce U's and C's on this chain and the new codons thus produced could be brought into use as the mechanism for protein synthesis evolved. Eventually G would be substituted for I. At no stage would the message become complete nonsense. The idea that the initial nucleic acid contained only two bases is thus a very plausible one. It remains to be seen whether primitive ribosomal RNA and primitive tRNA could be constructed using only two bases.

The Stereochemical Alternative

As stated earlier, it seems very unlikely that there is any stereochemical relationship between all the present amino acids and specific triplets of bases; but it is by no means ruled out that a few amino acids can interact in this way. If this were possible, it would certainly help in the initial stages of the evolution of the code. However, sooner or later a transition would have had to be made to the present type of system, involving tRNA's, ribosomes, etc. It seems to us that this could only happen easily if the code at that stage was fairly simple and only coded a rather small number of amino acids.

The Evolution of the Primitive Code

Whatever the early steps in the evolution of the code, it seems highly likely that it went through a stage when only a few amino acids were coded. At this stage either the mechanism was rather imprecise and thus could recognize most of the triplets, or only a few triplets were used, perhaps because the message contained only two types of base. We must now consider what would happen next.

A complication should be introduced into this simple picture. It could well be that at this stage the recognition mechanisms were not very precise and that any given codon corresponded to a *group* of amino acids (see Woese, 1965, who has stressed this point). Thus codons for alanine might also incorporate glycine, those for threonine might also code serine, etc. However, it is by no means certain that this happened. It seems highly likely that a "cavity" to accept threonine would also accept serine to some extent, but the converse mistake is less likely and could depend on the exact nature of the structure involved. Thus, though the early coding machinery probably produced errors, we can only guess at their extent.

We shall argue that by far the most likely step was that these primitive amino acids spread all over the code until almost all the triplets represented one or other of them. Our reasons for believing this are that too many nonsense triplets would certainly be selected against, so that most codons would quickly be brought into use (Sonneborn, 1965). In addition, it would be easier to produce a new tRNA, altered only in its anticodon, while still recognizing the amino acid, than to produce both a new anticodon and a new recognition system for attaching a new amino acid. Thus, we can reasonably expect that the intermediate code had two properties:

(i) few amino acids were coded, and

(ii) almost all the triplets could be read.

Moreover, because of the way this primitive code originated, the triplets standing for any one amino acid are likely to be related. At this stage the organism could only produce rather crudely made protein, since the number of amino acids it could use was small and the proteins had probably not evolved very extensively.

The final steps in the evolution of the code would involve an increase in the precision of recognition and the introduction of new amino acids. The cell would have to produce a new tRNA and a new activating enzyme to handle any new amino acid, or any minor amino acid already incorporated because of errors of recognition. This new tRNA would recognize certain triplets which were probably already being used for an existing amino acid. If so, these triplets would be ambiguous. To succeed, two conditions would have to be fulfilled.

(1) The new amino acid should not upset too much the proteins into which it was incorporated. This upset is least likely to happen if the old and the new amino acids are related.

(2) The new amino acid should be a positive advantage to the cell in at least one protein. This advantage should be greater than the disadvantages of introducing it elsewhere.

In short, the introduction of the new amino acid should, on balance, give the cell a reproductive advantage.

For the change to be consolidated we would expect many further mutations, replacing the ambiguous codons by other codons for the earlier amino acid when this was somewhat better for a protein than the later one. Thus, eventually the codons involved would cease to be ambiguous and would code only for the new amino acid.

There are several reasons why one might expect such a substitution of one amino acid for another to take place between structurally similar amino acids. First, as mentioned above, such a resemblance would diminish the bad effects of the initial substitution. Second, the new tRNA would probably start as a gene duplication of the existing tRNA for those codons. Moreover, the new activating enzyme might well be a modification of the existing activating enzyme. This again might be easier if the amino acids were related. Thus, the net effect of a whole series of such changes would be that *similar amino acids would tend to have similar codons*, which is just what we observe in the present code.

It is clear that such a mechanism for the introduction of new amino acids could only succeed if the genetic message of the cell coded for only a small number of proteins and especially proteins which were somewhat crudely constructed. As the process proceeded and the organism developed, more and more proteins would be coded and their design would become more sophisticated until eventually one would reach a point where no new amino acid could be introduced without disrupting too many proteins. At this stage the code would be frozen. Notice that it does not necessarily follow that the original codons, of the original primitive code (as opposed to the intermediate code) will necessarily keep their assignments to the primitive amino acids. In other words, the evolution of the code may well have wiped out all trace of the primitive code. For this reason arguments about which base-pair came into use first on the nucleic acid should not depend too heavily on the assignments of the present code.

The idea described above is crucial to the evolution of the code. It seems to me not to be the same as the idea, suggested by several authors (Sonneborn, 1965; Goldberg & Wittes, 1966), that the code is designed to minimize the effects of mutations. The implication is that the mutations are those occurring in the many proteins of the organism, and in fact are still occurring today. This is not quite the same as the idea that it is the situation produced by the introduction of a new amino acid to the *developing* code that we have to consider. Moreover, the disturbances had to be minimized not to the present day proteins but to the small number of more primitive proteins then existing. The minimizing of the effects of mutations is in any case likely to have only a small selective advantage even at the present time, and I think it unlikely that it could have had any appreciable effect in moulding the genetic code. Woese (1967) has made the same point.

An idea rather close to the one presented above has been developed by Woese (1965). He emphasizes in his discussion the fact that the early translation mechanism would probably be prone to errors. This is indeed an important idea and may well be what actually occurred but it is not identical to the idea suggested above, as can be easily seen by making the rather unlikely assumption that the early mechanism was rather accurate. In this case Woese's ideas are irrelevant and one is driven to the scheme outlined above. Nevertheless, Woese's discussion (Woese, 1967) follows much the same line as that presented here. However, he argues that by this mechanism it is unlikely that the code could reach the truly optimum code. There is no reason to believe, however, that the present code is the best possible, and it could have easily reached its present form by a sequence of happy accidents. In other words, it may not be the result of trying all possible codes and selecting the best. Instead, it may be frozen at a local minimum which it has reached by a rather random path.

On the other hand, the basic idea has been very clearly stated by Jukes (1966) in his book *Molecules and Evolution* (p. 70) though he does not give it any particular emphasis.

There is one feature of the process by which new amino acids were added to a primitive code which is far from clear. This is why several versions of the genetic code did not emerge. It is, of course, easy to say that in fact several did emerge and only the best one survived, but the argument is rather glib. A detailed discussion of what was likely to have happened at this period would involve the consideration of genetic recombination. Did it occur at a very early stage, perhaps even before the evolution of the cell, and, if so, what form did it take? Surprisingly enough, no writer on the evolution of the code seems to have raised this point. Naturally only rather simple processes would be expected, but the selective advantages of such a process would be very great. Perhaps a simple fusion process would suffice for the origin of the code (a suggestion made by Dr Sydney Brenner, personal communication). This would provide spare genes for further evolution and in as far as the code for the fusing organisms differed it would produce fruitful ambiguities. One might even argue that the population which defeated all its rivals and survived was the one which first evolved sex, a curious twist to the myth of the Garden of Eden.

General Features of the Code

We must now go back and ask whether we can explain the *general* features of the code in terms of the ideas sketched above.

The Four Distinct Bases

We have argued that originally there may have been only two bases in the nucleic acid. Why should there be four today? The likely answer seems to be that four were stereochemically possible (i.e. could fit into a double-helical structure) and that two was too restrictive a number. If only the first two bases of the triplet were originally distinguished, the mechanism could only code for four things (three amino acids and a space?), and even if the present "wobble" mechanism applied only a maximum of eight things could be coded. This could well be too few to construct really efficient proteins.

Whether six distinct base-pairs are stereochemically possible has been discussed elsewhere (Rich, 1962; Crick, 1964). It should be possible to settle this point experimentally.

Why a Triplet?

We have argued that the code must have been basically a triplet code from a very early stage, so that one is not entitled to use sophisticated arguments which would apply only to a later stage, although one could argue that early organisms with doublet or quadruplet codes actually existed but became extinct, only the triplet code surviving.

However, we are inclined to suspect that the reason in this case may be a structural one. If indeed there is no direct stereochemical relationship between an amino acid and a triplet, the problem of constructing an adaptor to recognize the codon may be a difficult one to solve. In effect, one wants to perform a rather complicated act of recognition *within a rather limited space*, since two adaptors need to lie side by side, and attached to adjacent codons on the mRNA, during the act of synthesis. This is probably very difficult to perform if protein is used for the adaptor. On the other hand, nucleic acid, by employing the base-pairing mechanism, can do a very neat job in a small space.

For various reasons the adaptor cannot be too simple a molecule. For example, the amino acids on adjacent adaptors need to be brought together—this is probably done at the present using the flexible . . . CCA tail. It must have, to some extent, a definite structure and this is likely to be based on stretches of double-helix. Thus the *diameter* of a double-helix (since two may have to lie side by side) may have dictated the *size of the codon*, in that a doublet-code (moving along two bases at a time) would present an impossible recognition problem.

The 20 Amino Acids

According to the theory sketched above, both the number 20 and the actual amino acids in the code are at least in part due to historical accident.

First note that if the wobble theory of the interaction between codon and anticodon is correct, then the maximum number of things which can be coded in a positive way is 32 (say 31 amino acids and a chain terminator) not 64. Thus, the multiple representation of eight of the amino acids is not excessive. On this view, only eight of the 21 things coded appear more than once. If the code evolved as I have suggested, it would in fact be surprising if each amino acid did occur only once. However, the theory of wobble must not be trusted too far, if only because it does not easily explain the fact that UGA codes differently from both UG_{C}^{U} and UGG.

Discussion of the actual amino acids used in the code may not be very profitable. Some less common amino acids, such as cysteine and histidine, would clearly seem to have an advantage because of their chemical reactivity; but whether, say, methionine could be justified in this way seems less obvious. It might be more useful to consider which amino acids are *not* used in the code. However, the answer, if this general scheme is correct, really depends upon very complicated considerations, partly accidental, during the early evolution of the code. In particular, it would depend on the exact nature of the primitive proteins. It seems unlikely that one could come to any firm conclusions by following this line of argument.

As already mentioned, the theory does explain in a general way why similar amino acids often use similar codons. This does not answer the question whether the allocation of particular amino acids is entirely due to chance. However, if it is assumed that the primitive code used tRNA molecules and that the recognition site for the amino acid was distinct from the anticodon, then even if activating enzyme did not exist at this stage and instead the amino acid fitted into a specific cage in the tRNA, the association between amino acid and anticodon *could* be due to pure chance. Thus, a code with this property is not outrageous. Always remember that the present tRNA molecules must necessarily have evolved at *some* time or another.

The Two Theories Contrasted

The evolution of the code sketched here has the property that it could produce a code in which the actual allocation of amino acid to codons is mainly accidental and yet related amino acids would be expected to have related codons. The theory seems plausible but as a theory it suffers from a major defect: it is too accommodating. In a loose sort of way it can explain anything. A second disadvantage is that the early steps needed to get the system going seem to require rather a lot of chance effect. A theory of this sort is not necessarily useless if one can get at the facts experimentally. Unfortunately, in this problem this is just what is so difficult to do. A theory involving stereochemical relationships between amino acids and triplets, on the other hand, not only makes it easier to see how the system could start but there is at least a reasonable chance that well-designed experiments could prove that such specific interactions are possible. It is therefore essential to pursue the stereochemical theory. However, vague models of such interactions are of little use. What is wanted is direct experimental proof that these interactions take place (expressed as binding constants) and some idea of their specificity.

REFERENCES

Crick, F. H. C. (1964). In *Proc. Plenary Sessions 6th Int. Cong. Biochem.* p. 109. *Int. Union Biochem.* vol. **33**. Federation of American Societies for Experimental Biology.
Crick, F. H. C. (1966). *Cold Spr. Harb. Symp. Quant. Biol.* **31**, 3.
Crick, F. H. C. (1967*a*). *Nature,* **213**, 119.
Crick, F. H. C. (1967*b*). *Nature,* **213**, 798.
Dunnill, P. (1966). *Nature,* **210**, 1267.
Epstein, C. J. (1966). *Nature,* **210**, 25.
Goldberg, A. L. & Wittes, R. E. (1966). *Science,* **153**, 420.
Goodman, H. M., Abelson, J., Landy, A., Brenner, S. & Smith, J. D. (1968). *Nature,* **217**, 1019.
Jukes, T. H. (1966). *Molecules and Evolution.* New York: Columbia University Press.

Madison, J. T., Everett, G. A. & King, H. (1966). *Science,* **153**, 531.
Pelc, S. R. & Welton, M. G. E. (1966). *Nature,* **209**, 868.
Rich, A. (1962). In *Horizons in Biochemistry,* ed. by A. Kasha & B. Pullman, p. 103. New York: Academic Press.
Sonneborn, T. M. (1965). In *Evolving Genes and Proteins,* ed. by V. Bryson & H. J. Vogel, p. 377. New York: Academic Press.
Welton, M. G. E. & Pelc, S. R. (1966). *Nature,* **209**, 870.
Woese, C. (1965). *Proc. Nat. Acad. Sci., Wash.* **54**, 1546.
Woese, C. R. (1967). *The Genetic Code.* New York: Harper & Row.

2 SOURCES OF VARIATION

These individual differences are of the highest importance for us, for they are often inherited, as must be familiar to everyone; and they thus afford materials for natural selection to act on and accumulate, in the same manner as man accumulates in any given direction individual differences in his domesticated productions.
Charles Darwin, The Origin of Species (1859, p. 51).

Hereditary variation has always been a central element in the Darwinian theory of evolution. Indeed, one of the most significant innovations by Darwin was his stress on individual variation within populations as a precondition for evolutionary change (Mayr, 1975). Although all of the early Darwinians accepted the importance of intrapopulation variation, the nature of these mutations was to remain mysterious for nearly a century after the publication of *The Origin of Species*.

The concept of mutation has undergone its own evolution from the time of its first use. The 19th century paleontologist Wilhelm Waagen first employed the term to denote sharp morphological discontinuities in fossil series of ammonites (Dobzhansky *et al.*, 1977); by the early 1900s Bateson and de Vries used the term mutation specifically for those hereditary discontinuities between generations that caused marked changes in morphology or pigmentation; by 1915 it had come to be employed by Morgan and his associates to denote any hereditary change of either large or small phenotypic effect. In 1923 Muller pointed out that the term mutation was being applied to three categories of genetic change--large scale chromosomal alterations, small scale changes within genes (point mutations), and certain special cases of recombinational events. He proposed that the term be restricted to the second class, point mutations, and this usage has since prevailed.

Since the beginnings of population genetics in the 1920s and 1930s, point mutations and chromosomal rearrangements have been accepted as the primary source of genetic novelty (a position that may require modification in light of the recent discovery of movable genes). The physical basis of point mutations was both unknown and unimaginable until the formulation of the Watson-Crick model of DNA structure. According to this model, a point mutation could be an event as subtle as the substitution of a single base by another through a replication error (Watson and Crick, 1953b). One of the triumphs of molecular genetics in the 1950s and 1960s was the validation of the Watson-Crick hypothesis of transition mutation, and its extension to include other kinds of base-sequence alterations that occur. Nucleotide additions and deletions (Crick *et al.*, 1961; Streisinger *et al.*, 1966) were also found.

These discoveries in molecular genetics raise a question pertinent to the evolution of genes--namely, whether all classes of point mutations contribute equally to gene diversification, or if some are more important than others. The study of mutational change in prokaryotes reveals that base substitutions usually have less drastic effects than single base additions or deletions ("frameshift" mutations); unless a stop codon is produced, a base substitution either has no effect (because of codon redundancy) or yields a single amino acid substitution, whereas an addition or deletion produces multiple alterations in the amino acid sequence through changes in the reading frame (Crick *et al.*, 1961). Thus, one might expect base-substitution mutations to occur more frequently in gene evolution. The first paper in this chapter, by Sankoff, Morel and Cedergren (1973), confirms this expectation. In this study the probable evolutionary pattern of mutational substitutions in 5S rRNA is reconstructed. The results show that even in this gene, which does not encode a protein, base substitutions are considerably more common in evolution than either base insertions or deletions. A more extensive treatment of the same problem can be found in Sankoff *et al.* (1976). A recent detailed analysis of part of the mitochondrial genome by Brown *et al.* (1982) has also shown that frameshift mutations are much less common, and has demonstrated that at shorter intervals after divergence, pyrimidine-pyrimidine and purine-purine substitutions (transitions) are significantly more frequent than pyrimidime-purine substitutions (transversions).

What are the biological consequences of the mutational changes that have been fixed during evolution of the 5S rRNA gene? Are they adaptive, reflecting functional differences between these genes in different organisms or are they neutral? In this case, for the majority of changes the answer is probably the latter, since there is no evidence of adaptive significance in terms of change of function or more efficient function for any of these sequence changes. Instead, the observed changes are probably a residuum after selective pressures have eliminated the functionally disruptive base-sequence changes. Recent studies show that among newly arising mutations, presumably before natural selection has had a significant effect, the frequency of deletions of one or more bases is much higher than is found differentiating the various 5S RNAs among species (de Jong and Ryden, 1981). This conclusion raises an interesting point: in analyses of gene evolution in which the gene has retained its function in the organisms under study, the surviving changes are probably unimportant for that function. One may begin to identify the adaptive changes primarily in the evolution of new functions from preexisting ones (see Chapter 5). Whatever the explanation, mutational studies and molecular phylogenies, such as those of Sankoff *et al.* are beginning to provide a fuller picture of the mutational sources of variation.

Another important source of hereditary variation is that which generates new combinations from preexisting variation through genetic recombination associated with sexual reproduction. The evolutionary significance of sexual reproduction and its attendant opportunity for genetic recombination has been recognized for over half a century. The basic argument, advanced by Fisher (1930) and Muller (1932), is that recombination speeds genetic change within populations by providing a greater range of variants. The arguments and evidence in favor of the proposition have been reviewed by Felsenstein (1974) and Maynard Smith

(1978), and an experimental confirmation of the Fisher-Muller hypothesis has been provided for bacteriophage populations by Malmberg (1977). However, the importance of conventional recombination in sexual reproduction as an agent for long-term macroevolutionary changes has been questioned by Stanley (1979), who based his argument on the great stability of lineages of many sexually reproducing taxa, as seen in the fossil record. These lineages may show long periods of apparent stasis that may be punctuated by periods of rapid diversification (punctuated equilibria). If conventional genetic recombination was an efficient generator of a greater range of variants (the argument goes), one should see more continuous patterns of change rather than the punctuated equilibria (Eldridge and Gould, 1972) seemingly shown by the fossil record. However, the Stanley argument is not convincing as it stands since it suggests that recombination should imply orthogenesis, the idea that internal forces cause slow continuous change (see references in Mayr and Provine, 1980). The kinds of genetic change that may lead to macroevolutionary diversification are only beginning to be understood and will be discussed again in Chapter 6.

In addition to standard meiotic recombination in which reciprocal exchanges occur between homologous chromosomes, another type of recombinational event has been observed in which blocks of DNA move from one region of a DNA molecule to another. This class of events has been revealed by molecular studies and is important in both the generation of new genetic functions and in the evolution of genome structure. The most important of these recombinational events is DNA sequence duplication, which was first postulated to have evolutionary importance from genetic and cytogenetic evidence (Bridges, 1935; Lewis, 1951). The precise mechanism of gene duplication is still unknown, but the process of tandem duplication of single sequences must certainly include some kind of unequal insertion event that follows an error in replication or alignment.

Molecular studies have extended the cytogenetic evidence and provided proof for the importance and ubiquity of DNA sequence duplication in eukaryotic genomes. These investigations have consisted of both comparative protein studies (Ingram, 1961, Chapter 1) and direct nucleic acid analysis (Britten and Kohne, 1968, Chapter 1), with the two approaches complementing one another. An example of this complementarity is provided by the discovery of nonexpressed pseudogenes. On the basis of the early findings on the proteins of the hemoglobin family, Zuckerkandl and Pauling (1962) suggested that duplication of a gene could occasionally be followed by a mutation inactivating one of the duplicates, thereby producing a "dormant" gene. These genes would not be expressed but, after further mutational alteration, might eventually be re-expressed with a modified function. Such a mechanism has also been put forward to explain certain features of tRNA evolution (Holmquist, Jukes, and Pangburn, 1973). The recent discovery of pseudogenes by the methods of DNA cloning has given substance to these ideas.

The paper by Lacy and Maniatis (1980, this chapter) presents an analysis of a pseudogene and illustrates the information about gene evolution that can be extracted from a detailed analysis of nucleotide sequence. The pseudogene sequence shows the mutation that prevents its expression and, based on molecular clock assumptions (see Chapter 4), allows one to deduce the approximate time of gene inactivation. The second part of the Zuckerkandl-Pauling hypothesis (the evolution of pseudogenes into functional genes with new properties) has yet to be established but seems possible. (Another mechanism for the evolution of new functions--direct mutation of duplicate genes to produce proteins with new properties--is discussed in Chapter 5.) For a general treatment of the role of gene duplication in the evolution of new functions, the reader should consult the book by Ohno (1970).

It is not yet clear that pseudogenes have a function, but their widespread occurrence is consistent with the possibility that they play an indirect role in gene regulation (Vanin *et al.*, 1980). Another explanation for the perpetuation of pseudogenes is that they are one form of "junk" DNA that is not eliminated because the energetic cost of its replication is only a small fraction of the total cost of cell replication. By this hypothesis, much of the eukaryotic genome may be junk, or "selfish" DNA (Doolittle and Sapienza, 1980; Orgel and Crick, 1980).

The process of sequence duplication, whether the sequences are adaptive or neutral, must be responsible for one of the distinctive characteristics of the eukaryotic genome--the presence of large amounts of repetitive DNA (Britten and Kohne, 1968, Chapter 1; see also Chapter 3). Unlike the hemoglobin gene family, a large number of these repetitive sequences appear to have no coding function. Repeated sequences can be classified into families of identical or very similar members and the organization of these families can be in either of two quite different arrangements: clustered (tandem) arrays, or widely dispersed throughout the genome. The third and fourth papers in this chapter deal with these two basic arrangements and the possible ways in which they arise. The paper by Smith (1974, this chapter; also, Smith, 1976; Ohta, 1980) is concerned with tandemly arranged repeated sequences and the hypothesis that once a sequence has duplicated to form a pair of tandem copies, subsequent unequal crossing-over between homologous but misaligned duplicates can generate further duplicates. Smith discusses how the process can amplify or contract the number of members in an array, and how, coupled with the occurrence of new mutations, it can lead to the evolution of new multiply-repeated sequences within the array. This proposed mechanism for sequence duplication does not depend on a particular arrangement of the sequences within the chromosome strand, but can generate arrays of sequences, starting from virtually any unit size. The postulated sequence of events may have played a major role in the formation, expansion, and evolution of such tandem arrays as the satellite sequences (Brutlag, 1980), the rRNA cistrons (Long and Dawid, 1980) and the histone-gene clusters (Kedes, 1979). Dover (1982) has elaborated on this theme by proposing that closely related repeated sequences, even if dispersed, may be subject to "homogenization" by a biased gene-conversion event. He points out that this "concerted" evolution, or molecular drive, would be a different mechanism for the fixation of variants than either normal selection or drift. However, selection may modulate either the rate of homogenization, if different sequences in the family have different selective values, or the increase in sequence number, if the numbers of different sequences have different selective effects. Dover has proposed that these processes of fixation may contribute to incipient speciation between populations.

Another suggested mechanism for duplication of a segment of DNA is that of Keyl (1965), who observed that certain polytene chromosome bands in *Chironomus* apparently arise by a series of sequential doubling events and proposed that in duplication chromosome loops undergo sister chromatid joining at their bases. Some recent evidence that eukaryotic chromosomes are organized as a series of large loops (Marsden and Laemmli, 1979; Pardoll *et al.*, 1980) lend some weight to this hypothesis. However, more must be known about the organization of chromosomal DNA replication before a more specific mechanism for DNA duplication is established.

The majority of middle repetitive sequences appear to be dispersed throughout the genome but the mode of dispersal of these sequences is unknown. Evidence is accumulating that suggests that members of such dispersed families may have much in common with the transposable elements first reported in maize by McClintock (1951) and later discovered in prokaryotes (reviewed by Calos and Miller, 1980). Recently, transposable elements have been discovered in *Drosophila* and yeast. The paper by Strobel, Dunsmuir and Rubin, (1979, this chapter) describes several transposable elements in *Drosophila*. These elements can move with high frequency and the movement (transposition) is often accompanied by an increase in the number of copies per genome. The possible relevance of transposable elements to the evolution of the genome is obvious; the fact that they can be mutagenic (Green, 1978; Thompson and Woodruff, 1980)--see Problem 9--further emphasizes the possibility of their genetic importance. It may yet prove to be the case that "jumping genes" are as important a source of genetic variation as conventional point mutations and recombination. In addition, certain classes, the so-called P elements and I elements, can create sterility barriers between populations, a potentially important step in speciation (Engeles, 1981).

Duplication of DNA sequences occurs both on a small scale, in which single genes and sub-gene segments are duplicated, and on a very large scale, in which multigene segments are duplicated. It has been hypothesized that entire prokaryotic (Riley and Anilionis, 1978; De Martelaere and Van Gool, 1981) and eukaryotic genomes (Sparrow and Nauman, 1976) may have arisen by processes of complete chromosomal doublings. Furthermore, many complexes of plant species, both domesticated and wild, exist that have resulted from either autopolyploidy (duplication of the genome of a normal member of a species one or more times) or allopolyploidy (duplication of the chromosomes of an interspecific hybrid (see Stebbins, 1971). However, beyond these clearcut cases, many plants and animals are known in which large differences in DNA content exist between phenotypically and genetically close species (the C-value paradox) and for which these differences cannot be explained by simple piecemeal sequence duplication. The possible functional significance of these large DNA differences has been discussed by Stebbins (1966) and Cavalier-Smith (1978).

Another source of genetic variation of great potential significance for the evolution of new functions was discovered in 1978--the shuffling and recombination of segments of eukaryotic genes. The discovery that the structural genes of many eukaryotes consist of coding sequences interrupted by stretches of noncoding DNA (intervening sequences or introns) has elicited a flood of speculation on the origin and functional consequences of this organization. One possible consequence is evolutionary: genes may trade sections at their intron boundaries to form new composite functions having new properties, an idea proposed by Gilbert (1978, this chapter). This hypothesis is given in more detail in Gilbert (1979, 1981) and more information may be found in Chapter 5.

The final reprinted paper in this chapter--that by Ferris, Brown, Davidson and Wilson (1981a)--is concerned with the amount of genetic variation (as fixed point mutations) in a population. For some time mutation rates in eukaryotes have been known to be under genetic control; furthermore, the flow of new mutations into populations is recognized to have evolutionary significance (Sturtevant, 1937). Discoveries of prokaryotic strains with altered substitution rates (reviewed by Drake, 1974, and Cox, 1976) have emphasized the generality of this phenomenon and raised the question whether observed mutation rates are at the minimum possible, or at some higher but optimal level that promotes evolutionary change (Cox, 1976; Maynard Smith, 1978). Two points must be considered in analyzing this question. First, selective pressure for increasing the fidelity of DNA synthesis has been strong and indeed the mistake rate during DNA replication may approach the minimum attainable (Alberts and Sternglanz, 1977). Evidently successful biological systems have minimized mutation rates. Second, under rapidly changing environmental conditions, in which rapid adaptation is at a premium, an increased rate of mutation may speed the acquisition of fitness (Ayala, 1969). This finding may explain the seeming paradox that *Escherichia coli* mutator strains outgrow wild-type strains when both are grown in a chemostat (Cox and Gibson, 1974), yet they do not prevail in natural populations. The prevalence of standard (nonmutator) rates may imply that, in general, the periods in which an increased mutation rate would be advantageous are not frequent enough to offset the increased genetic load that accompanies an increased mutation rate.

Mutation rates are just one of the determining factors in setting levels of population genetic variation, since newly occurring mutations might either accumulate (if neutral or advantageous) or be eliminated (if deleterious); the nature and strength of selective forces are the ultimate arbiters of these levels. One of the first triumphs of molecular evolution was the estimation of genetic variation in populations by detection of electrophoretic variants of enzymes (Hubby and Lewontin, 1966; Harris, 1966, Chapter 1). These experiments and the host of similar studies that have followed, revealed much higher levels of intrapopulational variation than had been previously suspected, and generated new debates about the significance of this variation (Lewontin, 1974; see Chapter 4). This approach measures variation at the protein level only, and underestimates the amount of genetic variation for a variety of reasons (including degeneracy of the code, synonymous substitutions, and possible higher levels of variability in noncoding regions of the genome).

Through the use of restriction-enzyme technology, variation of DNA sequences can be detected directly. With this technique well-defined DNA sequences from different individuals are isolated and each sample is digested with specific restriction enzymes; from the fragment sizes and number one can calculate the number of enzyme-specific sites present in the DNA of each individual. An estimate of base-pair differences between individuals can then be calculated. For reasons of size and convenience of isolation, the mitochondrial DNA molecule is particularly suitable for these measurements. The biochemical methodology and the procedure for calculating rates of base-sequence change in this molecule have been described by Brown *et al* (1979). By using this technique, Ferris *et al.* (1981a, this chapter), determined the extent of mitochondrial DNA polymorphism in higher primate populations. Their approach demonstrates that considerable genetic variation exists within populations (though in this respect humans are less variable than was expected from their population size). In addition, the paper demonstrates the use of molecular data to make deductions both about evolutionary history (phylogeny) and about ancestral population structure.

The large number of types of genetic variation, as well as the large amounts of variation within populations, was not suspected 20 years ago (see Lewontin, 1974). The discovery of this variation has been a major contribution of molecular biology to the study of evolution.

Evolution of 5S RNA and the Non-randomness of Base Replacement

SEQUENCES of 5S rRNA have been published for five widely divergent organisms: *Escherichia coli*, *Pseudomonas fluorescens*, yeast, human KB carcinoma, and *Xenopus laevis*[1-5]. The question arises whether sequences for the ancestors of these organisms, represented by X, Y, and Z in Fig. 1, can be reconstructed to any degree of confidence; were this possible, statistical analysis of mutation types would become feasible. Such reconstructions are well known in protein studies[6] and for tRNAs[7], but they are somewhat more difficult for 5S rRNA.

The main problem is to align the various sequences so that bases in corresponding position in different sequences are fairly certain to reflect a common term in the ancestral sequence. In protein studies, the fact that sequences are based on a twenty-symbol alphabet makes it unlikely that identical di- or tri-peptides in different sequences in approximately the same part of the molecule could have arisen by accident, and this is of great help in constructing alignments. With tRNA, the fixed position of certain modified bases in many sequences, plus base-pairing constraints in parts of the molecule, similarly ease the solution of the alignment problem. No such clues are available with 5S RNA. Alignment problems are primarily due to the confusing effects of base insertion or deletion mutations, which necessitate the use of gaps in alignment schemes. Since it is seldom clear in which sequences and between which positions gaps should be placed, the task of evaluating possible alignments becomes a combinatorial problem of great complexity. This problem has recently been solved[8] for the case where the evaluation is based on a minimal mutation criterion. The alignment algorithm developed for this solution is computationally impractical for large numbers of sequences, but for the case of three known sequences a, b, and c, and one unknown sequence x, in the configuration it is quite easy to implement a computer program for the method. (For example $a = E.\ coli$, $b = P.\ fluorescens$, $c = yeast$, $x =$ unknown proto-prokaryote).

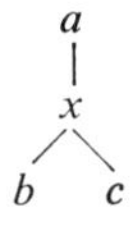

The algorithm is based on a recursion relationship which calculates, for three sequences, $a_1, a_2, \ldots, a_m$; $b_1, \ldots, b_n$; and $c_1, \ldots, c_p$, the minimal total number of mutations along branches ax, bx, and cx necessary to relate the observed sequences to any hypothetical or reconstructed sequence x. Once this total is known, a minimising sequence $x_1, \ldots, x_q$ can be constructed.

Weighting insertions, deletions, and replacements of single bases equally, and letting $d(i, j, k)$ be the minimal total number of mutations required for the subsequences $a_1, \ldots, a_i$; $b_1, \ldots, b_j$; and $c_1, \ldots, c_k$, we have

$$d(i,j,k) = \min \begin{cases} d(i-1, j-1, k-1) \begin{cases} +0 & \text{if } a_i = b_j = c_k \\ +1 & \text{if } a_i = b_j \neq c_k,\ a_i = c_k \neq b_j \\ & \text{or } b_j = c_k \neq a_i \\ +2 & \text{if } a_i \neq b_j \neq c_k \neq a_i \end{cases} \\ d(i-1, j-1, k) \begin{cases} +1 & \text{if } a_i = b_j \\ +2 & \text{if } a_i \neq b_j \end{cases} \\ d(i-1, j, k-1) \begin{cases} +1 & \text{if } a_i = c_k \\ +2 & \text{if } a_i \neq c_k \end{cases} \\ d(i, j-1, k-1) \begin{cases} +1 & \text{if } b_j = c_k \\ +2 & \text{if } b_j \neq c_k \end{cases} \\ d(i, j, k-1) \quad +1 \\ d(i, j-1, k) \quad +1 \\ d(i-1, j, k) \quad +1 \end{cases}$$

for $i = 1, \ldots, m$; $j = 1, \ldots, n$; and $k = 1, \ldots, p$, where $d(r, 0, 0) = d(0, r, 0) = d(0, 0, r) = r$ for $r = 1, 2, \ldots$ Once $d(m, n, p)$ has been calculated by systematically computing $d(1, 1, 1)$, $d(1, 1, 2)$, $d(1, 2, 1)$, $\ldots$, an optimal alignment can be constructed by reversing the calculation, that is, by seeing which of a_i, b_j and/or c_k contributed to the minimal choice in calculating each $d(i, j, k)$, starting with $d(m, n, p)$. These procedures are described in more detail and proofs of optimality given elsewhere[8].

After finding the alignment, the unknown sequence can be largely reconstructed as follows. If $a_i = b_j = c_k$ are aligned, the three of these must be aligned with an x_h, where $x_h = a_i = b_j = c_k$. If $a_i = b_j$ or $b_j = c_k$ or $c_k = a_i$ (or two gaps) are aligned, then they are also aligned with an identical x_h (or a gap). All other cases lead to uncertainty in the x sequence (for example, if $a_i \neq b_j \neq c_k \neq a_i$ are aligned).

We used this procedure to establish a preliminary reconstruction for X in Fig. 1, by setting $a = E.\ coli$, $b = P.\ fluorescens$, $c =$ yeast, and $x =$ X. Similarly, a tentative sequence for Y was constructed by setting $a =$ KB, $b = Xenopus$, $c =$ yeast, and $x =$ Y. In both cases, the uncertain positions were filled in by means of a random choice among the three different aligned bases (or a gap) at that position. An approximation to Z could then be derived by using $a =$ X, $b =$ Y, $c =$ yeast, and $x =$ Z.

A better version of X (according to the minimal mutation criterion over the whole tree of Fig. 1) could then be produced by setting $a = E.\ coli$, $b = P.\ fluorescens$, $c =$ Z and $x =$ X. Similar attempts to improve Y and Z produced no change. The five known sequences and the three reconstructed sequences are aligned in Fig. 2.

Once the alignments and reconstructed ancestral sequences are available, it becomes possible to infer which mutations have occurred along each of the branches X→*E. coli*, X→*P. fluorescens*, Z→yeast, Z→Y, Y→KB, Y→*Xenopus*, and X→Z, although for the latter there is no way of knowing whether X or Z retains the ancestral base. Total mutations along each branch are indicated in Fig. 1. In Table 1, the 117 mutations reconstructed by our method are broken up into deletions, insertions and base replacements of various types. This shows that each transition type of mutation (pyrimidine-pyrimidine or purine-purine replacement) occurs more frequently than any transversion (pyr-pur or pur-pyr). In additon, transitions involving C and U are by far the most frequent type of mutation.

Through a systematic study of protein sequences, Vogel[9] and others have found that the evolution of the DNA coding for these proteins similarly shows evidence for a predominance of

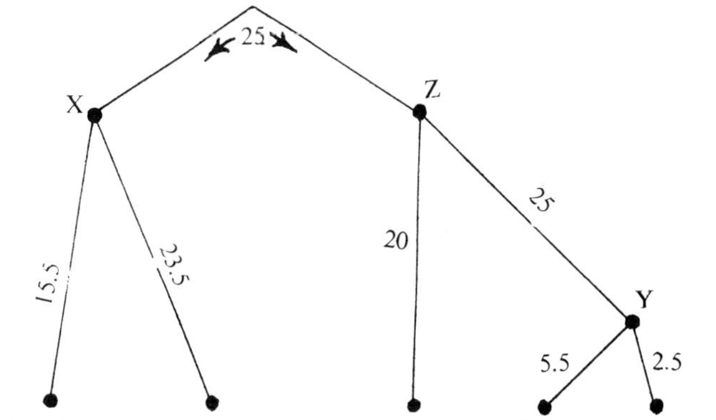

Fig. 1 Phylogenetic tree of known and ancestral organisms. X, Proto-prokaryote; Y, proto-vertebrate; Z, proto-eukaryote. Numbers represent estimated numbers of mutations along each branch.

transitions, particularly pyr-pyr mutations. The frequency of fixed mutations from these and our own studies is thus: pyr-pyr > pur-pur > pur-pyr, pyr-pur. In addition, the transitions > transversions effect is also observable in tRNA data[10].

Table 1 Estimated Number of Each Mutation Type Occurring in 5S RNA

Original base	New base A	C	G	U	None
A		5.0	8.0	6.3	3.3
C	3.7		6.2	11.3	4.3
G	7.3	4.0		6.3	4.0
U	3.0	14.0	4.2		3.0
None	6.3	4.3	6.0	6.3	

Bottom row represents insertions; extreme right column lists deletions. Transition mutation types underlined.

```
EC UGC CUGGCGGCCG UAGC GCGGUGGUCCCACCUGACC CCAUGCC
PP UG? CU GCGGCCA UAGCAGC?UUGGAA?CACCUGAUC CCAU?CC
PF UGUUCU UUGACGAGUAGUAGCAUUGGAA CACCUGAUC CCAUCCC
PE  G? CU GCGGCCA UACCA?C UUGAAAGCACC?GAUCUCCGU CC
YT  GG UU GCGGCCA UACCAUC UAGAAAGCACC GUUCUCCGU CC
KB  GU CU ACGGCCA UACCACC CUGAACGCGCCCGAUCU CGU CU
PV  G? CU ACGGCCA UACCACC CUGAAAGCGCCCGAUCU CGU CU
XL  GC CU ACGGCCA CACCACC CUGAAAGUGCCCGAUCU CGU CU
          |         |         |         |
         10        20        30        40

EC GA ACUCAGAAGUGAAACGC C GU AGCGCC GA UG GUA GUGU
PP GA ACUCAGAAGUGAAACGC U GU AGCGCC GA UG GUA GUGU
PF GA ACUCAGAGGUGAAACGA U GC AUCGCC GA UG GUA GUGU
PE GAUACUC?GAAGU UAA GC UGGU AG?GCCUGA UGAGUA GUGU
YT GAUAACCUGUAGU UAA GC UGGUAAGAGCCUGACCGAGUA GUGU
KB GAU CUCGGAAGC UAA GCAGGGU CGGGCCUGG UUAGUACUUGG
PV GAU CUCGGAAGC UAA GCAGGGU CGGGCCUGG UUAGUACUUGG
XL GAU CUCGGAAGC CAA GCAGGGU CGGGCCUGG UUAGUACUUGG
    |         |         |         |         |
   50        60        70        80        90

EC   GGGGUCUCCCCAUG CGAG A GUAGGGAAC U GCCAGGCAU
PP   GGGGU?UCCCCAUG CGAG A?CUA GGA?C U G CAGGCAU
PF   GGGGUUUCCCCAUGUCAAG AUCU  CGACCAUAG  A GCAU
PE A UGGGUGACC?CAUG CGAA ACCUA GGUGC U G CAGGC?U
YT AGUGGGUGACCAUACG CGAA ACCUA GGUGC U G CA AUCU
KB A UGGGAGACCGCCUG GGAAUACC G GGUGC U G UAGGCUU
PV A UGGGAGACCGCCUG GGAAUACC A GGUGC•U G UAGGCUU
XL A UGGGAGACCGCCUG GGAAUACC A GGUGU C G UAGGCUU
      |         |         |         |
     100       110       120       130
```

Fig. 2 Alignment of known and reconstructed 5S RNA sequences. From top to bottom: EC, *E. coli*; PP, proto-prokaryote; PF, *P. fluorescens*; PE, proto-eukaryote; YT, yeast; KB, human KB carcinoma; PV, proto-vertebrate; XL, *Xenopus*. Reconstructed sequences are shaded; ?, undetermined base or gap.

It could be argued that the protein data are explicable by some selective force controlling the type of amino acid substitution permitted. The resulting non-random protein mutations could then be related by the genetic code to non-random DNA mutations. One might also argue that the RNA data are the result of selection based on either function or three dimensional structure. That one finds identical patterns of non-random distributions in both protein and RNA data, however, indicates that the process responsible for this is independent of the particular type of molecule. Moreover, if selective forces are involved, this could only be at the level of DNA replication and/or RNA transcription. The basis of this selection could be structural, that is, the excess of transitions could imply that mispairing is permitted more often between a purine and a pyrimidine. This proposal is in agreement with the tautomerisation theory of mutations, in which G (enol form) pairs with T, and C (imino form) pairs with A. Another possibility is that known DNA repair enzymes correct pur-pur and pyr-pyr mismatches more often since they disrupt the DNA duplex more than pur-pyr mismatches. Neither tautomerisation nor selective repair, however, can explain the fact that pyr-pyr mutations are fixed more often that pur-pur.

Figure 2 suggests another general tendency in the evolutionary history of 5S RNA; that insertions and deletions have involved mainly single nucleotides. This contrasts with tRNA[7], and protein data[12], where simultaneous insertion or simultaneous deletion of several consecutive nucleotides has been shown.

This work was supported in part by le Ministère de l'Education du Québec.

DAVID SANKOFF
CRISTIANE MOREL
ROBERT J. CEDERGREN

Centre de Recherches Mathématiques and Département de Biochimie, Université de Montréal, Case Postale 6128, *Montréal* 101

Received June 21, 1973.

1 Brownlee, G. G., Sanger, F., and Barrell, B. G., *J. molec. Biol.*, **34**, 379 (1968).
2 DuBuy, B., and Weissman, S. M., *J. biol. Chem.*, **246**, 747 (1971).
3 Brownlee, G. G., Cartwright, E., McShane, T., and Williamson, R., *FEBS Lett.*, **25**, 8 (1972).
4 Forget, B. G., and Weissman, S. M., *Science, N.Y.*, **158**, 1695 (1967).
5 Hindley, J., and Page, S. M., *FEBS Lett.*, **26**, 157 (1972).
6 Dayhoff, M. O., *Scient. Am.*, **221**, 86 (1969).
7 Cedergren, R. J., Cordeau, J. R., and Robillard, P., *J. theor. Biol.*, **37**, 209 (1972).
8 Sankoff, D., *Publ. Cent. Rech. Math., Technical report 262* (1973).
9 Vogel, F., *J. mol. Evol.*, **1**, 334 (1972).
10 Dayhoff, M. O., in *Atlas of Protein Sequence and Structure*, **5**, 113 (Nat. Biomed. Res. Found., Washington, DC, 1972).
11 Watson, J. D., and Crick, F. H. C., *Cold Spring Harb. Symp. quant. Biol.*, **18**, 193 (1953).
12 Dayhoff, M. O., in *Atlas of Protein Sequence and Structure*, **5**, 44 (Nat. Biomed. Res. Found., Washington, DC, 1972).

Unequal Crossover and the Evolution of Multigene Families

George P. Smith

Departments of Genetics and Medical Genetics, University of Wisconsin, Madison, Wisconsin 53706

The 18 S and 28 S ribosomal RNAs in eukaryotes are coded by hundreds of identical or nearly identical genes which I will refer to collectively as the rRNA family. There are about 500 such genes in the frog *Xenopus laevis* (Buongiorno-Nardelli et al., 1972) which I will assume to form a single cluster of tendemly repeated genes. The repeated unit in this rRNA family contains not only the "gene" portion, from which the rRNA molecules are transcribed, but also a "spacer," which has no known function (Wensink and Brown, 1971; Miller and Beatty, 1969). Within one species of animal all the rRNA repeats are very similar or identical, but there are striking differences between the repeats in different species (Brown et al., 1972). Most of the interspecies differences reside in the spacer; the gene portion, which is presumably under more intense selective pressure, shows much less phylogenetic variation.

Although I will concentrate on the evolution of the rRNA family in this paper, I believe that the theory that I will develop for this family applies in all essentials to the evolution of the 5 S RNA (Brown et al., 1971; Brown and Sugimoto, this volume; Pardue, this volume) and satellite DNA families (Southern, 1970; Sutton and McCallum, 1972; Gall et al., this volume; Blumenfeld, this volume) as well.

The simplest view of the evolution of the rRNA family supposes that each repeat evolves *independently* of all the other repeats in the family. The most compelling objection to this scheme is raised by the spacer portions of the rRNA repeats. They accumulate mutations rapidly in evolutionary time, giving rise to large interspecies differences, yet remain very homogeneous within a species. To reconcile this with the simple evolutionary scheme seems to require a selective force with the extraordinary property of fixing distinct sets of multiple, exactly parallel mutations in hundreds of spacers in distinct phylogenetic lines. Several alternative evolutionary schemes for familes of tandem repeats have been proposed, including the master-slave hypothesis (Callan, 1967; Thomas, 1970), the saltatory replication theories (Britten and Kohne, 1968; Buongiorno-Nardelli et al., 1972; Amaldi et al., 1973), and the democratic gene conversion theory (Edelman and Gally, 1970; Edelman, 1970).

In this paper I join other authors at this symposium (Brown and Sugimoto, this volume; Tartof, this volume) in support of what I consider a much simpler and more plausible explanation of the homogeneity of the rRNA repeats than the aforementioned theories. This explanation is based on the fact that if an array of tandem repeats undergoes a sufficient number of homologous but unequal crossovers with itself, eventually the descendants of all but one of the starting repeats will be eliminated, being replaced in the process by the descendants of that one remaining repeat. When this happens, I will say that the surviving repeat has become "fixed," and the entire process I will call "crossover fixation." This process is illustrated in Figure 1. I emphasize that crossover fixation does not require the relatively radical event of eliminating all but a very few of the repeats and repopulating the entire multigene family with the descendants of those few survivors. It occurs even if the total number of repeats never varies much from some average number.

Crossover fixation has been proposed previously as a possible mechanism for the evolution of the rRNA family. For example, Edelman and Gally (1970) suggested "that these genes are held in a dynamic equilibrium in the population. The number of genes can either increase or decrease as a result of unequal crossing over. . . . In a dynamic system of this sort, an allele arising from a point mutation in an rRNA gene which confers selective advantage to an organism can increase in gene frequency, both in the population *and* in the tandem set of genes in one organism. This might well account for the coevolution of genes in a species." The question I will address myself to is whether unequal crossover is a quantitatively reasonable explanation of the facts.

The average time for one of the repeats, present at some starting time, to become fixed will be called the crossover fixation time. Obviously, the more crossovers that occur per unit time, the shorter will be the crossover fixation time. If crossover is frequent enough, the crossover fixation time will be too short for the repeats to have accumulated an appreciable number of mutations during their descent from their most recent common ancestor. The steady state of such a family will be such that

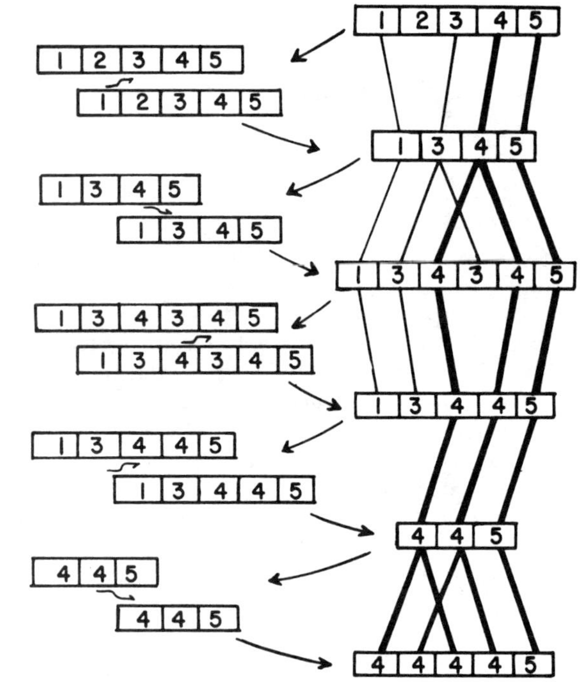

Figure 1. Schematic illustration of crossover fixation. Each box represents a repeat, and the number inside each box refers to the starting repeat from which the repeat represented by the box descends. A line connects each repeat with that from which it descends and that (or those) which descend from it. Bold lines represent the descent of the contemporary repeats, which are numbered underneath to show their correspondence to Figure 4.

the repeats are essentially homogeneous, as the repeats in the rRNA family in fact are.

The spacer and gene portions of the rRNA repeats would not be expected to differ markedly in their degree of homogeneity according to this theory. A short crossover fixation time leading to homogeneity of one would necessarily lead to homogeneity of the other as well. However, the two portions of the repeat might easily be expected to differ in their respective rates of acceptance of mutations. It is perfectly reasonable to suppose, on the one hand, that most mutations in the gene part would be deleterious. Natural selection would rapidly eliminate chromosomes in which a repeat harboring such a mutation is on the way to fixation. It is equally plausible to suppose, on the other hand, that many mutations in spacers are more or less selectively neutral. Natural selection would *not* eliminate chromosomes in which a repeat harboring a neutral mutation is on the way to fixation, and these mutations would therefore accumulate in time. If we make these two plausible suppositions, then the rapid accumulation of mutations in the spacer portions, the evolutionary conservation of the gene portions, and the intraspecies homogeneity of both portions are not at all puzzling observations.

Kinetics of Fixation in a Single Lineage of Chromosomes by Intrachromosomal Exchange

Crossover fixation might be brought about by two types of exchange, which I will call intrachromosomal and interchromosomal. By intrachromosomal exchange (Fig. 1) I mean crossover between two identical DNA molecules which are daughters of the same parental DNA molecule. Such crossovers might occur after any of the DNA replications in the germ line, either during meiossi or during one of the many germ line mitoses. By interchromosomal exchange I mean crossover between not necessarily identical homologues, each of which is derived from a different one of the organism's parents. Interchromosomal exchange includes ordinary meiotic crossover and also, at least occasionally, mitotic crossover between homologues.

I shall concentrate on intrachromosomal exchange as the mechanism for fixation because it is easy to compute. If we assume that *no* interchromosomal exchange occurs within the rRNA family, then the fixation of a starting repeat in a single lineage of chromosomes will not be affected by the other chromosomes in the population and can thereore be computed separately.

I know of no analytical treatment of this type of fixation, but an adequate idea of its kinetics can be obtained by averaging the results of many computer simulations. Figure 1 illustrates how fixation in a single lineage of chromosomes by intrachromosomal crossover can be simulated. The computer program performs a large number of successive unequal crossovers and keeps a record of the starting repeat from which each contemporary repeat descends. In Figure 1, for example, this information is represented by the numbers inside the boxes. Fixation would be complete when all the numbers inside the boxes are the same, indicating that all the contemporary repeats derive from the same starting repeat.

For each computer simulation, the starting number of repeats and the allowed latitude in repeat number were specified parameters. The program used two random numbers to generate each crossover. The first specified the number of repeats that would be contained in the chromosome resulting from the crossover—the "new" chromosome. This number had to be within the prescribed latitude in repeat number. Two identical replicas of the "old" chromosome were then aligned as required so that a crossover between them would give a new chromosome with the number of repeats specified by the first random number. The second random number then specified where the crossover

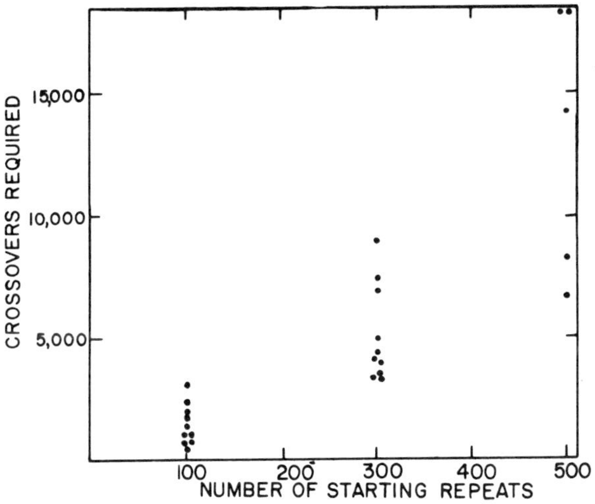

Figure 2. Fixation in a single lineage of chromosomes by intrachromosomal crossover: number of crossovers required for nearly complete fixation (clonality = 0.9; see footnote 1) in families with various numbers of starting repeats. Each point represents a separate simulation. In each case, the number of repeats in the family was allowed to vary within 10 percent of the starting number.

would occur within the overlapping repeats in the old chromosomes.

I allowed crossover to occur between, but not within, repeats. Crossover within the repeats would not alter the rate of fixation but would merely introduce the possibility that different starting repeats would become fixed for different portions of the final repeats. This is a possibility of no real consequence in the case of the homogeneous rRNA family, since the various starting repeats that might potentially become fixed do not differ perceptibly in sequence.

Figure 2 shows the number of intrachromosomal crossovers required for essentially complete fixation[1] in families with various numbers of starting repeats. In each simulation, the number of repeats per chromosome was allowed to vary within 10 percent of the starting number. It will be seen that the number of crossovers required for fixation increases roughly linearly with the number of repeats. Figure 3 shows that large differences in the allowed latitude in repeat number do not greatly affect the rate of crossover fixation: fixation does not occur significantly faster in a family allowed to vary between 75 and 125 repeats than in a family allowed to vary between 95 and 105 repeats, and only a few times faster than in a family allowed to vary only between 99 and 101 repeats.

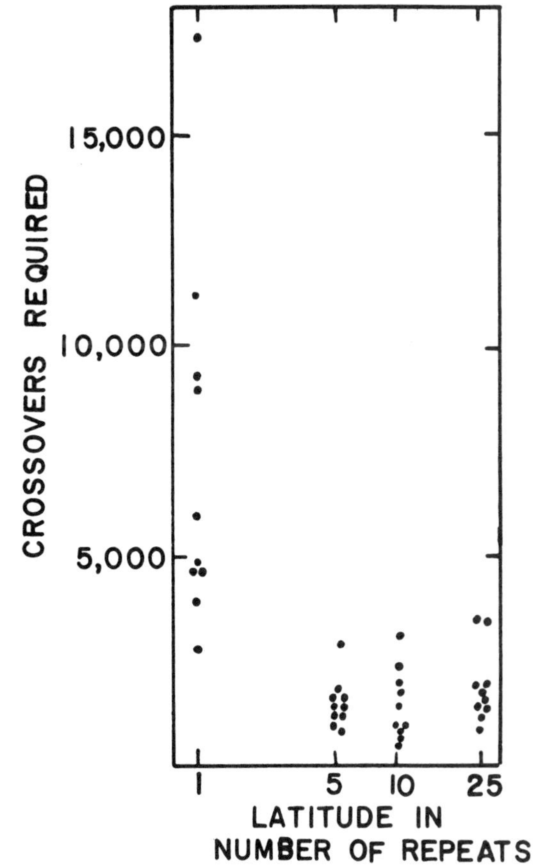

Figure 3. Fixation in a single lineage of chromosomes by intrachromosomal crossover: number of crossovers required for nearly complete fixation (clonality = 0.9; see footnote 1) in a family of 100 starting repeats allowed to vary within various latitudes from that number. Each point represents a separate simulation.

From Figure 2, we estimate that about 10^4 intrachromosomal crossovers would be required for fixation in a family of 500 rRNA repeats. If we assume a crossover rate of 1 per generation and a mutation rate of 5×10^{-9} mutations/nucleotide pair/generation, only about one in every 20,000 nucleotide pairs would have mutated during one crossover fixation time. Even if the heterogeneity were 100 times greater than this, it would be un-

[1] In order to guage the progress of crossover fixation, I used a measure I will call "clonality," which is the sum of the squares of the proportions of the contemporary repeats which are descendants of the various starting repeats. In the final array in Figure 1, for example, four of the five repeats descend from starting repeat 4 and one of the five repeats descends from starting repeat 5; the clonality is therefore $(4/5)^2 + (1/5)^2 = 17/25$. Clonality, which varies between 0 and 1, has the following characteristics: (1) if N starting repeats are equally represented by descendants, the clonality is $1/N$; (2) the clonality is 1 if, and only if, all the contemporary repeats descend from a single one of the starting repeats; and (3) as the clonality approaches 1, its square root approximates the proportion of the contemporary repeats which are descendants of the most highly represented starting repeat. At a clonality of 0.9, which I used as a cut-off point for simulations, about 95 percent of the contemporary repeats descend from a single one of the starting repeats. In a number of simulations I followed the increase in clonality with the number of crossovers; in these cases, clonality seemed to increase roughly linearly.

detected by ordinary methods, such as DNA reannealing or sequence analysis.

"Rectification" by unequal intrachromosomal crossover is even a plausible explanation for the homogeneity of very large families of tandem repeats, such as the satellite DNAs of *Drosophila virilis*, each family of which contains about 10^7 repeats (Gall et al., this volume; Blumenfeld, this volume). Assuming that the data in Figure 2 indeed extrapolate linearly to very large numbers of repeats, a family of 10^7 repeats would require about 2×10^8 crossovers for fixation. Assuming a mutation rate of 5×10^{-9} mutations/nucleotide pair/year, a crossover rate of 1 per generation, and 10 generations per year, about 10 percent of the nucleotide pairs would have mutated during one crossover fixation time. The actual degree of heterogeneity in these satellites is actually at least 10 times less than this (Gall et al., this volume; Blumenfeld, this volume), but the foregoing calculation did not take into account the possibility that selection maintains the repeats more homogeneous than they otherwise would be. Indeed, we know in the case of at least one *D. virilis* satellite that there must be strong selection against mutations, since there is a satellite with the same sequence in *D. americana* (Gall et al., this volume). Even a relatively weak selection against mutations would be able to maintain the repeats at least 10 times more homogeneous than they would be if mutations accumulated at random.

I emphasize that the assumed numbers on which the above calculations were based are reasonable, but not necessarily accurate; these calculations should not therefore be considered as predictions of the theory, but only as illustrations of its plausibility.

It should be borne in mind that all the time fixation is occurring within each chromosomal array by unequal crossover, the chromosomes themselves are undergoing allelic fixation by the ordinary processes of population dynamics. Thus, not only will the repeats within one chromosome tend to remain homogeneous, but also the repeats in different chromosomes in the organismal population will tend to resemble each other. The ordinary population genetics of tandem repeats is not at issue in this paper; in any case, fixation within natural populations of organisms must be rapid relative to the acceptance of mutations, since families of tandem repeats (such as the rRNA and satellite DNA families) do not, in fact, vary perceptibly from individual to individual within a species.

I have not been able to analyze fixation due to interchromosomal exchange satisfactorily. The difficulty arises from the fact that fixation by homologue exchange cannot be analyzed in one chromosome without at the same time analyzing it in the other chromosomes with which it is crossing-over. Consequently, a simulation of this process must keep track not only of the repeats within chromosomes, but also of the chromosomes within the population of organisms. A few simulations with very small numbers of organisms (up to 50) indicate that fixation requires several times more interchromosomal crossovers than intrachromosomal crossovers. I believe—but have not rigorously proved—that, given a certain rate of intrachromosomal crossover, additional interchromosomal exchanges will not slow the rate of crossover fixation.

Callan (1967) cited work of H.-G. Keyl which suggested that the numbers of repeats in some multigene families vary by factors of two. More recently, Buongiorno-Nardelli et al. (1972) and Amaldi et al. (1973) have presented more direct evidence indicating that the numbers of repeats in most or all multigene systems vary by this factor. These striking twofold variations would not be expected if the numbers of repeats changed solely by unequal crossover, and, to that extent, the present theory must be accounted an uneconomical explanation of the facts. Nevertheless, the crossover theory is not directly contradicted by this pattern of variation; measurements of repeat numbers are not nearly precise enough to exclude small fluctuations due to unequal crossover superimposed on the more obvious twofold variations noted by these authors.

Crossover Fixation in Immunoglobulin V-gene Families

The immunoglobulin polypeptides are coded by three complexes of closely linked genes: one complex each for kappa, lambda, and heavy chains. Each complex consists of a family of V genes coding for V regions (the amino-terminal, approximately 110 amino acids) and one or more C genes coding for C regions (the carboxy-terminal, approximately 110, 330, or 440 residues) (Smith et al., 1971; Smith, 1973). There is some controversy over the number of V genes per family, the germ line theory supposing that there are at least several hundreds, the somatic theories envisioning far fewer. If (as I believe) the former figure is more nearly correct, the pattern of variation in immunoglobulin V regions presents some evolutionary anomalies which (as I hope to show in this section) are elegantly resolved by a minor modification of the theory of crossover fixation.

The first anomaly is the existence of *subgroups* in each family of V regions. The sequences within one subgroup are much more similar to one another than they are to sequences in different subgroups. If V genes evolved independently of one another, it is difficult to understand how they could assort themselves into a few subgroups.

The second anomaly is that subgroups in distantly related mammalian species are species-specific and do not correspond to each other on a one-to-one basis. That is, first, all the sequences in one subgroup in one species are more closely related to one another than they are to even the most similar sequences in other species; and second, each species has subgroups which are absent in other species. These facts seem to imply that some of the genes in a V-gene family, having converged to form a subgroup in the first place, then coevolve to maintain that subgroup while at the same time accumulating many mutational differences from sequences in other species. Meanwhile, subgroups would have to be continually lost and gained in different phylogenetic lines.

These subgroups, which are so unexpected if each V gene evolved independently of all the others, would be an entirely natural result of crossover fixation. To show this it will be necessary to analyze the evolutionary relationships to be expected in a family of repeats undergoing unequal crossover. Such repeats will be related to each other by a branched evolutionary tree descending from their most recent common ancestor. For example, in Figure 1 the contemporary repeats are related by the branches printed in bold lines. The same relationships are shown again in Figure 4, abstracted from the other details of Figure 1. This evolutionary "tree" consists of two parts, one descending from starting repeat 4, the other from starting repeat 5.

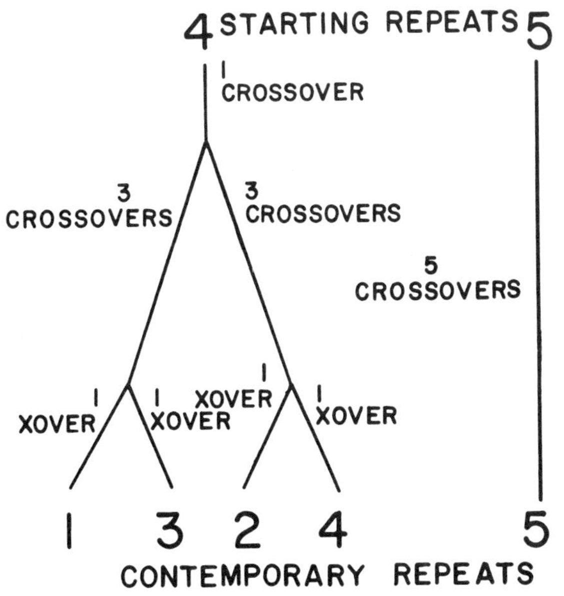

Figure 4. Evolutionary relationships of the contemporary repeats in the example of crossover fixation shown in Figure 1.

If fixation had been complete, there would have been a single branched structure descending from one of the starting repeats.

The lengths of the branches in Figure 4 are proportional to the number of crossovers occurring between the points of time they connect. Number of crossovers is at least approximately a measure of time; presumably, then, the length of an evolutionary branch in crossovers would be roughly proportional to its length in terms of the more directly observable measure of the passage of time in evolution—namely, the number of mutations that occurred in the evolutionary lineage represented by that branch.

I determined the structure of such an evolutionary tree in a computer simulation of intrachromosomal crossover fixation in a family of 45–55 repeats. Rather than keeping track of the starting repeat from which each contemporary repeat descends, as in the simulations described previously in this paper, this simulation kept track of the evolutionary relationships among the contemporary repeats. If this program had performed the series of crossovers shown in Figure 1, for example, its output would have been the evolutionary scheme shown in Figure 4. Figure 5 shows the tree obtained in this simulation. It will be seen that the 46 final repeats fall into nine subgroups. The general structure of this tree is strikingly similar to the structure of the trees reconstructed from actual V-region amino acid sequences (Smith et al., 1971; Barker et al., 1972; Smith, 1973).

Each number in Figure 5 gives the position of that contemporary repeat in the final array. The members of a given subgroup are widely (though not randomly) scattered along the array. Donald Brown (pers. commun.) has pointed out that crossover fixation is the only current theory consistent with such a scattered distribution of subgroups, whether in V genes or in other families of repeats.

Crossover fixation also readily explains the species-specific nature of V-region subgroups. To show this, let us take the hypothetical tree in Figure 5 as representing the descent of a mouse V-gene family. Let us assume that the arc drawn about a third of the way down the descent delineates the epoch of the most recent common ancestor of mouse and man—an assumption in accord with the evidence (Barker et al., 1972). To see why V genes within a mouse subgroup will be much more similar to each other than to any human V gene, it is merely necessary to note that no human gene can have diverged from a mouse gene more recently than the time designated by the arc, when the two species separated.

The mouse/man ancestor, like present-day mice and men, had many V genes, whose family tree

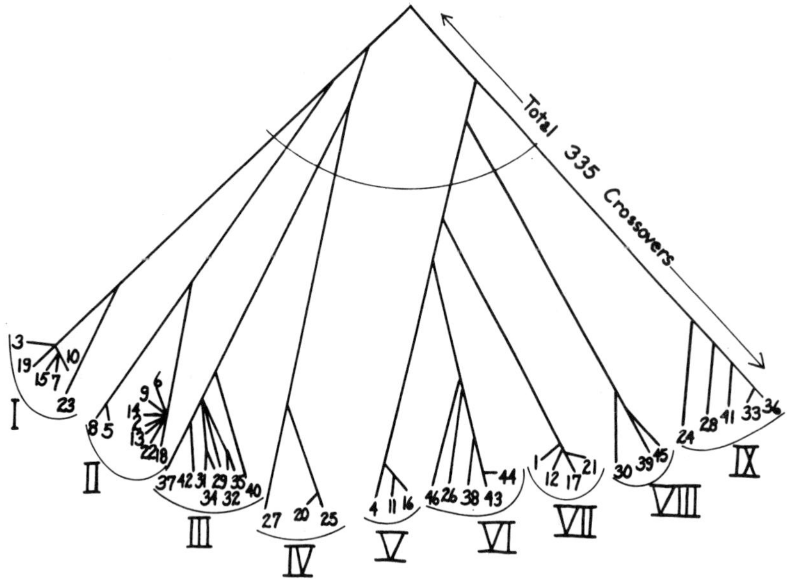

Figure 5. An evolutionary tree generated in a simulation in which 50 starting repeats were allowed to vary between 45 and 55 in a single chromosomal lineage by intrachromosomal crossover. There happened to be 46 repeats in the final array and they were related to each other by the evolutionary tree shown. The roman numerals designate subgroups. Branch lengths are proportional to number of crossovers. The arc near the top of the tree is discussed in the text.

(according to this view) looked much like that in Figure 5. Among this ancestor's subgroups, two have given rise to the subgroups in our hypothetical contemporary mouse, one to subgroups I–IV, the other to subgroups V–IX. The human V genes, on the other hand, would have undergone an independent series of unequal crossovers since man and mouse diverged. They might well, therefore, have descended from *different* subgroups in the ancestor. Moreover, even if some human and mouse subgroups descended from the same ancestral subgroup, there is no reason to believe that the *number* would correspond in the two species. This, then, is how the theory of crossover fixation accounts for the fact that subgroups in distantly related mammalian species do not correspond to one another on a one-to-one basis.

One aspect of the pattern of variation in V regions that is somewhat disturbing to the theory of crossover fixation is the absence of any clear recombinants (Smith et al., 1971; Smith, 1973). Assuming that the translated portion forms an appreciable fraction of the repeat, would not crossover be expected to occur within the translated portion and thus be perceivable as recombinant amino acid sequences? It is true, of course, that most of the recombinants generated during crossover fixation would not be fixed by the subsequent crossovers and thus will be lost. But a crude calculation suggests that enough recombinants should survive to have been seen among the many immunoglobulin V-region sequences that have been determined. One reasonable explanation of this discrepancy is that crossover occurs much more frequently between the very similar V genes within one subgroup than between the very different V genes in different subgroups. Such recombinants would seldom be detectable by sequence comparisons. Since (as I pointed out above) genes in the same subgroup would be widely scattered in the array, this restriction would have little effect on crossover fixation.

Another, less direct piece of evidence against crossover fixation in V genes is the apparent absence of rabbit chromosomes that are recombinant for heavy chain V-gene genetic markers. This subject is discussed elsewhere (Smith et al., 1971).

I feel that these two reservations are minor ones. Apart from them, the theory of crossover fixation provides a simple explanation of many salient features of the evolution of V genes.

Summary

Much of the theoretical interest in families of repeated DNA sequences has stemmed from the suspicion that their features imply radically new evolutionary processes. In this paper I have tried to allay that suspicion—though not, I hope, to detract from that interest—by showing that the evolutionary characteristics of two very different systems of multiple repeats, the rRNA and immunoglobulin V-gene families, can plausibly be attributed to a moderate amount of unequal crossover.

Acknowledgments

These ideas have been developed during, and in large measure as a result of, my association with my postdoctoral advisor, Oliver Smithies. I hereby forgive him for those many helpful suggestions which caused me to rewrite this paper so many times. I was first informally introduced to the idea of "expanding and contracting" genes by Leroy Hood at this Symposium in 1967, and he is responsible for many of the concepts on which this theory is based. I thank the Helen Hay Whitney Foundation for a generous postdoctoral fellowship. This is publication number 1668 from the Laboratory of Genetics, University of Wisconsin.

References

AMALDI, F., P. A. LAVA-SANCHEZ, and M. BUONGIORNO-NARDELLI. 1973. Nuclear DNA content variability in *Xenopus laevis*: A redundancy regulation common to all gene families. *Nature* **242:** 615.

BARKER, W. C., P. J. MCLAUGHLIN, and M. O. DAYHOFF. 1972. Evolution of a complex system: The immunoglobulins. In *Atlas of protein sequence and structure*, ed. M. O. Dayhoff, p. 31. National Biomedical Research Foundation, Silver Spring, Md.

BRITTEN, R. J. and D. E. KOHNE. 1968. Repeated sequences in DNA. *Science* **161:** 529.

BROWN, D. D., P. C. WENSINK, and E. JORDAN. 1971. Purification and some characteristics of 5S DNA from *Xenopus laevis*. *Proc. Nat. Acad. Sci.* **68:** 3175.

———. 1972. A comparison of the ribosomal DNA's of *Xenopus laevis* and *Xenopus mulleri*: The evolution of tandem genes. *J. Mol. Biol.* **63:** 57.

BUONGIORNO-NARDELLI, M., F. AMALDI, and P. A. LAVA-SANCHEZ. 1972. Amplification as a rectification mechanism for the redundant rRNA genes. *Nature New Biol.* **238:** 134.

CALLAN, H. G. 1967. The organization of genetic units in chromosomes. *J. Cell Sci.* **2:** 1.

EDELMAN, G. M. 1970. The structure and genetics of antibodies. In *The neurosciences: Second study program*, ed. F. O. Schmitt, p. 885. Rockefeller University Press, New York.

EDELMAN, G. M. and J. A. GALLY. 1970. Arrangement and evolution of eukaryotic genes. In *The neurosciences: Second study program*, ed. F. O. Schmitt, p. 962. Rockefeller University Press, New York.

MILLER, O. L. and B. R. BEATTY. 1969. Visualization of nucleolar genes. *Science* **164:** 955.

SMITH, G. P. 1973. *The variation and adaptive expression of antibodies* pp. 75–107. Harvard University Press, Cambridge, Mass.

SMITH, G. P., L. HOOD, and W. M. FITCH. 1971. Antibody diversity. *Ann. Rev. Biochem.* **40:** 969.

SOUTHERN, E. M. 1970. Base sequence and evolution of guinea-pig alpha-satellite DNA. *Nature* **227:** 794.

SUTTON, W. D. and M. MCCALLUM. 1972. Related satellite DNA's in the genus *Mus*. *J. Mol. Biol.* **71:** 633.

THOMAS, C. A. 1970. The theory of the master gene. In *The neurosciences: Second study program*, ed. F. O. Schmitt, p. 973. Rockefeller University Press, New York.

WENSINK, P. C. and D. D. BROWN. 1971. Denaturation map of the ribosomal DNA of *Xenopus laevis*. *J. Mol. Biol.* **60:** 235.

Reprinted from *Nature* **271**: 501.

Why genes in pieces?

from Walter Gilbert

OUR picture of the organisation of genes in higher organisms has recently undergone a revolution. Analyses of eukaryotic genes in many laboratories[1-10], studies of globin, ovalbumin, immunoglobulin, SV40 and polyoma, suggest that in general the coding sequences on DNA, the regions that will ultimately be translated into amino acid sequence, are not continuous but are interrupted by 'silent' DNA. Even for genes with no protein product such as the tRNA genes of yeast and the rRNA genes in *Drosophila*, and also for viral messages from adenovirus, Rous sarcoma virus and murine leukaemia virus, the primary RNA transcript contains internal regions that are excised during maturation, the final tRNA or messenger being a spliced product.

The notion of the cistron, the genetic unit of function that one thought corresponded to a polypeptide chain, now must be replaced by that of a transcription unit containing regions which will be lost from the mature messenger—which I suggest we call introns (for intragenic regions)—alternating with regions which will be expressed—exons. The gene is a mosaic: expressed sequences held in a matrix of silent DNA, an intronic matrix. The introns seen so far range from 10 to 10,000 bases in length; I expect the amount of DNA in introns will turn out to be five to ten times the amount in exons.

This model immediately accommodates two aspects of the genetic structure of higher cells. Heterogeneous nuclear RNA clearly is the long transcription products out of which the much smaller ultimate messengers for expressed polypeptide sequences are spliced. The unexpected extra DNA in higher cells, the excess of DNA over that needed to code for the number of products defined genetically, now is ascribed to the introns.

What are the benefits of this intronic/exonic structure for genes? For the sake of argument let us assume that the splicing mechanism is general and independent of the specific gene or the state of the cell, reflecting simply some secondary structure in the RNA. For example, base-pairing in the messenger could generate sites which would serve as signals for enzymes, such as those that excise tRNAs from their precursors, to cut out a section. The cut would be resealed by an RNA ligase. Even if RNA processing is general, the presence of infilling sequences can speed evolutiony

Single base changes, the elementar. mutational events, not only can change protein sequences by the alteration of single amino acids but now, if they occur at the boundaries of the regions to be spliced out, can change the splicing pattern, resulting in the deletion or addition of whole sequences of amino acids. During the course of evolution relatively rare single mutations can generate novel proteins much more rapidly than would be possible if no splicing occurred.

Furthermore, the splicing need not be a hundred per cent efficient; changes in sequence can alter the process so that base pairing and splicing occurs only some of the time. Even mutations in silent third base positions, could modify the joining so that the products of a single transcription unit can be both the original gene product and a new product, also synthesised at a high rate. Evolution can seek new solutions without destroying the old. A classic problem is resolved: the genetic material does not have to duplicate to provide a second copy of an essential gene in order to mutate to a new function. Rather than a special duplication, the extra material is scattered in the genome, to be called into action at any time. After a new gene function appears, if a higher level of product is needed, there will be selective pressure for gene duplication (as well as pressure for the loss of the introns in highly repeated genes). One consequence of the intronic model is that the dogma of one gene, one polypeptide chain disappears. A gene, a contiguous region of DNA, now corresponds to one transcription unit, but that transcription unit can correspond to many polypeptide chains, of related or differing functions.

Recombination now becomes more rapid. Since the gene is spread out over a larger region of DNA, recombination, which should be hampered in higher cells by the inability of DNA molecules to get together, will be enhanced. Furthermore, if exonic regions correspond to functions put together by splicing to form special combinations in the finished protein, then recombination within introns will assort these functions independently. Middle repetitious sequences within introns may create hot spots for recombination to rearrange the exonic sequences.

Recombination within introns will generate curious genetic structures for eukaryotic genes. Structural mutations should be clustered, separated by long distances from mutations in other exons. Mutations in different functions may be interspersed, when one product's intron becomes another's exon.

According to this view, introns are both frozen remnants of history and as the sites of future evolution. Nevertheless, they could also have other roles. Specific recombinations between introns can bring together exons into a transcription unit to make special differentiation products. Specific new splicing patterns could be turned on by special gene products. A differentiation pathway may be determined by the appearance of a new splicing enzyme, calling forth new proteins out of the heterogeneous nuclear RNA.

On this can be based a striking hypothesis to explain the behaviour of immunoglobulin heavy chains. At an early stage of the immune response a single lymphocyte can synthesise two different immunoglobulins, IgM and IgD, with the same idiotype; two different constant portions attached to the same V_H region. This may be the result of a V_H region translocating by recombination within an intron near the constant genes so that a trranscription unit is formed for a V_H-C_μ-C_δ message. Splicing can then create contiguous messenger sequences for V_HC_μ and V_HC_δ chains. The switch from IgM to IgG might be a new translocation of the V_H gene, but, alternatively, it may be a new enzyme that changes the processing of a V_H-$C_{\mu\gamma}$ C_δ-C_γ message to produce a V_HC-product. □

Walter Gilbert is American Cancer Society Professor of Molecular Biology at Harvard University.

1. *News and Views, Nature* **268**, 101 (1977).
2. Berget *et al. Proc. natn. Acad. Sci. U.S.A.* **74**, 3171 (1977).
3. Papers in *Cell* **12** (1) (1977).
4. Aloni *et al. Proc. natn. Acad. Sci. U.S.A.* **74**, 3686 (1977).
5. Leder *et al. Cold Spring Harbor Symp. quant. Biol.* (in the press).
6. *News and Views, Nature* **270**, 295 (1977).
7. Breathmark *et al. Nature* **270**, 314 (1977).
8. Deol *et al. Nucleic Acids Res.* **4**, 3701 (1977).
9. Jeffreys & Flavell *Cell* **12**, 1097 (1977).
10. Brack & Tonegawa *Proc. natn. Acad. Sci. U.S.A.* **74**, 5652 (1977).

Polymorphisms in the Chromosomal Locations of Elements of the *412*, *copia* and *297* Dispersed Repeated Gene Families in Drosophila

Edward Strobel, Pamela Dunsmuir and Gerald M. Rubin
Department of Tumor Biology
Sidney Farber Cancer Institute
and Department of Biological Chemistry
Harvard Medical School
Boston, Massachusetts 02115

Summary

The number and chromosomal locations of elements of the *412*, *copia* and *297* dispersed repeated gene families differ extensively when the genomes of four D. melanogaster strains are compared. Differences among individuals from the same laboratory stock in the arrangement of these elements are also observed. In contrast to these polymorphisms, the structures of the elements themselves are closely conserved. Our results indicate that *412*, *copia* and *297* are capable of evolutionarily rapid transpositions to new chromosomal sites.

Introduction

The elements of the *412*, *copia* and *297* dispersed repeated gene families of Drosophila melanogaster are generally more abundant in the genomes of tissue culture cell lines than in embryo DNA. Furthermore, the additional elements which occur in tissue culture cells are distributed to different locations in the genome rather than amplified locally (Potter et al., 1979). These data demonstrate that the elements of these dispersed repeated gene families are amplifiable and transposable in the Drosophila genome during the course of cell proliferation in culture.

We have asked whether this apparent plasticity in both the number and distribution of *412*, *copia* and *297* elements is a general property of these sequences. To do this, we compared the genomes of four different strains of D. melanogaster, using filter hybridization to restriction endonuclease digests of embryo DNA and in situ hybridization to salivary gland polytene chromosomes. The Oregon R, Canton S, Seto and Swedish C strains used in these experiments are derived from widely separated wild populations of D. melanogaster—they have been maintained in the laboratory for at least 30 years (Bridges and Brehme, 1944). The four strains are morphologically indistinguishable, fully reproductively compatible isolates of the same species; their polytene chromosome banding patterns are homosequential. We present data demonstrating variations in the number and distribution of *412*, *copia* and *297* elements among the genomes of these four strains. Moreover, differences in the arrangement of these elements occur among individuals from the same laboratory stock. These data indicate that the elements of dispersed repeated gene families undergo amplification, and/or diminution, together with transposition within the D. melanogaster genome.

Results

The Elements of the *412*, *copia* and *297* Dispersed Repeated Gene Families Differ in Number and Organization in the Genomes of Four D. melanogaster Strains

We have examined the pattern of restriction enzyme cleavage sites within those sequences homologous to the elements of the *412*, *copia* and *297* dispersed repeated gene families which occur in DNA isolated from Oregon R, Canton S, Seto and Swedish C embryos. In the Oregon R genome, most elements of a given gene family have been shown to be similar with respect to their restriction enzyme cleavage patterns (Rubin, Finnegan and Hogness, 1976; Finnegan et al., 1978; G. M. Rubin and B. Backner, unpublished results). An experiment demonstrating such conservation of restriction sites within the sequences homologous to *412* in DNA from each strain is shown in Figure 1. Three characteristic restriction fragments were generated from within the *412* element of the cloned Oregon R DNA segment Dm412 (Rubin et al., 1976); sequences homologous to each of these three restriction fragments were confined to fragments of the same size in embryo DNA of each strain. Similar results were obtained with fragments from within the *copia* and *297* elements (Figure 2), although for *copia* the strains appear to differ in the amount of these sequences. Thus the characteristic restriction sites within those sequences homologous to *412*, *copia* and *297* elements are conserved in the DNA from each of the four D. melanogaster strains considered.

We have examined the pattern of restriction sites in the DNA adjacent to elements of the *412*, *copia* and *297* gene families in D. melanogaster DNA isolated from embryos of the Oregon R, Canton S, Seto and Swedish C strains. Consider first the results obtained for the *copia* dispersed repeated gene family. DNA isolated from embryos of each strain was digested with Eco RI, and the sizes of the resultant fragments homologous to the Hpa I-Eco RI internal fragment of *copia* were determined (Figure 3). Numerous qualitative and quantitative differences exist among the hybridization patterns of these embryo DNAs, suggesting that there are different amounts and distributions of *copia* elements in the genomes of the four D. melanogaster strains. Analogous results were obtained when the patterns of restriction sites adjacent to either *297* elements (Figure 4) or *412* elements (data not shown) were compared among the four strains. Not all D. melanogaster dispersed repeated sequences exhibit such marked strain differences in

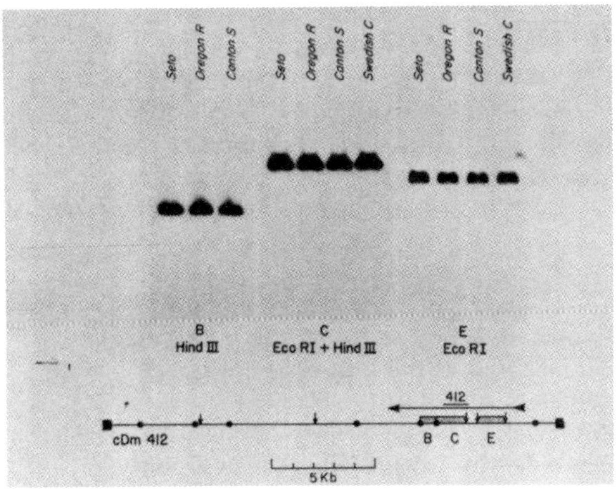

Figure 1. Comparison of Restriction Fragments in cDm412 with Restriction Fragments from Total DNA of Oregon R, Canton S, Seto and Swedish C Embryos Which Contain *412* Sequences

For each set of results, either Hind III, Eco RI or Hind III plus Eco RI digests of each total genome DNA were separated on 1.4% agarose gels. Each panel shows the autoradiograph obtained when these restriction fragments are subsequently hybridized according to the method of Southern (1975) with ^{32}P-labeled restriction fragments derived from cDm412 as indicated. A physical map of cDm412 (Rubin et al., 1976) is shown below the autoradiograms. The thin horizontal line represents Drosophila DNA. The thick blocks at the end of the map represent Col E1 DNA. Restriction enzyme cleavage sites for Eco RI (↓) and Hind III (●) are indicated. The shaded regions delineate the restriction fragments used as hybridization probes. The position of the *412* element is indicated above the map.

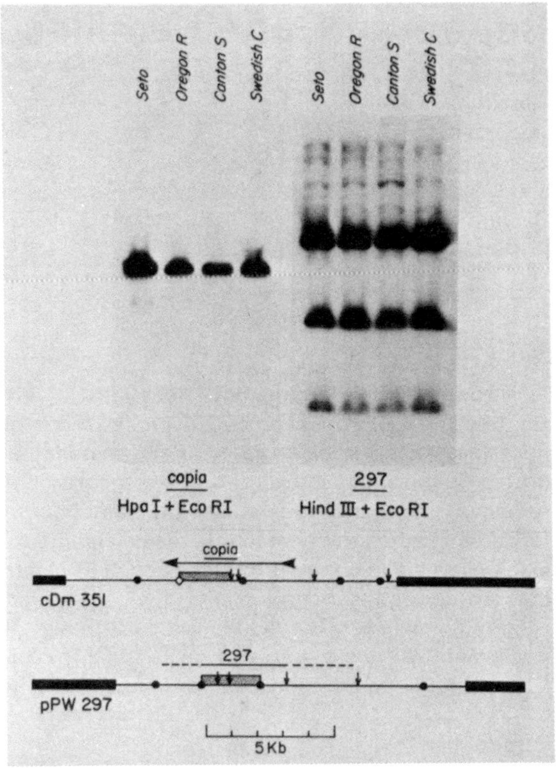

Figure 2. Comparison of Restriction Fragments in cDm351 and pPW297 with Restriction Fragments from Total DNA of Oregon R, Canton S, Seto and Swedish C Embryos Which Contain *copia* or *297* Sequences

For *copia* sequences, Hpa I plus Eco RI digests of each total genome DNA were separated on a 1% agarose gel and subsequently hybridized with the indicated fragments of cDm351 according to the method of Southern (1975). For *297* sequences, Hind III plus Eco RI digests were separated on a 1.4% agarose gel and hybridized with the indicated Hind III fragment of pPW297. Autoradiograms are shown. In the maps of cDm351 (D. J. Finnegan, G. M. Rubin and D. S. Hogness, unpublished results) and pPW297 (G. M. Rubin and B. Backner, unpublished results), the thin horizontal lines represent Drosophila DNA and the heavy lines represent DNA of the plasmid vector (Col E1 for cDm351 and pMB9 for pPW297). Restriction enzyme cleavage sites for Eco RI (↓), Hind III (●) and Hpa I (◇) are indicated. The shaded regions delineate the restriction fragments used as hybridization probes. The positions of the *copia* element in cDm351 and the *297* element in pPW297 are indicated above the maps. The precise end points of the *297* element have not been determined, but fall within the limits shown by the dashed portion of the line.

adjacent genomic sequences. In contrast to the results obtained with the large, conserved, terminally redundant elements (*412, copia* and *297*), restriction sites in sequences adjacent to Dm27 sequences, a set of smaller clustered repeat units present at over 70 chromosomal sites (D. J. Finnegan, G. M. Rubin, D. J. Bower and D. S. Hogness, manuscript in preparation), were nearly identical in the four strains (data not shown).

The data obtained from hybridization to restriction fragments of embryo DNA suggest that sequences homologous to the *412, copia* and *297* elements of the Oregon R strain are also conserved as intact elements in the genomes of three other D. melanogaster strains—Canton S, Seto and Swedish C. However, the number and genomic environments of these elements differ among these four strains.

The Elements of the Dispersed Repeated Gene Families Are Organized Differently in the Salivary Gland Chromosomes of Different Strains of D. melanogaster

The data from hybridization experiments to restriction digests of embryo DNA suggest that intact elements of the *412, copia* and *297* repeated gene families are distributed differently among the four strains of D. melanogaster examined. Direct demonstration of these differences in genomic organization was provided by hybridization of ^{3}H-labeled fragments of *412, copia* and *297* elements to polytene salivary gland chromosomes of individuals from each of the four D. melanogaster strains: Oregon R, Canton S, Seto and Swedish C. Differences in both the number and location of labeled sites were observed.

As an example of the differences among strains, the in situ hybridization patterns of *copia* sequences to Canton S and Seto chromosomes are shown in Figures 5 and 6. There are 43 regions labeled on the chromosome arms of Seto and only 19 in Canton S. The chromocenter (a region which results from the association of the centric heterochromatin of all chromosomes) in each strain also contains sequences

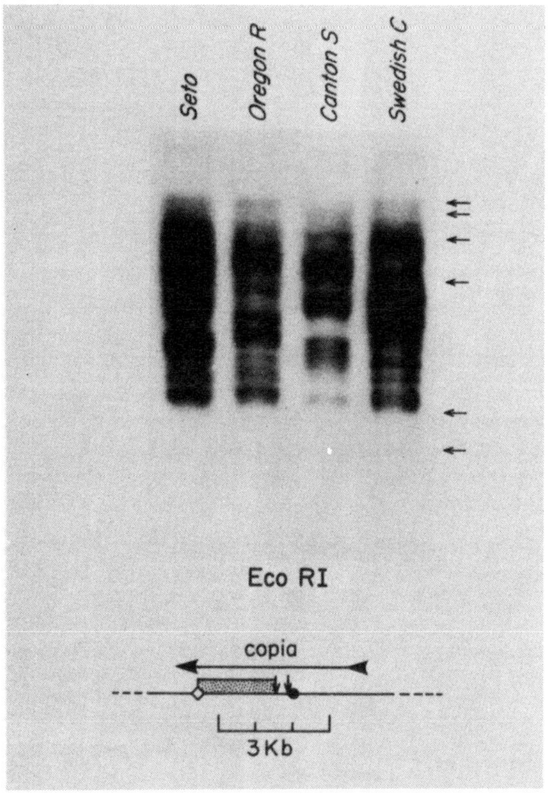

Figure 3. Distribution of Eco RI Restriction Sites in Sequences Adjacent to *copia* Elements in the Genomes of Oregon R, Canton S, Seto and Swedish C Embryos

Eco RI digests of each total genome DNA were separated on a 0.5% agarose gel. The autoradiograph obtained when these fragments were hybridized according to the procedure of Southern (1975) with the indicated ^{32}P-labeled Eco RI plus Hpa I restriction fragment of cDm351 is shown. A portion of the physical map of cDm351 is shown. The horizontal line represents Drosophila DNA, and restriction enzyme cleavage sites for Eco RI (↓), Hpa I (◇) and Hind III (●) are indicated. The shaded region delineates the restriction fragment used as a hybridization probe. The position of the *copia* element is indicated above the map. The arrows indicate the positions of Hind III fragments of bacteriophage λ DNA which are 23, 9.8, 6.6, 4.5, 2.5 and 2.2 kb long (Murray and Murray, 1975).

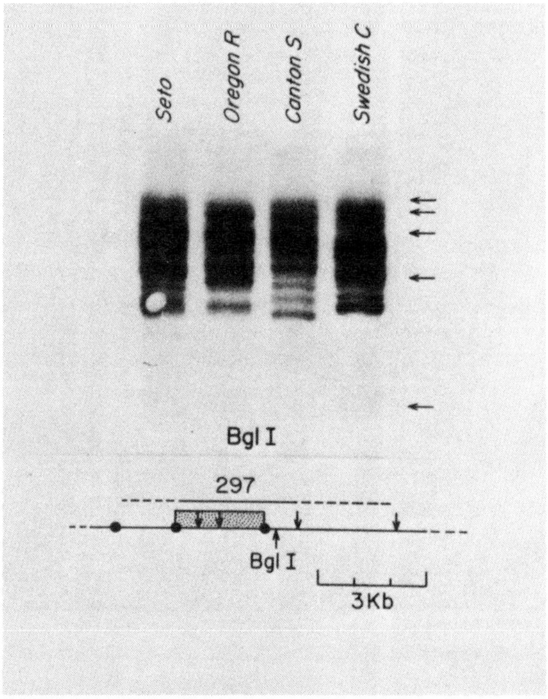

Figure 4. Distribution of Bgl I Restriction Sites in Sequences Adjacent to *297* Elements in the Genomes of Oregon R, Canton S, Seto and Swedish C Embryos

Bgl I digests of each total genome were separated on a 0.5% agarose gel. The autoradiograph obtained when these restriction fragments were hybridized according to the method of Southern (1975) with the indicated ^{32}P-labeled Hind III fragment of pPW297 is shown. A portion of the physical map of pPW297 is presented. The horizontal line represents Drosophila DNA, and restriction enzyme cleavage sites for Eco RI (↓), Hind III (●) and Bgl I (↑) are indicated. The shaded region delineates the restriction enzyme fragment used as a hybridization probe. The position of the *297* element is shown above the map. The arrows indicate the positions of Hind III fragments of bacteriophage λ DNA which are 23, 9.8, 6.6, 4.5 and 2.5 kb long (Murray and Murray, 1975).

homologous to *copia*. Some of these heterochromatic *copia* sequences are adjacent to satellite DNA (Carlson and Brutlag, 1978).

We have compared the number of in situ hybridization sites of *412*, *copia* and *297* in several individuals from each of the four strains. A numerical summary of these data is presented in Tables 1–3 for the sequences homologous to *412*, *copia* and *297* elements, respectively. Our determinations of the number of chromosomal sites are subject to several sources of variance inherent in the cytological techniques used: a number of closely associated sites may appear as a single band of hybridization unless the chromosomes are well stretched; regions which are distorted or highly constricted may be less accessible to hybridization with the labeled probe; and sites close to the centric heterochromatin may be obscured by extensive hybridization to the chromocenter. To minimize the contribution of these factors to the number of observed sites, we have combined the data from many nuclei of each individual. (The variability in the number of sites of hybridization among individuals of a population is real and is discussed in more detail in the following section.) There are clear differences in the number of sites labeled when each chromosome arm is compared among the strains. The right arm of the third chromosome exhibits the greatest variability irrespective of which dispersed repeated gene family is being considered.

The full extent of the plasticity in the genomic organization of the dispersed repeated gene families is apparent when we examine the chromosomal sites of the *412*, *copia* and *297* elements, rather than just their numbers. A convenient method for directly demonstrating these differences in the chromosomal distribution of elements is provided by in situ hybridization to the chromosomes of the F1 progeny from

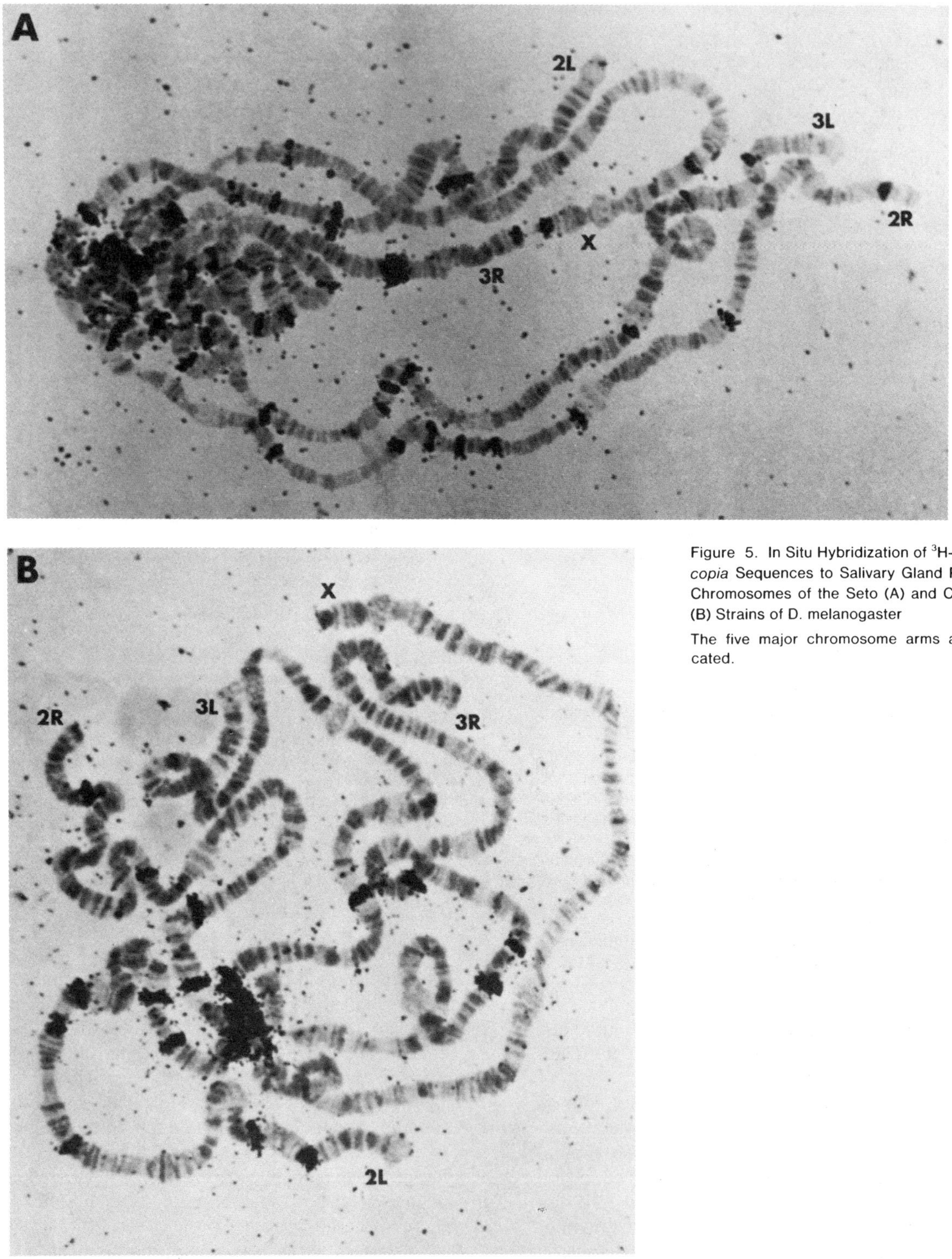

Figure 5. In Situ Hybridization of ^{3}H-Labeled *copia* Sequences to Salivary Gland Polytene Chromosomes of the Seto (A) and Canton S (B) Strains of D. melanogaster

The five major chromosome arms are indicated.

crosses between different strains. Homologous chromosomes normally remain paired in Drosophila salivary gland nuclei; each chromosome arm depicted in Figures 5 and 6 represents the intimate synapsis of both maternal and paternal homologues. Occasionally the homologues become separated as a result of the squashing procedure, and examination of such regions allows a direct comparison of the sites of a particular dispersed repeated gene family in the chromosomes of the two parent strains. Figure 7 shows *412* sites within such an asynapsed region on the 2L chromosome arm of an F1 Oregon R × Seto individ-

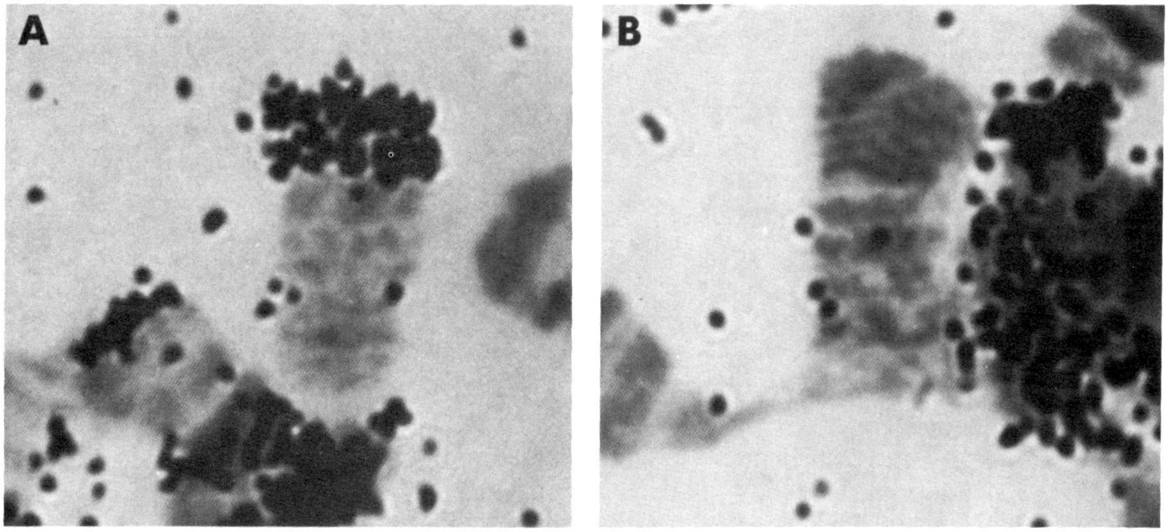

Figure 6. In Situ Hybridization of ^{3}H-Labeled *copia* Sequences to the Fourth Chromosomes of the Seto (A) and Canton S (B) Strains of D. melanogaster

Two *copia* sites occur on the Seto chromosome; there are none on the Canton S chromosome, although both strains show extensive labeling at the chromocenter.

Table 1. Number of *412* in Situ Hybridization Sites

	Chromosome Arm						
	Total Sites	X	2L	2R	3L	3R	4
Oregon R							
Individual 1	26.2 ± 1.7 (8)	4.8 ± 0.7 (8)	6.6 ± 0.5 (7)	5.0 ± 1.0 (7)	3.9 ± 0.7 (7)	4.7 ± 0.5 (7)	0 (6)
Individual 2	26.3 ± 2.0 (6)	4.4 ± 0.5 (9)	6.6 ± 0.5 (7)	4.9 ± 0.6 (9)	3.7 ± 0.5 (7)	4.8 ± 0.5 (8)	0 (7)
Overall	26.3 ± 1.8 (14)	4.6 ± 0.6 (17)	6.6 ± 0.5 (14)	4.9 ± 0.8 (16)	3.8 ± 0.6 (14)	4.7 ± 0.5 (15)	0 (13)
Canton S							
Individual 1	33.4 ± 1.7 (8)	2.9 ± 0.3 (10)	7.8 ± 0.5 (8)	7.0 ± 0.5 (8)	7.8 ± 0.7 (6)	7.2 ± 0.6 (10)	0 (8)
Individual 2	32.7 ± 1.7 (13)	2.8 ± 0.4 (12)	7.8 ± 0.4 (9)	7.4 ± 0.5 (9)	7.3 ± 0.6 (11)	7.1 ± 0.3 (10)	0 (10)
Overall	33.0 ± 1.7 (21)	2.9 ± 0.4 (22)	7.8 ± 0.4 (17)	7.2 ± 0.5 (17)	7.5 ± 0.7 (17)	7.2 ± 0.5 (20)	0 (18)
Seto							
Individual 1	29.9 ± 1.2 (7)	6.3 ± 0.6 (13)	4.0 ± 0.9 (11)	4.9 ± 0.7 (10)	6.4 ± 0.5 (10)	5.9 ± 0.6 (9)	0 (5)
Individual 2	33.2 ± 2.2 (5)	5.7 ± 1.0 (9)	3.9 ± 0.6 (9)	5.1 ± 0.7 (10)	6.7 ± 0.5 (9)	5.4 ± 0.6 (5)	0 (5)
Overall	31.2 ± 2.3 (12)	6.0 ± 0.8 (22)	4.0 ± 0.8 (20)	5.0 ± 0.7 (20)	6.5 ± 0.5 (19)	5.7 ± 0.6 (15)	0 (10)
Swedish C							
Individual 1	32.6 ± 2.5 (10)	5.8 ± 0.4 (11)	5.1 ± 0.8 (11)	5.3 ± 0.5 (13)	4.1 ± 0.7 (11)	11.0 ± 0.6 (11)	0 (12)
Individual 2	31.4 ± 2.3 (5)	5.8 ± 0.4 (11)	4.7 ± 0.7 (15)	5.4 ± 0.5 (16)	4.0 ± 0.6 (14)	10.8 ± 0.6 (15)	0 (10)
Overall	32.2 ± 2.5 (15)	5.8 ± 0.4 (22)	4.9 ± 0.8 (26)	5.3 ± 0.5 (29)	4.0 ± 0.6 (25)	10.9 ± 0.6 (26)	0 (22)

The numbers for each individual represent the mean and standard deviation of hybridization sites in nuclei from one salivary gland. The number of nuclei counted is indicated in the brackets. The overall mean and standard deviation were calculated by pooling the data from all nuclei. A single-factor analysis of variance was used to test for significant differences among individuals of a given strain. An asterisk denotes significance at the 0.1% level.

ual; none of three sites in this short asynaptic region is common to both homologues.

It is not possible to prepare chromosome squashes consistently with chromosomes totally asynapsed, yet when the chromosomes are closely synapsed it is not possible to determine reliably whether an observed hybridization site is due to the presence of the element on one or both homologues. However, a comparison of the number of labeled sites on each arm in two parent strains with that of the hybrid progeny allows one to estimate the number of sites, if any, that are common to the two strains. For example, we found that approximately 30% of the elements of each dispersed repeated gene family were located at the same

Table 2. Number of copia in Situ Hybridization Sites

	Chromosome Arm						
	Total Sites	X	2L	2R	3L	3R	4
Oregon R							
Individual 1	29.7 ± 2.5 (11)	2.0 ± 0.0 (7)	4.5 ± 0.5 (6)	6.7 ± 0.5 (7)	7.3 ± 1.0 (7)	9.7 ± 0.5 (7)	0 (8)
Individual 2	32.4 ± 1.7 (11)	2.8 ± 0.4 (12)	4.9 ± 0.4 (7)	6.8 ± 0.4 (6)	7.2 ± 0.7 (8)	9.9 ± 0.4 (8)	0 (7)
Individual 3	28.7 ± 2.1 (3)	0.9 ± 0.3 (9)	4.6 ± 0.5 (7)	6.1 ± 0.7 (14)	6.0 ± 0.9 (11)	9.1 ± 0.7 (7)	0 (5)
Individual 4	29.0 ± 3.6 (3)	2.8 ± 0.4 (10)	4.3 ± 0.6 (3)	6.2 ± 0.5 (4)	6.3 ± 0.8 (7)	9.3 ± 0.6 (3)	0 (5)
Individual 5	25.0 ± 0.0 (3)	3.0 ± 0.0 (3)	6.0 ± 0.0 (3)	6.0 ± 0.0 (3)	4.0 ± 0.0 (3)	6.0 ± 0.0 (3)	0 (3)
Individual 6	27.0 ± 0.0 (3)	2.0 ± 0.0 (3)	4.0 ± 0.0 (3)	6.0 ± 0.0 (3)	5.0 ± 0.0 (3)	10.0 ± 0.0 (3)	0 (3)
Individual 7	28.0 ± 0.0 (3)	2.0 ± 0.0 (3)	4.0 ± 0.0 (3)	7.0 ± 0.0 (3)	6.0 ± 0.0 (3)	9.0 ± 0.0 (3)	0 (3)
Overall	29.6 ± 2.9 (37)	2.2 ± 0.8 (47)*	4.6 ± 0.7 (32)*	6.4 ± 0.6 (40)	6.3 ± 1.2 (42)*	9.2 ± 1.2 (34)*	0 (34)
Canton S							
Individual 1	19.4 ± 1.3 (5)	0 (6)	7.1 ± 0.4 (15)	5.0 ± 0.8 (10)	2.9 ± 0.4 (8)	4.0 ± 0.0 (6)	0 (6)
Individual 2	19.4 ± 0.6 (5)	0 (7)	6.9 ± 0.4 (7)	3.7 ± 0.5 (10)	2.0 ± 0.5 (10)	3.8 ± 0.4 (5)	0 (5)
Individual 3	20.0 ± 1.2 (5)	0 (7)	7.0 ± 0.0 (6)	5.1 ± 0.4 (7)	2.2 ± 0.4 (6)	4.2 ± 0.4 (5)	0 (7)
Individual 4	19.9 ± 1.2 (8)	0 (9)	6.7 ± 0.5 (6)	5.3 ± 0.5 (6)	2.8 ± 0.8 (6)	5.0 ± 1.0 (7)	0 (7)
Overall	19.7 ± 1.1 (23)	0 (29)	7.0 ± 0.4 (34)	4.7 ± 0.9 (33)*	2.4 ± 0.6 (30)	4.3 ± 0.8 (23)	0 (25)
Seto							
Individual 1	42.2 ± 1.9 (5)	6.2 ± 0.4 (6)	7.2 ± 0.4 (6)	8.1 ± 0.9 (9)	8.2 ± 1.0 (8)	11.5 ± 0.9 (8)	2.0 ± 0.0 (8)
Individual 2	41.8 ± 1.4 (6)	6.0 ± 0.0 (5)	7.1 ± 0.4 (7)	8.7 ± 0.5 (8)	8.9 ± 0.7 (10)	11.7 ± 0.5 (7)	2.0 ± 0.0 (5)
Individual 3	42.6 ± 3.0 (8)	6.2 ± 0.5 (8)	6.2 ± 0.5 (8)	8.7 ± 0.8 (7)	10.4 ± 0.5 (7)	12.4 ± 0.8 (7)	2.0 ± 0.0 (10)
Individual 4	43.4 ± 2.2 (7)	5.9 ± 0.4 (7)	6.2 ± 0.4 (6)	8.8 ± 0.7 (8)	10.0 ± 0.6 (7)	13.6 ± 0.5 (9)	2.0 ± 0.0 (7)
Individual 5	44.2 ± 1.8 (6)	6.1 ± 0.4 (7)	5.6 ± 0.5 (8)	8.1 ± 0.8 (8)	10.8 ± 0.5 (8)	14.7 ± 0.5 (10)	2.0 ± 0.0 (8)
Overall	42.8 ± 2.2 (32)	6.2 ± 0.4 (32)	6.4 ± 0.7 (35)*	8.6 ± 0.8 (32)	9.3 ± 1.1 (40)*	12.9 ± 1.4 (40)*	2.0 ± 0.0 (38)
Swedish C							
Individual 1	34.1 ± 2.0 (9)	4.4 ± 0.5 (7)	6.9 ± 0.6 (9)	7.5 ± 0.7 (10)	6.8 ± 0.4 (9)	6.8 ± 0.6 (10)	0 (10)
Individual 2	35.6 ± 2.3 (5)	4.0 ± 0.0 (6)	7.0 ± 0.6 (12)	7.6 ± 0.5 (12)	8.0 ± 0.7 (12)	7.2 ± 0.8 (10)	0 (14)
Overall	34.6 ± 2.2 (14)	4.2 ± 0.4 (13)	7.0 ± 0.6 (21)	7.6 ± 0.6 (22)	7.5 ± 0.9 (21)*	7.0 ± 0.7 (20)	0 (24)

Legend as in Table 1.

chromosomal location when the strains Oregon R and Seto were compared in this way. These data suggest that there is extensive variability in the organization of dispersed gene family elements among strains, so much that one wonders whether any loci are common to all four D. melanogaster populations examined. We have studied the chromosome locations of *297* and *copia* elements in individuals from each of the four D. melanogaster strains. In each strain there are more than twenty sites at which *297* sequences hybridize; of those only four sites are common to all strains—12B on the X chromosome, and the 67C, 76D and 77D on the third chromosome. Of the nineteen or more sites where *copia* sequences hybridize in each strain, only one at 30A on chromosome 2 is common to all four strains. It may be that these regions are more stable; a direct test of this hypothesis is not possible, however, until we construct appropriate chromosome lines and monitor the sequence distributions over an extended period. It is interesting that one of the common "conserved" *297* sites, 76D, may correspond to one of the five sites of hybridization determined for *297* elements in the D. melanogaster sibling species, D. simulans. The "conserved" *copia* site at 30A does not overlap, however, with the three regions of *copia* hybridization observed in D. simulans (P. Dunsmuir, M. Schweber and M. Meselson, personal communication). Clearly a dramatic reshuffling of *412*, *copia* and *297* elements has occurred since the separation of these four strains of D. melanogaster.

There Are Differences in the Chromosomal Organization of Elements of the Dispersed Repeated Gene Families among Individuals of an Inbreeding Laboratory Strain

In the course of the in situ hybridization experiments to determine whether strain specific organizations of dispersed repeated genes occur, we noticed differences in element distributions among individuals within each strain. Evidence for population heterogeneity in the number of hybridization sites is contained

Table 3. Number of *297* in Situ Hybridization Sites

	Chromosome Arm						
	Total Sites	X	2L	2R	3L	3R	4
Oregon R							
Individual 1	19.5 ± 1.6 (6)	5.5 ± 0.6 (14)	3.0 ± 0.0 (9)	3.9 ± 0.3 (9)	4.0 ± 0.7 (5)	3.4 ± 0.7 (11)	1.0 ± 0.0 (5)
Individual 2	19.7 ± 1.1 (10)	5.4 ± 0.5 (8)	5.0 ± 0.5 (8)	4.1 ± 0.4 (8)	3.2 ± 0.4 (9)	2.4 ± 0.5 (7)	1.0 ± 0.0 (11)
Individual 3	20.2 ± 1.2 (8)	4.2 ± 0.4 (9)	4.8 ± 0.4 (9)	4.1 ± 0.4 (8)	3.1 ± 0.4 (8)	4.2 ± 0.4 (9)	1.0 ± 0.0 (10)
Individual 4	19.9 ± 1.4 (7)	3.4 ± 0.5 (9)	4.6 ± 0.5 (8)	4.2 ± 0.5 (8)	3.6 ± 0.5 (9)	3.9 ± 0.4 (7)	1.0 ± 0.0 (7)
Overall	19.8 ± 1.2 (31)	4.7 ± 1.0 (40)*	4.3 ± 0.9 (34)*	4.1 ± 0.3 (33)	3.4 ± 0.6 (31)	3.5 ± 0.8 (34)*	1.0 ± 0.0 (33)
Canton S							
Individual 1	21.6 ± 1.1 (9)	5.6 ± 0.5 (9)	4.0 ± 0.0 (7)	3.0 ± 0.0 (9)	5.9 ± 0.4 (8)	5.2 ± 0.4 (9)	1.0 ± 0.0 (6)
Individual 2	22.0 ± 1.2 (7)	5.9 ± 0.3 (10)	4.0 ± 0.0 (7)	2.2 ± 0.3 (13)	3.8 ± 0.4 (5)	6.2 ± 0.7 (9)	1.0 ± 0.0 (5)
Individual 3	22.6 ± 0.9 (5)	6.8 ± 0.4 (6)	4.0 ± 0.0 (8)	2.0 ± 0.0 (7)	5.0 ± 0.0 (7)	4.4 ± 0.6 (5)	1.0 ± 0.0 (5)
Individual 4	22.7 ± 1.2 (6)	5.6 ± 0.5 (11)	3.0 ± 0.0 (5)	2.9 ± 0.3 (11)	5.8 ± 0.4 (5)	5.0 ± 0.0 (7)	1.0 ± 0.0 (6)
Overall	22.1 ± 1.2 (27)	5.9 ± 0.6 (36)*	3.8 ± 0.4 (27)*	2.6 ± 0.5 (40)*	5.2 ± 0.9 (25)*	5.3 ± 0.8 (30)*	1.0 ± 0.0 (22)
Seto							
Individual 1	28.4 ± 1.4 (9)	6.4 ± 0.5 (10)	1.6 ± 0.5 (7)	3.0 ± 0.0 (9)	5.9 ± 0.4 (8)	9.3 ± 0.8 (12)	1.0 ± 0.0 (6)
Individual 2	30.3 ± 1.5 (7)	8.7 ± 0.7 (9)	1.9 ± 0.3 (8)	3.2 ± 0.4 (9)	6.9 ± 0.3 (10)	8.9 ± 0.4 (7)	1.0 ± 0.0 (6)
Individual 3	31.2 ± 0.8 (5)	8.5 ± 0.6 (4)	2.0 ± 0.0 (5)	4.0 ± 0.0 (5)	6.0 ± 0.0 (5)	9.0 ± 0.0 (5)	1.0 ± 0.0 (4)
Individual 4	30.8 ± 1.6 (5)	8.1 ± 0.5 (11)	2.8 ± 0.5 (8)	3.1 ± 0.4 (7)	7.3 ± 0.5 (7)	9.1 ± 0.4 (6)	1.0 ± 0.0 (6)
Individual 5	29.6 ± 1.8 (11)	8.7 ± 0.5 (9)	2.0 ± 0.5 (10)	3.1 ± 0.3 (9)	6.0 ± 0.0 (11)	9.4 ± 0.6 (13)	1.0 ± 0.0 (6)
Overall	29.8 ± 1.7 (37)	8.0 ± 1.1 (43)*	2.0 ± 0.6 (38)*	3.2 ± 0.4 (39)*	6.4 ± 0.7 (41)*	9.2 ± 0.6 (43)	1.0 ± 0.0 (28)
Swedish C							
Individual 1	20.0 ± 1.3 (11)	3.0 ± 0.0 (8)	2.8 ± 0.4 (9)	4.1 ± 0.3 (11)	6.3 ± 0.8 (10)	3.6 ± 0.5 (7)	1.0 ± 0.0 (6)
Individual 2	21.6 ± 1.3 (9)	4.0 ± 0.0 (6)	2.8 ± 0.4 (5)	3.8 ± 0.4 (6)	7.0 ± 0.0 (5)	4.8 ± 0.4 (12)	1.0 ± 0.0 (5)
Individual 3	20.7 ± 1.5 (9)	3.2 ± 0.4 (5)	2.0 ± 0.0 (6)	4.0 ± 0.0 (5)	6.2 ± 0.4 (5)	4.6 ± 0.5 (6)	1.0 ± 0.0 (5)
Individual 4	21.5 ± 1.0 (6)	3.8 ± 0.4 (5)	2.0 ± 0.0 (4)	3.5 ± 0.7 (2)	7.0 ± 0.0 (3)	4.5 ± 0.7 (2)	1.0 ± 0.0 (2)
Overall	20.7 ± 1.4 (35)	3.4 ± 0.5 (24)*	2.5 ± 0.5 (24)*	4.0 ± 0.4 (24)	6.5 ± 0.7 (23)	4.4 ± 0.7 (27)*	1.0 ± 0.0 (18)

Legend as in Table 1.

in the data presented in Tables 1–3 for *412*, *copia* and *297* elements, respectively. There are a number of instances where the variance in the number of sites on a particular chromosome arm is greater among the individuals considered than the variance among nuclei of a given individual. This statistical analysis suggests that there is variability in the number of locations for repeated gene family elements among individuals which cannot be attributed to artifacts of the in situ hybridization procedure.

We have identified the sites of *copia* hybridization in three individuals of the Oregon R strain; three nuclei from each individual were scored. Although the nuclei from each individual were identical, differences were observed in hybridization patterns among the individuals. The number of sites on the third chromosome ranged from 10–15; only six of these were shared by the three individuals. The sites on the second chromosome ranged from 10–13, of which nine were common; and of the two or three sites on the X chromosome, two were identical in all three individuals. The right arm of the third chromosome exhibited the most heterogeneity; only three of the 6 to 10 sites observed were common among individuals.

The full extent of variation in the positions of elements of the dispersed repeated gene families is revealed when the maternal and paternal homologues of each polytene chromosome are examined separately. This is possible only if asynaptic chromosome regions are generated as a result of the chromosome squashing procedure. In such asynaptic regions, we frequently observed that an element was present on only one of the two homologues. For example, Figure 8 shows in situ hybridization of *copia* sequences to three different asynaptic regions of the Oregon R third chromosome. Within the region 86–90 which constitutes about one fifth of the right arm of this chromosome, three *copia* sites are observed. Only one of these sites, however, is present on both homologues. Two additional examples of polymorphic *copia* loci along asynaptic regions of the third chromosome are also shown in Figure 8.

This detailed comparison of *copia* sequence elements in three individuals of the Oregon R strain

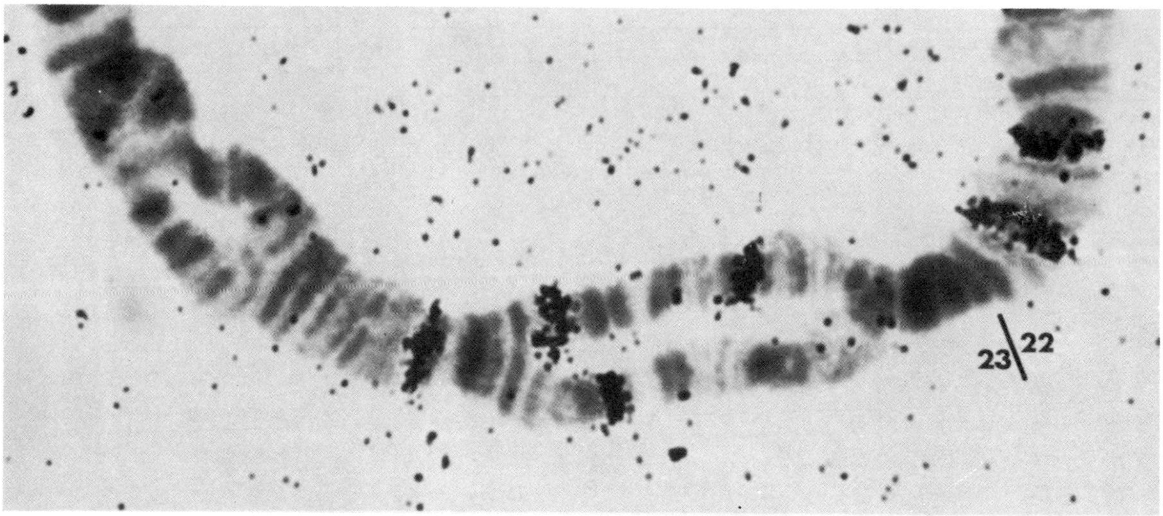

Figure 7. In Situ Hybridization of ^{3}H-Labeled *412* Sequences to an Asynaptic Region of the Second Chromosome in a F1 Oregon × Seto Individual The numbers refer to the standard Drosophila chromosome divisions (Lindsley and Grell, 1968).

clearly demonstrates the heterogeneity in chromosomal distribution of these dispersed repeated gene sequences which occurs within an inbreeding population. It suggests that transposition of these elements may be occurring during the relatively brief time that these stocks have been maintained in the laboratory.

Discussion

Our in situ hybridization studies show that elements of the *412, copia* and *297* gene families have different distributions in the salivary gland chromosomes of four D. melanogaster strains. The number and location of chromosomal sites observed for each dispersed repeated gene family differ dramatically among these strains. Few sites are common to all four strains. Analysis of restriction enzyme cleavage sites within genomic sequences homologous to these elements indicates that most, if not all, of these sequences occur as intact elements in the genomes of each of the four D. melanogaster strains. Analysis of the restriction enzyme cleavage sites in DNA flanking these elements confirms that the local genomic environments in which these elements occur differ among the strains. The results demonstrate that the recent evolution of the D. melanogaster genome has involved rearrangements of these elements to many alternative chromosomal sites.

Individuals from within a strain also exhibit differences in the arrangement of *412, copia* and *297* elements; homologous chromosomes within individual flies are often polymorphic in the organization of these elements. This variability within a strain is surprising, especially since these stocks have been maintained in the laboratory for at least 30 years (Bridges and Brehme, 1944). The genomes of Drosophila strains tend to become increasingly homozygous in time as a result of laboratory propagation (Lewontin, 1974); we would therefore expect an even greater diversity in the organization of these elements in natural populations. The results presented in this paper and similar studies carried out by Ilyin et al. (1978) document the most dramatic genetic differences detected to date among D. melanogaster strains, and as such represent a previously undetected type of genetic variability. It is interesting to note that the integration of dispersed repeated genes into new salivary gland chromosome sites does not noticeably alter the morphology of the chromosome at the site of integration (Figures 7 and 8). The pattern of chromomeres in the region containing the inserted element is identical to that of the homologous chromosome segment where the element is absent.

From our data, it is not possible to determine the rate at which dispersed repeated gene family elements transpose to new chromosomal sites. It is possible that homologous chromosomes with different dispersed gene family arrangements have been segregating in these strains for quite some time. The original source of variability in the chromosomal arrangements of these elements must, however, result from transposition of elements to new chromosomal sites. Once such transpositions have taken place, recombination between homologous chromosomes can generate an even greater variability in the arrangement of dispersed repeated gene family elements. Even taking this effect into account, we believe that the surprisingly high degree of variation observed among individuals in the same laboratory stock suggests that transpositions of these elements are occurring at a sufficient rate to measure in the laboratory. We are presently initiating experiments to determine the rate at which *412, copia* and *297* elements transpose to new chromosomal sites.

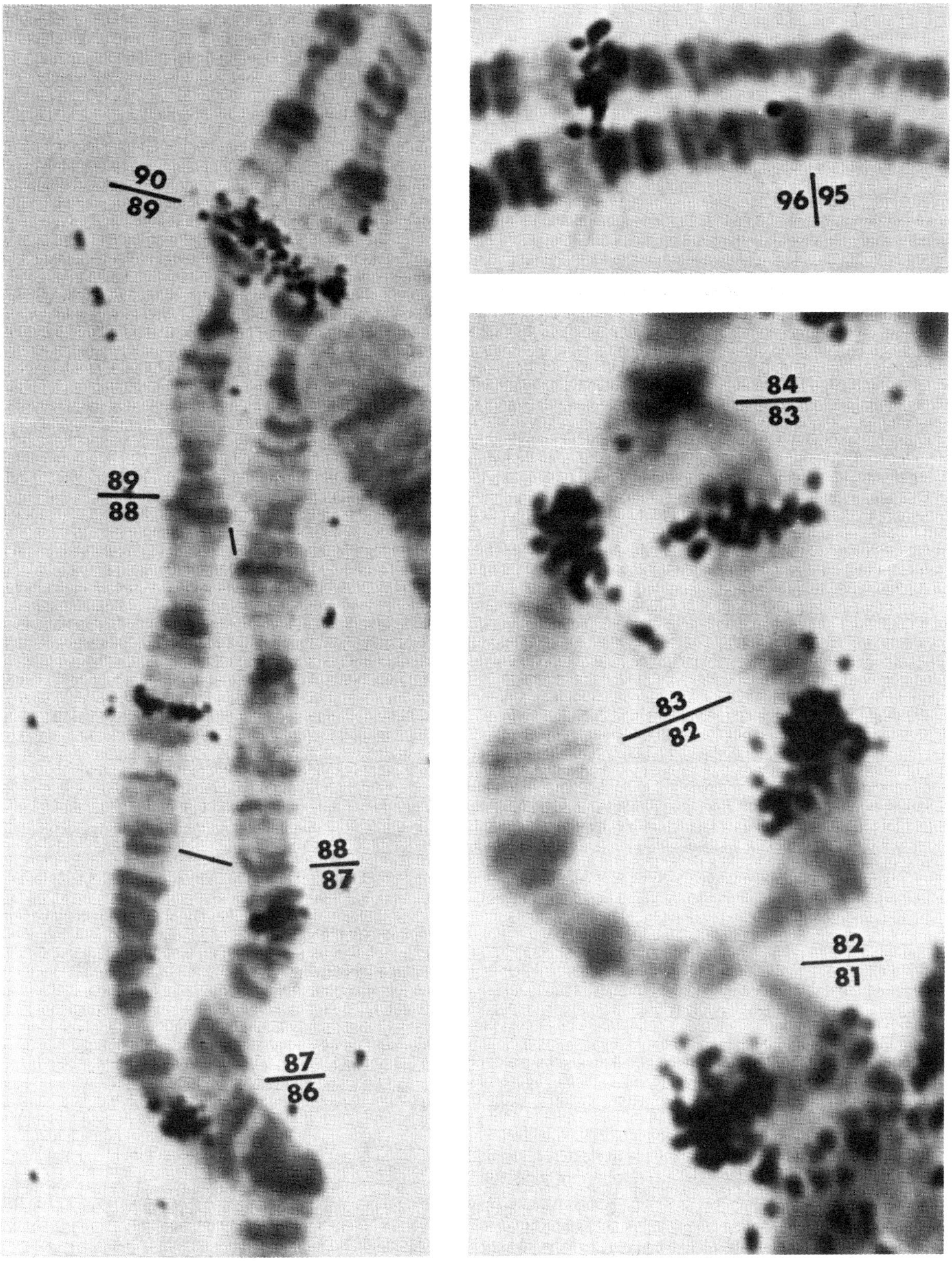

Figure 8. In Situ Hybridization of ^{3}H-Labeled *copia* Sequences to Regions of the Third Chromosome of the Oregon R Strain of D. melanogaster The numbers refer to the standard Drosophila chromosome divisions (Lindsley and Grell, 1968).

Dispersed repeated DNA sequences with elements of similar structure to those of *412, copia* and *297* gene families have recently been found in the genome of the yeast Saccharomyces cerevisiae (Cameron, Loh and Davis, 1979). Like its Drosophila counterpart, the yeast element is about 5 kb long, codes for an abundant poly(A)-containing RNA and is bordered by direct terminal repeats of about 0.25 kb. The yeast elements also appear to be capable of transposition and exhibit different genomic organizations in different yeast strains. The fact that such dispersed repeated gene families exist in both a unicellular eucaryote and a highly evolved metazoan suggests that such elements may be a common feature in all eucaryotic genomes. The Drosophila and yeast elements are similar in structure both to certain procaryotic transposable elements (Kleckner, 1977) and to the integrated DNA genomes of RNA viruses of the retrovirus group (Hughes et al., 1978). The evolutionary or functional significance of these structural similarities, if any, is unclear.

Elements such as *412, copia* and *297* which change their location in the genome will affect the expression of other genes as long as the movement of such elements is not limited to a small number of genetically inert sites. The *412, copia* and *297* elements do not appear to be limited in this way. Almost 100 different chromosomal sites are seen for each of these dispersed repeated gene families when only four laboratory stocks are compared. Studies of the transposition of these elements in Drosophila cell cultures suggest that the number of potential insertion sites may exceed several hundred (Potter et al., 1979). It is possible that the sole function of these elements is to promote genetic variability, and that their gene products may only be necessary for the maintenance and mobility of the elements themselves, rather than for other cellular processes. It is interesting to note that there are many parallels between dispersed repeated gene families in Drosophila and transposable controlling elements in maize (McClintock, 1957), and although detailed comparisons are premature, previous speculations on the molecular nature of maize controlling elements (Fincham and Sastry, 1974; Peterson, 1977; Nevers and Saedler, 1977) are strikingly similar to the properties we observe for *412, copia* and *297*.

There is considerable genetic evidence in support of the existence of insertion mutations in Drosophila (reviewed by Green, 1977). These presumptive insertion mutations are characterized by their high spontaneous reversion rates and, in some cases, their causal association with the production of specific deletions at the chromosome site to which they map. We believe it is probable that some of these mutations result from the local integration of dispersed repeated gene family elements. To test this hypothesis, we are hybridizing *412, copia* and *297* sequences in situ to polytene chromosomes carrying such insertion mutations and their revertants.

Experimental Procedures

D. melanogaster Strains

Stocks of Canton S, Seto and Swedish C were obtained from the Mid-American Drosophila Stock Center (Bowling Green, Ohio). The Oregon R stock came from the laboratory of M. Meselson. Flies were grown on cornmeal-agar medium at 18°C. Crosses between strains were performed using 0–4 hr virgin females of Oregon R with males of the other strains.

Nucleic Acid Preparations and Hybridizations

Methods for DNA isolations, agarose gel electrophoresis, hybridizations to restriction enzyme digests of genomic DNAs (Southern, 1975) and the preparation of labeled DNA by nick translation (Rigby et al., 1977) have been described (Potter et al., 1979). Individual restriction fragments were purified by electrophoresis of digests of the appropriate plasmid in horizontal agarose gels followed by elution of the desired fragment by a modification of the method of Thuring, Sanders and Borst (1975). The following purified restriction fragments were used as probes in the in situ hybridization experiments: the 4.9 kb Hind III fragment of cDm412 (Finnegan et al., 1978) which contains the right end of the *412* element (see map in Figure 1) including one terminal repeat sequence and approximately 0.5 kb of nonrepetitive DNA from the Dm412 chromosomal site; the 4.2 kb Hha I fragment of cDm351 which contains nearly the entire *copia* element; the 2.3 kb Eco RI fragment of pPW297 (see map in Figure 2).

Preparation of Salivary Gland Chromosomes for in Situ Hybridization

Salivary glands were prepared for in situ hybridization by a modification of the procedure of Pardue and Gall (1975). Third instar larvae were dissected in 45% acetic acid. The excised salivary glands were transferred to a 20 μl drop of 45% acetic acid on a subbed slide (Pardue and Gall, 1975) and fixed for 5–10 min. A siliconized coverslip was placed over the glands and tapped gently to allow lateral motion of the coverslip. Chromosome spreading was monitored under the phase-contrast microscope before the slide was placed on a warming plate at 45°C for several minutes. The chromosomes were flattened by exerting as much thumb pressure as possible on the coverslip; this was done rapidly, since sustained pressure results in lateral movement of the coverslip which distorts the chromosomes. After squashing, the slide was placed on a warming plate at 45°C until cytoplasmic flowing stopped and then immersed in liquid nitrogen for 2–5 min before removing the coverslip with a razor blade. Finally, the chromosomes were fixed in methanol:acetic acid (3:1) for 10 min and passed through an ethanol series (70, 70, 95, 95%) before air-drying. The slides can be stored at this stage for several weeks at room temperature.

In Situ Hybridization

Prior to hybridization, slides were treated with pancreatic RNAase (Worthington) at 100 μg/ml in 2 × SSC for 1 hr at 37°C. The RNAase was removed by washing in 2 × SSC (4 × 10 min); the slides were then dehydrated through an ethanol series and air-dried. The chromosomal DNA was denatured in 0.07 N NaOH for 3 min at room temperature, and the slides were washed in 2 × SSC and dried. In situ hybridization of ^{3}H-labeled nick-translated DNA was performed under 22 mm square coverslips in a total volume of 5–20 μl ($\sim 10^5$ cpm per slide; 10^7 cpm/μg). Hybridizations were either in 0.3 M NaCl, 0.05 M Na phosphate (pH 7.0), 5 mM $MgCl_2$, 0.02% ficoll, 0.02% polyvinyl pyrrolidone, 0.02% bovine serum albumin (Denhardt, 1966) at 65°C, or in 50% Formamide (Fluka), 0.4 M NaCl, 0.01 M PIPES (pH 6.5) at 37°C for 12–15 hr. After hybridization at 65°C, slides were washed in 2 × SSC, 0.02% ficoll, 0.02% polyvinyl pyrrolidone, 0.02% bovine serum albumin at 65°C; if hybridized at 37°C, slides were washed in 50% formamide, 0.4 M NaCl, 0.01 M PIPES (pH 6.5) at 37°C for 15 min and then in 2 × SSC at 25°C. Finally, slides were dehydrated through an ethanol series, air-dried, dipped in Kodak NTB2 nuclear emulsion and exposed at 4°C with a dessicant (Drierite). Development was for 2 min in Kodak D-19; chromosomes were stained with Giemsa stain (Fisher Scientific) diluted in 0.01 M Na phosphate (pH 7.0) (1:20).

Acknowledgments

This work was supported by a grant from the NIH. P. D. was supported by a grant from the Damon Runyon-Walter Winchell Cancer Fund, and E. S. was supported by an NIH postdoctoral fellowship.

The costs of publication of this article were defrayed in part by the payment of page charges. This article must therefore be hereby marked *"advertisement"* in accordance with 18 U.S.C. Section 1734 solely to indicate this fact.

Received February 1, 1979; revised March 26, 1979

References

Bridges, C. B. and Brehme, K. F. (1944). The Mutants of Drosophila melanogaster. Carnegie Inst. Washington Publication 552.

Cameron, J. R., Loh, E. Y. and Davis, R. W. (1979). Evidence for transposition of dispersed repetitive DNA families in yeast. Cell *16,* 739–751.

Carlson, M. and Brutlag, D. (1978). One of the *copia* genes is adjacent to satellite DNA in Drosophila melanogaster. Cell *15,* 733–742.

Denhardt, D. T. (1966). A membrane-filter technique for the detection of complementary DNA. Biochem. Biophys. Res. Commun. *23,* 641–646.

Fincham, J. R. S. and Sastry, G. R. F. (1974). Controlling elements in maize. Ann. Rev. Genet. *8,* 15–50.

Finnegan, D. J., Rubin, G. M., Young, M. W. and Hogness, D. S. (1978). Repeated gene families in *Drosophila melanogaster.* Cold Spring Harbor Symp. Quant. Biol. *42,* 1053–1063.

Green, M. M. (1977). The case for DNA insertion mutations in *Drosophila.* In DNA Insertion Elements, Plasmids, and Episomes, A. I. Bukhari, J. A. Shapiro and S. L. Adhya, eds. (New York: Cold Spring Harbor Press), pp. 437–445.

Hughes, S. H., Shank, P. R., Spector, D. H., Kung, H.-J., Bishop, J. M., Varmus, H. E., Vogt, P. K. and Breitman, M. L. (1978). Proviruses of avian sarcoma virus are terminally redundant, co-extensive with unintegrated linear DNA and integrated at many sites. Cell *15,* 1397–1410.

Ilyin, Y. V., Tchurikov, N. A., Ananiev, E. V., Ryskov, A. P., Yenikolopov, G. N., Limborska, S. A., Maleeva, N. E., Gvozdev, V. A. and Georgiev, G. P. (1978). Studies on the DNA fragments of mammals and Drosophila containing structural genes and adjacent sequences. Cold Spring Harbor Symp. Quant. Biol. *42,* 959–969.

Kleckner, N. (1977). Translocatable elements in procaryotes. Cell *11,* 11–23.

Lewontin, R. C. (1974). The Genetic Basis of Evolutionary Change (New York: Columbia University Press).

Lindsley, D. and Grell, R. (1968). Genetic Variations of Drosophila melanogaster. Carnegie Inst. Washington Publication 627.

McClintock, B. (1957). Controlling elements and the gene. Cold Spring Harbor Symp. Quant. Biol. *21,* 197–216.

Murray, K. and Murray, N. E. (1975). Phage lambda receptor chromosomes for DNA fragments made with restriction endonuclease III of *Haemophilus influenzae* and restriction endonuclease I of *E. coli.* J. Mol. Biol. *98,* 551–564.

Nevers, P. and Saedler, H. (1977). Transposable genetic elements as agents of gene instability and chromosomal rearrangements. Nature *268,* 109–115.

Pardue, M. L. and Gall, J. G. (1975). Nucleic acid hybridization to the DNA of cytological preparations. In Methods in Cell Biology, *10,* D. M. Prescott, ed. (New York: Academic Press), pp. 1–16.

Peterson, P. A. (1977). The position hypothesis for controlling elements in maize. In DNA Insertion Elements, Plasmids, and Episomes, A. I. Bukhari, J. A. Shapiro, and S. L. Adhya, eds. (New York: Cold Spring Harbor Press), pp. 429–435.

Potter, S. S., Brorein, W. J., Jr., Dunsmuir, P. and Rubin, G. M. (1979). Transposition of elements of the *412, copia* and *297* dispersed repeated gene families in Drosophila. Cell *17,* 415–427.

Rigby, P. W. J., Dieckmann, M., Rhodes, C. and Berg, P. (1977). Labeling deoxyribonucleic acid to high specific activity *in vitro* by nick translation with DNA polymerase I. J. Mol. Biol. *113,* 237–251.

Rubin, G. M., Finnegan, D. J. and Hogness, D. S. (1976). The chromosomal arrangement of coding sequences in a family of repeated genes. Prog. Nucl. Acid Res. Mol. Biol. *19,* 221–226.

Southern, E. M. (1975). Detection of specific sequences among DNA fragments separated by gel electrophoresis. J. Mol. Biol. *98,* 503–517.

Thuring, R. W. J., Sanders, J. P. M. and Borst, P. (1975). A freeze-squeeze method for recovering long DNA from agarose gels. Anal. Biochem. *66,* 213–220.

Reprinted from *Cell* **21**: 545–553.

The Nucleotide Sequence of a Rabbit β–Globin Pseudogene

Elizabeth Lacy* and Tom Maniatis
Division of Biology
California Institute of Technology
Pasadena, California 91125

Summary

We report the nucleotide sequence of a rabbit β-globin pseudogene, ψβ2. A comparison of the ψβ2 sequence with that of the rabbit adult β-globin gene, β1, reveals the presence of frameshift mutations and premature termination codons in the protein coding sequence which render ψβ2 unable to encode a functional β-globin polypeptide. ψβ2 contains two intervening sequences at the same locations in the globin protein coding sequence as β1 and all other sequenced β-globin genes. An examination of the DNA sequences at the intron/exon junctions suggests that a putative ψβ2 precursor mRNA could not be spliced normally. We compare the flanking and noncoding sequences of ψβ2 and β1 and discuss the evolutionary relationship between these two genes.

Introduction

A cluster of four different β-like globin gene sequences has been isolated from a bacteriophage λ library of rabbit chromosomal DNA in a set of overlapping clones which together contain 44 kilobase pairs (kb) of contiguous DNA (Maniatis et al., 1978; Lacy et al., 1979). The linkage arrangement of the four genes, β1–β4, is shown in Figure 1. Approximately 5–8 kb of DNA separate each gene pair and all four genes are transcribed from the same strand of DNA in the orientation 5′–β4–β3–β2–β1–3′ (Lacy et al., 1979).

The nucleotide sequence of gene β1 shows that it encodes the rabbit adult β-globin protein (Hardison et al., 1979). The presence of mature mRNA transcripts from β3 and β4 in nucleated reticulocytes from the blood islands of 12-day rabbit embryos suggests that these genes encode embryonic and/or fetal β-globin polypeptides. Furthermore, the β3 and β4 genes hybridize more efficiently to embryonic than to adult erythroid RNA (Hardison et al., 1979). Therefore, the β1, β3 and β4 genes appear to be expressed differentially during development.

Gene β2 hybridizes more efficiently to adult than to embryonic β-globin mRNA sequences. However, no β2 mRNA transcripts have been detected in either adult bone marrow and reticulocytes or in embryonic erythroid cells (Hardison et al., 1979). One possible explanation for the apparent lack of β2 transcripts is that gene β2 is a globin pseudogene, a β-like sequence that does not code for a functional β-globin polypeptide. α- and β-globin genes which cannot be identified with known polypeptides have been discovered in several other mammalian species including mouse (Nishioka, Leder and Leder, 1980; Jahn et al., 1980; Vanin et al., 1980), human (Proudfoot and Maniatis, 1980; Fritsch, Lawn and Maniatis, 1980; Lauer, Shen and Maniatis, 1980) and goat (Haynes et al., 1980). To determine whether β2 contains deletions, insertions or base changes that would generate a nonfunctional globin gene, we have determined its nucleotide sequence and compared it with that of a known functional β-globin gene, β1.

* Present address: Department of Zoology, University of Oxford, South Parks Road, Oxford OXI 3PS, England.

Results and Discussion

The Protein Coding Region

Figure 1 shows the strategy we used to determine the DNA sequence of β2 by the Maxam-Gilbert base-specific chemical degradation technique (Maxam and Gilbert, 1977). An alignment of the nucleotide sequences of β2 and β1 is presented in Figure 2. To make this alignment, deletions and insertions were included in the β2 sequence wherever necessary to maintain identical reading frames and DNA sequence homology in the protein coding regions.

An examination of the two coding sequences reveals that a base has been deleted in β2 at codon 20. This deletion would shift the translational reading frame of a putative β2 mRNA relative to β1 and result in an in-phase terminator, TGA, spanning codons 28 and 29 in β1. Consequently, a β2 mRNA would code for a protein that is only 27 amino acids long. Another terminator in phase with the protein sequence of β1 occurs in β2 at codon 125 and a second frameshift mutation (a deletion) occurs at codon 128. In addition, base changes in several codons have replaced amino acids which are conserved in many β-globin polypeptides (Dayhoff, 1972). Some of these altered amino acids are known to function in heme binding and in interactions with α-globin chains (Eaton, 1980). Thus the nucleotide sequence analysis of β2 has revealed the presence of mutations that have rendered it unable to produce a functional β-globin polypeptide. We will therefore refer to this gene as ψβ2 to indicate that it is a β-globin pseudogene.

Comparison of the Noncoding and Flanking Sequences of ψβ2 and β1

There are at least two classes of mutations that could potentially make a gene unable to code for a functional polypeptide. One class consists of insertions/deletions and base changes which result in frameshifts, missense mutations and premature termination codons in the protein coding sequence. As shown above, such alterations are found in the nucleotide sequence of ψβ2. A second class consists of mutations affecting the transcription of genes and the processing of nuclear mRNA precursors. In general, genes coding for defective polypeptides may also selectively accumulate mutations preventing the generation and accumulation of nonfunctional proteins. Sequences which may function in the regulation of transcription and processing have been identified in the noncoding and flanking regions of several eucaryotic genes on the basis of sequence conservation (Benoist et al., 1980). A comparison of the noncoding and flanking sequences of ψβ2 and β1 is presented in the following sections to determine whether ψβ2 has acquired base changes in the sequences which are conserved in mammalian globin and other eucaryotic genes.

5′ Flanking Region

Figure 3A shows an alignment of the ψβ2 sequence from base pair −101 to +1 (see Figure 2) with the sequences 5′ to the mRNA capping site in the rabbit β1, human β and mouse β^{maj}-globin genes. Counting each insertion or deletion as one mismatch, the β1 and ψβ2 5′ flanking sequences are 71% homologous. Most of this homology is found in three regions 5′ to a putative mRNA capping site in ψβ2: (1) between −12 and +1, (2) between −18 and −35 and (3) between −77 and −84 bp. Both regions 2 and 3 are highly conserved among functional adult β-globin genes (Figure 3A).

Region 2 contains an AT-rich sequence originally identified in Drosophila histone genes (the Goldberg-Hogness box) (Goldberg, 1979) and subsequently shown to begin 30 to 31 bp 5′ to the cap site in most eucaryotic structural genes (for references see Baker et al., 1979; Benoist et al., 1980). A comparison of a number of different β-like globin genes has revealed that the AT-rich sequence CATAAA is found in most of these genes, but that the only sequence shared by all the β-like globin genes is PyATAPu. This sequence was therefore designated the ATA box (Efstratiadis et al., 1980). The conserved position of the ATA box relative to the mRNA capping site and its similarity to the Pribnow box (TATPuATG) of procaryotic promoters (Pribnow, 1979) has led a number of investigators to propose that the ATA sequence is involved in the initiation or processing of transcripts from eucaryotic structural genes (for discussion see Baker et al., 1979; Grosschedl and Birnstiel, 1980; Efstratiadis et al., 1980).

Region 3 contains a sequence, CACCCT, which is found in all except one of the adult β-globin genes sequenced thus far (Efstratiadis et al., 1980). The fact that the CACCCT sequence is not conserved in the embryonic and fetal human β-like genes (Efstratiadis et al., 1980) or in other eucaryotic genes (Benoist et al., 1980), including the mouse and human α-globin genes (Nishioka and Leder, 1979; S. Liebhaber, M. Goosens and Y. W. Kan, manuscript submitted), suggests that this sequence is specific to adult β-globins.

A region of strong sequence homology found in the 5′ flanking region of all mammalian globin genes studied thus far is deleted in ψβ2. With the exception of the human δ-globin gene, which has the sequence CCAAC, all the β-like globin genes contain the sequence CCAAT (CCAAT box) 70–80 bp 5′ to the mRNA capping site. In addition, an identical sequence is found at a similar location in both the mouse (Nishioka and Leder, 1979) and human (S. Liebhaber et al., manuscript submitted) α-globin genes. The CCAAT box, however, does not appear to be unique to mammalian globin genes. Benoist et al. (1980) have observed a related sequence in a similar location in the 5′ flanking sequences of the chicken ovalbumin and conalbumin genes and the adenovirus 2 early 1A gene. The alignment in Figure 3A clearly shows that the CCAAT sequence is missing in ψβ2 and that this is the major difference between the 5′ flanking region of ψβ2 and that of β1 and other mammalian globin genes.

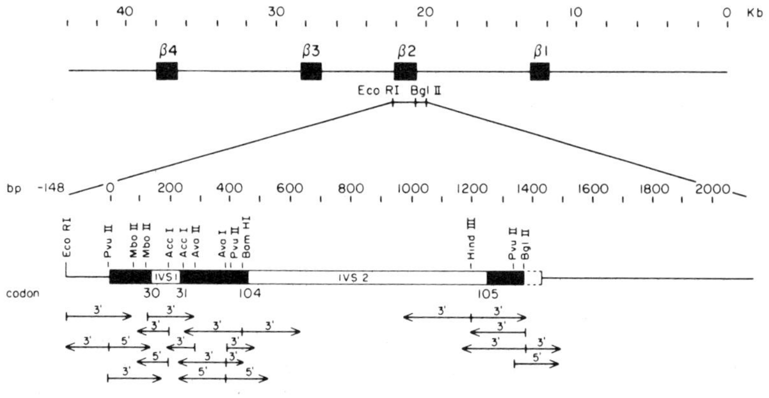

Figure 1. Linkage Arrangement of the Rabbit β-Like Globin Genes and the Strategy for Determining the Nucleotide Sequence of ψβ2

The top line shows the linkage arrangement of the four rabbit β-like globin genes. The direction of transcription of all four genes is 5′ → 3′ from left to right. Gene ψβ2 is contained within a 2.24 kb fragment generated by a limit Eco RI and partial Bgl II digestion. A plasmid subclone of this fragment (pRI·Bgl II 2.24) was used in the sequence analysis. A fine structure restriction enzyme map of this region is shown in the bottom half of the figure. Only those restriction enzyme sites used in deriving the sequence are indicated. Base pairs (bp) are numbered in both directions from the putative mRNA capping site (0 bp). The putative mRNA coding region of the gene (filled boxes) and intervening sequences (open boxes) are indicated. The open box bordered by a dotted line denotes the sequenced portion of the putative 3′ untranslated region. The amino acid codon numbers were assigned on the basis of a DNA sequence comparison between the β1 and ψβ2 genes. The horizontal arrows below the map denote the regions of the DNA that were sequenced. The arrows are labeled with a 3′ or 5′ to indicate whether the restriction fragments were radioactively labeled at their 3′ or 5′ ends.

5′ Noncoding Region

Figure 3B shows an alignment of the $\psi\beta2$ sequence from positions +1 to +51 (see Figure 2) with the sequences between the mRNA capping site and the initiator ATG in the rabbit $\beta1$, human β and mouse β^{maj}-globin genes. There are ten base changes between $\beta1$ and $\psi\beta2$ in the 5′ noncoding region, resulting in an overall homology of 81%. Thus the 5′ noncoding regions of $\beta1$ and $\psi\beta2$ are as conserved as those of the mouse adult β^{maj} and β^{min} genes, which have a homology of 80% (Konkel, Maizel and Leder, 1979), but less conserved than the 5′ noncoding regions of the two human adult genes, β and δ, which share a homology of 92% (Efstratiadis et al., 1980).

The sequence comparisons in Figures 3A and 3B identify a putative mRNA capping site in $\psi\beta2$ and suggest that the first three nucleotides of a mature $\beta2$ mRNA would be AUG. A second AUG is found in the putative 5′ noncoding region of $\psi\beta2$ at the site expected for the initiation of protein synthesis. Since translation initiates at the AUG closest to the 5′ end of an mRNA (Kozak, 1978), it is conceivable that a $\psi\beta2$ message beginning with a capped AUG might not be translated correctly or efficiently.

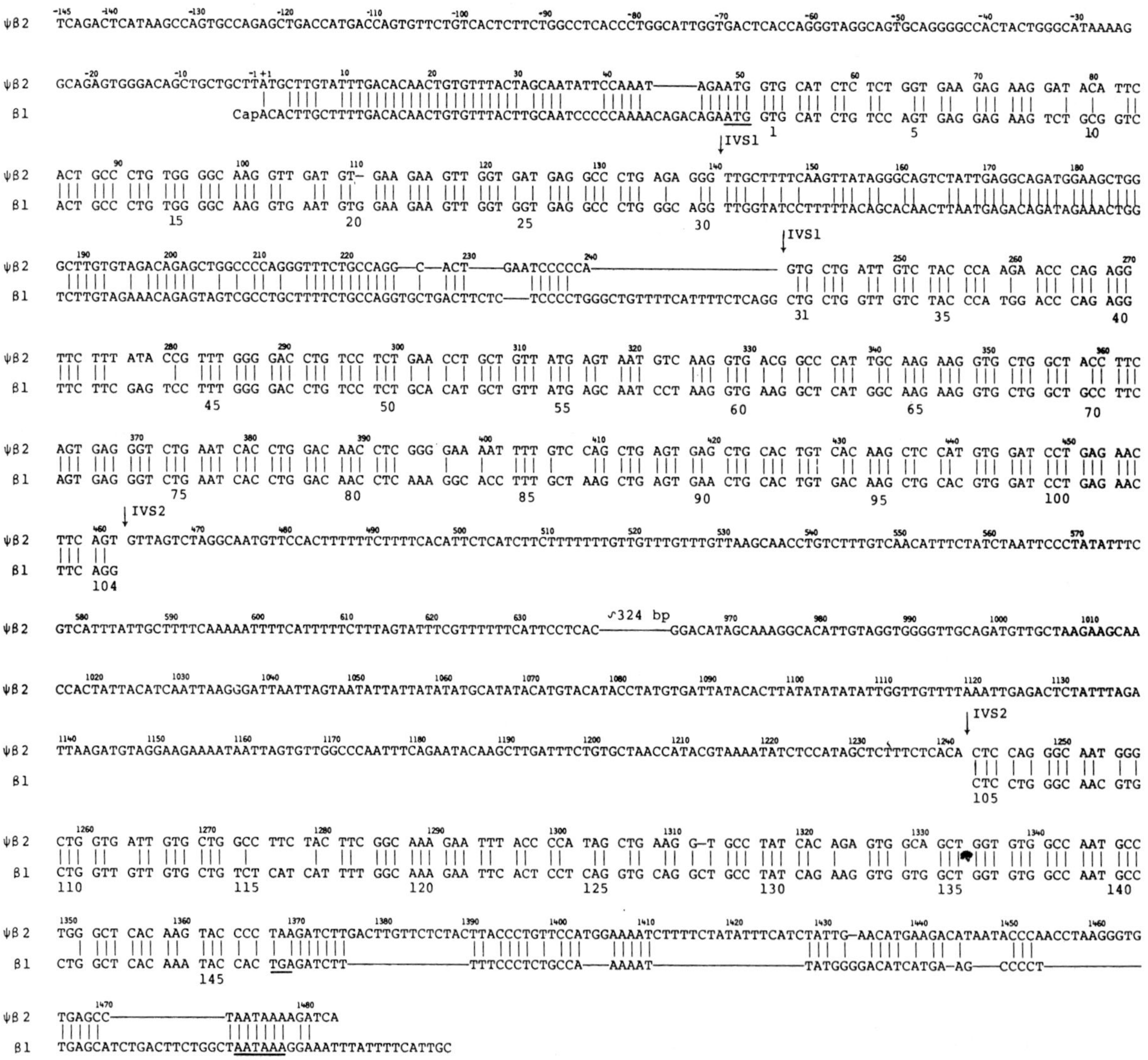

Figure 2. Comparison of the Nucleotide Sequences of the Rabbit $\psi\beta2$- and $\beta1$-Globin Genes

The nucleotide sequences of the mRNA synonymous strands of the $\psi\beta2$ and $\beta1$ (Hardison et al., 1979) genes are aligned. The $\beta1$ DNA sequence from the mRNA capping site to the poly(A) addition site, excluding IVS 2, is shown. Insertions/deletions are included in both the $\psi\beta2$ and $\beta1$ sequences wherever necessary to maintain maximum DNA sequence homology. Vertical lines indicate homologous bases in the two sequences. The $\psi\beta2$ sequence is numbered in both directions from the putative cap site. The numbers are placed above the $\psi\beta2$ sequence so that the first digit (or the minus sign) is directly above the numbered nucleotide. The numbers beneath the $\beta1$ sequence designate amino acid codons of the $\beta1$ gene. The boundaries of the intervening sequences (IVS) are designated by arrows. The initiator codon, ATG, the terminator, TGA, and the hexanucleotide, AATAAA, in the 3′ untranslated region are underlined in the $\beta1$ sequence.

A

5' Flanking Region

```
Rabbit ψβ2   TGTCACTCT-TCTGGCCTCACCCTGGCATTGG———TGAC—TCACCAGGG-TAGGCAGTGCAGGGGCCACTAC-TGGGCATAAAAGGCAGAG-TGGGACAGCTGCTGCTTA
Rabbit β1    CCTCACCCTGCAGAGCCACACCCTGGTGTTGGCCAATCTAC—ACA-CGGGG-TAGGGATTACATAGTTCAGGACTTGGGCATAAAAGGCAGAGCAGGG-CAGCTGCTGCTTA
Human β      CCTCACCCTGTGGAGCCACACCCTAGGGTTGGCCAATCTAC—TCC-CAGGAGCAGGGAGGGCAGGAGCCAGGGC-TGGGCATAAAAGTCAGGGCAGAGCCATCTATTGCTTA
Mouse β^maj  CCTGATTCCGTAGAGCCACACCCTGGTAAGGGCCAATCTGCTCACA-CAGGA-TAGAGAGGGCAGGAGCCAGGGC-AGAGCATATAAGGTGAGGTAGGATCAGTTGCTCCTCA
```

B

5' Noncoding Region

```
Rabbit ψβ2   ATGCTTGTATTTGACACAACTGTGTTTACTAGCAATATTCCAAAT———AGAATG
Rabbit β1    ACACTTGCTTTGACACAACTGTGTTTACTTGCAATCCCCAAAACAGACAGAATG
Human β      ACATTTGCTTCTGACACAACTGTGTTCACTAGCAACCTC——AAACAGACACCATG
Mouse β^maj  ACATTTGCTTCTGACATAGTTGTGTTGACTCACAACCCCAG-AAACAGACATCATG
```

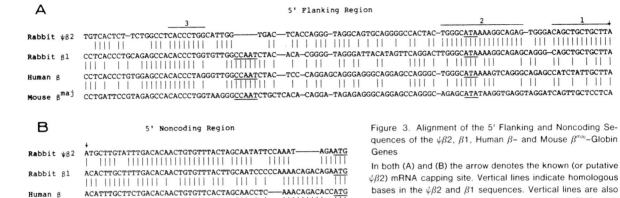

Figure 3. Alignment of the 5′ Flanking and Noncoding Sequences of the ψβ2, β1, Human β- and Mouse β^{maj}-Globin Genes

In both (A) and (B) the arrow denotes the known (or putative ψβ2) mRNA capping site. Vertical lines indicate homologous bases in the ψβ2 and β1 sequences. Vertical lines are also drawn wherever bases are common to the rabbit β1, human β and mouse β^{maj} gene sequences. (A) An alignment of the ψβ2 sequence from base pair −101 to +1 with the sequences 5′ to the mRNA capping site in the rabbit β1 (Mantei et al., 1979), human β (Lawn et al., 1980) and mouse β^{maj} (Konkel et al., 1978) genes. The ATA and CCAAT sequences are underlined. The lines above the ψβ2 sequence denote regions of extensive sequence homology between ψβ2 and β1. (B) An alignment of the ψβ2 sequence from base pair 1 to 51 with the sequences from the mRNA capping site through the initiator ATG in the rabbit β1, human β and mouse β^{maj} genes. The initiator ATG is underlined.

The sequence CUUPyUG, first noted by Baralle and Brownlee (1978), is found in the 5′ noncoding region of all mammalian α- and β-globin genes for which sequence information is available (Efstratiadis et al., 1980). Since the CUUPyUG sequence is complementary to a conserved purine-rich region at the 3′ end of eucaryotic 18S rRNAs, Hagenbüchle et al. (1978) have speculated that this sequence may function in the initiation of translation. ψβ2 contains a similar sequence (UAUUUG) which differs in two bases from the sequence in β1 (CUUUUG) (Figure 3B). An alignment of the β1 and ψβ2 sequences with the purine-rich sequence in 18S rRNAs, 3′ UAGGAAGGCGU 5′ (Hagenbüchle et al., 1978), indicates that the ψβ2 sequence would form a less stable hybrid than the β1 sequence.

Between the putative capping site and initiator ATG, ψβ2 contains 48 bp whereas β1 contains 53 bp (see Figure 3B). The difference in length can be accounted for by a deletion of 5 bp from β1, 3 bp before the ATG. An examination of the sequence in this region reveals that part of a direct repeat, CAGACAGA, has been lost in ψβ2. An analysis of the DNA sequence at hotspots for spontaneous mutations in the *laci* gene of E. coli suggests that direct repeats may be involved in the generation of deletions (Farabaugh and Miller, 1978). A similar loss of direct repeats in noncoding sequences has been noted by Efstratiadis et al. (1980) when making pairwise comparisons of the rabbit, human and mouse β-like globin genes. To obtain an alignment of a given gene pair, it was often necessary to assume that deletions had occurred in one of the two genes. They observed that deletion sites are flanked by short direct repeats and that a deletion removes one repeat completely and part or none of the other repeat. A mechanism based on models proposed for procaryotic systems (Streisinger et al., 1966; Farabaugh and Miller, 1978) is presented by Efstratiadis et al. (1980) to explain how deletions might be generated from the mispairing of repeats during replication.

Intervening Sequences

The mRNA coding region in all α- and β-globin genes studied thus far is interrupted by two intervening sequences (Jeffreys and Flavell, 1977; Konkel, Tilghman and Leder, 1978; Lawn et al., 1978; Leder et al., 1978; Smithies et al., 1978; Tilghman et al., 1978a; Proudfoot and Baralle, 1979; Dodgson, Strommer and Engel, 1979; Hardison et al., 1979; Konkel et al., 1979; Mantei et al., 1979; Efstratiadis et al., 1980; Lauer et al., 1980). DNA sequence analyses of β-globin genes have shown that in all cases one intron (IVS 1) is located between codons for amino acids 30 and 31, while the second, larger intron (IVS 2) is located between codons 104 and 105. Therefore, a globin mRNA sequence is encoded in the genome in three discontinuous blocks: exons I, II and III. We previously identified an intron of 700–900 bp in the ψβ2 gene by hybridization and restriction mapping experiments (Hardison et al., 1979). The existence of this large intron (IVS 2) is confirmed by comparing the nucleotide sequences of the ψβ2 and β1 genes (Figure 2). This comparison also shows that ψβ2 contains a second, smaller intervening sequence (IVS 1) and that the locations of both introns in ψβ2 are identical to those found in all other β-like globin genes.

IVS 1 of ψβ2 is 26 bp smaller than IVS 1 of β1 (100 versus 126 bp). Most of this difference in size can be accounted for by assuming that a 21 bp region immediately adjacent to the 3′ end of IVS 1 was deleted in ψβ2 (Figure 2). We note that the pentanucleotide GGCTG occurs near both endpoints of the putative deletion. In the previous section on the 5′ noncoding region of ψβ2 we discussed the possible involvement of short, direct repeats in the generation of deletions

(Marotta et al., 1977; Farabaugh and Miller, 1978; Efstratiadis et al., 1980). It is possible that this deletion was produced by the proposed mispairing mechanism.

IVS 2 of $\psi\beta 2$ is approximately 200 bp larger than IVS 2 of $\beta 1$. Only a portion (457 bp) of the second intron in $\psi\beta 2$ has been sequenced, including 176 bp from the 5′ junction and 281 bp from the 3′ junction. We reported previously that the large intron of the $\psi\beta 2$ gene does not hybridize to the $\beta 1$ gene (Hardison et al., 1979). This observation is consistent with the fact that, with the exception of a few nucleotides at the intron/exon junctions, we were unable to make any reasonable alignments between IVS 2 of $\beta 1$ and the available nucleotide sequence of IVS 2 in $\psi\beta 2$.

A consensus sequence has been derived from an analysis of the DNA sequences at the intron/exon junctions of several eucaryotic genes (Breathnach et al., 1978; Crick, 1979; Seif, Khoury and Dhar, 1979). The consensus sequence defines a common splicing frame which predicts, with three exceptions, that an intron begins with GT and ends with AG (Breathnach et al., 1978; Lerner et al., 1980). An examination of the four intron/exon junctions in $\psi\beta 2$ indicates that, if this pseudogene is transcribed, the introns could not be excised according to the splicing frames specified by the consensus sequence in the $\beta 1$ gene.

Figure 4 presents a comparison of the DNA sequences at the intron/exon boundaries of $\psi\beta 2$ and $\beta 1$. The 5′ junction of IVS 1 in $\psi\beta 2$ differs from the $\beta 1$ junction in 4 out of 15 bp. The presence of the dinucleotide GT in $\psi\beta 2$ suggests that the 5′ junction could serve as a substrate for splicing. However, the occurrence of a deletion at the 3′ end of IVS 1 makes it unlikely that a $\psi\beta 2$ transcript could be spliced normally. An AG dinucleotide is present in $\psi\beta 2$ near the IVS 1/exon II junction, but its location within exon II would lead to a splice which would alter the normal globin translation reading frame.

The sequence at the 5′ junction of IVS 2 in $\psi\beta 2$ shares 13 of 15 bp with the $\beta 1$ sequence and is consistent with a GT/AG splicing frame. The 3′ junction, on the other hand, can best be aligned with the $\beta 1$ sequence if a deletion is included in $\psi\beta 2$ at the G in the AG dinucleotide. However, if the splice is placed after the AG in the second codon of exon III, the $\psi\beta 2$ sequence lines up with the consensus sequence in Figure 4 in 9 of 11 bases. This splice would not disrupt the translation reading frame, but it would delete the codons for amino acids 105 and 106 in the resulting mRNA.

The mouse, rabbit and human β-globin genes are transcribed into 15–17S nuclear RNA precursors which contain both intron sequences. The intervening sequences are subsequently spliced out in steps to generate mature mRNAs (Kinniburg, Mertz and Ross, 1978; Tilghman et al., 1978b; Kinniburg and Ross, 1979; Flavell et al., 1979; Hardison et al., 1979; Kantor, Turner and Nienhuis, 1980; Maquat et al., 1980). The observation that a $\psi\beta 2$ transcript may not be a suitable substrate for RNA processing raises the possibility that $\psi\beta 2$ transcripts may accumulate in the nucleus as unspliced precursor mRNAs. Kantor et al. (1980) and Maquat et al. (1980) have reported that the deficiency of mature β-globin mRNAs in several β^{+}-thalassemia patients results from the inefficient processing and consequent accumulation of nuclear precursors. However, the nucleotide sequences have not been determined for the β^{+}-thalassemia genes in these studies, and thus it is not known whether there are base changes in the intron/exon junctions.

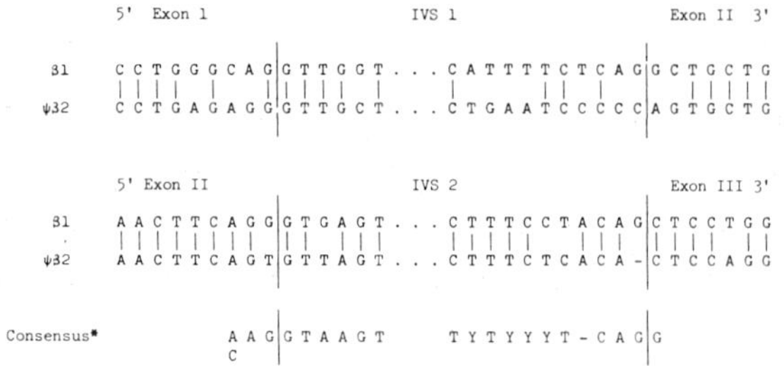

Figure 4. Alignment of Sequences Surrounding Intron/Exon Junctions of $\psi\beta 2$ and $\beta 1$

An alignment of the sequences surrounding the four intron/exon junctions of $\psi\beta 2$ and $\beta 1$. Vertical lines denote the splicing sites in $\beta 1$ which follow the GT/AG rule (Breathnach et al., 1978). Lines are drawn between bases common to the $\psi\beta 2$ and $\beta 1$ genes. *The consensus sequences for the 5′ and 3′ junctions are from Lerner et al. (1980).

Mature mRNA transcripts from $\psi\beta 2$ are not detected in anemic adult bone marrow and reticulocyte RNA or in RNA from 12-day embryos (Hardison et al., 1979). To determine whether $\psi\beta 2$ precursor mRNAs are present in rabbit erythroblasts, a gel blot of total and poly(A) RNA (Alwine, Kemp and Stark, 1977) from the bone marrow of an anemic rabbit was hybridized to DNA probes containing the second intron from either $\psi\beta 2$ or $\beta 1$. $\beta 1$, but not $\psi\beta 2$, precursor RNA molecules were detected (E. Lacy, unpublished results). It was estimated that, if $\psi\beta 2$ is transcribed, its mRNA precursors are present at less than a tenth the concentration of $\beta 1$ precursors in anemic adult rabbit bone marrow cells. The lack of $\psi\beta 2$ cytoplasmic and nuclear precursor mRNA sequences in embryonic and adult erythroid cells (Hardison et al., 1979) suggests that $\psi\beta 2$ may not be transcribed in vivo or, if it is, that its transcripts are degraded rapidly.

3′ Noncoding Region

A termination codon, TAA, is found in $\psi\beta 2$ at the end of the putative protein coding sequence (Figure 2). A different termination codon, TGA, is used in the $\beta 1$ sequence. Figure 2 shows a comparison of the noncoding sequences 3′ to the termination codon in $\psi\beta 2$ and $\beta 1$. The two sequences clearly do not share extensive homology in this region, since an alignment requires the introduction of several deletions or insertions. The only sequence common to ψβ2 and β1 is the hexanucleotide AATAAA, which precedes the poly(A) addition site in several eucaryotic mRNAs (Proudfoot and Brownlee, 1976).

This lack of significant sequence homology is consistent with the results of blot hybridization experiments which indicate that the 3′ end of ψβ2 cannot be detected with a labeled β1 probe (Lacy et al., 1979). It should be noted, however, that the divergence between the 3′ untranslated regions of ψβ2 and β1 is not necessarily a direct consequence of ψβ2 evolving as a peudogene. The 3′ noncoding regions of the human, mouse and rabbit β-like globin genes have sustained many deletions and/or insertions during their evolution (Efstratiadis et al., 1980).

The Evolution of ψβ2

To learn more about the evolutionary relationship between ψβ2 and β1, we have calculated the divergence between the protein coding sequences of these two genes. Over evolutionary time, the DNA and protein sequences of homologous genes accumulate changes and diverge. In general, two types of base changes occur: those that generate amino acid replacements (replacement substitutions) and those that produce synonymous codons (silent substitutions). The accumulation of replacement substitutions in a particular family of proteins is proportional to time, although different proteins accumulate replacement at different rates. This phenomenon, referred to as the evolutionary clock hypothesis, was first established by comparing the amino acid sequences of related proteins (for review see Wilson, Carlson and White, 1977). Perler et al. (1980) have recently developed an improved method for calculating the divergence between the protein coding sequences of homologous genes. Their analysis of the divergence of insulin and globin gene sequences also indicates that replacement substitutions provide a reliable evolutionary clock and that the rate of accumulation of silent substitutions, which is greater than that of replacements, cannot serve as a clock over a long time scale. However, silent substitutions may be able to serve as a clock when the sequences in question have diverged in recent evolutionary time (that is, within the last 85 million years) (Perler et al., 1980).

To confirm that ψβ2 is more closely related to the adult gene, β1, than to the embryonic genes, β3 and β4, we have estimated the amount of divergence between the protein coding sequences of the four rabbit β-like globin genes (Table 1). Our analysis includes only the second exons because the available nucleotide sequence for genes β3 and β4 (provided by R. Hardison and E. Butler, unpublished results) does not include the first and third exons. The percentages of divergence were calculated from pairwise comparisons following the procedures described by Perler et al. (1980). The resulting percentages were

Table 1. Corrected Percent Divergences of the Coding Sequences of β-Like Globin Gene Pairs

Exon II	Replacement Sites	Silent Sites
ψβ2/β1	14.8	33.3
ψβ2/β3	29	79.3
ψβ2/β4	26.9	74.1
Total Coding Region	Replacement Sites	Silent Sites
ψβ2/β1	16.9	44.3
Human δ/β	3.7	31.8

The corrected percent divergence between each pair of coding sequences was calculated as described by Perler et al. (1980) except that the codons UUA (Leu), UCG (Ser) and AGA (Arg) were included in the calculation. The sequence data for genes β1, β3 and β4 are from Hardison et al. (1979), R. Hardison (unpublished results) and E. Butler (unpublished results), respectively. The corrected percent divergence for the human δ and β gene pair is from Efstratiadis et al. (1980). The calculations for exon II include only those of the four rabbit β-like globin genes.

corrected for multiple events, assuming that base changes are Poisson-distributed (Salser, 1977; Perler et al., 1980). It is clear from the results for replacement site substitutions listed in Table 1 that ψβ2 shares more homology with β1 than with either β3 or β4.

The methodology of Perler et al. (1980) has been used to calibrate an evolutionary clock for globins (Efstratiadis et al., 1980). According to this clock, a 1% change in replacement sites requires 10 million years to become fixed in two initially identical genes, a UEP (unit evolutionary period) of 10. The approximate time of divergence for two β–globin genes can therefore be estimated from the percentage of divergence calculated from replacement site substitutions. A "percent divergence" value for a pair of genes actually represents the sum of the sequence differences each gene has accumulated as it diverged from the common ancestral gene. In addition, the time of divergence predicted for a pair of genes does not necessarily correspond to the time when the gene duplication event took place. If the initial products of the gene duplication were corrected against each other for an unknown period of time, the time of divergence would correspond to the time of the last gene correction event (Efstratiadis et al., 1980).

Table 1 also compares the percent divergence calculated from an alignment of the entire protein coding sequence of ψβ2 and β1 with the percent divergence calculated for the human δ– and β–globin genes (Efstratiadis et al., 1980). ψβ2 and β1 exhibit a greater degree of replacement site divergence (16.9%) than do δ and β (3.7%) although both pairs of genes have accumulated a similar number of silent substitutions. The percent divergence in silent substitutions for the δ/β gene pair predicts the same time of divergence as the replacement site substitutions, 40 million years (MY) ago (Efstratiadis et al., 1980). According to the clock constructed from silent substitutions, β1 and ψβ2 also began to diverge quite recently, that is, 55 MY ago. In contrast, the replacement substitutions predict that the rabbit adult β-globin genes diverged approximately 170 MY ago. One interpretation of the discrepancy in the predicted times of divergence is that the replacement sites in ψβ2 were not under selective pressure for the entire period of time since the β1-ψβ2 divergence. In this case, the percent divergence calculated from silent substitutions would more accurately predict the time of divergence of β1 and ψβ2.

If ψβ2 was a pseudogene from the time it started to diverge from β1 55 MY ago, we might expect replacement sites to be equivalent to silent sites. In other words, all base changes would be silent substitutions and replacement substitutions would accumulate at a similar rate to silent substitutions. Since the percent divergence in replacement sites is approximately twofold lower than in silent sites, ψβ2 probably diverged for a time from β1 as a functional gene before acquiring mutations that rendered it nonfunctional and unselected.

It is possible to estimate the time at which ψβ2 became a pseudogene if we consider the divergence between two related genes as the sum of the divergences of each gene from the common ancestor. The rate of change in replacement sites between two selected globin genes is 0.1%/MY, while the rate of change in silent sites is 0.8%/MY (Efstratiadis et al., 1980). Thus the 16.9% divergence between β1 and ψβ2 would be the sum of the percent divergence accumulated by β1 as a selected gene for the last 55 MY (55×0.05) and of the percent divergence accumulated by ψβ2 during N years under selection ($N \times 0.05$) and during 55−N years as an unselected globin gene ($55-N \times 0.4$). This calculation predicts that ψβ2 diverged as a functional globin gene for 22 MY. Approximately 33 MY ago, well after the mammalian radiation (85 MY ago), ψβ2 began to diverge as an unselected globin gene.

Concluding Remarks

The occurrence of pseudogenes in eucaryotic gene clusters was first reported in the oocyte 5S DNA repeat unit of Xenopus laevis (Jacq, Miller and Brownlee, 1977; Miller et al., 1978). In each repeat unit there are two 5S gene sequences. One gene encodes the complete oocyte 5S RNA while the second gene (the pseudogene) differs from the first by 10 base substitutions and a 3′-terminal deletion of 19 bp. Since pseudogene transcripts were not detected in vivo, the presence of a pseudogene in the 5S repeat unit may have no functional significance. Rather, its presence may simply be a consequence of the mechanisms of duplication and gene correction which are thought to act on multigene families (Miller et al., 1978).

The aberrant nucleotide sequence of ψβ2 and the apparent absence of ψβ2 transcripts in vivo indicate that this gene is also a pseudogene. Similar pseudogenes have been identified in other mammalian globin gene clusters. Nucleotide sequence analyses of a human α-globin pseudogene, ψα1 (Proudfoot and Maniatis, 1980; Lauer et al., 1980), and a mouse β-globin pseudogene, βh3 (Jahn et al., 1980), revealed the presence of deletions, insertions and base changes which altered the translational reading frame of the protein coding sequences. Human ψα1 contains two intervening sequences at positions identical to those of functional α-globin genes. However, as in the case of rabbit ψβ2, splicing cannot occur at the intron/exon junctions of ψα1 in accordance with the splicing frames specified by the consensus sequences of Lerner et al. (1980) and Breathnach et al. (1978). The DNA sequence of another α pseudogene in mouse (designated α30.5 by Vanin et al., 1980, and α-3 by Nishioka et al., 1980) demonstrated not only the presence of premature termination codons and frameshift mutations in the protein coding sequence but also the precise excision of both intervening sequences.

The location of the mouse α-globin pseudogene with respect to the functional mouse α-globin genes is unknown. However, the locations of ψβ2, βh3 and ψα1 in their respective globin gene clusters have been established. In each case, the pseudogene is found between the embryonic (or fetal) genes and the adult genes. A β-like sequence, ψβ1, which cannot be identified with a known globin polypeptide, is found in the human β-like globin gene cluster, also between the adult and fetal genes (Fritsch et al., 1980). The equivalent position occupied by pseudogenes in the different mammalian globin gene linkage groups suggests that pseudogenes may have some as yet unidentified function in the gene clusters. However, the observation that ψβ2 and β1 diverged well after the time at which adult and embryonic (fetal) specific globin sequences began to appear [approximately 200 MY ago (Efstratiadis et al., 1980)] indicates that the creation of the rabbit β pseudogene was not coincident with the formation of the gene cluster. In addition, the fact that the rabbit ψβ2 gene and the human α-globin gene, ψα1, diverged relatively recently from their functional counterparts suggests that pseudogenes arose independently in different gene clusters. Consequently, the pseudogenes that have been identified in different mammalian globin gene families cannot resemble each other in location or function as a result of a common evolutionary history.

As in the case of the 5S pseudogene, globin pseudogenes may have no function but may simply be products of gene duplication and subsequent sequence divergence. A common feature of globin gene clusters which reflects their evolution by gene duplication is the occurrence of two adjacent genes that are coordinately expressed during a given developmental stage. In humans, the fetal β-like and the adult α-like globin polypeptides are encoded in pairs of nearly identical and closely linked genes, $^{G}\gamma$-$^{A}\gamma$ and α1-α2, respectively (Slightom, Blechl and Smithies, 1980; Lauer et al., 1980). Both members of the α-globin gene pair are expressed at similar levels, whereas the ratio of $^{G}\gamma$ to $^{A}\gamma$ expression varies during the fetal to adult switch (Comi et al., 1980). In contrast, the similar but nonidentical adult β^{maj}- and β^{min}-globin genes of the mouse are expressed at quite different levels in some strains of mice (Russell and McFarland, 1974). In other strains, only one adult β-globin polypeptide is thought to be expressed, although two genes can be detected by hybridization. One possible explanation of this observation is that one of the two adult globin genes is defective (Weaver et al., 1979). A similar situation is observed in the human δ-β-globin gene pair. The δ protein comprises less than 2.5% of the adult β-globin polypeptides in erythrocytes (Bunn, Forget and Ranney, 1977). Although the δ-globin gene is found in many primate species, this gene appears to be silent in certain Old World monkeys (Martin et al., 1980). Thus there is a wide spectrum of divergence in the structure and levels of expression within different globin gene pairs. Perhaps globin pseudogenes represent the most extreme case of divergence of structure and function. In fact, there may be other pseudogenes in globin gene clusters which have diverged so extensively that they are no longer detected by globin gene hybridization probes.

In summary, the structural analysis of a number of different globin gene clusters suggests that globin gene families are in evolutionary flux (for discussion see Lauer et al., 1980). Perhaps pseudogenes are simply a natural consequence of the mechanisms by which multigene families evolve.

Experimental Procedures

The procedures used to construct plasmid subclones, to prepare plasmid DNAs and to derive restriction endonuclease cleavage maps have been described elsewhere (Lawn et al., 1978; Lacy et al., 1979). DNA fragments were sequenced using the procedure of Maxam and Gilbert (1977). Fragments were labeled with ^{32}P at their 5′ ends with T4 polynucleotide kinase (Boehringer-Mannheim) following dephosphorylation with bacterial alkaline phosphatase (Bethesda Research). Fragments were labeled at their 3′ ends with the Klenow fragment (Klenow and Henningston, 1970) of E. coli DNA polymerase 1. 1–20 μg of DNA were labeled with 50 μCi each of two α-^{32}P-dXTPs (2000–3000 Ci/mmole, Amersham) in the presence of 50 mM Tris-HCl (pH 8.0), 5 mM $MgCl_2$, 10 mM β-mercaptoethanol and 2 units of Klenow polymerase (Boehringer-Mannheim) for 15 min at 25°C. The reaction was chased for 10 min at 25°C with 0.1 mM of all four unlabeled deoxynucleoside triphosphates and an additional unit of enzyme when labeling fragments with 3′ recessive ends.

Acknowledgments

We thank B. Seed, P. F. R. Little and C.-K. J. Shen for helpful discussions, and J. Posakony for the computer analysis of the replacement site and silent site substitutions. E.L. and T.M. were supported by an NIH graduate training grant to the California Institute of Technology and the Rita Allen Foundation, respectively. This work was supported by a grant from the NSF.

The costs of publication of this article were defrayed in part by the payment of page charges. This article must therefore be hereby marked "*advertisement*" in accordance with 18 U.S.C. Section 1734 solely to indicate this fact.

Received June 11, 1980

References

Alwine, J. C., Kemp, D. J. and Stark, G. R. (1977). Method for detection of specific RNAs in agarose gels by transfer to diazobenzyloxymethyl-paper and hybridization with DNA probes. Proc. Nat. Acad. Sci. USA *74*, 5350–5354.

Baker, C. C., Herisse, J., Courtois, G., Galibert, F. and Ziff, E. (1979). Messenger RNA for the Ad2 DNA binding protein: DNA sequences encoding the first leader and heterogeneity at the mRNA 5′ end. Cell *18*, 569–580.

Baralle, F. E. and Brownlee, G. G. (1978). AUG is the only recognisable signal sequence in the 5′ non-coding regions of eukaryotic mRNA. Nature *274*, 84–87.

Benoist, C., O'Hare, K., Breathnach, R. and Chambon, P. (1980). The ovalbumin gene—sequence of putative control regions. Nucl. Acids Res. *8*, 127–142.

Breathnach, R., Benoist, C., O'Hare, K., Gannon, F. and Chambon, P. (1978). Ovalbumin gene: evidence for a leader sequence in mRNA and DNA sequences at the exon-intron boundaries. Proc. Nat. Acad. Sci. USA *75*, 4853–4857.

Bunn, H. F., Forget, B. G. and Ranney, H. M. (1977). Human Hemoglobins (Philadelphia: W. B. Saunders).

Comi, P., Giglioni, B., Ottolenghi, S., Gianni, A. M., Polli, E., Barba, P., Covelli, A., Migliaccio, G., Condorelli, M. and Peschle, C. (1980). Globin chain synthesis in single erythroid bursts from cord blood: studies on γ → β and $^{G}\gamma \rightarrow {}^{A}\gamma$ switches. Proc. Nat. Acad. Sci. USA *77*, 362–365.

Crick, F. (1979). Split genes and RNA splicing. Science *204*, 264–271.

Dayhoff, M. O. (1972). Atlas of Protein Sequence and Structure (Washington, D.C.: National Biomedical Research Foundation).

Dodgson, J. B., Strommer, J. and Engel, J. D. (1979). Isolation of the chicken β-globin gene and a linked embryonic β-like gene from a chicken DNA recombinant library. Cell *17*, 879–887.

Eaton, W. A. (1980). The relationship between coding sequences and function in hemoglobin. Nature *284*, 183–185.

Efstratiadis, A., Posakony, J. W., Maniatis, T., Lawn, R. M., O'Connell, C., Spritz, R. A., DeRiel, J. K., Forget, B., Weissman, S. M., Slightom, J. L., Blechl, A. E., Smithies, O., Baralle, F. E., Shoulders, C. C. and Proudfoot, N. J. (1980). The structure and evolution of the human β-globin gene family. Cell *21*, in press.

Farabaugh, P. J. and Miller, J. H. (1978). Genetic studies of the lac repressor. VII. On the molecular nature of spontaneous hotspots in the lac i gene of Escherichia coli. J. Mol. Biol. *126*, 847–863.

Flavell, R. A., Bernards, R., Grosveld, G. C., Hoeijmakers-Van-Dommelen, H. A. M., Kooter, J. M., De Boer, E. and Little, P. F. R. (1979). The structure and expression of globin genes in rabbit and man. In Eukaryotic Gene Regulation, ICN-UCLA Symposium on Molecular and Cellular Bioogy, *XIV*, R. Axel, T. Maniatis and C. F. Fox, eds. (New York: Academic Press), pp. 335–354.

Fritsch, E. F., Lawn, R. M. and Maniatis, T. (1980). Molecular cloning and characterization of the human β-like globin gene cluster. Cell *19*, 959–972.

Goldberg, M. (1979). Ph.D. thesis, Stanford University, Stanford, California.

Grosschedl, R. and Birnstiel, M. L. (1980). Identification of regulatory sequences in the prelude sequences of an H2A histone gene by the study of specific deletion mutants in vivo. Proc. Nat. Acad. Sci. USA *77*, 1432–1436.

Hagenbüchle, O., Santer, M., Steitz, J. A. and Mans, R. J. (1978). Conservation of the primary structure at the 3′ end of 18S rRNA from eucaryotic cells. Cell *13*, 551–563.

Hardison, R. C., Butler, E. T., III, Lacy, E., Maniatis, T., Rosenthal, N. and Efstratiadis, A. (1979). The structure and transcription of four linked rabbit β-like globin genes. Cell *18*, 1285–1297.

Haynes, J. R., Smith, K., Rosteck, P., Schon, E. A., Gallagher, P. M., Burks, D. J. and Lingrel, J. B. (1980). The isolation of the β^A, β^C, and γ globin genes and a presumptive embryonic globin gene from a goat DNA recombinant library. J. Biol. Chem., in press.

Jacq, C., Miller, J. R. and Brownlee, G. G. (1977). A pseudogene structure in 5S DNA of Xenopus laevis. Cell *12*, 109–120.

Jahn, C. L., Hutchinson, C. A., III, Phillips, S. J., Weaver, S., Haigwood, N. L., Voliva, C. F. and Edgell, M. H. (1980). DNA sequence organization of the β-globin complex in the BALB/c mouse. Cell *21*, 159–168.

Jeffreys, A. J. and Flavell, R. A. (1977). The rabbit β-globin gene contains a large insert in the coding sequence. Cell *12*, 1097–1108.

Kantor, J. A., Turner, P. H. and Nienhuis, A. W. (1980). Beta thalassemia: mutations which affect processing of the β-globin mRNA precursor. Cell *21*, 149–157.

Kinniburgh, A. J. and Ross, J. (1979). Processing of the mouse β-globin mRNA precursor: at least two cleavage-ligation reactions are necessary to excise the large intervening sequence. Cell *17*, 915–921.

Kinniburgh, A. J., Mertz, J. E. and Ross, J. (1978). The precursor of mouse β-globin messenger RNA contains two intervening RNA sequences. Cell *14*, 681–693.

Klenow, H. and Henningston, I. (1970). Selective elimination of the exonuclease activity of the DNA polymerase from E. coli B by limited proteolysis. Proc. Nat. Acad. Sci. USA *65*, 168–175.

Konkel, D. A., Tilghman, S. M. and Leder, P. (1978). The sequence of the chromosomal mouse β-globin major gene: homologies in capping, splicing and poly(A) sites. Cell *15*, 1125–1132.

Konkel, D. A., Maizel, J. V., Jr. and Leder, P. (1979). The evolution and sequence comparison of two recently diverged mouse chromosomal β-globin genes. Cell *18*, 865–873.

Kozak, M. (1978). How do eucaryotic ribosomes select initiation regions in messenger RNA? Cell *15*, 1109–1123.

Lacy, E., Hardison, R. C., Quon, D. and Maniatis, T. (1979). The linkage arrangement of four rabbit β-like globin genes. Cell *18*, 1273–1283.

Lauer, J., Shen, C.-K. J. and Maniatis, T. (1980). The chromosomal arrangement of human α-like globin genes: sequence homology and α-globin gene deletions. Cell *20*, 119–130.

Lawn, R. M., Fritsch, E. F., Parker, R. C., Blake, G. and Maniatis, T. (1978). The isolation and characterization of linked δ- and β-globin genes from a cloned library of human DNA. Cell *15*, 1157–1174.

Lawn, R. M., Efstratiadis, A., O'Connell, C. and Maniatis, T. (1980). The nucleotide sequence of the human β-globin gene. Cell *21*, in press.

Leder, A., Miller, H. I., Hamer, D. H., Seidman, J. G., Norman, B., Sullivan, M. and Leder, P. (1978). Comparison of cloned mouse α- and β-globin genes: conservation of intervening sequence locations and extragenic homology. Proc. Nat. Acad. Sci. USA *75*, 6187–6191.

Lerner, M., Boyle, J., Mount, S., Wolin, S. and Steitz, J. (1980). Are snRNPs involved in splicing? Nature *283*, 220–224.

Maniatis, T., Hardison, R. C., Lacy, E., Lauer, J., O'Connell, C., Quon, D., Sim, G. K. and Efstratiadis, A. (1978). The isolation of structural genes from libraries of eucaryotic DNA. Cell *15*, 687–701.

Mantei, N., van Ooyen, A., van den Berg, J., Beggs, J. D., Boll, W., Weaver, R. F. and Weissmann, C. (1979). Synthesis of rabbit β-globin-specific RNA in mouse L cells and yeast transformed with cloned rabbit chromosomal β-globin DNA. In Eucaryotic Gene Regulation, ICN-UCLA Symposium on Molecular and Cellular Biology, *XIV*, R. Axel, T. Maniatis and C. F. Fox, eds. (New York: Academic Press), pp. 477–498.

Maquat, L. E., Kinniburgh, A. J., Beach, L. R., Honig, G. R., Lazerson, J., Ershler, W. B. and Ross, J. (1980). Processing of the human β-globin mRNA precursor to mRNA is defective in three patients with β^+ thalassemia. Proc. Nat. Acad. Sci. USA, in press.

Marotta, C. A., Wilson, J. T., Forget, B. G. and Weissman, S. M. (1977). Human β-globin messenger RNA. III. Nucleotide sequences derived from complementary DNA. J. Biol. Chem. *252*, 5040–5053.

Martin, S. L., Zimmer, E. A., Kan, Y. W. and Wilson, A. C. (1980). Silent δ globin gene in Old World monkeys. Proc. Nat. Acad. Sci. USA *77*, 3563–3566.

Maxam, A. M. and Gilbert, W. (1977). A new method for sequencing DNA. Proc. Nat. Acad. Sci. USA *74*, 560–564.

Miller, J. R., Cartwright, E. M., Brownlee, G. G., Federoff, N. V. and Brown, D. D. (1978). The nucleotide sequence of oocyte 5S DNA in Xenopus laevis. II. The GC-rich region. Cell *13*, 717–725.

Nishioka, Y. and Leder, P. (1979). The complete sequence of a chromosomal mouse α-globin gene reveals elements conserved throughout vertebrate evolution. Cell *18*, 875–882.

Nishioka, Y., Leder, A. and Leder, P. (1980). An unusual alpha globin-like gene that has cleanly lost both globin intervening sequences. Proc. Nat. Acad. Sci. USA *77*, 2806–2809.

Perler, F., Efstratiadis, A., Lomedico, P., Gilbert, W., Kolodner, R. and Dodgson, J. (1980). The evolution of genes: the chicken preproinsulin gene. Cell *21*, 555–565.

Pribnow, D. (1979). Genetic control signals in DNA. In Biological Regulation and Development, *1*, R. Goldberger, ed. (New York: Plenum Press), pp. 219–277.

Proudfoot, N. J. and Brownlee, G. G. (1976). 3′ non-coding region sequences in eukaryotic messenger RNA. Nature *263*, 211–214.

Proudfoot, N. J. and Baralle, F. E. (1979). Molecular cloning of the human ε-globin gene. Proc. Nat. Acad. Sci. USA *76*, 5435–5439.

Proudfoot, N. J. and Maniatis, T. (1980). The structure of a human α-globin pseudogene and its relationship to α-globin gene duplication. Cell *21*, 537–544.

Russell, E. S. and McFarland, E. C. (1974). Genetics of mouse hemoglobins. Ann. NY Acad. Sci. *244*, 25–38.

Salser, W. (1977). Globin mRNA sequences: analysis of base pairing and evolutionary implications. Cold Spring Harbor Symp. Quant. Biol. *42*, 985–1002.

Seif, I., Khoury, G. and Dhar, R. (1979). BKV splice sequences based on analysis of preferred donor and acceptor sites. Nucl. Acids Res. *6*, 3387–3398.

Slightom, J. L., Blechl, A. E., and Smithies, O. (1980). Human fetal $^G\gamma$- and $^A\gamma$-globin genes: complete nucleotide sequences suggest that DNA can be exchanged between these duplicated genes. Cell *21*, in press.

Smithies, O., Blechl, A., Denniston-Thompson, K., Newell, N., Richards, J., Slightom, J., Tucker, D. and Blattner, F. (1978). Cloning human fetal γ-globin and mouse α-type globin DNA: characterization and partial sequencing. Science *202*, 1284–1289.

Streisinger, G., Okada, Y., Emrich, J., Newton, J., Tsugita, A., Terzaghi, E. and Inouye, M. (1966). Frameshift mutations and the genetic code. Cold Spring Harbor Symp. Quant. Biol. *31*, 77–84.

Tilghman, S. M., Tiemeier, D. C., Seidman, J. G., Peterlin, B. M., Sullivan, M., Maizel, J. and Leder, P. (1978a). Intervening sequence of DNA identified in the structural portion of a mouse β-globin gene. Proc. Nat. Acad. Sci. USA *75*, 725–729.

Tilghman, S. M., Curtis, P. J., Tiemeier, D. C., Leder, P. and Weissman, C. (1978b). The intervening sequence of a mouse β-globin gene is transcribed within the 15S β-globin mRNA precursor. Proc. Nat. Acad. Sci. USA *75*, 1309–1313.

Vanin, E. F., Goldberg, G. I., Tucker, P. W. and Smithies, O. (1980). A mouse alpha globin-related pseudogene (ψα30.5) lacking intervening sequences. Nature *286*, 222–226.

Weaver, S., Haigwood, N. L., Hutchison, C. A., III and Edgell, M. H. (1979). DNA fragments of the Mus musculus β globin haplotypes Hbb^s and Hbb^d. Proc. Nat. Acad. Sci. USA *76*, 1385–1389.

Wilson, A. C., Carlson, S. S. and White, T. J. (1977). Biochemical evolution. Ann. Rev. Biochem. *46*, 573–639.

Extensive polymorphism in the mitochondrial DNA of apes

(cleavage map/ribosomal gene/ape population/human evolution)

STEPHEN D. FERRIS*, WESLEY M. BROWN†, WILLIAM S. DAVIDSON‡, AND ALLAN C. WILSON

Department of Biochemistry, University of California, Berkeley, California 94720

Communicated by Sherwood L. Washburn, June 18, 1981

ABSTRACT Ape species are 2–10 times more variable than the human species with respect to the nucleotide sequence of mtDNA, even though ape populations have been smaller than the human population for at least 10,000 years. This finding was made by comparing purified mtDNAs from 27 individuals with the aid of 25 restriction endonucleases; for an additional 59 individuals, comparisons were made with fewer enzymes by using the blot hybridization method. The amount of intraspecific sequence divergence was greatest between orangutans of Borneo and Sumatra. Among common chimpanzees, a large component of the variation is due to two highly distinct forms of mtDNA that may reflect a major geographic subdivision. The least amount of sequence variation occurred among lowland gorillas, which exhibit only twice as much sequence variation as humans. The large intraspecific differences among apes, together with the geological and protein evidence, leads us to propose that each ape species is the remnant of an ancient and widespread population that became subdivided geographically and reduced in size and range, perhaps by hominid competition. The low variation among human mtDNAs is consistent with geological evidence that the human species is young. The distribution of site changes within the mitochondrial genome was also examined. Comparison of closely related mtDNAs shows that the ribosomal RNA genes have diverged more slowly than the rest of the genome.

The human species has an unusually low level of genetic variation in mtDNA. Two humans picked at random are expected to possess mtDNAs that differ by only 0.36% in nucleotide sequence (1). By contrast, the level of intraspecific variation reported for other mammals is 3–30 times higher (2–4).

The research described below was aimed at finding out whether the low variation in mtDNA is unique to our species or shared by our closest relatives, the great apes. By using many restriction endonucleases, which provide a fast method of estimating the amount of difference in nucleotide sequence among mtDNAs (5), we have obtained estimates of intraspecific variation in chimpanzees, gorillas, and orangutans. These estimates contrast with those for humans and shed light on the evolutionary history of ape and human populations. Our intraspecific comparisons also provide a perspective on the susceptibility of the region coding for rRNA to evolutionary change.

MATERIALS AND METHODS

Tissues and Cell Lines. mtDNA was purified (5) from 5 orangutans (*Pongo pygmaeus*), 10 common chimpanzees (*Pan troglodytes*), 2 pygmy chimpanzees (*Pan paniscus*), and 4 gorillas (*Gorilla gorilla*). Records show no immediate kinship among these individuals. The six human mtDNAs analyzed correspond to individuals 3, 6, 9, 10, 15, and 21 of Brown (1). Total cellular DNA was prepared from one additional pygmy chimpanzee and 59 additional common chimpanzees. The tissues, blood samples, and cell lines used were supplied by zoos and primate research centers.

Restriction Endonucleases and Electrophoresis. Nineteen restriction endonucleases (New England BioLabs) were used to digest purified mtDNAs. Digestions were performed according to the supplier's directions. Fragments were radioactively labelled, subjected to electrophoresis, and detected as described by Brown (1). The smallest routinely scored fragment was 150 base pairs (bp) in 1.2% agarose and 60 bp in 3.5% acrylamide gels.

Cleavage Maps and Fragment Patterns. The location of restriction sites in mtDNA was determined for 19 endonucleases, using as a reference the published map for a single representative of each species (6). Six other endonucleases made too many fragments to be mapped conveniently. For these, the fragment patterns were determined for individual chimpanzees and gorillas as Brown (1) has done for humans.

Blot Hybridization. Total cellular DNA from blood samples was prepared as described in Zimmer *et al.* (7). A radioactive hybridization probe (specific activity, 10^8 cpm of $^{32}P/\mu g$ DNA) was made by nick translation of 50 ng of purified pygmy chimpanzee mtDNA (8). The mtDNA fragments present in a restriction endonuclease digest of 2–7 μg of cellular DNA were detected by hybridization with a labelled probe (10^7 cpm per filter) after the fragments had been separated by electrophoresis in 0.8% agarose gels and transferred to nitrocellulose filters (7).

Estimation of Sequence Divergence from Cleavage Maps. The percentage divergence in base sequence among mtDNAs was estimated from comparison of cleavage maps by using equation 15 of Nei and Li (9). An assumption of this method is that the cleavage site differences are due to base substitution. Empirical evidence justifying this assumption has been presented (1, 5).

RESULTS

Orangutans. Fig. 1 shows the mtDNA cleavage maps for 29 variable sites in five orangutans. Although differing from one another at many sites, the maps are identical in length and in the arrangement of 33 invariant sites. The biggest differences are between Bornean and Sumatran orangutans, which are considered to be distinct subspecies (10). The most similar mtDNAs come from the same island.

A tree relating the mtDNA maps was constructed (Fig. 2). All Sumatran orangutans cluster together, as do those from Borneo. The mean difference in nucleotide sequence between

The publication costs of this article were defrayed in part by page charge payment. This article must therefore be hereby marked "*advertisement*" in accordance with 18 U. S. C. §1734 solely to indicate this fact.

Abbreviation: bp, base pair(s).

* Present address: Department of Genetics, Stanford University, Stanford, CA 94305.

† Present address: Division of Biological Sciences, University of Michigan, Ann Arbor, MI 48109.

‡ Present address: Department of Biochemistry, Memorial University, St. Johns, Newfoundland A1B 3X9.

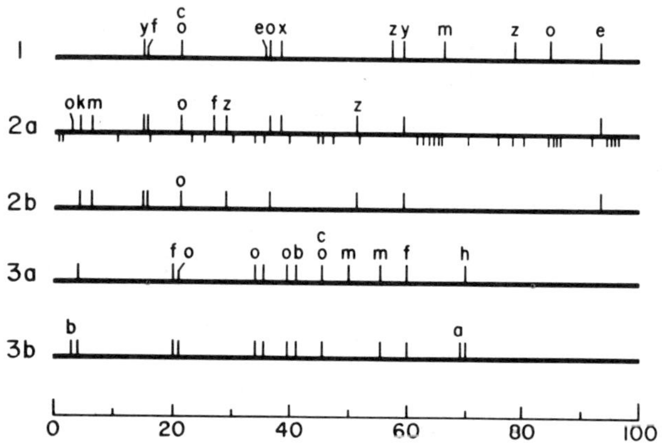

FIG. 1. Location of 29 variable restriction sites in mtDNAs of three Sumatran (maps 1, 2a, and 2b) and two Bornean (maps 3a and 3b) orangutans. Vertical lines represent variable cleavage sites. a, *Eco*RI; b, *Hin*dIII; c, *Hpa* I; d, *Bgl* II; e, *Xba* I; f, *Bam*HI; g, *Pst* I; h, *Pvu* II; i, *Sal* I; j, *Sac* I; k, *Kpn* I; l, *Xho* I; m, *Ava* I; n, *Sma* I; o, *Hin*cII; w, *Bst*EII; x, *Bcl* I; y, *Bgl* I; z, *Fnu*DII. Vertical lines that lack letters are homologous in position to one on a map above. The scale is in map units with the origin of replication at 0 and the direction of replication to the right. The 33 invariant sites appear below map 2a (cf. ref. 6).

the Bornean and Sumatran branches of the tree is estimated from the maps to be 5%.

Chimpanzees. *Maps.* Chimpanzee mtDNA exhibits a lower degree of variability than orangutan mtDNA. The 31 variable positions in the maps of 13 individual mtDNAs are shown in Fig. 3. From the fraction of sites in common, we estimated the percentage difference in nucleotide sequence for all possible pairs of these 13 individuals (Table 1). The biggest difference (3.7%) is between common and pygmy chimpanzees. Among all common chimpanzees, the mean pairwise difference is 1.3% and among all pygmy chimpanzees it is 1.0%.

An evolutionary tree for the mtDNA maps (Fig. 4) shows the existence of three major types of mtDNA in chimpanzees. One type is characteristic of pygmy chimpanzees and the other two occur exclusively in the common chimpanzees. The sequence difference between the latter two types is 2.0%.

Fragments. Ten of the common chimpanzee mtDNAs and one of the pygmy chimpanzee mtDNAs that had been analyzed by cleavage mapping were digested with six restriction enzymes that recognize four or five base sites. The average number of fragments detected electrophoretically was 140 per mtDNA—i.e., with this method, we were able to detect ≈140 restriction sites

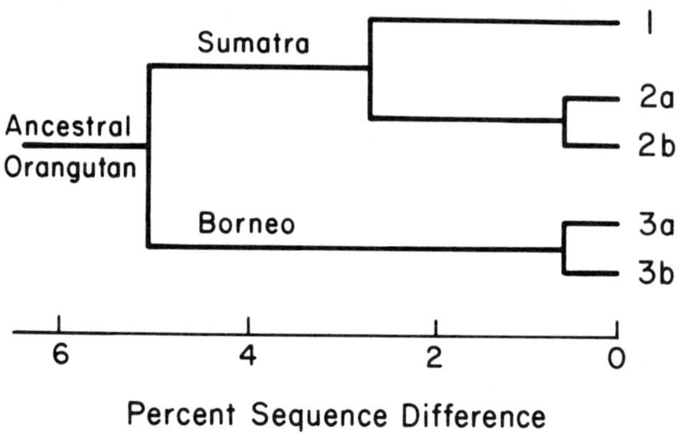

FIG. 2. Evolutionary tree for the mtDNAs of five orangutans. This tree was obtained by the parsimony method (6). The percent sequence differences between branches of the tree are calculated from the maps by using equation 15 of Nei and Li (9).

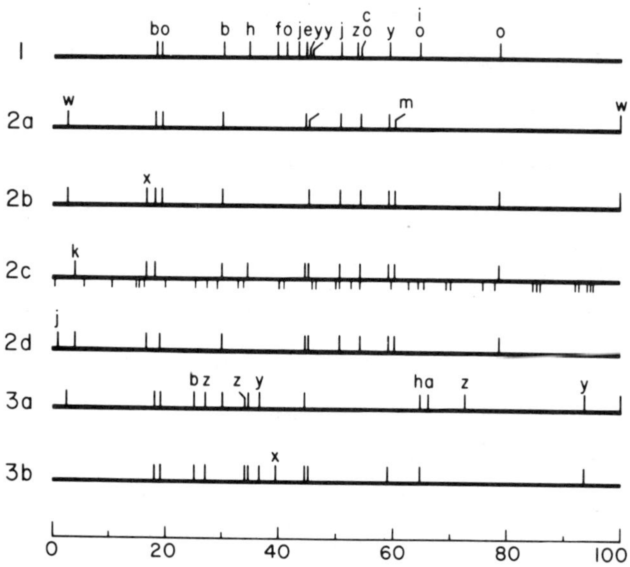

FIG. 3. Locations of 31 variable restriction sites in mtDNAs of 10 common chimpanzees [map 1 (individuals 4, 5, and 8), map 2a (9), map 2b (individuals 1 and 6), map 2c (individuals 2, 3, and 10), and map 2d (individual 7)] and three pygmy chimpanzees [map 3a (individuals 1 and 3) and map 3b (individual 2)]. Restriction sites are as in the legend to Fig. 1. The mtDNA of pygmy chimpanzee 3 was mapped by the blot hybridization method (see ref. 6). The 36 invariant sites appear below map 2c (see ref. 6). *Bam*HI sites occur at 15, 20, and 40 map units.

in addition to those mapped. As is evident from the fragment patterns listed in Table 2, this method has great sensitivity. Several of the individuals having identical cleavage maps for six base-recognition sites were readily differentiated on the basis of fragment patterns produced by digestion with enzymes recognizing four base sites (e.g., individuals 4, 5, and 8, Table 2). Furthermore, the pygmy chimpanzee shares no mtDNA fragment patterns with common chimpanzees and the two major types found within the common chimpanzees also share no patterns. Thus, the fragment-pattern comparisons confirm and extend the inferences based on the mapping approach.

Screening by the blot hybridization method. We screened mtDNA from 59 additional common chimpanzees by examining *Bam*HI digests of cellular DNA with the blot hybridization method. This enzyme was chosen because it distinguished readily between the two major types of common chimpanzee mtDNA, as is evident from the maps in Fig. 3. In chimpanzees exhibiting type 1 mtDNA, *Bam*HI produced fragments of 12,400, 3,200, and 870 bp; in those with type 2, a site loss has resulted in the fusion of the two largest fragments into a 15,600-bp fragment. According to this test, 10 mtDNAs were of type 1, 48 were of type 2, and one had a pattern that can be derived from type 1 by the gain of a *Bam*HI site within the 3200-bp fragment. Our typing of the 59 common chimpanzee mtDNAs was confirmed by a similar, but more limited, study in which *Sac*

Table 1. Comparison of mtDNA maps among chimpanzees

	Percent sequence difference					
Map no.	1	2a	2b	2c	2d	3a
2a	2.1					
2b	2.3	0.5				
2c	1.7	1.3	1.1			
2d	2.3	1.3	1.1	0.7		
3a	4.3	3.0	3.6	4.1	4.5	
3b	3.5	2.9	3.6	3.2	3.6	1.5

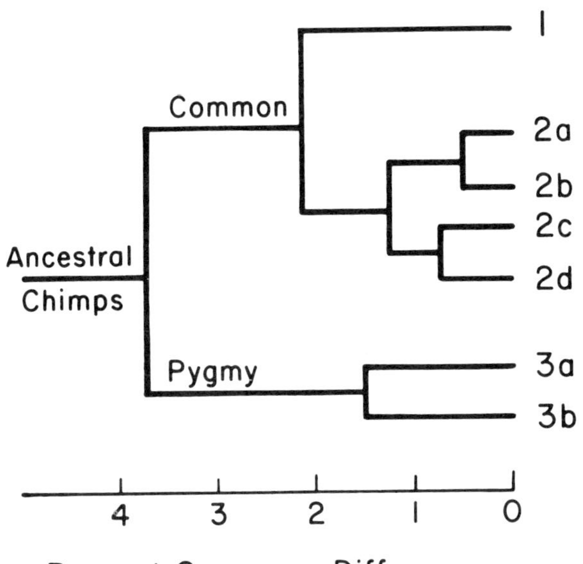

FIG. 4. Evolutionary tree for the seven types of chimpanzee mtDNA shown in Fig. 3. The tree was obtained by the parsimony method (6). The percent sequence differences between branches of the tree are taken from Table 1. Gorilla mtDNA was used to root the tree.

I digests of cellular DNA were analyzed with a mtDNA probe (B. Chapman, unpublished results).

Table 3 gives the distribution of the two types of mtDNA among 49 chimpanzees that have been tentatively classified as to subspecies on the basis of external morphology. Type 1 mtDNA occurs almost exclusively in the eastern subspecies (*P.t. schweinfurthi*) and type 2 mtDNA predominates in the western subspecies (*P.t. verus*).

Gorillas. Gorillas were the least variable for mtDNA of the great apes. The mtDNAs from four lowland gorillas were variable at five positions (Fig. 5). The mean pairwise sequence difference was 0.55% and the two most divergent branches in the gorilla tree differ by 0.9%. Digests of gorilla mtDNA with *Hpa* II, *Mbo* I, *Hinf* I, *Mbo* II, *Taq* I, and *Ava* II, which provided ≈140 additional sites, showed no differences between individuals 1 and 4, which also had identical cleavage maps.

Humans. Humans were the least variable of the hominoids for mtDNA. Only five positions were variable in a sample comprised of two Blacks, two Whites, and two Asians. No variation beyond that reported by Brown (1) was observed when four

Table 2. Cleavage patterns for restriction endonucleases recognizing four- and five-base sites in mtDNA from 10 common chimpanzees

Map no.*	Individual	Fragment pattern: *Hpa* II	*Mbo* I	*Taq* I	*Hinf*I	*Mbo* II	*Ava* II
1	4	A	A	A	A	A	A
1	5	B	B	A	A	A	A
1	8	A	B	A	A	A	A
2a	9	C	C	B	B	B	B
2b	1	C	D	C	B	C	C
2b	6	C	D	D	B	C	C
2c	2, 3 and 10	D	E	E	C	C	C
2d	7	E	F	F	D	D	C

Fragment patterns were obtained from mtDNAs by digestion with restriction enzyme and then labeling the fragments at the ends, separating them electrophoretically, and autoradiographing them.
* See Fig. 3.

Table 3. Correspondence between subspecies designation and type of mtDNA in common chimpanzees at the Holloman Primate Center

Subspecies	Number of individuals: Type 1	Type 2	Correspondence, %	Geographic region*
verus	1	29	97	A
troglodytes	0	3	100	B
schweinfurthi	8	6	57	C
koolakamba	1†	1	50	D

Subspecies designations are based on external morphology (11). One of the identifications is inconsistent with records concerning geographic origin—although identified as *schweinfurthi*, records indicate that this chimpanzee came from Sierra Leone (region A); consistent with this geographic origin, its mtDNA is of type 2. In addition, some of the other specimens designated as *schweinfurthi* probably do not belong to this subspecies (see ref. 18).
* Region in which the subspecies is supposed to occur (11): A, West African countries; B, region from eastern Nigeria to Congo Brazzaville; C, Central and East African countries; D, southern Congo, Brazzaville, and southern Gabon.
† *Bam*HI fragment pattern is derived from type 1 by a single site gain.

additional restriction endonucleases were used (*Bgl* I, *Fnu*DII, *Bcl* I, and *Bst*EII). The mean pairwise difference, 0.30%, was calculated from maps based on the 19 endonucleases used for the ape species. This value is in reasonable agreement with the value, 0.36%, found in studies using much larger sample sizes and comparing many more cleavage sites (ref. 1; R. L. Cann, personal communication.).

Distribution of Site Changes. Besides giving information about genetic variability among individuals, our mtDNA comparisons extend knowledge of the distribution of sites at which variation has occurred within this genome. It is expected that comparisons of closely related mtDNAs will provide the most sensitive way of detecting differences in rates of evolutionary change between different regions. Only 1 of the 51 site changes observed in intraspecific comparisons of hominoid mtDNAs that differ by less than 3% has occurred in the ribosomal region [i.e., at 81–97 map units (6)]. If the changes were distributed at random, six would be expected in this region. The deviation from expectation is statistically significant.

DISCUSSION

Ape Species May Be Old. The big differences among ape individuals shown by mtDNA comparisons can be reconciled with the view that ape species are older than the human species. A modern ape species typically consists of fewer than 10,000 individuals distributed over a small geographic area. The ex-

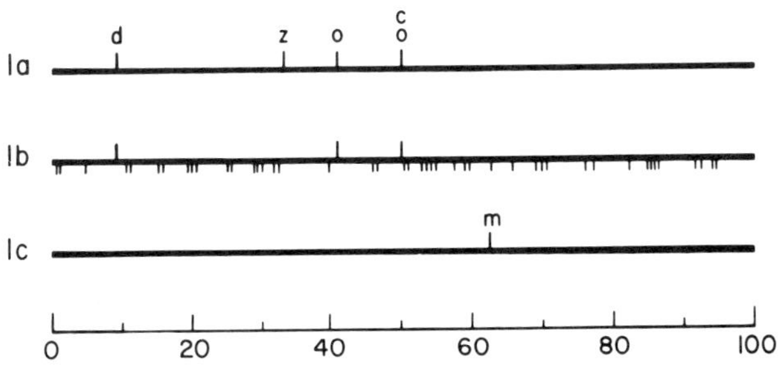

FIG. 5. Locations of five variable restriction sites in the mtDNAs of four lowland gorillas. Restriction sites are as in the legend to Fig. 1. Two individuals, 1 and 4, had identical maps (1a); the other two individuals, 2 and 3, had maps 1b and 1c, respectively. The 45 invariant sites are shown below map 1b (see ref. 6).

istence of highly divergent mtDNA lineages within a species prompts us to think of it as the remnant of a formerly widespread and large population that became subdivided hundreds of thousands or even millions of years ago by geographic barriers and, perhaps, by hominid competition.

From mtDNA map comparisons, we can calculate the divergence time (t) of two populations (X and Y) by using Eq. 1:

$$t = 0.5\,[\delta_{xy} - 0.5\,(\delta_x + \delta_y)] \qquad [1]$$

where δ_{xy} is the mean percent sequence divergence between the two populations and δ_x (or δ_y) is the mean percent sequence divergence for all possible pairs of individuals within population X (or Y). This equation is based on equation 25 of Nei and Li (9) and the assumption that mtDNA evolution in apes has occurred at the rate of 2 substitutions per 100 bp per million years. This is the approximate rate inferred from studies of other mammals (5).

The divergence time of the Bornean and Sumatran orangutans is estimated to be at least 1.5 million years from the maps, and this estimate is consistent with geological and protein evidence. Although now confined to two islands, orangutans occurred widely in the Pleistocene from China to Indonesia (10). Their wide distribution would have been facilitated by the frequent existence of land bridges between the Asian mainland and the islands of Indonesia during the past 2 million years (12). Fossil evidence for the presence of hominids in Java 1.9 million years ago (13) raises the possibility that they reduced the range of orangutans and confined them to Borneo and Sumatra. Contact between the two populations of orangutans may also have been hindered by the presence of large rivers in the Sunda land mass between Borneo and Sumatra (14). The idea of an ancient divergence time for these populations is also supported by protein evidence (15).

The mtDNA sequence difference between common and pygmy chimpanzees is also suggestive of an ancient divergence time. The estimated time, based on the map comparisons, is 1.3 million years ago. This agrees with geological evidence that the Zaire River, which separates the two species, has maintained its course for the past 1.5 million years (16). This time is consistent with protein evidence (15, 17).

Among common chimpanzees, the geographic range of the two major types (1 and 2) of mtDNA remains to be determined rigorously. Our studies were conducted using captive chimpanzees whose geographic origins are, in most cases, uncertain (see Table 3). As is evident from the careful electrophoretic study of polymorphic proteins by Goodman and Tashian (18), authentic eastern representatives of the common chimpanzee differ markedly from western representatives. The differences in allelic frequencies are suggestive of an ancient separation between these populations, and we calculate a divergence time of 1 million years between the two types of mtDNA. It is also evident from their study that the morphological classification of captive chimpanzees may sometimes have been inaccurate.[§] Hence, a mtDNA study of chimpanzees in the wild will be required to ascertain the geographical distribution of the two types of mtDNA.

Table 4. Site variability in two regions of mtDNA

Region	Changes per site: Closely related mtDNAs	Changes per site: Distantly related mtDNAs
Ribosomal RNA	0.08	1.31
Nonribosomal	0.45	1.09
Whole genome	0.41	1.11

Changes per site were calculated by dividing the number of site changes (inferred phylogenetically) by the number of positions surveyed (13 for intraspecific and interspecific comparisons in the ribosomal RNA region and 110 for intraspecific and 119 for interspecific comparisons in the nonribosomal regions). Interspecific results are from Ref. 6. Closely related mtDNAs are those differing by <3% in nucleotide sequence.

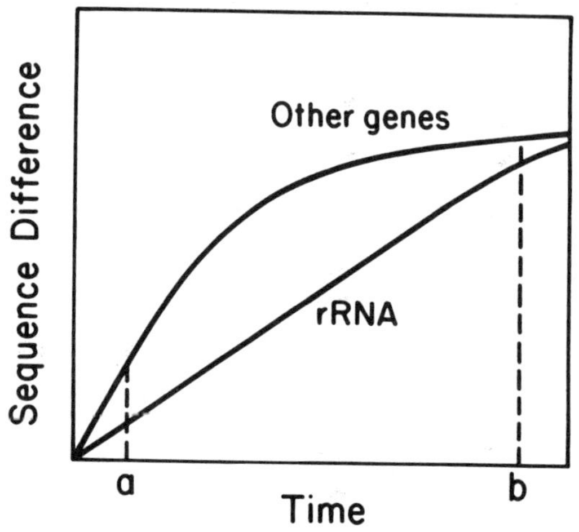

FIG. 6. Suggested dependence of sequence divergence on time of divergence for ribosomal RNA genes and other genes of mtDNA.

Gorillas were the least variable of the apes for mtDNA sequences. We attribute this to the sampling of only one subspecies, the lowland gorilla, which occupies a small geographic area (21). Protein divergence between lowland and mountain gorillas indicates a divergence time as old as that between the two species of chimpanzees (17). So, we expect the mtDNA divergence between the two gorilla subspecies to be large. If this is true, then every great ape species has a level of mtDNA diversity exceeding that in the human species by a factor of three or more.

The Human Species Is Young. The low level of mtDNA variation in humans fits with geological evidence that our species is a young one. Fossil studies indicate that the transformation of *Homo erectus* into *H. sapiens* was in progress 200,000 years ago and may not have been completed until within the past 100,000 years (22–24). Thus, we conclude, tentatively, that mtDNA diversity is related to species age in hominoids.[¶]

Implications for Primate Management. Our demonstration of large genetic differences within ape species points to the need for renewed attention to the problem of managing breeding colonies of both captive and wild apes.

A rational approach to the preservation of the genetic diver-

§ There is further evidence for a great deal of nuclear gene polymorphism among common chimpanzees. Morphologically defined eastern and western subspecies showed significant differences at four out of six blood group loci (19). A high level of globin gene polymorphism also exists in chimpanzees (ref. 7; unpublished data). There are other reports, however, indicating a low level of protein polymorphism in chimpanzees (15, 20). In our opinion, there is need for a more rigorous and extensive comparison of the level of nuclear polymorphism in chimpanzees and other hominoids.

¶ Another factor could contribute to the low level of mtDNA variability among humans. Tree analysis of both map and nucleotide sequence data shows that the lineage leading to human mtDNA is slightly shorter than those leading to chimpanzee and gorilla mtDNA (unpublished). Such an effect can be explained by back mutation, parallelism, or an evolutionary slowdown in the hominid lineage. A slowdown, if in effect during the period of human diversification, would contribute to the low level of variability in the human population. Evidence that this contribution is unlikely to be large will be presented elsewhere.

sity present in apes will require molecular screening of wild populations. Additional screening of captive populations will also be important, to avoid inadvertent mixing of genetically very distinct lineages of apes.

Slow Divergence of Ribosomal DNA. Comparison of the distribution of site changes in mtDNA of closely and distantly related organisms gives a perspective on the rate at which ribosomal RNA genes diverge. Our intraspecific comparisons in hominoids suggest that the rate of divergence in ribosomal RNA genes is several times lower than that in the rest of the mitochondrial genome. A similar observation has been made with rodents (ref. 2; unpublished) and by extensive studies of variation among humans (R. L. Cann, personal communication).

The intraspecific observations appear to contrast with previous observations based on interspecific comparisons, which indicate that the extent of divergence in the ribosomal RNA genes is about the same as for other regions of the mitochondrial genome (Table 4; refs. 5 and 25). It may be possible to reconcile the two observations with the aid of Fig. 6, which draws attention to the nonlinear dependence of sequence difference on time of divergence for the mitochondrial genome as a whole (5). This nonlinear dependence is probably due to the existence of two classes of sites (i.e., silent and replacement) in the sequences coding for proteins. These sequences, which account for >65% of the genome (6), have experienced a high rate of substitution at silent sites (not causing amino acid substitutions) and a low rate at replacement sites (causing amino acid substitutions) (E. M. Prager, personal communication). Our suggestion is that ribosomal genes diverge at a rather steady rate that is intermediate between the silent and replacement rates. When closely related mtDNAs are examined (Fig. 6, time a), the ribosomal genes appear less divergent than the rest of the genome. By contrast, when distantly related mtDNAs are compared (Fig. 6, time b), the ribosomal genes seem to be about as divergent as the rest of the genome.

For providing primate materials, we thank the following institutions—Cincinnati Zoo, Holloman Primate Center, Houston Zoo, National Institutes of Health, San Diego Zoo, Stanford University, W. Reid Hospital, and Yerkes Primate Center—and the following individuals—J. Cronin, C. Graham, P. Grant, D. Ledbetter, D. Premack, P. Russell, O. Ryder, B. Swenson, and G. Todaro. We also thank R. Cann, E. Prager, S. Washburn, and E. Zimmer for discussion, B. Chapman for collaboration, and P. McCutchan for preparation of the manuscript. This research was supported by a grant from the National Science Foundation and fellowships from the Miller Institute at Berkeley (to S.D.F.) and the Medical Research Council of Canada (to W.S.D.).

1. Brown, W. M. (1980) *Proc. Natl. Acad. Sci. USA* **77**, 3605–3609.
2. Brown, G. G. & Simpson, M. V. (1981) *Genetics*, **97**, 125–143.
3. Avise, J. C., Lansman, R. A. & Shade, R. O. (1979) *Genetics* **92**, 279–295.
4. Avise, J. C., Giblin-Davidson, C., Laerm, J., Patton, J. C. & Lansman, R. A. (1979) *Proc. Natl. Acad. Sci. USA* **76**, 6694–6698.
5. Brown, W. M., George, M. & Wilson, A. C. (1979) *Proc. Natl. Acad. Sci. USA* **76**, 1967–1971.
6. Ferris, S. D., Wilson, A. C. & Brown, W. M. (1981) *Proc. Natl. Acad. Sci. USA* **78**, 2432–2436.
7. Zimmer, E. A., Martin, S. L., Beverley, S. M., Kan, Y. W. & Wilson, A. C. (1980) *Proc. Natl. Acad. Sci. USA* **77**, 2158–2162.
8. Maniatis, T., Jeffrey, A. & Kleid, D. G. (1975) *Proc. Natl. Acad. Sci. USA* **72**, 1184–1188.
9. Nei, M. & Li, W-H. (1979) *Proc. Natl. Acad. Sci. USA* **76**, 5269–5273.
10. Smith, R. J. & Pilbeam, D. R. (1980) *Nature (London)* **284**, 447–448.
11. Hill, O. (1969) in *The Chimpanzee*, ed. Bourne, G. (Karger, Basel), pp. 22–49.
12. Ashton, P. S. (1972) in *The Quaternary Era in Malaysia*, eds. Ashton, P. S. & Ashton, M. (University of Hull, Hull, England), Dept. Geography Misc. Ser. no. 13, pp. 35–49.
13. Ninkovich, D. & Burckle, L. H. (1978) *Nature (London)* **275**, 306–307.
14. Haile, N. S. (1971) *Quaternaria* **15**, 333–343.
15. Bruce, E. J. & Ayala, F. J. (1979) *Evolution* **33**, 1040–1056.
16. Horn, A. D. (1979) *Am. J. Phys. Anthrop.* **51**, 273–282.
17. Sarich, V. M. (1977) *Nature (London)* **265**, 24–28.
18. Goodman, M. & Tashian, R. E. (1969) *Human Biol.* **91**, 237–249.
19. Moor-Jankowski, J. & Wiener, A. (1972) in *Pathology of Simian Primates*, ed. Fiennes, T.-W. (Karger, Basel), pp. 270–317.
20. King, M-C. & Wilson, A. C. (1975) *Science* **188**, 107–116.
21. Dorst, J. & Dandelot, P. (1970) *A Field Guide to the Larger Mammals of Africa* (Houghton Mifflin, Boston).
22. Kennedy, G. (1980) *Nature (London)* **284**, 11–12.
23. Cronin, J. E., Boaz, N. T., Stringer, C. B. & Rak, Y. (1981) *Nature (London)* **292**, 113–122.
24. Hennig, G. J., Herr, W., Weber, E. & Xirotiris, N. I. (1981) *Nature (London)* **292**, 533–536.
25. Borst, P. & Grivell, L. A. (1981) *Nature (London)* **290**, 443–444.

PROBLEMS

1. Define the following: Goldberg-Hogness box, Pribnow box, capping site, Shine-Dalgarno sequence, domain, intron-exon splicing junction, initiator codon, termination codon, coding sequence, noncoding sequence, flanking sequence, pseudogene, transition, transversion, structural gene, mutation rate, transposition, illegitimate recombination, plasmid, Charon phage, restriction enzyme, ligase, nick translation, Southern blot, in situ hybridization, polytene chromosome.

2. The following is the sense strand surrounding the initiation codon (ATG) of a structural gene:

   GTAATACACAATGTCTATTC
   -9 1 10

   CAGAAACTCAAAAAGCTGA
   20

   (a) What are the likely consequences of a transversion at position -5? a transition at 1, a single base addition at 21, and a deletion at 19?
   (b) In the first three codons, what are the silent and replacement transition mutations that can be generated by base substitution? Are there any possible transversion or termination mutations in the sequence given? Of the missense mutations in the first three codons, how many are changes to a different electrophoretic class of amino acid?

3. What are the features of the globin genes that make them (1) an usually favorable subject for genetic and molecular studies, (2) of special interest to the molecular evolutionist, and (3) of significance to the geneticist interested in the regulation of gene expression in development? You may want to refer to Efstratiadis *et al.* (1980).

4. Topal and Fresco (1976) have predicted, on stereoisomeric grounds, that transversions will only occur 1/10th as often as transitions in spontaneous base-substitution mutations.
   (a) If single base substitution were a completely random event, what relative frequency of newly arising transversions and transitions would be expected?
   (b) Do the relative frequencies of transversions and transitions in the data of Sankoff *et al.* (1973, this chapter) match either prediction?
   (c) What might the reasons for the discrepancy be?
   (d) For an essential polypeptide-coding gene sequence, would one necessarily expect comparable frequencies of the various mutation events calculated by Sankoff *et al.*? Why?

5. Most of the contemporary analysis of gene duplications involves Southern blotting with cloned gene fragments. Successful hybridization on Southern blots requires a high degree of homology between the DNA probe and the electrophoresed restriction fragments that have been transferred to nitrocellulose filters.
   (a) Evaluate, in light of the homology

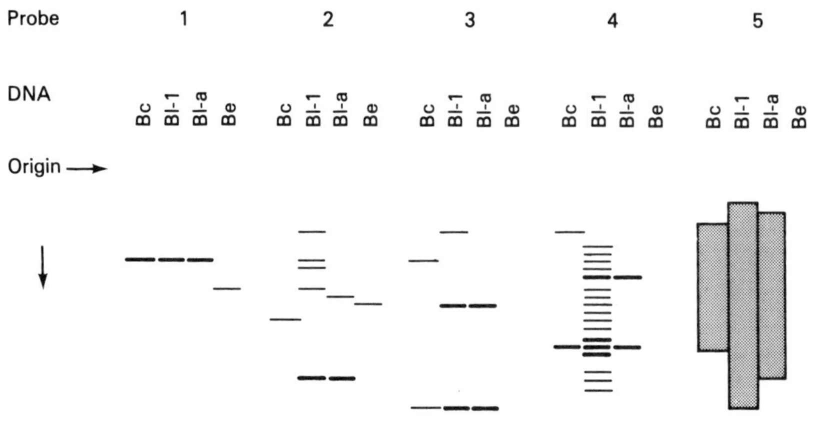

Figure 2-5

requirement, the data in Figure 2-5, showing the ability of five cloned DNA fragments (1-5) of the round worm *Caenorhabditis elegans* strain Bristol to hybridize with DNA fragments from several related sources. The DNA molecules from *C. elegans* strain Bergerac larvae (Bc), strain Bristol larvae (B1-1) and adults (B1-a) and *C. briggsae* adults (Be) were isolated, digested with Bam HI, electrophoresed on agarose gels, transferred to nitrocellulose filters, and probed with the labelled fragments. In the figure an idealized autoradiogram of the results is shown, in which the size of bars denote the intensity of the spots. (Data adapted from Hirsh *et al.*, 1979).
(b) Why might one get multiple bands hybridizing to single cloned and labelled fragments, as seen with probes 2-5?
(c) Suggest an explanation for the differences in patterns between Bristol larvae and adults when probed with fragments 2-4.

6. In the primates, alpha-hemoglobin sequences are normally present in duplicate copies, spaced about 3 kb apart. Zimmer *et al.* (1980) report that, not uncommonly, individual chimpanzees and humans are found to have either single or triple linked copies. Despite the existence of two or more copies in most individuals, there is substantially less intraspecific variation for alpha-globin than interspecific variation, as revealed by protein sequence studies. Thus, within a species, there seems to be concerted evolution of the multiple alpha-globin loci. Discuss these observations in light of the ideas of Smith (1974, this chapter) and of Dover (1982).

7. The histone genes and rRNA genes are archetypal examples of tandemly clustered, repeated gene families which exist outside the main clusters as isolated copies, or orphons (Childs *et al.*, 1981). Recently, it has been found that isolated histone and rRNA genes exist outside the main clusters as isolated copies, or orphons (Childs *et al.*, 1981). What does the finding say about the distinction between tandemly repeated and dispersed gene families? What does it seem to reveal about the requirements for sequence specificity in the dispersal of genes?

8. Gilbert (1978, this chapter) has discussed the possible selective advantages of introns in the eukaryotic genome. (A somewhat different view of the extra DNA has been given by Orgel and Crick (1980).) Can you think of possible selective disadvantages that might serve to keep the number of introns in check? Does the Gilbert proposal satisfactorily explain why eukaryotes put up with introns in their DNA?

9. Male recombination (MR) elements have been found in many natural populations of *Drosophila melanogaster*. These genetic elements promote recombination in males, and increase the frequency of mutation, chromosome breakage, hybrid sterility, transposition of genes and distortion of segregation. It is now clear that a large part, if not all, of these genetic consequences are due to the presence in MR strains of certain transposable elements. In the experiment shown in Table 2-9, flies from different populations were crossed to yield hybrid males and the frequency of visible mutation on the X chromosome in these males was assayed. (The hybrid and control males were mated to attached X females; the male progeny in such crosses will each have the paternal X chromosome).

TABLE 2-9.
Mutation frequencies in populations of wild and control lines and among the progeny of interpopulation crosses

Cross		*Mutations in hybrid males*		
Female parent	*Male parent*	*Number of mutants*	*Total*	*%*
OK1	OK1	0	13,355	0
M-4	M-4	13	6,319	0.21
N-34	N-34	3	5,259	0.06
Canton-S	Canton-S	0	8,590	0
Oregon-R	Oregon-R	0	3,170	0
N-34	M-4	4	4,386	0.09
OK1	N-34	18	6,314	0.28
N34	OK1	38	2,762	1.38
OK1	M-4	8	3,886	0.20
M-4	OK1	14	2,635	0.53
OK1	Canton-S	5	4,342	0.12
Canton-S	OK1	61	2,982	2.04
Oregon-R	Canton-S	0	3,654	0
Canton-S	Oregon-R	1	3,339	0.03

Strain OK1 is one of the original MR strains, isolated from the wild in Oklahoma; N-34 was isolated from a different, natural Oklahoma population; M-4 is from Australia; Canton-S and Oregon-R are standard lab strains lacking MR activity. (Data from Thompson and Woodruff, 1980)

(a) What general conclusion do the data suggest about the mutator activity in intrapopulational crosses versus outcrossings?
(b) In Table 2-9 are there any apparent exceptions to this generalization?
(c) Is there a difference between reciprocal crosses? Note particularly the crosses utilizing a wild isolate and a standard laboratory strain.
(d) What general conclusions might you draw about the control of variability induced by MR elements in *Drosophila* populations, and what might the evolutionary significance be?

10. The discovery of high mutator activity associated with transposable elements in wild strains of *D. melanogaster* raises the question of how much the mutation occurring in natural populations is due to these elements and how much to more conventional point mutations, small deletions, etc. Let us say you have a cloned structural gene, for a locus whose mutants are comparatively easy to select from natural populations. The alcohol dehydrogenase (Adh) locus is one possibility; see O'Donnell *et al.* (1975) and Benyajati *et al.* (1981). How might you go about assessing the relative contributions of transposable element-mediated mutations and point mutations in the spontaneous mutability of this gene?

3 GENOME ORGANIZATION AND CHANGE

...an explanation of the scale on which significant evolutionary invention has taken place may require in addition whole new classes of mechanisms, such as selection in favor of genomes with favorable types of sequence organization (e.g. suitable sources of regulatory genes) or with optimal rates of reorganization.
R.J. Britten and *E.H. Davidson* (1971).

The Watson-Crick model for the structure of DNA, published in 1953, has as its central feature two complementary strands held together by many weak noncovalent bonds. Subsequently, much experimental attention was given to the physical chemistry of strand separation and reassociation of DNA (melting and reannealing). After the publication of the paper of Hoyer et al. (1964) it became evident that this approach could be used for examining the general properties of genomic DNA and its evolution. Since then, a more detailed examination of nucleotide sequences and organization has been made possible by the recombinant DNA techniques resulting from the collaboration of microbial geneticists with biochemists. The geneticists have provided the autonomously replicating units such as viruses and plasmids, which are employed as vectors for the amplification and maintenance of particular DNA fragments (cloning): biochemists provided the enzymes that permit the controlled *in vitro* manipulation of nucleic acids required for the construction of the vector-fragment hybrid molecules and also the techniques of nucleotide-sequence analysis (for an overview, see Sinsheimer, 1977; Wu, 1978).

A casual inspection of the distribution of genome size across a broad spectrum of organisms suggests a crude correlation between the minimum genome size in a taxonomic group, and the level of biochemical and morphological complexity of that group (Sparrow and Nauman, 1976). However, among eukaryotes, anomalies are immediately apparent. For example, even within a reasonably homogeneous group of organisms (the plant genus *Lilium*) there can be a fivefold difference in genome size, and between groups at distinctly different levels of biological complexity (primates and salamanders), the less complex can possess an order of magnitude more genomic DNA. This problem has been termed the C-value paradox (Cavalier-Smith, 1978) and raises the possibility that large amounts of genomic DNA may not be involved in direct coding functions, as pointed out by King and Jukes (1969). Britten and Davidson (1971) and Davidson and Britten (1979) suggested that noncoding DNA may be utilized in complex regulatory circuits, while Cavalier-Smith (1978) proposed that sheer bulk of DNA may have a generalized regulatory function. Orgel and Crick (1980), and Doolittle and Sapienza (1980) have even suggested that the eukaryotic genome carries an indeterminate amount of "parasitic" or junk DNA. In this chapter we shall focus primarily upon the larger eukaryotic genome, and discuss only very briefly the smaller prokaryotic and organelle genomes.

Britten and Kohne (1968, Chapter 1) investigated the reannealing of DNA as a tool for examining genome structure. They proposed a simple graphical representation of the kinetics of reannealing of DNA in which the proportion of remaining single-stranded DNA is plotted against Cot, the product of the DNA concentration and time of DNA annealing. Britten and Kohne showed that prokaryotic DNA reannealed with kinetics that conform to a simple second-order reaction, whereas eukaryotic DNA yielded a complex Cot curve. The curve for eukaryotic DNA was interpreted as the summation of several second-order reactions, each of which reflected the reannealing of a different frequency class of nucleotide sequences. In fact, no absolutely distinct frequency classes are seen, but for convenience of description three classes of sequences have been defined: highly repetitive sequences, which occur up to a million times; middle repetitive sequences, which are found in the range of 10^2–10^5 copies per genome; and unique or single-copy sequences, for which 1—10 copies are present.

Detailed characterization of specific classes of nucleotide sequences had led to important insights into the evolutionary constraints upon sequence alterations in these classes. In particular, it has been found that some identifiable classes of sequences appear to be free to change at a rate apparently limited only by the fidelity of DNA replication and drift, while others show a high degree of conservatism. Sufficient data have been collected to begin to discern some characteristic differences between the frequency classes.

Highly repetitive (rapidly reannealing) DNA has long been recognized as a distinct class termed satellite DNA. It accounts for as much as 20 percent of the genome and was originally distinguished from bulk DNA by its buoyant density (base composition), and by its almost exclusive association with heterochromatic regions of the chromosome. This class of DNA has been shown to consist usually of very short blocks of 2 to 10 nucleotides, tandemly repeated many thousands of times, and with total repetition frequencies numbering in the millions. Initially interspecific hybridization studies indicated that there is rapid evolutionary change in satellite DNA sequences. However, more recent work with sequences of cloned fragments has shown that often the basic repeat units are similar in closely related species, but that different variants are amplified in different species. Furthermore, within species this type of analysis has revealed interesting patterns of both regular and random variations in sequences of the repeated units, as well as interruptions of long tandem arrays by nonhomologous sequences. Cytological studies employing in situ hybridization have shown that blocks of identical or closely related sequences may be found at different locations either on the same chromosome or on nonhomologous chromosomes. The patterns of repetitive variation and dispersion are consistent with the model of unequal recombination proposed by Smith (1974, Chapter 2), when combined with some form of transposition (Dover, 1982). This class of DNA sequence has been reviewed by Brutlag (1980), Singer (1982), and Jelinek and Schmid (1982); its function still remains obscure.

The middle repetitive sequences have

proved to be very diverse with respect to their genomic distribution and to their evolution both within and between taxa; accordingly, they defy simple description and classification. Despite an enormous amount of data from a broad spectrum of organisms, the functional, organizational, and evolutionary significance of most of this DNA sequence class remains mysterious (Bouchard, 1982). Early studies of both purified sequences of known function (such as rRNA genes), and bulk (hence heterogeneous) middle repetitive DNA indicated that these sequences could exist as either tandemly repeated or dispersed elements. More recent work using specific cloned fragments of unknown function has shown that in addition, at least some of the middle repetitive sequences are highly mobile or "nomadic" (Young, 1979).

Graham et al. (1974) observed a regular pattern of interspersion of single copy and repetitive DNA in the eukaryotic genome. The blocks of repetitive sequences were found after using short reannealing times (which meant that only highly repetitive sequences had time to reanneal) and digesting the remaining single-stranded DNA with S1 nuclease (an enzyme active only against single-stranded DNA segments). These repetitive blocks were shown to be adjacent to longer blocks of single-copy sequences by using DNA sheared to varying extents and subjected to short term reannealing with short labeled fragments. Single-stranded, and hence, low-copy-number sequences, were found to be covalently attached to a high proportion of the reannealed middle repetitive blocks. The interspersion patterns of a variety of animals are described in the paper of Davidson, Galau, Angerer and Britten (1975, this chapter), while Flavell, Rimpau and Smith (1977, this chapter) were able to demonstrate that this general pattern of sequence organization is present in plants as well. The generally observed short-period interspersion pattern has blocks of unique sequence several thousand nucleotides long, alternating with blocks of repetitive sequence ranging from several hundred to several thousand nucleotide pairs. In contrast, *Drosophila* DNA has a long-period interspersion pattern in which unique sequence blocks containing many thousands of base pairs alternate with repetitive sequences having several thousand base pairs. Indeed a critical evaluation of the voluminous literature in this area (Bouchard, 1982) leads to the conclusion that a wide variety of interspersion patterns exist even between closely related taxa. Moyzis et al. (1981) suggest that at least some of the observed variability may be artifactual due to variation in stringency of annealing. Finally, examination of the relatedness of the dispersed middle repetitive sequences reveals that in some organisms a few sequences predominate, whereas in others a wide variety of different sequences are found.

A great deal of work has been done on the interspersed middle repetitive sequences, stimulated in part by the original proposal of Britten and Davidson (1971) that this class of nucleotide sequences may be involved in complex regulatory cascades presumed to distinguish eukaryotes from prokaryotes. It was recognized relatively early (Britten and Davidson, 1976) that for any particular species remarkable sequence homology is present within a repeated family, whereas between species, divergence occurs that is similar to that observed for bulk low-copy-number DNA. This result suggested that either specific sequences are amplified in particular lines of descent or a mechanism of sequence homogenization is present (Dover, 1982). However, more recent studies with sea urchins, using cloned members of repeated sequence families, has shown that sequences exist that are highly conserved between species, but whose repetition frequencies vary widely between these species (Moore et al, 1978). In a complementary series of studies on the interspersed repetitive sequences of a group of cereals (grasses) Flavell et al. (1977c, this chapter) were able to identify families of sequences that expanded and diverged in particular branches of the phylogenetic tree. Application of the techniques of recombinant DNA analysis permitted Bedbrook, Jones, and Flavell (1981c, this chapter) to extend this study to a direct examination of the distribution of specific repetitive sequences throughout the genome. Smith (1974, Chapter 2; 1976), recognizing the significance of repetitive sequences in the evolution of the eukaryotic genome, has provided a useful theoretical model, invoking unequal recombination, for the origin, expansion, and diversification of these sequences. Dover (1982) has considered the problem of the apparent maintenance of sequence identity, even following dispersion of a proliferated sequence. He has described a possible homogenization mechanism in the form of a biased gene conversion, occurring in the course of a putative recombination event between scattered repetitive sequences.

By isolating both bulk interspersed repetitive sequences as well as cloned fragments of specific interspersed sequences of sea urchin DNA, Costantini et al. (1978) were able to show that a large fraction of the sequences of both complementary strands is transcribed, though none of the transcripts become associated with ribosomes. Although the genomic frequency of these particular sequences varied widely in related species, the composition of the RNA transcript pool of the specific sequences was quite stable (Moore et al., 1980). In the human genome a single family of sequences, the Alu family, dominates the interspersed repetitive class. The Alu sequences have been recovered as inverted repeat sequences in heterogeneous nuclear RNA (Jelinek et al., 1980). The sequence conservation, transcription without apparent translation, and the conserved level of transcription all suggest that these elements may play some sort of regulatory role (Moore et al., 1980).

The middle repetitive sequences of known function have been classified by Long and Dawid (1980) into "dosage" and "variant" repetitive genes. The former are those precisely repeated sequences whose products (such as rRNA and histones) are required in large numbers; the latter are those families of somewhat related sequences (such as globins or immunoglobulins and their pseudogenes) that are in the process of diverging from each other, and, accordingly, merge imperceptibly into the unique sequence class. For the obvious reason of the ready availability of a hybridization probe, the ribosomal RNA genes were among the first of the middle repetitive sequences to receive close scrutiny and thus have provided probably the best comparative database for evolutionary inferences.

Brown and Sugimoto (1974, this chapter) give an early description of the structure of the tandemly repeated region coding for the 28S, 18S and 5S RNAs of two species of *Xenopus* and conclude that while those sequences that are matured into structural RNA molecules exhibit strong evolutionary conservation, the sequence and length of the transcribed and nontranscribed spacer regions are much less restrained. Some evidence suggests that the spacer regions consist in part of repeated short sequences and are related in some way to the highly repetitive satellite DNA. In contrast, Tartof (1979) found that in *Drosophila*, aside from a short nonconserved transcribed spacer region, the 18S and 28S spacers (nontranscribed) are at least as highly conserved as the sequences that are matured into rRNA. In all of the systems studied, the rRNA sequence appears to be highly conserved and, thus, it has been pointed out (Woese et al., 1975) that such sequences should provide a useful measure for distant relationships, a point examined more thoroughly in Chapter 8. The sequences of tRNA appear to be even more conserved and perhaps allow a glimpse back into precellular evolution, as we discuss in Chapter 9.

"Variant" repetitive sequences encompass multigene families in which, by duplication and subsequent divergence, presumed ancestral protein sequences have spawned families of sequences of greatly varying degrees of relatedness, both within a species and between species. This sequence variability makes the group very difficult to study by techniques that rely on DNA hybridization, because the precise conditions of reannealing impose an arbitrary limit on the detection of homology (Bonner et al., 1973). Thus, some low-copy-number sequences may represent members of repetitive families that have escaped the forces of homogenization and have diverged beyond recognition at the selected level of hybridization. Britten and Davidson (1976) estimated that much of this nonrepetitive DNA might not encode proteins and showed that the rate of base substitutions in this class of nucleotide sequence is only limited by the accuracy of DNA replication. Measurements of evolutionary relatedness by means of sequence divergence of bulk low-copy-number DNA in eukaryotes (Kohne et al., 1972) and prokaryotes (DeLey, 1974, Chapter 8) have yielded results generally consistent with those from other methods. A detailed examination of sequences in terms of rates of amino acid or nucleotide substitutions and functional constraints upon such changes are the primary topics of Chapter 4.

A final class of repetitive elements having unknown function but intriguing properties and possible evolutionary significance are the movable elements described in *Drosophila* by

Strobel et al. (1979, Chapter 2). Indeed, in a survey of the *Drosophila* genome, Young (1979) suggested that much of the dispersed middle repetitive DNA is similarly mobile. As pointed out by Strobel et al. (Chapter 2), similar elements have been described in both yeast and maize, and their structures are reminiscent of both prokaryotic transposable elements and eukaryotic retroviruses.

With certain interesting exceptions (see Chapter 8), the genomes of prokaryotes and of eukaryotic organelles reanneal with simple second-order kinetics; when taken together with independent measures of genome size, these kinetics indicate that most of the sequences are contained in single-copy DNA. Few or no generalizations can yet be made about the evolution of these genomes, though they manifest interesting features that may yield future insights. The complete sequence determination of bacteriophage øX174 (Sanger et al. 1978) has revealed a remarkable economy of coding for genetic information in that several of the genes overlap. Borst and Grivell (1981) and Wallace (1982) have briefly reviewed the peculiarities of mitochondrial genomes. Not only are there differences in size and efficiency of sequence utilization in the mitochondrial genome of different organisms but also departures from the nuclear universal code, and, at least among higher vertebrates, a high mutation rate is present (Brown et al., 1979; 1982). More information about these points can be found in Chapter 7. Palmer (1982) and Wallace (1982) have briefly summarized the data on chloroplast genomes, which indicate that in the majority of species studied, genome size and structure are conserved, at least in terms of the approximate location of identifiable genes.

In summary, direct analysis of genomic DNA by a variety of techniques has revealed a number of interesting features. The eukaryotic genome differs from the standard prokaryotic genome in having large amounts of noncoding and repetitive DNA. This repetitive DNA appears to amplify and diminish and to transpose at quite a high frequency, and endows the eukaryotic genome with an unexpected plasticity. In addition, much of the low-copy-number DNA is interspersed with repetitive sequences having a long- or short-period pattern. Both prokaryotes and eukaryotes have been shown to harbor a variety of transposable elements that provide an additional source of plasticity. Aside from a few known families of repetitive sequences, the function of this DNA is unknown, and the significance of the short- versus long- period interspersion is not understood. Detailed comparative studies, such as those of Barrie et al. (1981) on the beta-globin gene cluster should lead ultimately to a deeper understanding of both the function of the interspersd repetitive sequences and the significance of the interspersion. The mechanism of sequence dispersion and of the maintenance of sequence homogeneity (Dover, 1982) is likewise not understood and requires further attention. For a broad coverage of the topic of genome evolution, the reader is referred to a recent book edited by Dover and Flavell (1982).

Reprinted from *Cold Spring Harbor Symp. Quant. Biol.* **38**: 501–505.

The Structure and Evolution of Ribosomal and 5 S DNAs in *Xenopus laevis* and *Xenopus mulleri*

Donald D. Brown and Kazunori Sugimoto
Carnegie Institution of Washington, Department of Embryology, Baltimore, Maryland 21210

The isolation of genes of known structure has provided information on their arrangement and evolution. This analysis is most advanced for the genes for ribosomal RNAs due to the ease with which these RNAs can be purified for use in hybridization assays of the genes. In addition, the genes are present in many copies in the genome and have unusual physical and chemical characteristics that have facilitated their purification (Wallace and Birnstiel, 1966; Brown and Weber, 1968a).

The genes coding for the structure of the 18 S and 28 S ribosomal RNAs (rDNA) (Birnstiel et al., 1966; Brown and Dawid, 1968; Brown et al., 1972) and those for the low molecular weight 5 S ribosomal RNA (5 S DNA) (Brown et al., 1971; Brown and Sugimoto, 1973) have been purified from the DNA of *Xenopus laevis* and *Xenopus mulleri*. These genes are highly repetitive and occur in one or more clusters within the genome. The genes are separated by what has been called, for want of a better term, "spacer" DNA (Brown and Weber, 1968b; Birnstiel et al., 1968; Miller and Beatty, 1969). These spacer regions resemble satellite DNAs in several ways. They are very different from the average DNA in the genome, since they do not hybridize with other sequences in the DNA. They are highly repetitive like the genes which they separate, but unlike these genes, they evolve rapidly with little restriction on their exact nucleotide sequence (Brown et al., 1972; Brown and Sugimoto, 1973).

The ribosomal RNA genes and their spacers have evolved together as a family of sequences—a phenomenon we have referred to as "horizontal" evolution (Brown et al., 1972). In this paper, we summarize the structure of these reiterated genes (in the two species of frogs), their chromosomal location, the similarities and differences that exist in the DNA sequences within a species and between the two species, and finally experiments designed to understand the genetic mechanisms that might cause this "horizontal" evolution.

Ribosomal DNA

Structure. Figure 1 and Table 1 summarize what we know about the arrangement of sequences in ribosomal DNA of Xenopus. The 18 S and 28 S RNAs of *X. laevis* and *X. mulleri* are the same size, have the same base composition, and hybridize with homologous and heterologous rDNA with equal efficiency and stability (Brown et al., 1972). By these criteria 18 S and 28 S genes are indistinguishable in the two species. Together the genes for 18 S and 28 S rRNA comprise about half of each repeat length in rDNA (Birnstiel et al., 1968; Gall, 1969). The spacer regions are about the same length in the two species but very different in sequence. Although both spacer sequences are high in GC content, *X. mulleri* spacers are 4% lower than those of *X. laevis* rDNA. Complementary RNA transcripts from these spacers hybridize only about 20% as efficiently to heterologous spacer DNA as to the homologous rDNA. Furthermore, these heterologous hybrids have lower thermal stability compared to homologous cRNA–rDNA hybrids (Brown et al., 1972). The spacer sequences of *X. laevis* and *X. mulleri* rDNA must differ by at least 1000 of their approximately 7000 base pairs.

The 18 S and 28 S genes are diagrammed as close together in Figure 1 but separated by a short region. We now believe that part of the "transcribed" spacer (that region of rDNA transcribed as a precursor molecule but discarded during maturation) is located between the two genes. Evidence for this is twofold. RNA secondary structure maps (Wellauer and Dawid, this volume) show these sequences are divided equally on either side of the 18 S gene region. Second, heteroduplex molecules between *X. laevis* and *X. mulleri* rDNA show a small "blister" between two fully duplexed regions of

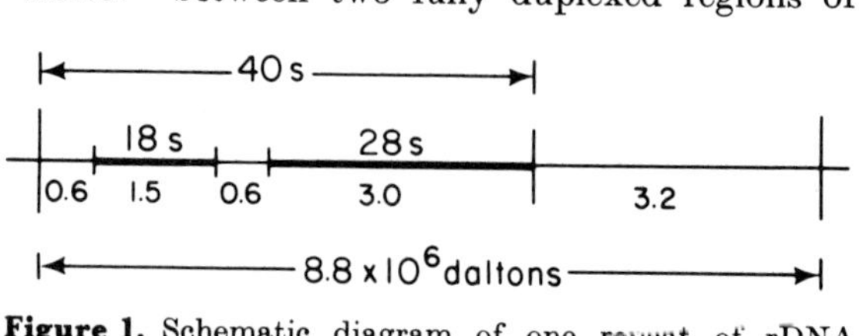

Figure 1. Schematic diagram of one repeat of rDNA from either species of Xenopus. The 40 S RNA precursor is transcribed from the region shown. The 18 S and 28 S gene regions are denoted by heavy lines. The small regions on either side of the 18 S gene are the "transcribed spacer" sequences. Their position has been determined by secondary structure maps of the *X. laevis* 40 S precursor RNA (Wellauer and Dawid, this volume and pers. commun.).

Table 1. Characteristics of Xenopus rDNA

Percent of total DNA	0.2
Average number of repeats	450
Chromosomal location	nucleolar organizer (one chromosome)
Repeat length	8.7×10^6
Gene: spacer	1:1
Spacer base composition (% GC)	*X. laevis*, 73 *X. mulleri*, 69

DNA which have the proper lengths for 18 S and 28 S genes (Forsheit and Davidson, pers. commun.). This mismatched region must be part of the "transcribed" spacer which has diverged in the two genomes like the rest of the transcribed and nontranscribed spacer sequences.

Homogeneity of rDNA within a species. Within each species the several hundred repeats of rDNA are very similar, if not identical, both in length and nucleotide sequence. This is true for the spacer regions as well as for the gene regions. Two kinds of experiments have demonstrated this fact. When rDNA is denatured and the strands reassociated, the duplex is reformed faithfully so that its thermal stability is as great as the original duplex. There is no difference in the stability of reassociated rDNA and phage T2 and T7 DNAs, which are known to be unique sequences (Dawid et al., 1970; Brown et al., 1972). More recent evidence which corroborates these melting experiments comes from electron microscopic studies of reassociated homoduplexes of rDNA. Reassociated molecules of rDNA appear to be very faithfully rematched. These molecules are duplexed completely even when spread at high formamide concentrations (Forsheit and Davidson, pers. commun.).

Denaturation maps of both kinds of rDNA molecules are highly reproducible, demonstrating a regular repeating arrangement of denatured regions every 8.7×10^6 daltons along the molecule (Wensink and Brown, 1971; Brown et al., 1972). As many as 15 repeats have been visualized along a single rDNA molecule. Both the length of the repeats and the reproducibility of their denaturation profile are very uniform. Molecules of rDNAs from the two species are distinguished easily by this technique, since the spacer region of *X. laevis* rDNA is uniformly high in GC content, whereas half of the spacer region of *X. mulleri* rDNA has on the average a much higher GC content than the other half. This is visualized by a denatured region of DNA adjacent to a native region within each spacer sequence. The regularity of rDNA repeats supports the notion that they are not interspersed with other kinds of DNA sequences.

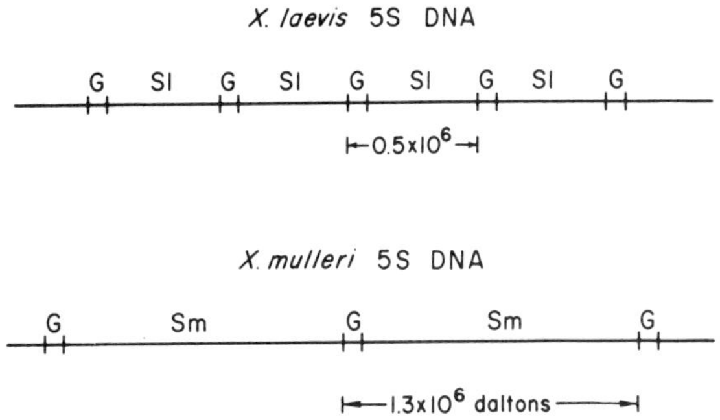

Figure 2. Schematic diagram of 5 S DNA from the two species of Xenopus. The *average* repeat lengths are given. Denaturation map of *X. laevis* 5 S DNA shows that the repeat length is very regular (Brown et al., 1971). However the same experiment has not yet been carried out for *X. mulleri* 5 S DNA. S = spacer; G = 5 S gene region.

Chromosomal localization of rDNA. Both species of Xenopus have 18 haploid chromosomes. One nucleolar organizer is visualized in *X. laevis*, and two chromosomes with secondary constrictions have been reported in the karyotype of *X. mulleri* (Tymowska and Kobel, 1972). Previous analyses of the nucleolar mutant in *X. laevis* established that all its rDNA is at the nucleolar organizer location, since this deletion removed greater than 99% of the several hundred repeating genes (Wallace and Birnstiel, 1966). No such mutant has been described as yet for *X. mulleri*. However, in situ hybridization of *X. mulleri* chromosomes shows that rDNA is present on a single pair of homologues just as it is in *X. laevis* (Pardue and Brown, unpublished).

The Structure of 5 S DNA

Figure 2 and Table 2 summarize the structure and some characteristics of 5 S DNA from *X. laevis* and *X. mulleri*. Saturation hybridization and other methods show that the 5 S gene sequences comprise about one-seventh (Brown et al., 1971) and one-eighteenth (Brown and Sugimoto, 1973)

Table 2. Characteristics of Xenopus 5 S DNA

	laevis	*mulleri*
Percent of total DNA	0.7	0.7
Average number of repeats	24,000	9000
Chromosomal location	most telomeres	most telomeres
Average repeat length $\times 10^6$	0.5 to 0.6	1.2 to 1.5
Gene:spacer	1:6	1:15
Spacer base composition (% GC)*	35	43

* Calculated from nearest neighbor values of the [^{32}P]cRNA synthesized from 5 S DNA by *E. coli* polymerase (Brown and Sugimoto, 1973). These values have been corrected for a 5 S RNA base composition of 57% GC.

of the base pairs of the two purified 5 S DNAs, respectively. Since the 5 S RNAs produced by both animals have the same length, the spacers must be different in length. Sequences of the two spacers have been compared by reassociating complementary strands and hybridizing cRNA to the DNAs. The majority, if not all, of the complementarity between the two 5 S DNAs resides in the 5 S RNA gene regions. No spacer complementarity has been detected by these methods (Brown and Sugimoto, 1973).

Despite the differences between these spacers, they nevertheless resemble each other in certain respects. Both kinds of spacers have similar dinucleotide frequencies characterized by greater than random abundance of ApA and TpT, whereas the alternating sequences ApT and TpA are rare. In addition, the individual strands differ in buoyant density by 23 mg/cm^3 in alkaline CsCl. The asymmetry is caused by spacer sequences in both DNAs, which have clusters of A residues on one strand and T residues on the other. The 5 S RNA is transcribed from the light strand of both DNAs (Brown and Sugimoto, 1973).

Heterogeneity of 5 S DNA within a species. In contrast to rDNA, reassociation of denatured 5 S DNA does not produce perfect duplexes. Only about 70–80% hyperchromicity is regained after renaturation, and the reassociated duplexes are displaced by 2–3° from the native melt (Brown and Sugimoto, 1973). Since most of each DNA is spacer sequence, we can assume that the melting profile is largely a consequence of the spacer region. Therefore by applying the formula of 1.5% mismatching for a 1° decrease in T_m (Laird et al., 1969), we presume that 70% of the sequences are mismatched by about 3–5% of their nucleotides, and the remainder of the spacer sequences fail to reassociate altogether.

5 S RNA heterogeneity. Whereas the multiple 5 S genes in mammalian genomes appear to produce only a single homogeneous 5 S RNA species, *Xenopus laevis* synthesizes at least two 5 S RNAs. Cultured somatic cells of *X. laevis* synthesize one kind of 5 S RNA, which has been sequenced in two laboratories (Brownlee et al., 1972; Wegnez et al., 1972). It contains 120 nucleotides, eight of which differ from mammalian 5 S RNA. Recently it has been shown that young oocytes synthesize another 5 S RNA of the same size, which is seven nucleotides different from the somatic cell type (Wegnez et al., 1972; Ford and Southern, 1973). There is evidence that there may be additional minor 5 S RNA components synthesized by oocytes. Therefore the 5S RNA genes themselves are heterogeneous within the *X. laevis* genome. Preliminary sequencing experiments have indicated that somatic 5 S RNA from *X. mulleri* cultured cells is the same as that from *X. laevis* (Brown and Williamson, unpublished). The oocyte 5 S RNA of *X. mulleri* has not been sequenced. However, it has been compared with the somatic 5 S RNA on MAK columns, which are known to separate *X. laevis* somatic and oocyte 5 S RNAs (Denis et al., 1972). The same kind of separation is affected for oocyte and somatic 5 S RNA from *X. mulleri*, suggesting that similar, and perhaps identical, nucleotide differences exist between these two 5 S RNAs as those which distinguish the two kinds of 5 S RNAs in *X. laevis*. If this is true, then this particular divergence in 5 S RNA genes must have occurred before divergence of the two Xenopus species.

In terms of gene arrangement, this heterogeneity poses two problems. First, what is the relative abundance of the two kinds of sequences in the purified 5 S DNA and within the genome? Second, how are the two kinds of RNA genes arranged with respect to each other in the genome, and how similar are their respective spacer sequences?

We have some preliminary evidence on the first question. George Brownlee (pers. commun.) has begun sequencing cRNA transcribed from 5 S DNA by *E. coli* polymerase. Oligonucleotides from T1 and pancreatic RNase digests have been found which are copies of either strand of the 5 S RNA gene region. Searching for the expected nucleotides from gene and antigene transcripts, he has found only oocyte type sequences. If the *E. coli* polymerase transcripts are faithful representations of the DNA, then the 5 S DNA of *X. laevis* must be primarily, and perhaps entirely, of the oocyte type. It is entirely reasonable that most of the 24,000 5 S genes should encode for oocyte type 5 S RNA, since the production of 5 S RNA in early oocytes probably proceeds at synthetic rates higher than at any time during the life of the animal. If oocyte type 5 S RNA is only transcribed in oocytes and in no other cell type in vivo, we have purified a differentiated group of genes that are normally repressed in somatic cells and therefore must be regulated by an additional control mechanism. At present, we know nothing about the number and possible arrangement of the somatic-type 5 S genes. It is conceivable that somatic cells, which make 5 S RNA at rates similar to rRNA, need about the same number of genes—about 450. If this is the case, it will be difficult to detect these genes among the 24,000 oocyte type genes.

5 S DNA spacer heterogeneity. Aside from 5 S RNA gene heterogeneity, there is detectable sequence heterogeneity in the spacers of 5 S DNA. This heterogeneity is detected by the thermal denaturation studies described earlier and has also been documented by the sequencing studies of George Brownlee (pers. commun.). A group of spacer oligonucleotides of about 15 nucleotides, which differ from each other by one or two bases, has been found. The yield of one of these oligonucleotides has been compared to oligonucleotides derived from the 5 S gene region. If each repeat within the DNA is a unique sequence with no internal repetitions, then the mole ratio of these two oligonucleotides should be one. Surprisingly, however, Brownlee (pers. commun.) has found a 5 to 1 ratio of this spacer oligonucleotide compared to a gene oligonucleotide. This means that the spacer in *X. laevis* 5 S DNA has some internal repetition. The arrangement and extent of this repetition within each spacer is not yet known. However, it complicates the interpretation of experiments that analyze the reassociated DNA, since the internal repetition conceivably would permit the strands to reassociate out of register even if all the spacers were exactly the same.

The 5 S DNA from a single animal. The 5 S DNA has been purified from the DNA of a single *X. laevis* adult. The melt of this reassociated DNA is identical to that of 5 S DNA normally isolated from the nucleated erythrocytes pooled from many animals (Brown and Sugimoto, 1973). We conclude that heterogeneity is not introduced into 5 S DNA by using DNA from many animals; it is present in the 5 S DNA of a single frog.

Chromosomal location of 5 S DNA. By in situ hybridization the 5 S DNA has been localized at the telomeres of the long arms of the majority of chromosomes in both species of frogs. At least 15 out of the 18 haploid chromosomes hybridize with cRNA transcribed from 5 S DNA (Pardue et al., 1973). The distribution of these sequences on many chromosomes suggests another source of heterogeneity for 5 S DNA. It is possible that the 5 S DNA repeats on each chromosome are identical to each other but different from those on other chromosomes. This would account for heterogeneity in reassociated duplexes, since a strand from one class would be expected to duplex only rarely with its complement. There is enough similarity between the majority of all 5 S DNA sequences to cause some cross-hybridization between them. The essential question in this regard then is to determine experimentally whether adjacent sequences are always identical but different from those on other DNA molecules. Put another way, are repeats with different nucleotide sequences ever adjacent on the same molecule? The question is unanswered at present but is being studied by a number of electron microscopic techniques. In our laboratory, Dana Carroll is measuring the spacing of the 5 S gene regions along the DNA. In addition, he is studying size classes of DNA fragments produced by restriction enzymes. As will be discussed, the mapping of 5 S DNA heterogeneity is crucial to distinguishing between models that could account for the "horizontal" evolution of the 5 S genes.

Mechanisms of "Horizontal" Evolution

The 450 apparently identical rDNA repeats are adjacent on a single chromosome, and about 24,000 similar but not identical 5 S DNA spacers are located in clusters of hundreds to thousands at the ends of most of the chromosomes. Possible mechanisms for the "horizontal" evolution of each of these kinds of genes will not be detailed here, but they are divided into two groups, each of which makes certain experimental predictions. [See the article by Smith (this volume) which details some of these mechanisms.]

Sudden correction mechanisms. These mechanisms include the "master-slave," "expansion-contraction," and "saltatory expansion." They imply a sudden change in the sequences of a series of adjacent repeats. This change need not happen very often to maintain tandem repeat homogeneity. For example, it could occur once every fifty generations, in which case the event will be very difficult to detect experimentally. The important feature of these mechanisms experimentally is that adjacent sequences will always be identical or more closely related than more distant sequences.

Gradual correction mechanisms. The major mechanism in this category is unequal crossing-over at meiosis between homologues and sister chromatid exchange at meiosis or during mitosis of germ cells. The homogeneity of multiple sequences is related to the number of sequences, the frequency with which exchange occurs, the size of the breeding population, and the mutation rate. If correction occurs by unequal crossing-over, the interesting prediction is made that a family of related, but not identical, sequences need not have their most closely related members in juxtaposition (Smith, this volume). In the case of 5 S DNA, unequal crossing-over would be assumed to occur often enough to maintain the various repeats as a family. However, in contrast to rDNA, the mutation rate within 5 S DNA exceeds the correcting influence of crossing-over so that the multiple sequences are not identical. Experimentally we should be able to show nonidentity of nucleotide sequences or repeat lengths of adjacent repeats and a high rate of exchange between sequences on homologous chromosomes and presumably also on nonhomologous chromosomes. Exchange must play some part in the spread of mutations. At least it presumably accounts for the maintenance of sequence relatedness between 5 S gene loci on nonhomologous chromosomes.

Conclusions

The structures of ribosomal and 5 S DNAs from two species of Xenopus have been summarized. These DNAs represent the first genes of known function to be purified and characterized. The emphasis in this paper has been on the value of these repeating sequences for the understanding of how multiple repeating sequences evolve together as a family. This phenomenon of "horizontal" evolution is being studied in these genes by mapping the location of the heterogeneity within the DNA. Various "correction" theories predict different arrangements of this heterogeneity. The answer is not yet known, but 5 S DNA, which has been shown to contain sequence heterogeneity, is especially suited for such experiments.

References

Birnstiel, M., J. Speirs, I. Purdom, K. Jones, and U. E. Loening. 1968. Properties and composition of the isolated ribosomal DNA satellite of *Xenopus laevis. Nature* **219:** 454.

Birnstiel, M. L., H. Wallace, J. L. Sirlin, and M. Fischberg. 1966. Localization of the ribosomal DNA complements in the nucleolar organizer region of *Xenopus laevis. Nat. Cancer Inst. Monogr.* **23:** 431.

Brown, D. D. and I. B. Dawid. 1968. Specific gene amplification in oocytes. *Science* **160:** 272.

Brown, D. D. and K. Sugimoto. 1973. The 5S DNAs of *Xenopus laevis* and *Xenopus mulleri:* The evolution of a gene family. *J. Mol. Biol.* **78:** 397.

Brown, D. D. and C. S. Weber. 1968a. Gene linkage by RNA-DNA hybridization. I. Unique DNA sequences homologous to 4S RNA, 5S RNA and ribosomal RNA. *J. Mol. Biol.* **34:** 661.

———. 1968b. Gene linkage by RNA-DNA hybridization. II. Arrangement of the redundant gene sequences for 28S and 18S ribosomal RNA. *J. Mol. Biol.* **34:** 681.

Brown, D. D., P. C. Wensink, and E. Jordan. 1971. Purification and some characteristics of 5S DNA from *Xenopus laevis. Proc. Nat. Acad. Sci.* **68:** 3175.

———. 1972. Comparison of the ribosomal DNA's of *Xenopus laevis* and *Xenopus mulleri:* The evolution of tandem genes. *J. Mol. Biol.* **63:** 57.

Brownlee, G. G., E. Cartwright, T. McShane, and R. Williamson. 1972. The nucleotide sequence of somatic 5S RNA from *Xenopus laevis. FEBS Letters* **25:** 8.

Dawid, I. B., D. D. Brown, and R. H. Reeder. 1970. Composition and structure of chromosomal and amplified ribosomal DNA's of *Xenopus laevis. J. Mol. Biol.* **51:** 341.

Denis, H., M. Wegnez, and R. Willem. 1972. Recherches biochimiques sur l'oogenese. V. Comparaison entre le RNA 5S somatique et le RNA 5S des oocytes de *Xenopus laevis. Biochimie* **54:** 1189.

Ford, P. J. and E. M. Southern. 1973. Different sequences for 5S RNA in the kidney cells and ovaries of *Xenopus laevis. Nature New Biol.* **241:** 7.

Gall, J. G. 1969. The genes for ribosomal RNA during oogenesis. *Genetics* (Suppl. 1) **61:** 121.

Laird, C. D., B. L. McConaughy, and B. J. McCarthy. 1969. Rate of fixation of nucleotide substitutions in evolution. *Nature* **224:** 149.

Miller, O. L. and B. R. Beatty. 1969. Visualization of nucleolar genes. *Science* **164:** 955.

Pardue, M. L., D. D. Brown, and M. L. Birnstiel. 1973. Localization of the genes for 5S ribosomal RNA in *Xenopus laevis. Chromosoma.* **42:** 191.

Tymowska, J. and H. R. Kobel. 1972. Karyotype analysis of *Xenopus mulleri* (Peters) and *Xenopus laevis* (Daudin), Pipidae. *Cytogenetics* **11:** 270.

Wallace, H. and M. L. Birnstiel. 1966. Ribosomal cistrons and the nucleolar organizer. *Biochim. Biophys. Acta* **114:** 296.

Wegnez, M., R. Monier, and H. Denis. 1972. Sequence heterogeneity of 5S RNA in *Xenopus laevis. FEBS Letters* **25:** 13.

Wensink, P. C. and D. D. Brown. 1971. Denaturation map of the ribosomal DNA of *Xenopus laevis. J. Mol. Biol.* **60:** 235.

Comparative Aspects of DNA Organization in Metazoa

Eric H. Davidson, Glenn A. Galau, Robert C. Angerer, and Roy J. Britten*

Division of Biology, California Institute of Technology, Pasadena, Calif. 91125, U.S.A.

Abstract. Data on sequence organization in metazoa are reviewed and tabulated. It is shown that the features of sequence organization previously observed in *Xenopus* DNA are extremely widespread. At least 70% of DNA fragments 2,000–3,000 nucleotides long contain both single copy and repetitive sequence in all the organisms examined except *Drosophila*.

Introduction

Transcription level control in animal genomes probably depends on the organization of sequences in the DNA. Study of sequence organization in animal DNA has been stimulated by the discovery of highly ordered patterns of sequence arrangement. Sufficient data are now available to permit significant comparisons across a wide phylogenetic area.

We are now aware of two distinct forms of DNA sequence organization. One of these is typified by the alternating interspersion of repetitive and nonrepetitive sequences demonstrated first for *Xenopus* DNA. For ease of communication we refer to this as "the *Xenopus* pattern". This form of sequence organization is characterized by single copy sequence lengths of about 800 to several thousand nucleotides, with a large fraction of the single copy sequence less than 1,500 nucleotides in length. The single copy sequences are terminated by repetitive sequence elements which themselves are typically only about 300 nucleotides long. Many diverse experimental approaches have been utilized in deriving the *Xenopus* pattern, including hydroxyapatite binding experiments, Sl nuclease and hyperchromicity measurements, kinetic studies and electron microscope observations on renatured DNA (Davidson *et al.*, 1973, 1974; Chamberlin *et al.*, 1975).

A distinctly different pattern of sequence organization has been found in *Drosophila* DNA. Here middle repetitive sequences are of an average length of 5,600 nucleotides, though about 10% (only 2–3% of the total DNA) are 500 nucleotides or less. The most fundamental difference, however, is in the length of single copy sequence elements, which appear to extend for at least 10,000 nucleotides on the average without interruption by repetitive sequences (Manning *et al.*, 1975). These conclusions are based primarily on electron microscope data but are supported as well by hydroxyapatite binding studies which confirm the relative absence of sequence interspersion within a distance less than 2,500 nucleotides. Pearson and Bonner (unpublished data) have carried out hyperchromicity measurements on renatured middle repetitive *Drosophila* DNA, and obtained results in accordance with expectation for repetitive sequences which are much longer than a few hundred nucleotides. Furthermore only long repetitive sequences (>2,000 nucleotides) lacking internal repetition have so far been identified in the experiments of Wensink *et al.* (1975). These workers have clonally replicated and characterized randomly chosen fragments of *Drosophila* DNA. Additional corroboration of the absence or relative scarcity of short interspersed repeats in *Drosophila* DNA derives from Sl nuclease studies carried out in our laboratory (Eden, Davidson, and Britten, unpublished data). An earlier report to the effect that *Drosophila* DNA has essentially the same form of sequence organization as *Xenopus* DNA (Wu *et al.*, 1972), appears to have overestimated the quantitative frequency of short repetitive sequence elements interspersed at distances of 1–2,000 nucleotides. Though it remains possible that such repeats exist in *Drosophila* DNA they would have to be unusually divergent, so as to have escaped

* Also staff member, Carnegie Institution of Washington.

detection by the usual procedures. We are left with the probability that the *Drosophila* and *Xenopus* models of genomic organization are in fact basically different.

Phylogenetic Considerations

A major object of the work described in the preceding paper (Goldberg *et al.*, 1975) has been expansion of present knowledge of sequence organization in order that the phylogenetic distribution of each form of sequence organization may be appreciated. This object has been accomplished, since it is now clear that the *Xenopus* form of sequence organization occurs in most major branches of the phylogenetic tree.

In Fig. 1 we present a phylogenetic tree, showing the evolutionary position of animals whose genomes have been studied. Organisms whose DNA has been subjected to relatively extensive investigation are underlined. For the other organisms listed partial data exist, in each case suggesting the presence of interspersed repetitive sequence elements only a few hundred nucleotides long. References are given in the legend to Fig. 1, as space does not permit a further review of these data here.

The inference suggested by Fig. 1 is that the *Xenopus* pattern of sequence organization stems from a remote evolutionary stage antedating the divergence of the metazoa. Thus the DNA of a coelenterate (*Aurelia*) is organized in the same way as are the DNAs of higher animals. Similarly the genomes of both proto-

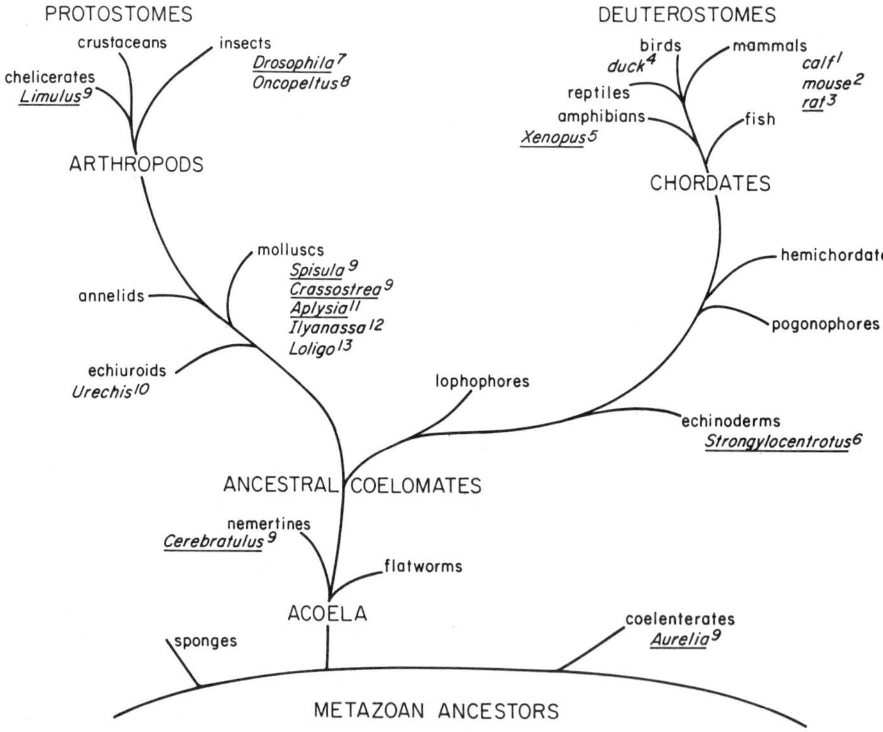

Fig. 1. Phylogenetic tree, essentially after Hyman (1940), showing positions of organisms for which information on genomic sequence organization exists. DNAs studied at least as extensively as those in the accompanying paper (*9*) are underlined. Only generic or common names are given. References: *1* Britten and Smith, 1970; Britten, unpublished data. *2* Rice, 1971; Britten, 1972. *3* Rice, 1971; Bonner *et al.*, 1974; Pearson, Wilkes and Bonner, unpublished data. *4* Bishop and Freeman, 1974; Bishop, personal communication. *5* Davidson *et al.*, 1973; Chamberlin *et al.*, 1975. *6* Graham *et al.*, 1974. *7* Manning *et al.*, 1975; Wu *et al.*, 1972; Wensink *et al.*, 1975. *8* Lagowski *et al.*, 1973. *9* Goldberg *et al.*, 1975. *10* Neufeld, Smith, Davidson, and Britten, unpublished data. *11* Angerer, Davidson, and Britten, in preparation. *12* Davidson *et al.*, 1971. *13* Galau, unpublished data

stomial and deuterostomial animals, including an extremely primitive acoelomate protostome (*Cerebratulus*), are organized along the lines of the *Xenopus* pattern. The protostomial and deuterostomial evolutionary lines diverged before the beginning of the Cambrian fossil record. A possible interpretation is that interspersed sequence organization of the *Xenopus* type provided part of the basis for the evolution of multicellular forms. This would be true, for example, if sequence interspersion is required for the regulation of gene activity which underlies cell differentiation, as suggested earlier (Britten and Davidson, 1969, 1971).

Summary of Sequence Organization Data and Some Generalizations

Table 1 collates key data from sequence organization studies on organisms whose generic names are underlined in Fig. 1. Several general conclusions can be drawn from this table:

1. Genomic Complexity. There is no simple correlation between the sequence complexity of the genome and the apparent biological complexity of the animal. We have made this point earlier (Britten and Davidson, 1971) in comparing complexities of amphibian genomes of diverse sizes. However it is now clear that the true genomic sequence complexity cannot be appreciated until the sequence interspersion pattern of the DNA is known so that the actual content of single copy sequence (rather than the fraction of fragments bearing only single copy sequence) can be measured. As can be seen in Table 1 the fraction of the genome which is single copy ranges around 70%. However, the complexity of the oyster genome, 3.8×10^8 nucleotide pairs, is only 46% of that of the surf clam genome, 8.2×10^8 nucleotide pairs, yet both belong to the same molluscan class Bivalvia. The complexity of the *Aurelia* genome is almost six times that of *Drosophila*, and yet *Aurelia* totally lacks organ systems. The complexity of the *Cerebratulus* genome is greater than that of *Aplysia* (a gastropod mollusc), or *Drosophila*, and yet the acoelomate nemertean would appear to be a less biologically complex animal in terms of construction, organ systems, developmental pathways and capabilities. Single copy complexity is thus to be regarded as a measure of the *potential*, rather than the actual (*i.e.*, utilized) genomic information content.

2. Frequency of Occurrence of Repetitive Sequences. A striking fact emerging from Table 1 is that most of the organisms listed possess a repetitive sequence class in which the sequences are present between 30 and 200 times per genome. Such components may also exist in the organisms which are apparent exceptions. In addition most genomes include sequences present 1,000–4,000 times. Much more highly repetitive frequency classes such as those found in some mammalian genomes, for example the 66,000 repeat class studied extensively in cow DNA by Britten and Smith (1970), occur only sporadically, or if present generally are in very small quantity. Low and moderate repetition frequency sequences could thus be regarded as a basic requirement though present data suggest that some exceptions such as cow DNA, may exist. There is also evidence for the interspersion with single copy DNA of either or both classes of repetitive sequence in various genomes. The usual interspersion experiments tend to detect only the more highly repetitive interspersed sequence since if less repetitive sequence elements are present on the same fragments they will be carried along at the faster rate.

3. Repetitive Sequence Length. Though all organisms listed contain some repetitive sequence in the 200–400 nucleotide range the proportion varies sharply. Except for *Drosophila* all show a peak in the repetitive sequence length distribution at about 300 nucleotides. The quantity of short repetitive sequence in the genome always suffices to account for the amount of interspersion of single copy DNA observed, as pointed out in the previous paper (Goldberg *et al.*, 1975). The longer repetitive sequences terminate eventually and thus are also interspersed with single copy DNA providing one considers a great enough length. However it is obvious that the far more numerous 300 nucleotide repetitive sequences account for most of the interspersed repeats. The fraction of the repetitive DNA in long middle repetitive sequence is not correlated in any simple way with the

Table 1. Approximate parameters of sequence organization for ten animals DNAs

1 Organism	2 Genome size (pg per haploid genome)	3 Single copy fraction of DNA	4 Single copy complexity (nucleotide pairs)	5 Repetitive sequence classes (average occurrence of each sequence per haploid genome)	6 Are repetitive sequences interspersed at length given in column 9 ?	7 Fraction of repetitive DNA in 200–400 nucleotide elements	8 Fraction of single copy sequence interspersed with repetitive sequence at length given in column 9	9 Fragment length (nucleotides)	10 Reference
Spisula solidissima	1.2	0.75	8.2×10^8	30 3,700	yes ?	0.60	>0.70	2,300	Goldberg *et al.*, 1975
Crassostrea virginica	0.69	0.60	3.8×10^8	40	yes	0.35	>0.75	3,000	Goldberg *et al.*, 1975
Aplysia californica	1.8	~0.40	10.7×10^8	85 4,600	? yes	0.60	>0.80	2,500	Angerer, Davidson, and Britten, in prep.
Limulus polyphemus	2.8	0.70	17.9×10^8	50 ~2,000	yes yes	0.75	>0.70	2,000	Goldberg *et al.*, 1975
Cerebratulus lacteus	1.4	0.60	7.7×10^8	40 1,000	yes probably	0.55	>0.70	2,800	Goldberg *et al.*, 1975
Aurelia aurita	0.73	0.70	4.7×10^8	180	yes	0.60	>0.80	2,000	Goldberg *et al.*, 1975
Strongylocentrotus purpuratus	0.89	0.75	6.1×10^8	100 1,500	? yes	0.75	0.7	3,300	Graham *et al.*, 1974
Xenopus laevis	2.7	0.75	18.5×10^8	100 2,000	? yes	0.75	0.7	3,700	Davidson *et al.*, 1973, 1974
Rattus norvegicus	3.2	0.75	22.3×10^8	low frequency ? 1,800	? yes	?	>0.65	3,200	Bonner *et al.*, 1974; Holmes and Bonner, 1974; Pearson, Wilkes and Bonner, unpublished data
Drosophila melanogaster	0.12	0.75	0.82×10^8	35	no	0.1	none observed	2,500	Manning *et al.*, 1975

genome size. For example in the small oyster genome about 65% of the repetitive sequence is found in long sequence elements, while in *Spisula* DNA, which has a genome almost twice as large, only 40% of the repetitive sequences are long. However, the absolute quantity of short interspersed repetitive sequence appears to be correlated with the quantity (complexity) of single copy DNA.

4. Correlation between Mismatch and Repetitive Sequence Length. In the DNAs of *Xenopus* (Davidson *et al.*, 1974), *Spisula* (Goldberg *et al.*, 1975), calf (Britten, unpublished data), and sea urchin (Britten, Graham, Eden, and Davidson, in preparation), it has been found that longer repetitive sequences melt with higher T_m's. This is true even after correction for the effect of duplex length on melting temperature. We conclude that there is a correlation, as yet not understood, between repetitive sequence organization and sequence divergence during evolution.

5. Length of Interspersed Single Copy Sequence Elements. As Table 1 shows, in all of the organisms studied (except *Drosophila*) over 70% of the single copy DNA is included in sequence elements of 3,000 nucleotides or less, terminated by repetitive sequence. In no case are we able to show that all the non-repetitive sequence is included in this category, though of course it is possible that this is actually the case and technical problems have simply interfered with the demonstration. Such problems might include strand scission, and interspersion of very low repetition frequency sequences or of highly divergent repetitive sequences. We have recently reported that in the sea urchin genome the single copy sequence contiguous to the interspersed repetitive sequence includes most or all of those structural genes which are active in embryogenesis (Davidson *et al.*, 1975). This may also be true for other organisms whose DNA is organized similarly. The observed single copy sequence element length distribution would fit well with this view, since a small fraction of structural genes are longer than 3,000 nucleotides while most are shorter (the modal size is probably about 1,200 nucleotides). Rationalization of sequence organization in such terms, however, awaits another crucial item of evidence. This is knowledge of the fraction of the total single copy sequence which is present in functional structural genes.

Acknowledgements. This work was supported by NIH Training Grant No. HD-00014 to the Marine Biological Laboratory's Embryology Course and by NIH Grant No. GM-20927. R.C.A. holds a postdoctoral fellowship from the American Cancer Society (No. PF-925).

References

Bishop, J. O., Freeman, K. B.: DNA sequences neighboring the duck hemoglobin genes. Cold Spr. Harb. Symp. quant. Biol. **38**, 707–716 (1974)

Bonner, J., Garrard, W. T., Gottesfeld, J., Holmes, D. S., Sevall, J. S., Wilkes, M.: Functional organization of the mammalian genome. Cold Spr. Harb. Symp. quant. Biol. **38**, 303–310 (1974)

Britten, R. J.: DNA sequence interspersion and a speculation about evolution. In: Evolution of genetic systems (H. H. Smith, ed.), p. 80–94. New York: Gordon and Breach 1972

Britten, R. J., Davidson, E. H.: Gene regulation for higher cells: a theory. Science **165**, 349–357 (1969)

Britten, R. J., Davidson, E. H.: Repetitive and nonrepetitive DNA sequences and a speculation on the origins of evolutionary novelty. Quart. Rev. Biol. **46**, 111–138 (1971)

Britten, R. J., Smith, J.: A bovine genome. Carnegie Inst. Yearbook **68**, 378–386 (1970)

Chamberlin, M. E., Britten, R. J., Davidson, E. H.: Sequence organization in Xenopus DNA studied by the electron microscope. J. molec. Biol. (in press, 1975)

Davidson, E. H., Graham, D. E., Neufeld, B. R., Chamberlin, M. E., Amenson, C. S., Hough, B. R., Britten, R. J.: Arrangement and characterization of repetitive sequence elements in animal DNAs. Cold Spr. Harb. Symp. quant. Biol. **38**, 295–301 (1974)

Davidson, E. H., Hough, B. R., Amenson, C. S., Britten, R. J.: General interspersion of repetitive with nonrepetitive sequence elements in the DNA of Xenopus. J. molec. Biol. **77**, 1–23 (1973)

Davidson, E. H., Hough, B. R., Chamberlin, M., Britten, R. J.: Sequence repetition in the DNA of Nassaria (Ilyanassa) obsoleta. Develop. Biol. **25**, 445–463 (1971)

Davidson, E. H., Hough, B. R., Klein, W. H., Britten, R. J.: Structural genes adjacent to interspersed repetitive DNA sequences. Cell **4**, 217–238 (1975)

Goldberg, R. B., Crain, W. R., Ruderman, J. V., Moore, G. P., Barnett, T. R., Higgins, R. C.,

Gelfand, R. A., Galau, G. A., Britten, R. J., Davidson, E. H.: Sequence organization in the genomes of five marine invertebrates. Chromosoma (Berl.) **51**, 225–251 (1975)
Graham, D. E., Neufeld, B. R., Davidson, E. H., Britten, R. J.: Interspersion of repetitive and nonrepetitive DNA sequences in the sea urchin genome. Cell **1**, 127–137 (1974)
Holmes, D. S., Bonner, J.: Sequence composition of rat nuclear DNA and high molecular weight nuclear RNA. Biochemistry **13**, 841–848 (1974)
Hyman, L. H.: The invertebrates: Protozoa through Ctenophora, vol. 1, ch. 2. New York: McGraw-Hill Book Co. 1940
Lagowski, J. M., Yu, M. Y. W., Forrest, H. S., Laird, C. D.: Dispersity of repeat DNA sequences in Oncopeltus fasciatus, an organism with diffuse centromeres. Chromosoma (Berl.) **43**, 349–373 (1973)
Manning, J. E., Schmid, C. W., Davidson, N.: Interspersion of repetitive and nonrepetitive DNA sequences in the Drosophila melanogaster genome. Cell **4**, 141–155 (1975)
Rice, N.: Thermal stability of reassociated repeated DNA from rodents. Carnegie Inst. Yearbook **69**, 472–479 (1971)
Wensink, P. C., Finnegan, D. J., Donelson, J. E., Hogness, D. S.: A system for mapping DNA sequences in the chromosomes of Drosophila melanogaster. Cell **3**, 315–325 (1975)
Wu, J.-R., Hurn, J., Bonner, J.: Size and distribution of the repetitive segments of the Drosophila genome. J. molec. Biol. **64**, 211–219 (1972)

Received and accepted April 11, 1975 by J. G. Gall
Ready for press April 17, 1975

Repeated Sequence DNA Relationships in Four Cereal Genomes

Richard B. Flavell, Jürgen Rimpau[1] and Derek B. Smith

Department of Cytogenetics, Plant Breeding Institute, Trumpington, Cambridge CB2 2LQ, England; [1]present address: Institut für Pflanzenzüchtung der Universität Göttingen, D-3400 Göttingen, Federal Republic of Germany

Abstract. The effect of DNA fragment size on the extent of hybridisation that occurs between repeated sequence DNAs from oats, barley, wheat and rye has been investigated. The extent of hybridisation is very dependent on fragment size, at least over the range of 200 to 1000 nucleotides. This is because only a fraction of each fragment forms duplex DNA during renaturation. From these results estimates of the proportions of repeated sequences of each of the cereal genomes that are homologous with repeated sequences in the other species have been determined and a phylogenetic tree of cereal evolution constructed on the basis of the repeated sequence DNA homologies. It is proposed that wheat and rye diverged after their common ancestor had diverged from the ancestor of barley. This was preceded by the divergence of the common ancestor of wheat, rye and barley and the ancestor of oats. Once introduced in Gramineae evolution most families of repeated sequences appear to have been maintained in all subsequently diverging species. – The repeated sequences of oats, barley, wheat and rye have been divided into Groups based upon their presence or absence in different species. Repeated sequences of related families are more closely related to one another within a species than between species. It is suggested that this is because repeated sequences have been involved in many rounds of amplification or quantitative change via unequal crossing over during species divergence in cereal evolution.

Introduction

The chromosomes of higher organisms contain repeated nucleotide sequences (Britten and Kohne, 1968). These kinds of sequences can account for a considerable proportion of the total nuclear DNA, especially in plants and amphibians with higher nuclear DNA contents (Flavell et al., 1974; Straus, 1971). The repeated sequence complements of genomes can change considerably during evolution and there are often large differences between the repeated sequences of related species (Rice, 1972; Goldberg et al., 1972; Rice and Esposito, 1973; Straus, 1972; Mizuno et al., 1976). They are therefore useful in determining taxonomic and phylogenetic relationships of species.

In earlier DNA/DNA hybridisation studies Bendich and McCarthy (1970) and Smith and Flavell (1974) investigated the relationships between the repeated sequence DNA complements of the wheat, barley, oats and rye genomes. However, these studies and virtually all those in other laboratories on other species (e.g. Goldberg et al., 1972; Rice and Esposito, 1973; Stein and Thompson, 1975; Chooi, 1971; Rice, 1972) have failed to consider the effect of fine interspersion in the genome of relatively short repeated sequences common to two species, species-specific repeated sequences and non-repeated sequences. These genome organisation patterns would lead to overestimates of repeated sequence homologies between species unless extremely small DNA fragments were used. We have therefore studied the effect of fragment size on estimates of repeated sequence DNA homology between the cereal genomes and have provided better quantitative estimates of these repeated sequence homologies. Furthermore, from studies on related repeated sequences within and between species we have speculated on the evolution of the families of repeated sequences in the present-day cereal genomes.

Materials and Methods

a) Isolation of Unlabelled and Tritium Labelled DNAs. Unlabelled DNAs were isolated from young leaves from many plants of *Triticum aestivum* (wheat), variety "Chinese Spring", *Secale cereale* (rye) variety "Petkus", *Hordeum vulgare* (barley), variety "Sultan" and from *Avena sativa* (oats) variety "Maris Titan". The extraction and purification procedures were as described in detail previously (Smith and Flavell, 1974; Smith and Flavell, 1975). At least three different preparations of DNA from each species were used in the work reported here. The DNA preparations had $(OD_{260}/OD_{230}) \geqq 2.0$ and $(OD_{260}/OD_{280}) \geqq 1.9$ and more than 98% of their OD_{260} material was retained on hydroxyapatite at 60° C in 0.12 M phosphate buffer.

Tritium labelled DNAs from the four species were extracted from 3 day old seedlings germinated under sterile conditions in the presence of ^3H thymidine (CH_3 labelled, 52 Ci/m mol). At least two preparations from each species were used in this work. The purification method was exactly as described previously (Smith and Flavell, 1974). The specific activities of the four DNAs determined after drying on glass fibre filters, were between 28,500 cpm/μg and 51,400 cpm/μg.

b) Fractionation of ^3H Labelled DNAs and Fragment Size Determinations of Labelled and Unlabelled DNAs. The tritium labelled DNAs were sedimented after brief sonication in preparative 5 to 11% (w/w) linear sucrose gradients in 0.1 N NaOH for 19 and 40 h at 24,000 rpm (80,000 × g) at 20° C. The gradients were fractionated and the fractions used as sources of DNA of different fragment sizes. The weight average denatured DNA fragment lengths in each fraction were subsequently determined by sedimentation in identical gradients together with unlabelled markers as described by Burgi and Hershey (1963). The marker DNAs were sonicated wheat or calf thymus DNAs and their weight average denatured fragment lengths were determined by boundary velocity sedimentation using an MSE Centriscan analytical ultracentrifuge. The order of fractions in the original gradient agreed with the order of sedimentation coefficients subsequently determined for each fraction separately. The values used for average fragment sizes in each fraction were taken from the best-fit line through a plot of fraction number against sedimentation coefficient.

All unlabelled DNA average denatured fragment lengths were determined by sedimentation in 0.9 M NaCl 0.1 N NaOH in the analytical ultracentrifuge using the equation of Studier (1965).

c) Renaturation in vitro and Hydroxyapatite Chromatography of Renaturation Products. For the determination of unlabelled DNA renaturation kinetics (see Fig. 1), different concentrations of unlabelled DNAs were denatured by boiling at 100° C for 5 min and incubated at 60° C in 0.12 M phosphate buffer (0.18 M Na^+) for different times. The renatured DNAs were estimated quantitatively by hydroxyapatite chromatography (see below).

In most experiments involving duplex formation between labelled and unlabelled DNA fragments, ^3H labelled DNAs of appropriate average fragment size were added to a large excess (ratio > 1:8000) of unlabelled DNA (see details of individual experiments in Results) previously sonicated to an average fragment size of 300 to 400 nucleotides. DNAs in 0.12 M phosphate buffer were denatured by boiling for 5–10 minutes before incubation at 50 or 60° C to the appropriate C_0t value. (C_0t = DNA concentration in moles nucleotide per litre × incubation time in sec). The samples were then diluted with 0.12 M phosphate buffer and applied to hydroxyapatite columns (BioRad H.T.P.) previously equilibrated to either 50° or 56° C. DNA eluting at these temperatures was considered denatured and DNA in the renatured fraction was recovered by elution with the same buffer at 95° C.

DNAs in 62% formamide 0.69 M NaCl were denatured by heating to 85° C for 5 min and incubated at 42.5° C to the chosen C_0t value. These conditions are approximately equivalent to 80° C in 0.12 M phosphate buffer (McConaughty et al., 1969). Renaturation was terminated by rapidly cooling the samples and diluting them with cold phosphate buffer to a final concentration of 0.03 M and a formamide concentration of < 5%. The samples were then passed over hydroxyapatite columns equilibrated to 0.03 M phosphate buffer at room temperature. After washing the columns with 0.03 M and 0.12 M phosphate buffers at room temperature the remaining DNA was eluted with 0.12 M phosphate buffer at 80° C and 95° C. The DNA eluting at room temperature and 80° C was considered denatured and that eluting at 95° C renatured. Control experiments have shown that hydroxyapatite retains well matched duplexes at 80° C as well as at 60° C (Flavell and Smith, 1976).

Unlabelled DNA concentrations in the eluates were determined from OD_{260} after brief centrifugation to sediment any hydroxyapatite. They were corrected for any OD_{260} appearing in eluates from columns to which no DNA had been added. Labelled DNA was estimated by precipitating the DNA with trichloracetic acid (final concentration of 5%) in the cold in the presence of 100 μg bovine serum albumin and collection of the DNA on glass fibre filters as described elsewhere (Smith and Flavell, 1974; Flavell and Smith, 1976). The recovery and quantitative estimations of labelled DNA by precipitation and collection on glass fibre filters before counting in a scintillation mixture are not dependent on the DNA fragment size over the range used in this work (Flavell and Smith, 1976).

d) DNA Melting Curves. The duplex DNAs were adsorbed on to a hydroxyapatite column in 0.12 M phosphate buffer at 56° C. The temperature was raised in 5° C steps up to 100° C. At each temperature any melted DNA was eluted with three, 3 ml washes of 0.12 M phosphate buffer. The total DNA or labelled DNA content of each eluate was determined and the temperature (T_m) at which 50% of the duplexes under study had been eluted from the column was calculated.

Results

1. Renaturation Kinetics of Cereal DNAs

The renaturation kinetics of denatured wheat, oats, barley and rye DNAs were determined at 60° C in 0.12 M phosphate buffer using DNA fragments with a mean length of around 300 nucleotides. The formation of double stranded DNA was determined by hydroxyapatite chromatography. The renaturation "C_0t curves" (Britten and Kohne, 1968) for all four DNAs were very similar (unpublished results). The curve for wheat is shown in Figure 1 as an example. The kinetics of wheat DNA renaturation have been presented in more detail in a previous paper (Smith and Flavell, 1975).

After incubation of denatured DNAs to C_0t values below 10^{-3} less than 10% of the DNA was double stranded (Fig. 1) and some of this is due to extremely rapid intrastrand reassociation (Wilson and Thomas, 1974; Bonner, 1974; Smith and Flavell, 1975; Flavell and Smith, 1976; Huguet et al., 1975). We have previously reported detailed studies on this very rapidly reannealing fraction in wheat (Smith and Flavell, 1975; Flavell and Smith, 1976) and rye (Smith and Flavell, 1977) and how the proportion of the genome included in this fraction increases with increasing average DNA fragment size.

Most of the DNA of the cereal genomes reanneals between C_0t 10^{-3} and C_0t 50 (Fig. 1). After C_0t 50 very little more DNA reanneals until after C_0t 500 where single or few copy sequences would be expected to reanneal. The presence of single or very few copy sequences per haploid genome has been demonstrated for wheat (Smith and Flavell, 1975). Thus the 70% or more of each of the cereal genomes renaturing between C_0t 10^{-2} and C_0t 10^2 consists mostly of repeated DNA sequences. The renaturation curves for these fractions are not second order.

The marked kinetic distinction between the renaturation of repeated and non-repeated sequences in the cereal genomes (see Fig. 1 and Smith and Flavell, 1974, 1975) enables the repeated sequences of these species to be separated from most of the non-repeated sequences (Smith and Flavell, 1974, 1975) and facilitates comparison of essentially the whole complement of repeated sequences in interspecies hybridisation experiments.

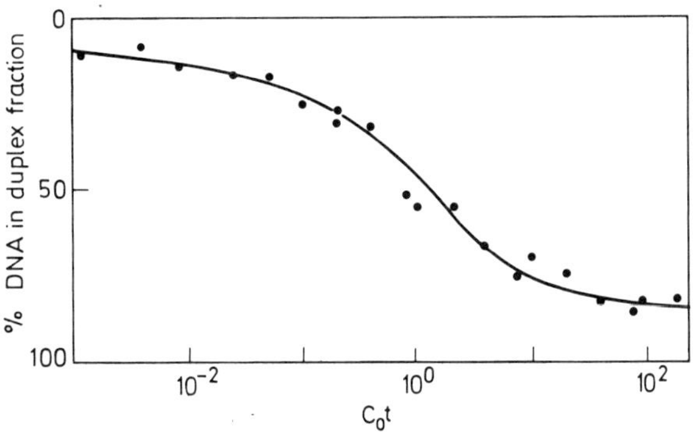

Fig. 1. Renaturation kinetics of wheat DNA. Wheat DNA with an average single stranded fragment size of 300 nucleotides at 50 and 500 μg/ml was denatured by boiling and incubated at 60° C in 0.12 M phosphate buffer. Samples were withdrawn at appropriate C_0t intervals and the proportion of renatured DNA determined by hydroxyapatite chromatography

2. *The Extent of Heteroduplex Formation between Cereal DNAs*

a) The Effect of Fragment Size

Tritium labelled DNA of each species (wheat, barley, rye and oats) was incubated with an 8000 or more fold excess of unlabelled DNA of each of the other species in turn at 60° C in 0.12 M phosphate buffer to C_0t 120. This excess of unlabelled DNA ensures that essentially all labelled DNA incorporated into duplex DNA is hybridised to an unlabelled fragment even when only a small proportion of the labelled DNA is able to hybridise with the unlabelled DNA. Incubation to C_0t 120 is sufficient to allow all repeated sequences in cereal genomes to be included in the duplex fraction (see Fig. 1) and also allows for the slower renaturation of related but more diverged sequences in interspecies hybridisations (Bonner, 1973). A range of labelled DNA fragment sizes were used while the unlabelled DNA sizes were kept constant. The proportions of labelled DNA of each species and of each fragment size in the duplex fractions were determined by hydroxyapatite chromatography and are illustrated in Figure 2. All results have been corrected for the very rapidly reannealing DNA which may renature in a monomolecular reaction of intrastrand reassociation (see legend to Fig. 2 and Flavell and Smith, 1976; Smith and Flavell, 1977).

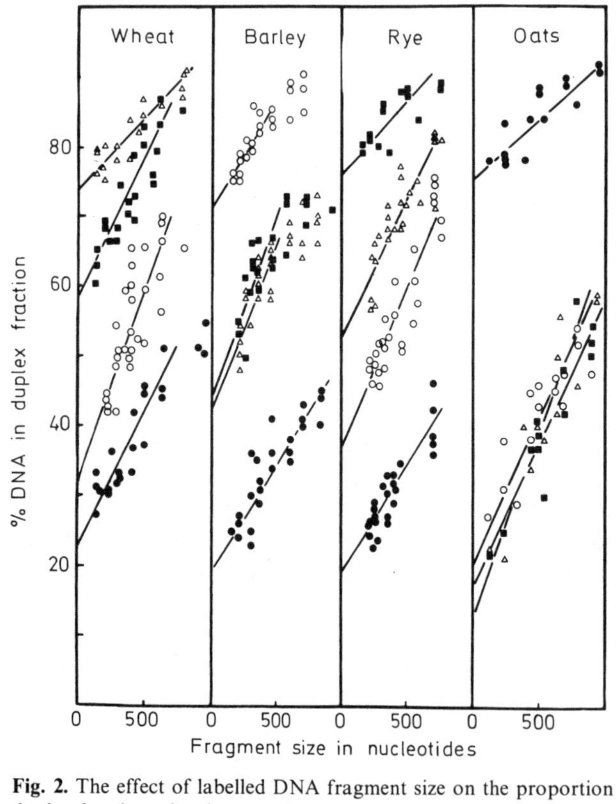

Fig. 2. The effect of labelled DNA fragment size on the proportion of DNA in the hydroxyapatite duplex fraction after intraspecies and interspecies DNA hybridisation. Aliquots of labelled wheat, barley, rye and oats DNAs of various average fragment sizes were mixed with unlabelled wheat, rye, barley or oats DNA sheared to an average single stranded fragment size of 300 to 400 nucleotides. The concentration of unlabelled DNA was 500 µg/ml in 0.12 M phosphate buffer and the ratio of unlabelled to labelled DNA exceeded 8000:1 in every case. After denaturation and incubation to a C_0t of approximately 120 the DNAs were fractionated on hydroxyapatite. The proportion of the labelled DNA in the duplex fraction was determined in each DNA mixture. The labelled DNAs used are indicated at the top of the appropriate section of the figure. The unlabelled DNAs are as follows: (△) wheat, (■) rye, (○) barley, (●) oats. The straight-line extrapolations are drawn to primarily fit the points for fragment sizes shorter than 500 nucleotides since in some experiments (e.g. with labelled barley DNA), reductions in slope commence around 500 nucleotides reaching a plateau with fragments over 1000 nucleotides (Rimpau et al., in preparation)

In all combinations of DNAs the proportion of labelled DNA in the duplex fraction increased markedly with increasing fragment size, especially in the interspecies combinations. This clearly demonstrates that DNA/DNA hybridisation values within and between these species depend considerably on fragment size.

The scatter of points at each fragment size in Figure 2 is not atypical for these kinds of experiments (see for example Davidson et al., 1973). The proportion of labelled DNA retained in the duplex fraction on hydroxyapatite is sensitive to, amongst other things, the extent of DNA strand scission during denaturation, incubation and chromatography, the batch properties of the hydroxyapatite, and variation in counting efficiency of TCA precipitated DNA on filters. The points within each curve in Figure 2 were combined from several experiments carried out many months apart.

The increasing proportion of DNA in the duplex fraction with increasing labelled fragment size in Figure 2 could be due to (1) some imperfect duplexes being too short with small fragments to bind to hydroxyapatite and/or (2) longer labelled fragments associating with two short unlabelled DNA fragments, thereby forming sufficiently long regions of duplex to bind to hydroxyapatite. In a series of experiments reported in detail elsewhere (Smith et al., 1976) we have found no evidence to suggest that either of these possibilities is responsible for much of the increased labelled DNA in the duplex fraction with longer fragment lengths. The additional DNA in the duplex fraction when long labelled fragments are used, is not renatured DNA but single stranded. Some small single stranded "tails" may remain on duplexes because the duplex structures interfere with renaturation of the single stranded tails (Smith et al., 1976; Britten and Davidson, 1976). However, most of the single stranded tail DNA, especially when long labelled fragments are involved, is single or few copy DNA which is unable to reanneal at C_0t 100 and in the interspecies hybridisations, repeated sequences which are not amplified in both DNAs (Smith et al., 1976).

Since duplexes formed from randomly sheared fragments frequently have single stranded tails, better estimates of the proportions of the labelled DNAs in the renatured conformation come from extrapolations of the curves in Figure 2 to the ordinate (Davidson et al., 1973). From the intraspecies hybridisations to C_0t 120 these extrapolations provide estimates of the proportion of repeated sequences in the genome (Davidson et al., 1973; Flavell and Smith, 1976) while in the interspecies hybridisations, they provide estimates of the repeated sequences in common between the species (Smith et al., 1976). The extrapolated hybridisation percentages are summarised in Table 1 for all the wheat, rye, oats and barley hybridisations.

b) The Effect of the Renaturation Temperature

It is now well-established that the extent to which duplexes form in vitro is dependent upon the salt concentration and temperature during incubation (McCarthy and Farquhar, 1972). We investigated the effect of temperature on the extent of hybridisation between cereal DNAs by carrying out the renaturation of labelled wheat fragments, of various average sizes, to unlabelled wheat,

Table 1. Percentage homologies between cereal DNAs
Percentages were obtained from extrapolations to the ordinates in Figure 2

Labelled DNA	Unlabelled DNA			
	Wheat	Rye	Barley	Oats
Wheat	74	58	32	22
Rye	52	74	38	19
Barley	42	44	71	20
Oats	14	17	20	75

barley, rye and oats DNAs at 50° C in 0.18 M Na^+ instead of 60° C in 0.18 M Na^+ as described above. The percentage labelled DNA in the duplex fractions was between 7 and 10% higher at 50° C than at 60° C after renaturation and fractionation of short fragments. This is due to more partially homologous repeated sequences being able to form stable duplexes at 50° C. These results, to be presented in detail elsewhere (Rimpau et al., in preparation), illustrate that estimates of percentage homology between different DNAs are dependent upon the incubation conditions under which renaturation occurs.

3. Repeated Sequences and Cereal Evolution

After establishing the proportions of repeated sequences in each genome that can renature with repeated sequences of the other genomes at 60° C in 0.12 M phosphate buffer, we investigated for each genome the relationship between the smaller proportion of sequences hybridising to one species and the larger proportion hybridising to another species. We wished to test, for example, whether the wheat repeated sequences homologous to repeated sequences in oats are also included in the wheat DNA homologous to barley and rye DNAs.

Labelled wheat, rye, barley and oats DNAs, with average fragment sizes of 400, 350, 360 and 345 nucleotides respectively were therefore mixed with combinations of unlabelled DNAs and 1:1 mixtures of two unlabelled DNAs. The specific hybridisation combinations chosen and the results for duplicate experiments are shown in Table 2.

In the hybridisation comparisons involving labelled wheat and rye DNAs, in no case was the percentage labelled DNA hybridisation greater to the mixture of unlabelled DNAs than to the component of the mixture with which the labelled DNA has higher homology (see Table 2). Thus the sequences in wheat which hybridise to barley DNA include those which hybridise to oats DNA. Similarly the sequences which hybridise to tye include those which hybridise to barley DNA. In the rye genome, those sequences which hybridise to barley DNA also include those which hybridise to oats DNA and those sequences which hybridise to wheat DNA include those which hybridise to barley DNA.

The results for the labelled barley and oats DNAs suggest that small proportions of sequences in these DNAs that hybridise to each other may not be amplified in the wheat and rye genomes. However, these proportions are small compared with the total heteroduplex formed between these DNAs. Thus it is also reasonable to conclude that most of the sequences in the barley genome that hybridise to wheat and rye DNAs include those which hybridise to oats

Table 2. Hybridisation of labelled cereal DNAs to mixtures of unlabelled DNAs
Labelled wheat, rye, barley and oats DNAs, with mean fragment lengths between 345 and 400 nucleotides were mixed with unlabelled DNAs (mean fragment lengths 340 to 450 nucleotides) in the combinations shown such that the labelled: unlabelled DNA ratios were in excess of 1:6250. The DNAs were denatured and incubated in 0.12 M phosphate buffer at 60° C to a C_0t of 90. The proportions of labelled DNA in the hydroxyapatite duplex fractions are shown in the table for two separate experiments. B = Barley, O = Oats, R = Rye and W = Wheat

Labelled DNA		Percentage hybridisation to DNAs from									
		W	R	B	O	R+B	R+O	W+B	W+O	B+O	W+R
Wheat	(i)	82.0	75.0	61.0	–	74.5	–	–	–	53.1	79.3
	(ii)	79.5	80.6	59.8	–	75.7	–	–	–	50.2	78.0
Rye	(i)	71.9	87.4	50.7	–	–	–	73.7	–	48.4	84.6
	(ii)	68.9	–	57.6	–	–	–	71.9	–	53.2	84.1
Barley	(i)	55.6	51.1	84.2	–	79.4	58.7	73.3	59.5	–	–
	(ii)	56.1	50.5	81.1	–	83.2	62.1	82.4	59.1	–	–
Oats	(i)	27.3	29.5	36.0	76.8	36.6	80.9	31.7	–	–	–
	(ii)	28.2	28.5	29.5	81.5	34.7	81.5	32.3	–	–	–

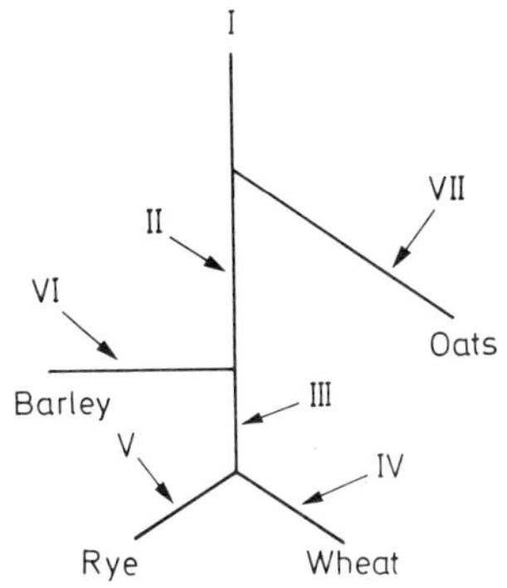

Fig. 3. The evolution of cereal species based on repeated sequence homology. The scheme is drawn approximately to scale, the distances between species representing the proportion of repeated sequences that are unable to form stable hybrids at 60° C in 0.18 M Na^+. These values are derived from table 1 and each value is the mean of the two reciprocal hybridisations involving each pair of species. For example, the value used for the relationship between oats and wheat is the mean of 74–22 and 75–14 = 56.5 The Roman numbers indicate different Groups of repeated sequences distinguished by the interval in which they first became amplified in the lineages of cereal evolution

DNA and most of the sequences in oats DNA which hybridise to barley DNA are the same sequences which hybridise to wheat and rye DNAs.

These results provide important information for the construction of a phylogenetic scheme of the evolution of cereal DNAs and for the division of the repeated sequences in cereal genomes into different groups.

A phylogenetic scheme for the evolution of cereal genomes based on percentage heteroduplex formation summarised in Table 1 is shown in Figure 3. The scheme is drawn approximately to scale with the distance between any two species representing the mean proportion of repeated sequence DNA not sufficiently homologous to form heteroduplexes at 60° C in 0.18 M Na^+. These values were derived from Table 1 as described in the legend to Figure 3. The scheme substantiates the evolutionary tree derived from genetic and taxonomic considerations (Bell, 1965).

After each branch point, new repeated sequences are present and these account for the increased sequence homology between more closely related species. When families of repeated sequences introduced during the same internode are combined into a Group, seven different Groups of repeated sequences can be recognised in the four cereal species, as illustrated in Figure 3.

The conclusions derived from Table 2 indicate that for most families of repeated sequences once they have been introduced into the cereal lineage, they are present in all species subsequently emerging from the lineage. Therefore Group I repeated sequences are present in oats, barley, wheat and rye. Group II sequences are in barley, wheat and rye but not in oats. Group III sequences are in wheat and rye but not in barley or oats and sequences in Groups IV, V, VI and VII are specific to the wheat, rye, barley and oats genomes respectively. The proportions of each of the Groups of sequences in the four cereal genomes can therefore be derived from the extrapolations to the ordinate in Figure 2 (see Table 1 and the legend to Table 3) and are given in Table 3 together with the absolute amounts of DNA in picograms per 1C genome in each of the Groups.

4. Sequence Heterogeneity in Groups of Repeated Sequences Within and Between Species

Estimates of heterogeneity among families of related sequences can be gained from assaying the thermal stabilities of duplexes formed in vitro at random between the related sequences. The thermal stabilities of the duplexes formed between tritium labelled cereal DNAs and large excesses of unlabelled homologous DNAs have been published previously (Smith and Flavell, 1974).

Renatured cereal DNAs have much broader melting profiles than native DNAs, melting at reduced temperatures. The mean melting temperatures (T_m) of the renatured repeated sequences of cereal DNAs are approximately 10° lower than that of the native DNAs. These results suggest that many if not all the families of repeated sequences in the cereal genomes are heterogeneous.

Table 3. Division of cereal repeated sequence DNAs into Groups, their proportions and amounts in each genome

The Groups I to VII are defined in the text. The proportions of each genome in each Group are derived from the ordinate intercept values in figure 2. For example, the intercept value when labelled wheat was hybridised to unlabelled oats DNA is 22%. Group I in wheat is therefore 22%. The intercept value in the labelled wheat+unlabelled barley DNA curve is 32%. Group II in wheat is therefore 32–22=10%. The intercept in the labelled wheat+rye DNA curve is 58%. Group III in wheat is therefore 58–32=26%. The intercept in the labelled wheat+unlabelled wheat DNA curve is 74%. Group IV in wheat is therefore 74–58=16%. The proportions of each of the Groups in the other species were determined similarly. – This method of estimating the DNA content of each of the Groups of a species is permissible because all the labelled sequences hybridising to DNA from a more distantly related species also hybridise to DNA from a more closely related species (see Table 2). – Group II in barley DNA was determined from the mean of the intercepts of the curves with unlabelled wheat and rye DNAs. Group I in oats DNA is the mean of the intercepts of the curves with unlabelled wheat, barley and rye. – The absolute amounts of DNA in each Group per 1 C genome (shown in brackets) were determined from wheat, rye, barley and oats DNAs having 1 C DNA contents of 17.3, 8.3, 5.5 and 13.2 pg respectively (Bennett and Smith, 1976). Wheat and oats are hexaploids, rye and barley diploids. – denotes Group absent from the species

Species	Group						
	I	II	III	IV	V	VI	VII
Wheat	22 (3.8)	10 (1.7)	26 (4.5)	16 (2.8)	–	–	–
Rye	19 (1.6)	19 (1.6)	14 (1.15)	–	22 (1.8)	–	–
Barley	20 (1.1)	23 (1.3)	–	–	–	28 (1.5)	–
Oats	17 (2.2)	–	–	–	–	–	58 (7.7)

Further experiments have now been carried out on wheat and rye DNAs to determine if all of the Groups contain heterogeneous families of repeated sequences. Initially, repeated sequences able to form near perfect duplexes in vitro, with very little mismatching, were selected from wheat and rye DNAs by renaturation under very stringent conditions and then the proportions of Groups I, II, III, IV or V in these repeated sequences were determined by separate hybridisation to oats, barley, wheat and rye DNAs. The experimental scheme and results are shown in Figures 4 and 5. Tritium labelled rye or wheat DNA (average fragment size around 100 nucleotides) was mixed with a large excess of sheared unlabelled rye or wheat DNA, denatured and incubated at 42.5° C in 62% formamide 0.69 M NaCl to a C_0t of 90. These incubation conditions are approximately equivalent to 80° C at 0.18 M Na^+ (McConaughty et al., 1969; Flavell and Smith, 1976; Smith and Flavell, 1977), and thus allow only essentially identical repeated sequences to form duplexes. Incubation to a C_0t of 90 is sufficient to allow essentially all of the repeated sequences to reanneal which are able to reanneal under these conditions (unpublished results). Approximately 45% of the labelled rye DNA and 35% of the labelled wheat DNA was isolated in the renatured fraction by hydroxyapatite chromatography (see Figs. 4 and 5). These results are in agreement with other studies published elsewhere (Flavell and Smith, 1976; Smith and Flavell, 1977). Sheared unlabelled oats, barley, wheat and rye DNAs were then added to separate aliquots of the renatured and non-renatured fractions such that the newly added unlabelled DNA was present in a 500 to 1000 fold excess over the *unlabelled* DNA already present. After denaturing, the DNA mixtures were incubated at 60° C in 0.12 M phosphate buffer to a C_0t of 95.

The proportions of the labelled wheat and rye DNA fractions from the first incubations that were able to form duplexes with oats, barley, wheat and rye DNAs in the second incubation are shown in Figures 4 and 5 for two separate experiments. The proportions of Groups I, II, III and IV or V in each of the wheat and rye fractions were derived from these proportions as indicated in the legend to Table 4.

The results from these experiments clearly show that all of the Groups of sequences in the wheat and rye genomes contain sequences which will form near-perfect duplexes under very stringent incubation conditions as well as other related but non-identical sequences which cannot form stable duplexes at very stringent conditions but can under moderately stringent conditions (60° C at 0.18 M Na^+). Thus each of the Groups in the wheat and rye genomes contain families of repeated sequences within which there is, on average, a similar degree of sequence heterogeneity.

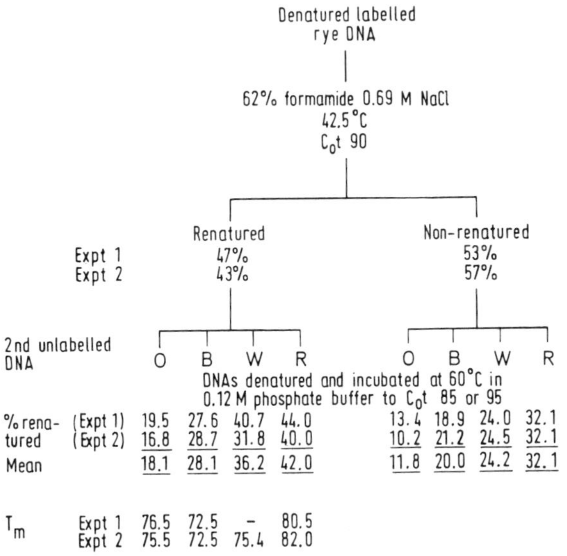

Fig. 4. Proportions of Groups of repeated sequences in rye that form duplexes under high stringency incubation conditions. Duplicate labelled rye samples (average fragment size = 100 nucleotides) were melted and incubated at 42.5° C in 62% formamide 0.69 M NaCl, to a C_0t of 90. The renatured and non-renatured DNAs were isolated by hydroxyapatite chromatography as described in Materials and Methods. Small aliquots of the renatured and non-renatured fractions were mixed with aliquots of unlabelled oats, wheat and rye DNAs (500 μg/ml in 0.12 M phosphate buffer; average fragment sizes 360 to 500 nucleotides) such that the ratios of labelled to unlabelled DNAs were at least 1:500. These DNA mixtures were melted and incubated at 60° C to a C_0t of between 80 and 95. The renatured DNAs were then isolated by hydroxyapatite chromatography. Melting curves of these renatured DNAs were obtained as described in Materials and Methods. *O* Oats DNA, *B* Barley DNA, *R* Rye DNA, *W* Wheat DNA

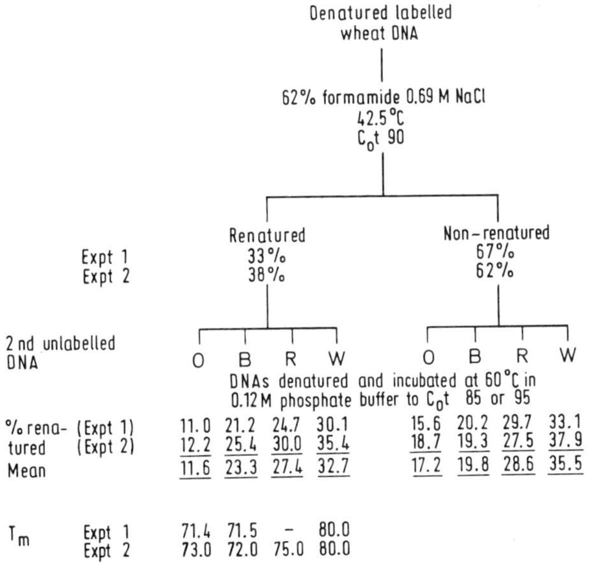

Fig. 5. Proportions of Groups of repeated sequences in wheat that form duplexes under high stringency incubation conditions. The method was as described in the legend to Figure 4 except labelled wheat DNA, average fragment size = 100 nucleotides, was used in the initial incubations. *O* Oats DNA, *B* Barley DNA, *R* Rye DNA, *W* Wheat DNA

Table 4. The proportion of each Group of repeated sequences in wheat and rye that can form duplexes in high stringency incubation conditions

Group	Proportion (%) of DNA of whole genome[a] in		Total repeated sequence DNA in both fractions
	Renatured fraction	Non-renatured fraction	
	rye		
I	14	12	26
II	10	8	18
III	8	4	12
V	6	8	14
	wheat		
I	8	17	25
II	11	3	14
III	4	9	13
IV	6	7	13

[a] The proportions were derived from the mean hybridisation percentages shown in Figures 4 and 5, by the same method as described in the legend to Table 3. Four per cent of extremely rapidly renaturing DNA, that is probably not heteroduplex (Smith and Flavell, 1977; Flavell and Smith, 1976) was initially subtracted from the hybridisation percentages in Figures 4 and 5 before the Group percentages were calculated. Because these proportions are derived directly from the proportion of DNA binding to hydroxyapatite and not from extrapolations to zero fragment length as in Figure 2 and Table 3 estimates for Groups I and II will be overestimates and III, IV and V underestimates compared with the values in Table 3. The total repeated sequence DNA gained by summing the individual Group values is lower than the value expected from Table 3, especially for wheat. This is probably because some fragments were damaged during the multiple denaturation, incubation and fractionation procedures

In the same experiments, the thermal stabilities were measured of the duplexes formed at 60° C in 0.18 M Na^+ between those wheat and rye repeated sequences able to renature in 62% formamide 0.69 M NaCl at 42.5° C and the repeated sequences of oats, barley, wheat and rye DNAs. The T_ms are given in Figures 4 and 5 under the appropriate duplexes.

The repeated sequences selected as being able to form very stable duplexes in 62% formamide 0.69 M NaCl at 42.5° C formed duplexes with T_ms equal to or above 80° C when hybridised again with DNA from the same species. However, when hybridised to DNA from other species the T_ms of the duplexes were at least 5° lower in all cases. Similar results are estimated when essentially all the repeated sequences of wheat and rye are hybridised to oats, barley, wheat and rye DNAs at 60° C in 0.12 M phosphate buffer (See Smith and Flavell, 1974).

These results provide the important conclusion that repeated sequences in these cereal genomes are on average more closely related to other sequences in the same Group of the same species than they are to the sequences in the same Group of a related species.

Discussion

1. Repeated Nucleotide Sequences and Cereal Phylogeny

The proportion of one cereal DNA that is in the renatured DNA hydroxyapatite-bound fraction after hybridisation to another cereal DNA, as well as to DNA from the same species, is very dependent upon fragment size (Fig. 2). This is almost certainly because a high proportion of the repeated sequences common to different cereal genomes are short and are interspersed with different sequences (non-repeated or repeated) throughout large fractions of the genome.

This arrangement of sequences which is discussed extensively elsewhere (Smith et al., 1976; Rimpau et al., in preparation), results in overestimates of the repeated sequence proportion of a genome that will hybridise to another genome. Thus it is particularly important to recognise the consequences of interspersion of different short sequences in experiments designed to estimate the extent of homology between two DNAs.

Our revised estimates of homology between the repeated sequences in the oats, barley, wheat and rye genomes (Table 1) indicate the same phylogenetic relationships for cereal evolution as our previous results, which did not allow for sequence interspersion (Smith and Flavell, 1974). However, the proportion of repeated sequences which are related to repeated sequences in the other genomes are clearly considerably less than previously estimated (Smith and Flavell, 1974).

The evolutionary scheme in Figure 3 derived from repeated sequence homologies and the scheme based on other morphological and genetic considerations (Bell, 1965) both indicate that wheat and rye diverged relatively recently in Gramineae evolution, that their common ancestor diverged from the ancestor of barley sometime previous while the ancestor of oats diverged from the ancestor of barley, wheat and rye even earlier.

The molecular events during speciation in Gramineae evolution are unknown but the chromosomes of the related species now possess different families of repeated sequences. New families of repeated sequences appear to have arisen at or after each of the evolutionary branch points (see Fig. 3) and, in general, all families of repeated sequences present before a branch point appear to have been retained during subsequent evolution (see Table 2). This has enabled us to divide the repeated sequences of cereal genomes into Groups, each Group consisting of repeated sequences introduced at or after a different evolutionary branch point (see Fig. 3). This is a useful way of fractionating repeated sequences but we do wish to emphasize that each Group does not necessarily consist entirely of identical or closely related sequences. Each Group may contain hundreds of unrelated families of sequences but each family consists of hundreds or thousands of identical or related sequences. Furthermore, the proportion of each genome in each Group is to some extent defined by the hybridisation conditions. For example lowering the renaturation temperature increases the proportion of the genome in Group I (see Results and Rimpau et al., in preparation).

2. The Evolution of Families of Repeated Sequences

Families of identical or closely related repeated sequences have their origin in some kind of amplification event in which, possibly through some kind of error in replication, a chromosomal region is replicated hundreds, thousands or even millions of times. Britten and Kohne (1968) have called this "saltatory" amplification. The mechanism of saltatory amplification is not known. Where the sequence length is not very short, then a rolling circle mechanism now shown to be responsible for amplifying rDNA during oogenesis in *Xenopus laevis* (Hourcade et al., 1973) is attractive (Southern, 1975; Botcham, 1974). For very short sequences, Wells et al. (1967) have suggested a mechanism involving "slippage" replication.

When a completely different family of repeated sequences appears in a genome, it presumably is the result of amplification of a sequence previously present in only one or a few copies per genome. Thus the different groups of repeated sequences in the cereal genomes consist of families of repeated sequences amplified at different times from different, essentially non-repeated sequences. We infer from the evolutionary tree, that the sequences related to existing Group I sequences were amplified first in the ancestral genome before the ancestor of oats and the ancestor of barley, wheat and rye diverged. The sequences closely related to existing Group II sequences were presumably first amplified after these ancestral species had diverged while Group III sequences were presumably amplified after the ancestor of barley had diverged from the ancestor of wheat and rye. Groups IV, V, VI and VII probably consist of families of sequences amplified from effectively non-repeated sequences after the four species diverged from their closest relatives among the four species studied here.

When a newly amplified family of repeated sequences is first stabilised in a genome, all the sequences are probably identical. However, repeated sequences seem able to tolerate the introduction of base changes fairly readily rendering the families of repeated sequences heterogeneous. Related repeated sequences within the cereal genomes differ on average by about 6 to 10%. We have inferred this from the thermal stabilities of the duplexes formed between related sequences in vitro (Smith and Flavell, 1974). Assuming that families of repeated sequences are able to tolerate considerable heterogeneity, it would be expected that the oldest families would have the most heterogeneity while very recently amplified families would be very homogeneous. However, Group I sequences within each of the cereal genomes do not appear to have diverged significantly more than the sequences in Groups II, III, IV or V, which appear to have been first amplified considerably later in evolution (Table 4). Furthermore, if mutations accumulate in families of repeated sequences at random, then the distribution of sequence heterogeneity in a given family of sequences ought to be the same between species as within species. Our measurements on sequence differences within and between cereal genomes clearly show this is not the case. The repeated sequences of a family are more closely related to each other within a species than they are to the repeated sequences of the corresponding family in another species (Figs. 4 and 5). These results strongly argue against a single amplification of each family of repeated sequences in evolution. Instead, we suggest that further amplification of already amplified sequences has been common in cereal chromosome evolution. If different members of a family of related but not identical repeated sequences were amplified (or deleted) in diverging species then the repeated sequences of this family would be more closely related within species than between species. The hypothesis for multiple amplifications in diverging species is illustrated in Figure 6.

An alternative hypothesis to account for the amplification of individual members of a family of diverged repeated sequences has been proposed by Smith (1973, 1976). By computer simulation studies he has shown it is possible, by expansion and contraction of a set of diverged repeated sequences by unequal crossing over, to form families of identical repeated sequences from one diverged member of an older family of repeated sequences. If different diverged sequences of the same family were "selected" in this way during species divergence, then this could also explain how repeated sequences of a family are more closely related within a species than between species. It is quite possible that both saltatory amplification and unequal crossing over have been instrumental in establishing the spectrum of repeated sequences now observed in plant chromosomes.

Our results on repeated sequence homologies within and between cereal species and the explanations that could account for them are similar in principle to those which have come from detailed studies of satellite DNA sequences in related animal species. These latter studies provide very strong evidence that in the evolution of families of repeated sequences initial amplification to form a family of tandemly repeated sequences is followed by cycles of sequence divergence and further amplification of diverged sequences, with unequal crossing over probably also playing a role (Cooke, 1975; Southern, 1975; Rice and Straus, 1972).

With cycles of amplification of diverged repeated sequences occurring within species throughout evolution, it would be expected that diverged species would frequently have different numbers of copies of sequences in closely related families. This may be the explanation behind the different amounts of DNA

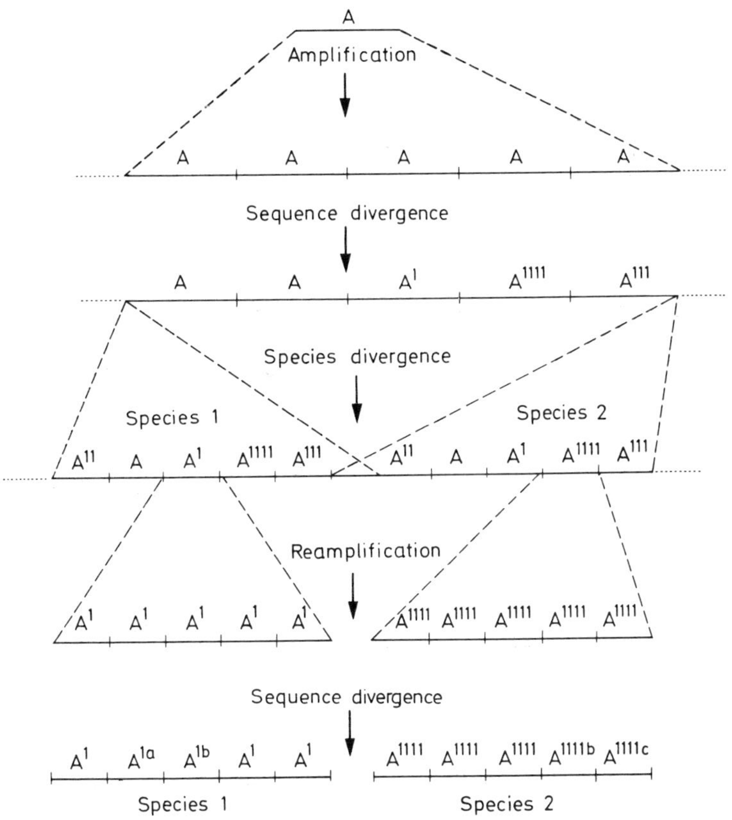

Fig. 6. A simplified scheme for the evolution of related families of repeated sequences in different cereal genomes. A nucleotide sequence A in an ancestral genome is initially amplified to form a family of identical repeated sequences. Over a period of time "mutations" occur at random in these sequences. The "mutations" could be base substitutions, additions or deletions. The repeated sequences gradually diverge. Let us assume that at this time two species diverge from each other and this family of diverged sequences is present in both diverging species. Then, in species 1, an amplification of sequence A^1 occurs to produce a high proportion of A^1 sequences in the family of repeated A sequences. At a different time, separate amplification of sequences A^{1111} occurs in species 2 to produce high proportions of A^{1111} sequences in the family of repeated A sequences. The repeated sequences are now more closely related within each species than they are between the two species as is found in cereal genomes today. This remains so even though more mutations may accumulate in the reamplified sequences. Deletions have not been illustrated but may also occur and widen the differences in related sequences between species.

The mechanism of "amplification" could be of the "rolling circle" type (Hourcade et al., 1973) where a sequence is replicated many times more than other chromosomal sequences or a large number of unequal recombination events at meiosis (Smith, 1973, 1976)

in the same Group in the four cereal species (Table 3).

In the foregoing discussion of the evolution of present day families of repeated sequences in cereal chromosomes, we have emphasized amplification events. These would cause increases in the total DNA content of the genomes. Very early ancestors of wheat, barley, oats and rye probably had lower amounts of repeated sequence DNA in which case increases in total DNA content have occurred from those times. However, throughout evolution, loss of DNA as well as addition of new families of repeated sequences has probably also occurred.

Acknowledgements. We wish to thank Michael O'Dell for his excelllent technical assistance. J.R. was on leave from Institut für Pflanzenzüchtung der Universität Göttingen and supported by Deutsche Forschungsgemeinschaft. We thank Professor R. Riley and Dr. C.N. Law for critically reading the manuscript.

References

Bell, G.D.H.: The comparative phylogeny of the temperate cereal. In: Essays in crop plant evolution (J. Hutchinson, ed.), pp. 70–102. Cambridge: University Press, Cambridge, England 1965

Bendich, A.J., McCarthy, B.J.: DNA comparisons among barley, oats, rye and wheat. Genetics **65**, 545–566 (1970)

Bennett, M.D., Smith, J.B.: Nuclear DNA amounts in Angiosperms. Phil. Trans. roy. Soc. (Lond.) B **274**, 227–274 (1976)

Bonner, T.I.: Reduction in rate of DNA reassociation by sequence divergence. J. molec. Biol. **81**, 123–135 (1973)

Bonner, T.I.: Hairpin-forming sequences in mammalian DNA. Carnegie Inst. Wash. Year Book **73**, 1079–1088 (1974)

Botcham, M.R.: Bovine satellite 1 DNA consists of repetitive units 1,400 base pairs in length. Nature (Lond.) **251**, 288–292 (1974)

Britten, R.J., Davidson, E.H.: Studies on nucleic acid reassociation kinetics: Empirical equations describing DNA reassociation. Proc. nat. Acad. Sci. (Wash.) **73**, 415–419 (1976)

Britten, R.J., Kohne, D.E.: Repeated sequences in DNA. Science **161**, 529–540 (1968)

Burgi, E., Hershey, A.D.: Sedimentation rate as a measure of molecular weight of DNA. Biophys. J. **3**, 309–321 (1963)

Chooi, W.Y.: Comparison of the DNA of six Vicia species by the method of DNA-DNA hybridisation. Genetics **68**, 213–230 (1971)

Cooke, H.J.: Evolution of the long range structure of satellite DNAs in the genus Apodemus. J. molec. Biol. **94**, 87–99 (1975)

Davidson, E.H., Hough, B.R., Amenson, C.S., Britten, R.J.: General interspersion of repetitive with non-repetitive sequence elements in the DNA of Xenopus. J. molec. Biol. **77**, 1–23 (1973)

Flavell, R.B., Bennett, M.D., Smith, J.B., Smith, D.B.: Genome size and the proportion of repeated sequence DNA in plants. Biochem. Genet. **12**, 257–269 (1974)

Flavell, R.B., Smith, D.B.: Nucleotide sequence organisation in the wheat genome. Heredity **37**, 231–252 (1976)

Goldberg, R.B., Bemis, W.P., Siegel, A.: Nucleic acid hybridisation studies within the genus Cucurbita. Genetics **72**, 253–266 (1972)

Hourcade, D., Dressler, P., Wolfson, J.: The amplification of ribosomal RNA genes involves a rolling circle intermediate. Proc. nat. Acad. Sci. (Wash.) **70**, 2926–2930 (1973)

Huguet, T., Jouanin, L., Bazetoux, S.: Occurrence of palindromic sequences in wheat DNA. Plant Sci. Letters **5**, 379–385 (1975)

Mizuno, S., Andrews, C., Macgregor, H.C.: Interspecific "common" repetitive DNA sequences in salamanders of the genus Plethodon. Chromosoma (Berl.) **58**, 1–31 (1976)

McCarthy, B.J., Farquhar, M.N.: The rate of change of DNA in evolution. Brookhaven Symp. Biol. **23**, 1–44 (1972)

McConaughty, B.C., Laird, C.D., McCarthy, B.J.: Nucleic Acid Reassociation in Formamide. Biochemistry **8**, 3289–3295 (1969)

Rice, N.R.: Change in repeated DNA in evolution. Brookhaven Symp. Biol. **23**, 44–79 (1972)

Rice, N. and Esposito, P.: Relatedness among several hamsters. Carnegie Inst. Year Book **72**, 200–204 (1973)

Rice, N.R. and Straus, N.A.: Evolution of repeated sequences in the genus Mus. Carnegie Inst. Year Book **71**, 264–266 (1972)

Smith, D.B., Flavell, R.B.: The relatedness and evolution of repeated nucleotide sequences in the DNA of some Gramineae species. Biochem. Genet. **12**, 243–256 (1974)

Smith, D.B., Flavell, R.B.: Characterisation of the wheat genome by renaturation kinetics. Chromosoma (Berl.) **50**, 223–242 (1975)

Smith, D.B., Flavell, R.B.: Nucleotide sequence organisation in the rye genome. Biochim. biophys Acta (Amst.) **474**, 82–97 (1977)

Smith, D.B., Rimpau, J., Flavell, R.B.: Interspersion of different repeated sequences in the wheat genome revealed by interspecies DNA/DNA hybridisation. Nucleic Acid Res. **3**, 2811–2825 (1976)

Smith, G.P.: Unequal crossover and the evolution of multigene families. Cold Spr. Harb. Symp. quantitative Biol. **38**, 507–514 (1973)

Smith, G.P.: Evolution of repeated DNA sequences by unequal crossover. Science **191**, 528–535 (1976)

Smith, M.J., Britten, R.J., Davidson, E.H.: Studies on nucleic acid reassociation kinetics. Reactivity of single stranded tails in DNA-DNA renaturation. Proc. nat. Acad. Sci. (Wash) **72**, 4805–4809 (1975)

Southern, E.M.: Long range periodicity in mouse satellite DNA. J. molec. Biol. **94**, 51–69 (1975)

Stein, D.B., Thompson, W.F.: DNA hybridisation and evolutionary relationships in three Osmunda species. Science **189**, 888–890 (1975)

Straus, N.A.: Comparative DNA renaturation kinetics in amphibians. Proc. nat. Acad. Sci. (Wash.) **68**, 799–802 (1971)

Straus, N.A.: Reassociation of Bean DNA. Carnegie Inst. Wash. Year Book **71**, 257–259 (1972)

Studier, F.W.: Sedimentation studies of the size and shape of DNA. J. molec. Biol. **11**, 373–390 (1965)

Ullman, J.S., McCarthy, B.J.: The relationship between mismatched base pairs and the thermal stability of DNA duplexes. Biochim. biophys. Acta (Amst.) **294**, 405–415 (1973)

Wells, R.D., Buchi, H., Kossel, H., Outsuka, E., Khorana, H.G.: Studies on polynucleotides. LXX: Synthetic deoxyribopolynucleotides as templates for the DNA polymerase of Escherichia coli: DNA-like polymers containing repeating tetranucleotide sequences. J. molec. Biol. **27**, 265–272 (1967)

Wilson, D.A., Thomas, C.A.: Palindromes in Chromosomes. J. molec. Biol. **84**, 115–144 (1974)

Received February 9–April 21, 1977 / Accepted April 24, 1977 by H.C. Macgregor
Ready for press April 30, 1977

 Reprinted from *Cold Spring Harbor Symp. Quant. Biol.* **45**: 755–760.

Evidence for the Involvement of Recombination and Amplification Events in the Evolution of *Secale* Chromosomes

J. BEDBROOK,* J. JONES,† AND R. FLAVELL†

* *Department of Genetics, Division of Plant Industry, CSIRO, Canberra, A.C.T. Australia;* † *Department of Cytogenetics, Plant Breeding Institute, Trumpington, Cambridge CB2 2LQ, England*

In addition to cultivated rye, the genus *Secale* includes 11 other species or taxa (Jain 1960; Stutz 1972). All the various taxa are diploids with 14 chromosomes and can be intercrossed to give partially fertile hybrids. Within the genus there is a 20% variation in nuclear DNA content that is correlated with variation in the size of blocks of telomeric heterochromatin (Bennett et al. 1977). More than 70% of the DNA in *Secale* chromosomes is highly repeated (Rimpau et al. 1978). The arrangements and chromosomal distribution of specific repeated sequences in this genus and other cereal species have been compared (Bedbrook et al. 1980a,b). On the basis of these experiments, we propose here that (1) much of the nuclear DNA of *Secale* species consists of a few different sequence-divergent elements that are present in many copies and capable of recombining to form various permutations of linear order; (2) such permutation can undergo amplification at specific chromosomal locations and be dispersed to other chromosomal locations; and (3) on the basis of cytological observations of wheat-rye crosses, four specific recombination-amplification events could have contributed to the speciation of *Secale*.

The 120 Family—An Example of a Divergent, Distributed, Repeat Element in *Secale*

In support of the first two proposals mentioned above, we use information derived from analysis of a specific-repeat family, the 120 family (Bedbrook et al. 1980b). More general analysis (Flavell et al. 1980) as well as analysis of other specific-repeat families from rye and wheat (see Flavell et al., this volume) would suggest that the arrangements of the 120 family are typical of repeated sequences in cereals. The 120 family was first isolated by screening clone banks of *Secale cereale* DNA with probes enriched for the DNA of rye telomeric heterochromatin. Renaturation analysis in 0.18 M Na^+ and at 60 °C showed that this cloned sequence was related to more than 2% by weight of *S. cereale* DNA. This sequence was found to be present in high copy number in all *Secale* taxa tested (J. Jones, unpubl.) and also in wheat (Bedbrook et al. 1980b).

Sheared, radiolabeled, cloned DNA of the 120 family was hybridized with a vast excess of sheared *S. cereale* DNA to a C_0t of 0.3, and the duplex fraction was collected on hydroxyapatite columns. The DNA was eluted in 0.18 M Na^+ by increasing the temperature. As an internal control, the elution of the total DNA was followed by measuring the absorbance at 260 nm, and the elution of the cloned probe was measured by radioactivity. The elution of sheared native rye DNA was also measured. Figure 1 gives the results of such an experiment and shows that the thermal elution profile of native rye DNA is a sharp biphasic transition curve with a T_m of approximately 90 °C. For renatured rye DNA, the thermal elution profile is broader than that for native DNA, with a T_m of about 82 °C. The 120 family also shows a very broad profile with a low T_m, indicating that the cloned sequence is representative of a highly divergent repeated-sequence family.

In situ hybridization analysis with the 120-family probe (Fig. 2) shows that the sequence is present on all seven pairs of rye chromosomes. Although the sequence is present in very high copy number at the telomeres, it is also distributed throughout both arms of all chromosomes.

The 120 family consists of tandem arrays of the 120-bp element plus a myriad of complex arrangements with other sequences. Hybridization of the cloned 120 family to blots (Southern 1975) of rye DNA digested with various restriction endonucleases is illustrated in Figure 3. *Hae*III digests (lane 1) show that much of the 120 family exists as a simple tandem-repeat sequence, since hybridization reveals a progression of fragments with a 120-bp-length interval. Such a pattern is expected for a family of tandemly arranged repeated sequences in which each repeated-unit sequence contains a *Hae*III site that is occasionally lost by random mutation to give fragments of dimer, trimer, etc. length. More-detailed hybridization analysis (Bedbrook et al. 1980b) to *Hae*III digests showed that the tandem repeat contains both a 120-bp and a 117-bp unit. The two size variants have essentially the same sequence and are interspersed in the tandem arrays. Figure 3 also illustrates that the restriction endonucleases *Bam*HI (lane 2), *Bgl*II (lane 3), *Hind*III (lane 4), and *Eco*RI (lane 5) cut sequences that hybridize the 120-family probe less frequently than *Hae*III. After digestion with these enzymes, there is a smear of hybridization throughout the size classes of fragments generated by the enzymes. *Bam*HI and *Eco*RI produce, in addition to the smear, an integral series of bands with a periodicity of 120 bp. Such frag-

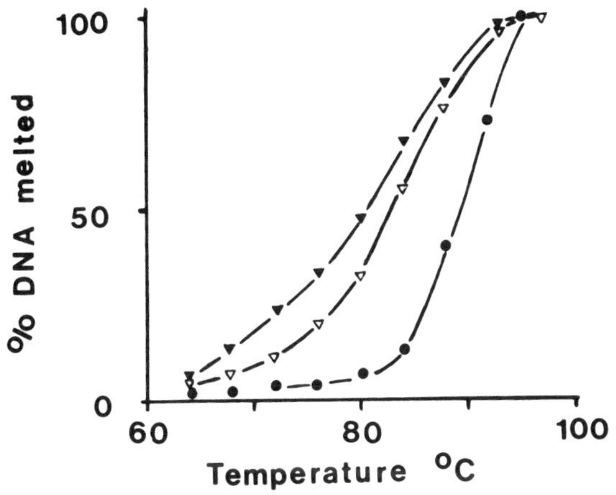

Figure 1. Thermal elution profiles of duplexes formed between a cloned example of the 120-bp family and total *S. cereale* DNA. (●) Elution profile of sheared native rye DNA (measured by OD_{260}); (▽) elution profile of total renatured rye repeated DNA; (▼) elution profile of the 120-family rye hybrids.

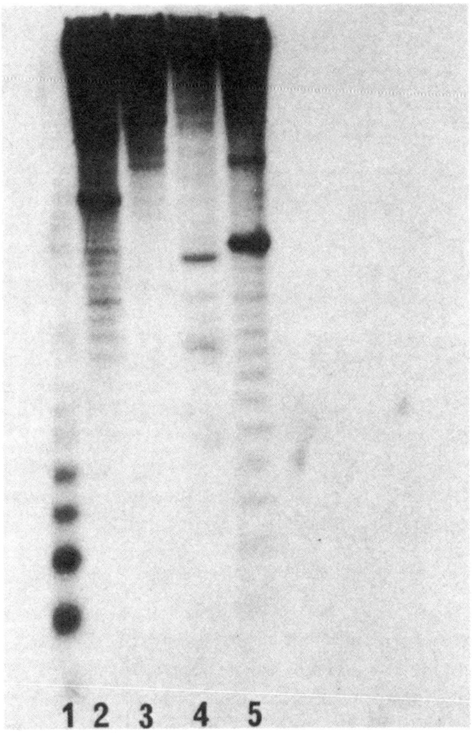

Figure 3. Hybridization of a cloned example of the 120 family to rye DNA digested with various restriction endonucleases. Hybridization to: (*1*) *Hae*III-digested DNA; (*2*) *Bam*HI; (*3*) *Bgl*II; (*4*) *Hin*dIII; (*5*) *Eco*RI.

ments presumably arise by the creation, by mutation, of sites for these enzymes in the tandem arrays of the 120-bp repeating unit. *Bgl*II and *Hin*dIII do not generate such integral-length series of fragments. Therefore, no sequence close to the recognition sequence for *Bgl*II or *Hin*dIII can exist in the 120-bp repeating unit.

The presence of the smear of background hybridization and hybridization to bands in *Hin*dIII and *Bgl*II digests that are not length multiples of 120 bp is indicative that the 120-family sequence is often found interfaced with other sequences in the chromosomes. Direct evidence that the 120 family exists in forms other than tandem repeats of 120-bp units was obtained by analysis of cloned sequences. Clones that hybridized the cloned representative of the 120-bp family but that were not integral multiples of 120 bp were obtained. The structure of one cloned fragment is given in Figure 4.

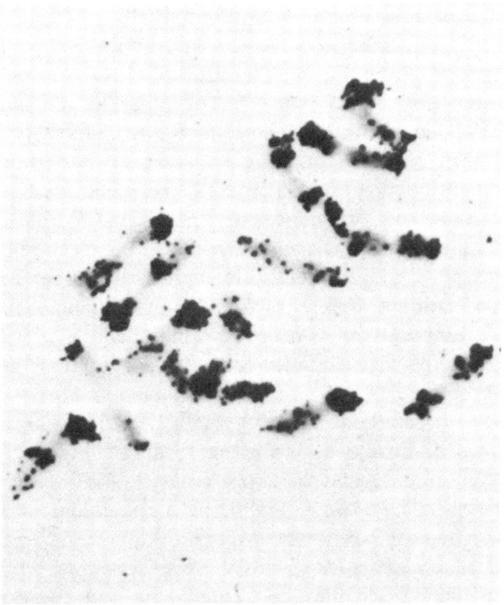

Figure 2. In situ hybridization to rye metaphase chromosome of a radiolabeled probe from a cloned example of the 120 family.

The fragment 2.2 kb in length contains *Eco*RI sites at its termini and single *Pst*I and *Sal*I sites. The 120-family clone hybridizes only to the right-hand 1200-bp *Sal*I-*Eco*RI fragment, and detailed mapping locates the hybridization within the region of the thick line in Figure 4; that is, within this 2.2-kb fragment, the 120-bp family is not a simple tandem array but is interfaced with nonhomologous sequences. The 2.2-kb sequence comigrates with a hybridization band in *Eco*RI digests arrowed in Figure 3.

For ease of discussion, we call forms of the 120 family that are of the type represented by the 2.2-kb sequence "complex arrangements of the 120 family."

Complex arrangements of the 120 family are occasionally amplified. The blot experiment shown in Figure 3 was done under conditions that would detect, using a 1-kb probe, a minimum of 50 to 100 copies per nucleus in a single band; that is, the various complex forms detected in *Hin*dIII and *Bgl*II digests must be repeated in the chromosomes.

The 2.2-kb complex repeat discussed above can be seen as a hybridization band of high stoichiometry (see lane 5 of Fig. 3).

It is possible to measure the relative stoichiometry of

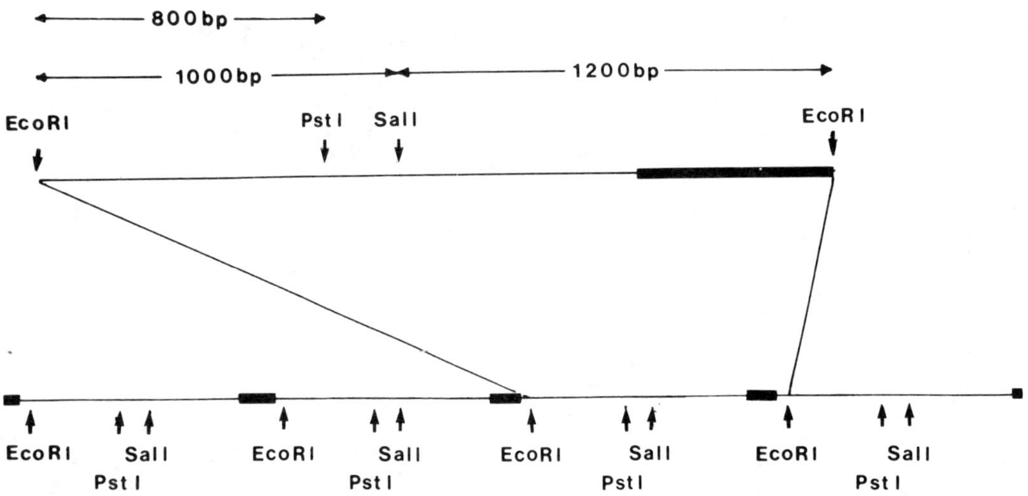

Figure 4. Structure of a 2.2-kb complex repeat of the 120 family. (—) The location of the sequence hybridizing the 120-family probe.

this repeat by using relative fluorescence in ethidium-bromide-stained gels. This complex form of the 120 family is present in approximately 4000 copies per haploid genome. Hybridization analysis to partial *Eco*RI digests shows that at least some of this 2.2-kb repeated sequence is in the form of tandem repeats. Hybridization to bands up to pentamer length can be detected. The tandem structure of the 2.2 kb is illustrated in Figure 4.

Since multiple and independent events leading to the formation of complex repeats such as the 2.2-kb repeat seems an unlikely possibility, we assume that the multiple copies of the repeat are a consequence of amplification events.

Specific complex arrangements of the 120 family are present on one to all of the seven rye chromosomes. Wheat-rye addition lines (Riley and Chapman 1958) make it possible to investigate the chromosomal distribution of amplified complex forms of the 120 family. Each wheat-rye addition line has a full complement of wheat chromosomes and a single pair of rye chromosomes. The presence of rye chromosomes in one such addition line is illustrated in Figure 5A by in situ hybridization with a probe specific to the telomeres of rye chromosomes. Figure 5B illustrates blot experiments in which the 120-family probe was hybridized to *Eco*RI-digested rye DNA (lane 1), *Secale silvestre* DNA (lane 2), the seven wheat-rye addition lines (lanes 3–9), and wheat DNA (lane 10).

The 2.2-kb complex repeat (arrowed in Fig. 5B) is present in *S. silvestre* DNA but absent or in very low copy number in wheat. Hybridization to the addition lines shows that this particular complex form is present on all seven rye chromosomes. Two other amplified complex forms revealed by *Eco*RI digestion are present only in three of the seven chromosomes (lanes 4, 7, and 9). *Bam*HI digests (not shown) reveal complex forms of the 120 family present in one chromosome only.

The finding that some amplified complex forms of the 120 family are chromosome-specific suggests that amplification of the sequence may occur initially at a single location and be subsequently distributed to other chromosomes by recombination.

Complex arrangements of the 120 family contain sequences that are amplified independently of the 120-bp sequence. As discussed above, the 2.2-kb complex form of the 120 family contains sequences that are not related to the 120-bp sequence itself. Figure 6 shows an experiment in which a probe prepared from the left-hand, non-120-homologous *Eco*RI-*Pst*I fragment (Fig. 4) was hybridized to rye DNA digested with *Hin*dIII (lane 2). The hybridization pattern is compared with that given by the 120-family probe (lane 1). It is possible to detect relatively high copy-number bands that are hybridized by the 120-family probe but are not hybridized by the 2.2-kb-specific probe. The converse is also true; that is, hybridization bands indicative of repeated sequences are revealed by the 2.2-kb-specific probe and not by the 120-family probe.

This result suggests that the 2.2-kb repeat can be regarded as a set of sequence elements, including the 120-bp sequence itself, that are capable of being amplified together or independently in various arrangements. Amplification of different variants must take place relatively frequently in evolutionary time since it is possible to detect repeated complex forms specific to closely related, crossable species. Figure 5 illustrates an experiment in which the 2.2-kb complex-repeat-specific probe (described above) was hybridized to *Eco*RI digests of *S. cereale* (lane 3) and *S. silvestre* DNA (lane 4). This probe reveals at least six high-copy-number complex repeats common to the two species and one sequence unique to *S. silvestre.*

On the basis of data presented here and elsewhere (Bedbrook et al. 1980a; Flavell et al. 1980 and this volume), we propose that much of the DNA in the chromosomes of *Secale* species and other cereals is the

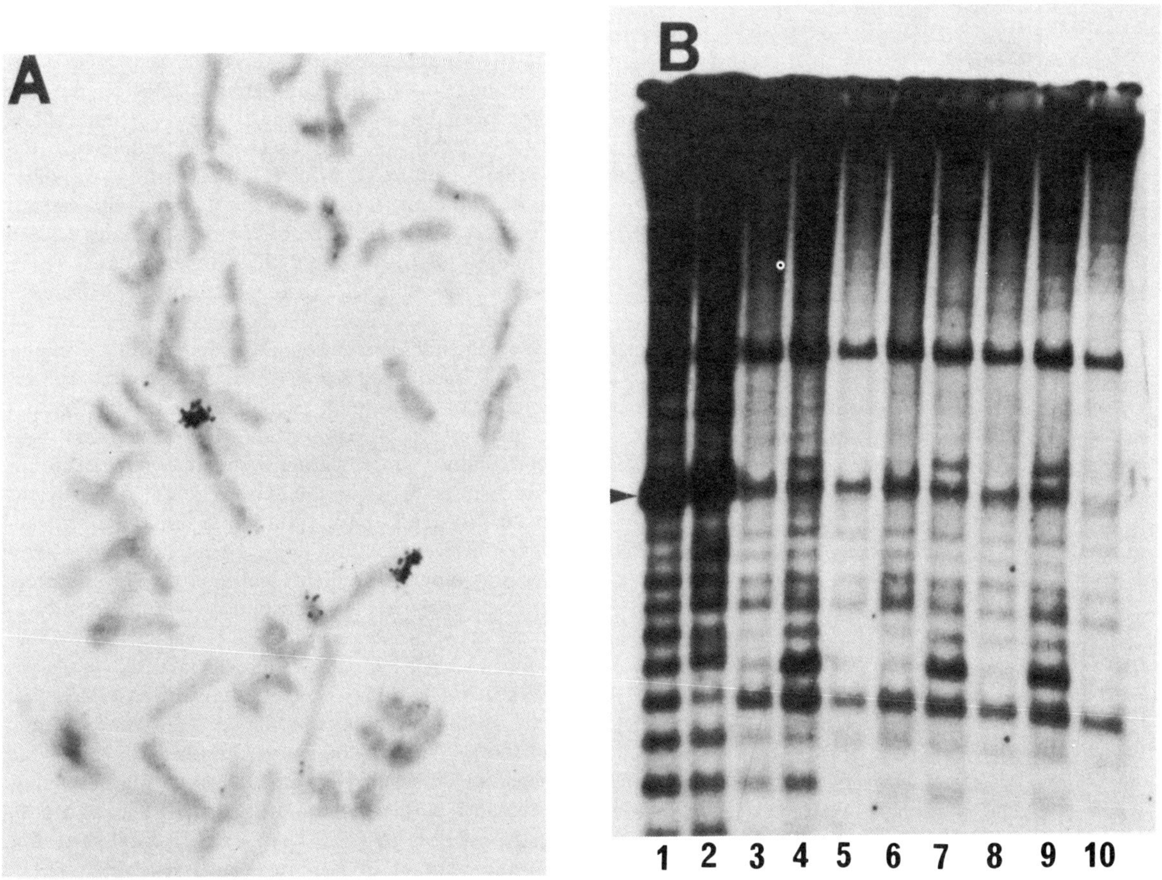

Figure 5. Hybridization of the 120-family probe to wheat-rye addition lines. (*A*) In situ hybridization to metaphase chromosomes of a wheat-rye addition line with a probe derived from a rye-DNA-specific sequence. (*B*) Hybridization of the 120-family probe to (*1*) rye DNA digested with *Eco*RI, (*2*) *S. silvestre* DNA, (*3-9*) wheat-rye addition line DNA, and (*10*) wheat DNA.

product of recombination of relatively few diverged sequence families and that various rearrangements are amplified to form high copy-number repeats. After new permutations are amplified at one location, they are distributed by recombination to multiple chromosomal locations.

We have no evidence for a mechanism for the recombination or amplification processes.

Specific Recombination-Amplification Events in Relation to Speciation in *Secale*

There is quantitative variation in the amounts of telomeric heterochromatin in the various taxa of the genus *Secale* (Bennett et al. 1977).

We (Bedbrook et al. 1980b) have shown that most of this intertaxa variation is due to the presence of four, complex, highly repeated sequences. The lengths and copy numbers for these repeats in *S. cereale* are given in Table 1. All the various *Secale* taxa can be distinguished by in situ hybridization of cloned examples of the four repeats to chromosomes of the various taxa (J. Jones, unpubl.).

We have proposed that these repeats can be envisaged as arising by an insertion recombination-amplification mode, illustrated in Figure 7; that is, a sequence of relatively low copy number is inserted within a preexisting simple tandem-repeat sequence. A portion of the tandem repeat plus the inserted element are then amplified together to form a complex repeat containing a subrepeating unit and an unrelated unique sequence per repeat unit. Evidence for this model has been obtained by comparing the structures of DNA sequences from the complex repeats in different *Secale* taxa. Specifically, we have compared arrangements in *S. silvestre,* considered representative of a primitive member of the genus, and *S. cereale,* representative of an advanced member (Evans 1976).

Table 1. Sizes and Amounts of Repeats in *S. cereale* Telomeric Heterochromatin

Repeat-unit length (bp)	Percentage weight of *S. cereale* DNA	Copy number per haploid genome
480	6.1	1×10^6
610	2.7	3.5×10^5
630 } 356 }	0.6	7.4×10^4

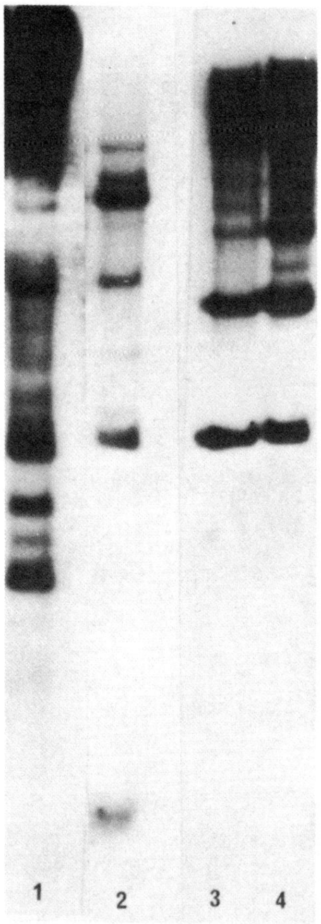

Figure 6. Hybridization of a complex form of the 120 family to *S. cereale* and *S. silvestre* DNAs. (*1*) Hybridization of the 120-family probe to *S. cereale* DNA digested with *Hin*dIII; (*2*) hybridization of the 2.2-kb complex-repeat-specific probe to *S. cereale* DNA; (*3*) hybridization as in *2* to *Eco*RI-digested *S. cereale* DNA; (*4*) hybridization to *Eco*RI-digested *S. silvestre* DNA.

The 356–630 family, for example, represents two complex forms of a 120-bp simple tandem-repeat unit. In *S. silvestre* DNA, only the simple repeat can be detected; in *S. cereale,* the complex forms are present in high copy number (Bedbrook et al. 1980b).

As indicated from Table 1 the 610 family is a very high copy-number repeat in *S. cereale.* Sequences complementary to the 610 family are not detected in *S. silvestre* DNA by solution hybridization. In blot experiments capable of detecting single sequences, the 610-family probe hybridizes 10–20 single-copy-size classes of fragments in *S. silvestre* digested with various restriction enzymes. None of these size classes have a periodicity of 610 bp, and they do not appear to be linked to the subrepeating unit found in the 610-bp repeat unit.

We conclude, therefore, that sometime in the evolution of *S. cereale* from a *S. silvestre*–like ancestor, a recombination event such as that illustrated in Figure 7 has given rise to the 610 repeat. We speculate that these recombination-amplification events may have been important in speciation in the *Secale* since (1) these events have led to quantitative variation in the amounts of telomeric heterochromatin in the *Secale,* and (2) the presence of telomeric heterochromatin on rye chromosomes has been proposed to contribute to kernal sterility in crosses between rye and wheat (a cereal species without major blocks of telomeric heterochromatin).

Wheat-rye hybrids or triticale (Gustafson 1976) suffer from meiotic irregularity (Merker 1976; Thomas and Kaltsikes 1976), leading most frequently to polyploid nuclei, which results in kernel sterility and shriveling. Cytological studies have indicated that there is a positive correlation between the degree of kernal shriveling and the number of abnormal nuclei. Bennett (1973, 1977) has proposed that most of the aberrant nuclei arise as the consequence of anaphase bridges that are not resolved in the endosperm, and he suggests that the large amount of late-replicating terminal heterochromatin in rye chromosomes that is absent in wheat is the factor in the formation of these anaphase bridges. Univalent formation in triticale also seems to be related to the telomeric heterochromatin of the rye chromosomes; that is, quantitative variation in the amount of telomeric heterochromatin can lead to sterility in crosses between closely related cereals. It does not seem unreasonable, then, to speculate that variation in the amount of telomeric heterochromatin as the consequence of the above-mentioned sequence amplifications has led to partial reproductive barriers in the genus *Secale.*

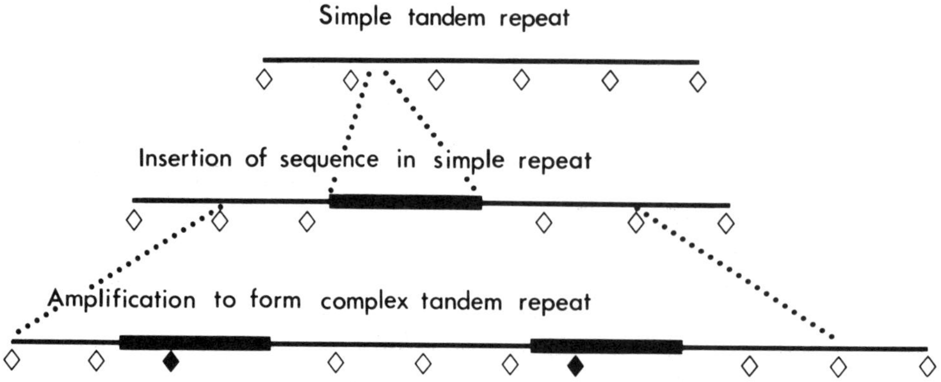

Figure 7. Scheme for the evolution of the telomeric-specific repeats of *S. cereale.*

REFERENCES

Bedbrook, J. R., M. O'Dell, and R. B. Flavell. 1980a. Amplification of rearranged repeated DNA sequences in cereal plants. *Nature* **288:**133.

Bedbrook, J. R., J. Jones, M. O'Dell, R. D. Thompson, and R. B. Flavell. 1980b. A molecular description of telomeric heterochromatin in *Secale* species. *Cell* **19:** 545.

Bennett, M. D. 1973. Meiotic, gametophytic and early endosperm development in triticale. In *Triticale* (ed. R. MacIntyre and M. Campbell), p. 137. International Development Research Centre, Ottawa.

———. 1977. Heterochromatin, aberrant endosperm nuclei and grain shrivelling in wheat-rye genotypes. *Heredity* **39:** 411.

Bennett, M. D., J. P. Gustafson, and J. B. Smith. 1977. Variation in nuclear DNA in the genus *Secale. Chromosoma* **61:** 149.

Evans, G. M. 1976. Rye. In *Evolution of crop plants* (ed. N. W. Simmonds), p. 108. Longmans, London.

Flavell, R. B., J. Bedbrook, J. Jones, M. O'Dell, W. L. Gerlach, T. A. Dyer, and R. D. Thompson. 1980. Molecular events in the evolution of cereal chromosomes. In *Proceedings of the Fourth John Innes Symposium: The plant genome.* (ed. D. R. Davies and D. A. Hopwood.), p. 15. The John Innes Charity, Norwich, England.

Gustafson, J. P. 1976. The evolutionary development of triticale: The wheat-rye hybrid. *Evol. Biol.* **9:** 107.

Jain, S. K. 1960. Cytogenetics of rye (*Secale* spp.). *Bibliogr. Genet.* **19:** 1.

Merker, A. 1976. The cytogenetic effect of heterochromatin in hexaploid triticale. *Hereditas* **83:** 215.

Riley, R. and V. Chapman. 1958. The production and phenotypes of wheat-rye chromosome addition lines. *Heredity* **12:** 301.

Rimpau, J., D. B. Smith, and R. B. Flavell. 1978. Sequence organisation analysis of the wheat and rye genomes by interspecies DNA/DNA hybridization. *J. Mol. Biol.* **123:** 327.

Southern, E. M. 1975. Detection of specific sequences among DNA fragments separated by gel electrophoresis. *J. Mol. Biol.* **98:** 503.

Stutz, H. C. 1972. On the origin of cultivated rye. *Am. J. Bot.* **59:** 59.

Thomas, J. B. and P. J. Kaltsikes. 1976. The genomic origin of unpaired chromosomes in triticale. *Can. J. Genet. Cytol.* **18:** 687.

PROBLEMS

1. Briefly define or describe the following terms in the context of genome structure: first-order reaction, second-order reaction, hypochromicity, hyperchromicity, nitrocellulose filters, hydroxyapatite column, S1 nuclease, high stringency, low stringeny, Cot curve and $Cot_{1/2}$, T_m (melting temperature).

2. Exercises in the qualitative interpretation of Cot curves (see Figure 3-2):
(a) Suggest a structure for the material yielding curve A.
(b) Curve B was observed for a eukaryote. What are the approximate proportions of highly repetitive, middle repetitive and low- or single-copy DNA?

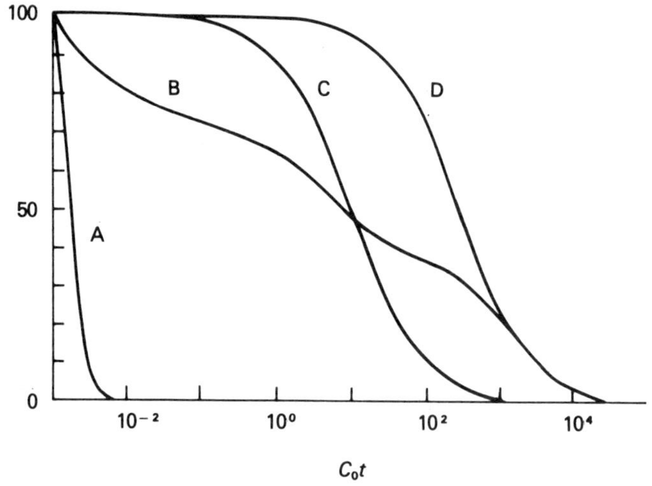

Figure 3-2

(c) Curve C is for a prokaryote whose genome size is 5 x 10^6 base pairs and curve D is for the reisolated single-copy fraction from curve B. Why is the $Cot_{1/2}$ of the reisolated fraction shifted to the left, relative to the $Cot_{1/2}$ of the corresponding part of the complex curve B? What is the total genome size of the eukaryote B?

3. Estimates of the proportions of various classes of DNA sequences depends strongly on the stringency of the hybridization conditions. Describe briefly, and in general terms, the expected effect of increasing stringency upon:
(a) A standard eukaryotic Cot curve.
(b) A standard interspersion analysis.
(c) Melting curve of the duplex molecules resulting from curve D?

4. This is an exercise in the qualitative interpretation of thermal denaturation curves (Figure 3-4).
(a) Panel A shows melting curves for both the main peak DNA (dashed line) and a satellite peak DNA (solid line). Account for the differences in both the melting temperatures (T_m) and the slopes of the curves.
(b) Panel B shows the melting curves for a sample of native eukaryotic DNA (thin solid line), the same DNA sheared to 200 base-pair fragments and melted and reannealed under high stringency conditions to an intermediate Cot value (heavy solid line), and the same sheared preparation

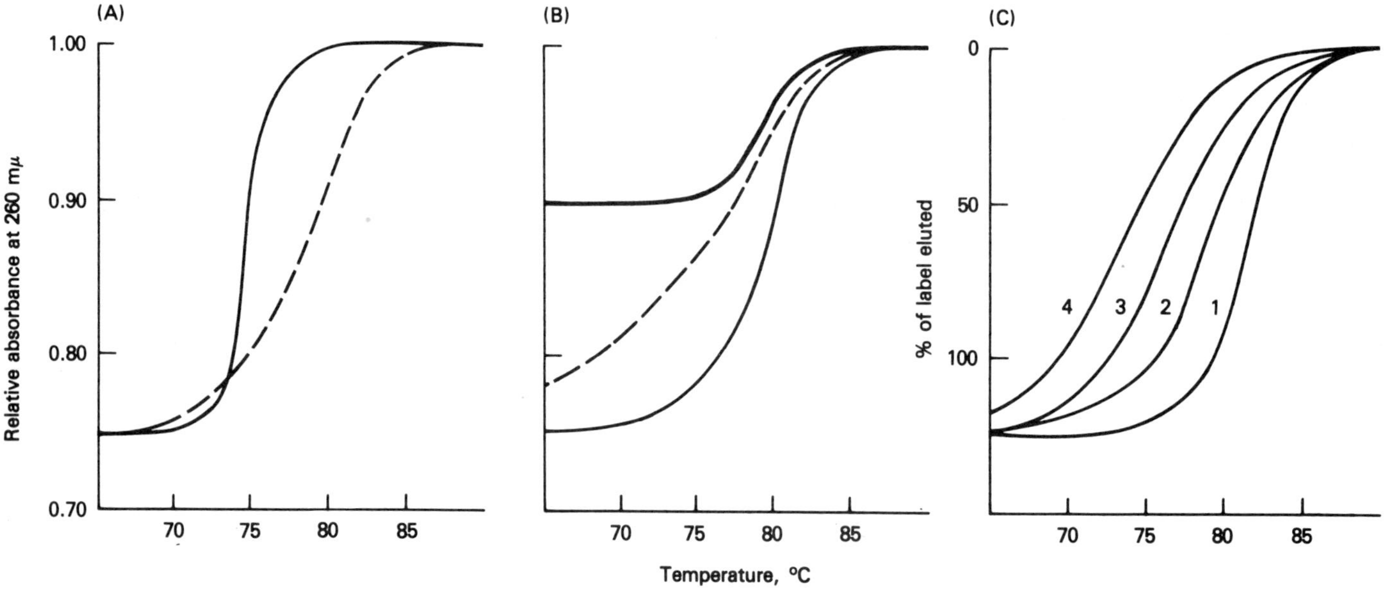

Figure 3-4

reannealed under low stringency conditions followed by purification by hydroxyapatite chromatography (dashed line). Account for the differences observed in these three curves.
(c) In panel C are shown the hydroxyapatite thermal elution profiles for the heteroduplexes formed between the labelled single-copy DNA of species 1 and an excess of homologous DNA (curve 1, control), of species-2 DNA (curve 2), of species-3 DNA (curve 3), and of species-4 DNA (curve). What can you say about the overall single-copy sequence divergence between these four species, and, hence, of the degree of relatedness?

5. Britten and Davidson (1976) reannealed sea urchin DNA to a low Cot value, treated the DNA with S1 nuclease, and isolated the resulting double-stranded fragments by hydroxyapatite chromatography. These were fractionated into long (more than 1500 base pairs) and short (approximately 300 base pairs) fragments, and melting profiles were obtained (Figure 3-5). What do these profiles say about the sequence homogeneity of the long and short segments and what does this suggest about their evolution?

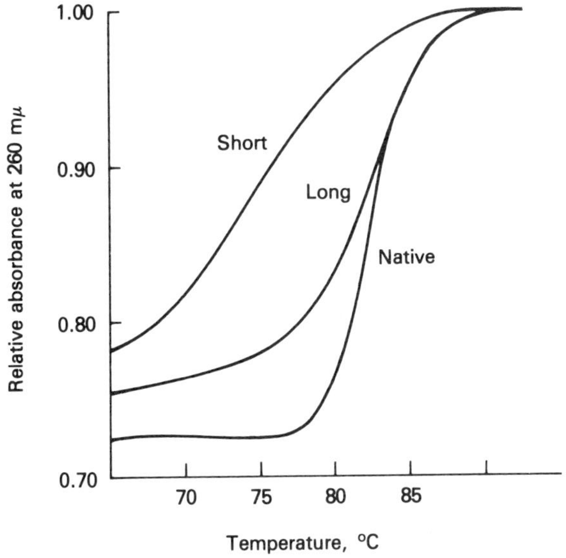

Figure 3-5

6. In order to study the interspersion pattern of repetitive and single-copy DNA, an excess of highly fragmented DNA can be hybridized with labeled DNA fragments of various (measured) length, to a Cot value at which only repetitive DNA anneals. For each of the following patterns of interspersion (obviously oversimplified), draw an expected profile of the percentage of label in duplex DNA as a function of the size of the labeled fragments, if
(a) The entire genome consists of alternating blocks of 200-bp repetitive DNA and 300-bp nonrepetitive DNA.
(b) The entire genome consists of alternating blocks of 200-bp repetitive DNA and 800-bp nonrepetitive DNA.
(c) 50 percent of the genome is arranged as in (a) and 50 percent of the genome as in (b).
(d) 50 percent of the genome is arranged as in (a) and 50 percent of the genome is continuous repetitive DNA
(e) 50 percent of the genome is arranged as in (b) and 50 percent of the genome is continuous nonrepetitive DNA.

7. The human (haploid) genome consists of approximately 2.5×10^9 base pairs of which 40 percent is S1 resistant following reannealing to a $Cot_{1/2}$ of 68 (only middle repetitive and highly repetitive DNA have reannealed). Houck *et al.* (1979) reported that approximately 12 percent of this S1-resistant fraction is in 300-bp fragments and that more than 60 percent of these 300-bp fragments are digestible by AluI (and by none of a dozen other restriction enzymes tested) to yield 120-bp and 180-bp fragments. This class of 300-bp fragments is termed the Alu family.
(a) What fraction of the genome does this sequence (or sequence family) represent and what must its repetition frequency be? Why are these estimates likely to be too low?
(b) How might you investigate whether this family of sequences is found primarily interspersed with single-copy sequences? Alternatively, how might you check if there is any appreciable amount involved in extensive tandem repetition (either precise or diverged)? If the former is the case, how might you check whether the sequences with which the Alu family are interspersed are repetitive sequences or single-copy sequences?

8. Total genomic DNA is digested with a restriction enzyme; the resulting fragments are separated on agarose gels and transferred to nitrocellulose filters. When the filter is finally probed with a

labeled known repeated element, often a "ladder" results in which the "rungs" are separated by a fixed increment. What is the most straightforward explanation for such a pattern? If use of a different restriction enzyme yielded a single higher molecular weight band and a very high molecular weight smear, what would this suggest about the original repeated element?

9. Flavell *et al.* (1977, this chapter) described seven groups of *Graminae* repetitive-sequence DNA that are defined by their relative frequency of occurrence in wheat, oat, barley, and rye genomes. In further work (Rimpau *et al.*, 1980), they have asked more detailed questions about the patterns of interspersion of these groups of sequences in the genomes of the four representative cereals. In a particular experiment, labeled barley DNA fragments of lengths varying from 200 bp to 11,000 bp were annealed with an 800-fold excess of short fragments of barley, wheat, rye, or oat DNA to a Cot of 120. The amount of labeled DNA involved in hybridization (either directly by formation of double-stranded DNA or indirectly by virtue of covalent linkage to regions that form double-stranded stretches) is measured by hydroxyapatite chromatography. The results are presented in Figure 3-9.

(a) From the three curves given, what proportion of the repetitive DNA of the barley genome may be attributed to group I (common to all four species), group II (common to barley, wheat, and rye) and to group VI (specific to barley)? What percentage of the unique sequence DNA appears to be in runs of 10,000 bp or greater?
(b) What proportion of the barley genome contains group-VI repetitive sequences exclusively, and what must be the pattern of interspersion in that region?
(c) What proportion of the genome contains interspersed group-I sequences and what is the pattern of interspersion? How might you assess the proportion of this region that is actually nonrepeated sequence?

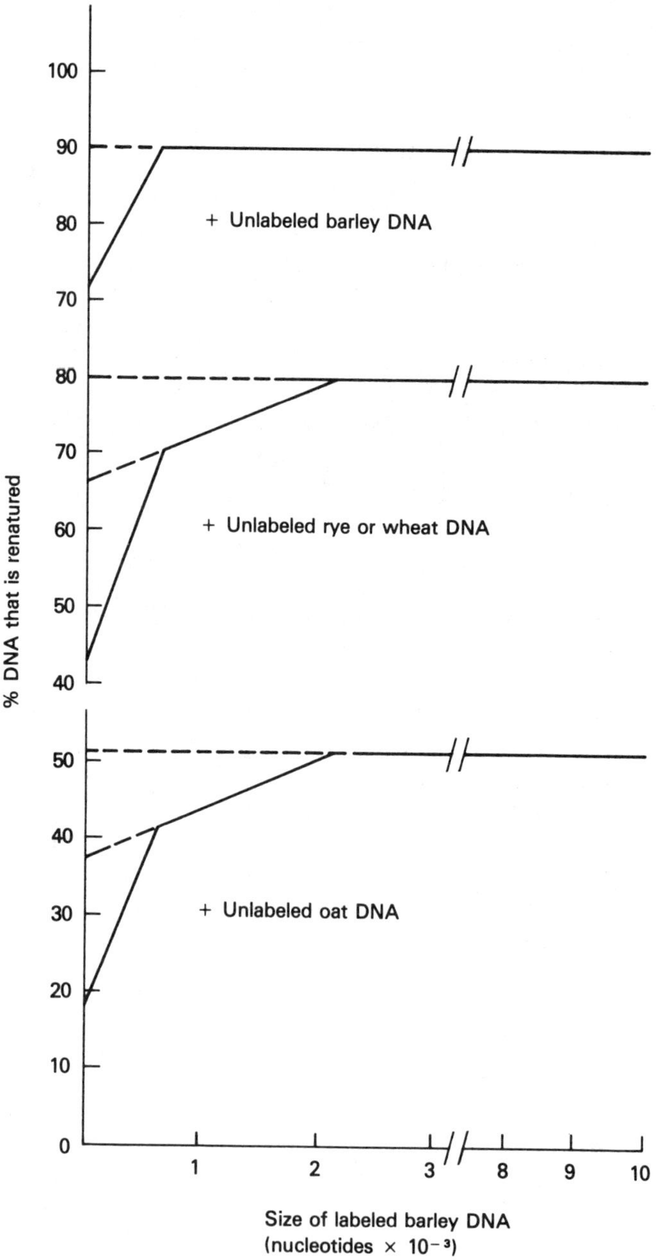

Figure 3-9

10. What specific features of a cultivar, like wheat, makes possible the experiment reported by Bedbrook *et al.* (1981, this chapter) that led to the conclusion that the 2.2-kb complex, containing the 120-bp repeat element, is carried on all chromosomes? By extension, what is the advantage of working with a system such as the cultivated wheats (see Feldman and Sears, 1981), if one wishes to examine genomic events that occur on a relatively short time scale?

4 CODING SEQUENCES—DESCENT WITH MODIFICATION

...before long we shall have a subject which might be called "protein taxonomy"—the study of the amino acid sequences of the proteins of an organism and the comparison of them between species. It can be argued that these sequences are the most delicate expression possible of the phenotype of an organism and that vase amounts of evolutionary information may be hidden away within them.
F. Crick (1958)

A cornerstone of molecular evolution is the proposition that an organism carries a record of its history inscribed in its nucleotide sequence. It follows that the degree of evolutionary relatedness between two organisms should be reflected in the degree of homology in their macromolecular sequences (Crick, 1958; Zuckerkandl and Pauling, 1962). This prediction has been strongly supported during the last 25 years by many comparative studies of both protein and nucleic acid sequences: cytochrome c (Fitch, 1976), myoglobins (Romero-Herrera *et al.*, 1978), hemoglobins (Goodman *et al.*, 1979), globins (Lesk and Chothia, 1980), pancreatic ribonuclease (Beintema *et al.*, 1977), and snake venoms (Hseu *et al.*, 1977). In this chapter we will consider the evolution of proteins whose homology may be surmised by identity of function.

In the Introduction a clear distinction was made between the treatment of evolution as history and the study of the mechanisms of evolutionary change. In considering the evolution of coding sequences we shall take up these two aspects of evolutionary studies in that order.

With the assumption that relative sequence divergence is a function of the time elapsed since the splitting of a population into two or more isolated populations and that this divergence is neutral as far as gene function is concerned, sequence data provide a unique opportunity for the objective reconstruction of trees of descent.

An early systematic approach to the use of sequence data was done by Eck and Dayhoff (1966b), and aided by the rapid development of computer technology an entire mini-discipline of "tree construction" has arisen. Perhaps because the data are both precise and biologically meaningful, the problem of tree construction has attracted the more formal attention of mathematicians interested in graph theory (for example, Waterman and Smith, 1978; Hendy *et al.*, (1980). It is easy to construct evolutionary trees from sequence data, as superficially discussed in Hood *et al.* (1975), but difficult to do it well. Special problems arise when dealing with widely divergent sequences and when little or no reliable independent information is available about the historical time or phylogenetic location of the divergence of specific sequences. In particular, it is difficult to distinguish between homologous sequences, whose divergence reflects directly the organismal phylogeny, and paralogous sequences, which reflect a divergence predating that of the organisms. Additional complications are posed by the great variation in conservatism of specific sequences, even at the sub-gene level, and by the various nonclassical modes of genetic change and exchange, as discussed in several other chapters. Discussions of tree-building methods are in the papers of Boulter *et al.* (1972, Chapter 7) and De Ley (1974, Chapter 8); Cedergren *et al.* (1981) and Felsenstein (1982) present excellent reviews of this topic. The comparison of trees has answered criticisms concerning the supposed inability of evolutionary theory to make falsifiable predictions by showing that different proteins from the same array of organisms yield similar trees (Penny *et al.*, 1982), thereby confirming the ideas of Crick (1958) and Zuckerkandl and Pauling (1962).

One of the first conclusions from molecular studies coupled with dating of divergence times from the fossil record was that over long periods of time a particular protein (even in different lines of descent) had similar rates of amino-acid substitutions. From this observation Zuckerkandl and Pauling (1965) suggested the existence of an evolutionary molecular clock, a term now widely used to describe the observation that similar rates of change are observed with homologous proteins from different organisms. The central idea was that any protein with a fixed function (Dickerson, 1971) will have a characteristic rate of evolution, but that this rate is not the same for all proteins. Examples are the similar rates of change for hemoglobin in different branches of the vertebrate evolutionary tree, or for cytochrome c in vertebrates and flowering plants (Ramshaw *et al.*, 1972). Rates for different proteins can be found in Dayhoff (1972) and in the supplements to that work. The question as to why proteins evolve at different rates has been a less contentious issue. Dickerson (1971) and Kimura and Ohta (1974, this chapter) have provided qualitatively useful guidelines for understanding the different rates of change; these include constraints imposed by the conservation of active sites and of points of interaction with other macromolecules as well as the maintenance of the overall tertiary conformation of the protein. Fitch and Markowitz (1970, this chapter) discuss their model of covarions (concomitantly variable codons), which proposes that only some amino acid sites are free to change at any one time, but that the ensemble of permissible changes will alter as substitutions occur in the protein. Thus, differences in rates of amino acid substitution between proteins may largely disappear if attention is restricted to those positions that are unconstrained at the time of substitution. This view of shifting internal constraints on change is conceptually appealing and useful for understanding such things as parallel changes in different lines of descent or certain details of amino acid replacement. The main difficulty has been that it has proved difficult to identify the sites that are free to accept changes at a particular point in time.

The existence of a molecular clock was unexpected in the context of classical population genetics and naturally raised questions about the factors that shape structural gene evolution. Evolutionists had assumed by analogy with morphological studies that sequences in some lines of descent would be very constant (in living fossils), whereas other lines would have shown rapid changes (after adaptive radiation). However, there were several early suggestions from protein chemists that the amino acid positions that change most frequently may not be very important in protein structure (Doolittle and Blomback, 1964; Zuckerkandl and Pauling, 1965). In consequence, protein (gene) evolution might proceed largely independently of morpholoigcal evaluation.

Kimura (1968, Chapter 1) explicitly recognized that the existence of neutral mutations (those neither selected for nor selected against) would lead to a stochastic (probabilistic) molecular clock based on relatively constant opportunities for random nucleotide substitutions. He showed that the expected rate of substitution of neutral mutations (selective advantage s = 0) is independent of population size. The larger the population, the higher the chance of a mutation, but the increase in the number of mutations is counterbalanced by the lower chance of any particular mutation being fixed in the population. Thus, the rate of amino acid substitution is proportional only to the mutation rate. This neutralist hypothesis entails the existence of a molecular clock and, contrary to the expectations of classical population geneticists, requires that many gene loci be heterozygous, as had been found by Harris (1966, Chapter 1) and Hubby and Lewontin (1966). Lewontin (1974) and Kimura and Ohta (1974, this chapter) discuss some of the implications of the high levels of heterozygosity.

A significant controversy has existed between exponents of this neutralist view and advocates of panselectionism, the view that virtually all changes in DNA sequence require positive selection. Both groups recognize that many mutations that alter a protein sequence would be deleterious and therefore would be eliminated by selection; they differ in the relative frequency of mutations for which there is positive selection, compared to those that are neutral. In practice, opinions may range continuously from the view that every substitution is neutral to the view that no substitution is neutral.

At least two factors have added confusion to the debate. The first is the human tendency to ovestate the opposing case. The initial neutralist statement was that "many, if not the majority" of substitutions would be neutral. However, the statement that often is tested is that "all substitutions are neutral," because it is easiest to take as the null hypothesis that all mutations are neutral, and then calculate expected rates of change or levels of hetrozygosity (see for example Fitch and Langley, 1976). The other point of confusion has been a semantic one. The use of the terms "Darwinian" for panselectionism and "non-Darwinian" for neutralism reflects a misunderstanding of Darwin who pointed out that "many structures are now of no direct use to their possessors, and may never have been of any use to their progenitors" ("Origin of Species", Chapter 6).

Several selectionist theories have attempted to explain the molecular clock. Van Valen (1974) has based one on his Red Queen hypothesis, which points out that with finite availability of resources, a species must keep improving or lose its share of resources, as other species become more competitive (see also Zuckerkandl, 1976). However, in general, advocates of selectionism have concentrated on looking for irregularities in the rate of sequence evolution (Goodman, 1981; Baba *et al.*, 1981), which may be interpreted as a burst of positive selection following, for example, gene duplication. This approach may not be decisive in placing a protein on the continuum from no changes being neutral to all changes being neutral, because an increase in the rate of substitution may result from a reduction of functional constraints following gene duplication. This issue is considered further in problem 7.

Certainly, some specific amino acid substitutions can alter the physiological properties of proteins. For example, there are 39 differences between human adult beta globin and the fetal gamma sequences. Of these differences, one (histidine at position 143 in beta globin is replaced by serine in the gamma chain), may account for the different affinities for oxygen of the adult and fetal hemoglobins (Harris, 1976). As Harris comments, "the score for natural selection is apparently one out of 39, with 38 residues to go". However, it is not sufficient just to show that an amino acid substitution changes the physiological properties of the protein; it is also necessary to show that this change alters the probability of survival of the organism in its environment. The question of the relative importance of neutral and selected changes remains unresolved, but in the view that there is a continuum of selective values, the controversy may be fading out.

An alternative approach to using sequence

similarities to detect possible evolutionary relationships among proteins is to compare three-dimensional structures. Such comparisons involve a broader, and perhaps more meaningful approach to the problem of sequence alignment. Examination of an extended series of globins (ranging from the alpha and beta chains of human globin to the leghemoglobins of legume root nodules) reveals nearly identical backbone conformations for all of the molecules, despite a sequence homology that may be as low as 16 percent (Lesk and Chothia, 1980). Almassy and Dickerson (1978, this chapter), similarly focusing on tertiary structure, substantially expanded the cytochrome c family of molecules. Furthermore, the lysozymes of bacteriophage T4 and hen egg white (Grutter, Weaver, and Matthews, 1983, this chapter), which have no significant amino acid sequence homology, are nearly identical with respect to backbone conformation, mode of substrate binding, and mechanism of catalysis. This general line of inquiry has important implications not only for those interested in "descent with modification", but also for those interested in understanding the details of the relationship between protein function and primary, secondary, and tertiary structure.

The lysozyme data raise the important question whether, in the absence of any discernible sequence homology, divergence from a common ancestor can be distinguished from independent convergence from different ancestors. Indeed, Richardson (1977) has pointed out that among the large number of sterically possible beta-sheet topologies, a relatively small number are observed. This apparent conservatism may be attributed either to an early evolutionary establishment and widespread use of this protein element, or to convergence towards a particularly stable conformation. Thus, even though similiarity of backbone conformation and similarity of function and mechanism, seem to argue strongly for common evolutionary descent, the apparent restriction of conformational possibilities in apparently unrelated proteins dictates caution in interpretation. There may well be cases that cannot be decided on the basis of present evidence (Doolittle, 1981).

A different approach to the question of sequence relations and the constraints on DNA nucleotide sequence change is provided by the technology of gene cloning and nucleotide-sequence determination (Gilbert, 1981). The study of van Ooyen, van den Berg, Martel, and Weissmann, (1979, this chapter) reveals a variety of constraints on nucleotide-sequence alterations in third positions of codons or in noncoding intervening sequences. This observation suggests that selective constraints operate at a level other than that of protein function and may explain the conservation of amino acids at specific sites that do not appear to be otherwise functionally constrained.

The approaches explored in this chapter focus on genes that have retained their function or sequence during evolution and therefore cast little light on the problem of the origins of new functions. Only with difficulty can these approaches detect relations between parts of otherwise unrelated proteins. The recent recognition of functional domains in proteins and their correlation with exons (Gilbert, 1978, Chapter 2) provide one means by which homologous regions may be recognized in widely divergent proteins. This point will be examined in the next chapter, along with a more explicit treatment of the problem of derivation of new functions.

An Improved Method for Determining Codon Variability in a Gene and Its Application to the Rate of Fixation of Mutations in Evolution

Walter M. Fitch[1] **and Etan Markowitz**[2]

Received 15 Jan. 1970—Final 1 April 1970

If one has the amino acid sequences of a set of homologous proteins as well as their phylogenetic relationships, one can easily determine the minimum number of mutations (nucleotide replacements) which must have been fixed in each codon since their common ancestor. It is found that for 29 species of cytochrome c *the data fit the assumption that there is a group of approximately 32 invariant codons and that the remainder compose two Poisson-distributed groups of size 65 and 16 codons, the latter smaller group fixing mutations at about 3.2 times the rate of the larger. It is further found that the size of the invariant group increases as the range of species is narrowed. Extrapolation suggests that less than 10% of the codons in a given mammalian cytochrome* c *gene are capable of accepting a mutation. This is consistent with the view that at any one point in time only a very restricted number of positions can fix mutations but that as mutations are fixed the positions capable of accepting mutations also change so that examination of a wide range of species reveals a wide range of altered positions. We define this restricted group as the concomitantly variable codons. Given this restriction, the fixation rates for mutations in concomitantly variable codons in cytochrome* c *and fibrinopeptide A are not very different, a result which should be the case if most of these mutations are in fact selectively neutral as Kimura suggests.*

INTRODUCTION

Kimura (1968) has suggested that during the evolution of cytochromes *c*, hemoglobins, and triosephosphate dehydrogenase, the vast majority of mutations fixed were selectively neutral mutations. King and Jukes (1969) agree but for different reasons. Corbin and Uzzell (1970) also agree and go on to calculate the absolute mutation rate on the assumption that all mutations are neutral in the codons for the most variable positions of fibrinopeptides A and B. The rate they found is approximately the same as the rate of mispairing of bases during replication as computed by Watson (1965). Maynard Smith (1968) takes issue with Kimura by choosing to dispute the validity of using Haldane's "cost of natural selection" (1957). Maynard Smith then devises a model which permits selection at many loci simultaneously and thereby escapes the dilemma. O'Donald (1969) agrees with Smith's conclusion regarding Haldane but also regards Maynard Smith's model as biologically unrealistic. O'Donald then proposes his own model and comes to the same ultimate conclusion as Kimura regarding the cost.

But regardless of how the cost is accounted, the conclusion that most of the mutations fixed in the structural genes for these proteins were selectively neutral may nevertheless be correct. However, if the vast majority of those mutations fixed are selectively neutral, then in those positions where mutations may be fixed, they should be fixed with equal probability and therefore at equal rates.[3] This should be a strong consistency criterion for the validity of the neutrality of most fixed mutations. The difficulty lies in a proper estimate of the number of those positions. This paper will

---

Paper number 1382 from the Laboratory of Genetics. Work performed in part at the University of Iowa, Department of Preventive Medicine and Environmental Health and Department of Statistics, Iowa City, Iowa. Computing supported by the Graduate College, University of Iowa.

[1] Department of Physiological Chemistry, University of Wisconsin, Madison, Wisconsin.

[2] Departments of Physiological Chemistry, Medical Genetics, and Statistics, University of Wisconsin, Madison, Wisconsin.

[3] We recognize that there may exist more acceptable alternatives for some codons than for others at any one point in time. For the present study, we assume that this variability averages out when a collection of codons is examined over long periods of time.

first show how to calculate the number of codon positions capable of fixing nondeleterious mutations and then, using this number, show that mutations are indeed being fixed at approximately equal rates in different genes.

DETERMINATION OF CODON VARIABILITY

A previously published procedure (Fitch and Margoliash, 1967*a*) showed that, provided a few recognizable coding positions were excluded from the data, the remainder of the mutations found in cytochrome *c* distributed themselves as if there were a number of invariant positions and the rest were equally likely to be the position fixing the next mutation; that is, the number of mutations fixed in the variable positions followed a Poisson distribution. That procedure is improved upon here. First of all, no coding positions are arbitrarily excluded. Secondly, an iterative procedure obtains the maximum likelihood estimates for the parameters. This leads to an expected distribution which may be compared to the observed distribution by the Pearson chi-square test. The procedure examines three possible models. Model 1 assumes that all positions are equally variable. Model 2 assumes that there are some invariant positions and that the rest are equally variable. Model 3 assumes that there are some invariant positions and that the rest constitute two classes of variability. Details are presented in the accompanying paper (Markowitz, 1970).

CODON VARIABILITY IN CYTOCHROME *c*

Figure 1 shows the phylogeny of 29 species based upon their cytochrome *c* sequences using the method of Fitch and Margoliash (1967*b*). When this particular phylogeny is assumed, the 29 present-day sequences may be accounted for by a minimum of 366 fixations [i.e., nucleotide replacements; *cf.* Fitch (1970) for method]. Their distribution over the 113 codons is shown by group A in Table I, which gives the number of codons that were found to have fixed the number of nucleotide replacements indicated at the top of each column. Models 1, 2, and 3 immediately below group A are the best-fitting solutions according to the three possible models, respectively, outlined in the preceding paragraph. The χ^2 values are also shown in Table I for each of the three possible models. It can be seen that only the third model gives a reasonable fit to the data. For this distribution, the number of invariant codons is estimated at 32, with the 81 remaining variable codons being divided into two groups of size 65 and 16 with Poisson parameters of 3.2 and 10.1, respectively, which is to say that mutations, on a per codon basis, are being fixed in the latter small group at 3.2 times the rate of fixation in the former large group.

This substantiates the earlier conclusion (Fitch and Margoliash, 1967*a*) that there is a small group of codons which, relative to the others, is hypervariable,[4] and contradicts the conclusion of King and Jukes (1969) that the "so-called 'hypermutable sites'. . . are predictable in terms of [a single] Poisson distribution." Since we are using the same basic data, the discrepancy needs to be accounted for. King and Jukes found no need to postulate a second Poisson distribution because they used a less sensitive procedure in determining the number of mutations in each coding position. They simply examined all the amino acids in each position and determined the number of nucleotide replacements necessary to accomplish their interconversion [their method is more explicitly given in Jukes (1969)]. This fails to take proper account of the total number of replacements required as a consequence of the phylogeny. For example, in position 47 of cytochrome *c* of the 20 species examined in the earlier study (Fitch and Margoliash, 1968) there is an alanine in *Neurospora*, a threonine in the screw worm fly, the kangaroo, and the horse, and a serine in two fungi, one insect, one fish, two reptiles, four birds, and six other mammals. There is no reasonable explanation of these data

[4] "Hypervariable" refers to the fact that, relative to the majority of the codons fixing mutations, this group is fixing them more rapidly. Since one cannot assume that these codons are more mutable, the term "hypermutable" (Fitch and Margoliash, 1967*a*) has been replaced by "hypervariable" (Fitch and Margoliash, 1968).

other than to postulate that on three separate occasions, a mutation from a codon for serine to one for threonine has been fixed, and we would thus count it as three fixations rather than one. King and Jukes have not followed this practice. This is of some importance since, of the 230 mutations in that paper (Fitch and Margoliash, 1968), there were 41 such parallel mutations as well as three back-mutations comprising 20% of the total mutations. These would not have been assessed by King and Jukes.

A model with precisely two classes of variable codons is not as biologically attractive as a model with only one. If the variable positions are not all equally variable, and this is clear from the data, then it is more reasonable to assume that there is a range of variability. Our using only two classes of variability is a simplification that merely reflects a combination of the limited sensitivity of the method, the limited amount of sequence data, and the generally well-behaved nature of the property we are measuring. A model that predicted how the range of variability was distributed could be tested,

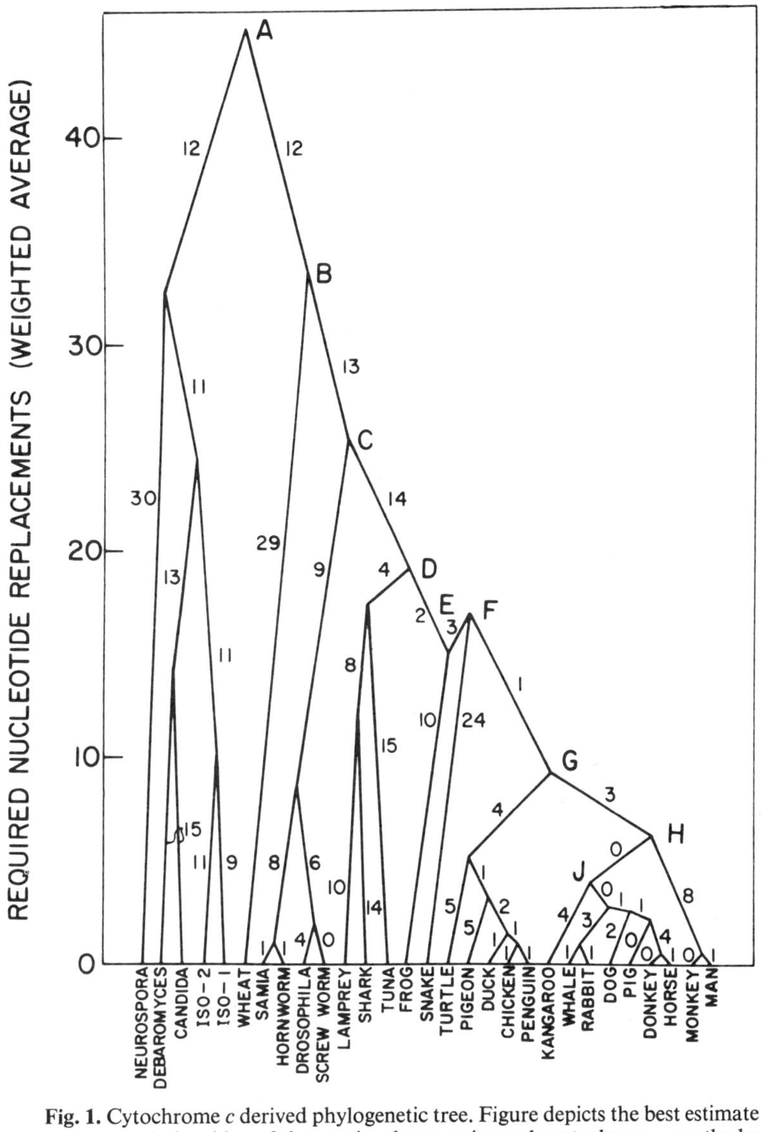

Fig. 1. Cytochrome *c* derived phylogenetic tree. Figure depicts the best estimate of ancestral relationships of the species shown using only cytochrome *c* as the basis for that estimate. Numbers on segments are the actual number of nucleotide replacements required to account for the descent of the present sequences. The following cytochromes *c*, in the order they appear in the figure, were used in constructing the tree: *Neurospora crassa*; *Debaromyces kloeckeri*; *Candida krusei*; *Saccharomyces cerevisiae* iso-1; *Saccharomyces* iso-2; wheat; *Protogarce sexta*; *Samia cynthia* moth; *Drosophila melanogaster*; *Haematobia irritans*; *Entosphenus tridentatus*; *Squalus sucklii*; tuna; frog; *Crotalus adamanteus*; *Chelydra serpentina*; pigeon; *Anas platyrhyncos*; chicken; *Aptenodytes patagonica*; *Macropus canguru*; *Rachianectes glaucus*; rabbit; dog; pig; donkey; horse; *Macacus mulatta*; man.

Table I. Distributions of Fixations According to Sites with Various Numbers of Fixations[a]

F/S	0	1	2	3	4	5	6	7	8	9	10	11	12	13	14	15	χ^2	df	p
Group A	35	8	17	10	15	7	3	4	4	3	0	0	2	1	3	1			
Model 1	4.4	14.3	23.2	25.1	20.3	13.2	7.1	3.3	1.3	0.5	0.2	0.05	< 0.02				> 453*	> 112	$< 10^{-26}$
Model 2	35	0.8	8.2	12.6	14.7	13.6	10.6	7.0	4.1	2.1	1.0	0.4	< 0.25				> 100	> 10	$< 10^{-14}$
Model 3	35	8.7	13.8	14.5	11.4	7.2	4.7	3.1	2.4	2.2	2.0	1.9	1.6	1.2	0.8	1.4	15	11	0.18
Group J	102	6	4	1															
Model 1	97.2	14.6	1.1	0.06													189*	112	$< 10^{-6}$
Model 2	102	6.6	3.1	1.2													0.3	1	0.58
Group K	2	1	4	2	0	3	3	2	1	0	1								
Model 3	2	1.9	2.2	2.3	2.3	2.2	2.0	1.6	1.1	0.7	0.78						5.8	5	0.45

[a] For each of three groups of amino acid sequences, the number of codon sites found to have fixed zero, one, two, etc., mutations is given under the column whose heading is the number of fixations per site (F/S). For each model is indicated the expected distribution of mutations in the group on the assumption that there are no invariant codons (model 1), that there are invariant codons (model 2), and that there are not only invariant codons but also that the variable codons are a mixture of two Poisson-distributed classes with different variability (model 3). Group A are the cytochromes *c* from the 29 species shown in Fig. 1. A total of 366 fixations is distributed over 113 codons. In determining χ^2 values, the rightmost expected value of the distributions contains all of the residual expectation lumped together. Model 1 χ^2 values, shown with an asterisk, are based upon Fisher's variance test or index of dispersion (*cf.* Markowitz, 1970). If all the data for fixations/site > 9 are lumped together, model 2 still gives a $\chi^2 < 51$ which, for 8 degrees of freedom, gives a probability $< 10^{-6}$ of occurring by chance. However, the same lumping of the data for model 3 reduces χ^2 from 15 down to 5.6 with an associated probability of 0.36. The Poisson parameters for the two variable classes of model 3 are 3.2 and 10.1. The 35 codons in model 3 that have never varied are composed of two groups: 32 invariant codons plus three variable but unvaried codons. Group J are the data from the cytochromes *c* of the nonprimate mammals of Fig. 1. A total of 17 fixations is distributed over 113 codons. The 102 codons of model 2 that have never varied are composed of two groups: 95 invariant codons plus seven variable but unvaried codons. No computation was performed for model 3 since there are insufficient degrees of freedom. Group **K** are data based upon the fibrinopeptides A from 25 mammalian species. A total of 80 fixations is distributed over 19 codons. The estimate of the number of invariant codons for group **K** is 1.1, with an estimated standard deviation of 1.9.

but for the present the limited approximation of only two degrees of variability will suffice.

CONCOMITANTLY VARIABLE CODONS

In what was to have been an attempt to show that the same distribution of variability among the codons is obtained regardless of the range of organisms examined, the fungi were omitted from group A to form group B and the computations repeated. As a result, the number of invariant codons increased from 32 to 51. The computation was successively repeated, each time using a new group which was formed by omitting the most remote members of the previous group. These groups are shown in Fig. 1 by a letter at the appropriate apex, the members of a particular group being all descendants from the apex bearing that group's letter designation. The reduction in range of species continued until, for group J, only nonprimate mammals remained. The 17 mutations required to account for their differences were distributed as shown in Table I by Group J. Also shown are the best-fitting distributions for models 1 and 2 (there are too few degrees of freedom to fit model 3). For model 2, the estimate of the number of invariant codons is 95, a sharp contrast to the 32 calculated for trial group A. The percentage of the gene which is estimated to be invariant increases as the range of species narrows. That range may be represented by the height of the apex of the group shown in Fig. 1. Figure 2 shows that the inverse relationship between the range of

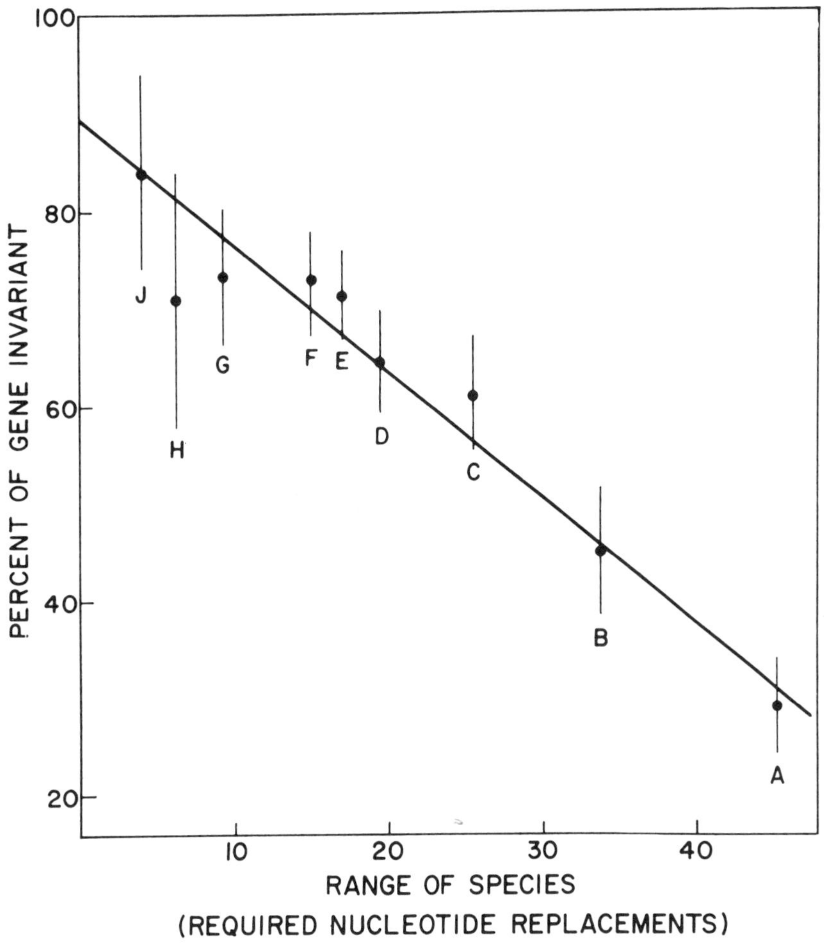

Fig. 2. Concomitantly variable codons. The percent of the gene found to be invariant (y) is plotted as a function of the range of species examined (i.e., peak height in Fig. 1). A least-squares fit to the results is extrapolated to the ordinate, which estimates the fraction of the gene for which all mutations are lethal or malefic. The rightmost and leftmost values plotted are for the groups A and J, respectively, of Table I. Each vertical line extends one estimated standard deviation above and below its y value.

species examined and the percentage of the gene that is invariant is linear in the region for which we have data.[5] The line shown is a weighted least-squares fit; each point's weight is the reciprocal of its approximate variance. It is vital to recognize that the invariant codons are only a subset of the codons that have not varied. As the range of species narrows, the number of codons that have not varied necessarily increases but the same is not necessarily true for the invariant codons.

The line in Fig. 2 may be extrapolated upward to the y axis. We interpret this intercept to mean that when the range of species being examined is reduced to zero (i.e., when only one species' cytochrome *c* gene is considered), over 90% of the codons in the mammalian cytochrome *c* gene are invariant. Although this appears to contradict our experience that, if a broad enough range of species is examined, most of the codons will have varied, there is a rational biological explanation available. This explanation asserts that because of the structural restraints imposed by functional requirements, mutations that will not be selected against are available only for a very limited number of positions. We shall use the term acceptable for such mutations. However, as such acceptable mutations are fixed they alter the positions in which other acceptable mutations may be fixed. Thus, only about 10 codons, on the average, in any cytochrome *c* may have acceptable mutations available to them but the particular codons will vary from one species to another. We shall term those codons at any one instant in time and in any given gene for which an acceptable mutation is available as the *concomitantly variable codons*.[6]

King and Jukes (1969) state that "74 to 81 residues are variable" and that "there is very little restriction on the type of amino acid that can be accommodated at most of the variable sites." We would interpret our results as indicating that, contrariwise, there is much restriction in the cytochrome *c* gene most of the time with very few of the codons having acceptable mutations available and we see no reason to assume otherwise than that the range of acceptable mutations in those codons may be similarly limited.

SPATIAL CORRELATIONS IN AMINO ACID REPLACEMENTS

The implication that events in one coding position may be dependent upon events in other positions may be related to the interesting observations of Wyckoff (1968) who noted a spatial relationship between many pairs of the substitutions observed between rat and bovine ribonucleases. This was possible because the three-dimensional structure of bovine RNase is known. For example, the amino acids in positions 38 and 39 of rat RNase are glycine and serine, respectively. Now glycine could mutate to aspartate but, as Wyckoff points out, presumably this would be damaging because it could interact with lysine 41 and pull this necessary residue out of the active site. Also, the serine could mutate to arginine, with there being no particular reason to suspect that this might not be acceptable. In bovine RNase, the groups are indeed aspartate and arginine but the positively charged arginine neutralizes the negatively charged aspartate and, perhaps, prevents any deleterious effect of the aspartate on the critical lysine 41. If this is true, we presume that the fixation of arginine at 39 must have preceded the fixation of aspartate. This illustrates well how the positions belonging to the group of concomitantly variable codons may change since before the arginine fixation position 38 might not tolerate an aspartate, whereas after both fixations position 39 might not tolerate the return of arginine to a neutral residue.

Another example is provided by positions 57 and 79, which in bovine RNase are

[5] Although the curve is linear in this region, we would assume that it must, as the range of species increases, eventually approach, as a lower limit, an asymptote parallel to the x axis at a height indicative of the truly invariant positions. If prokaryotic cytochromes *c* of *Pseudomonas fluorescens* (Ambler, 1963) and *Rhodospirilum rubrum* (Dus *et al.*, 1968) are considered to be homologous to the eukaryotic horse cytochrome *c* as shown in Table II, there are only seven of the 75 common positions identical so that we may put an upper limit for the asymptote at 9.4%.

[6] The phrase *co*ncomitantly *vari*able cod*ons* is awkward and we would suggest the term "covarions" to describe this particular set of codons.

Table II. Comparison of Prokaryotic and Eukaryotic Cytochromes *c*[a]

Ps. fluor.						GLU	asp	pro	GLU	val	leu	PHE
Rh. rubrum		glu	GLY	ASP	ala	ala	ala	GLY	GLU	LYS	–	–
Horse			GLY	ASP	val	GLU	lys	GLY	lys	LYS	ile	PHE
			•					•				•
	lys	asn	LYS	gly	CYS	val	ALA	CYS	HIS	ala	ile	ASP
	VAL	ser	LYS	lys	CYS	leu	ALA	CYS	HIS	THR	phe	ASP
	VAL	gln	LYS	–	CYS	ala	gln	CYS	HIS	THR	val	glu
	–	–	–	thr	lys	met	VAL	GLY	PRO	ala	tyr	lys
	gln	GLY	GLY	ala	asn	LYS	VAL	GLY	PRO	ASN	LEU	phe
	lys	GLY	GLY	lys	his	LYS	thr	GLY	PRO	ASN	LEU	his
						•					•	
	asp	VAL	ala	ala	lys	phe	ALA	GLY	GLN	ALA	gly	ala
	GLY	VAL	PHE	glu	asn	thr	ALA	ala	his	lys	asp	asn
	GLY	leu	PHE	gly	arg	lys	thr	GLY	GLN	ALA	pro	gly
	•				•			•				•
	–	–	–	–	–	–	–	–	–	–	–	–
	tyr	ala	TYR	ser	glu	ser	tyr	thr	glu	met	LYS	ala
	phe	thr	TYR	thr	asp	ala	asn	–	–	–	LYS	asn
			•			•	•					
	–	–	–	–	–	–	GLU	ALA	glu	LEU	ALA	gln
	LYS	GLY	leu	THR	TRP	thr	GLU	ALA	asn	LEU	ALA	ala
	LYS	GLY	ile	THR	TRP	lys	GLU	glu	thr	LEU	met	glu
					•							
	arg	ile	LYS	ASN	gly	ser	gln	gly	VAL	trp	GLY	pro
	TYR	val	LYS	ASN	PRO	LYS	ala	phe	VAL	leu	glu	lys
	TYR	leu	glu	ASN	PRO	LYS	lys	tyr	ile	pro	GLY	–
	•	•			•	•	•	•	•	•	•	
	ile	pro	–	–	–	–	–	–	–	MET	pro	pro
	ser	gly	asp	pro	lys	ala	lys	ser	LYS	MET	thr	PHE
	–	–	–	–	–	–	–	thr	LYS	MET	ile	PHE
								•	•			
	asn	ala	val	ser	–	ASP	ASP	GLU	ala	gln	thr	LEU
	–	lys	leu	thr	LYS	ASP	ASP	GLU	ile	GLU	asn	val
	ala	gly	ile	lys	LYS	lys	thr	GLU	arg	GLU	asp	LEU
	ala	lys	trp	val	leu	–	–	ser	gln	LYS		
	ILE	ALA	TYR	LEU	LYS	–	–	THR	leu	LYS		
	ILE	ALA	TYR	LEU	LYS	lys	ala	THR	asn	glu		

[a] A dash (–) indicates a gap inserted to maintain the homologous alignment. Underlining indicates those 7 positions which are still invariant. Except for the invariant asparagine (ASN), all of these residues interact with the heme or with residues which do in bovine cytochrome *c* (R. E. Dickerson, personal communication). Positions with a dot under them are invariant among the eukaryotic cytochromes *c* so far examined.

valine and methionine and in the rat are isoleucine and leucine, respectively. Although the individual amino acids are of differing volumes, the pairs have the same total length. If there is no further room in the site filled by these residues, we would have to assume that in the ancestral pair, the longer of the amino acids (say isoleucine) had to be replaced by a shorter descendant (say valine) before the other ancestral amino acid (say leucine) could tolerate a mutation changing it to a longer descendant (in this illustration, methionine). This avoids trying to fit both methionine and isoleucine into the restricted space at the same time. For this to be a neutral process, it must be assumed that there is tolerance for the loss of a methylene group in this region. Wyckoff presents evidence for such tolerance at another site inR Nase.

At position 14 is an aspartate that is charge-bonded to an arginine in position 33 of the cow and position 32 of the rat. Presumably, in the intermediate form, aspartate 14 has an option between the two arginines (rather than none) but since only one was needed, it was acceptable for one to be replaced subsequently and it was. A summary of these three cases is shown in Fig. 3.

This discussion of paired replacements in RNase is extended because the very specific nature of the replacements observed necessarily focuses one's attention upon the selective requirements of the positions. Nevertheless, as we hope the preceding indicates, it is possible to provide a rational explanation for these events in terms of neutral mutations even though the latitude permitted to what is acceptably neutral may be severely limited by selective forces and even though the neutrality of the intermediates has not been demonstrated. It also shows that we are not yet forced to postulate any deleterious intermediates to account for such paired fixations. And finally, it shows with a biological system how the positions that can accept mutations may change as mutations are fixed. These views encompass assertions regarding the order of the mutational events. As a consequence of knowledge of phylogeny, some orderings have already been postulated for cytochrome *c* (Fitch and Margoliash, 1968) without recourse to such considerations as presented here. It will be interesting to see if similar predictions based upon the forthcoming three-dimensional structure of cytochrome *c* will agree with the phylogenetic orderings.

There is corroborative evidence for the view that the number of concomitantly variable codons is very restricted for cytochrome *c*. It was previously noted that the number of codons that had fixed nucleotide replacements in two of its three nucleotides in the interval between two successive nodes of the phylogenetic tree was greatly in excess of expectation if the mutations were being fixed at random (Fitch and Margoliash, 1968). These so-called double mutations appeared to imply a non-randomness in the fixations observed and therefore to suggest the possible operation of selective forces. However, that computation was performed assuming 82 variable codons so that if four mutations were fixed between two successive nodes in the tree, the probability that two of them would have occurred in the same codon was about 0.04. When this computation was carried out for each of the 26 internodal segments containing two or more fixations in that tree for 20 species of cytochrome *c*, there was an expectation of only 11.3 "double" mutations. In fact, there were 32. The number of

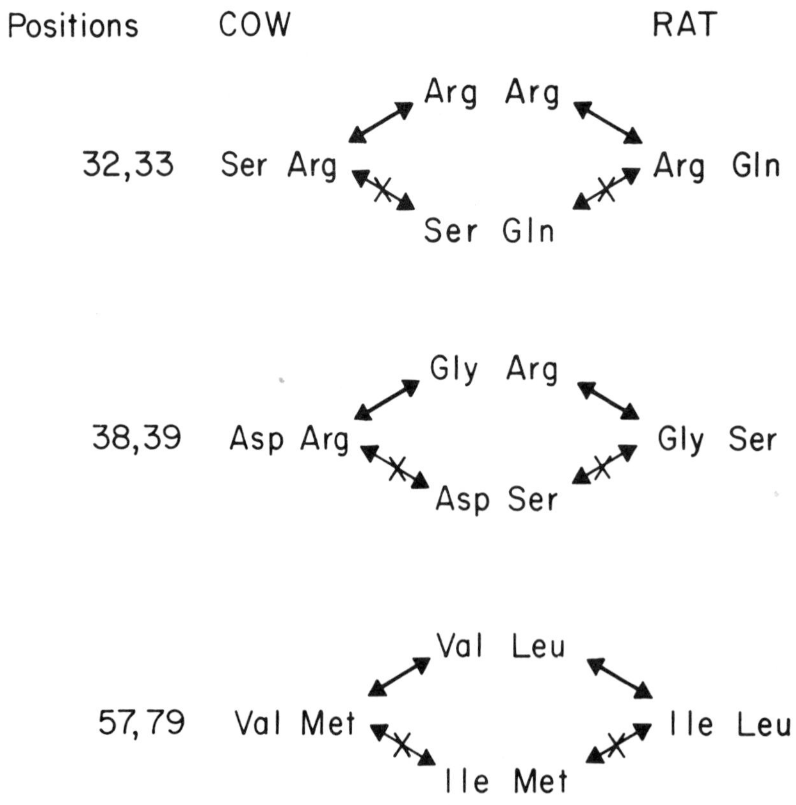

Fig. 3. Transition states in ribonuclease. The figure shows the bovine and rat ribonuclease residues in three selected pairs of positions together with two alternative intermediate forms. In each case, the upper alternative is predicted to have been the true ancestral intermediate on the basis of information supplied by Wyckoff (1968) and elaborated on in the text.

expected double mutations would, of course, be increased if the number of concomitantly variable codons were considerably less. A repeat of that calculation assuming 10 rather than 82 concomitantly variable codons yields a value of 32 expected "double" mutations, the number actually found. This means first of all that there is an independent basis (the larger number of "double" mutations) for concluding that there is a restricted number of concomitantly variable codons in the gene for cytochrome *c*, and the agreement of the two methods at a value of 10 such codons is encouraging. Secondly, it means that the "double" mutation data can now be accounted for strictly on the basis of random events and thus selective forces are not necessarily involved.

It should be noticed that the agreement between the expected and found number of "double" mutations is too good in that the calculation of the expected number is inadequate in two ways. One involves the absence of a correction for the possibility that the second fixation in the same codon may be in the same nucleotide. The second involves the absence of a correction for the fact that the 10 concomitantly variable codons represent a sort of moving average and, after a series of fixations, are not the same codons that were concomitantly variable at the beginning. That this second effect is surely present can be seen from Table III, where the preceding data are divided into two groups according to the internodal segment length as determined by the number of mutations thereon. We underestimate the number of "double" mutations in the segments with only a few mutations and overestimate that number in the segments with many mutations. This is a result that presumably would be corrected by using fewer concomitantly variable codons while simultaneously correcting for the change in the codons that are concomitantly variable. Work designed to accomplish these corrections is currently in progress and, in addition to increasing the reliability of this estimate of the number of concomitantly variable codons, it should also give information on the degree to which a fixation affects other positions. This, of course, is subject to the qualification that fixations in other genes with gene products that interact with cytochrome *c* (e.g., cytochrome oxidase) may also affect which codons will belong to the group of concomitantly variable codons.

The number of invariant residues of fibrinopeptide A was determined using the data from 25 mammalian sequences and is shown in Table I as group K. The estimate of the number of invariant positions is only one. Further restriction of the range, unlike the case for cytochrome *c*, did not change the estimated number of invariant codons. Presumably, that one invariant codon is the carboxyl-terminal arginine that is necessary for the specificity of thrombin which acts on fibrinogen to form fibrin in the blood-clotting mechanism. In any event, extrapolation to a zero range of species gives a

Table III. Distribution of "Double" Fixations by Segment Length[a]

Internodal mutations (m)	No. of such internodal segments	Number of "double" fixations		$\frac{(E-F)^2}{E}$
		Expected (E)	Found (F)	
$2 \leq m \leq 7$	16	5	10	5.00
$9 \leq m \leq 30$	10	27	22	0.93
Total	26	32	32	5.93

[a] Data are from Fitch and Margoliash (1967*a*), which gives the codon position and the location on the phylogenetic tree of all postulated mutations in the gene for cytochrome *c* during the evolution of 20 eukaryotic species. "Double" mutations are nucleotide replacements in two of the three nucleotides of a single codon on a tree segment between two successive ancestral nodes or between the present-day species and its most recent ancestor. The table shows that while the total expected and found "double" mutations agree, their distribution by segment length does not. χ^2 is 5.93 for 1 degree of freedom. Such a departure from expectation would occur by chance less than 2% of the time if the underlying assumptions about the distribution of "double" mutations were correct.

single invariant codon and therefore 18 concomitantly variable codons, a result consistent with the view that this fragment is important simply by virtue of its existence as a blocking group and that its specific structural configuration may be of little importance.

EVOLUTIONARY RATES

The cytochromes *c* and the fibrinopeptides A were chosen in part because of the difference in the rate at which they appear to be evolving. Both groups contain sequences for the horse and pig and, since their common ancestor, there have been five fixations in the 104 codons for cytochrome *c* and 13 fixations in the 19 codons for fibrinopeptide A. This gives rates of 0.048 and 0.684 fixations/codon in the two lines since their divergence from a common ancestor. Thus it would appear that fixations are occurring in the part of the gene encoding fibrinopeptide A at more than 14 times the rate they occur in the cytochrome *c* gene. If, however, we assume that the rate should be calculated on the basis of the number of codons that can accept a mutation, the rates become, respectively, $5/10 = 0.50$ and $13/18 = 0.72$ fixations/concomitantly variable codon. Given the limits of the estimate of the number of concomitantly variable codons, these two fixation rates cannot be said to differ. Furthermore, if the number of concomitantly variable codons in cytochrome *c* is in fact less than 10, as suggested above by the "double" mutation data, then the agreement may be even better.

The calculation of the number of concomitantly variable codons does not presuppose the nature of any selective forces involved. It only requires that a mutation, to be acceptable, must not be selected against, which leaves open the question of its possible neutrality. However, if the mutations are indeed neutral, then the fixation rates in concomitantly variable codons should be the same.[3] Since the rates are the same, then either the mutations are, as Kimura suggested, mostly neutral or else the agreement is fortuitous. There is a third possibility, namely, that selection proceeds to fix mutations at essentially similar rates in all genes in all species at all times. But to contrive a selective theory so as to mimic the results expected of a neutral mutation theory is philosophically unpalatable. In view of the diverse nature of these two polypeptides, we would reject the explanation based upon fortuity. In any event, when similar rates can be estimated for other gene products, we will know whether the agreement is fortuitous.[7]

There is additional reason to believe that the accumulation of mutations has been of a random rather than a selective nature. If selective forces were greatly affecting the nature of cytochrome *c*, one would expect that, say, wheat, *Drosophila*, shark, duck, and human cytochromes *c* would show greatly varying degrees of differentness when compared to *Neurospora* cytochrome *c*. In fact, when the number of nucleotide replacements required to account for the differences in these five pairs is determined, they are found to be 69, 60, 71, 63, and 64, respectively. Is it really reasonable to suggest that the proper interpretation of these data is that, since the time of their common ancestor, the shark's cytochrome *c* has been evolving under selection more than 20% faster[8] than the human cytochrome *c*? Or do these data simply reflect sampling variability? Nolan and Margoliash (1968) have given many other examples of how, for any one gene, approximately the same number of mutations separate the members of two groups having the same common ancestor. This multiplication of such examples lends further support to the idea that most of the mutations fixed in the genes for cytochromes *c*, fibrinopeptides, and hemoglobins have been selectively

[7] The most obvious next choice for examination would be hemoglobin, but only about a half dozen species of each of the α and β chains have been completely sequenced whereas at least a dozen for any one orthologous gene should be available. This emphasizes the importance of doing further sequences on orthologous gene products from closely related species.

[8] The 20% comes from the fact that 71 is 10% greater than 64, that the difference must necessarily be relatable to mutations which occurred since the common ancestor of the shark and man, and that presumably not more than half (32) of the mutations separating the vertebrate and fungal lines have occurred since the time of the common ancestor of the shark and man.

neutral. This conclusion, regarding fibrinopeptide A, stands in sharp contrast to that of King and Jukes (1969), who state, "within the short fibrinopeptide A fragment, however, some positions are notably less changeable than others. It is quite likely that only a minority of the changes that occur in this portion of the fibrinogen gene are selectively neutral." Group K of Table I clearly depicts the changeable character the authors speak of, but the row below also clearly shows that the most variable positions "are predictable in terms of the Poisson distributions," to use the authors' own words from another context.

The difficulty perhaps lies in an unwillingness of King and Jukes to assume the existence of a group of codons more variable than another. We might ask therefore how a second, more variable group might arise. Possibilities include: (1) their codons are hot spots, i.e., they are hypermutable; (2) a greater variety of alternatives is acceptable at these positions; (3) an alternative is acceptable over a greater portion of the total time span.

While we are not as yet prepared to choose among these alternatives,[9] they appear sufficiently reasonable to permit acceptance of the possibility of detecting two groups of codons with differing degrees of variability.

If we assume that the fixation rate is 0.72 and that the common ancestor of the horse and pig existed 80 million years ago (Romer, 1966), then the fixation rate would appear to be 4.5 fixations/concomitantly variable codon/10^9 years in any one line of descent. The fixation rate should be relatable to the mutation rate, but the calculation is complicated by the fact that not all mutations in a concomitantly variable codon are necessarily acceptable.

Finally, we would make a point of noting the importance of verifying the idea that most mutations fixed in structural genes have been neutral and that on the basis of the number of concomitantly variable codons, fixation rates are uniform because, given one good paleontological reference point we shall then have for the first time a very excellent evolutionary time scale at our disposal. With luck, this could extend back into pre-Cambrian times and permit the temporal location of many events such as speciation and gene duplications which now can only be guessed at.

ADDENDUM

Since this paper was submitted and reviewed, the sequence of porcine ribonuclease has appeared (Jackson and Hirs, 1970). The amino acids at positions 57 and 79 are identical to those in the cow. The amino acids at positions 32, 33 and 38, 39 are ArgArg and GlyArg, respectively, and are the intermediates between the bovine and rat forms predicted on the basis of Wyckoff's data and shown in Fig. 3.

ACKNOWLEDGMENTS

This project received support from National Science Foundation grant GB-7486. The University of Wisconsin Computing Center, whose facilities were used, also receives support from NSF and other government agencies.

REFERENCES

Ambler, R. P. (1963). The amino acid sequence of *Pseudomonas* cytochrome c-551. *Biochem. J.* **89:** 349.

Corbin, K. W., and Uzzell, T. (1970). Natural selection and mutation rates in mammals. *Amer. Nat.* **104:** 37.

Dus, K., Sletten, K., and Kamen, M. K. (1968). Cytochrome c_2 of *Rhodospirillum rubrum*. *J. Biol. Chem.* **243:** 5507.

Fitch, W. M. (1970). Distinguishing homologous from analogous proteins. *Systematic Zool.* **19:** 99.

[9] The third possibility cannot apply to fibrinopeptide A since the number of concomitantly variable codons is constant at 18 regardless of the range of species examined and hence all 18 are always variable.

Fitch, W. M., and Margoliash, E. (1967*a*). A method for estimating the number of invariant amino acid coding positions in a gene using cytochrome *c* as a model case. *Biochem. Genet.* **1:** 65.
Fitch, W. M., and Margoliash, E. (1967*b*). The construction of phylogenetic trees. *Science* **155:** 279.
Fitch, W. M., and Margoliash, E. (1968). The construction of Phylogenetic trees. II. How well do they reflect past history? *Brookhaven Symp. Biol.* **21:** 217.
Haldane, J. B. S. (1957). The cost of natural selection. *J. Genet.* **55:** 511.
Jackson, R. L., and Hirs, C. H. W. (1970). The primary structure of porcine pancreatic ribonuclease. *J. Biol. Chem.* **245:** 637.
Jukes, T. H. (1969). Evolutionary pattern of specificity regions in light chains of immunoglobins. *Biochem. Genet.* **3:** 109.
Kimura, M. (1968). Evolutionary rate at the molecular level. *Nature* **217:** 624.
King, J. L., and Jukes, T. H. (1969). Non-Darwinian evolution. *Science* **164:** 788.
Markowitz, E. (1970). Estimation and testing goodness-of-fit for some models of codon fixation variability. *Biochem. Genet.* **4:** 595.
Maynard Smith, J. (1968). "Haldane's dilemma" and the rate of evolution. *Nature* **219:** 1114.
Nolan, C., and Margoliash, E. (1968). Comparative aspects of primary structures of proteins. *Ann. Rev. Biochem.* **37:** 727.
O'Donald, P. (1969). "Haldane's dilemma" and the rate of natural selection. *Nature* **221:** 815.
Romer, A. S. (1966). *Vertebrate Paleontology*, 3rd ed., University of Chicago Press, Chicago.
Watson, J. D. (1965). *Molecular Biology of the Gene*, Benjamin, New York.
Wyckoff, H. W. (1968). Discussion. *Brookhaven Symp. Biol.* **21:** 252.

Reprinted from *Proc. Natl. Acad. Sci. USA* **71**: 2848–2852.

On Some Principles Governing Molecular Evolution*

(population genetics/mutational pressure/negative selection/random drift)

MOTOO KIMURA† AND TOMOKO OHTA†

National Institute of Genetics, Mishima, Japan

Contributed by Motoo Kimura, May 1, 1974

ABSTRACT **The following five principles were deduced from the accumulated evidence on molecular evolution and theoretical considerations of the population dynamics of mutant substitutions: (*i*) for each protein, the rate of evolution in terms of amino acid substitutions is approximately constant/site per year for various lines, as long as the function and tertiary structure of the molecule remain essentially unaltered. (*ii*) Functionally less important molecules or parts of a molecule evolve (in terms of mutant substitutions) faster than more important ones. (*iii*) Those mutant substitutions that disrupt less the existing structure and function of a molecule (conservative substitutions) occur more frequently in evolution than more disruptive ones. (*iv*) Gene duplication must always precede the emergence of a gene having a new function. (*v*) Selective elimination of definitely deleterious mutants and random fixation of selectively neutral or very slightly deleterious mutants occur far more frequently in evolution than positive Darwinian selection of definitely advantageous mutants.**

Recent development of molecular genetics has added a new dimension to the studies of evolution. Its impact is comparable to that of Mendelism and cytogenetics in the past. Accumulated evidence suggests (1–8) that, as causes of evolutionary changes at the molecular (genic) level, mutational pressure and random gene frequency drift in Mendelian populations play a much more important role than the orthodox view of neo-Darwinism could lead us to believe.

In the present paper, we intend to enumerate some basic principles that have emerged from recent evolutionary studies of informational macromolecules. Of these, the first four are empirical, while the last one, which is theoretical, enables us to interpret the four empirical principles in a unified way.

(i) *For each protein, the rate of evolution in terms of amino acid substitutions is approximately constant per year per site for various lines, as long as the function and tertiary structure of the molecule remain essentially unaltered.* In their influential paper on the evolution of "informational macromolecules," Zuckerkandl and Pauling (9), noting that the mean evolutionary rates of globins are approximately equal per year among different lineages, suggested the existence of a molecular evolutionary clock. Actually, the idea of such a clock was implicit in the earlier writings of Ingram (10) and Jukes (11). The approximate constancy of the evolutionary rate in globins has since been confirmed by a number of authors (2, 12–14). For example, the number of observed amino acid differences between the α and β hemoglobin chains of man is approximately equal to that between the α chain of the carp and the β chain of man (12). Table 1 lists the numbers of amino acid sites in these two sets of comparisons that can be interpreted from the code table as due to a minimum of 0, 1, and 2 nucleotide substitutions. Also, the number of gaps due to insertion and/or deletion is listed. Since the human and carp α chains differ from each other at roughly 50% of the amino acid sites, the data suggest that the two structural genes coding for the α and β chains of hemoglobin have diverged independently of each other and to the same extent in the two lines since their origin by duplication which occurred possibly at the end of the Ordovician period. It is remarkable that mutant substitutions at gene loci coding for the α and β chains have occurred at practically the same average rates in the two separate lines that have evolved independently over nearly a half billion years. From these comparisons, the rate of amino acid substitution/site per year turns out to be about 0.9×10^{-9}. On the other hand, from comparisons of the α hemoglobin chains among various mammalian species, we obtain roughly the rate 10^{-9} site per year which is in good agreement with the above estimate. Although local fluctuations no doubt occur, constancy rather than variation of the evolutionary rate distinguishes the process of molecular evolution. This is

TABLE 1. *Comparison of amino acid differences between α and β hemoglobins*

Type of change*	Human α vs. human β	Carp α vs. human β
0	63	61
1	53	49
2	22	29
Gap	9	10
Total	147	149

* The numbers of amino acid sites that can be interpreted from the code table as due to a minimum of 0, 1, 2 nucleotide substitutions in two sets of comparisons involving the α and β hemoglobin chains. The number of gaps is also listed for each comparison.

* Contribution no. 1000 from the National Institute of Genetics, Mishima, Shizuoka-ken, 411, Japan.

† *We dedicate this paper to Dr. Hitoshi Kihara the former director of our institute in honor of his 80th birthday anniversary. He was really far-sighted when he wrote, as early as 1947, in relation to his outstanding cytogenetical work on the origin of cultivated wheat, "The history of the earth is recorded in the layers of its crust; the history of all organisms is inscribed in the chromosomes" (original in Japanese, ref. 42). With this paper we also celebrate the 25th anniversary of the National Institute of Genetics.*

particularly noteworthy since it is well-known (15) that there are enormous differences among evolutionary rates at the organism level; some forms have evolved very rapidly while others have stayed essentially unchanged over hundreds of millions of years (especially in organisms known as living fossils). Approximate constancy of the evolutionary rate per year has also been noted in cytochrome *c* and fibrinopeptides (4, 16, 17) although each has its characteristic rate; i.e., the evolutionary rate of cytochrome *c* is about $^1/_3$ while that of fibrinopeptides is roughly 4 to about 9 times that of hemoglobin. Constancy of the evolutionary rate per year has also been noted in albumin evolution of primates (18).

Recently, some authors have questioned the concept of a molecular clock by emphasizing local variation of evolutionary rates. For example, Goodman and his associates (19) emphasize that the evolutionary rates of the hemoglobin α chain slowed down in higher primates. Their method is based on estimating hidden mutant substitutions with the so-called "maximum parsimony" method, and accepts time spans from paleontological studies. In our opinion, the validity of their method (particularly the maximum parsimony principle) has to be tested in several cases rather than being taken for granted. More recently, Langley and Fitch (20) performed a somewhat more reliable analysis on the variation of evolutionary rates among the branches of a phylogenetic tree involving simultaneously the evolution of the α and β hemoglobins, cytochrome *c*, and fibrinopeptide A. They found that variation of evolutionary rates among branches ("legs") over proteins is significantly higher than expected by pure chance, with a χ^2 value about 2.5 times its degree of freedom. Since the expected value of χ^2 is equal to its degrees of freedom, their results mean that variation of evolutionary rate in terms of mutant substitutions among lines is about 2.5 times as large as that expected from chance fluctuations. Their estimation of the number of mutant substitutions is based on the assumption of minimum evolution, and it is likely that the estimation is biased in such a direction that lineages with more branches tend to show more hidden mutant substitutions. Yet, their results essentially agree with our previous analysis of the variation of evolutionary rates among lines using data on hemoglobins and cytochrome *c* (21). Namely, the observed variance of evolutionary rates among mammalian lines is roughly 1.5 to about 2.5 times the expected variance. However, the existing data indicate that, when averaged over a long period, the rate of evolution is remarkably uniform among different lineages, even though local fluctuations do occur.

We conclude, therefore, that constancy of evolutionary rate per year is valid as a first approximation. Such a constancy can be explained by the neutral mutation-random drift hypothesis if we assume that the rate of occurrence of neutral mutants is constant per year (4–6). Highly complicated and arbitrary sets of assumptions must be invoked regarding mutation, gene interaction, and ecological conditions as well as population size in order to explain the approximate constancy solely from the neo-Darwinian viewpoint. As predicted by one of us (12), it is likely that genes of "living fossils" in general have undergone essentially as many DNA base substitutions as corresponding genes in more rapidly evolving species. It is this constancy which makes the molecular data so useful and of such great potential value in constructing phylogenetic trees. Eventually, it will be possible to go far back into the history of life to clarify the early stage of evolution far beyond the capability of the traditional methods based on phenotypes.

(ii) *Functionally less important molecules or parts of a molecule evolve (in terms of mutant substitutions) faster than more important ones.* The rate of amino acid substitution has been estimated (with differing degrees of accuracy) for more than twenty different proteins as shown in Table 6-1 of Dayhoff (22). The highest rate is represented by fibrinopeptides (9×10^{-9}/amino acid per year according to their estimation) while the lowest rate is that of histone IV (0.006×10^{-9}). From this table it turns out that the median rate is 1.3×10^{-9}/amino acid per year. (represented by myoglobin in the table). This is not very different from 1.6×10^{-9}/amino acid per year which was estimated earlier by King and Jukes (2) as the average rate for seven proteins. Thus, hemoglobins show an evolutionary rate typical of those proteins that have been studied.

It is interesting to note that fibrinopeptides, the most rapidly evolving molecules, have little known function after they become separated from fibrinogen in the blood clot. The relationship between the functional importance (or more strictly, functional constraint) and the evolutionary rate has been beautifully explained by Dickerson (17) as follows. In fibrinopeptides, virtually any amino acid change (mutant substitution) that permits the peptides to be removed is "acceptable" to the species. Thus, the rate of evolutionary substitution of amino acids may be very near to the actual mutation rate. Hemoglobins, because they have a definite function of carrying oxygen and, so specifications for them are more restrictive than for fibrinopeptides, have a lower evolutionary rate. Cytochrome *c* interacts with cytochrome oxidase and reductase, both of which are much larger than it, and there is more functional constraint in cytochrome *c* than in hemoglobins. Thus cytochrome *c* has a lower evolutionary rate than hemoglobins. Histone IV binds to DNA in the nucleus, and is believed to control the expression of genetic information. It is quite probable that a protein so close to the genetic information storage system is highly specified with little evolutionary change over a billion years. Boyer *et al.* (23) reported that the δ chain of hemoglobin A_2 ($\alpha_2\delta_2$), which forms the minor component of adult hemoglobin, shows higher evolutionary rates and a higher level of polymorphism than the β chain which forms the major component A ($\alpha_2\beta_2$). This appears to agree with the present principle that less constraint enables more rapid change.

The evolutionary rate differs not only between different molecules but also between different parts of one molecule. For example, in both the α and β hemoglobin chains, the surface part of the molecule evolves nearly 10 times as fast as the functionally important heme pocket (7). In addition, two histidines binding to the heme are absolutely invariant throughout the entire history of vertebrate evolution extending nearly a half billion years (13). The Perutz model of hemoglobins (24) helps us greatly to interpret such observations in terms of structure and function of these molecules. More generally, if we consider the oil drop model of globular proteins (25), the inside of a molecule is filled with nonpolar (hydrophobic) amino acids, while the surface parts are occupied by polar (hydrophilic) amino acids. The functionally vital "active center" is located inside a crevice, and the rate of evolutionary substitutions of amino acids in this part is expected to be very low. On the other hand, the surface parts are usually not very critical in maintaining the function or the tertiary structure, and the evolutionary rates in these parts are expected to be much higher. Another interesting example is the middle segment (C) of the proinsulin molecule. This part is removed when the active insulin is formed, and it is now known that this part evolves at the rate 4.4×10^{-9}/ amino acid per year, which is roughly 10 times as fast as that of insulin (6, 17). An additional example is afforded by the recent report of Barnard *et al.* (26). According to them, sequence 15–24 of pancreatic ribonucleases evolves at a very high rate comparable to rapidly evolving parts of fibrinopeptides, and this "hypervariability" can be correlated with a lack of any contribution of this part either to the enzymatic activity or to the maintenance of structure required for the activity. Incidentally, their Table 3 listing frequencies of amino acids in hypervariable segments suggests that in such regions there might still exist some selective constraint in amino acid substitutions, so that not all of the mutations are tolerated.

All the observations in this section allow a very simple interpretation from the neutral mutation-random drift hypothesis. Namely, in a molecule or a part of a molecule which is functionally less important, the chance of a mutant being selectively neutral (or very slightly deleterious) is higher, and therefore it has a higher chance of being fixed in the population by random drift. On the other hand, from the neo-Darwinian view-point, we must assume that a rapidly evolving part has an important functional role and is undergoing very rapid adaptive improvements by accumulating many advantageous mutations. It may be argued that the smaller the effect of a mutational change, the higher the chance of it being beneficial as Fisher (27) said, and therefore observations in this section can also be explained by positive natural selection. However, if the selective advantage of a mutant becomes small, then the chance of its fixation in the population becomes correspondingly small. Thus, apart from the problem of validity of Fisher's statement when applied to molecular data, it may not necessarily follow that the smaller the effect, the higher the rate of mutant substitution by natural selection.

(iii) *Those mutant substitutions that disrupt less the existing structure and function of a molecule (conservative substitutions) occur more frequently in evolution than more disruptive ones.* The conservative nature of amino acid substitutions was earlier noted by Zuckerkandl and Pauling (9). They also noted that the code table itself is conservative in that single base substitution often leads to both the substitution of a similar amino acid as well as a synonymous substitution. Since then, the conservative nature of substitutions has been amply documented in evolutionary studies of proteins (17, 22, 28, 29). Clarke (30) treated this problem in quantitative terms by using Sneath's (31) measure of chemical similarity of amino acids and by considering the regression of the relative frequency of evolutionary substitutions on the similarity. His results confirm the well-known fact that chemically similar substitutions occur more frequently than dissimilar ones.

The principle of conservative substitution holds also for nucleotide substitutions. In their extensive study on the evolution of transfer RNA, Holmquist *et al.* (32) found that, among the mispairings in the helical regions, G·U or U·G pairs that do not interfere with helicity occur much more frequently than other forms of mispairing; of 68 observed "non-Watson-Crick pairs," 43 turned out to be either G·U or U·G. For each transfer RNA molecule, the total number of mispairings in helical regions is limited to one or two, suggesting that beyond such a small number, a mutation leading to an additional mispairing becomes highly deleterious and rejected (it is likely that even the mutation causing the first mispairing is deleterious, but it can be fixed by random drift due to very small effect, see ref. 8); only when one of the existing mispairings is closed by a mutant substitution, is the molecule ready to accept a new mutation through random drift and/or selection. This offers an excellent model of Fitch's concept of concomitantly variable codons or "covarions" (33); according to him, only 10% of codons in cytochrome *c* can accept mutations at any moment in the course of evolution. He also found (34) that the proportion of covarions is about 35% in the hemoglobin α, but nearly 100% in fibrinopeptide A. A remarkable fact emerging from his analyses is that if the rate of amino acid substitution is calculated on the bases of covarions, cytochrome *c*, hemoglobin α, and fibrinopeptide A are all evolving at about the same rate. Fitch's covarion idea, we believe, has a clearer meaning now in the light of selective constraints involved in the secondary and tertiary structure necessary for the function of the molecule.

Similarly, one might expect that synonymous substitutions causing no change in amino acids would occur more frequently in evolution than nucleotide substitutions leading to amino acid change. From studies of amino acid sequences of tryptophan synthetase A-chains of three bacterial species, *Escherichia coli*, *Salmonella typhimurium*, and *Aerobacter aerogenes*, in conjunction with the estimated nucleotide sequence differences among the corresponding structural genes (determined by mRNA·DNA hybridization, Li *et al.* (35) obtained results suggesting that synonymous codon differences in the gene for tryptophan synthetase A chain are quite common. According to their estimate, there are about as many base differences that do not alter the amino acid sequences as those that alter the sequences. It is possible, as Li *et al.* point out, that not every synonymous substitution is completely neutral with respect to natural selection. Some of them might be subject to selective elimination based on structural requirement (such as the one involved in forming the secondary structure of the RNA molecule). However, because synonymous substitutions, in general, must have a higher chance of being selectively neutral or only very slightly deleterious (other things being equal) than mis-sense substitutions, they have a greater chance of becoming fixed in the population by random drift. One prediction that we could therefore make is that the slower the evolutionary rate of a protein molecule, the higher the ratio of synonymous to mis-sense substitutions.

(iv) *Gene duplication must always precede the emergence of a gene having a new function.* The importance of gene duplication in evolution has been noted earlier by the great Drosophila workers of the Morgan school (see ref. 5). The crucial point pertinent here is that the existence of two copies of the same gene enables one of the copies to accumulate mutations and to eventually emerge as a new gene, while another copy retains the old function required by the species for survival through the transitional period. Shielded by the normal counterpart in the corresponding site of the duplicated DNA segment, mutations that would have been rejected before duplication can now accumulate, and through their accumulation, a stage is set for emergence of a new gene. The creative role which gene duplication plays in evolution has been much clarified by Ohno (36) in his stimulating book in which he considers new evidence based on modern molecular, cytological, and paleontological researches. Together with his recent paper (37), Ohno has made an important contribution to the modern evolutionary theory by bringing to light the remarkably conservative nature of mutant substitutions in evolution. Gene duplication, at the same time, must have caused a great deal of degeneration in duplicated DNA segments. This is because many mutations, which would have been definitely deleterious before duplication, become neutral or only very slightly deleterious after duplication, thus enabling them to spread in the population by random drift (38, 39).

(v) *Selective elimination of definitely deleterious mutants and random fixation of selectively neutral or very slightly deleterious mutants occur far more frequently in evolution than positive Darwinian selection of definitely advantageous mutants.* This is an extended form of the neutral mutation-random drift hypothesis, and is based on the thesis put forward by one of us (8) which argues that very slightly deleterious mutations as well as selectively neutral mutations play an important role in molecular evolution. Adaptive changes due to positive Darwinian selection no doubt occur at the molecular level, but we believe that definitely advantageous mutant substitutions are a minority when compared with a relatively large number of "non-Darwinian" type mutant substitutions, that is, fixations of mutant alleles in the population through the process of random drift of gene frequency. We emphasize that neutral or nearly neutral mutations should be considered not as a limit of selectively advantageous mutants but as a limit of deleterious mutants when the effect of mutation on fitness becomes small. In other words, mutational pressure causes evolutionary change whenever the negative-selection barrier is lifted. As an application of this principle, let us consider the evolutionary change of guinea pig insulin. Although the insulin (A and B segments) in general has a very low evolutionary rate (about 0.33×10^{-9}/amino acid per year), guinea pig insulin is exceptional in that it diverged very rapidly with the estimated rate of 5.3×10^{-9}/amino acid site per year (2). From the neo-Darwinian point of view, one might naturally consider such a rapid evolutionary change the result of adaptive change by natural selection. In fact, even King and Jukes (2) in their paper "Non-Darwinian Evolution" invoked "positive natural selection" to explain the rapid change. We suggest that guinea pig insulin lost its original selective constraint in the process of speciation. This allowed the accumulation of mutations which before would have been rejected. This inference is supported by a recent report of Blundell *et al.* (40) who studied the three-dimensional structure of insulin molecules. According to them, guinea pig insulin is accompanied by the loss of zinc in the islet cells (coinciding with the loss of usually invariant histidine B10). This suggests a drastic change in the tertiary structure. It is assumed then that, with the loss of the zinc constraint, mutations in guinea pig insulin started to accumulate at a very high rate approaching the rate in fibrinopeptides (the rate that might be called the fibrinopeptide limit).

When we consider the action of natural selection at the molecular level, we must keep in mind that higher order (i.e., secondary, tertiary, and quaternary) structures rather than the primary structure (i.e., amino acid sequence) are subject to selective constraint, usually in the form of negative selection, that is, elimination of functionally deleterious changes. The existence of selective constraint, often inferred from nonrandomness in amino acid or nucleotide sequences, does not contradict the neutral mutation-random drift hypothesis. Incidentally, it is interesting to note that the fibrinopeptide rate, when expressed in terms of nucleotide substitutions, is roughly equal to the rate of nucleotide substitution in the DNA of the mammalian genome (39). We note also that accumulation of very slightly deleterious mutants by random drift is essentially equivalent to the deterioration of environment, and definitely adaptive gene substitutions must occur from time to time to save the species from extinction.

Although clearly documented cases at the genic level are rather scarce, there is not a slightest doubt that the marvellous adaptations of all the living forms to their environments have been brought about by positive Darwinian selection. It is likely, however, that the ways in which mutations become advantageous are so opportunistic that no simple rules could be formulated to describe them. On the whole, mutations are disadvantageous, and, when a mutant is advantageous, it can be advantageous only under restricted conditions (41). We note also that difference in function at the molecular level, does not necessarily lead to effective natural selection at the level of individuals within a population.

In the past half century, with the rise of neo-Darwinism or more precisely, the synthetic theory of evolution, the claim that mutation is the main cause of evolution has completely been rejected. Instead, the orthodox view has been formed which maintains that the rate and direction of evolution are almost exclusively determined by positive natural selection. We believe that such a view has to be re-examined, particularly regarding evolutionary changes at the molecular level. We think that evolution by mutational pressure is a reality.

We thank Drs. T. H. Jukes and J. L. King for stimulating discussions which helped greatly to compose the manuscript. Especially, we are indebted to Dr. Jukes for critically reviewing the first draft and offering many suggestions for improvement. Thanks are also due to Drs. J. F. Crow and E. R. Dempster for reading the manuscript and offering suggestions for improving the presentation.

1. Kimura, M. (1968) *Nature* **217**, 624–626.
2. King, J. L. & Jukes, T. H. (1969) *Science* **164**, 788–798.
3. Crow, J. F. (1969) *Proc. XII Intern. Congr. Genetics (Tokyo)* **3**, 105–113.
4. Kimura, M. & Ohta, T. (1971) *J. Mol. Evolut.* **1**, 1–17.
5. Kimura, M. & Ohta, T. (1971) *Theoretical Aspects of Population Genetics* (Princeton University Press, Princeton, N.J.).
6. Kimura, M. & Ohta, T. (1972) *Proc. 6th Berkeley Symp. on Math. Stat. and Probability* **5**, 43–68.
7. Kimura, M. & Ohta, T. (1973) *Genetics (Sup.)* **73**, 19–35.
8. Ohta, T. (1973) *Nature* **246**, 96–98.
9. Zuckerkandl, E. & Pauling, L. (1965) in *Evolving Genes and Proteins*, eds. Bryson, V. & Vogel, H. J. (Academic Press, New York), pp. 97–166.
10. Ingram, V. M. (1961) *Nature* **189**, 704–708.
11. Jukes, T. H. (1963) *Advan. Biol. Med. Phys.* **9**, 1–41.
12. Kimura, M. (1969) *Proc. Nat. Acad. Sci. USA* **63**, 1181–1188.
13. Jukes, T. H. (1971) *J. Mol. Evolut.* **1**, 46–62.
14. Air, G. M., Thompson, E. O. P., Richardson, B. J. & Sharman, G. B. (1971) *Nature* **229**, 391–394.

15. Simpson, G. G. (1944) *Tempo and Mode in Evolution* (Columbia Univ. Press, New York).
16. Margoliash, E., Fitch, W. M. & Dickerson, R. E. (1968) *Brookhaven Symp. Biol.* **21**, 259–305.
17. Dickerson, R. E. (1971) *J. Mol. Evolut.* **1**, 26–45.
18. Sarich, V. M. & Wilson, A. C. (1967) *Proc. Nat. Acad. Sci. USA* **58**, 142–148.
19. Goodman, M., Barnabas, J., Matsuda, G. & Moore, G. W. (1971) *Nature* **233**, 604–613.
20. Langley, C. H. & Fitch, W. M. (1973) in *Genetic Structure of Populations* ed. Morton, N. E. (Univ. Press of Hawaii, Honolulu), pp. 246–262.
21. Ohta, T. & Kimura, M. (1971) *J. Mol. Evolut.* **1**, 18–25.
22. Dayhoff, M. O. (1972) *Atlas of Protein Sequence and Structure 1972* (National Biomedical Research Foundation, Silver Spring, Md.).
23. Boyer, S. H., Crosby, E. F., Thurmon, T. F., Noyes, A. N., Fuller, G. F., Leslie, S. E., Shepard, M. K. & Herndon, C. N. (1969) *Science* **166**, 1428–1431.
24. Perutz, M. F. & Lehman, H. (1968) *Nature* **219**, 902–909.
25. Dickerson, R. E. & Geis, I. (1969) *The Structure and Action of Proteins* (Harper & Row, New York, Evanston, London).
26. Barnard, E. A., Cohen, M. S., Gold, M. H. & Kim, Jae-Kyoung (1972) *Nature* **240**, 395–398.
27. Fisher, R. A. (1930) *The Genetical Theory of Natural Selection* (Clarendon Press, Oxford).
28. Epstein, C. J. (1967) *Nature* **215**, 355–359.
29. Lanks, K. W. & Kitchin, F. D. (1972) *Nature* **226**, 753–754.
30. Clarke, B. (1970) *Nature* **228**, 159–160.
31. Sneath, P. H. A. (1966) *Theoret. Biol.* **12**, 157–193.
32. Holmquist, R., Jukes, T. H. & Pangburn, S. (1973) *J. Mol. Biol.* **78**, 91–116.
33. Fitch, W. M. & Markowitz, E. (1970) *Biochem. Genet.* **4**, 579–593.
34. Fitch, W. M. (1972) in *Haematologie und Bluttransfusion* ed. Martin, H. (J. F. Lehmanns Verlag, Munich, Germany), pp. 199–215.
35. Li, S. L., Denney, R. M. & Yanofsky, C. (1973) *Proc. Nat. Acad. Sci. USA* **70**, 1112–1116.
36. Ohno, S. (1970) *Evolution by Gene Duplication* (Springer-Verlag, Berlin).
37. Ohno, S. (1973) *Nature* **244**, 259–262.
38. Nei, M. (1969) *Nature* **221**, 40–42.
39. Ohta, T. & Kimura, M. (1971) *Nature* **233**, 118–119.
40. Blundell, T. L., Cutfield, J. F., Cutfield, S. M., Dodson, E. J., Dodson, G. G., Hodgkin, D. C., Mercola, D. A. & Vijayan, M. (1971) *Nature* **231**, 506–511.
41. Ohta, T. (1972) *J. Mol. Evolut.* **1**, 305–314.
42. Kihara, H. (1947) *Ancestors of Common Wheat* (in Japanese) (Sōgensha, Tokyo).

Reprinted from *Proc. Natl. Acad. Sci. USA* **75**: 2674–2678.

Pseudomonas cytochrome c_{551} at 2.0 Å resolution: Enlargement of the cytochrome *c* family

(bacterial metabolism/protein evolution)

ROBERT J. ALMASSY AND RICHARD E. DICKERSON

Norman W. Church Laboratory of Chemical Biology, California Institute of Technology, Pasadena, California 91125

Communicated by John D. Baldeschwieler, March 20, 1978

ABSTRACT The structure of respiratory cytochrome c_{551} of *Pseudomonas aeruginosa,* with 82 amino acids, has been solved by x-ray analysis and refined to a crystallographic *R* factor of 16.2%. It has the same basic folding pattern and hydrophobic heme environment as cytochromes *c*, c_2, and c_{550}, except for a large deletion at the bottom of the heme crevice. This same "cytochrome fold" appears to be present in photosynthetic cytochromes *c* of green and purple sulfur bacteria, and algal cytochromes *f*, suggesting a common evolutionary origin for electron transport chains in photosynthesis and respiration.

Cytochrome c_{551} is found in various Pseudomonads and in *Azotobacter vinelandii,* where it plays a respiratory role analogous to mitochondrial cytochrome *c* in eukaryotes and cytochrome c_{550} in *Paracoccus denitrificans* (1–4). It is significantly smaller than these latter proteins, however, having only 82 amino acids instead of 103–134. Amino acid sequence comparisons (table 2 of ref. 4) have suggested a similarity of folding between c_{551} and the larger cytochromes, but in the absence of x-ray data it was difficult to decide where the "deletions" in the c_{551} chain should be placed to make the proper sequence alignment (5–8). The preliminary low-resolution x-ray analysis of *Pseudomonas aeruginosa* cytochrome c_{551} (discussed in ref. 4) showed that the folding patterns in *c*, c_2, c_{550}, and c_{551} were indeed the same. This has now been confirmed by x-ray analysis and constrained difference map refinement at 2.0 Å resolution.

METHODS

The original cytochrome c_{551} was the gift of Henry Harbury, who then trained one of us (R.J.A.) in the techniques of growth of culture of *P. aeruginosa* and purification of the cytochrome (ref. 2; H. Harbury, personal communication). Crystals were grown in 40–50% saturated ammonium sulfate solutions, with 1 M NaCl and 0.01 M ammonium phosphate buffer, pH 5.6–5.9. The crystals are space group $P2_12_12_1$ with unit cell dimensions: a = 29.43 Å, b = 49.00 Å, c = 49.66 Å. Three heavy-atom derivatives were prepared by soaking crystals in stock solutions: K_2PtCl_4, $UO_2(NO_3)_2$, and $NaAu(CN)_2$. All high-resolution data were collected on a modified General Electric XRD-490 x-ray diffractometer, to a resolution of 2.0 Å for native protein crystals and 2.4 Å for the isomorphous derivatives.

Full details of the structure analysis will be reported elsewhere, but the strategy may be outlined here. The mean figure of merit for multiple isomorphous replacement phase refinement (9) at 2.4 Å resolution was 0.924 for centric reflections and 0.763 for all data. Atomic coordinates were measured from a

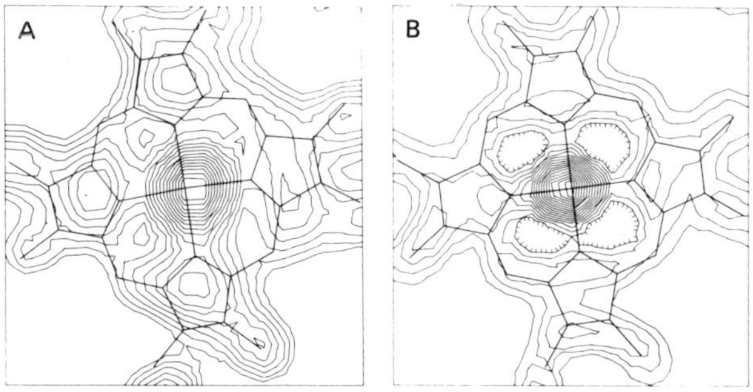

FIG. 1. (*A*) Section through the heme group, from the multiple isomorphous replacement electron-density map of *P. aeruginosa* cytochrome c_{551}. Heme coordinates as obtained from the multiple isomorphous replacement map are superimposed. (*B*) Section through the heme group in the final ($2F_o - F_c$) map from the refined structure, with the refined coordinates superimposed. The five-membered rings appear as flat plates, with their substituent groups clearly defined; the six-membered rings including the heme iron have deep negative centers. The cysteine 12 attachment is at the left side of the lower pyrrole ring; the cysteine 15 attachment is at the top of the left pyrrole ring. The two propionic acid groups extend out of the sectioning plane at the upper right corner.

Kendrew wire model built in a Richards box. These were used as the starting point for refinement on a minicomputer, alternating cycles of (*a*) automated and occasionally manual shifts in atomic position based on Fourier difference maps, and (*b*) adjustment of bond distances, angles, and torsion angles toward standard values followed by calculation of a new difference map (10). At present, with stereochemically acceptable bond lengths and angles, a planar heme, and the addition of 42 water molecules per cytochrome molecule, the conventional crystallographic *R* factor* is 16.2% at 2.0 Å resolution.

As a rough indication of the effect of refinement, two sections through the plane of the heme group are compared in Fig. 1, with heme skeletons superimposed. Fig. 1*A* shows the multiple isomorphous replacement map at 2.4 Å resolution, and Fig. 1*B* the refined ($2F_o - F_c$) exp $i\phi_c$ map at 2.0 Å.

RESULTS

The α-carbon atoms, heme group, and a few key side chains of cytochrome c_{551} are shown in the stereo drawing of Fig. 2; the folding of c_{551} is compared with that of tuna *c* (11) in the ribbon drawings of Fig. 3. Among the common structural features are the amino-terminal α helix, cysteine and histidine attachments to the heme, the 20's loop at the right, the 60's helix (40's helix in c_{551}), methionine ligand to the heme, and the carboxy-ter-

The costs of publication of this article were defrayed in part by the payment of page charges. This article must therefore be hereby marked "*advertisement*" in accordance with 18 U. S. C. §1734 solely to indicate this fact.

$$* R \equiv \frac{\sum_{hkl} \| F_o| - |F_c\|}{\sum_{hkl} |F_o|} \times 100.$$

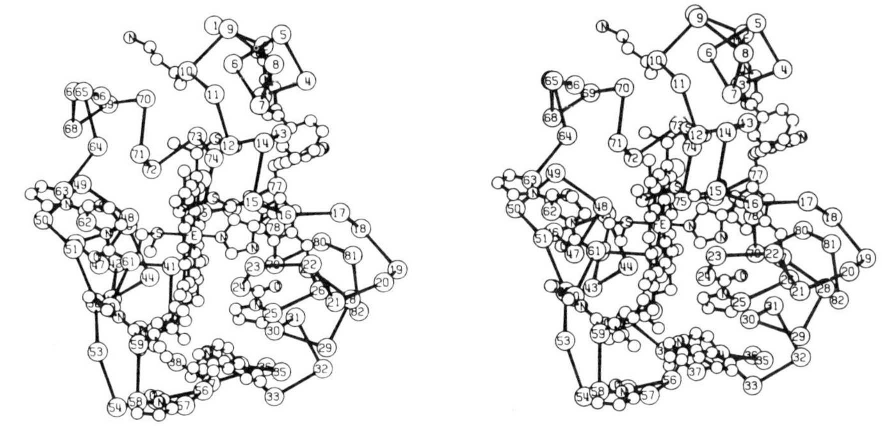

FIG. 2. Stereo drawing of the main chain backbone, heme, and key side chains in c_{551}. Similar drawings of tuna *c*, *Paracoccus* c_{550}, and *Rhodospirillum rubrum* c_2 can be found in ref. 4.

minal α helix. The principal differences are the deletion of the bottom of the *c* molecule in c_{551}, its replacement by a 30's helix at the lower rear, and the pulling downward of the 70's loop of *c* to close off the bottom of the *c* molecule. The deletion in cytochrome *c* required to produce the c_{551} molecule can be described approximately as a joining of residue 40 to 56, and removal of residues 41–55. Residues 58–63, with the amino acid sequence Pro-Ile-Pro-Met-Pro-Pro, are now seen to form an almost ideal polyproline threefold helix.

One interesting aspect, an apparent consequence of the deletion of chain at the bottom of the molecule, is the tilting of the heme within its polypeptide cage relative to that in cytochromes *c*, c_2, and c_{550}. If the polypeptide chains are made to coincide by rotating 44 α-carbon atoms of c_{551} onto corresponding atoms in homologous regions of tuna cytochrome *c* using a least-squares fitting program, then the heme in c_{551} is found to be rotated approximately 11° to the left about a vertical axis in Fig. 2, and tilted forward 16° about a horizontal axis, in comparison with its orientation in tuna *c*. (A similar tilt of the heme has subsequently been observed in cytochrome c_{555}; see *Discussion*.)

These structure comparisons now enable a precise alignment of amino acid sequences to be made, as shown in Fig. 4 for tuna *c* and *Pseudomonas* c_{551}.

The hydrophobic environment around the heme group is almost identical in the two proteins. Table 1 shows the comparison of structurally and sequentially equivalent hydrophobic side chains packed around the heme in 60 eukaryotic *c* sequences and 6 prokaryotic c_{551}. At only three places along the sequences is a hydrophobic heme contact in one protein not matched in the other. At two of these places the residue even remains hydrophobic, but is only turned so as to be in less obvious contact with the heme plane. This near-identity of heme environments is striking support for the evolutionary relatedness of the two proteins in spite of their size differences.

Aromatic side chains are less well conserved (Table 2). A cytochrome heme seems to demand the nearby presence of aromaticity, but with the placement in three dimensions being less critical. Only the two rings in what has been called the "right channel" (12), positions 10 and 97 in *c* or 7 and 77 in c_{551}, are invariant in all species. None of the aromatic rings that once were invoked in the electron transfer mechanism—59, 67, 74, 82—has an exact equivalent in c_{551}. However, the role of tryptophan 59, which is hydrogen bonded to the inner propionic

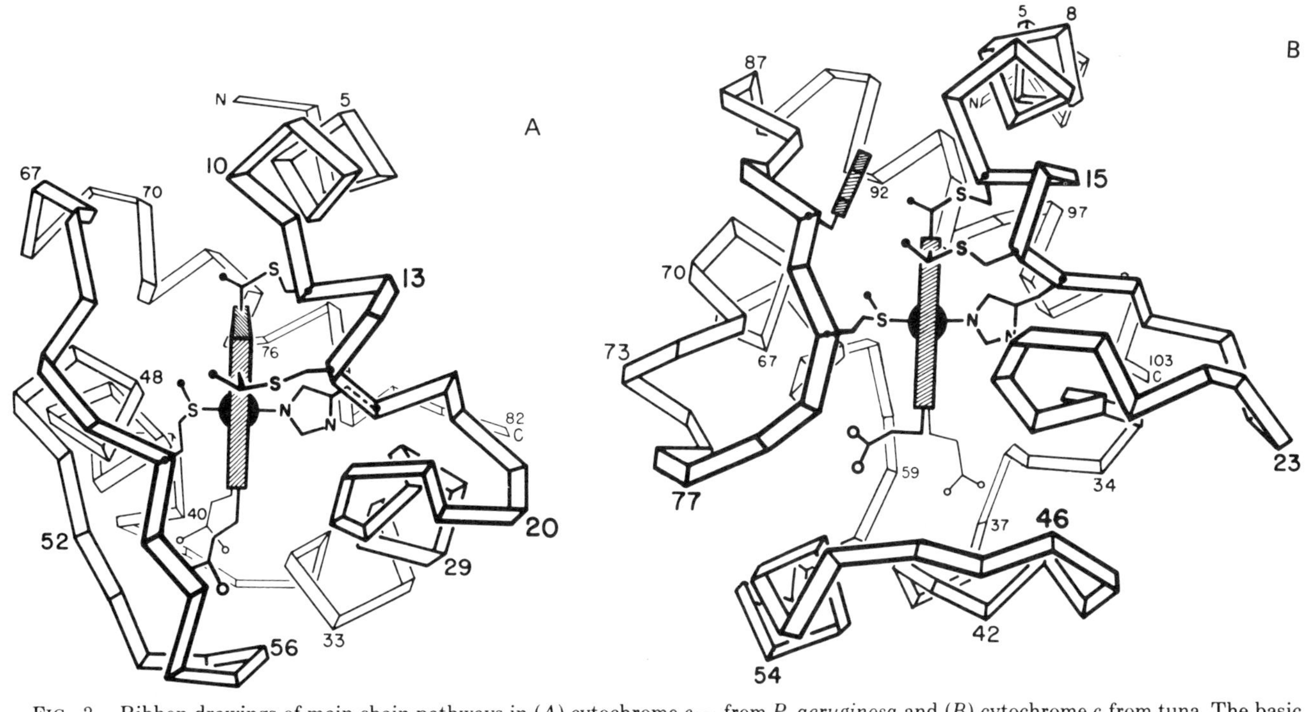

FIG. 3. Ribbon drawings of main chain pathways in (*A*) cytochrome c_{551} from *P. aeruginosa* and (*B*) cytochrome *c* from tuna. The basic molecular folding is the same. The heme group with its iron is represented by a cross-hatched slab with a central black ball. Covalent cysteine attachments, the methionine and histidine iron ligands, and the inner and outer heme propionic acid groups (at the bottom) are also shown.

```
                                         α
                  1  ┌──────────────────────────────────────┐
                                        10                        20
Tuna c:           Gly-Asp-Val-Ala-Lys-Gly-Lys-Lys-Thr-Phe-Val-Gln-Lys- x -Cys-Ala-Gln-Cys-His-Thr-Val-Glu-Asn-Gly-

Pseudomonas c551:            Glu-Asp-Pro-Glu-Val-Leu-Phe-Lys-Asn-Lys-Gly-Cys-Val-Ala-Cys-His-Ala-Ile-Asp-Thr- x -
                             1    └──────────────────────10┘                      └──┘  20
                                              α                                   α

                                                                                             α
                     30                               40                                ┌ 50
-Gly-Lys-His-Lys-Val-Gly-Pro-Asn-Leu-Trp-Gly-Leu-Phe-Gly-Arg-Lys-Thr-Gly-Gln-Ala-Glu-Gly-Tyr-Ser-Tyr-Thr-Asp-Ala-

- x - x -Lys-Met-Val-Gly-Pro-Ala-Tyr-Lys-Asp-Val-Ala-Ala-Lys-Phe-Ala- x - x - x - x - x - x - x - x - x - x - x -
                        └──────────────30──────────┘
                                       α

    α                                α                      α
──────┐                    ┌60────────────────────────70┐──────────────────┐
-Asn-Lys-Ser-Lys-Gly-Ile-Val-Trp-Asn-Asn-Asp-Thr-Leu-Met-Glu-Tyr-Leu-Glu-Asn-Pro-Lys-Lys-Tyr-Ile- x -Pro-Gly-Thr-

- x - x - x - x -Gly-Gln-Ala-Gly-Ala-Glu-Ala-Glu-Leu-Ala-Gln-Arg-Ile-Lys-Asn-Gly-Ser-Gln-Gly-Val-Trp-Gly-Pro-Ile-
                        └──40──────────────────────────────50┘
                                           α

                                                         α
    80                        ┌──90──────────────────────────────────────────────────┐
-Lys-Met-Ile-Phe-Ala-Gly-Ile-Lys-Lys-Lys-Gly-Glu-Arg-Gln-Asp-Leu-Val-Ala-Tyr-Leu-Lys-Ser-Ala-Thr-Ser.

-Pro-Met-Pro-Pro-Asn-Ala-Val- x -Ser-Asp-Asp-Glu-Ala-Gln-Thr-Leu-Ala-Lys-Trp-Val-Leu-Ser-Gln-Lys.
 60                             └──────70──────────────────────────┘
                                              α
```

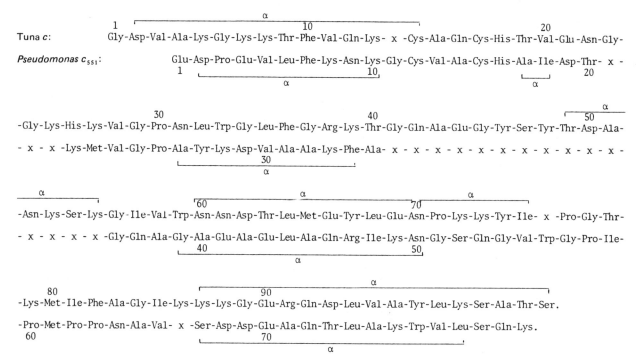

FIG. 4. Amino acid sequence alignments for tuna cytochrome *c* and *P. aeruginosa* c_{551}, made on the basis of the x-ray analyses. Brackets with α indicate regions of α helix. Ala 17-Ile 18 in c_{551} are considered α-helical because they make the proper hydrogen bonds to the beginning of helix 26–34.

acid group of the heme in all molecules of *c*, c_{550}, and c_2, is played by the sequentially unrelated tryptophan 56 of c_{551}. This appears to be an example of evolutionary convergence, as if the molecule, lacking an essential tryptophan at one position, developed a compensating residue somewhere else and made the requisite main chain adjustments to bring the tryptophan to its former position.

The environments of the buried and exposed propionic acid groups on the heme are similar in *c* and c_{551}, although the results are achieved in different ways from a sequence standpoint. The aromatic rings of invariant tyrosine 48 and tryptophan 59 in *c* are matched by invariant tryptophan 56 and semi-invariant phenylalanine/tyrosine/asparagine 34 around the buried propionic group in c_{551}. Serine/threonine 49 and invariant threonine 78 in *c*, hydrogen bonded to the outer propionic group, are paralleled by a semi-invariant serine/threonine at position 52 or 53 in c_{551}, also hydrogen bonded to the outer propionic group. In all known cytochrome *c* structures there seems to be a need for (*i*) a tryptophan residue hydrogen bonded to the buried heme propionic acid (*ii*) another nearby aromatic ring, (*iii*) a serine/threonine hydrogen bond to the outer propionate, and (*iv*) a pair of aromatic side chains to the right of the heme.

The highly asymmetric distribution of positively and negatively charged side chains that has been remarked for other cytochromes *c* (3, 12, 13) is present in cytochrome c_{551} also. The molecule has eight lysines, one arginine, five aspartic acids, and five glutamic acids, for a net charge count among side chains of −1. However, if a plane is drawn parallel to the page in Fig. 2 through the center of gravity of the molecule, then six positive charges (residues 8, 10, 28, 33, 47, and 49) and one negative

Table 1. Hydrophobic heme contacts in cytochromes *c* and c_{551}

Cytochrome *c*		Cytochrome c_{551}	
Tuna	60 Eukaryotes	*P. aeruginosa*	6 Prokaryotes
Phe^{10}	60 Phe	Phe^{7}	5 Phe, 1 Tyr
Pro^{30}	60 Pro	Pro^{25}	6 Pro
Leu^{32}	60 Leu	Tyr^{27}	3 Phe, 2 Leu, 1 Tyr
Leu^{35}	51 Leu, 5 Ile, 2 Val, 2 Phe	Val^{30}	6 Val
Leu^{64}	57 Leu, 2 Met, 1 Phe	Leu^{44}	5 Leu, 1 Ile
Tyr^{67}	59 Tyr, 1 Phe	No heme contact in c_{551}	
Leu^{68}	60 Leu	Ile^{48}	6 Ile
Pro^{71}	60 Pro	Gly^{51}	6 Gly
No heme contact in *c*		Pro^{62}	6 Pro
Phe^{82}	60 Phe	No heme contact in c_{551}	
Ile^{85}	36 Leu, 24 Ile	Val^{66}	6 Val
Leu^{94}	58 Leu, 2 Ile	Leu^{74}	6 Leu
Val^{95}	55 Ile, 4 Val, 1 Leu	Ala^{75}	6 Ala
Leu^{98}	57 Leu, 3 Met	Val^{78}	4 Val, 2 Ile

Eukaryotic cytochrome *c* sequences from tables IX and X of ref. 18; prokaryotic c_{551} sequences from table 2 of ref. 4.

Table 2. Aromatic side chains in cytochromes *c* and c_{551}

Cytochrome *c*		Cytochrome c_{551}	
Tuna	60 Eukaryotes	*P. aeruginosa*	6 Prokaryotes
Phe^{10}	60 Phe	Phe^{7}	5 Phe, 1 Tyr
Not aromatic		Tyr^{27}	3 Phe, 2 Leu, 1 Tyr
Phe^{36}	53 Phe, 3 Ile, 2 Val, 2 Tyr	Not aromatic	
Not aromatic		Phe^{34}	3 Tyr, 2 Asn, 1 Phe
Tyr^{46}	33 Tyr, 27 Phe	Deletion region	
Tyr^{48}	60 Tyr	Deletion region	
Trp^{59}*	60 Trp	Trp^{56}*	6 Trp
Tyr^{67}	59 Tyr, 1 Phe	Not aromatic	
Tyr^{74}	58 Tyr, 2 Phe	Not aromatic	
Phe^{82}	60 Phe	Not aromatic	
Tyr^{97}	59 Tyr, 1 Phe	Trp^{77}	6 Trp

* Molecular convergence: same position in three dimensions, although different positions along polypeptide chains.

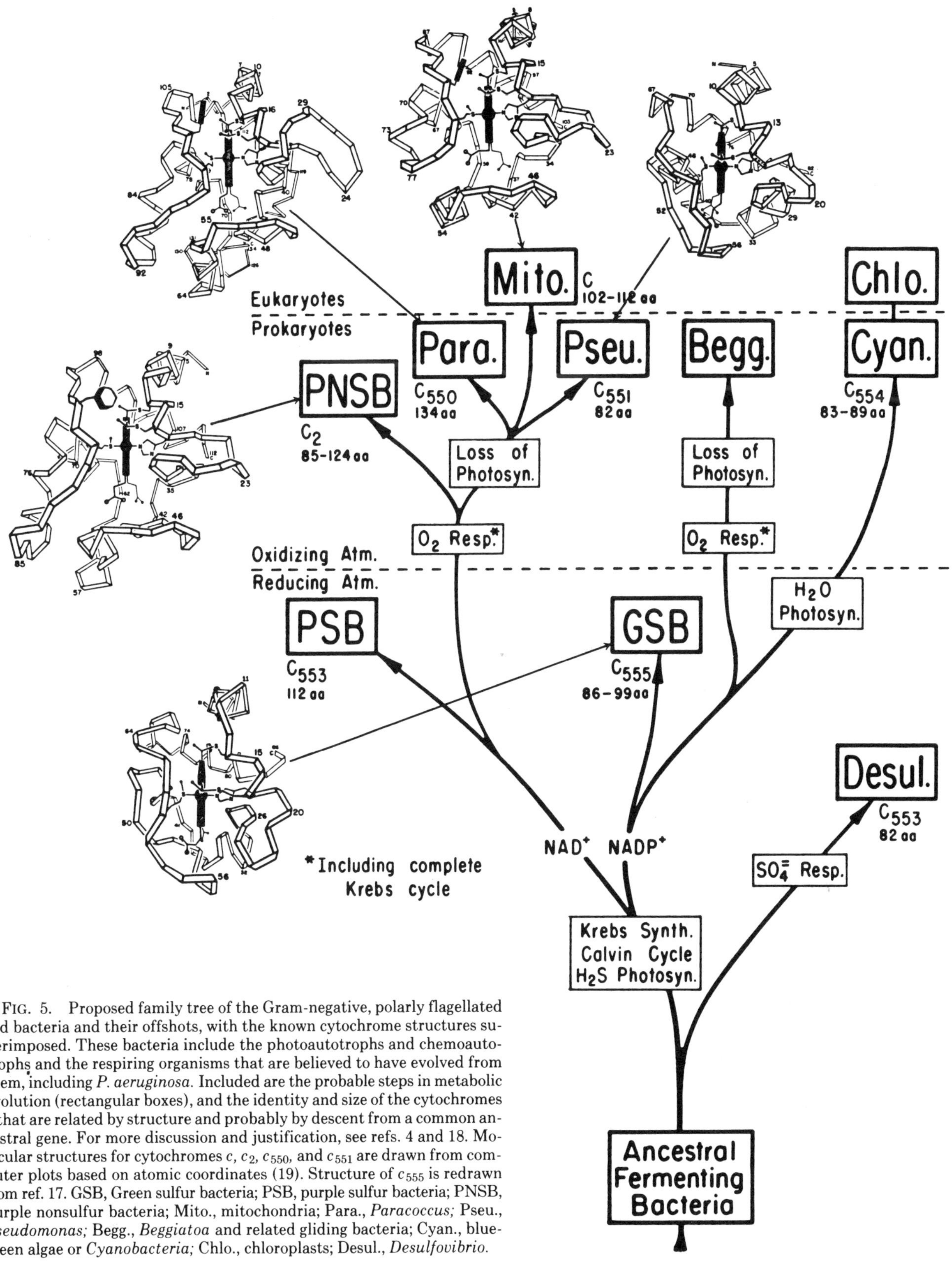

FIG. 5. Proposed family tree of the Gram-negative, polarly flagellated rod bacteria and their offshots, with the known cytochrome structures superimposed. These bacteria include the photoautotrophs and chemoautotrophs and the respiring organisms that are believed to have evolved from them, including *P. aeruginosa*. Included are the probable steps in metabolic evolution (rectangular boxes), and the identity and size of the cytochromes *c* that are related by structure and probably by descent from a common ancestral gene. For more discussion and justification, see refs. 4 and 18. Molecular structures for cytochromes *c*, c_2, c_{550}, and c_{551} are drawn from computer plots based on atomic coordinates (19). Structure of c_{555} is redrawn from ref. 17. GSB, Green sulfur bacteria; PSB, purple sulfur bacteria; PNSB, purple nonsulfur bacteria; Mito., mitochondria; Para., *Paracoccus;* Pseu., *Pseudomonas;* Begg., *Beggiatoa* and related gliding bacteria; Cyan., blue-green algae or *Cyanobacteria;* Chlo., chloroplasts; Desul., *Desulfovibrio.*

charge (residue 29) lie on this bisecting perimeter, eight negative charges (residues 1, 2, 4, 41, 43, 68, 69, and 70) and two positive (residues 76 and 82) lie on the back half of the molecular surface, and only one side chain of each charge (aspartic 19 and lysine 21) is found on the front hemisphere. The cytochrome c_{551} molecule is essentially a sphere with the heme crevice opening to a hydrophobic front hemisphere, diametrically opposed to a negatively charged back hemisphere, with a belt of positive charges separating the two. The positive charges around the perimeter of the heme face of the molecule, which have been commented upon in other cytochromes *c*, are present here also. Lysines 10 at the top of the crevice, 21 at the right, and 47 at the left correspond to lysines 13, 25/27, and 72/73 in eukaryotes. These residues are evolutionarily invariant among the six c_{551} sequences known (as are most of the other lysines). The acidic side chains are more variable, as was also the case for the eukaryotic cytochromes *c*.

DISCUSSION

The x-ray crystal structure shows clearly that cytochrome *c* belongs in the same evolutionary family with eukaryotic, mitochondrial *c*, *Paracoccus* c_{550}, and c_2 from purple nonsulfur photosynthetic/respiratory bacteria such as *Rhodospirillum*. Possible evolutionary relationships between these proteins, and between their host organisms, have been discussed in ref. 4. At the time of that paper it appeared as if cytochromes *c*, c_2, and c_{550} were the most closely related group and that the c_{551}s with their chain deletion at the bottom of the molecule were more distantly related. Among the cytochromes c_2 from various purple nonsulfur photosynthetic bacteria whose sequences had been determined by Ambler and others (14, 15) were molecules that were as small as eukaryotic *c* or as large as c_{550}, but with nothing in the size range of c_{551}. Recent work has changed this picture (R. Ambler, private communication). Sequences have been determined for at least one c_2 from every formal species listed in *Bergey's Manual* (16). In at least two of the three genera, *Rhodospirillum* and *Rhodopseudomonas*, Ambler found species containing cytochromes c_2 that in both size and amino acid sequence appear to be homologous with *Pseudomonas* c_{551}. It appears that all of these small cytochromes, from c_{551} with 82 amino acids to c_{550} with 134, should be considered as one evolutionary family, with a standard pattern of acceptable insertions and deletions that is followed in diverse organisms: respirers and photosynthesizers, prokaryotes and eukaryotes.

Other members of this family probably include cytochrome c_{553} of purple sulfur bacteria (Chromatiaceae) with 112 amino acids, c_{555} of green sulfur bacteria (Chlorobiaceae) with 86 amino acids, cytochromes *f* or c_{554} of prokaryotic and eukaryotic algae with 83–89 amino acids, and perhaps even the single-heme cytochrome c_{553} of the sulfate-respiring *Desulfovibrio* with 82 amino acids. Korszun and Salemme have recently reported the folding of the polypeptide chain backbone in cytochrome c_{555} of *Chlorobium thiosulfatophilum* (17) as obtained from an unrefined multiple isomorphous replacement analysis at 2.7 Å resolution, and it is apparent that this backbone is virtually identical with that reported for *Pseudomonas* c_{551} in ref. 4 and this paper. The same large deletion is observed at the bottom of the molecule, and from the ribbon drawing of the c_{555} chain path, the heme appears to be tilted forward as in c_{551}. One interesting difference observed in c_{555} is that the tryptophan that appears to be hydrogen bonded to the buried propionic acid (17) is present in the *same* position along the chain in eukaryotic cytochromes *c* and all published c_{555} sequences, even though these two proteins probably are the most distantly related in an evolutionary sense. The shifted tryptophan observed in c_{551} may represent a special adaptation in the Pseudomonads.

The probable evolutionary relationships between the cytochrome *c*-containing metabolic pathways in bacteria and eukaryotes are outlined in Fig. 5, which is an extension of an earlier metabolic tree in ref. 4. At least the main chain pathway is now known for five of the cytochromes indicated in this figure; c_{555}, c_2, c_{551}, c_{550}, and *c*, in the probable order of evolution. The "small" form of cytochrome *c* with the bottom chain deletion is found in green sulfur and purple nonsulfur photosynethetic bacteria, respiratory bacteria (c_{551}), and cyanobacteria or blue-green algae. The "large" form of cytochrome *c*, in contrast, has been observed so far only in the branch of Fig. 5 leading to the purple sulfur and nonsulfur bacteria. Hence, it seems more likely that the small cytochrome is the ancestral form and that one branch has seen the *addition* of more chain at the bottom of the heme crevice.

All of the bacteria discussed in this paper and shown in Fig. 5 are similar enough in general morphology to be grouped by Brock (20) into one category: the Gram-negative, polarly flagellated rods. It is interesting to find this morphological similarity matched by a biochemical similarity: all members perform either photosynthesis or respiration or both, all possess electron transport chains with a cytochrome *c* near the high-potential end, and all of these cytochromes *c* appear to be structurally or sequentially homologous. This class of cytochromes arose soon after bacteria developed the ability to trap light and use it for chemical purposes, and has been retained in the divergent evolution of present-day photosynthesis and respiration.

We thank Dr. Richard P. Ambler for permission to mention unpublished results prior to publication. This investigation was supported by the National Institutes of Health under Grant GM-12121 and the National Science Foundation under Grant PCM75-05586. This is contribution no. 5632 of the Norman W. Church Laboratory of Chemical Biology.

1. Horio, T., Higashi, T., Sasagawa, M., Kusai, K., Nakai, M. & Okunuki, K. (1960) *Biochem. J.* **77**, 194–201.
2. Ambler, R. (1963) *Biochem. J.* **89**, 341–349 and 349–378.
3. Timkovich, R. & Dickerson, R. E. (1976) *J. Biol. Chem.* **251**, 4033–4046.
4. Dickerson, R. E., Timkovich, R. & Almassy, R. J. (1976) *J. Mol. Biol.* **100**, 473–491.
5. Needleman, S. B. & Blair, T. T. (1969) *Proc. Natl. Acad. Sci. USA* **63**, 1227–1233.
6. Dickerson, R. E. (1971) *J. Mol. Biol.* **57**, 1–15.
7. McLachlan, A. D. (1971) *J. Mol. Biol.* **61**, 409–424.
8. Dayhoff, M. O. (1976) *Atlas of Protein Sequence* (Georgetown Univ., Washington, DC), Vol. 5, Suppl. 2, pp. 26–27.
9. Dickerson, R. E., Weinzierl, J. E. & Palmer, R. A. (1968) *Acta Crystallogr. Sect. B* **24**, 997–1003.
10. Chambers, J. L. & Stroud, R. M. (1977) *Acta Crystallogr. Sect. B* **33**, 1824–1837.
11. Mandel, N., Mandel, G., Trus, B. L., Rosenberg, J., Carlson, G. & Dickerson, R. E. (1977) *J. Biol. Chem.* **252**, 4619–4636.
12. Dickerson, R. E., Takano, T., Eisenberg, D., Kallai, O. B., Samson, L., Cooper, A. & Margoliash, E. (1971) *J. Biol. Chem.* **246**, 1511–1535.
13. Salemme, F. R., Freer, S. T., Xuong, Ng. H., Alden, R. A. & Kraut, J. (1973) *J. Biol. Chem.* **248**, 3910–3921.
14. Ambler, R. P. (1976) in *CRC Handbook of Biochemistry and Molecular Biology*, ed. Fasman, G. D. (CRC Press, Cleveland, OH), 3rd Ed., Proteins Section, Vol. III, pp. 294–298.
15. Ambler, R. P., Meyer, T. E. & Kamen, M. D. (1976) *Proc. Natl. Acad. Sci. USA* **73**, 472–475.
16. Buchanan, R. E. & Gibbons, N. E., eds. (1974) *Bergey's Manual of Determinative Bacteriology* (Williams and Wilkins, Baltimore, MD), 8th Ed.
17. Korszun, Z. R. & Salemme, F. R. (1978) *Proc. Natl. Acad. Sci. USA* **74**, 5244–5247.
18. Dickerson, R. E. & Timkovich, R. (1975) in *The Enzymes*, ed. Boyer, P. (Academic Press, New York), Vol. 11, pp. 397–547.
19. Dickerson, R. E. (1978) in *International Symposium on Biomolecular Structure, Conformation, Function and Evolution Proceedings*, Madras, 4–7 January 1978, ed. Srinivasan, R. (Pergamon Press, New York).
20. Brock, T. D. (1970) *Biology of Microorganisms* (Prentice Hall, New York), pp. 579–606.

Comparison of Total Sequence of a Cloned Rabbit β-Globin Gene and Its Flanking Regions with a Homologous Mouse Sequence

Abstract. *The nucleotide sequence of a cloned rabbit chromosomal DNA segment of 1620 nucleotides length which contains a β-globin gene is presented. The coding regions are separated into three blocks by two intervening sequences of 126 and 573 base pairs, respectively. The rabbit sequence was compared with a homologous mouse sequence. The segments flanking the rabbit gene, as well as the coding regions, the 5′ noncoding and part of the 3′ noncoding messenger RNA sequences are similar to those of the mouse gene; the homologous introns, despite identical location, are distinctly dissimilar except for the junction regions. Homologous introns may be derived from common ancestral introns by large insertions and deletions rather than by multiple point mutations.*

We have recently described the cloning and characterization of a 5100-base pair (bp) Kpn I fragment of rabbit DNA containing a β-globin gene (*1*). The coding sequences were arranged in three blocks, separated by two intervening sequences or introns, a smaller one of 126 and a larger one of 573 base pairs. The positions of both introns relative to the coding sequences were identical to those found in a mouse β-globin major gene cloned by Tilghman *et al.* (*2*). Although the corresponding mouse and rabbit β-globin introns had very similar sequences in the vicinity of the junctions to the coding sequences, the similarities within the introns diminished rapidly with increasing distance from the junctions, as far as the sequences were determined.

We now report the complete sequence of a 1620-bp rabbit DNA segment extending from 223 nucleotides before the start of the sequence coding for β-globin messenger RNA (mRNA) to 109 nucleotides beyond its terminus. Moreover, we have determined the sequence of most of the mouse β-globin chromosomal gene β-G2 isolated by Tilghman *et al.* (*2*); our findings agree in all but 16 positions with the sequence determined by Konkel *et al.* (*3*). The rabbit and mouse sequences show homology, except for the introns and part of the 3′ noncoding sequence. It seems that, although the introns have common ancestral sequences, they have been subject to considerable genetic drift, which suggests that no sequence specific function is associated with most of the intron. Conversely, the homologies retained in other regions, in particular those preceding the beginning

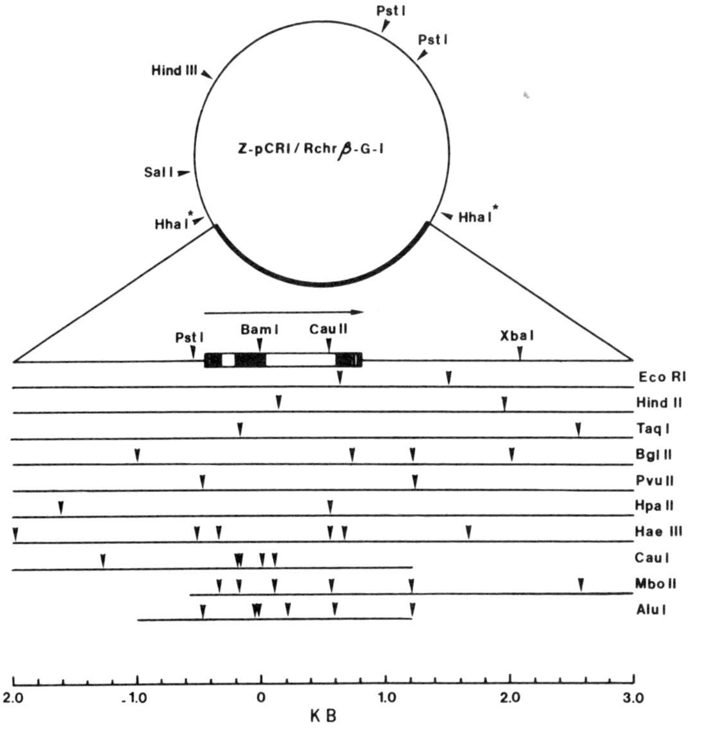

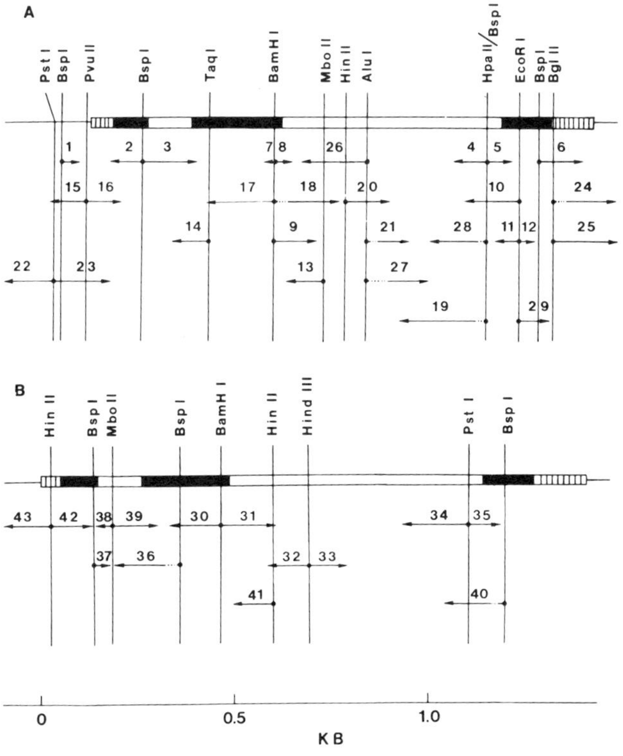

Fig. 1. (left). Restriction site map of a cloned rabbit DNA Kpn I fragment containing a β-globin gene and its flanking regions. The hybrid plasmid Z-pCRI/RchrβG-1 (*1*) was cleaved with Bam HI and the two resulting 5′ termini were labeled with [γ-^{32}P]ATP and polynucleotide kinase (*5*). The labeled DNA was further cleaved with Sal I (or in some cases with Eco RI), and the two labeled fragments were separated by agarose gel electrophoresis. Each ^{32}P-labeled fragment was subjected to partial cleavage with the restriction enzymes indicated, and the products were analyzed by polyacrylamide gel electrophoresis (*4*). The restriction sites between −0.7 and 1.2 kbp were confirmed by sequence analysis. The restriction sites of the pCRI moiety are taken from (*35*); only the two Hha I sites closest to the insert are indicated. The top line of the map shows the position of restriction sites present only once within the insert, and the location of the coding regions (black boxes), intervening sequences (white boxes), and 5′ and 3′ noncoding sequences (hatched boxes). Fig. 2 (right). Strategy for sequencing the β-globin gene and its flanking regions. (A) Rabbit β-globin DNA of Z-pCRI-RchrβG-1 (*1*). (B) Mouse β-globin major DNA β-G2 (*2*) recloned in pBR322 (*1*). The nucleotide sequence was determined (*5*) from the restriction sites shown [vertical lines; see Fig. 1 and (*1*)]. The arrows originating at the dots on the vertical lines indicate the direction (5′ to 3′) and extent of the readout; the discontinuous horizontal lines show the regions in which the sequence was not determined. The numbers above the arrows refer to the preparation of the fragment described in Table 1. Distances are indicated relative to the beginning of the RNA sequence. The different regions of the gene are indicated as in Fig. 1.

of the mRNA sequence, suggest a functional role for these segments.

The restriction map [data obtained as in (*4*)] of the 5100-bp Kpn I fragment of rabbit DNA, which is joined to the plasmid pCRI by AT-linkers (A, adenine; T, thymine) at the Eco RI site, is shown in Fig. 1. It is of practical interest that the rabbit DNA insert, perhaps due to the characteristic dearth of CG (C, cytosine; G, guanine) doublets in vertebrate DNA, contains no sites for endonuclease HhaI (*GCGC*) and can be excised intact from the hybrid DNA by this enzyme.

The nucleotide sequence of the β-globin gene and that of its flanking regions was determined, using the fragments indicated in Fig. 2, A and B. The approach used to generate, label, and purify each fragment is shown (Table 1). In many instances the sequences of the two complementary DNA strands were determined to preclude errors that may arise when a DNA strand containing methylated C residues is sequenced. The second C in the Eco RII recognition sequence, C-C-(A/T)-G-G, gives rise to only a very weak band, or more often a gap, in the C + T lane of the Maxam-Gilbert ladder; the correct sequence can be obtained from the opposite strand (*5*). Moreover, in most cases additional sequencing was carried out across the 5′ termini, which served as origins for sequencing because we have not been able to determine unambiguously the first 5′ proximal one to three nucleotides from a Maxam-Gilbert sequencing gel.

In the complete sequence of the rabbit gene (Fig. 3), five positions, marked by asterisks, have not been reliably established. Seven positions (marked by dots) were not deduced by sequencing, but were established from the known sequence of a restriction site and the amino acid sequence in that region. The sequence of the coding part of the gene

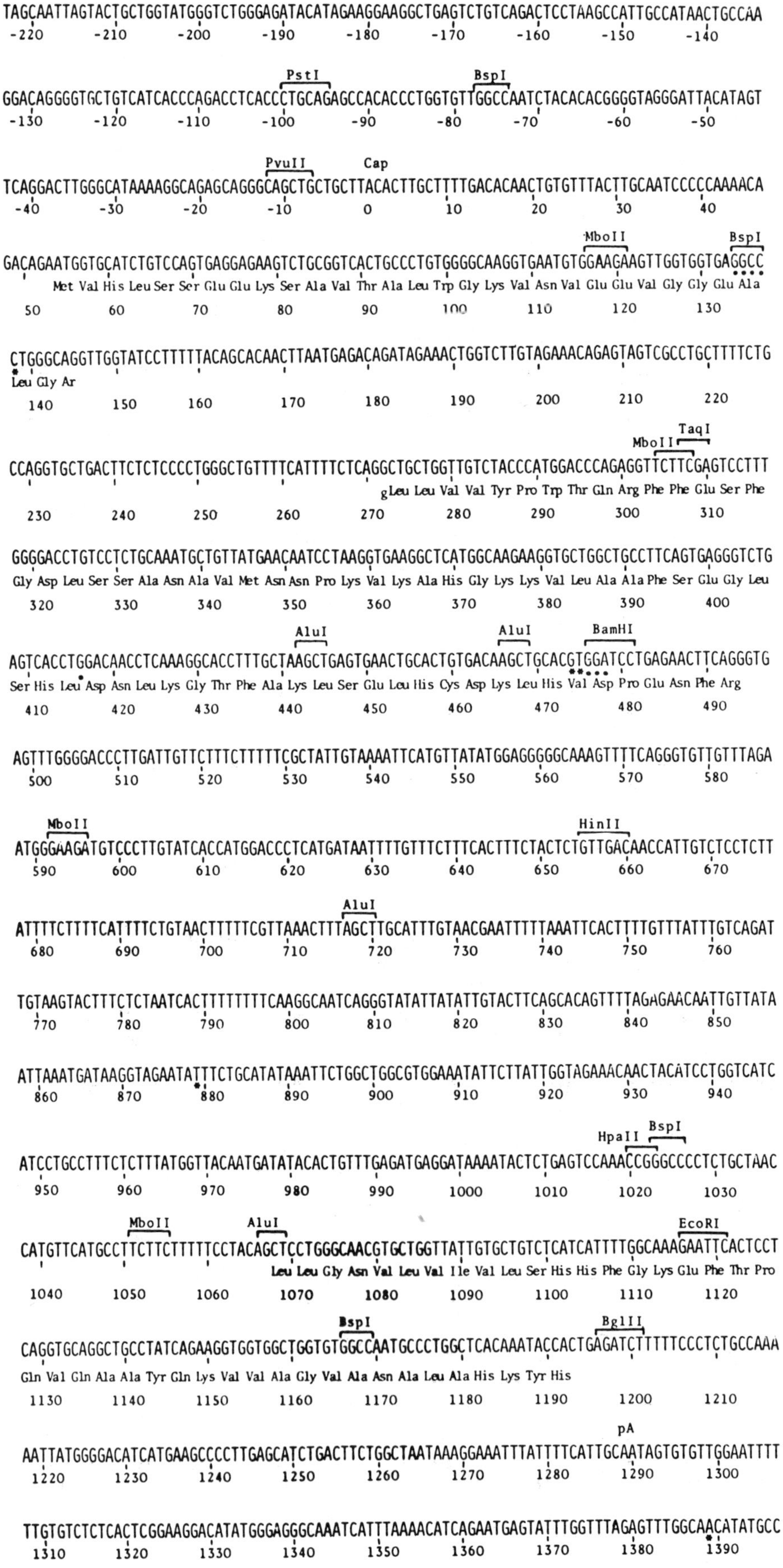

Fig. 3. The complete nucleotide sequence of a rabbit β-globin gene and its flanking regions. Position 1 corresponds to the capped nucleotide of the mRNA (*36*). Nucleotides marked by a dot have been deduced from the recognition sequence of a restriction site *and* the amino acid sequence in that region. Five nucleotides marked by an asterisk have not been reliably determined. Cap and pA designate the positions of the cap and poly(A) tail, respectively. The borders of the introns have not been determined experimentally; they have been placed at the positions predicted by the Chambon rule (*31*).

agrees with that of a rabbit β-globin complementary DNA (cDNA) as established by Efstratiadis *et al.* (*6*). As determined in our laboratory, several nucleotide positions of the mouse β-globin gene, in particular in the region from 1090 to 1120 [numbering as in (*3*)], do not agree with the sequence given by Konkel *et al.* In Fig. 4, we have indicated by asterisks the discrepant positions, and by dashed lines the nucleotide sequences not determined by us, but taken from (*3*).

The rabbit sequence studied may be subdivided into three major sections: the middle portion (1288 bp), corresponding to the sequence transcribed into the 15*S* β-globin mRNA precursor (see *7*), and two flanking sequences. The middle portion comprises the 5′ noncoding sequence (53 bp), followed by three coding sequences (93, 222, and 129 bp, including initiation and termination triplets), intermingled with two introns (126 and 573 bp) and the 3′ noncoding sequence (92 bp). The mouse β-globin major gene has a similar general structure (*3*), except for differences in the lengths of the noncoding regions, mainly the large intron—646 bp according to (*3*) or 650 bp if our corrections are taken into account—and the 3′ noncoding sequence (130 bp).

The nearest-neighbor frequency was determined for various DNA segments and expressed as the ratio of the value found to that expected for a random sequence of the same base composition (*8*). The values for the coding and the noncoding segments of the rabbit β-globin sequence plus strand (Fig. 5, A and B), and those calculated for the double-stranded DNA (Fig. 5, C and D) were compared with those of total DNA of rabbit liver (Fig. 5E) (*9*). In all cases the value for CG (that is, C + G) is strikingly low, ranging from 0.13 in the coding regions to 0.17 in the noncoding regions and 0.25 for total rabbit liver DNA; the corresponding value for the sequenced mouse globin DNA fragment is 0.1 (*3*). A deficit in CG has been described as a general and distinctive feature of vertebrate DNA (*8*). Russell *et al.* (*8*) have suggested that this feature is characteristic for protein coding sequences and that therefore the bulk of the nuclear DNA shows the general design of DNA coding for polypeptides. Our data on the fragments that contain the rabbit β-globin gene, however, show that the CG deficit is common to all segments, whether they be coding or not. In addition, the overall pattern of the nearest neighbor distribution of total liver DNA of the rabbit closely resembles that of the noncoding regions of the β-globin DNA, rather than that of the coding regions. The deficit of CG in noncoding regions is also apparent in mouse DNA fragments containing the β-globin (*3*) and immunoglobulin light chain genes (*10*). The CG deficit is thus not restricted to coding regions; in fact, some eukaryotic mRNA's are quite rich in this doublet (*11*, *12*). Therefore the CG deficit requires a different explanation. Heindell *et al.* (*13*) propose that the CpG sequence is a mutational "hot spot" because it is a major methylation site, and methylated C, once deaminated, is not subject to the repair pathway involving uracil *N*-glycosidase, which removes uracil residues from DNA (*14*). This would lead to a depletion of CpG sequences whenever there is no selective pressure to conserve them and to a concomitant enrichment of TpG and CpA doublets. In fact, there is an overrepresentation (Fig. 5) of TpG and CpA in all segments of the rabbit sequence.

There is no general, simple method of determining the degree of relatedness of two nucleotide sequences. In the simplest approach, two sequences of equal length are lined up, the number of positions containing the same nucleotides are scored and compared to the values given by random sequences. If the two sequences to be compared are related, but differ as a consequence of deletions or insertions (or both), homology may be detected only in part or not at all by such a simple alignment. Only by introducing gaps (or insertions) in appropriate positions can the homologous sequences be aligned (*15*). But by introducing a sufficient number of gaps, any two heteropolymeric sequences (including random ones) can be adjusted to give substantial correspondence of nucleotides (*15*). Thus, if the introduction of a pair of gaps permits the alignment of 20 previously unmatched, adjacent nucleotides, a significant homology has been uncovered. If only one or two nucleotides can be lined up by introducing a gap, then it is most likely not a meaningful result since this is readily obtained also with random sequences.

We therefore use the following rule in lining up two sequences. For each gap inserted, a penalty of N points is levied, while each matched nucleotide pair is credited with one point; in order for the introduction of a gap to be permissible, the net gain in points (calculated over the entire sequence) must be ≥ 0. Using this scoring system on eight pairs of random sequences of 100 nucleotides each, we determined that, on average, for $N = 4$, 0.9 gap could be introduced per pair, leading to an increase of matched nucleotides from 27 to 31 percent, while for $N = 5$, 0.5 gap could be introduced, raising the percentage of matched nucleotides to 29.

This rule, with $N = 4$, was applied in aligning the rabbit, mouse, and human (*16*) β-globin sequences and surrounding regions, as far as they were known (Fig. 4). Although it is, in principle, very difficult to optimize the alignment because of the enormous number of combinations that would have to be tested (which certainly also surpass current computing capacity), the degree of matching attained in practice was not very different when carried out by different investigators. The similarities of the different segments, expressed as number of matching nucleotides per number of positions compared (including gaps) is given in Table 2. In Fig. 6, the percentage of matching nucleotides of the complete rabbit and mouse sequences (determined for overlapping blocks of 20 nucleotides) is plotted along the length of the sequences. The greatest similarity is found among the coding sequences (81 percent), the 5′ noncoding mRNA sequence (75 percent), the 5′ flanking sequence (68 percent), and the last 50 nucleotides of the 3′ noncoding sequence (72 percent). Both the large and small introns show very little similarity (average, 53 percent for the small and 40 percent for the large intron) except at the junctions with the coding sequences (Fig. 4) and a few stretches of about 12 to 15 nucleotides in the middle region of the large introns. The similarity of the large introns is only slightly higher than that of random sequences (Table 2).

In the case of the coding sequences one may distinguish three classes of sites, namely (i) replacement sites, where each nucleotide substitution leads to an amino acid replacement, (ii) totally silent sites, where no nucleotide substitution gives rise to an amino acid change, and (iii) mixed sites, in which only some nucleotide substitutions cause an amino acid replacement. The similarity (Table 2) among replacement sites (88 percent) is distinctly higher than that among totally silent (70 percent) or mixed sites (67 percent). This is also true for the rabbit-human and mouse-human pairs. It seems reasonable to postulate that conservation of sequences reflects an evolutionary constraint due to some functional significance. Constraint seems to be exercised preferentially at the protein level inasmuch as nucleotide changes in replacement sites are less frequent than in totally silent sites. However, in a comparison of human and rabbit β-globin mRNA, Kafatos *et al.* (*17*) pointed out that even silent sites are more strongly conserved than the "variable regions" of fibrinopeptides, which are considered to be under little or no constraint and are used as "neutrality standard." Furthermore, they note that silent and nonsilent substitutions tend to be clustered, suggesting that evolutionary constraints may operate not only at the protein level, but also at that of the mRNA.

Comparison of the sequences coding for the human, mouse, and rabbit β-globin mRNA's reveals fewer differences between human and rabbit sequences than between human and mouse or rabbit and mouse. The 98 positions in which nucleotide differences occur are scattered more or less uniformly over the entire length (444 nucleotides) of the coding sequence, except for two regions of 38 and 51 nucleotides, respectively, in each

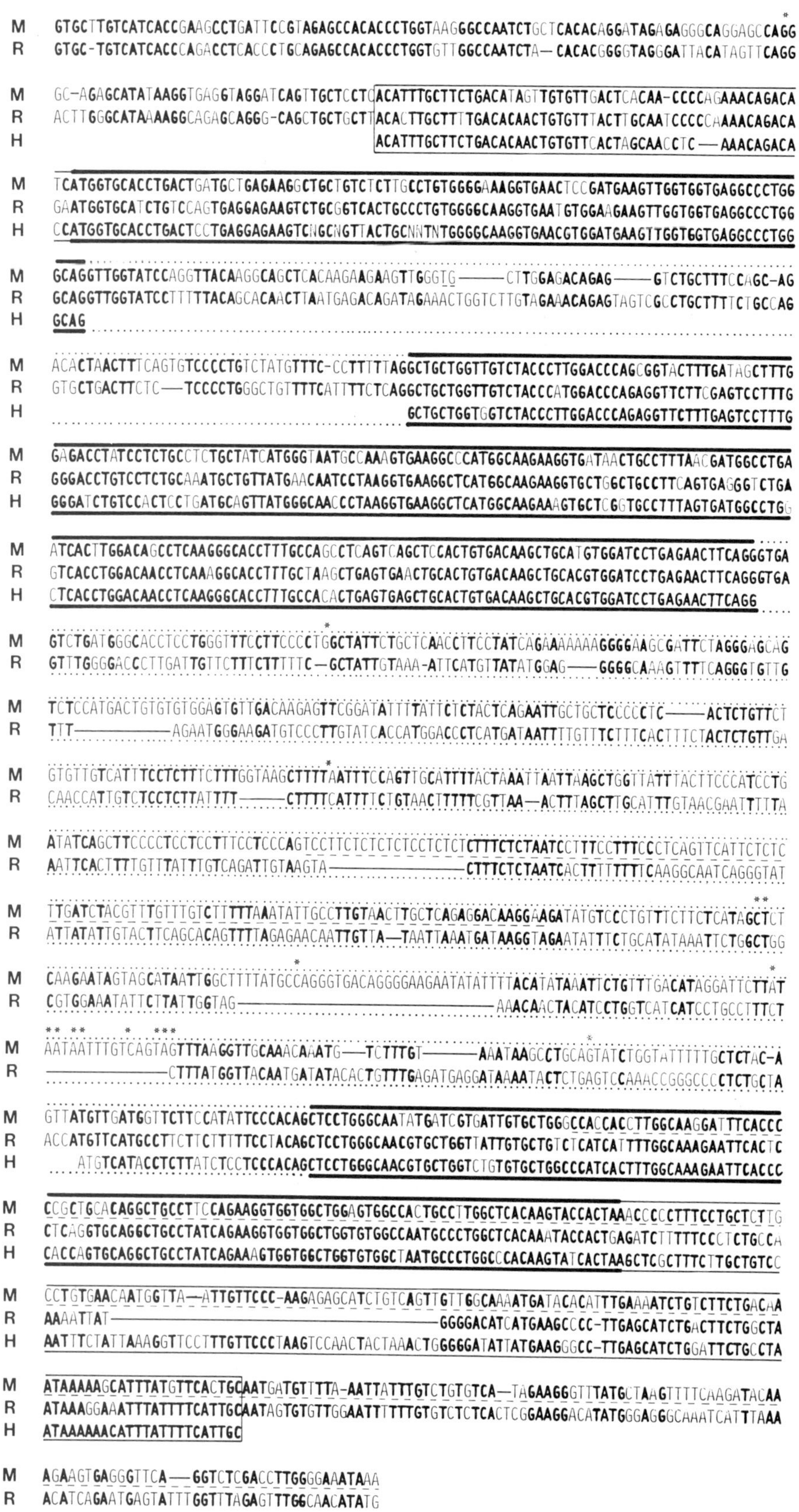

Fig. 4. Comparison of the nucleotide sequences of β-globin genes of mouse (M), rabbit (R), and human (H). The sequences were aligned as described. The nucleotide sequence of the rabbit β-globin gene is that shown in Fig. 3; that of the mouse β-globin DNA [the fragment βG-2 cloned by Tilghman *et al.* (*2*)] was determined in our laboratory, except for the regions indicated by a dashed line, which are from Konkel *et al.* (*3*). The following discrepancies (indicated by asterisks) were noted between our sequence and that of Konkel *et al.* (*3*) (numbering is according to Konkel *et al.*): Konkel's sequence lacks a G residue each between nucleotides 38 and 39 and between 598 and 599; a C residue between 1037 and 1038; a CT sequence between 1006 and 1007 and TAG sequence between 1111 and 1112. The T residue at 772, the C residue at 1108, and the G residue at 1153 were not found in our analysis. At positions 1096, 1098, 1099, 1101, and 1102 there should be an A rather than a G. The first 49 nucleotides of the sequences shown were determined only in our laboratory. The primary structure of the human coding sequences are from (*37*), and the sequence at the edge of the large intron are from (*28*). Heavy type indicates positions identical in two or more sequences. The 5′ and 3′ noncoding regions of the mRNA are framed with a thin line, the coding sequences with a thick line, and the introns with a dotted line.

Table 1. Preparation of ^{32}P-labeled fragments of β-globin DNA for nucleotide sequence determination. The 5′ terminal labeling was carried out as described by Maxam and Gilbert (*5*). Fragments were isolated on 5 percent polyacrylamide gels in 50 m*M* tris-borate (*p*H 8.3), 1 m*M* EDTA; or on 1 percent agarose gels in 2 m*M* EDTA, 50 m*M* tris-acetate, 20 m*M* sodium acetate (adjusted to *p*H 7.8 with acetic acid). The asterisks preceding the endonucleases indicate the 5′ labeled restriction site. The number following the endonuclease represents the length of the fragment (in nucleotides, and not including overhanging ends); the numbers in parentheses refer to the arrows in Fig. 2.

Starting material	First enzymatic cleavage + labeling	Second enzymatic cleavage	Fragments isolated
		Rabbit β-globin DNA	
Total plasmid	Bam HI	Eco RI	*Bam HI-Eco Ri 17′000 (7) and *Bam HI-Eco RI 636 (8, 9)
	Eco RI	Bgl II	*Eco RI-Bgl II 1700 (10, 11) and *Eco RI-Bgl II 76 (12)
	Pst I	Bgl II	*Pst I-Bgl II 400 (22) and *Pst I-Bgl II 1299 (23)
Hha I* fragment 6000 bp	Bsp I	Pvu II†	*Bsp I-Pvu II 144 (2) and *Bsp I-Pvu II 66 (1)
		Bam HI†	*Bsp I-Bam HI 341 (3) and *Bsp I-Bam HI 546 (4)
		Eco RI†	*Bsp I-Eco RI 600 (6) and *Bsp I-Eco RI 102 (5)
	Mbo II	Bam HI	*Mbo II-Bam HI 123 (13)
	Taq I	Hpa II	*Taq I-Hpa II 1400 (14)
	Pvu II	Hpa II	*Pvu II-Hpa II 1100 (15) and *Pvu II-Hpa II 1030 (16)
Eco RI‡ fragment 900 bp	§	Pvu II	*Eco RI-Pvu II 500 (29)
Bam HI-Eco RI fragment‖ 636 bp	Hpa II	Alu I	*Bam HI-Alu I 238 (18) and *Hpa II-Alu I 302 (19)
	Hin II	Bsp I	*Hin II-Bsp I 367 (20)
	Alu I	Bsp I	*Alu I-Bsp I 306 (21)
Bam HI-Eco RI fragment 17000 bp	§	Bsp I	*Bam HI-Bsp I 341 (17)
Bgl II fragment 400 bp	§	Alu I	*Bgl II-Alu 400 (24, 25)
Bgl II fragment 1700 bp	Alu I	Bam HI¶	*Alu I-Bam HI 238 (26)
		Bsp I¶	*Alu I-Bsp I 306 (27)
	Hpa II	Alu I	*Hpa II-Alu I 302 (28)
		Mouse β-globin DNA	
Total plasmid	Bam HI	Eco RI	*Bam HI-Eco RI 1800 (30) and *Bam HI-Eco RI 5000 (31)
	Hind III	Bam HI	*Hind III-Bam HI 220 (32) and *Hind III-Bam HI 8500 (33)
	Pst I	Bam HI + Eco RI	*Pst I-Bam HI 609 (34) *Pst I-*Pst I 4500
*Pst I-*Pst I 4500 bp	§	Bsp I	*Pst I-Bsp I 87 (35)
Bam HI-Eco RI 1800 bp	Bsp I	Mbo II	*Bsp I-Mbo II 167 (36) *Bsp I-Mbo II 53 (37)
	Mbo II	Bsp I	*Mbo II-Bsp I 53 (38) *Mbo II-Bsp I 167 (39)
Bsp I-Bsp I 810 bp	§	Bam HI	*Bsp I-Bam HI 696 (40)
Hin II-Hin II 577 bp	§	Bam HI	*Hin II-Bam HI 133 (41) *Hin II-Bam HI 439 (42)
Hin II-Hin II 600 bp	§	Alu I	*Hin II-Alu I 200 (43)

*Total plasmid DNA was cleaved with Hha I and the largest Hha I fragment was isolated by sucrose gradient centrifugation. †Triple digestion with Pvu II, Bam HI and Eco RI. ‡Total plasmid was cleaved with Eco RI and the 900 bp fragment was isolated by sucrose gradient centrifugation. §The starting material was labeled directly. ‖The 17500 bp Eco RI-Eco RI fragment isolated by sucrose gradient centrifugation was cleaved with Bam HI and the Bam HI-Eco RI 17000 bp and Bam HI-Eco Ri 638 bp fragments were isolated. ¶Double digestion with Bsp I and Bam HI.

of which only one position is variable (Fig. 4). These highly conserved sequences are located around the positions corresponding to amino acids 30 and 104 (or 105), where the introns are located. No other β-globin RNA sequences are known at present; however, inspection of amino acid sequences shows that these are also most stringently conserved in the same two regions (23 to 38 and 88 to 108) for various species including chicken and frog (*18*). Whether the conservation in these two regions is due to functional requirements at the level of the hemoglobin or whether they reflect requirements of the splicing mechanism remains to be determined.

We have pointed out that sequences of 11 nucleotides, identical except for one site, flank the positions of both the large and the small introns in rabbit and mouse:

```
                 30   31   32
                 arg  leu  leu
Small intron ────────┐
               CAGGCTGCTGG*
               CAGGCTCCTGG*
Large intron ────────┘
                 arg  leu  leu
                 104  105  106
```

*(human, mouse, rabbit)

This sequence occurs also in the human β-globin gene, and the corresponding amino acid sequence (Arg or Lys)-Leu-Leu (Arg, arginine; Lys, lysine; Leu, leucine) is common to all known β-globin sequences at positions 30 to 32 and 104 to 106 (*18*).

The strong similarities of β-globin mRNA 5′ noncoding sequences from human, rabbit, and mouse have already been discussed (*19*). With respect to the 3′ noncoding segment of the β-globin mRNA, Proudfoot (*20*) has pointed out that the rabbit and the human sequences are extensively homologous, except that the human sequence has a stretch of 39 additional nucleotides. Proudfoot suggests that part of this DNA segment arose by a duplication of a segment of 31 nucleotides following the termination codon. We note that the mouse 3′ noncoding sequence also possesses the "additional" sequence, which shows some homology to the corresponding human sequence. If the "additional" sequence indeed arose by reduplication, then we must conclude that, in the course of evolution, the rabbit line diverged before a common ancestor of man and mouse developed the reduplication. This is in contrast to the conclusion reached by comparing amino acid (*18*) or nucleotide sequences (Fig. 4), where rabbit and human are more closely related in regard to the β-globin gene. It thus seems more likely that the length differences in the 3′ noncoding sequences are due to a deletion in the rabbit sequence; a less likely alternative would be independent reduplication or insertion at the same positions in mouse and human. If the deletion (or insertion) is disregarded, there is again

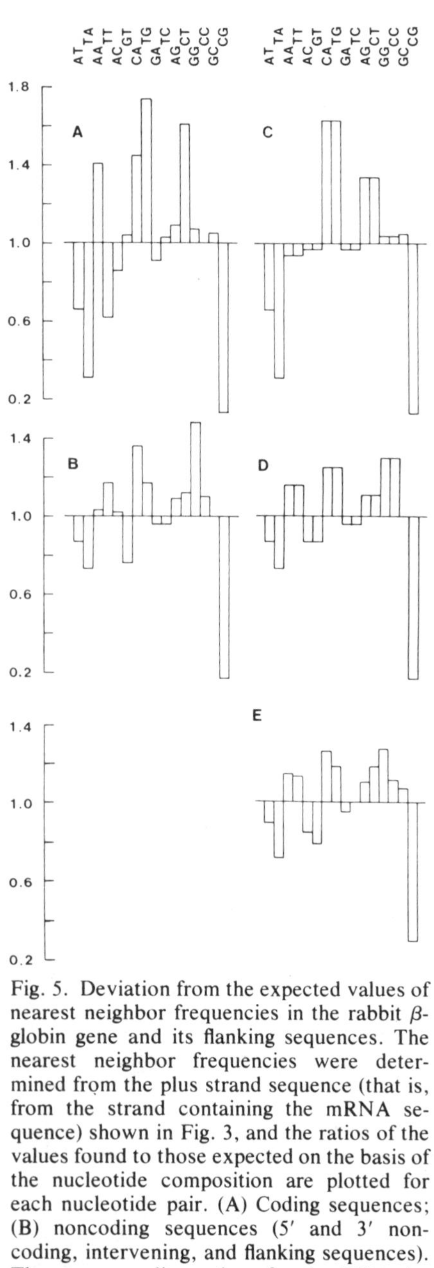

Fig. 5. Deviation from the expected values of nearest neighbor frequencies in the rabbit β-globin gene and its flanking sequences. The nearest neighbor frequencies were determined from the plus strand sequence (that is, from the strand containing the mRNA sequence) shown in Fig. 3, and the ratios of the values found to those expected on the basis of the nucleotide composition are plotted for each nucleotide pair. (A) Coding sequences; (B) noncoding sequences (5′ and 3′ noncoding, intervening, and flanking sequences). The corresponding values for the DNA duplex are given in (C) and (D), and are compared with those calculated for total rabbit DNA (E) (*11*).

more similarity between human and rabbit β-globin than between any other pair of 3′ noncoding sequences.

What constraint is responsible for the conservation of the last 60 nucleotides of the 3′ noncoding sequences? Experiments by Kronenberg et al. (*21*) have shown that the 3′ terminal region of the rabbit β-globin mRNA is not required for translation in a wheat germ system. Inasmuch as these results reflect the situation in vivo, this region would have a different role, perhaps in RNA processing, interactions with proteins (formation of ribonucleoproteins), termination of transcription, or polyadenylation.

The extensive homology between rabbit and mouse DNA in the region preceding the beginning of the mRNA may be related to the initiation of transcription and to its regulation. We have found that the 15*S* β-globin precursor and the mature β-globin mRNA of the mouse have the same 5′ terminal sequence (*22*) and the same cap structure (*23*). Ziff and Evans (*24*) have shown that the adenovirus major mRNA is initiated with the nucleotide which is subsequently capped; since we have no evidence to the contrary, we tentatively assume that the situation is similar in the case of the β-globin mRNA of the rabbit (*7*). If longer precursors than the 15*S* RNA exist, as proposed for mouse β-globin (*25*), they may extend beyond the 3′ terminal region of the mature mRNA. Hogness (*26*) has noted that in a number of cases a sequence of eight nucleotides or a variant thereof precedes the postulated transcription initiation site by 23 ± 1 positions (counted from the first nucleotide following the "box," and including the first nucleotide of the mRNA). The canonical structure is TATAAATA; however, the last two nucleotides show less constancy than the others. In the case of the β-globin genes of rabbit and mouse the following sequences, compatible with Hogness' observation, were found:

TTGGGCATAAAAGGCA ....20.... ACA
rabbit β-globin

CAGAGCATATAAGGTG ....21.... ACA
mouse β-globin

At least one sequence of the type described by Hogness (CTGCA*TATAAAT*-*T*CTGG) occurs in the large intron (between positions 880 and 900, as in Fig. 3); it is not known whether any initiation occurs in that region. In mouse and rabbit, several identical regions, of 9 to 16 positions, precede the Hogness sequence; conceivably, such regions may contribute to a recognition sequence involved in control or initiation (or both) of RNA synthesis, which is perhaps specific for globin genes.

We have argued (*1*) that the introns in mouse and rabbit were homologous because they occurred in the same positions relative to the coding sequence, because they had similar lengths, and because the similarity in sequence (at least at the edges) exceeded that expected statistically. We concluded that corresponding introns were derived from a common ancestral sequence, becoming separated when the evolutionary lines leading to mouse and rabbit diverged about 70 million years ago (*18*). Recent data on the structure of the human β-globin gene show that the position of the introns is the same as in rabbit and mouse (*27, 28*) except that the large intron is almost 900 nucleotides in length (*28*), that is, about 50 percent longer. That there is a strong conservation of the amino acid sequence in β-globins around the positions corresponding to the intron locations in rabbit, mouse, and human, suggests that β-globins of all higher organisms will prove to contain introns at similar positions. Moreover, Leder and his colleagues (*29*) have found that the mouse α-globin gene also contains two introns, located at the corresponding positions as in the β-globin major [and β-globin minor, (*30*)] gene. The common ancestral intron sequence must therefore be older than about 500 million years, which is when α- and β-globins are thought to have arisen from a common globin ancestor (*29*). It will be of great interest to examine the myoglobin gene in regard to possible introns, since the common ancestor of myoglobin and the hemoglobins is more than 10^9 years old. If the myoglobin gene lacked one or both of the introns, this would suggest that introns were introduced into uninterrupted genes in the course of evolution, rather than being present in the DNA segment from the onset of its expression.

Breathnach *et al.* (*31*) have compared the flanking regions of the seven ovalbumin introns, as well as of some other introns. The prototype sequences deduced by them for the 5′ (TCAG↓GTA) and 3′ (TXCAG↓G) junctions of the introns agree moderately well with those of the

Table 2. Similarity between the various parts of mouse and rabbit chromosomal β-globin genes. The data are from Fig. 4; M, mouse, R, rabbit.

	1 Nucleotides (No.) compared (M/R)*	2 Gaps* No.	2 Gaps* Total length	3 Matching nucleo tides†	4 Transi-tions†	5 Trans-versions†	6 Transitions / Transversions	7 Adjusted total length*	8 Similar-ity‡ (percent)
5′ Flanking sequence	128/125	4	5	88	25	11	2.3	129	68
mRNA 5′ noncoding sequence	52/53	1	1	40	8	4	2	53	75
Coding sequence	444/444	0	0	358	46	40	1.2	444	81
Silent	76 .			53	13	10	1.3		70
Mixed	90			60	19	11	1.7		67
Replacement	278			245	14	19	0.7		88
Small intron	116/126	5	6	68	27	18	1.5	129	53
Large intron	650/573	14	109	265	113	179	0.63	666	40
mRNA 3′ noncoding sequence	92/130	2	38	55	17	20	0.85	130	42
3′ Flanking sequence	101/107	3	6	56	16	29	0.55	107	52
Random sequences§	800	7	14	247	183	363	0.50	807	31

*When two sequences of different length were compared, gaps were introduced so as to render both sequences of equal length "adjusted total length" and to optimize the matching of the two sequences, following the rules explained in the text. Total gap length is expressed as the number of nucleotides spanned by the gaps. †Number of matching nucleotides after optimizing alignment of the sequences. Nonmatching pairs of nucleotides are classified as transitions or transversions, the underlying assumption being that the sequences are related. ‡(Number of matching nucleotides/total length) × 100. §Eight pairs of random sequences of equimolar base composition and 100 nucleotides length were aligned and compared, with the same rules applied to the globin sequences. The percent of matching nucleotides ranged from 20 to 33 prior to, and from 24 to 37 following, alignment.

rabbit β-globin and the mouse β-globin major genes. Interestingly, the sequence occurring at the 5′ terminal junction of large intron (TTCAGG↓GTG: the arrow indicates the presumed splice site) is repeated within the large intron (position 570 to 578, Fig. 3), and a sequence from the 3′ terminal junction of the small intron, CAG↓GCTGC, is found in the third coding segment, from position 1134 to 1141. If a six- to eight-nucleotide sequence sufficed to induce RNA cleavage or splicing, we might expect to find aberrant β-globin-specific RNA sequences as side products of splicing. No such molecules have been identified so far; however, they may be generated only at low levels or may have a short half-life, thereby escaping detection. Alternatively, a more complex signal (including, for example, specific secondary and tertiary structures) may be required to induce splicing.

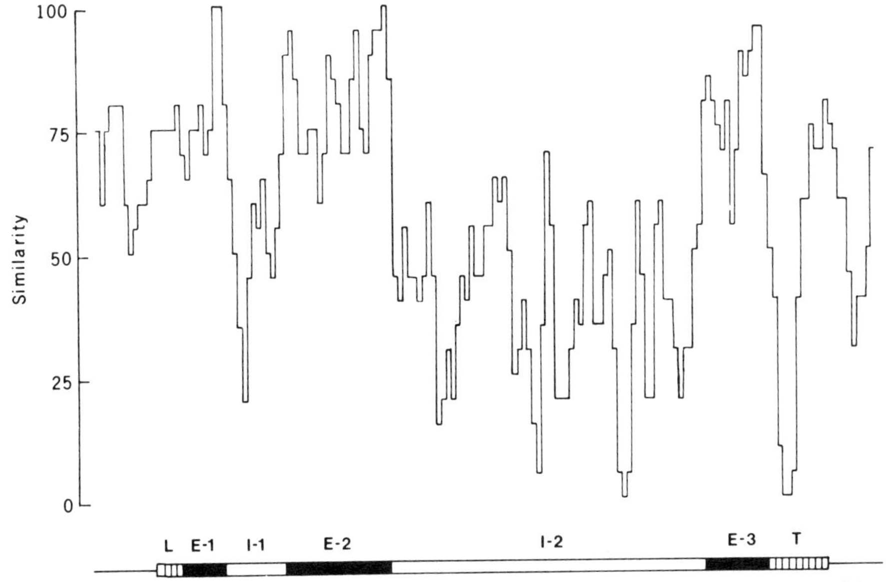

Fig. 6. Similarity of sequences along the rabbit and mouse β-globin genes and their flanking regions. The rabbit and mouse sequences were aligned as shown in Fig. 4. The matching nucleotides were scored within blocks of 20 positions (including nucleotides and gaps) and expressed as percentages. Each block overlaps the neighboring ones by ten nucleotides. The ordinate represents a hypothetical ancestral sequence (comprising the gaps introduced into both mouse and rabbit sequences). Since the gaps are included, the map is distorted, especially in the region of the intervening sequences. *L* and *T*, 5′ and 3′ noncoding sequence of the mRNA. *E-1*, *E-2*, and *E-3*; First, second, and third coding segment; *I-1* and *I-2*, small and large intervening sequences, respectively.

The large introns of rabbit and mouse are almost as different as two random sequences. If the divergence were due to point mutations, the mutation rate within introns (0.4 to 1.5 × 10^{-8}) (*32*) would be at least 2 to 6 times higher than that of the β-globin silent sites (2.7 × 10^{-9}) or the variable regions of fibrinopeptides (2 to 4 × 10^{-9}) (*33*). An alternative explanation is that the internal part of the introns, whatever the genesis of introns may be, are subject to frequent or massive insertions and deletions; this would account not only for the unexpectedly strong sequence divergence of homologous introns, but also for the striking differences in their size. In the case of clearly related sequences (the 5′ flanking, 5′ noncoding, and coding—with the exception of replacement sites—mRNA sequence) the ratio of transitions to transversions is between 1.2 and 2.3. Comparison of two random sequences gives a value of 0.5, as would be expected statistically. We propose that, in eukaryotic DNA, as in the case of Qβ RNA (*34*), transitions are more frequent than transversions, but that selection at the protein (or RNA) level may lead to a modification of the ratio of transition to transversion. The finding that this ratio for the large introns is 0.63 could mean that the difference in sequence arises as a consequence of large insertions and deletions rather than multiple point mutations.

A. van Ooyen, J. van den Berg*
N. Mantei, C. Weissmann
Institut für Molekularbiologie I, Universität Zürich, 8093 Zurich, Switzerland

References and Notes

1. J. van den Berg, A. van Ooyen, N. Mantei, A. Schamböck, G. Grosveld, R. A. Flavell, C. Weissmann, *Nature (London)* **275**, 37 (1978).
2. S. M. Tilghman, D. C. Tiemeier, F. Polsky, M. H. Edgell, J. G. Seidman, A. Leder, L. W. Enquist, B. Norman, P. Leder, *Proc. Natl. Acad. Sci. U.S.A.* **74**, 4406 (1977).
3. D. A. Konkel, S. M. Tilghman, P. Leder, *Cell* **15**, 1125 (1978).
4. H. O. Smith and M. L. Birnstiel, *Nucl. Acids Res.* **3**, 2387 (1976).
5. A. M. Maxam and W. Gilbert, *Proc. Natl. Acad. Sci. U.S.A.* **74**, 560 (1977); H. Ohmori, J. Tomizawa, A. M. Maxam, *Nucl. Acids Res.* **5**, 1479 (1978).
6. A. Efstratiadis, F. C. Kafatos, T. Maniatis, *Cell* **10**, 571 (1977).
7. R. A. Flavell, G. C. Grosveld, F. G. Grosveld, E. De Boer, J. M. Kooter, 11th Miami Winter Symp. (1979) in press.
8. G. J. Russell, P. M. B Walker, R. A. Elton, J. H. Subak-Sharpe, *J. Mol. Biol.* **108**, 1 (1976).
9. M. N. Swartz, T. A. Trautner, A. Kornberg, *J. Biol. Chem.* **237**, 1961 (1962).
10. O. Bernard, N. Hozumi, S. Tonegawa, *Cell* **15**, 1133 (1978).
11. S. Nakanishi, A. Inoue, T. Kita, M. Nakamura, A. C. Y. Chang, S. N. Cohen, S. Numa, *Nature (London)* **278**, 423 (1979).
12. W. Salser, *Cold Spring Harbor Symp.* **42** (2), 985 (1977).
13. H. C. Heindell, A. Liu, G. V. Paddock, G. M. Studnicka, W. A. Salser, *Cell* **15**,43 (1978).
14. T. Lindahl, *Proc. Natl. Acad. Sci. U.S.A.* **71**, 3649 (1974).
15. T. H. Jukes and C. R. Cantor, In *Mammalian Protein Metabolism*, H. N. Munro, Ed. (Academic Press New York, 1969), p. 21.
16. F. E. Baralle, *Cell* **12**, 1085 (1977).
17. F. C. Kafatos *et al.*, *Proc. Natl. Acad. Sci. U.S.A.* **74**, 5618 (1977).
18. M. O. Dayhoff, *Atlas of Protein Sequence and Structure*, (National Biomedical Research Foundation, Washington, D.C., 1972), vol 5, p. D371.
19. F. E. Baralle and G. G. Brownlee, *Nature (London)* **274**, 84 (1978).
20. N. J. Proudfoot, *Cell* **10**, 559 (1977).
21. H. M. Kronenberg, B. E. Roberts, A. Efstratiadis, *Nucl. Acids Res.* **6**, 153 (1979).
22. R. Weaver, W. Boll, C. Weissmann, *Experientia*, **35**, 983 (1979).
23. P. J. Curtis, N. Mantei, C. Weissmann, *Cold Spring Harbor Symp. Quant. Biol.* **42**, 971 (1977).
24. E. B. Ziff and R. M. Evans, *Cell* **15**, 1463 (1978).
25. R. N. Bastos and H. Aviv, *ibid.* **11**, 641 (1977).
26. D. Hogness, personal communication.
27. O. Smithies, A. E. Blechl, K. Denniston-Thompson, N. Newell, J. E. Richards, J. L. Slightom, P. W. Tucker, F. R. Blattner, *Science* **202**, 1284 (1978).
28. R. M. Lawn, E. F. Fritsch, R. C. Parker, G. Blake, T. Maniatis, *Cell* **15**, 1157 (1978).
29. A. Leder, H. I. Miller, D. H. Hamer, J. G. Seidman, B. Norman, M. Sullivan, P. Leder, *Proc. Natl. Acad. Sci. U.S.A.* **75**, 6187 (1978).
30. D. C. Tiemeier, S. M. Tilghman, F. I. Polsky, J. G. Seidman, A. Leder, M. M. Edgell, P. Leder, *Cell* **14**, 237 (1978).
31. R. Breathnach, C. Benoist, K. O'Hare, F. Gannon, P. Chambon, *Proc. Natl. Acad. Sci. U.S.A.* **75**, 4853 (1978).
32. The calculation was carried out by the formula of Kimura (*33*), $k_{nuc} = -3/4 \ln(1 - 4/3\lambda)/2T$, where λ is the fraction of sites by which two homologous sequences differ from each other, T is the time in years since the divergence of the two lineages (70 × 10^6 years) (*18*) and k_{nuc} is the rate of nucleotide substitution per site per year. We have carried out the calculations for the large introns of rabbit and mouse in different ways, either counting the gaps or not in computing the number of nucleotides compared or using values of 2/3 and 3/2, respectively, in the formula given above, to account for the fact that random sequences, after introduction of gaps, differ in about 2/3, rather than 3/4 of their nucleotides. In the text we indicate the extreme values, which differ by a factor of four. The values of λ used in comparing mouse and rabbit sequences were 0.3 for silent sites (Table 2) and 0.42 or 0.5 for the two introns, depending on whether the gaps are counted or not.
33. M. Kimura, *Nature (London)* **267**, 275 (1977).
34. E. Domingo, D. Sabo, T. Taniguchi, C. Weissmann, *Cell* **13**, 735 (1978).
35. K. A. Armstrong, V. Hershfield, D. R. Helinski, *Science* **196**, 172 (1977).
36. R. E. Lockard and U. L. RajBhandary, *Cell* **9**, 747 (1976).
37. C. A. Marotta, J. T. Wilson, B. G. Forget, S. M. Weissman, *J. Biol. Chem.* **252**, 5040 (1977).
38. Supported by the Schweizerische Nationalfonds (No. 3.114.77) and the Kanton of Zürich. Supported by grants (to A.v.O.) from EMBO and the Netherlands Organization for the Advancement of Pure Research (ZWO), and grants (to J.v.d.B.) from EMBO and Koningin Wilhelmina Fonds.
* Present address: Gist Brocades, Postbus 1, Delft, Netherlands.

14 May 1979; revised 11 July 1979

Goose lysozyme structure: an evolutionary link between hen and bacteriophage lysozymes?

M. G. Grütter*, L. H. Weaver & B. W. Matthews

Institute of Molecular Biology and Department of Physics, University of Oregon, Eugene, Oregon 97403, USA

During evolution, the amino acid sequence of a protein is much more variable and changes more rapidly than its tertiary structure. Given sufficient time, the amino acid sequences of proteins derived from a common precursor may alter to the point that they are no longer demonstrably homologous. The ability to make meaningful comparisons between such distantly related proteins must therefore come primarily from structural homology, and only secondarily (if at all) from sequence homology[1]. On the other hand, structural homology in the absence of sequence homology might be attributed to convergent rather than divergent evolution. (A common fold might be dictated by functional or folding requirements.) We have previously argued, on the basis of structural and functional similarities, that the lysozymes of hen egg-white and bacteriophage T4 have a common evolutionary precursor, even though their amino acid sequences have no detectable similarity[2,3]. Here we report the structure of the lysozyme from Embden goose, a representative of a third class of lysozymes[4–10] that has no sequence homology[11,12] (or perhaps very weak homology[13]) with either the hen egg-white or the phage enzyme. The structure of goose egg-white lysozyme has striking similarities to the lysozymes from hen egg-white and bacteriophage T4. However, some parts of goose lysozyme resemble hen lysozyme while other parts correspond only to the phage enzyme. The nature of the structural correspondence strongly suggests that all three lysozymes evolved from a common precursor.

The structure determination of goose lysozyme[14] by isomorphous replacement[1,15] was based on four heavy-atom replacements supplemented by an additional data set collected[16,17] for one of the derivatives at lower substitution. Electron density maps calculated at 2.8 Å and 3.2 Å resolution showed several helices and allowed extended segments of the polypeptide backbone to be followed, but also contained many ambiguous regions. In order to trace the complete backbone of the molecule we alternated model building[18,19] with cycles of crystallographic refinement. Details will be given elsewhere. The model of the structure used for the structure comparisons described here was refined to a crystallographic residual of 0.28 at 2.8 Å resolution. Subsequent refinement has reduced the residual to 0.25 at 2.1 Å resolution.

The polypeptide backbone of goose egg-white lysozyme (GEWL) was compared with those of hen egg-white lysozyme[20,21] (HEWL) and T4 phage lysozyme (ref. 22, and L.H.W., T. Gray & B.W.M., unpublished results) (T4L) by the methods of Rossmann and Argos[23,24] and Remington and Matthews[25,26]. Application of these methods to HEWL and T4L have been described previously[2,3,23,25].

The three-dimensional structure of GEWL has striking similarities to both HEWL and T4L (Table 1, Figs 1, 2). Apart from the amino-terminal region, residues 1–46, essentially every residue of GEWL has a counterpart either in HEWL or T4L or both (Fig. 1). There are 90 spatially equivalent α-carbons in GEWL and HEWL, and 91 in GEWL and T4L. For the residues that structurally correspond, the minimum base change per codon is, in each case, close to the random-sequence value of 1.5, confirming the overall lack of amino acid sequence homology between any pair of the proteins.

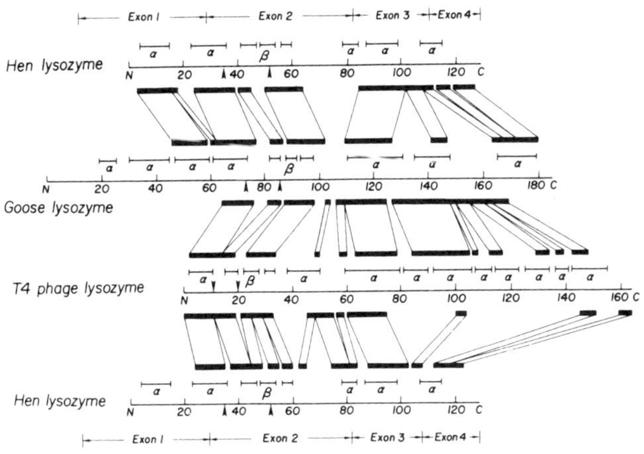

Fig. 1 Structural correspondence between goose, hen and phage lysozymes. The connected solid bars indicate parts of the polypeptide backbones that structurally correspond when the lysozymes are compared in pairs by the method of Rossmann and Argos[23,24]. The locations of the α-helices and β-strands are shown as are the exons of hen egg-white lysozyme. The arrowheads show the locations of the residues that are presumed to be involved in catalysis, *viz* Glu 35 and Asp 52 of HEWL, Glu 11 and Asp 20 of T4L, and Glu 73 and Asp 86 of GEWL.

There are two amino acids in HEWL that are thought to be of prime importance in catalysis, namely Glu 35 and Asp 52[27,28]. The respective counterparts in T4L are Glu 11 and Asp 20[29,30]. In the superposition of HEWL on GEWL (Figs 1, 2*a*) the α-carbon of Glu 35 (HEWL) is 0.8 Å from the α-carbon of Glu 73 (GEWL). Also, for T4L aligned with GEWL (Figs 1, 2*b*) the α-carbons of Glu 11 (T4L) and Glu 73 (GEWL) are 1.7 Å apart. Obviously, Glu 73 of GEWL is a counterpart to Glu 35 in HEWL and Glu 11 in T4L. Similarly, Asp 86 of GEWL corresponds to Asp 52 of HEWL and Asp 20 of T4L, although here the agreement is not so precise (see also ref. 31). According to the Rossmann–Argos procedure, Asp 86 of GEWL is spatially equivalent to Lys 19 rather than Asp 20 of T4L. In the case of HEWL, Asp 86 (GEWL) is topologically equivalent to Asn 44 (Fig. 1) rather than Asp 52, but its α-carbon is only 5.7 Å from that of the latter residue. All three lysozymes have their catalytic aspartates within their common three β-strand unit. For GEWL and T4L, the aspartate is in

Table 1 Lysozyme structural homologies

	GEWL and HEWL	GEWL and T4L	T4L and HEWL
No. of equivalent residues (N_e)	90	91	74
100 N_e/N_1	70%	55%	57%
100 N_e/N_2	49%	49%	45%
R.m.s. discrepancy	3.2 Å	3.2 Å	3.8 Å
Minimum base change per codon for equivalent residues	1.41	1.44	1.46

The table summarizes the comparisons of the three lysozymes by the method of Rossmann and Argos[23,24]. The procedure uses a complicated algorithm that can result in slightly different sets of equivalent residues, depending on the starting parameters. For this reason, little significance should be attached to small differences in the number of equivalent residues. N_1 and N_2 are, respectively, the number of amino acids in the smaller and larger protein being compared. HEWL has 129 residues, T4L 164 residues and GEWL 185 residues.

* Present address: Abt. Biophysikalische Chemie, Biozentrum der Universität Basel, CH-4056 Basel, Switzerland.

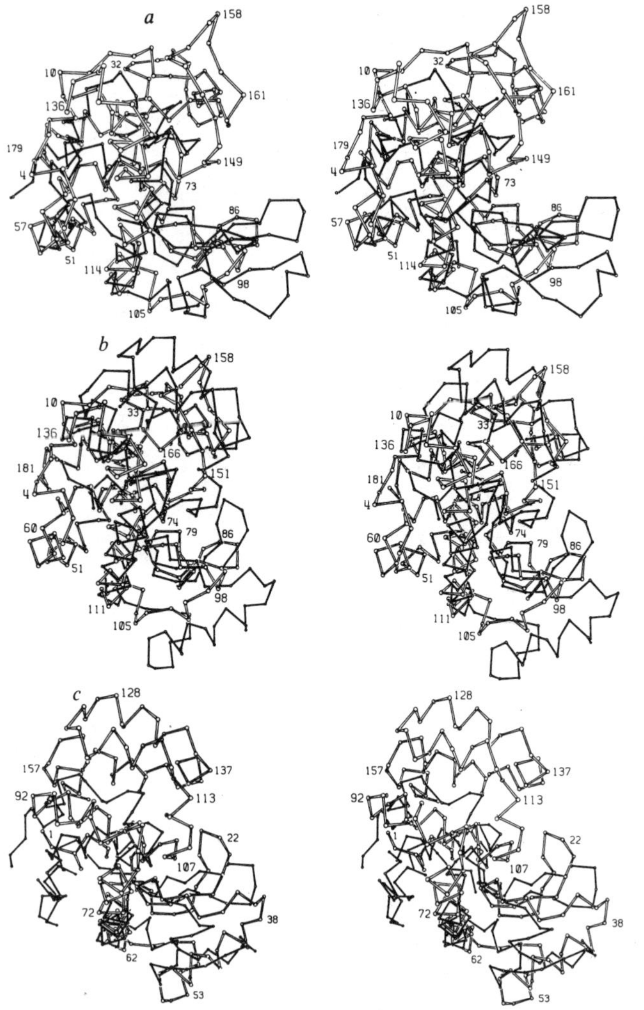

Fig. 2 Stereo drawings showing the superposition of different lysozyme structures. Essentially the same view is used throughout. *a*, GEWL (open connections) and HEWL (solid connections). Numbering for GEWL. *b*, GEWL (open connections) and T4L. Numbering for GEWL. *c*, T4L (open connections) and HEWL (solid connections). Numbering for T4L. (Following Rossmann and Argos[23].)

the first β-strand, whereas Asp 52 of HEWL is in the middle strand. Glu 86 and Asp 73 of GEWL are located on opposite sides of the presumptive active-site cleft. Their carboxyls are about 8 Å apart in the present mode (7–8 Å in HEWL and T4L). Supporting evidence for the location of the active site comes from (*hk*0) and difference Fourier projections (data not shown) indicating that *N*-acetylglucosamine binds to the crystalline enzyme in this region.

When comparing two proteins that have similarities in folding but dissimilar amino acid sequences, there is always a potential ambiguity. The similar conformation might reflect a common evolutionary precursor or, on the other hand, might be a conformation dictated by some functional or folding requirement. This ambiguity occurs quite often (ref. 2 cites a number of instances), and has led to a great deal of controversy concerning divergent versus convergent evolution. Divergent evolution is typified by the well-known families of proteins such as the globins[32], the serine proteases[33] and the cytochromes *c* (ref. 34). At the other extreme, subtilisin and α-chymotrypsin[35], and thermolysin and carboxypeptidase A[36] are normally accepted as examples of convergent evolution. However, in cases such as HEWL and T4L, where the amino acid sequences are not homologous, it is difficult, if not impossible, to choose between divergence and convergence. Now, with three classes of lysozyme structure available, we are in a stronger position. There are parts of four α-helices, together with the three-strand β-sheet, that occur in all three lysozymes, and presumably include the catalytically essential elements of the respective structures. These common residues constitute the core of the molecule and form both sides of the active-site cleft (Fig. 3). All other residues can be regarded as catalytically 'non-essential'. Included among such 'non-essential' residues are 128–141 of GEWL, residues that correspond closely to 85–98 of T4L (average discrepancy 3.0 Å) yet have no structural counterpart in HEWL. Conversely, and perhaps more striking, residues 48–60 of GEWL coincide with the amino-terminal helix, residues 4–16, of HEWL (average discrepancy 2.8 Å), whereas T4L lacks this helix entirely (Figs 1 and 2). It seems unlikely that each of the similarities between the catalytically 'non-essential' parts of the respective lysozymes occurred independently. Rather, these similarities suggest that all three lysozymes diverged from a common precursor, consistent with the accepted principle that secondary structure changes more slowly during evolution than does amino acid sequence[1,32–34].

The essence of the structural relation between the three lysozymes is shown in Fig. 4. The triangles, circles and squares are not intended to represent distinct structural domains. Rather, the triangle, for example, illustrates the parts of lysozyme structure that correspond in HEWL and GEWL, but are absent from T4L. This is predominantly the α-helix at the amino-terminus of HEWL, but also includes a segment at the carboxy-terminus (Fig. 1). The three lysozyme types might have arisen by a number of possible evolutionary pathways. One could imagine that a hen-type lysozyme might evolve to (or from) a goose-type lysozyme by appropriate insertions and deletions. The same is true for the phage-type and goose-type enzymes. However, direct evolution from a phage-type to a hen-type lysozyme (or vice versa) seems very unlikely, since it would be necessary to delete some goose-like features from

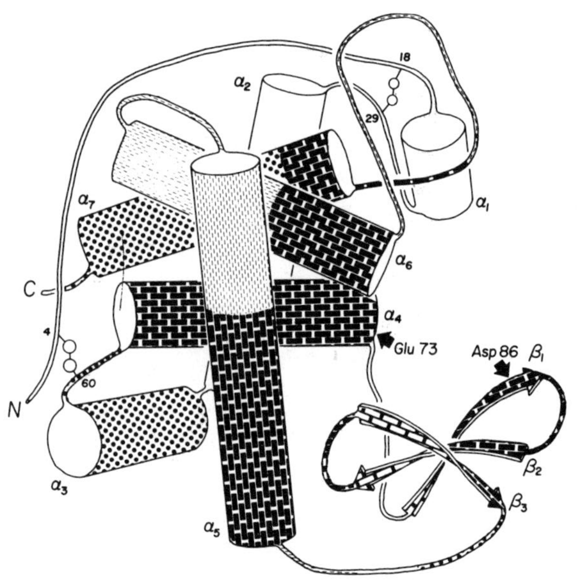

Fig. 3 Simplified drawing of goose egg-white lysozyme showing those parts of the structure that are common to hen and to phage lysozymes. Parts that are common to all three lysozymes are shown as bricks, parts common to GEWL and HEWL are shown dotted, parts common to GEWL and T4L are shown dashed, and parts that occur only in GEWL are shown as open.

Fig. 4 Formalized representation of the relation between the structures of hen-type, goose-type and phage-type lysozymes together with possible transitions that could have occurred between precursors of these lysozymes during the course of evolution. The symbols are not meant to represent distinct structural domains. Rather, the rectangles represent those structural elements that are common to all three lysozymes, the triangles represent features unique to GEWL and HEWL, the circles represent parts of the lysozyme structures that occur only in GEWL and T4L and the star represents the polypeptide segment that occurs only in GEWL (see Fig. 3).

the starting protein and introduce other goose-like features into the product. One possibility is that the three lysozymes diverged from a common precursor that was most akin to a goose-like molecule. It is also possible that the common precursor was essentially hen-like, and gave rise to a goose-type descendent that in turn led to the present phage enzyme. The lineage phage-type to goose-type is also possible, but pathways such as phage-type to hen-type to goose-type lysozyme seem very unlikely.

The previous structural comparison of HEWL and T4L[2] was taken as evidence in support of the view that the four exons of the gene of hen lysozyme[37] (Fig. 1) correspond to functional[37], and, to some extent, structural units of HEWL[38–40]. As will be discussed in more detail elsewhere, the present comparisons (Fig. 1) do not strongly suggest that exon units are conserved in the three structures.

We thank N. Isaacs for suggesting the use of a neutral platinum heavy-atom derivative, based on its application to crystals of black swan lysozyme[41], and, together with R. J. Simpson and F. J. Morgan, for communicating results on the sequences of black swan lysozyme and goose lysozyme in advance of publication; and S. J. Remington, T. Gray, G. H. Arscott and A. Hollister for their help. This work was supported in part by grants from the NIH (GM20066, GM21967), the NSF (PCM-8014311) and from the M. J. Murdock Charitable Trust.

Received 13 December 1982; accepted 15 April 1983.

1. Matthews, B. W. in *The Proteins* Vol. 3, 3rd edn (eds Neurath, H. & Hill, R. L.) 403–590 (Academic, New York, 1977).
2. Matthews, B. W., Grütter, M. G., Anderson, W. F. & Remington, S. J. *Nature* **290,** 334–335 (1981).
3. Matthews, B. W., Remington, S. J., Grütter, M. G. & Anderson, W. F. *J. molec. Biol.* **147,** 545–588 (1981).
4. Canfield, R. E. & McMurry, S. *Biochem. biophys. Res. Commun.* **26,** 38–42 (1967).
5. Canfield, R. E., Kammesman, S., Sobel, J. H. & Morgan, F. J. *Nature* **232,** 16–17 (1971).
6. Dianoux, A-C. & Jollès, P. *Biochim. biophys. Acta* **144,** 472–479 (1967).
7. Prager, E. M., Wilson, A. C. & Arnheim, N. *J. biol. Chem.* **249,** 7295–7297 (1974).
8. Arnheim, N., Inouye, M., Law, L. & Laudin, A. *J. biol. Chem.* **248,** 233–236 (1973).
9. Hindenburg, A., Spitznagel, J. & Arnheim, N. *Proc. natn. Acad. Sci. U.S.A.* **71,** 1653–1657 (1974).
10. Hindenburg, A., Spitznagel, J. & Arnheim, N. *Proc. natn. Acad. Sci. U.S.A.* **71,** 1653–1657 (1974).
11. Simpson, R. J., Begg, G. S., Dorow, D. S. & Morgan, F. J. *Biochemistry* **19,** 1814–1819 (1980).
12. Simpson, R. J. & Morgan, F. J. *FEBS Lett.* (submitted).
13. Schoentgen, F., Jollès, J. & Jollès, P. *Eur. J. Biochem.* **123,** 489–497 (1982).
14. Grütter, M. G., Rine, K. L. & Matthews, B. W. *J. molec. Biol.* **135,** 1029–1032 (1979).
15. Green, D. W., Ingram, V. M. & Perutz, M. F. *Proc. R. Soc.* **A225,** 287–307 (1954).
16. Rossmann, M. G. *J. appl. Crystallogr.* **12,** 225–238 (1979).
17. Schmid, M. F. *et al. Acta crystallogr.* **A37,** 701–710 (1981).
18. Molnar, C. E., Barry, C. D. & Rosenberger, F. U. *Tech. Memo.* No. 229 (Computer Systems Laboratory, Washington University, St Louis, 1976).
19. Jones, T. A. in *Crystallographic Computing* (ed. Sayre, D.) 303–317 (Oxford University Press, 1982).
20. Diamond, R. *J. molec. Biol.* **82,** 371–391 (1974).
21. Imoto, I., Johnson, L. N., North, A. C. T., Phillips, D. C. & Rupley, J. in *The Enzymes* Vol. 7, 3rd edn (ed. Boyer, P.) 665–868 (Academic, New York, 1972).
22. Remington, S. J., Ten Eyck, L. F. & Matthews, B. W. *Biochem. biophys. Res. Commun.* **75,** 265–269 (1977).
23. Rossmann, M. G. & Argos, P. *J. molec. Biol.* **105,** 75–96 (1976).
24. Rossmann, M. G. & Argos, P. *J. molec. Biol.* **109,** 99–129 (1977).
25. Remington, S. J. & Matthews, B. W. *Proc. natn. Acad. Sci. U.S.A.* **75,** 2180–2184 (1978).
26. Remington, S. J. & Matthews, B. W. *J. molec. Biol.* **140,** 77–99 (1980).
27. Blake, C. C. F. *et al. Nature* **206,** 757–761 (1965).
28. Ford, L. O., Johnson, L. N., Machin, P. A., Phillips, D. C. & Tjian, R. *J. molec. Biol.* **88,** 349–371 (1974).
29. Matthews, B. W. & Remington, S. J. *Proc. natn. Acad. Sci. U.S.A.* **71,** 4178–4182 (1974).
30. Anderson, W. F., Grütter, M. G., Remington, S. J. & Matthews, B. W. *J. molec. Biol.* **147,** 523–543 (1981).
31. Jollès, J., Schoentgen, F. & Jollès, P. *C.r. hebd. Séanc. Acad. Sci., Paris* **292,** 891–892 (1982).
32. Hendrickson, W. A. & Love, W. E. *Nature* **232,** 197–203 (1971).
33. James, M. N. G., Delbaere, L. T. J. & Brayer, G. D. *Can. J. Biochem.* **56,** 396–402 (1978).
34. Almassy, R. J. & Dickerson, R. E. *Proc. natn. Acad. Sci. U.S.A.* **75,** 2674–2678 (1978).
35. Robertus, J. D. *et al. Biochemistry* **11,** 2439–2449 (1972).
36. Kester, W. R. & Matthews, B. W. *J. biol. Chem.* **252,** 7704–7710 (1977).
37. Jung, A., Sippel, A. E., Grez, M. & Schultz, G. *Proc. natn. Acad. Sci. U.S.A.* **77,** 5759–5763 (1980).
38. Gilbert, W. *Nature* **271,** 501 (1978).
39. Blake, C. C. F. *Nature* **273,** 267 (1978).
40. Artymiuk, P. J., Blake, C. C. F. & Sippel, A. E. *Nature* **290,** 287–288 (1981).
41. Masakuni, M., Simpson, R. J. & Isaacs, N. W. *J. molec. Biol.* **135,** 313–314 (1979).

PROBLEMS

1. Define the following: orthologous and paralogous sequences; alpha-, beta-, gamma- and delta-globin chains; sigmoid oxygenation curve of hemoglobin; electrophoretic variants; polymorphism; synonymous, conservative and disruptive substitutions; covarion; molecular clock; neutralism and panselectionism; stochastic process.

2. A neutrality standard is the expected rate of amino acid substitution in proteins, or base-pair substitution in nucleic acids when there are no constraints on substitutions; it can be estimated by examining the rate of amino-acid change in fibrinopeptides or selected third-codon positions. Innumerable examples of conformation to or departure from this standard have been reported in the literature. Provide justifications for each of the following observations.
(a) An amino acid in the active centre of an enzyme is highly conserved.
(b) An amino acid in an interdomain loop is highly conserved.
(c) Portions of mammalian mitochondrial cytochrome b are highly variable.
(d) A portion of the nucleotide sequence of a homologous series of introns is highly variable.
(e) A 5' upstream site is highly conserved.
(f) The overall rate of change of mitochondrial DNA is greater than that of nuclear DNA.
You might also like to consider how you could test your explanations.

3. Lesk and Chothia (1980) observed that over the entire range of molecules that are identifiable as homologous heme-containing globins, only 4 percent of the amino acid positions are invariant and, between the most distantly related representatives of this group, only 16 percent of the two sequences are identical. Does this conclusion disagree in any fundamental way with the conclusion of Fitch and Markowitz (1970, this chapter) that only 10 percent of the cytochrome c molecule can accept mutations at any given time? How does the observation that a number of parallel substitutions are present in different lines (of both cytochrome c and globin) bear on this question? How could you test the covarion hypothesis?

4. With only amino acid sequences available, the correct alignment of distantly related protein molecules was difficult to resolve. However, with increasing ease of determining both tertiary structures (by x-ray crystallography) and gene nucleotide sequences, what additional clues are now available to assist in the alignment of amino acid sequences? Do any of these technical advances contribute to resolving the problem of whether two molecules serving the same function in two very different organisms (lysozyme of T4 and hen egg white, for example) are related by descent or convergence, or do they merely make the situation worse?

5. Kimura (1968, Chapter 1) has presented a formula for the probability of the fixation of neutral mutations (equation 2) and for the probability of an individual being heterozygous at a given locus (He). Consider a population of feral mice. What is the frequency of fixation of mutations in the "coding" portion of a beta-globin pseudogene and what is the expected frequency of heterozygosis, assuming a mutation rate (u) per gene of 10^6 and an effective population size (Ne) of, respectively, 10^2, 10^3, 10^4, and 10^5? While this approach has yielded useful insight into possible mechanisms of genetic change in populations, the biological parameters Ne and s are extremely difficult to measure. Describe very briefly what you perceive to be the difficulties. How might you explain levels of heterozygosity below those you have calculated?

6. Human pathogenic viruses offer a unique opportunity to study evolution in action in that they have quite a simple structure, are subject to constant pressure from the host immune system, are capable of high rates of reproduction, and are well documented (at least in recent years with the advent of a worldwide public health monitoring system). The influenza virus has been marked by periodic worldwide pandemics interspersed with more frequent and local epidemics. The significant alterations to the virulence of this virus are usually associated with changes of the

hemagglutinin (a protein in the viral capsid) and accordingly a lot of attention has been focused on this molecule. Gething *et al.* (1980) have shown that the primary structure of the hemagglutinin of the viral strains responsible for the last two pandemics are no more (nor less) homologous than the hemagglutinin from a related avian virus. However, the viruses responsible for several epidemics since the last pandemic have highly homologous hemagglutinins, with each variant distinguishable by virtue of amino acid replacements in four very narrowly proscribed regions at one end of the tertiary structure (Wiley *et al.*, 1981). On the basis of this sketchy summary, what would you surmise about the molecular basis of antigenic drift (responsible for epidemics) and antigenic shift (responsible for pandemics) in relation to the functional anatomy of the haemagglutinin molecule.

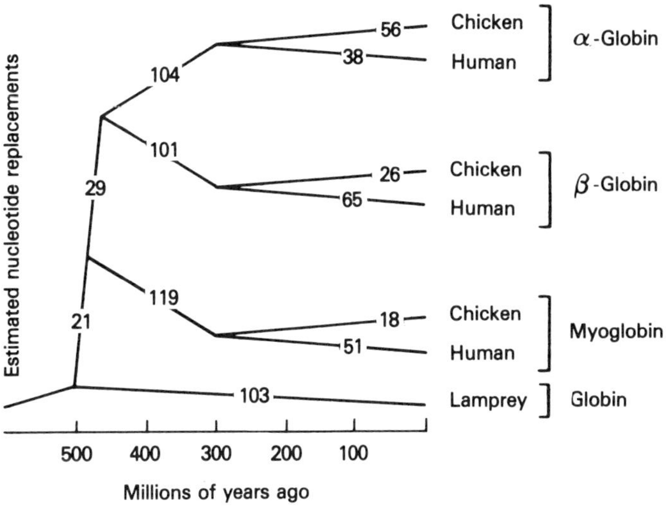

Figure 4-7

7. Goodman (1981) has proposed the following reconstruction for the descent of the globin family of molecules (see Figure 4-7). He came to two related conclusions: (1) the rate of globin evolution was significantly faster during the period of gene duplication and functional modification; and (2) those sites that seem to be involved in the interactions responsible for tetramer formation and the cooperative oxygen binding changed faster than the average rate for the whole molecule and much faster than for residues engaged in the binding of the heme group. Accepting these observations (that is, tree topology and allocation of changes), construct both a selectionist and a neutralist explanation that would account for these observations. (You may wish to read Dickerson, 1971.) Remembering that the fossil record does not show when a gene duplication occurs, suggest possible ways for eliminating the largest rate inhomogeneities and a potential method of testing these suggestions. What would you expect to happen to the rates of nucleotide change if one of the duplicated copies temporarily became a silent pseudogene, as suggested very early by Zuckerkandl and Pauling (1962).

8. Gilbert (1978, Chapter 2) proposed that introns provide opportunities for evolutionary experimentation by "exchanging" portions of proteins to generate new proteins. That introns often correspond to functional domains of proteins makes the proposal even more attractive. What do you make of the observation of Jensen *et al.*, (1981) that leghemoglobin (involved in leguminous nitrogen-fixation) has a third intron that splits the middle heme-binding domain in addition to the two introns common to other globins?

9. An objective criterion in support of the historical fact of evolution is the congruence of evolutionary trees calculated from several different proteins. Would you consider the discovery of a gross nonconcordance to be a falsification of an otherwise coherent and consistent scheme? Suggest explanations for the possible noncongruence of trees.

5 CODING SEQUENCES —ACQUISITION OF NEW FUNCTIONS

...natural selection...works like a tinkerer...who does not know exactly what he is going to produce... who uses everything at his disposal to produce some kind of workable object...what he ultimately produces is generally related to no special project, and it results from a series of contingent events, of all the opportunities he had to enrich his stock with leftovers.
F. Jacob (1977)

The acquisition of new functions, whether it be at the level of proteins, cell types or organs, is certainly one of the central problems to be explained by evolutionary theory. It is an area in which molecular approaches have been particularly helpful, not only in providing information about past events that have resulted in functional alterations to proteins, but also in suggesting mechanisms that lead to the development of new molecular functions.

A priori it can be hypothesized that an increase in biochemical or morphological complexity must be based on an increase in genetic information. Indeed, in 1918 Bridges (1935) pointed out that newly discovered duplications of a locus (for example, the Bar locus in *Drosophila*) could offer an explanation for an increase in the genetic information of an organism. Horowitz (1945, Chapter 1) later extended this idea to biochemical pathways and to the early stages of the origin of life. No direct evidence about coding sequences was available to these early workers, but their hypotheses can be summarized as follows: (1) Genes can be duplicated, giving at least two copies of the original. (2) One copy can diverge from its original function and may give rise to a new function.

In modern evolutionary studies, two approaches have been taken in testing these hypotheses: (1) the analysis of primary and tertiary structure of proteins provides a basis for the reconstruction of those past events that have led to functional divergence; (2) laboratory microevolutionary studies, particularly with bacteria, have provided insights into the kinds of mutations selected by novel nutritional opportunities.

The first method makes inferences about events in the evolution of genes by comparing either primary sequences or tertiary structures. Direct evidence for the existence of duplications first came from the sequencing of a number of similar proteins. Ingram (1961, Chapter 1) was able to develop an evolutionary tree of the globin family of sequences that included both myoglobin and the alpha-, beta-, gamma- and delta-hemoglobins and this finding provided strong evidence to support the hypothesis of gene duplication and divergence in evolution. The relations have been confirmed with many additional sequences, including cloned DNA sequences for the hemoglobin genes, which also provide information complementary to the protein studies--for instance, Efstratiadis *et al*. (1980) examined the nucleotide sequences of the cluster of human beta-globin genes and showed that a high degree of sequence homology is present and that introns occupy the same positions.

A duplication may lead to either two copies of the protein, or to an elongated polypeptide chain (Doolittle, 1981). Several possible events can follow a duplication and lead to the production of two separate proteins. If the duplication occurs at a locus for which heterozygosity is advantageous, then an appropriate recombinational event could fix the advantageous heterozygosity (Gottlieb, 1974). If there is a selective advantage in having more of a particular protein molecule, then a duplication may become fixed by selection (Hartley, Altosaar, Dothie and Neuberger, 1976, this chapter; Schimke *et al*., 1978). In either case, one or the other copy of the duplicated gene is free to acquire new properties or to lose its original function. In the latter case, the nonexpressed pseudogenes may become established in the population by genetic drift, as have those of the hemoglobins (see Lacy and Maniatis, 1980, Chapter 2). Clearly, if gene duplication is a reasonably frequent occurrence, and the genetic load occasioned by degeneration of one or the other copy is insignificant, then a pool of such "junk" DNA will accumulate at some equilibrium level in the genome.

Many good examples of functionally diverging families of proteins that appear to be descended from a common ancestor are now known. The best understood is still the hemoglobin-myglobin complex (Ingram, 1961, Chapter 1; Goodman *et al*., 1974; Efstratiadis *et al*., 1980). Another early discovery was the trypsin - chymotrypsin - elastase (serine proteases) group of enzymes (Hartley, 1966); more recently, several families of hormones have been recognized, representative members of which are insulin (Bradshaw, 1978), glucagon, somatostatin and the pancreatic polypeptides (Blundell and Humbel, 1980). The phylogenetic relationships of the proteins troponin C, parvalbumin, and light myosin chains, all of which bind calcium ions and appear to have a common origin, have also aroused interest (Barker and Dayhoff, 1979). One of the most complex examples of multiple duplication and specialization is the vertebrate immunoglobulin family, which has three interesting features: (1) "new functions" develop during generation of antibody diversity within an organism (Gough, 1982); (2) the gene organization is clearly related to protein functional units (Sakano *et al*., 1979); (3) sequence comparisons give information about the evolution of this large family (Honjo *et al*., 1981).

A comprehensive listing of proteins believed to have diverged from common ancestors may be found in Doolittle (1981). This last includes cases in which it is not yet clear whether a pair of proteins has diverged from a common ancestor or converged from different ancestors towards a similar function (the problem referred to earlier in Chapter 4 in connection with bacteriophage T4 and eggwhite lysozymes).

The other mode of duplication that leads to structural and functional divergence is the elongation of a protein chain by a partial (internal) duplication. This mechanism was suggested for ferredoxin (Tanaka *et al*., 1964; Eck and Dayhoff, 1966a, Chapter 1) and now all published sequences are routinely tested for evidence of internal duplications (Barker *et al*., 1978). The latter authors have reported evidence for internal duplications in 20 of the 116 superfamilies. The cases were divided into two types: (1) the most common class had two (and occasionally more) fairly large internal regions of homology, as found in immunoglobulins, serum albumins and the previously mentioned ferredoxin (Eck and Dayhoff, 1966a, Chapter 1); (2) in the second class repeated duplications of small segments of DNA had occurred (Ycas, 1972; 1973), giving proteins such as alpha-collagen that have a repeating unit over most of their length (Vogeli *et al.*, 1981).

It is important to remember that "new functions" include new regulatory mechanisms (Baumberg, 1981) as well as new biochemical reactions. In principle, regulation can occur by controlling either protein function or gene expression. Engel (1973, this chapter) identified an internal duplication in glutamate dehydrogenase, an enzyme whose activity is modulated by one of its cofactors. He was one of the first workers to discuss the evolution of regulator sites and suggested that the early enzyme had only catalytic activity and that regulatory control developed subsequently by means of a duplication of the catalytic binding site. This idea remains attractive and the data is compelling, but the regulatory binding site has not yet been identified unambiguously (Rossman *et al.*, 1974).

It had long been assumed that duplications occur at random with respect to a protein sequence. However, Gilbert (1978, Chapter 2; 1979) has pointed out that the existence of introns between coding sequences could mean that many duplications and/or recombinational events would involve the domains or functional subunits of a protein. This idea has been extended by Blake (1979), who suggested that an exon codes for a functional unit or domain of a protein that could be catalytic, structural, or regulatory. The shuffling of exons between genes would increase the chance that any new protein would form a stable three-dimensional structure and that this structure would have some enzymatic or regulatory function.

The idea of accretion of domains has been used to suggest a more detailed history of heme-binding proteins, including globins (myoglobin and hemoglobin) and cytochromes b and c (Argos and Rossman, 1979; 1980). Eaton (1980) correlated several aspects of function and regulation with exonic regions of the genes for alpha and beta chains. Peroxidase, catalase, and other cytochromes may eventually be included in this family of genes. A plant leghemoglobin gene has an additional intron in the middle of the heme-binding domain (Jensen *et al.*, 1981); this fact is particularly interesting because Go (1981) predicted that the heme-binding domain of the hemoglobins was originally two separate regions that had become fused into one exon. This latter interpretation suggests that the leghemoglobin condition is the primitive state and supports the idea that introns may be lost during evolution (Doolittle, 1978). Indeed, in recent studies a class of pseudogenes that have lost the normal introns--possibly by reverse transcription of a processed mRNA back to DNA—has been observed (Wilde *et al.*, 1982).

The analysis of three-dimensional structures of many proteins permits more distantly related proteins (or protein domains) to be recognized, as discussed by Argos and Rossman (1979) and Grutter *et al.* (1983, Chapter 4). From such data Rossman *et al.* (1974) have argued that the mononucleotide binding fold (mnbf) of a number of widely distributed enzymes are evolutionarily related to one another and to the ancestor of bacterial flavodoxin. Levine, Muirhead, Stammers, and Stuart, (1978, this chapter) have studied similarities between pyruvate kinase and triose phosphate isomerase and argue that the similarities are caused by shuffling of domains and not by independent convergence of different proteins to a particularly stable and catalytically active structure (see the introduction to Chapter 4). It is worth repeating at this point the caution offered by Richardson (1977) that the number of stable three-dimensional structures is small and that demonstrated similarities can reflect simply a convergence towards a given structure with stability being one of the primary criteria for selection. Indeed, Tang *et al.* (1978) pointed out that acid proteases are constructed from two conformationally similar halves and thus appear to be the result of gene duplication; however, the serine proteases, which have a similar two-fold symmetry, have opposite peptide polarities in the two halves and hence could not have arisen from a simple duplication.

Perhaps we may eventually see a complete phylogeny of proteins, with the order and timing of appearance of all proteins being known. A start has been made by simply classifying sequences into families and superfamilies (Dayhoff, 1976). Two groups of workers have independently estimated that there may be about 1000 superfamilies of related proteins (Dayhoff *et al.*, 1975; Zuckerkandl, 1975). The existence of these superfamilies has been deduced from sequence similarities, but it should be possible to start clustering super families into higher units with the aid of three-dimensional structures. Doolittle (1981) has suggested that the term protein family be used for all those clusters of sequences with demonstrable similarity and that a superfamily be comprised of all those families that intersect by virtue of at least one identifiable sequence similarity. With a sufficient number of sequences and with guidance from three-dimensional structures, one may expect eventually to discern a relatively small number of primeval polypeptides.

The second approach to exploring the origin of new functions utilized laboratory studies with selection for new functions in cell cultures and microorganisms. Such experiments can directly test the possibility that new functions arise by gene duplication and modification and can suggest other ways that organisms may respond to an altered biochemical environment. This work has been stimulated in part by the discovery of organisms that are resistant to antibiotics or are able to metabolize novel man-made chemicals. Schimke *et al.* (1978) have shown that cultured mouse cells can become resistant to the drug methotrexate by multiplication of a dihydrofolate reductase gene, which results in a higher concentration of the enzyme. The paper of Hartley, Altosaar, Dothie, and Neuberger (1976) reprinted in this chapter is one of a series of papers stemming from the work of Lerner *et al.* (1964) on mutants of the bacterium *Klebsiella* that were selected for the ability to grow on the sugar alcohol, xylitol. In the reprinted paper several ways in which a new activity for a substrate can arise are described. Enzyme specificity is rarely absolute and when an enzyme does have a low activity against a novel substrate, as is the case with ribitol dehydrogenase acting on xylitol, then simply increasing the amount of enzyme may increase substrate utilization sufficiently to allow growth. Hartley *et al.* (1976, this chapter) report that this end can be attained either by increasing the amount of enzyme (by gene duplication, or possibly by affecting a promoter so that transcription rate is increased), or by changes to an amino acid sequence that increase the activity of the enzyme for the new substrate. Wu *et al.* (1968) showed that in addition to changes directly affecting the structural gene of ribitol dehydrogenase, constitutive mutations of this gene and of arabitol permease (which transports xylitol) lead to improved growth on xylitol. These mechanisms by themselves do not strictly create new functions, but they show that changes in the amino acid sequence can lead to an increase in the specific activity against a particular substrate. Additional information on this system will be found in Hartley (1979).

It now appears that it is a simpler matter than previously thought to alter the specificity of an enzyme. Originally, Zuckerkandl and Pauling (1962) and Hartley (1966) suggested that first a duplicated gene would have to become silent (a pseudogene) and then to accumulate several mutations that could become established by genetic drift; then, if the gene became re-expressed, the new protein might possess a useful new function. The experimental observations are that only one or two specific amino acid substitutions are sufficient to change the activity of an enzyme. An example (Ala-196 to Pro-196 in ribitol dehydrogenase) is discussed in Hartley *et al.* (1976, this chapter). Similarly, in serine proteases specificity differences have been shown to be the result of only one or two amino acid differences in the active site (Hartley, 1979). Other cases are cited in Hall and Clarke (1977), and Wills and Jornvall (1979). These observations seem to contradict the conclusion from the previous chapter that a very large proportion of the amino acid sequence can be changed without destroying the function of the protein. However, both statements are true because many different sequences can yield a protein with the same three-dimensional structure, whereas certain substitutions can change enzymatic activity without greatly changing the three-dimensional structure.

The second reprinted paper concerning microevolutionary studies, that of Hall and Zuzel (1980, this chapter), is one of a series of papers on the *ebg* (evolved beta-galactosidase) gene, which has been found in *E. coli* strains carrying a deletion of the *lacZ* (beta-galactosidase) gene. The important features of this paper are that a possible mechanism for the origin of a new function (by recombination between two parental mutant genes) is illustrated and a class of genes of potential evolutionary significance is described. Gilbert (1981) has argued that in eukaryotes the existence of large introns greatly increases the probability of recombination between mutations in different exons, thereby increasing the chance of obtaining altered functions by recombination. A different mechanism for the generation of diversity by recombination, and perhaps for the generation of new features, is gene conversion involving linked, unlinked, or even silent members of a multigene family (Dover, 1982; Gough, 1982).

In conclusion, the papers by Hartley *et al.* (1976, this chapter) and Hall and Zuzel (1980, this chapter) demonstrate that the ability to use a novel substrate can result from gene duplications, changes in gene regulation, changes in enzyme specificity resulting from amino acid substitutions, and from recombination. Additional examples of the mutational derivation of new or modified enzyme activities can be found in Mortlock (1976, 1982), Hartley (1979), Clarke (1974, 1981), Betz *et al.* (1974), and Carlile *et al.* (1981).

Evidently, several mechanisms can lead to the development of new functions. Many further examples of these processes will certainly be found and possibly some new mechanisms will be uncovered, but a foundation has been laid for thinking about the ways biochemical complexity of organisms may increase.

Reprinted from *Nature* **241**: 118–120.

Evolution of Enzyme Regulator Sites: Evidence for Partial Gene Duplication from Amino-acid Sequence of Bovine Glutamate Dehydrogenase

MODULATION of the biological activity of proteins by metabolite levels is a widespread and well-documented phenomenon. As Monod, Wyman and Changeux[1] pointed out, most regulatory protein molecules are composed of several subunits. The regulation of some oligomeric proteins, notably haemoglobin[2], involves interaction of similar sites on separate similar subunits. In other cases such as that of aspartate transcarbamylase[3] there are separate catalytic and regulatory subunits. Some proteins, however, possess within individual subunits both active sites and regulatory sites capable of mediating homotropic and/or heterotropic interaction.

Glutamate dehydrogenase (GDH) from the liver of vertebrates is such a protein. Its catalytic activity is modulated by the coenzymes of the reaction and by such purine nucleotides as ADP and GTP[4–7], and yet the hexameric enzyme molecule contains only one type of polypeptide chain[8]. The precise relation between the purine nucleotide sites and the coenzyme binding sites remains uncertain, but it has been shown clearly that the enzyme can bind more than one molecule of NADH per subunit[4,9–12]. It appears that binding at the second site may account for inhibition by high levels of NADH[13].

If one assumes that in the course of enzyme evolution catalytic activity preceded the fine control imposed by regulatory sites, the GDH found in vertebrate liver must be descended from an ancestral monomeric dehydrogenase possessing the active but not the regulatory site. It is interesting, therefore, to consider the possible evolutionary origin of the regulatory site.

There are two basic possibilities:

(1) Gradual mutational modification of the genome resulting ultimately in the independent formation of a second, regulatory site, a process analogous to the original evolution of the active site. (2) The incorporation, by unequal crossing-over, for example, of part of a pre-existing gene. In the special case where the regulator is identical with or structurally similar to one of the substrates this would be most simply achieved through partial duplication of the structural gene for the enzyme itself. This would result (Fig. 1) in a protein containing a repeated sequence of amino-acid residues in its molecule.

The disadvantage of the first possibility is its inherently low probability[1]. The second is open to the objection that neither of the duplicated sequences would be in an intramolecular environment identical to that of the ancestral sequence. The protein might therefore fail to fold properly, resulting in obliteration rather than duplication of function. Such a protein would, of course, be discarded by natural selection. Even if the duplicated sequences folded correctly, full function might depend upon participation of other residues which would be accessible to only one of the duplicated sites (Fig. 1). Nevertheless, even if one site retained only a substrate-binding capacity, it could evolve a regulatory function, provided that the other site retained full functional activity. This method offers an evolutionary short-cut and appears more likely than the other.

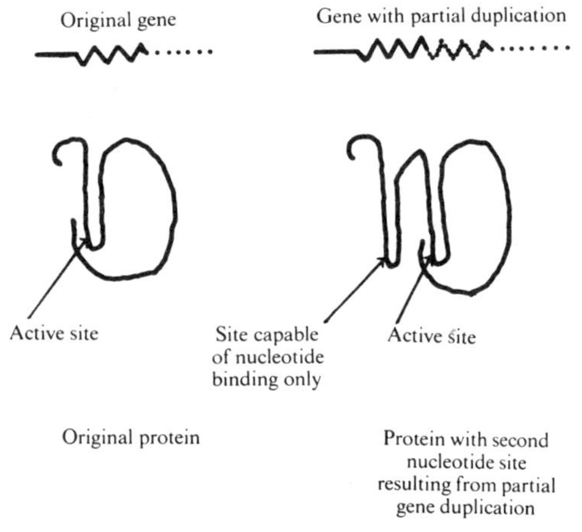

Fig. 1 Schematic diagram illustrating the possible partial gene duplication and its consequences in terms of amino-acid sequence, protein structure and enzyme function.

Although regulatory proteins have not been examined from this standpoint, there are documented cases of duplication having occurred within a single structural gene. Bacterial ferredoxin molecules consist of two clearly related halves[14–17], while those of the heavy chain of gamma G immunoglobulin contain three similar portions[18]. There is also evidence for two gene duplications in the evolution of papain[19].

The nucleotide sequence coding for the potential regulatory site, once established, would mutate independently and probably not conservatively. "Paralogous" sequences as defined by Fitch and Margoliash[20] (human myoglobin and α haemoglobin) tend to diverge more than "orthologous" sequences (human α and β haemoglobins). The duplicated sequences in regulatory proteins envisaged above may be regarded as a special case of paralogous sequences within the same protein. In this special case one might expect selective pressures to produce considerable divergences in those parts of the duplicated sequences concerned only with maintaining the structure of the protein.

It was of great interest to search the sequence of bovine liver GDH[8,21] for the vestigial homologies predicted on the basis of the hypothesis above. Two extended sequences in the protein subunit (similar in chicken GDH)[21] show considerable homology. If residues 114–163 are aligned with residues 269–318, no less than 12 of the 50 residues are in identical positions (Fig. 2). The homology includes a Lys-Cys sequence (114, 115) and a Pro-Lys sequence (134, 135) and is greatest in the vicinity of these two sequences. The intervening twelve residues show only one homology. This stretch includes Lys-126 which reacts with pyridoxal phosphate and is thought to be essential for catalytic activity[22]. In the homologous sequence this essential lysine is replaced by tryptophan and there are several other markedly non-conservative changes.

In addition to 24% of their residues in identical positions, the two sequences contain a further 38% that could have resulted from single base changes in the nucleotide sequence.

If one takes two completely random sequences each n residues in length and composed of the twenty common amino-acids, the probability of the occurrence of x and only x homologies is

$$\frac{n!\ 19^{n-x}}{x!\ (n-x)!\ 20^{n}}$$

From this it can be calculated that the probability of 1, 2 or 3 homologies is quite high (0.21, 0.27 and 0.23) but the probability dwindles rapidly for greater numbers of homologies and is 4.3×10^{-6} for 12 homologies. In the GDH sequence, however, there are 352 possible sequences of 50 residues with which to match the sequence in question, if overlapping pairs of sequences are ignored. The probability that one of these sequences will contain 12 homologies with the chosen sequence is therefore $352\times4.3\times10^{-6}$ $(=1.5\times10^{-3})$, as $4.3\times10^{-6}\ll1$. This is an approximation, as no account has been taken of the differing frequencies of different amino-acids.

Fig. 2 Comparison of suitably aligned partial sequences of beef liver GDH and pig muscle glyceraldehyde-3-phosphate dehydrogenase. The uppermost of the three sequences shows residues 198–261 of the porcine dehydrogenase. The middle sequence covers residues 267–330 of beef liver GDH, designated the "regulator" sequence in the text. The lower sequence is that of residues 112–175 of the bovine enzyme and includes the active site lysine, Lys 126. The homologies are indicated in capital letters. The GDH sequences are taken from the results of Moon, Piszkiewicz and Smith[21], while the glyceraldehyde-3-phosphate dehydrogenase sequence is that reported by Harris and Perham[23].

G3PDH	GLY	ALA	Ala	Glu	Asn	Ile	Ile	Pro	Ala	SER	Thr	GLY	Ala
GDH2	GLY	ALA	LYS	CYS	Val	Ala	VAL	Gly	Glu	SER	Asp	GLY	Ser
GDH1	Thr	Tyr	LYS	CYS	Ala	Val	VAL	Asp	Val	Pro	Phe	GLY	Gly
G3PDH	ALA	LYS	ALA	Val	Gly	LYS	Val	Ile	PRO	Glu	Leu	Asp	Gly
GDH2	Ile	Trp	Asn	Pro	Asp	Gly	ILE	Asp	PRO	LYS	Glu	Leu	Glu
GDH1	ALA	LYS	ALA	Gly	Val	LYS	ILE	Asn	PRO	LYS	Asn	Tyr	Thr
G3PDH	Lys	Leu	Thr	Gly	Met	Ala	Phe	Arg	Val	Pro	Thr	Pro	Asn
GDH2	ASP	Phe	Lys	LEU	Gln	His	Gly	THR	Ile	Leu	Gly	Phe	Pro
GDH1	ASP	Glu	Asp	LEU	Glu	Lys	Ile	THR	Arg	Arg	Phe	Thr	**Met**
G3PDH	Val	Ser	Val	Val	Asp	Leu	Thr	Cys	Arg	LEU	GLU	Lys	Pro
GDH2	Lys	Ala	Lys	Ile	Tyr	Glu	Gly	Ser	Ile	LEU	GLU	VAL	ASP
GDH1	Glu	Leu	Ala	Lys	Lys	Gly	Phe	Ile	Gly	Pro	Gly	VAL	ASP
G3PDH	Ala	Lys	Tyr	Asp	Asp	Ile	Lys	Lys	Val	Val	LYS	GLN	
GDH2	Cys	Asp	Ile	Leu	Ile	Pro	Ala	Ala	Ser	GLU	LYS	GLN	
GDH1	Val	Pro	Ala	Pro	Asn	Met	Ser	Thr	Gly	GLU	Arg	Glu	

It seems likely therefore that the homology between residues 114–163 and 269–318 of beef liver GDH reflects common ancestry. Convergent evolution seems most unlikely in this case in view of the marked differences between sequences 119–130 and 274–285. These differences probably reflect the divergence predicted above. Convergent evolution might in any case be expected to produce homologies less mathematically precise and more qualitative in nature.

Smith *et al.*[8] have compared the sequence of bovine GDH with that of glyceraldehyde-3-phosphate dehydrogenase from pig muscle[23], and conclude that there is very little homology, with the exception of the sequence immediately surrounding the active site lysine (Lys 126) of GDH. Residues 209–219 of glyceraldehyde-3-phosphate dehydrogenase and 123–133 of GDH have six residues in identical positions in both sequences. It seemed possible that the "regulator" sequence of GDH might retain homologies with the glyceraldehyde-3-phosphate dehydrogenase sequence that had been lost in the "active site" sequence. Accordingly in Fig. 2 the glyceraldehyde-3-phosphate dehydrogenase sequence is also aligned for comparison. The comparison, summarized in Table 1, shows that outside the region comprising residues 209–219 the glyceraldehyde-3-phosphate dehydrogenase sequence actually shows considerably greater homology with the regulator sequence than with the active site sequence of GDH. Twenty of the 64 residues in the regulator sequence are identical with the corresponding residues in at least one of the other sequences.

The evidence for an evolutionary relationship between the three sequences is persuasive. Divergence from an ancestral enzyme probably led separately to glyceraldehyde-3-phosphate dehydrogenase and to a primitive GDH lacking the regulatory site and possessing a subunit considerably smaller than that of beef liver GDH. Later, a partial gene duplication probably resulted in a protein in which a sequence of about 155 residues containing at least part of the active site was repeated. One of the identical sequences must have remained catalytically functional. If the other sequence retained its nucleotide-binding capacity it could then evolve as a regulatory site.

At present no other GDH sequences are available. It is significant, however, that most vertebrate GDHs studied so far are regulated by nucleotides and have molecular weights of about 330,000; the corresponding NADP-linked enzymes from yeast, *E. coli* and *Neurospora* are not regulated by nucleotides and available values for the molecular weights of GDH from lower organisms are all in the region of 250,000[4]. One may predict that when sequences become available for other higher organisms they will further document the partial gene duplication discussed above. The non-regulatory GDH from lower organisms, on the other hand, may well lack the duplicated sequence of 155 residues.

In view of the importance of Lys 126 in the active site sequence and its absence from the corresponding position of the presumed regulator sequence it is most interesting to note that Wallis and Holbrook[24] have shown that the chemical modification of this residue abolishes the binding of 2-oxoglutarate but not the binding of coenzymes. It is likely therefore, as required by the basic hypothesis developed in the present paper, that the replacement of lysine by tryptophan in the corresponding position of the regulator sequence, while abolishing any possible catalytic activity due to that sequence, would leave the nucleotide binding capacity intact.

The present study appears to represent the first evidence bearing upon the evolution of regulatory sites from ancestral catalytic sites. It remains to be seen whether similar mechanisms have operated in the evolution of other regulatory enzymes.

Table 1 Comparison of Homology of Glyceraldehyde-3-phosphate Dehydrogenase with "Active Site" and "Regulator" Sequences of Beef Liver GDH

	Identical residues	Alterations explicable by single base changes	Alterations requiring 2 or 3 base changes
G3PDH sequence and matching GDH "active site" sequence	6	30	28
G3PDH sequence and GDH "regulator" sequence	9	21	34
GDH "active" site sequence and GDH "regulator" sequence	12	28	24

The sequences used in this comparison are those shown in Fig. 2.

P. C. Engel

Department of Biochemistry,
University of Sheffield, S10 2TN

Received August 2; revised October 3, 1972.

1 Monod, J., Wyman, J., and Changeux, J.-P., *J. Mol. Biol.*, **12**, 88 (1965).
2 Pauling, L., *Proc. US Nat. Acad. Sci.*, **21**, 186 (1935).
3 Changeux, J.-P., and Gerhart, J. C., in *The Regulation of Enzyme Activity and Allosteric Interactions* (edit. by Kvamme, E., and Pihl, A.), 13 (Universitetsforlaget, Oslo, 1968).
4 Goldin, B. R., and Frieden, C., in *Current Topics in Cellular Regulation* (edit. by Horecker, B. L., and Stadtman, E. R.), **4**, 77 (Academic Press, New York, 1972).
5 Frieden, C., *J. Biol. Chem.*, **234**, 809 (1959).
6 Frieden, C., *J. Biol. Chem.*, **234**, 815 (1959).
7 Engel, P. C., and Dalziel, K., *Biochem. J.*, **115**, 621 (1969).

[8] Smith, E. L., Landon, M., Piszkiewicz, D., Brattin, W. J., Langley, T. J., and Melamed, M. D., *Proc. US Nat. Acad. Sci.*, **67**, 724 (1970).
[9] Jallon, J. M., and Iwatsubo, M., *Biochem. Biophys. Res. Commun.*, **45**, 964 (1971).
[10] di Prisco, G., *Biochemistry*, **10**, 585 (1971).
[11] Koberstein, R., and Sund, H., *FEBS Lett.*, **19**, 149 (1971).
[12] Melzi d'Eril, G., and Dalziel, K., *Biochem. J.*, **130**, 3P (1972).
[13] Goldin, B. R., and Frieden, C., *Biochemistry*, **10**, 3527 (1971).
[14] Eck, R. V., and Dayhoff, M. O., *Science*, **152**, 363 (1966).
[15] Tanaka, M., Nakashima, T., Benson, A. M., Mower, H. F., and Yasunobu, K. T., *Biochemistry*, **5**, 1666 (1966).
[16] Jukes, T. H., *Molecules and Evolution* (Columbia University Press. New York, 1966).
[17] Fitch, W. M., *J. Mol. Biol.*, **16**, 17 (1966).
[18] Ritishauser, U., Cunningham, B. A., Bennett, C., Konigsberg, W. H., and Edelman, G. M., *Proc. US Nat. Acad. Sci.*, **61**, 1414 (1968).
[19] Weinstein, B., *Biochem. Biophys. Res. Commun.*, **41**, 441 (1970).
[20] Fitch, W. M., and Margoliash, E., in *Evolutionary Biology* (edit. by Dobzhansky, T., Hecht, M. K., and Steere, W. C.), **4**, 67 (Appleton-Century-Crofts, New York, 1970).
[21] Moon, K., Piszkiewicz, D., and Smith, E. L., *Proc. US Nat. Acad. Sci.*, **69**, 1380 (1972).
[22] Piszkiewicz, D., Landon, M., and Smith, E. L., *J. Biol. Chem.*, **245**, 2622 (1970).
[23] Harris, J. I., and Perham, R. N., *Nature*, **219**, 1025 (1968).
[24] Wallis, R. B., thesis, University of Bristol (1972).

EXPERIMENTAL EVOLUTION OF A XYLITOL DEHYDROGENASE

B. Hartley, I. Altosaar, J. M. Dothie and M. S. Neuberger

Department of Biochemistry

Imperial College of Science and Technology

London SW7 2AZ

Previous experiments in growing *Klebsiella aerogenes* on xylitol in chemostats suggested that superproduction of ribitol dehydrogenase was the inevitable spontaneous response to this selective pressure, but recently we have obtained several improved xylitol dehydrogenases in this way. Sequence studies of one improved xylitol dehydrogenase have revealed only one amino acid change from its parent. The gene for ribitol dehydrogenase has been moved from *K. aerogenes* into the genome of *E. coli* K12 and subsequently into coliphage λ. Evolution of this *E. coli* K12 on xylitol gives both superproducing and specificity mutants. Preliminary sequence studies of a ribitol dehydrogenase from natural strains of *E. coli* C show 95% identity with the *K. aerogenes* enzyme.

A system for experimental enzyme evolution

The primary and tertiary structures of serine proteases tell us that they have evolved from a common ancestor whilst conserving the architecture of their polypeptide chains to an amazing degree. One or two amino acid changes in the substrate-binding site are sufficient to account for profound changes in the side-chain specificity. Yet 60% of the sequences are different, and the differences occur mostly in the surface away from the active centre, so that 90% of the surface residues differ. However even in the hydrophobic interior there are clusters of side chains that differ radically without affecting the architecture of the chain in the slightest.

One can speculate how these changes might have come about (e.g. Hartley, 1966), but hard evidence is likely to arise only from model experiments in which the sequence changes in an enzyme are investigated during the process of evolution towards a new substrate specificity.

We have tried to devise such a model by studying the ribitol dehydrogenase of *Klebsiella aerogenes* during prolonged growth on xylitol in chemostats. This organism grows well on ribitol and D-arabitol — which are relatively abundant in nature — by the pathways shown in Figure 1 (Mortlock, Fossitt and Wood, 1965; Lerner, Wu and Lin, 1964).

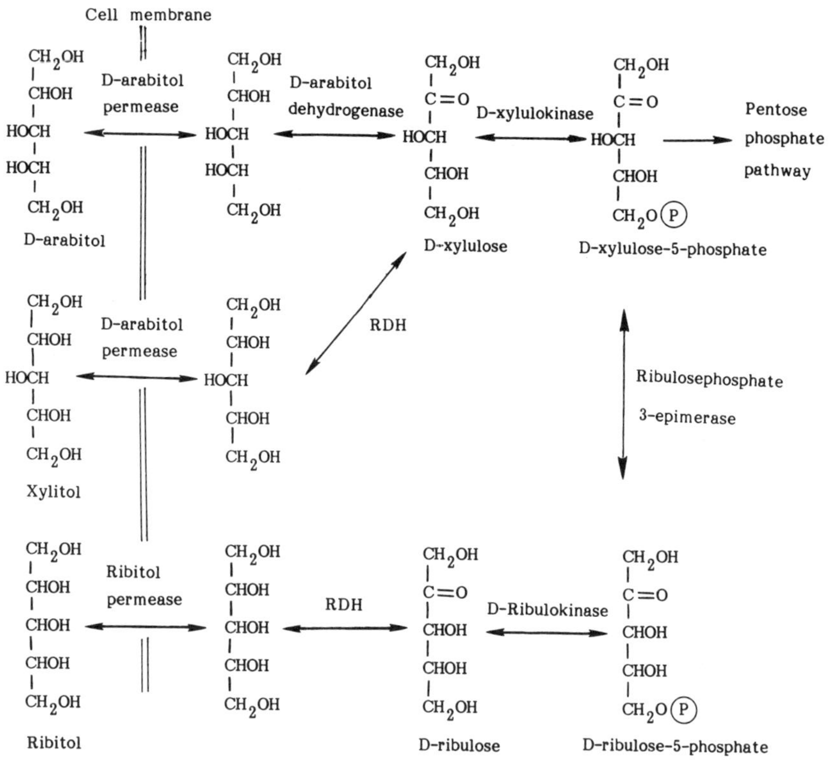

Fig.1
Pentitol metabolism in Klebsiella aerogenes.
RDH = ribitol dehydrogenase

Xylitol and L-arabitol are not metabolised, but mutants constitutive for ribitol dehydrogenase grow poorly on xylitol by using a side-specificity of this enzyme to convert xylitol to D-xylulose (Wu, Lin and Tanaka, 1968). At physiological concentrations of NAD (1 mM), the Km (app.) for ribitol is about 20 mM and kcat is $300s^{-1}$, whereas for xylitol Km (app.) is 1.2 M and kcat is $36s^{-1}$ (Burleigh, Rigby and Hartley, 1974). Hence the ribitol dehydrogenase (RHD) acts as a very poor xylitol dehydrogenase (XDH), and its activity limits the growth rate of the organism.

Therefore when strains of *K. aerogenes* that are constitutive for ribitol dehydrogenase are grown on xylitol in a chemostat, the steady-state biomass is well below that found for growth on ribitol, and there is a high concentration of residual xylitol in the effluent. The system can be used to monitor evolutionary steps that lead to adaptation to this new foodstuff, since faster-growth mutants take over from the ancestral population and such events are signalled by a rise in biomass and a fall in the effluent xylitol concentration (Rigby, Burleigh and Hartley, 1974).

Superproducers of ribitol dehydrogenase in K. aerogenes

We have described previously (Rigby et al., 1974) extensive experiments carried out in this way in the M.R.C. Laboratory of Molecular Biology in Cambridge. In brief, we found that initial takeover events occurred spontaneously after about 50-150 generations (1—3 weeks) and of 5 such events that we investigated, all the mutants proved to have enzymes with unchanged specificity but with 3-5 times higher specific activity in extracts. Having purified and investigated the kinetics of the enzyme (RDH-A) of the original constitutive mutant (A) (Burleigh et al., 1974), we can classify these events as a rise in the concentration of RDH-A from about 1% of the total protein in extracts of the ancestral strain A to about 5% in the first stage "evolvants" (A1, A3, A4 and A5). These initial takeovers were followed by second spontaneous takeovers (A11, A12 and A13) after a similar number of generations, and these proved to contain enzyme of unchanged specificity amounting to about 20% of the total soluble protein.

To speed up our studies we used a regime of mild UV mutagenesis of organisms *in situ* in the chemostats, and in this case also we observed the same phenomena of superproduction of RDH rather than evolution of a strain with improved xylitol specificity. In two such experiments the organism evolved in three or four distinct steps from a strain with maxiumum growth rate (µm) of 0.40 hr^{-1} containing 1.2% of its protein as RDH-A to strains with µm of 0.98 hr^{-1} containing 16% RDH-A or another with µm of 1.6 hr^{-1} containing over 20% of RDH-A.

By this time we had concluded that superproduction of RDH was the inevitable response of the organism to growth on xylitol under mild mutagenesis. We had obtained the same result with weak mutagenesis by N-methyl-N'-nitro-N-nitrosoguanidine (MNNG) or by investigating the further evolution of strain B — a "specificity mutant" obtained by Wu et al., (1968) after powerful MNNG mutagenesis (Hartley 1974). During these studies we had screened chemostats containing around 10^{11} organisms for over 3000 generations and detected over 20 takeover events, all of which appeared to be "superproducers" rather than "specificity mutants." If the latter could arise by single point mutations one would expect them to exist at frequencies greater than 10^{-10}. We therefore concluded that a minimum of two replacements was necessary to produce an efficient enzyme with better specificity for xylitol.

What is the mechanism for RDH superproduction in these strains? The original strain A produces RDH constitutively as over 1% of its total soluble protein, so one must suspect "up-promoter" mutations or gene-duplication as explanations. Lack of a well-characterised genetic system for *K. aerogenes* makes it difficult to answer such questions, but we analysed wild-type and three superproducers (AI, AII and A2II) by "gene-dosage" and "segregation" experiments and concluded that strain AI was *not* gene-duplicated whereas strain AII was. Strain A2II appeared to have three copies of the gene for RDH-A (*rbtD*) (Rigby et al., 1974).

Improved xylitol dehydrogenases

Our conclusion that improved xylitol dehydrogenases require multiple amino acid changes was apparently fortified by studies of strains subjected to more powerful MNNG mutagenesis. A culture subjected to severe MNNG mutagenesis was screened in a chemostat and samples from this run were inoculated into two further chemostats. Each run resulted in the isolation of a separate mutant with improved specificity for xylitol. In a separate experiment, powerful MNNG mutagenesis gave yet another specificity mutant, strain G. The properties of these strains are shown in Table 1.

Table 1
Mutants with altered ribitol dehydrogenases

Strain	μm (hr^{-1})	RDH S.A. (units/mg protein)	XDH/RDH activity ratio 500mM X / 5mM R	500 mM X / 50 mM R	Rm	$t_{\frac{1}{2}}$ at 55° (min)
A	0.53	7.3	0.10	0.04	0.39	20
B	0.61	1.3	0.50	0.25	0.39	6
D	-	24.0	0.41	0.15	-	-
D1	1.10	31.0	0.46	0.17	0.38	30
E	1.00	21.4	0.28	0.12	0.41	20
F	0.99	9.2	0.29	0.15	0.39	20
G	0.86	13.5	0.46	0.15	0.46	20

μm, the maximum specific growth rate on xylitol was measured by the washout method (Tempest, 1970). Ribitol dehydrogenase (RDH and xylitol dehydrogenase (XDH) activities in crude extracts were measured at 1 mM NAD and 5 mM or 50 mM ribitol or 500 mM xylitol (Burleigh et al, 1974). Rm is the relative electrophoretic mobility of the RDH activity on 7.5% acrylamide gels (Davis, 1964). The time for 50% inactivation ($t_{\frac{1}{2}}$) of the enzyme (0.05-0.10 mg/ml) in crude extracts were measured throughout incubation at 55°C in 0.1 M potassium phosphate-4mM NAD, pH7.0.

Enzyme RDH-H is particularly stable, and has been purified and examined in detail. Its kinetics resembles that of RDH-B in that the increased xylitol activity is exhibited in better binding of NAD and of xylitol, and in increased turnover rate. Moreover, as with RDH-B the increased activity towards xylitol is not at the expense of the activity towards ribitol (Burleigh et al, 1974).

Having completed the amino acid sequence of RDH-A (Morris, Williams, Midwinter and Hartley, 1974); Moore. C. H., Taylor, S. S., Smith, M. J. and

Hartley, B. S. - unpublished evidence) we have spent considerable effort in identifying the mutations in RDH-D. We had hoped that these might be revealed by peptide mapping, but in fact a rather exhaustive sequence study was necessary. A change of Ala-196 in RDH-A to Pro-196 in RDH-D was detected relatively easily, but over 90% of the rest of the D sequence has now been screened without finding any other change. Our argument that multiple changes are necessary for improved xylitol specificity therefore looks in peril|

We are being forced to the conclusion that improved xylitol dehydrogenases *can* arise by single point mutations, and the results of our chemostat studies since moving to Imperial College in 1974 support this. We performed six chemostat runs with strain "A" (the original arginine auxotroph XI of Wu et al., 1968) on xylitol at low dilution rates (D = 0.06 hr^{-1}) lasting 1-2 months each. In five of these there was no change in the XDH/RDH activity ratio throughout the run, and in the two cases that we investigated spontaneous takeovers by superproducers had occurred.

However, during another similar experiment strains were isolated from the chemostat wherein the XDH/RDH activity ratio had increased from 0.04 to 0.14 or 0.67. The latter finally took over completely. Separate chemostat runs of these "improved" strains gave superproducers in each case. Moreover in a run at normal dilution rates (ca. 0.2 hr^{-1}) three consecutive takeovers by "specificity mutants" were observed with XDH/RDH of 0.12, 0.47 and 1.00 respectively (Fig. 2).

Hence the surprising result of our evolutionary studies with *K. aerogenes* so far is as follows. In Cambridge, between 1969 and 1974, we studied 20 "evolutionary steps", after spontaneous or mild mutagenesis all of which were "superproducers" of wild-type RDH-A. In London since then, however, five specificity mutants of *K. aerogenes* have evolved spontaneously in response to the selective pressure for growth on xylitol, out of a total of about twelve takeover events. Furthermore the chemostat studies with *E. coli*, described below, have given five "specificity mutants" of RDH-A versus 4 "superproducers" in the nine takeovers studied so far. We are therefore now in a position to study the stepwise changes in sequence that are responsible for these spontaneous evolutionary events. We cannot yet explain our failure to observe such events in our previous studies, though we suspect that we may have inadvertently propagated a spontaneously arising superproducing mutant during sub-culturing of our ancestral strain A. Our recent experiments have been with a fresh isolate of the XI strain of Wu et al, (1968).

We must, however, correct our previous conclusion (Rigby et al., 1974) that improved xylitol dehydrogenases do not arise spontaneously and also, therefore, the argument that single amino acid changes cannot provide a significant increase in specificity of RDH for xylitol.

Fluctuating selective pressure

Nevertheless, our experiments show that multiplication of the gene is a

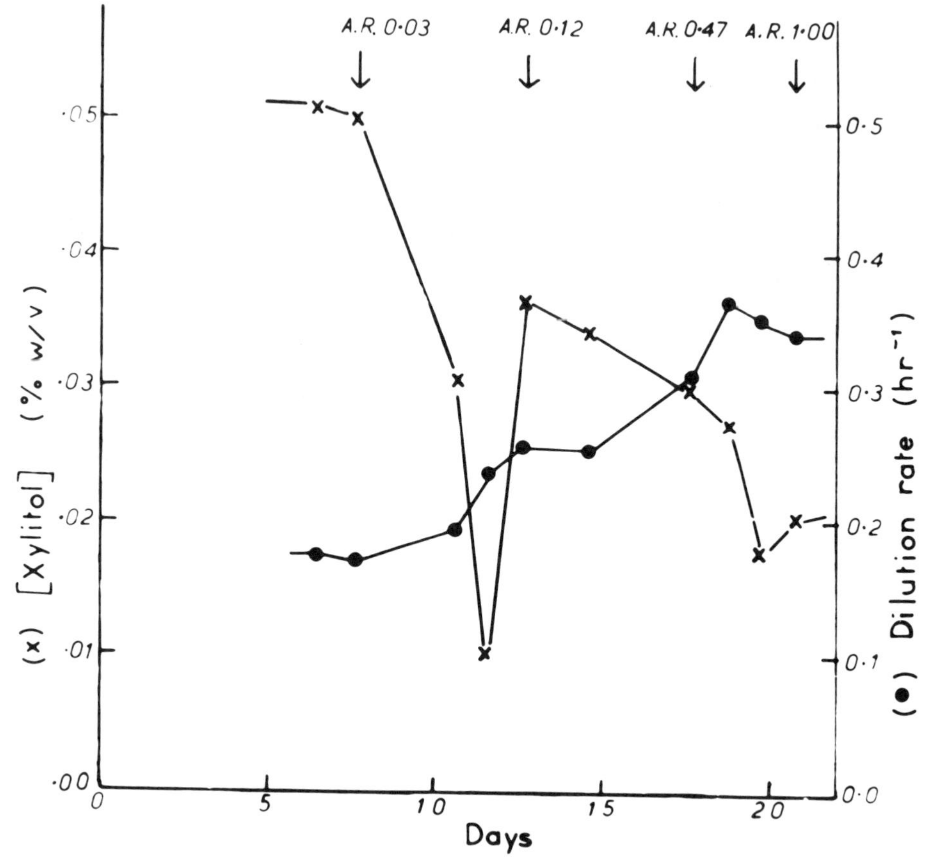

Fig. 2

Spontaneous evolution of xylitol dehydrogenase in *K. aerogenes* grown in xylitol in a chemostat.

Dilution rates were adjusted whenever a takeover was signalled by a drop in the effluent xylitol concentration. The starting strain A was an arginine auxotrophic strain XI of Wu et al (1968)--so media were supplemented with arginine. The "evolvant" strains indicated had taken over the chemostat at the times shown by the arrows, and A.R. indicates the ratio of dehydrogenase activities of extracts of these on 50 mM xylitol or 50 mM ribitol, 1 mM DAN, pH7.

frequent response to the selective pressure for growth on xylitol. This would be the necessary first step in the evolution of an enzyme family. We have argued (Hartley, 1966) that the second step might be the accumulation of mutations in one of these copies once the selective pressure for growth on the new foodstuff was removed. This implies that the rate of elimination of the second copy of the gene is low, and Koch (1972) has produced theoretical arguments that this may be so.

We have tried to test these hypotheses by growing *K. aerogenes* strain AII — which contains two copies of the *rbtD* gene (Rigby et al., 1974) on non-selective carbon sources for a large number of generations with repeated UV mutagenesis. If multiple mutations could thus accumulate in one of the copies of the RDH gene, a greatly improved xylitol dehydrogenase might then emerge when the population was switched back to growth on xylitol.

Table 2
Growth of a gene-doubled strain of K. aerogenes on changing substrates in chemostats

First substrate	Generations	Second substrate	Generations	A.R.	S.A.	Strain
-	-	-	-	0.04	18	A11
Ribitol	750 (2 u.v.)	Xylitol	750	0.03	26	A111
Ribitol	340 (2 u.v.)	Xylitol	120	0.04	25	A113
Ribitol	100 (MNNG)	Xylitol	70	0.05	33	A114
Glucose	700	Xylitol	400	0.04	28	A112
Inositol	470 (7 u.v.)	Xylitol	120	0.03	9	A11S
Inositol	300	-	-	0.04	5	A11S

Abbreviations as in Table 1. Chemostats were grown on 0.2%(w/v) substrate at dilution rates of 0.6 hr^{-1} with weak mutagenesis by u.v. irradiations or by treatment with N-methyl-N'-nitro-N-nitrosoguanidine (MNNG) as indicated.

Table 2 summarises the results of such experiments. With either ribitol or glucose as the first substrate, strains with an increased RDH specific activity but unchanged xylitol specificity took over the chemostats when growth was switched back to xylitol. We suspect that these are gene-tripled strains that would arise from AII by unequal crossover at a frequency of 1.4 x 10^{-3} (Rigby et al., 1974).

Glucose catabolite represses the synthesis of RDH, so the selective pressure to eliminate a second constitutive gene might be neglibible. Catabolite repression is, however, much less with inositol as sole carbon source, and here we found evidence for rapid selection of a single-gene strain during the growth on inositol.

We examined xylitol plates of samples taken from the chemostat throughout the nonselective phase, and found a rapid fall in the proportion of large colonies (gene doubled) to small colonies (putatively single gene segregants) between days 8 and 15. This clearly indicates selective pressure against strains that make a superfluous amount of temporarily unnecessary ribitol dehydrogenase, and argues against conclusions that strains duplicated for constitutive operons might persist for long periods after the selective pressure that elicited them has been removed. (Koch, 1972).

Transfer of the genes for pentitol metabolism from K. aerogenes *to* E. coli *K12 and coliphage* λ

Lack of information about the genetics of *K. aerogenes* limits the depth in which we can analyse our evolutionary model. We therefore decided to try to incorporate a fragment of the *K. aerogenes* genome containing the genes for pentitol metabolism into the chromosome of *E. coli K12*. This was before the days of "cut-and-stitch" genetic engineering, but we achieved our end by isolating strains of *K. aerogenes* that were sensitive to the transducing coliphage PICM *clr 100*, which confers chloramphenicol resistance on its host.

Lysates induced from such lysogens contain generalised transducing particles; thus after subsequent infection of the PICM *clr 100* lysate into *E. coli K12*, an *E. coli K12* construct (EA) capable of growth on both ribitol and D-arabitol was isolated. The *K. aerogenes* pentitol catabolic genes in this strain mapped at about 40 min on the *E. coli* chromosome (Rigby, Gething and Hartley, 1976).

It became clear that a specialised lambda transducing phage carrying the ribitol operon (λ*rbt*) would be extremely useful — both so as to allow very severe *in vitro* mutagenesis of the ribitol genes (and hence the isolation of a much improved xylitol dehydrogenase) as well as enabling the genetic manipulation and detailed investigation of the various mutant operons which caused very high synthesis of the RDH.

The construction of the λ*rbt* phage was achieved by the transduction of the ribitol genes into a strain deleted for the λ attachment site. Lysogenisation of this strain by a suitable λ (kindly provided by Dr. N. Murray) yielded a population of lysogens with λ integrated at the many secondary attachment sites dispersed over the *E. coli* chromosome (Shimada, K., Weisberg, R. A. and Gottesman, M. E. 1972). Subsequent induction yielded a lysate containing a variety of specialised transducing phage. Infection of this lysate into E. coli allowed the screening for the λ *rbt* phage. This isolate was found to confer on wild-type *E. coli* the ability to grow on ribitol but *not* D-arabitol. Such lysogens grow very weakly on xylitol, but can be evolved to better growth in chemostats.

Evolution of ribitol dehydrogenase in E. coli

Having integrated the gene for *K. aerogenes* ribitol dehydrogenase into *E. coli*, we can compare its evolution towards a xylitol dehydrogenase in two different organisms under identical selective pressures. Figure 3 summarises the takeover events that we have observed with our original *E. coli K12* construct (strain EA). At first the strain grows weakly on xylitol in a chemostat although it contains RDH-A as about 4% of its total protein, but soon reaches a higher steady state biomass. We suspect that such events may involve improvement of a xylitol transport system. As mentioned above, takeover events occur spontaneously and strains with improved xylitol specificity have been isolated in this way.

After we had constructed this *E.coli K12/K.aerogenes* hybrid, we learnt that Reiner (1975) had discovered natural strains of *E.coli C* which grow on ribitol and D-arabitol. The genes for the enzymes responsible for this metabolism are closely linked and map at about the same position (40 min.) in *E. coli C* as in our artificial *E. coli* K12 construct. However, Reiner was unable to find any natural *E. coli* B or *E. coli* K12 strains that grew on

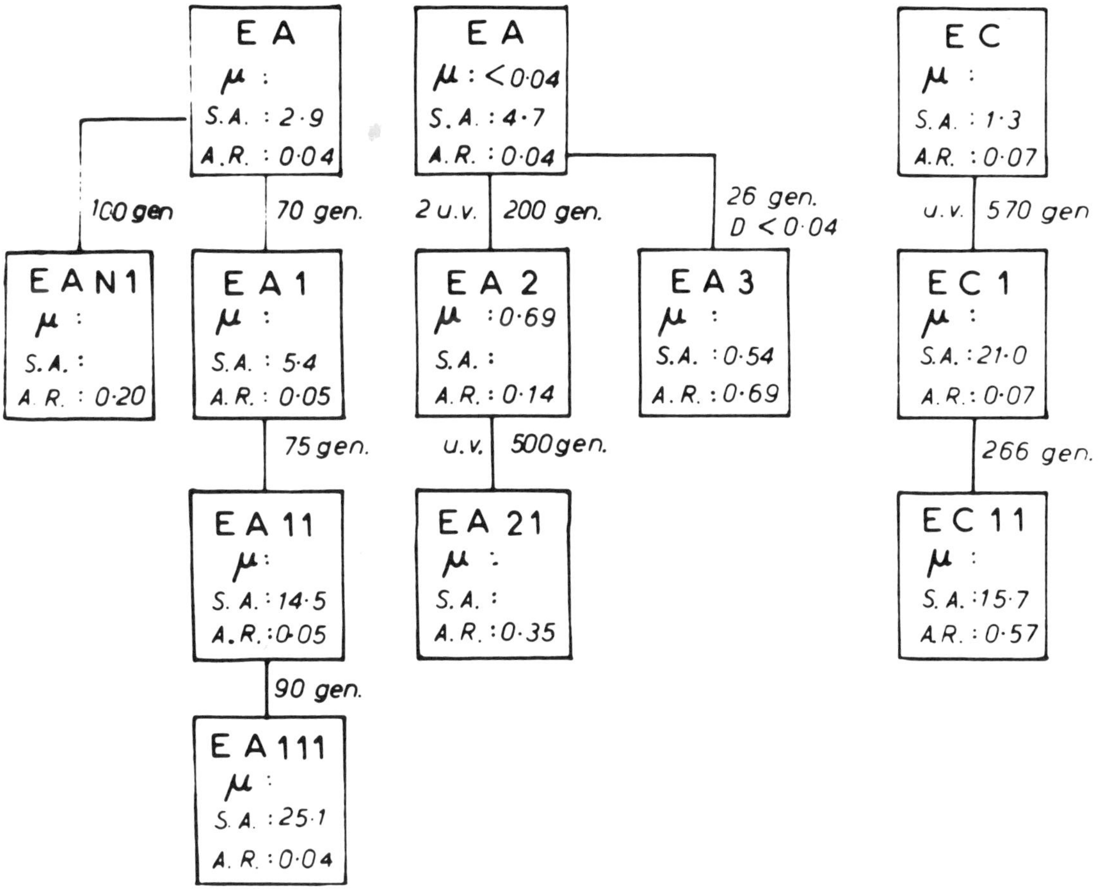

Fig. 3

Evolution of strains of *E. coli* during chemostat growth on xylitol

EA is the *E. coli* K12-*K. aerogenes* hybrid of Rigby, Gething and Hartley (1976). EC is the *E. coli* C strain of Reiner (1975). The "evolvant" strains indicated had taken over the chemostat after the number of generations shown. The dilution rate (D) was 0.2-0.8 except as indicated. EANI arose after mild mutagenesis by MNNG. Mild ultraviolet irradiations of the chemostat were performed as indicated. μ is the maximum specific growth rate (Tempest, 1970); S.A., the specific dehydrogenase activity of cell extracts with 50 mM ribitol (units/mg protein), and A.R. is the ratio of the activity with 500 mM xylitol to that with 50 mM ribitol.

these pentitols, nor does *E. coli* K12.

Charnetzky and Mortlock (1974) have shown that in *K. aerogenes* the genes for pentitol metabolism are closely clustered in the order:

dalK-dalD-dalC-rbtB-dalB-rbtC-rbtD-rbtK

where *B* and *C* are control genes (possibly repressor and operator respectively) for the ribitol (*rbt*) or D-arabitol (*dal*) pathways. *rbtD* codes for ribitol dehydrogenase and *dal K* for xylulokinase. It is therefore not surprising that we have been able to transduce the whole of this region into

E. coli with coliphage-PI; it almost looks as though nature has clustered these genes to make interspecies transfer easy!

Both Reiner and we suspect that *E. coli C* may have gained its pentitol genes by interspecies gene transfer from a *Klebsiella* -like species, and that this may have been a relatively recent evolutionary event. To test this hypothesis we are studying the sequence homology in the respective ribitol dehydrogenases. We have evolved superproducers of ribitol dehydrogenase (strain EC1) from Reiner's *E. coli C* strain (EC), by growth on xylitol in a chemostat. Further growth gives spontaneous takeover by an improved xylitol dehydrogenase. (Figure 3).

Using strain EC1 we can readily purify several grams of *E.coli C* ribitol dehydrogenase for sequence study. Our preliminary sequence results show that there are only 4 differences in 85 residues screened so far. A sequence identity of about 95% for the ribitol dehydrogenases from *K. aerogenes* and *E. coli C* is indicated. This is consistent with a relatively recent common ancestor for both these enzymes.

Prospects

We can now study the evolution of a xylitol dehydrogenase from ribitol dehydrogenase in three strains of microorganisms; *K. aerogenes* , *E. coli K12* and *E.coli C*. We have many such improved enzymes, and can relate the changes in enzyme properties and sequence to their genealogy.

Moreover, having isolated the genes responsible in a bacteriophage, we can study these evolutionary events at the DNA level. *In vitro* mutagenesis of this bacteriophage will be a powerful tool for envolving enzymes with yet higher xylitol activity.

The techniques used in this work are in principle applicable to any enzyme whose activity is, or can be made to be, growth limiting for a microorganism. The prospect of engineering enzymes with new specificities is therefore a step nearer.

REFERENCES

BURLEIGH, B. D. Jr., RIGBY, P. W. J. and HARTLEY, B. S. (1974). *Biochemical Journal* **143,** 341-352.

CHARNETSKY, W. T. and MORTLOCK, R. P.(1974). *Journal of Bacteriology* **119,** 176-182.

DAVIS, B. J. (1964). *Annals of the New York Academy of Science* **121,** 404-427.

HARTLEY, B. S. (1966). *The Advancement of Science* , May 1966 pp. 47-54.

HARTLEY, B. S. (1974). *Symposium of the Society of General Microbiology XXIV,* 151-182.

KOCH, A. L. (1972). *Genetics* **72**, 297-316.
LERNER, S. A., WUY, T.T. and LIN, E.C.C. (1964) *Science* **146**, 1313-1315.
MORRIS, H. R. WILLIAMS, D. H. MIDWINTER, G. G. and HARTLEY, B.S. (1974). *Biochemical Journal* **141**, 701-713.
MORTLOCK, R. P., FOSSITT, D.D. and WOOD, W.A. (1965). *Proceedings of the National Academy of Science, U.S.A.* **54**, 72-77.
REINER, A. M. (1975). *Journal of Bacteriology*, **123**, 530-536.
RIGBY, P. W. J., BURLEIGH, B. D. Jr., and HARTLEY, B. S. (1974). *Nature* **251**, 200-204.
RIGBY, P.W.J., GETHING, M.J. and HARTLEY, B.S. (1976). *Journal of Bacteriology* **125**, 728-738.
SHIMADA, K., WEISBERG, R.A. and GOTTESMAN, M.E. (1972). *Journal of Molecular Biology* **63**, 483-503.
TEMPEST, D. W. (1970). In Methods in Microbiology, Vol. 2. Eds. J. R. Norris and D. W. Ribbons. Academic Press, London 1972, pp. 259-276.
WU, T. T., LIN, E. C. C. and TANAKA, S. (1968). *Journal of Bacteriology* **96**, 447-456.

Reprinted from *Nature* **271**: 626–630.

Structure of pyruvate kinase and similarities with other enzymes: possible implications for protein taxonomy and evolution

Michael Levine*, Hilary Muirhead†, David K. Stammers‡ & David I. Stuart

Molecular Enzymology Laboratory, Department of Biochemistry, University of Bristol, Bristol, UK

The structure determination of pyruvate kinase shows that each subunit of the tetrameric molecule consists of three domains. The largest of these domains has a remarkable similarity to the structure of triosephosphate isomerase. Another domain shows similarities to many other nucleotide binding proteins. We discuss these similarities and their implications for current arguments on protein taxonomy and evolution.

We present here results from our X-ray determination of the three-dimensional structure of cat muscle pyruvate kinase (PK) at 2.6 Å resolution which may bear on questions of protein taxonomy and evolution. The X-ray data were collected photographically using an Arndt–Wonacott rotation camera[1], and phase angles were estimated by the method of isomorphous replacement with anomalous scattering measurements using three derivatives. Full experimental details will be published elsewhere.

The high resolution structures are known for about 50 proteins and certain patterns of structural similarity have appeared in a large proportion of these. Their close correlation with functional similarities has led some workers to propose evolutionary schemes for these proteins[2,3], although others have suggested that there are only a limited number of stable structures from which proteins are constructed[4,5]. In the globins and some serine proteases close correspondence in tertiary and primary structure, as well as function, strongly suggests divergence from an ancestral gene[2,3]. Amongst the dehydrogenases the sequences differ, but the dinucleotide binding domain forms a common structural feature[6,7] and divergence has been suggested here too. The observation that the 'mononucleotide binding fold' (mnbf) defined as one half of the dehydrogenase NAD binding domain, namely three parallel strands of β sheet and two interconnecting α helices (see Fig. 2*e*), is common to many other nucleotide binding proteins led Eventoff and Rossmann to propose that this may be an important evolutionarily conserved unit in the kinases[8]. (We shall use the term mnbf to designate this structure without implying any binding properties).

Our studies on PK have shown that the part of the structure involved in the binding of substrates and cofactors bears a striking similarity to the structure of triosephosphate isomerase (TIM)[9,10], an enzyme which does not require nucleotides, and is quite unlike the known structures of the other kinases[11–14]. Another part of the molecule resembles the mnbf and nucleotides bind here but in a novel fashion.

*Present address: Department of Physiology, University of Bristol, Bristol, UK.
†To whom reprint requests should be addressed.
‡Present address: The Wellcome Research Laboratories, Beckenham, Kent, UK.

Structure of pyruvate kinase

Details of the structure of one subunit of the PK molecule are shown in Fig. 1. Each subunit comprises some 500 amino acid residues and the molecular weight of the complete molecule, consisting of four identical subunits related by twofold axes as shown, is about 240,000. The polypeptide chain is folded into three distinct domains which are shown schematically in Fig. 1. Domain A is the largest, composed of about 220 residues, and contains a cylindrical β sheet of eight parallel strands. Adjacent strands are connected by α helices which form an outer cylinder coaxial with the first. This is shown in stereo in Fig. 2*a* and differs from the interpretation of the 3.1-Å map in that an extra strand of β sheet has been located[15]. This folding pattern has been observed previously in TIM[9]. To compare these two structures a computer search procedure similar to that of Rossmann and Argos[16] was carried out (Table 1 columns 5–10). A detailed comparison of the arrangement of β sheet and α helix in the two structures is shown in Fig. 2 *a–d*. To represent this on a two dimensional surface the α-carbon coordinates have been expressed in cylindrical polar form thereby effectively 'unwrapping' the barrel (Fig. 2 *c*, *d*). Between the third strand of β sheet and third helix of domain A the chain folds up into domain B, where about 100 residues have been located so far. As can be seen in Fig. 2, domain B occurs in the same position as the largest loop in TIM. Domain B does not seem to consist of α/β structure[17], the main secondary structural feature being anti-parallel β sheet. In TIM the chain terminates at the end of the final helix of the barrel whereas in PK the chain continues for about another 120 residues forming domain C. The first section of domain C comprises two long anti-parallel α helices. The rest of the chain is folded up into a five-stranded β sheet flanked by α helices. The first three strands of sheet and two inter-connecting α helices form the well known mnbf. The third helix leads to the fourth strand of sheet and the last strand lies between strands 1 and 4 and anti-parallel to them. The remarkable similarity of part of domain C to the first mnbf of lactate dehydrogenase (LDH) is illustrated in Fig. 2*e*.

Comparisons of structure and function

The comparisons of these folding patterns were quantified using the computer search procedure referred to above and the results are given in Table 1 columns 5–10. This shows that the similarity in fold between TIM and domain A of PK is more significant than that between the dehydrogenases. A part of domain C agrees as well with the mnbf of LDH as does phosphoglycerate kinase (PGK), but, the mnbf's of the dehydrogenases LDH and glyceraldehyde-3-phosphate dehydrogenase (GAPDH) agree rather better amongst themselves.

Substrate and cofactor difference maps have been published at 6-Å resolution previously for PK[18]. Phosphoenolpyruvate (PEP) lies at the carboxyl end of the barrel close to its axis and interacts with several large side chains. The

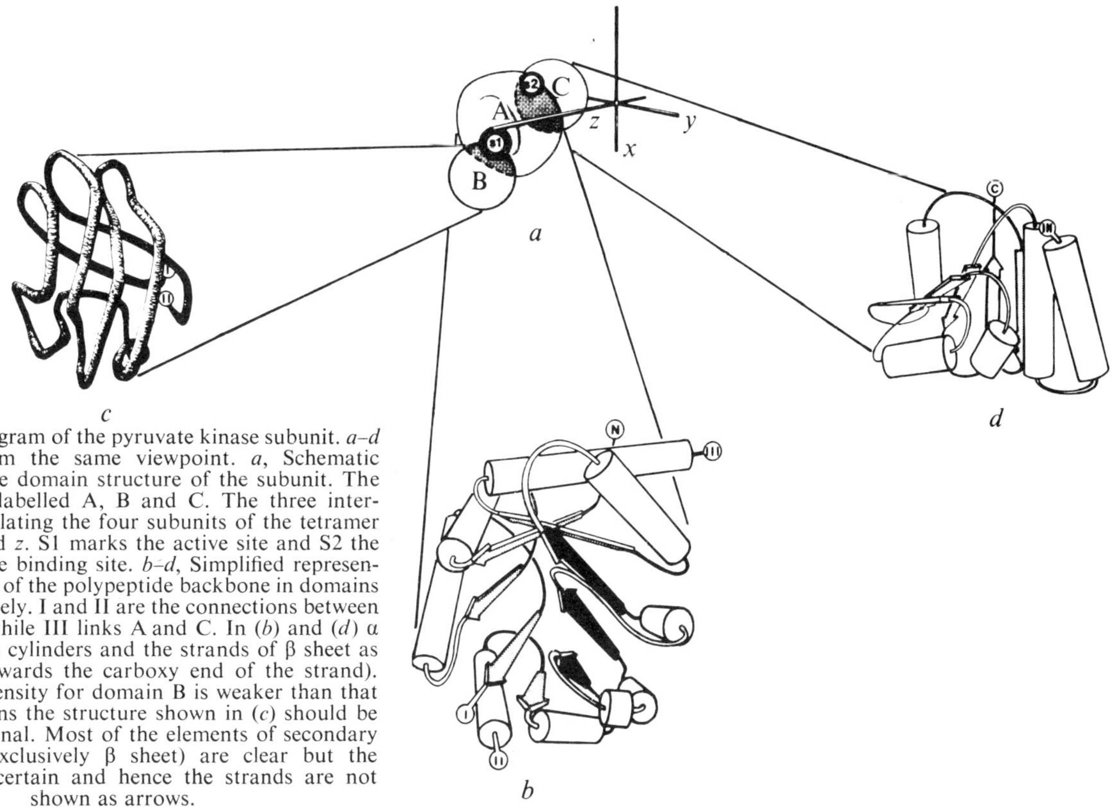

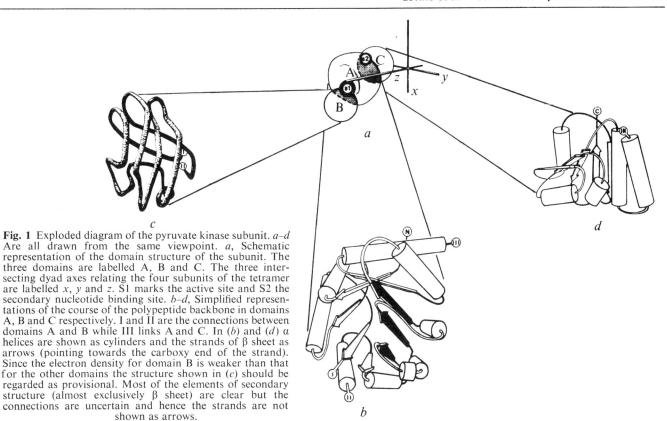

Fig. 1 Exploded diagram of the pyruvate kinase subunit. *a–d* Are all drawn from the same viewpoint. *a*, Schematic representation of the domain structure of the subunit. The three domains are labelled A, B and C. The three intersecting dyad axes relating the four subunits of the tetramer are labelled *x*, *y* and *z*. S1 marks the active site and S2 the secondary nucleotide binding site. *b–d*, Simplified representations of the course of the polypeptide backbone in domains A, B and C respectively. I and II are the connections between domains A and B while III links A and C. In (*b*) and (*d*) α helices are shown as cylinders and the strands of β sheet as arrows (pointing towards the carboxy end of the strand). Since the electron density for domain B is weaker than that for the other domains the structure shown in (*c*) should be regarded as provisional. Most of the elements of secondary structure (almost exclusively β sheet) are clear but the connections are uncertain and hence the strands are not shown as arrows.

Table 1 Results of structural comparisons between domains A and C of PK and other enzymes

(1) Structure 1	(2) Structure 2	(3) Figure of topological relatedness	(4)	(5) No. of equivalent residues between 1 and 2	(6) E_1	(7) E_2	(8) % of equivalent residues in structure 1	(9) % of equivalent residues in structure 2	(10) r.m.s. deviation for these equivalents (Å)
PK domain A	TIM	2,580,480	5,040	160	2.7	6.0	160/216 (74%)	160/247 (65%)	3.0
LDH dinucleotide binding fold	GAPDH	11,520	442	83	2.7	6.0	83/164 (51%)	83/162 (51%)	3.1
PK domain C mnbf	LDH	12	3	54	2.4	6.0	54/ 67 (81%)	54/ 76 (71%)	2.7
PGK mnbf	LDH	12	3	58	2.4	6.0	58/128 (45%)	58/ 76 (76%)	2.9
GAPDH mnbf	LDH	12	3	59	2.4	6.0	59/ 90 (66%)	59/ 76 (78%)	2.1

Columns (3) and (4): these measures are limited by being based on the extant structures and do not consider the significance that, for instance, 2 structures are eight-stranded α/β barrels, whatever the connections. However, since this measure has been used before in assessing the significance of structural similarities the numbers allow our results to be related to previous discussions, on the basis of simple, explicit assumptions. Col. (3), Schulz & Schirmer[22] calculated the number of distinct topologies which could be obtained by re-connecting the helices and β sheet strands of a given α/β structure. In their treatment the 'topology' is fully described by strand sequence in the sheet and the above below pattern of connections. For n strands of sheet there are $n!$ arrangements and 2^{n-1} ways of arranging the interconnecting helices above or below the sheet. Of these topologies half are superimposable on the other half by twofold rotation. Thus there are $n! \times 2^{n-2}$ distinct topologies. Col. (4), Sternberg and Thornton[5], and Richardson[4] have pointed out that sheet and helix in α/β structure are nearly always connected in such a way as to form a right-handed spiral. This reduces the number to approximately $n!/2$. For an eightfold barrel, assuming that the helices, for steric reasons, cannot be on the inside, we can choose the first strand anywhere. There are then seven choices for the second and so on. There are thus $(n-1)!$ distinct topologies. Since the barrel is a closed structure the restriction of Sternberg and Thornton does not affect this total. Cols (5)–(10), structure 2 is rotated through all possible angles and compared to structure 1 for each position (Rossmann and Argos[16]). The likelihood that residue i in structure 1 is equivalent to residue j in structure 2 is expressed as the probability $P_{ij} = \exp(d_{ij}^2/E_1^2)\exp(S_{ij}^2/E_2^2)$. d_{ij} = the distance between $c_{\alpha(i)}$ and $c_{\alpha(j)}$ in the current orientatoins of the molecules $S_{ij}^2 = (d_{ij}-d_{i+1,j+1})^2+(d_{ij}-d_{i-1,j-1})^2$ and gives a measure of how similar the shapes of the chains are on either side of $c_{\alpha(i)}$ and $c_{\alpha(j)}$. By varying the values of E_1 and E_2 we can adjust the relative weights attached to each of these two measures in assessing the overall similarity of two stretches of polypeptide chain. P_{ij} is calculated for all values of i and j (ref. 23). We choose that set of (i,j) for which the sum ΣP_{ij} $(P > 0.05)$ is maximum and i and j both increase from one pair to the next along the chains. Equivalent residues are then those residues which have $P_{ij} > 0.05$, the physical significance of this cut off being determined by the particular values that are used for E_1 and E_2. Orientation and translation parameters are found for which the number of equivalences is largest[24]. These numbers are in Col. (5) and expressed as percentages of each structure in Col. (8) and (9). The rotation matrix which brings TIM (protein data bank coordinates) into the same orientation as PK is:

−0.7376	0.5956	−0.3182
−0.5214	−0.2029	0.8288
0.4291	0.7773	0.4602

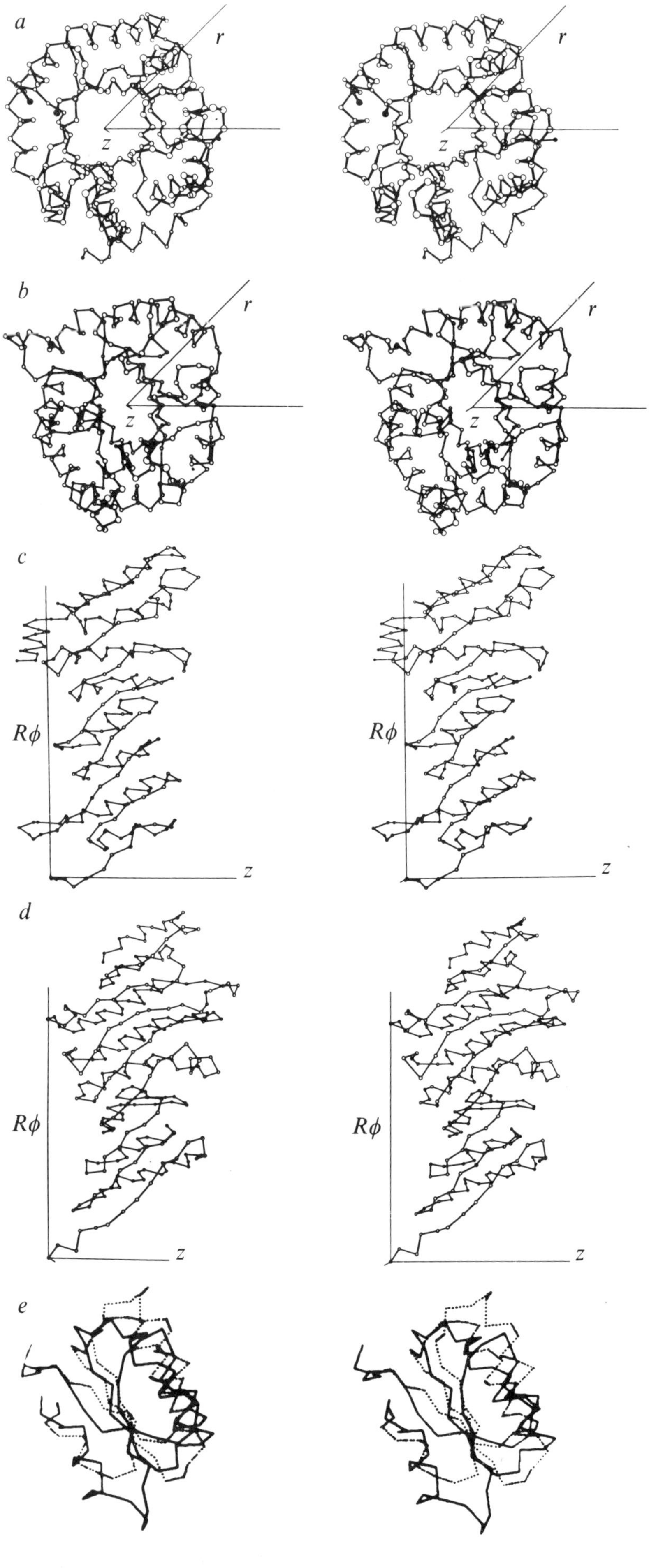

Fig. 2 Stereo diagrams showing the results of structural comparisons of pyruvate kinase with other enzymes. The α-carbon backbone is represented. The pyruvate kinase coordinates are as measured from a 1 cm/Å skeletal model. *a*, Domain A of PK. *b*, TIM, rotated so as to give the overall 'best fit' as defined in Table 1 when superimposed on domain A of PK. Note that the cross section of the barrel is more elliptical than in PK. *c*, Domain A of PK unwrapped. The cartesian coordinates of the α carbons have been transformed into cylindrical polar coordinates *r*, *z* and Φ where the cylindrical polar axes are as shown in (*a*). *r* Is the distance along the azimuthal vector, Φ is the azimuthal angle and *z* is the distance along the cylinder axis. *R* is the approximate radius of the β sheet cylinder. $R = 8.5$ Å. *d*, TIM unwrapped as in (*c*). $R = 8.0$ Å. This shows clearly any small differences in the orientation of the structure elements. *e*, The mnbf LDH (dotted line) compared to part of domain C of PK (full line).

density for Mn–ATP lies radially perpendicular to the barrel axis so that the end of the density interpreted as the terminal phosphate overlaps the PEP binding site and the other end is between strands 3 and 4 of the barrel β sheet. This is unlike PGK where the 3-phosphoglycerate (N. Walker, personal communication) and nucleotide[11,12] lie along the end of the β sheet. However, the PEP density is in a similar position to that of the substrate dihydroxyacetone phosphate (DHAP) in TIM[9] (see Fig. 4). In suitable conditions ADP can be bound at a site between domains A and C in such a position that it interacts with the amino ends of β strands of the barrel and the carboxy end of the second helix of the mnbf in domain C. Surprisingly it is not at the carboxy end of the β sheet of this structure (Fig. 1).

Since kinases are grouped together on the basis of phosphoryl transfer from ATP the similarity between domain A of PK and TIM is at first sight unexpected. However, there are in fact similarities in the reactions of the two enzymes if we consider enolisation of the 2-keto-triose substrates. In PK, PEP is produced by deprotonating and phosphorylating pyruvate, while in TIM the formation of glyceraldehyde-3-phosphate is accomplished through deprotonation of dihydroxyacetone phosphate. It has been proposed that in each case a base on the enzyme participates in the deprotonation reactions. The existence of a *cis*-ene-diol intermediate has been demonstrated for TIM[19,20]. Rose has shown that in the reaction catalysed by PK, enolisation of pyruvate may be decoupled from phosphorylation[21]. Figure 3 illustrates the similarity in this part of the proposed mechanisms.

We have already noted that the substrate binding sites are similarly located on the axis of the barrel in both PK and TIM. The next question to ask is are the active sites of the two proteins similar? Since the primary structure of PK is unknown we cannot identify the side chains with any certainty. We present the side chains which appear in the electron density map in the region where substrates bind, together with the corresponding region of the TIM model (Fig. 4). From this diagram it seems that there are side chains in the active sites of both PK and TIM that are associated with corresponding β-sheet elements. As far as can be seen the sizes and orientations are similar. Thus, as well as comparing PK with the other kinases, it may be useful to include consideration of the fine differences between PK and TIM in studies of the mechanisms of these two enzymes.

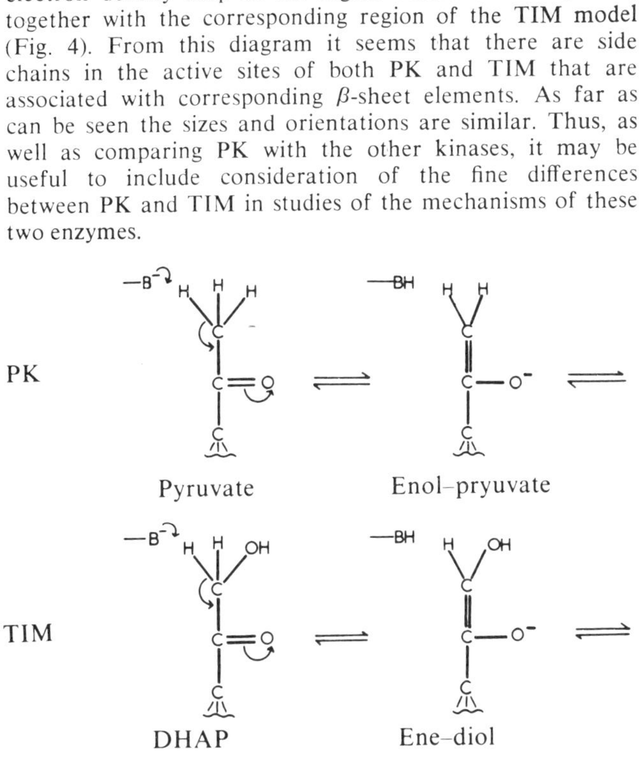

Fig. 3 The proton transfer reactions catalysed by pyruvate kinase and triose phosphate isomerase. In each case B is a base on the enzyme. Enol-pyruvate can be considered to be an enzyme-bound intermediate in the PK reaction[25] and a *cis*-ene-diol intermediate has been implicated in the TIM reaction[20].

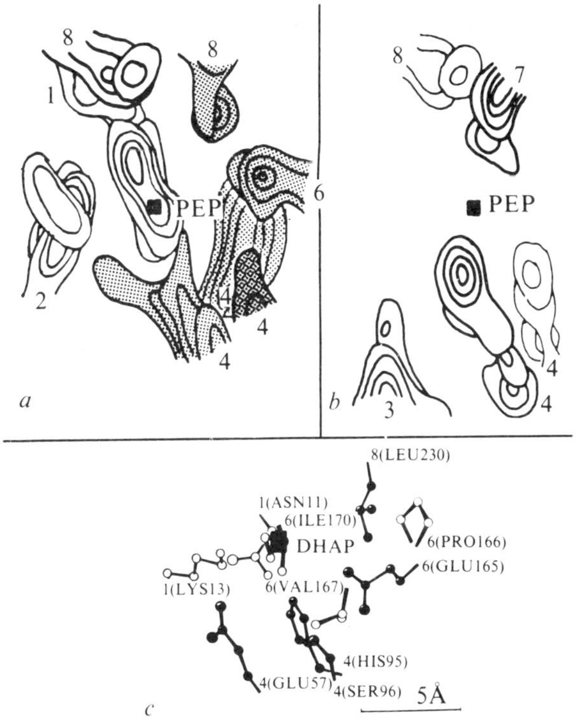

Fig. 4 *a*, *b*, Pyruvate kinase active site. The electron density corresponding to side chains close to the binding site of PEP is shown, this is the area marked S1 in Fig. 1. The direction of view is along the axis marked *y* in Fig. 1 with the *z* axis horizontal. Each side chain is numbered to indicate which element of the β-sheet barrel it is attached to. The *n*th element of the barrel is defined as containing all residues between the beginning of the *n*th strand and the end of the *n*th helix. In the electron density map, the unit cell of the crystal is sectioned perpendicular to the *y*-axis. The thickness of each section is 1.25 Å. The PEP site is centred on section 38 and is marked ■. (*a*) Shows a composite of sections 32 to 38 and *b*, of sections 38 to 42. The side chains only are shown, the main chain having been left out for clarity. *c*, Triosephosphate isomerase active site. The orientation corresponds to that of (*a*) and (*b*) and the atomic coordinates for the side chains of the residues close to the DHAP position are shown. C_2 of DHAP is marked ■. The labels show the element of barrel to which each residue is attached with the residue type and number given in parenthesis. The scale of this diagram is the same as (*a*) and (*b*). The residues shaded in (*a*) seem to occupy analagous positions to those with filled atoms in (*c*) and it will be seen that the secondary structural elements that they are attached to correspond. Phillips *et al.*[26] have shown that the proposal, based on the X-ray diffraction results, that Glu 165 is the base implicated in the mechanism (Fig. 3) is consistent with other biochemical evidence. In comparing the two diagrams note that the barrels are not exactly the same shape in TIM and PK (Fig. 2*a*, *b*) and consequently groups attached to different strands are displaced relative to each other. Note also that three of the residues of the fourth sheet-helix element seem to be in similar relative positions in the two structures. These three residues in TIM[26], His 95, Ser 96, Glu 97, are all members of the group of residues implicated in the active site.

Pyruvate kinase and evolution

Our results suggest that domain A of PK and TIM may have evolved from a common ancestor but contradict the idea[5] that the appearance of the mnbf in an enzyme implies the binding of a nucleotide in the same fashion as in the dehydrogenases and its evolutionary conservation for this function alone. Thus if one considers that any three consecutive strands of the eightfold barrel comprise a mnbf[9] then the two nucleotide binding sites in PK are both associated with a mnbf. However, the manner of binding is quite different to that found in the dehydrogenases. At the active site the nucleotide is bound at the carboxy end of the β sheet as in the dehydrogenases but is orientated along a radius instead of along the edge of the sheet. At the secondary site it is not even near to the β sheet. It could be argued[4,5] that a structure like the mnbf is likely to form in PK if only because it is so highly probable in such a large α/β structure. We feel that although this may be true it tends to divert attention from interesting considerations of structure and function and their possible relation to evolution. It may be that proposals for simple evolutionary relationships between structures should at present be confined to functional domains as in the dehydrogenases or domain A of PK and not applied to stable substructures such as the mnbf which could conceivably be dissected out from them. If the similarities between domain A and TIM do imply a divergent evolutionary relationship rather than convergent evolution then it would seem that either PK is a composite of parts which had separate evolutionary histories before their joining together to make the modern enzyme or TIM is derived from a fragment of an ancestral protein. We would still have to account for the derivations of domains B and C in PK. It is also apparent that the kinases do not form such a closely related group of structures as the dehydrogenases. All known kinase structures consist of α/β units[11–14]. PGK[11,12] possesses the 'dinucleotide binding fold' of LDH and apparently binds the nucleotide in a similar manner. PK, hexokinase[14] and adenyl kinase[13] all differ.

The basic difference between the PK/TIM homology and those observed previously is that whereas for, say, the dehydrogenases the function of the postulated conserved gene is characteristic of a group of enzymes accepted as belonging to a single class, in the PK/TIM case, the function conserved would cut across the two categories 'kinase' and 'isomerase'. As more protein structures are revealed the present categories may therefore be found inadequate and other groupings may emerge which reflect both similarities of domain structure and the different aspects of the function. The problem posed by the structural diversity of the enzymes now classed as kinases may then disappear.

We thank Dr John Williams for helpful discussions, Dr I. A. Wilson and Professor D. C. Phillips for use of unpublished results, the Protein Data Bank for atomic coordinates of proteins and the SRC for support.

Received 26 September; accepted 8 December 1977.

1. Arndt, U. W., Champness, J. N., Phizackerley, R. P. & Wonacott, A. J. *J. appl. Crystallogr.* **6**, 457–463 (1973).
2. Dayhoff, M. O., Hunt, L. T., McLaughlin, P. J. & Jones, D. D. *Atlas of Protein Sequence and Structure* **5**, 17–30 (1972).
3. Hartley, B. S. *Phil. Trans. R. Soc.* **B257**, 77–78 (1970).
4. Richardson, J. S. *Proc. natn. Acad. Sci., U.S.A.* **73**, 2619–2623 (1976).
5. Sternberg, M. J. E. & Thornton, J. M. *J. molec. Biol.* **105**, 367–382 (1976).
6. Rossmann, M. G., Moras, D. & Olsen, K. W. *Nature* **250**, 194–199 (1974).
7. Ohlsson, I., Nordström, B. & Brändén, C-I. *J. molec. Biol.* **89**, 339–354 (1974).
8. Eventoff, W. & Rossmann, M. G. in *CRC Critical Reviews of Biochemistry* (ed. Fasman, G. D.), vol. 3, 111–140 (CRC Press, Cleveland, 1975).
9. Banner, D. W. *et al. Nature* **255**, 609–614 (1975).
10. Banner, D. W., Bloomer, A. C., Petsko, G. A., Phillips, D. C. & Wilson, I. A. *Biochem. biophys. Res. Commun.* **72**, 146–155 (1976).
11. Blake, C. C. F. & Evans, P. R. *J. molec. Biol.* **84**, 585–601 (1974).
12. Bryant, T. N., Watson, H. C. & Wendell, P. L. *Nature* **247**, 14–17 (1974).
13. Schulz, G. E., Elzinga, M., Marx, F. & Schirmer, R. H. *Nature* **250**, 120–123 (1974).
14. Steitz, T. A., Fletterick, R. J., Anderson, W. F. & Anderson, C. M. *J. molec. Biol.* **104**, 197–222 (1976).
15. Stammers, D. K. & Muirhead, H. *J. molec. Biol.* **112**, 309–316 (1977).
16. Rossmann, M. G. & Argos, P. *J. molec. Biol.* **105**, 75–95 (1976).
17. Levitt, M. & Chothia, C. *Nature* **261**, 552–558 (1976).
18. Stammers, D. K. & Muirhead, H. *J. molec. Biol.* **95**, 213–225 (1975).
19. Rose, I. A. & Rieder, S. V. *J. biol. Chem.* **234**, 1007–1010 (1959).
20. Rose, I. A. *Brookhaven Synp. Biol.* **15**, 293–309 (1962).
21. Rose, I. A. *J. biol. Chem.* **235**, 1170–1177 (1960).
22. Schulz, G. E. & Schirmer, R. H. *Nature* **250**, 142–144 (1974).
23. Rossmann, M. G. & Argos, P. *J. biol. Chem.* **250**, 7525–7532 (1975).
24. Rao, S. T. & Rossmann, M. G. *J. molec. Biol.* **76**, 241–256 (1973).
25. Robinson, J. L. & Rose, I. A. *J. biol. Chem.* **247**, 1096–1105 (1972).
26. Phillips, D. C., Rivers, P. S., Sternberg, M. J. E., Thornton, J. M. & Wilson, I. A. *Biochem. Soc. Trans.* **5**, 642–647 (1977).

Evolution of a new enzymatic function by recombination within a gene

[evolved β-galactosidase (ebg)/β-galactosidase/experimental evolution/*Escherichia coli*]

BARRY G. HALL AND TIMOTHY ZUZEL

Microbiology Section, U-44, Biological Sciences, University of Connecticut, Storrs, Connecticut 06268

Communicated by Herschel L. Roman, March 10, 1980

ABSTRACT **Mutations that alter the *ebgA* gene so that the evolved β-galactosidase (ebg) enzyme of *Escherichia coli* can hydrolyze lactose fall into two classes: class I mutants use only lactose, whereas class II mutants use lactulose as well as lactose. Neither class uses galactosylarabinose effectively. In this paper we show that when both a class I and a class II mutation are present in the same *ebgA* gene, ebg enzyme acquires a specificity for galactosylarabinose. Although galactosylarabinose utilization can evolve as the consequence of sequential spontaneous mutations, it can also evolve via intragenic recombination in crosses between class I and class II *ebgA*⁺ mutant strains. We show that the sites for class I and class II mutations lie about 1 kilobase, or about a third of the gene, apart in *ebgA*. Implications of these findings with respect to the evolution of new metabolic functions are discussed.**

The *ebg* (evolved β-galactosidase) system of *Escherichia coli* is being employed as a model system to study the evolution of new metabolic functions. This model system has demonstrated, in the laboratory, the requirement for both structural and regulatory gene mutations for the evolution of a particular new metabolic function (1–3). It has also been used to demonstrate the existence of an evolutionary pathway, in that three sequentially selected spontaneous mutations in a structural gene were required for the evolution of a particular new metabolic function (4). We now turn our attention to the role of intragenic recombination in the evolution of new metabolic functions.

Strains of *E. coli* that bear deletions within the *lacZ* (β-galactosidase) gene, but in which the *lacY* (lactose permease) gene is intact, are unable to utilize lactose or other β-galactoside sugars as sole carbon and energy sources. A second β-galactosidase, enzyme ebg^o, is the product of the wild-type allele of the *ebgA* gene located at 66 min on the *E. coli* K-12 map (5, 6). Expression of the *ebgA* gene is under control of the tightly linked *ebgR* gene, which specifies a repressor (7). The wild-type enzyme, ebg^o, has little activity toward natural β-galactoside compounds, and even *ebgR*$^-$ strains, which synthesize about 5% of their soluble protein as ebg^o enzyme, are unable to utilize lactose, lactulose, or galactosylarabinose (Gal-Ara) (2, 4). Spontaneous single-point mutations can evolve the wild-type allele, *ebgA*^{o}, to *ebgA*$^+$, resulting in enzyme with greatly increased activity toward lactose (8).

Previous studies (2, 4) have shown that a number of different mutations in the *ebgA* gene can lead to enzyme with increased β-galactosidase activity. Selection for lactose utilization results in two classes of *ebgA*$^+$ mutants: class I strains grow rapidly on lactose but are unable to utilize lactulose; class II strains grow more slowly on lactose and grow at a moderate rate on lactulose (4). Table 2 shows the first-order growth rate constants for representative members of each class. Studies of growth rates, coupled with kinetic analyses of purified enzymes, have shown that a number of nonidentical mutations can occur within each class (2, 4). Selection for lactulose, rather than lactose, utilization results in strain previously designated class III (4). Because these *ebgA*$^+$ strains are phenotypically indistinguishable from class II, they will be referred to as class II strains in this and subsequent articles.

Class I and II strains carry single mutations within the *ebgA* gene (8), and all grow extremely slowly, and sometimes not at all, on Gal-Ara. When growth occurs, doubling times of 23–36 hr are typical (4). When class I strains (lactose-positive, lactulose-negative) are subjected to selection for lactulose utilization, class IV strains arise. Class IV strains carry *two* mutations in the *ebgA* gene. They differ from the previous classes in that (*i*) they grow faster on lactose than on lactulose, and (*ii*) they grow at a significant rate (doubling times less than 7 hr) on Gal-Ara. Class IV strains are of particular interest because they seem to be obligatory intermediates on the pathway to evolving *ebg* enzyme so that it can hydrolyze lactobionic acid (4).

We previously described an "obligatory" pathway for the evolution of lactobionate utilization consisting of a class I mutation, followed by a second mutation to give a class IV strain, followed by a third mutation to give lactobionate utilization (4). Class II strains were considered evolutionary dead ends simply because we did not know what selective pressure to apply in order to evolve further substrate specificities (4). We recently decided to apply selective pressure for Gal-Ara utilization to class II strains in order to determine whether Gal-Ara$^+$ strains would resemble class IV strains.

MATERIALS AND METHODS

Bacterial and Phage Strains. All bacterial strains are *E. coli* K-12 and bear the *lacZ* deletion W4680. Unless otherwise indicated in Table 1, all strains are *ebgR*$^-$, and thus synthesize ebg enzyme constitutively. The *ebgR*$^-$ allele in strain 5A11 was selected with phenyl-β-galactoside as in ref. 7. SJ-20 is a recombinant from a mating between 1B1 and SJ-7. SJ480 is a recombinant from a mating between D2 and SJ-7. Spontaneous *thyA*$^-$ mutants were selected by growth for 20 generations in minimal medium containing thymidine and trimethoprim. Genotypes are given in Table 1. Bacteriophage P1 cam ts100 was used for transductions as described (7).

Media. Minimal medium was described previously (1). As required, methionine and arginine were employed at 100 mg/liter and streptomycin sulfate was employed at 300 mg/liter. Lactose (4-*O*-β-D-galactopyranosyl-D-glucose), lactulose (4-*O*-β-D-galactopyranosyl-D-fructose), and Gal-Ara (3-*O*-β-D-galactopyranosyl-D-arabinose) were used at 1 g/liter. All media employing a β-galactoside also contained 0.2 mM isopropyl β-D-thiogalactopyranoside for the sole purpose of inducing synthesis of the lactose permease (1). MacConkey in-

The publication costs of this article were defrayed in part by page charge payment. This article must therefore be hereby marked "*advertisement*" in accordance with 18 U. S. C. §1734 solely to indicate this fact.

Abbreviation: ebg, evolved β-galactosidase.

Table 1. Relevant genotypes of bacterial strains

Strain	Relevant genotype	Ref.
1B1	*ebgA°*, HfrC, *spc* (= *rpsE*)	7
A2	*ebgA2*, HfrC, *spc*	1
D2	*ebgA168*, HfrC, *spc*	4
5A1	*ebgA51*, HfrC, *spc ebgR*$^+$	7
5A11	*ebgA51*, HfrC, *spc ebgR*$^-$ mutant of 5A1	This study
5A2	*ebgA52*, HfrC, *spc*	7
C1	*ebgA139*, HfrC, *spc*	4
C2	*ebgA141*, HfrC, *spc*	4
A23	*ebgA134*, HfrC, *spc*	4
5A11GA	*ebgA204*, HfrC, *spc*	This study
5A2GA	*ebgA203*, HfrC, *spc*	This study
C1GA	*ebgA201*, HfrC, *spc*	This study
C2GA	*ebgA202*, HfrC, *spc*	This study
SJ-7	*ebgA°*, F$^-$, *strA*, *metC*, *argG*, *tolC*, *ebgR*$^+$	7
SJ-8	*ebgA2*, F$^-$, *strA*, *metC*, *argG*	7
SJ-8T	*ebgA2*, F$^-$, *strA*, *metC*, *argG*, *thyA*	This study
SJ-12	*ebgA52*, F$^-$, *strA*, *metC*, *argG*	7
SJ-12T	*ebgA52*, F$^-$, *strA*, *metC*, *argG*, *thyA*	This study
SJ-20	*ebgA°*, F$^-$, *strA*, *tolC*	This study
SJ-480	*ebgA168*, F$^-$, *strA*, *metC*, *argG*	This study
SJ-480T	*ebgA168*, F$^-$, *strA*, *metC*, *argG*, *thyA*	This study
R41	*ebgA2*, F$^-$, *strA*	This study
R42	*ebgA143*, F$^-$, *strA*	This study
R61	*ebgA144*, F$^-$, *strA*	This study
R62	*ebgA52*, F$^-$, *strA*	This study
RT512	*ebgA108*, F$^-$, *strA*	This study
RT522	*ebgA107*, F$^-$, *strA*	This study
RT51168	*ebgA105*, F$^-$, *strA*	This study
RT52168	*ebgA106*, F$^-$, *strA*	This study
SJ12R/F′A2	F′122, *ebgA2*, *ebgR*$^-$/*ebgA52*, *metC*, *argG*, *recA*	This study
SJ8R/F′5A2	F′122, *ebgA52*, *ebgR*$^-$/*ebgA2*, *metC*, *argG*, *recA*	This study

dicator medium was prepared from MacConkey agar base (Difco), and contained the indicated fermentable β-galactoside at 10 g/liter. MacConkey agar kills *tolC*$^-$ cells (7), as does sodium deoxycholate, which was added to minimal medium at a concentration of 1 g/liter in some selective plates.

Growth Rates. Rates were measured as described in ref. 4.

RESULTS

Selection of Gal-Ara$^+$ mutants from class II ebgA$^+$ strains

Initial attempts to select spontaneous Gal-Ara$^+$ mutants on Gal-Ara minimal medium were hampered by significant background growth when more than 10^8 cells were spread per plate. That background growth was attributable to the marginal growth of class II strains on Gal-Ara. As an alternative, Gal-Ara$^+$ mutants were selected by a serial transfer method. The class II strains 5A2, 5A11, C1, and C2 were grown overnight in glycerol minimal medium containing 0.2 mM isopropyl thiogalactoside to fully induce the lactose permease, and inoculated into Gal-Ara minimal medium at an initial density of 10^8 cells per ml. The cultures were shaken at 37°C, and the optical density was monitored daily until the density had risen to 5×10^8 cells per ml. This required 6 days for strains C2 and 5A11, and 7 days for strains C1 and 5A2. The cultures were diluted 1:50 into the same medium and shaken for 3 days, by which time the densities of all cultures exceeded 5×10^8 cells per ml. The cultures were again diluted 1:50 and grown for 2 days to a density in excess of 5×10^8 cells per ml. Each culture was streaked onto a MacConkey Gal-Ara plate. The plates exhibited a preponderence of red (Gal-Ara$^+$) colonies, although a number of white (Gal-Ara$^-$) colonies were present on each plate. A single Gal-Ara$^+$ colony was isolated from each culture. The new strains were named 5A2GA, 5A11GA, C1GA, and C2GA. These strains, like their parents, were sensitive to the male-specific bacteriophage R17 and were resistant to the antibiotic spectinomycin.

To be sure that the Gal-Ara$^+$ phenotype was the result of mutation within the *ebgA* gene, the new mutations were mapped by transduction with bacteriophage P1. Strain SJ-20 (*ebgA°*, *ebgR*$^-$, *tolC*$^-$) was transduced with phage grown on each of the new strains, and *tolC*$^+$ transductants were selected on MacConkey lactose medium *ebgA*$^+$ cotransductants were obtained at the following frequencies: 5A2GA, 37%; 5A11GA, 43%; C1GA, 38%; and C2GA, 40%. These values are in good agreement with the previously reported value of 43.3% cotransduction between *ebgA* and *tolC* (7). All of the *ebgA*$^+$ cotransductants were replicated to MacConkey Gal-Ara plates were found to be Gal-Ara$^+$. Thus, the mutation that allows class II strains to utilize Gal-Ara effectively is in the *ebgA* gene.

Characteristics of Gal-Ara$^+$ strains

Table 2 shows the growth rates of the Gal-Ara$^+$ strains on lactose, lactulose, and Gal-Ara minimal medium. Comparison of these strains with their parental class II strains shows that in each case there was a great increase in the rate of Gal-Ara utilization such that all doubled in less than 7 hr. Likewise, in each case the growth rate on lactose increased by a factor of at least

Table 2. Growth rates on three β-galactoside sugars

		Growth rate constant, hr^{-1}		
Strain	Class	Lactose	Lactulose	Gal-Ara
1B1	WT	0	0	0
A2*	I	0.446 ± 0.033	0	0.029 ± 0.001
SJ8	I	0.365 ± 0.009	0	ND
D2*	I	0.451 ± 0.042	0	0.032 ± 0.005
5A2*	II	0.118 ± 0.016	0.255 ± 0.015	0.027 ± 0.001
SJ-12	II	0.128 ± 0.006	0.252 ± 0.003	ND
5A11	II	0.171 ± 0.004	0.233 ± 0.012	0
C1	II	0.181 ± 0.029	0.235 ± 0.010	0
C2*	II	0.227 ± 0.008	0.312 ± 0.006	0.034 ± 0.001
A23*	IV	0.360 ± 0.026	0.196 ± 0.005	0.139 ± 0.009
5A2GA	IV′	0.355 ± 0.025	0.145 ± 0.048	0.102 ± 0.022
5A11GA	IV′	0.352 ± 0.012	0.159 ± 0.018	0.136 ± 0.040
C1GA	IV′	0.376 ± 0.035	0.181 ± 0.025	0.145 ± 0.022
C2GA	IV′	0.360 ± 0.048	0.147 ± 0.024	0.102 ± 0.040
		Recombinant strains		
R41	I	0.377 ± 0.009	0	0
R42	II	0.140 ± 0.003	0.167 ± 0.004	0.031 ± 0.006
R61	I	0.389 ± 0.005	0	0
R62	II	0.162 ± 0.004	0.212 ± 0.005	0
RT512	IV	0.201 ± 0.004	0.092 ± 0.002	0.165 ± 0.012
RT522	IV	0.220 ± 0.006	0.090 ± 0.004	0.112 ± 0.002
RT51168	IV	0.230 ± 0.002	0.070 ± 0.008	0.101 ± 0.006
RT52168	IV	0.222 ± 0.003	0.069 ± 0.002	0.100 ± 0.011
		Diploid strains		
SJ12R/F′A2	II/F′I	0.312 ± 0.002	0.140 ± 0.004	0.041 ± 0.004
SJ8R/F′5A2	I/F′II	0.300 ± 0.001	0.182 ± 0.003	0.045 ± 0.011

Values shown are the first-order growth rate constants ± the 95% confidence interval. WT, wild type, ND, not determined.
* Data taken from ref. 4.

1.6, while the growth rate on lactulose decreased by at least a factor of 1.3. Class II strains all grow faster on lactulose than on lactose (4). Apparently mutations that permit significant growth rates on Gal-Ara reverse this relationship and result in more rapid growth on lactose than on lactulose. An examination of Table 2 shows that there is a striking similarity between the growth rates of the new Gal-Ara$^+$ strains and class IV strains on lactose, lactulose, and Gal-Ara. The Gal-Ara$^+$ strains, like class IV strains, carry two point mutations within the *ebgA* gene. The strikingly similar growth rates suggested that the *second* mutation in these strains might be equivalent to a class I mutation and that likewise the second mutation in a class IV strain might be equivalent to a class II mutation. On the basis of the similarity in growth rates the new strains are considered to be class IV; however, for the purposes of discussion they will be designated class IV′ to distinguish them from previously obtained class IV strains.

Recovery of single-point mutations from recombination within *ebgA*

If classes IV and IV′ are simply the result of a class I plus a class II mutation, it should be possible to recover *both* class I and class II type strains from crosses between a class IV or IV′ gene and a wild-type (*ebgA^o*) gene.

The class IV strain A23 and the class IV′ strain 5A2GA were each mated with the F$^-$, *tolC*$^-$, *ebgA^o*, *ebgR*$^-$, *strA*$^-$ strain SJ-20, and the mating mixture was plated onto MacConkey lactose agar containing streptomycin. The *tolC*$^+$ *strA*$^-$ recombinants that arose (about 1000 per plate) were replicated to MacConkey lactulose and MacConkey Gal-Ara plates. The replicated plates were screened for colonies that were Gal-Ara$^-$ but both lactose$^+$ and lactulose$^+$ (class II) and for colonies that were Gal-Ara$^-$ and lactulose$^-$ but lactose$^+$ (class I). Appropriate colonies were reisolated and the phenotypes were confirmed. Both of the expected recombinant types were recovered from each cross. From the A23 × SJ-20 cross the class I recombinant was designated R41, and the class II recombinant was designated R42. Likewise, from the 5A2GA × SJ-20 cross the class I recombinant was designated R61, and the class II recombinant was designated R62. Growth rates of these recombinant strains are shown in Table 2. The allele present in strain R41 should be identical with *ebgA2*, present (in the same genetic background) in strain SJ-8. The allele present in strain R62 should be identical to *ebgA52*, which is present in strain SJ-12. The alleles present in strains R42 and R61 did not previously exist, and were created by recombination.

Recovery of both class I and class II *ebgA*$^+$ alleles from these crosses is strong evidence that both class IV and class IV′ alleles are simply the sum of class I and a class II mutation in the *ebgA* gene.

Mapping the order of the class I and class II sites

To map the order of the class I and class II sites within the *ebgA* gene, strains A23 and 5A2GA were mated with strain S-J7. The mating mixture was plated onto lactose minimal medium containing sodium deoxycholate and streptomycin. The *tolC*$^+$ *strA*$^-$ recombinants that arose were replicated to MacConkey Gal-Ara plates to score for Gal-Ara$^-$ colonies, which would arise as the result of recombination between the class I and class II sites within the *ebgA* gene. In these crosses *all* of the Gal-Ara$^-$ recombinants are expected to be of that class whose site is closest to the *ebgR* gene. This is because *ebgR*$^+$ strains are lactose$^-$ regardless of the *ebgA* allele present (3) (see Fig. 1). In the A23 × SJ-7 cross an aliquot of the diluted mating mixture was also plated onto MacConkey lactose streptomycin plates to measure the recombination frequency between *ebgA* and *tolC* directly.

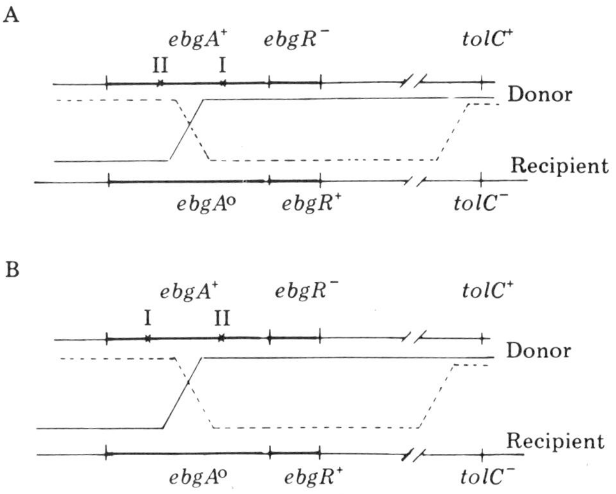

FIG. 1. Diagram of crosses between class IV *ebgA*$^+$ and *ebgA^o* alleles. The order *ebgA-ebgR-tolC* is as given in ref. 7 and confirmed by ref. 9. The distance between *ebgA* and *ebgR* is about 1.6% recombination (7). Colonies on the selective medium have received *tolC*$^+$ from the donor strain. A single crossover between *ebgR* and *tolC* results in an *ebgA^o ebgR*$^+$ strain, which is lactose-negative and thus does not form colonies on the selective medium (7). A single crossover between *ebgR* and *ebgA* results in an *ebgR*$^-$ *ebgA^o* strain, also lactose-negative (3). A single crossover to the left of *ebgA* yields a lactose-positive class IV or IV′ donor parental type, which will form colonies. A crossover between *tolC* and *ebgR* accompanied by a crossover within *ebgA* between the class I and class II sites (broken line) results in an *ebgR*$^+$ strain that cannot utilize lactose (3). A single crossover between the class I and class II sites within the *ebgA* gene (solid line) yields *ebgR*$^-$ nonparental lactose-positive recombinants. Alternative *A* shows that if the order is class II site-class I site-*ebgR*, all the nonparental colonies will be class I. Alternative *B* shows that if the order is class I site-class II site-*ebgR*, all nonparental colonies will be class II. Because of the tight linkage between *ebgA* and *ebgR* (7, 9) it was considered unlikely that double recombinants with one crossover between *ebgA* and *ebgR*, and the other crossover between the class I and the class II sites, would be detected. In fact no such double recombinants were observed.

From the cross between strain A23 and strain SJ-7, 1190 *tolC*$^+$, *ebgA*$^+$, *strA*$^-$ recombinants were recovered. Of these 18 were Gal-Ara$^-$, and all 18 were class I strains. This gives a recombination frequency of 1.5% between the class I and class II sites within the *ebgA* gene, and the order is class II site-class I site-*ebgR* gene. In that cross the recombination frequency between *ebgA* and *tolC* was 34.4%. Hartl and Dykhuizen (9) have carefully mapped the region around *ebgA*, and they estimate the distance between *ebgA* and *tolC* to be 0.6 min, or about 24.6 kilobases of DNA (6). If 34.4% recombination is equivalent to 24.6 kilobases, then 1.5% recombination is equivalent to about 1 kilobase.

That distance and order were confirmed by the cross between strain 5A2GA and SJ-7. Of 235 *tolC*$^+$, *ebgA*$^+$, *strA*$^-$ recombinants, 3, or 1.3%, were Gal-Ara$^-$. All three were class I strains.

Generation of Gal-Ara-utilizing mutants by recombination

If the class I and class II sites are about 1 kilobase apart, it should be relatively easy to obtain Gal-Ara$^+$ mutants from crosses between class I and class II *ebgA*$^+$ strains. Such recombinants would be expected to exhibit growth rates typical of class IV strains.

Two particular class I and two class II alleles were chosen because the enzymes specified by these alleles had been previously purified, characterized, and shown to have different

Table 3. Generation of Gal-Ara$^+$ strains by intragenic recombination

Donor	Recipient: Strain SJ480T *ebgA168* Class I		Recipient: Strain SJ8T *ebgA2* Class I		Recipient: Strain SJ12T *ebgA52* Class II	
Strain 5A11	*thyA*$^+$	13,700	*thyA*$^+$	10,600	*thyA*$^+$	4800
ebgA51	Gal-Ara$^+$	132	Gal-Ara$^+$	106	Gal-Ara$^+$	0
Class II	Recombination freq. 0.96%		Recombination freq. 1.0%			
	Recombinant strain name: RT51168		Recombinant strain name: RT512			
Strain 5A2	*thyA*$^+$	11,300	*thyA*$^+$	11,800		
ebgA52	Gal-Ara$^+$	110	Gal-Ara$^+$	118		
Class II	Recombination freq. 0.97%		Recombination freq. 1.0%			
	Recombinant strain name: RT52168		Recombinant strain name: RT522			
Strain A2	*thyA*$^+$	12,000				
ebgA2	Gal-Ara$^+$	0				
Class I						

catalytic constants (K_m and V_{max}) with respect to hydrolysis of lactose and *o*-nitrophenyl *β*-D-galactoside (2, 4). Three of the *ebgA*$^+$ alleles were introduced into F$^-$ strains that were subsequently made *thyA*$^-$. In each cross shown in Table 3 an Hfr strain carrying one *ebgA*$^+$ allele was crossed with an F$^-$ *thyA*$^-$ strain carrying a different *ebgA*$^+$ allele (*thyA* is distal to *ebgA* in these crosses). The mating mixture was plated onto Gal-Ara minimal agar containing arginine, methionine, and streptomycin, but lacking thymidine, to select Gal-Ara$^+$ *thyA*$^+$ recombinants. Each colony must arise from a zygote that has received the *thyA*$^+$ gene from the donor and has also undergone recombination within the *ebgA* gene. To determine the number of *thyA*$^+$ zygotes present, a 1:10 dilution of the mating mixture was plated on glucose minimal medium containing arginine, methionine, and streptomycin but lacking thymidine.

Table 3 shows that all crosses of a class I with a class II allele yielded Gal-Ara$^+$ recombinants at a frequency of about 1%. On the other hand, crosses between two different class I alleles, or two different class II alleles, failed to yield any Gal-Ara+ recombinants. This may mean that double class I or double class II mutants are Gal-Ara$^-$, or it may mean that recombination is too rare to have been observed. Twenty-five Gal-Ara$^+$ recombinants were picked from each cross and tested on MacConkey Gal-Ara plates. All were confirmed to be Gal-Ara$^+$ by this test. One recombinant from each cross was saved, and, in order to facilitate comparisons, the *ebgA*$^+$ gene was transduced into strain SJ-20. The growth rates of these isogenic strains (designated RT) are shown in Table 2. The results clearly demonstrate that the class IV phenotype can arise from recombination between a class I *ebgA*$^+$ gene and a class II *ebgA*$^+$ gene.

Complementation between class I and class II

F′ episomes were constructed that carried the class I allele *ebgA2* and the class II allele *ebgA52*. Details of the construction of these episomes will be presented elsewhere. The F′ *ebgA2* was introduced into an F$^-$ *recA*$^-$ strain carrying *ebgA52*, and the F′ *ebgA52* was introduced into an F$^-$ *recA*$^-$ strain carrying *ebgA2*. In neither case did the diploid strain grow at a rate expected for a class IV strain (Table 2). These results suggest that although the class IV phenotype can arise by recombination it is not generated by complementation.

DISCUSSION

The results presented here show that significant growth on Gal-Ara arises as a consequence of two sequential mutations within the *ebgA* gene. One of the mutations must be a class I mutation, which by itself permits utilization of lactose but not lactulose. The other mutation must be a class II mutation, which by itself permits utilization of both lactose and lactulose. The two mutations may be selected in either order, depending upon the selection pressure applied.

Two lines of evidence show that class IV alleles are simply the sum of a class I and a class II mutation: (*i*) Both class I and class II alleles can be recovered as a result of recombination between a class IV allele and a wild-type (*ebgA*^{o}) allele. This is independent of the method of selection of the class IV allele. (*ii*) Class IV alleles can be generated by recombination between a class I allele and a class II allele.

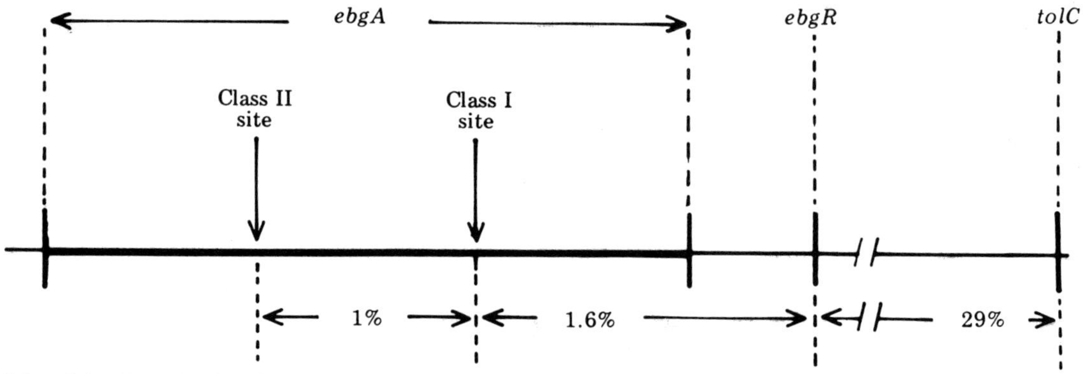

FIG. 2. Map of the *ebg* region based upon data presented here and in ref. 7. Distances are given in percent recombination. The class I site is placed 1.6% recombination from *ebgR* because the original mapping of *ebgR* with respect to *ebgA* employed the class I allele *ebgA2* (7).

The class I and class II sites within the *ebgA* gene were ordered with respect to the *ebgR* gene. The distance between the sites is about 650–1000 nucleotides, based upon recombination frequencies between 1% and 1.5%. As estimated on the basis of a subunit molecular weight of 120,000 for ebg enzyme (2), the sites are separated by 25–35% of the length of the gene. Fig. 2 shows a map of the *ebg* region.

These results show that a new enzymatic function, Gal-Ara hydrolysis, can evolve via recombination within the *ebgA* gene. Class I and class II *ebgA*$^+$ strains grow extremely slowly, and often not at all, on Gal-Ara as a sole carbon and energy source (Table 2). Because Gal-Ara-utilizing strains can evolve via recombination in the laboratory, we suggest that the same thing may happen in nature. We can envision a situation in which different alleles of the same gene have, under different selective pressures, diverged to give enzymes with somewhat different substrate specificities. Remixing of the populations would afford opportunities for intragenic recombination to generate a new allele with a substrate specificity present in neither parent. The new substrate specificity could permit the recombinant to exploit resources, or a new ecological niche, unavailable to either parent. Mechanisms of this sort at the intergenic recombination level have been discussed as mechanisms of speciation (10), but here we have directly demonstrated the evolution of a new metabolic function by intragenic recombination.

The existence of these alleles derived by recombination within the *ebgA* gene provides a unique opportunity for studying and comparing *functional* enzymes that carry different substitutions singly and together in the same molecule. We can, for instance, compare the wild-type enzyme ebg^o with the class I enzyme ebg^{+a}, with the class II enzyme ebg^{+b}, and with the recombinant class IV enzyme ebg^{+ab}. These studies should be of particular interest in that all of these mutations act to increase, rather than to impair, the activity of the enzyme.

This work was supported by Grant AI 14766 from the National Institute of Allergy and Infectious Diseases.

1. Hall, B. G. & Hartl, D. L. (1974) *Genetics* **76,** 391–400.
2. Hall, B. G. (1976) *J. Mol. Biol.* **107,** 71–84.
3. Hall, B. G. & Clarke, N. D. (1977) *Genetics* **85,** 193–201.
4. Hall, B. G. (1978) *Genetics* **89,** 453–465.
5. Hartl, D. L. & Hall, B. G. (1974) *Nature (London)* **248,** 152–153.
6. Bachman, B. J., Low, K. B. & Taylor, A. L. (1976) *Bacteriol. Rev.* **40,** 116–167.
7. Hall, B. G. & Hartl, D. L. (1975) *Genetics* **81,** 427–435.
8. Hall, B. G. (1977) *J. Bacteriol.* **129,** 540–543.
9. Hartl, D. L. & Dykhuizen, D. (1979) *Genetics* **91,** S45 (abstr.).
10. Lewontin, R. C. (1974) *The Genetic Basis of Evolutionary Change* (Columbia University Press, New York).

PROBLEMS

1. Define the following: historical reconstruction; acid protease; microevolutionary studies; leghemoglobin; fixed heterozygosity; evolved beta-galactosidase; ferredoxin; glutamate dehydrogenase; collagen; mononucleotide binding fold (mnbf); serine protease.

2. Using the data presented by Engel (1973, this chapter), determine how many amino acid identities there are in each of the following pairs of sequences:

   (a) G3PDH: 198-221, GDH: 112-135
   (b) G3PDH: 198-221, GDH: 267-290
   (c) GDH : 112-135, GDH: 112-135

   If these were random sequences of 23 residues, what would be the probability of observing these numbers of identities? Taking these differences in amino acid sequence at face value, what might you be tempted to say about the relative antiquity of the divergence of the two internal GDH sequences and the G3PDH sequence? Now, consult Rossman *et al.* (1974) and prepare a short, constructive critique of the Engel paper.

3. Eaton (1980) has pointed out that in the globin molecule there is a striking partition of function between the blocks of amino acids coded for by the three globin gene exons, with the middle exon being involved with heme binding, alpha (1)-beta (2) contacts, and cooperativity between dimers. The amino acids contributing to the other structural and functional refinements of the hemoglobin molecule encoded in the two "outside" exons. Argos and Rossman (1980) have further pointed out that *Pseudomonas aeruginosa* cytochrome c551 seems entirely involved in heme binding while bovine cytochrome b5 seems to have extra N- and C-terminal regions, the latter being involved with membrane binding. After looking again at Almassy and Dickerson (1978, Chapter 4), suggest two possible evolutionary pathways for bovine cytochrome b5 and a piece of evidence that might help you decide between these possibilities.

4. In the collagens so far analyzed (of which a number of chemically distinct species exist), the triple helix is formed by a staggered overlapping of three molecules at an interval of exactly 234 amino acids, denoted D. The basic repeating unit in all collagen is Gly-X-Y, but higher-order repeats also exist, the most interesting of which is one occurring at a frequency of D/13 (Hofmann *et al.*, 1980). Interestingly, this value corresponds exactly to the most common collagen exon size of 54 base pairs found in the gene of at least one of the collagens (Yamada *et al.*, 1981). If a detailed comparison of the structure of the repeats in several different collagens is made, homologous repeats in different collagens are found to be more similar than are the different repeats in the same collagen. On the basis of this information, suggest an evolutionary history for collagen starting from a primordial nonanucleotide (9 base pairs) sequence.

5. One of the criticisms that has been directed towards the stepwise retrograde scheme of Horowitz (1945, Chapter 1) for the development of metabolic pathways has been that often different classes of enzymes (for example, kinases versus isomerases) follow one another in actual pathways, and thus are presumed to be catalyzed by quite unrelated enzymes. In what way do the results of Levine *et al.* (1978, this chapter) reduce the strength of this argument?

6. Given that domain stability is one of the criteria by which a protein is likely to be judged by natural selection, it is perhaps not too surprising that a limited number of beta-sheet topologies are found in nature (Richardson, 1977). In comparing the tertiary structure of proteins with similar beta-sheet topologies but very different functions, she concludes that they are not necessarily evolutionarily related and that ".....it seems likely from overall statistics that the bulk of them are coincidences produced by the fact that the number of highly favourable chain-folding patterns is actually quite limited." Can you suggest an alternative evolutionary interpretation and list the kind of evidence that might bear on your alternative interpretation?

7. Given the opportunism of evolution alluded

to by Jacob in the quote at the beginning of this chapter and also from what you know about the pentitol-metabolizing system described in Hartley *et al.*, (1976, this chapter), list the kinds of mutations that might be helpful, either alone or in combination, in overcoming the difficulty of growing *Klebsiella aerogenes* with xylitol as a carbon source.

8. Specify and justify each step that was required in order to derive a strain of *Escherichia coli*, deprived of a functional *lacZ* gene, that was capable of metabolizing the disaccharide galactosylarabinose (Hall and Zuzel, 1980, this chapter).

9. The presence of duplicate, but functionally diverged genes, such as *ebg* in *E. coli* was an unexpected finding. Can you suggest a general method for examining *E. coli* for other similar duplicated and functionally altered or nonexpressed genes?

10. By pursuit of the appropriate references given in the introduction to this chapter and in the reprinted papers, assemble as complete a listing as possible of cases in which a single (or a defined small number of) amino acid substitutions appear to bring about an important quantum change in the function of a protein (not simply destruction of a function). What do you consider to be the minimal evidence required to demonstrate a mutationally altered enzyme function? What do these observations say about the early belief that the appearance of a new function should be preceded by duplication and a silent (pseudogene) phase?

6 THE REGULATION OF GENE EXPRESSION AND ORGANISMAL EVOLUTION

The fact that the macromolecules of most important structural genes have remained so similar, from bacteria to the highest organism, can be much better understood if we ascribe to the regulatory genes a major role in evolution. Since they strongly affect the viability of the individual they will be major targets of natural selection.... The day will come when much of the population genetics will have to be rewritten in terms of the interaction between regulator and structural genes.
E. Mayr, Populations, Species and Evolution (1970, p. 183)

The development of markedly different forms of animals and plants from ancestral stocks has been a major problem for our understanding of evolution. Darwin titled his major work *The Origin of Species* but his deepest interest was in the evolution of those larger changes of type that separate genera and higher taxonomic groupings. These changes are those designated as transspecific evolution (Dobzhansky *et al.*, 1977) or macroevolution (Goldschmidt, 1940; Simpson, 1944). Darwin himself felt that the emergence of different higher taxa was explicable as the cumulative effect of natural selection acting over long periods of time on small heritable differences ("individual variations") in natural populations. His belief in the efficacy of small individual differences as the source material for all biological evolution was not, as we have noted, shared by all evolutionists and led to some of the principal evolutionary disputes of the time.

In retrospect, it is not difficult to see why the debate between gradualists (Darwinians) and saltationists was unresolvable during the nineteenth century. Macroevolution often entails the evolutionary change in developmental programs over periods of time (whether long or short continues to be a contentious issue). Three requisite components for a successful macroevolutionary theory must be:

1. an understanding of genetic inheritance,

2. a theory of the genetic programming of development (the timing and sequence of gene expression),

3. an understanding of how functional novelty is introduced into developmental programs.

Early evolutionary theory lacked all three elements.

Seen in this light, it is not surprising that the rediscovery of Mendelian genetics was insufficient to resolve the problem. Indeed, the renewal of experimental genetics in this century deepened the divisions between saltationists and gradualists, because the work of the early followers of Mendel concentrated on readily detectable genetic differences to the exclusion of many other heritable characters. On the one hand, the experimental geneticists and paleontologists generally favored non-Darwinian, saltationist, or other modes of macroevolution, while, on the other hand, naturalist-systematists, with their attention to fine-grained intrapopulational differences, primarily supported Darwinian gradualism (Mayr and Provine, 1980). It was only with the emergence of the contemporary synthetic theory of evolution in the 1930s and 1940s that a consensus was finally achieved, and it was an agreement framed in twentieth-century ideas, but very close to the solution of Darwin: that the processes of microevolution--natural selection acting on small differences within populations--are sufficient to account for the evolutionary divergence of species and higher-order taxa. No appeal to unknown or metaphysical forces was felt to be required. Nevertheless, it was a consensus achieved largely without the participation of those who potentially had much to contribute on this issue--namely the embryologists and developmental biologists (see the discussion by Hamburger in Mayr and Provine, 1980).

In the last decade, the question of whether macroevolutionary trends can be explained by microevolutionary processes has again come to the fore, as evolutionists have become more interested in developmental mechanisms and the ways in which new forms and structures might result from changes in the genetic programming of development (Stebbins, 1973, 1974; Valentine and Campbell, 1975; Wallace, 1975; Gould, 1977; Stanley, 1979). This renewal of interest has accompanied and been stimulated by the rise of molecular biology and the consequent precision that can be brought to the analysis of the mechanisms of developmental programming in complex organisms. The critical scientific developments that have provided the impetus for this re-evaluation have been the formulation (and confirmation) of precise genetic regulatory models in prokaryotes (Jacob and Monod, 1961) and the elucidation of the characteristics of eukaryotic DNA and RNA sequences (Britten and Kohne, 1968, Chapter 1; Davidson *et al.*, 1975, Chapter 3). The latter studies have prompted various modifications of the prokaryotic regulatory models to fit the organizational complexities of the eukaryotic genome (Britten and Davidson, 1969; Davidson and Britten, 1979). Although we still lack a general theory of biological development, it is possible to speculate with increased confidence about the genetic basis of development and the effects of genetic change on developmental processes. The new consensus that is emerging is that regulatory gene changes have played a key role in macroevolutionary divergence (King and Wilson, 1975, this chapter), as they no doubt have in subspecific variability as well.

In thinking about regulatory mutations and macroevolution, it is helpful to consider the kinds of developmental transformations that require explanation. One major class of morphological change seems to result from the retardation or acceleration of relative growth rates during development, the phenomenon of heterochrony (de Beer, 1930; Huxley, 1942; Gould, 1977). These changes of rate can affect either the embryo as a whole or particular regions or primordia relative to others. The potential importance of such changes in evolution is attested by the dramatic, recognizable changes of molluscan shell-type that can be mimicked by computer simulations of differential growth rate changes (Raup and Michelson, 1965) and the simulated transformations between chimpanzee and human skulls that can be achieved in the same way (Sneath, 1967). Such growth-rate changes can explain the comparable geometrical transformations between different but related morphological groups, first noted by D'Arcy Thompson (1917).

The genetic changes responsible for such differential growth changes are unknown. At the morphological level, they can certainly be classified as regulatory, but it is not certain that all such changes include changes in gene expression, as opposed to changes in structural genes for particular growth-regulating proteins. Nevertheless, the general changes affecting many organ systems and embryonic fields seem likely to include altered programs of gene expression. Differential changes within individual embryonic fields (e.g. limb buds) often reflect different cellular readings of positional information (Wolpert, 1969), but this process may also entail changes in gene expression. In any event, it is important to be cautious when equating morphological regulatory change to regulatory changes at the genetic level.

The second major group of morphological changes within evolutionary series may be described as pattern-element changes--those affecting the numbers or arrangements of elements within standard patterns (for example, numbers of segments within arthropod bodies, numbers or placement of digits or bristles). In plants, certain single Mendelian genes are known that produce striking changes of form by changing the pattern of meristematic growth (Stebbins, 1973). In one simple animal group (nematodes) the importance of key regulatory cells in the morphogenesis of organ systems has been documented, and the possibility of producing saltatory evolution in morphogenesis through mutational alteration of the behaviour of the critical cells has been discussed by Sternberg and Horvitz (1981). In insects many gene mutations ("homoeotic" mutations) are known that markedly alter development through precisely regulated transformations of body segments (Lewis, 1978; Garcia-Bellido, 1977; Marx, 1981). Several of these genes act within compartments of imaginal disc primordia; the mutants appear to be defective in their cellular interpretation of positional information within certain embryonic fields, forcing the mobilization of a suite of genes normally expressed in other regions (Garcia-Bellido, 1977). Such "selector genes" may well be the eukaryotic analogues of prokaryotic regulator genes.

Developmental genetics, informed by the advances in molecular biology, is beginning to identify the kinds of genetic change that might play a role in macroevolutionary change. In addition, comparative macromolecular analyses are beginning to furnish the evidence that regulatory gene changes have played a major role in macroevolution. One principal kind of evidence concerns the rate of structural gene divergence in different groups of animals--as estimated from protein sequence differences relative to the rate of morphological change. These findings lead to the conclusion that the extent of structural gene divergence is not related in a fixed way to the rate of morphological change, though it is possible that the wrong comparisons are being made. Many changes in the sequence of structural genes may be neutral (see Chapter 4), so perhaps the rate of functional changes in proteins should be compared with the rate of morphological change. The existence of the molecular clock permits the earlier conclusion to be drawn: the rate of gene-sequence change has been essentially the same in evolutionarily conservative animals as it has been in more rapidly diversifying groups (Wilson *et al.* 1977). One such explicit comparison, using the albumin molecular clock, has been presented by Wallace *et al.* (1971), who compared a slowly diversifying amphibian group (frogs) with a rapidly diversifying group (mammals). King and Wilson (1975, this chapter), have examined the rates of structural gene divergence in chimpanzees and humans. These two species show over 99 percent sequence homology of selected proteins, yet, from our point of view, exhibit many developmental and morphological differences. Thus, the divergence of chimpanzees and humans must have involved a relatively small proportion of the genome, presumably those genes that regulate developmental sequences and, in particular, those that govern differential growth rates (Sneath, 1967).

The general importance of regulatory gene changes as a factor in mammalian divergence is further supported by studies on hybrid viability. The ability of two related organisms to form viable hybrids is often taken as a measure of the compatibility of their genetic regulatory systems: mammals show a much lower incidence of viable hybrid production--as a function of their time of divergence--than a reference group of slowly diverging vertebrates, the frogs (Wilson *et al.*, 1974a).

The extent of genetic variation of

regulatory elements within populations and the potential selective significance of this variation for macroevolution are two unresolved issues. The matter has been approached from two different perspectives, those of paleontology and of molecular biology. On paleontological grounds, it has been argued that normal intrapopulation genetic variability has played only a small role in macroevolution and that this process occurs in "bursts of speciation" possibly associated with karyotypic changes that are a source of genetic regulatory change through position effects (Eldredge and Gould, 1972; Stanley, 1979). Although karyotypic change has been implicated as one major source of genome regulatory change (Wilson *et al.*, 1974b), the correlation between rate or extent of karyotype change and organismal divergence is far from perfect (see White, 1978 and problem 8, this chapter). The alternative molecular biological view, which is in the mainstream of the evolutionary synthesis, is that the rate of evolution associated with regulatory gene change is a function of the normal intrapopulation variability of these genes. Dickinson (1980, this chapter), presents an experimental investigation of the nature and extent of regulatory gene variation within and between populations of the Hawaiian picture-wing *Drosophila* species, which have undergone an unusually rapid and extensive diversification. Dickinson documents the existence within these species of a high level of regulatory gene variation for five enzyme systems. The possible biological and evolutionary significance of these activity-level differences are discussed at the end of his paper. The results show that in this case considerable regulatory gene variation is present. This study and others of its kind may be the beginning of a whole new chapter in the modern evolutionary synthesis--a population genetics of regulatory and eventually developmental gene differences, a field anticipated by Mayr over a decade ago (see introductory quote to this chapter).

The final paper in this chapter deals with a topic that has received comparatively little attention--the evolution of regulation itself and, in particular, the change in regulatory networks during the evolution first of single and then of multi cellular eukaryotic cells from prokaryote-grade precursors. The paper by Tomkins (1975) is necessarily speculative but of great potential significance in view of the explosion of evolutionary diversity that followed the formation of eukaryotic cells. He discusses possible relationships between general physiological regulatory mechanisms and the evolution of intercellular communication, and stresses the potential importance of certain small ubiquitous molecules in the evolution of new regulatory relationships. These small regulatory molecules are very different from the macromolecules that have been the focus of attention in this book in that they leave no direct trace in the genetic record. However, since these molecules are important as mediators of various key intermolecular and intercellular reactions, their structure is strongly conserved and they may thus be considered to be molecular fossils in their own right (Zuckerkandl and Pauling, 1965; White, 1976). The evolutionary importance of such cofactors and coeffectors is discussed further in Eck and Dayhoff (1966a, Chapter 1); an illustration of the Tomkins thesis that small molecules present in simple cells may be employed in increasingly sophisticated ways during evolutionary advance is provided by the case of insulin (Le Roith, *et al.*, 1980).

An account of macroevolution must include some explanation of the selective forces that promote these evolutionary changes (Waddington, 1957; Stebbins, 1973; 1974; Gould, 1977). Each case obviously possesses its own special features, but one general principle has been emphasized: that often what is selected is not the final adult morphological form, but an initial developmental flexibility to respond to certain environmental challenges (Waddington, 1957; Stebbins, 1973), the end result of which is morphological change. Individuals with new developmental capabilities can be selected; with repeated selection for the appropriate modifiers the response may become canalized and assimilated, producing the new developmental pathway in the absence of the initial evocatory stimulus (Waddington, 1957). A particularly significant analysis of the importance of selection in macroevolution, from the point of view of population genetics, is embodied in the "shifting balance theory" of Wright; the most recent statement of this theory can be found in Wright (1982).

Since the number of unknowns and imponderables in macroevolution is large and those adaptive shifts that produce evolutionary novelty (Stebbins, 1974) are rare, it is safe to predict that many of the questions that intrigued Darwin and his contemporaries will continue to be of interest, and dispute, for some years to come. Furthermore, this field will also yield insights into the genetic basis of microevolutionary change, those smaller shifts of development, form, and behavior that contribute to species divergence in complex multi-cellular organisms.

Reprinted from *Science* **188**: 107–116.

Evolution at Two Levels in Humans and Chimpanzees

Their macromolecules are so alike that regulatory mutations may account for their biological differences.

Mary-Claire King and A. C. Wilson

Soon after the expansion of molecular biology in the 1950's, it became evident that by comparing the proteins and nucleic acids of one species with those of another, one could hope to obtain a quantitative and objective estimate of the "genetic distance" between species. Until then, there was no common yardstick for measuring the degree of genetic difference among species. The characters used to distinguish among bacterial species, for example, were entirely different from those used for distinguishing among mammals. The hope was to use molecular biology to measure the differences in the DNA base sequences of various species. This would be the common yardstick for studies of organismal diversity.

During the past decade, many workers have participated in the development and application of biochemical methods for estimating genetic distance. These methods include the comparison of proteins by electrophoretic, immunological, and sequencing techniques, as well as the comparison of nucleic acids by annealing techniques. The only two species which have been compared by all of these methods are chimpanzees (*Pan troglodytes*) and humans (*Homo sapiens*). This pair of species is also unique because of the thoroughness with which they have been compared at the organismal level—that is, at the level of anatomy, physiology, behavior, and ecology. A good opportunity is therefore presented for finding out whether the molecular and organismal estimates of distance agree.

Dr. King, formerly a graduate student in the Departments of Genetics and Biochemistry, University of California, Berkeley, is now a research geneticist at the Hooper Foundation and Department of International Health, University of California, San Francisco 94143. Dr. Wilson is a professor of biochemistry at the University of California, Berkeley 94720.

The intriguing result, documented in this article, is that all the biochemical methods agree in showing that the genetic distance between humans and the chimpanzee is probably too small to account for their substantial organismal differences.

Indications of such a paradox already existed long ago. By 1963, it appeared that some of the blood proteins of humans were virtually identical in amino acid sequence with those of apes such as the chimpanzee or gorilla (*1*). In the intervening years, comparisons between humans and chimpanzees were made with many additional proteins and with DNA. These results, reported herein, are consistent with the early results. Moreover, they tell us that the genes of the human and the chimpanzee are as similar as those of sibling species of other organisms (*2*). So, the paradox remains. In order to explain how species which have such similar genes can differ so substantially in anatomy and way of life, we review evidence concerning the molecular basis of evolution at the organismal level. We suggest that evolutionary changes in anatomy and way of life are more often based on changes in the mechanisms controlling the expression of genes than on sequence changes in proteins. We therefore propose that regulatory mutations account for the major biological differences between humans and chimpanzees.

Similarity of Human and Chimpanzee Genes

To compare human and chimpanzee genes, one compares either homologous proteins or nucleic acids. At the protein level, one way of measuring the degree of genetic similarity of two taxa is to determine the average number of amino acid differences between homologous polypeptides from each population. The most direct method for determining this difference is to compare the amino acid sequences of the homologous proteins. A second method is microcomplement fixation, which provides immunological distances linearly correlated with amino acid sequence difference. A third method is electrophoresis, which is useful in analyzing taxa sufficiently closely related that they share many alleles. For the human-chimpanzee comparison all three methods are appropriate, and thus many human and chimpanzee proteins have now been compared by each method. We can therefore estimate the degree of genetic similarity between humans and chimpanzees by each of these techniques.

Sequence and immunological comparisons of proteins. During the last decade, amino acid sequence studies have been published on several human and chimpanzee proteins. As Table 1 indicates, the two species seem to have identical fibrinopeptides (*3*), cytochromes c (*4*), and hemoglobin chains [alpha (*4*), beta (*4*), and gamma (*5, 6*)]. The structural genes for these proteins may therefore be identical in humans and chimpanzees. In other cases, for example, myoglobin (*7*) and the delta chain of hemoglobin (*5, 8*), the human polypeptide chain differs from that of the chimpanzee by a single amino acid replacement. The amino acid replacement in each case is consistent with a single base replacement in the corresponding structural gene.

Owing to the limitations of conventional sequencing methods, exactly comparable information is not available for larger proteins. Indeed, the sequence information available for the proteins already mentioned is not yet complete. By applying the microcomplement fixation method to large proteins, however, one can obtain an approximate measure of the degree of amino acid sequence difference between related proteins (*9*). This method indicates that the sequences of human and chimpanzee albumins (*10*), transferrins (*11*), and carbonic anhydrases (*4, 12*) differ slightly, but that lysozyme (*13*) is identical in the two species (Table 1) (*14*). Based on the proteins listed in Table 1, the average degree of difference between human and chimpanzee proteins is

$$\frac{19 \times 1000}{2633} = 7.2 \qquad (1)$$

amino acid substitutions per 1000 sites. That is, the sequences of human and chimpanzee polypeptides examined to date are, on the average, more than 99 percent identical.

Electrophoretic comparison of proteins. Electrophoresis can provide an independent estimate of the average amino acid sequence difference between closely related species. We have compared the human and chimpanzee polypeptide products of 44 different structural genes. Table 2 indicates the allelic frequencies and the estimated probability of identity at each locus. The symbol S_i represents the probability that human and chimpanzee alleles will be electrophoretically identical at a particular locus *i*, or

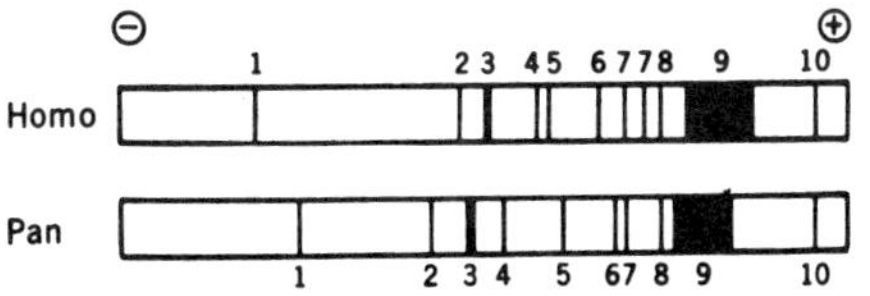

Fig. 1. Separation of human and chimpanzee plasma proteins by acrylamide electrophoresis at *p*H 8.9. The proteins are: 1, α_2-macroglobulin; 2, third component of complement; 3, transferrin; 4, haptoglobin; 5, ceruloplasmin; 6, α_{2HS}-glycoprotein; 7, Gc-globulins; 8, α_1-antitrypsin; 9, albumin; and 10, α_1-acid glycoprotein. The chimpanzee plasma has transferrin genotype *Pan* CC; the human plasma has transferrin genotype *Homo* CC and haptoglobin genotype 1-1. The direction of migration is from left to right.

$$S_i = \sum_{j=1}^{A_i} x_{ij} y_{ij} \qquad (2)$$

where x_{ij} is the frequency of the *j*th allele at the *i*th locus in human populations, and y_{ij} the frequency of the *j*th allele at the *i*th locus in chimpanzee populations for all A_i alleles at that locus. For example, Table 2 indicates the frequencies of the three alleles (AP^a, AP^b, and AP^c) found at the acid phosphatase locus for human and chimpanzee populations. The probability of identity of human and chimpanzee alleles at this locus, that is, S_i is $(0.29 \times 0) + (0.68 \times 1.00) + (0.03 \times 0)$, or 0.68.

Of the loci in Table 2, 31 code for intracellular proteins; 13 code for secreted or extracellular proteins. In general, the intracellular proteins were analyzed by starch gel electrophoresis of red blood cell lysates, with the buffer systems indicated in the table and stains specific for the enzymatic activity of each protein. For a few intracellular proteins (cytochrome c, the hemoglobin chains, and myoglobin), amino acid sequences have been published for both species, so that direct sequence comparison is also possible.

Most of the secreted proteins were compared by acrylamide gel electrophoresis of human and chimpanzee plasma (*15*). The electrode chamber contained tris(hydroxymethyl)aminomethane (tris) borate buffer, *p*H 8.9; acrylamide gel slabs were made with tris-sulfate buffer, *p*H 8.9. Gels were stained with amido black, a general protein dye. The identification of bands on a gel stained with this dye poses a problem, since it is not obvious, particularly for less concentrated proteins, which protein each band represents. We determined the electrophoretic mobilities of the plasma proteins by applying the same sample to several slots of the same gel, staining the outside columns, and cutting horizontal slices across the unstained portion of the gel at the position of each band. The protein was eluted separately from each band in 0.1 to 0.2 milliliter of an appropriate isotonic tris buffer (*9*) and tested for reactivity with a series of rabbit antiserums, each specific for a particular human plasma protein, by means of immunoelectrophoresis and immunodiffusion in agar (*15, 16*). The results of this analysis are shown in Fig. 1.

Some of the secreted proteins were compared by means of other electrophoretic methods as well. Albumin and transferrin were surveyed by cellulose acetate electrophoresis; and α_1-antitrypsin, Gc-globulin (group-specific component), the haptoglobin chains, lysozyme, and plasma cholinesterase were analyzed on starch gels, with the buffers indicated in Table 2.

The results of all electrophoretic comparisons are summarized in Fig. 2. About half of the proteins in this survey are electrophoretically identical for the two species, and about half of them are different. Only a few loci are highly polymorphic in both species (see *17*).

The proportion of alleles at an "average" locus that are electrophoretically identical in human and chimpanzee populations can be calculated from Table 2 and Eq. 3, where *L* is the number of loci observed:

$$\bar{S} = \frac{1}{L}(S_1 + S_2 + \ldots + S_L) = 0.52 \qquad (3)$$

Table 1. Differences in amino acid sequences of human and chimpanzee polypeptides. Lysozyme, carbonic anhydrase, albumin, and transferrin have been compared immunologically by the microcomplement fixation technique. Amino acid sequences have been determined for the other proteins. Numbers in parentheses indicate references for each protein.

Protein	Amino acid differences	Amino acid sites
Fibrinopeptides A and B (*3*)	0	30
Cytochrome c (*4*)	0	104
Lysozyme (*13*)	~0	130
Hemoglobin α (*4*)	0	141
Hemoglobin β (*4*)	0	146
Hemoglobin $^A\gamma$ (*5, 6*)	0	146
Hemoglobin $^G\gamma$ (*5, 6*)	0	146
Hemoglobin δ (*5, 8*)	1	146
Myoglobin (*7*)	1	153
Carbonic anhydrase (*4, 12*)	~3	264
Serum albumin (*10*)	~6	580
Transferrin (*11*)	~8	647
Total	~19	2633

In other words, the probability that human and chimpanzee alleles will be electrophoretically identical at a particular locus is about one-half.

Agreement between electrophoresis and protein sequencing. The results of electrophoretic analysis can be used to estimate the average number of amino acid differences per polypeptide chain for humans and chimpanzees, for comparison with the estimate based on amino acid sequences and immunological data. To calculate the average amino acid sequence difference between human and chimpanzee proteins, we need first an estimate of the proportion (c) of amino acid substitutions detectable by electrophoresis. Electrophoretic techniques detect only amino acid substitutions that change the net charge of the protein observed. Four amino acid side chains are charged at *p*H 8.6: arginine, lysine, glutamic acid, and aspartic acid. The side chain of histidine is positively charged below approximately *p*H 6. The proportion of accepted point mutations that would be detectable by the buffer

Table 2. Electrophoretic comparison of chimpanzee and human proteins. In the first column, Enzyme Commission numbers are given in parentheses; N is the number of chimpanzees analyzed, both in this study and by other investigators. Abbreviations: MW, molecular weight; aa, amino acids; tris, tris(hydroxymethyl)aninomethane; EDTA, ethylenediaminetetraacetate. Secreted proteins differ more frequently for the two species than intracellular proteins (*93*).

Locus (i) and allele (j)	Allele frequency: Human* (x_{ij})	Allele frequency: Chimpanzee (y_{ij})	Probability of identity† (S_i)	Comments and references‡
			Intracellular proteins	
Acid phosphatase (3.1.3.2); $N = 86$				
AP^a	0.29	0	0.68	Red cells; 15,000 MW; 110 aa; citrate-phosphate, *p*H 5.9, starch electrophoresis (*54, 55*)
AP^b	0.68	1.00		
AP^c	0.03	0		
Adenosine deaminase (3.5.4.4); $N = 22$				
ADA^1	0.96	0	0	Red cells; 35,000 MW; 300 aa; chimpanzee protein faster on starch electrophoresis (*54*); polymorphism in human populations (*16*)
ADA^2	0.04	0		
$ADA^{ape=5}$	0	1.00		
Adenylate kinase (2.7.4.3); $N = 86$				
AK^1	0.98	1.00	0.98	Red cells; 21,500 MW; 190 aa; well buffer is citrate-NaOH, *p*H 7.0; gel buffer is histidine-NaOH, pH 7.0, starch electrophoresis (*54, 56, 57*)
AK^2	0.02	0		
Carbonic anhydrase 1 or B (4.2.1.1); $N = 111$	1.00	1.00	1.00	Red cells; 28,000 MW; 264 aa; well buffer is borate-NaOH, *p*H 8.0; gel buffer is borate-NaOH, *p*H 8.6, starch electrophoresis (*56, 58*)
Cytochrome c	1.00	1.00	1.00	Mitochondria; 12,400 MW; 104 aa; sequence identity based on amino acid analysis (*5*); possible heterogeneity in man (*59*)
Esterase A_1 (3.1.1.6); $N = 111$	1.00	0	0	Red cells; well buffer is lithium borate, *p*H 8.2; gel buffer is lithium-borate and tris-citrate, *p*H 7.3, starch electrophoresis (*58, 60*)
	0	1.00		
Esterase A_2§ (3.1.1.6); $N = 111$	1.00	Absent		See esterase A_1
Esterase A_3 (3.1.1.6); $N = 111$				
$EstA_3{}^a$	1.00	0	0	See esterase A_1
$EstA_3{}^b$	0	1.00		
Esterase B (3.1.1.1); $N = 111$	1.00	1.00	1.00	See esterase A_1
Glucose-6-phosphate dehydrogenase (1.1.1.49); $N = 86$				
Gd^A	0.01	0	0.99	Red cells; six subunits, each 43,000 MW; ~ 370 aa; phosphate, *p*H 7.0, starch electrophoresis (*56*); A and B variants identical by microcomplement fixation (*61*); sequences differ by one amino acid, aspartic acid in A variant, asparagine in B variant (*61*)
Gd^B	0.99	1.00		
Glutamate-oxalacetate transaminase (soluble form) (2.6.1.1); $N = 63$				
$sGOT^1$	1.00	0	0	Red cells; two subunits, each 50,000 MW; ~ 430 aa; tris-citrate, *p*H 7.0, starch electrophoresis (*62*); chimpanzee protein faster (*63*)
$sGOT^2$	0	1.00		
Glutathione reductase (1.6.4.2); $N = 64$				
GSR^2 and GSR^3	0.97	1.00	0.97	Red cells; tris-EDTA, *p*H 9.6, starch electrophoresis; polymorphism in human populations (*64*), possibly associated with gout; GSR^2 and GSR^3 not distinguishable at *p*H 9.6
GSR^5	0.01	0		
GSR^6	0.02	0		
Hemoglobin α chain; $N = 108$				
$Hb_\alpha{}^A$	1.00	0.99	0.99	Red cells; 15,100 MW; 141 aa; tris-glycine, *p*H 8.4, cellulose acetate electrophoresis (*15*); tryptic peptides of human and chimpanzee α chains identical (*65*); chimpanzee α chain variant is electrophoretically identical to human Hb^J (*66*)
$Hb_\alpha{}^J$	< 0.01	0.01		

(Table 2 is continued on pages 223 and 224

Locus (i) and allele (j)	Allele frequency: Human* (x_{ij})	Allele frequency: Chimpanzee (y_{ij})	Probability of identity† (S_i)	Comments and references‡
Hemoglobin β chain; $N = 108$				
Hb_{β}^{A}	0.99	0.99	0.98	Red cells; 16,000 MW; 146 aa; tris-glycine, *p*H 8.4, cellulose acetate electrophoresis (*15*); amino acid sequences of β^A chains identical (*65*); chimpanzee Hb^B electrophoretically identical to human Hb^S (*66*)
$Hb_{\beta}^{S=B}$	0.01	0.01		
Hemoglobin $^{A}\gamma$ chain	1.00	1.00	1.00	Fetal red cells; 16,000 MW; 146 aa; amino acid sequence of human and chimpanzee γ chains identical; $^{A}\gamma$ and $^{G}\gamma$ arep roducts of different structural genes, differ at residue 136; A, alanine; G, glycine (*67*)
Hemoglobin $^{G}\gamma$ chain	1.00	1.00	1.00	See hemoglobin $^{A}\gamma$
Hemoglobin δ chain	1.00	1.00	1.00	Red cells; 16,000 MW; 146 aa; human and chimpanzee electrophoretic mobilities identical, but one amino acid difference at position 125: humane δ, methionine; chimpanzee δ, valine (*8*)
Lactate dehydrogenase H (1.1.1.27); $N = 74$	1.00	1.00	1.00	Red cells; H and M subunits each 34,000 MW; 330 aa; citrate-phosphate, *p*H 6.0, starch electrophoresis (*69*); three intermediate bands of five-band, tetrameric electrophoretic pattern have different mobilities for humans and chimpanzees, because of difference in M polypeptide (*70*)
Lactate dehydrogenase M (1.1.1.27); $N = 74$				
ldh M^{a}	1.00	0	0	See lactate dehydrogenase H
ldh M^{b}	0	1.00		
Malate dehydrogenase (cytoplasmic) (1.1.1.37); $N = 88$	1.00	1.00	1.00	Red cells; two subunits, each 34,000 MW; 330 aa; see LDH for procedures; polymorphic in some human populations (*71*)
Methemglobin reductase (1.6.99); $N = 86$				
MR^1	1.00	0	0	Red cells; tris-citrate, *p*H 6.8, starch electrophoresis (*72*) distinguishes human and chimpanzee enzymes, no difference with tris-EDTA, *p*H 9.3, electrophoresis (*56, 73*)
MR^2	0	1.00		
Myoglobin	1.00	1.00	1.00	Muscle; 16,900 MW; 153 aa; tryptic and chymotryptic peptides of cyanmethemoglobin electrophoretically identical at *p*H 8.6 (*74*), but at position 116, human has glutamine, chimpanzee has histidine (*7*)
Peptidase A (3.4.3.2); $N = 63$				
$PepA^1$ and $PepA^8$	0.99	1.00	0.99	Red cells; two subunits, each 46,000 MW; ~ 400 aa; tris-maleate, *p*H 7.4 starch electrophoresis, leucyl-glycine substrate (*65*); $PepA^1$ and $PepA^8$ not distinguishable in red blood cell lysates (*75*)
$PepA^2$	0.01	0		
Peptidase C (3.4.3.2); $N = 63$				
$PepC^1$	0.99	1.00	0.99	Red cells; 65,000 MW; ~ 565 aa; see peptidase A for procedures; polymorphism in human populations (*76*)
$PepC^4$	0.01	0		
Phosphoglucomutase 1 ‖ (2.7.5.1); $N = 168$				
PGM_1^1	0.77	0.26	0.20	Red cells; subunits PGM_1 and PGM_2 each 62,000 MW; ~ 540 aa; tris-maleate-EDTA, *p*H 7.4, starch electrophoresis (*16, 55, 61, 77*)
PGM_1^2	0.23	0		
PGM_1^{Pan}	0	0.74		
Phosphoglucomutase 2 (2.7.5.1); $N = 168$				
PGM_2^1	1.00	1.00	1.00	See phosphoglucomutase 1
PGM_2^3	$\ll 0.01$	< 0.01		
6-Phosphogluconate dehydrogenase (1.1.1.44); $N = 86$				
PGD^A	0.96	0	0.04	Red cells; two subunits, each 40,000 MW; 350 aa; see G6PD for procedures; chimpanzee allele electrophoretically identical to human "Canning" variant (*55*)
PGD^C	0.04	1.00		
Phosphohexose isomerase (5.3.1.9); $N = 86$				
PHI^1	1.00	0	0	Red cells; two subunits, each 66,000 MW; 580 aa; tris-citrate, *p*H 8.0, starch electrophoresis (*56*); chimpanzee protein has slower mobility, both cathodally migrating (*78*)
PHI^B	0	1.00		
Superoxide dismutase A (indophenol oxidase) (1.15.1.1); $N = 64$	1.00	1.00	1.00	Red blood cells; two subunits, each 16,300 MW; 158 aa (*68*); see phosphoglucomutase for procedure
Triosephosphate isomerase A (5.3.1.1)	1.00	1.00	1.00	Fibroblasts; dimers 48,000 MW; each polypeptide 248 aa (*79*); β polypeptide found only in hominoids.
Triosephosphate isomerase B (5.3.1.1)	1.00	1.00	1.00	See triosephosphate isomerase A

Locus (i) and allele (j)	Allele frequency: Human* (x_{ij})	Allele frequency: Chimpanzee (y_{ij})	Probability of identity† (S_i)	Comments and references‡
		Secreted proteins		
α_1-Acid glycoprotein (orosomucoid); $N = 123$				
Or^S	0.32	0	0.68	Glycoprotein in plasma; carbohydrate > 50 percent; 44,100 MW; 181 aa; acrylamide electrophoresis, *p*H 8.9 (see text); polymorphism in human populations detectable at *p*H 2.9 (*80*); isoelectric point is 1.82 for human and chimpanzee proteins, but proteins differ by quantitative precipitin analysis (*81*)
Or^F	0.68	1.00		
Albumin; $N = 123$				
Alb^A	0	1.00	0	Plasma; 69,000 MW, ~ 580 aa; tris-citrate, *p*H 5.5, cellulose acetate electrophoresis; acrylamide electrophoresis, *p*H 8.9; chimpanzee protein slower mobility, immunological difference detected by microcomplement fixation (*10, 42*); rare polymorphic alleles in human populations (*82*)
Alb^{Pan}	1.00	0		
α_1-Antitrypsin; $N = 123$				
Pi^M	0.95	0	0	Plasma; 49,000 MW; ~ 380 aa; anodal well buffer is citrate-phosphate, *p*H 4.5; cathodal well buffer is borate-NaOH, *p*H 9.0; gel buffer is tris-citrate, *p*H 4.8; starch electrophoresis (*56*); acrylamide electrophoresis, *p*H 8.9; polymorphism in human populations (*83*)
Pi^F	0.03	0		
Pi^S	0.02	0		
Pi^{Pan}	0	1.00		
Ceruloplasmin; $N = 123$				
Cp^A and Cp^{Pan}	0.01	1.00	0.01	Plasma; eight subunits, each 17,000 MW; ~ 150 aa; acrylamide electrophoresis, *p*H 8.9; possible adaptive significance of polymorphism in human populations (*84*)
Cp^B	0.98	0		
Cp^C	0.01	0		
Third component of complement; $N = 123$				
$C'3^{1=F}$	0.12	0	0	Plasma; total MW 240,000; acrylamide electrophoresis, *p*H 8.9; polymorphism in human populations detectable by high voltage electrophoresis (*85*)
$C'3^{2=S}$	0.87	0		
$C'3^8$	0.01	0		
$C'3^{Pan}$	0	1.00		
Group-specific component; $N = 206$				
Gc^1	0.74	0	0	Plasma; two subunits, each 25,000 MW; ~ 220 aa; acrylamide electrophoresis, *p*H 8.9; human Gc 2-2 and chimpanzee protein similar on acrylamide, chimpanzee slightly faster on starch or immunoelectrophoresis (*86*)
Gc^2	0.26	0		
Gc^{Pan}	0	1.00		
α_{2HS}-Glycoprotein; $N = 123$				
Gly^A	1.00	0	0	Plasma; 49,000 MW; ~ 400 aa; acrylamide electrophoresis, *p*H 8.9 (*15*)
Gly^B	0	1.00		
Haptoglobin α chain; $N = 300$				
$Hp_\alpha{}^1$	0.36	0	0	Plasma; α^1 chain is 8,900 MW, 83 aa; α^2 chain is 16,000 MW, 142 aa; β chain is 36,000 to 40,000 MW; ~ 330 aa; acrylamide electrophoresis, *p*H 8.9; borate-NaOH well buffer and tris-citrate gel buffer, *p*H 8.6, starch electrophoresis (*56*); chimpanzee Hp shares six human Hp 1-1 and eight Hp 2-2 antigenic determinants; Hp^2 evolved since human-chimpanzee divergence (*87*)
$Hp_\alpha{}^2$	0.64	0		
$Hp_\alpha{}^{Pan}$	0	1.00		
Haptoglobin β chain; $N = 300$	1.00	1.00	1.00	See haptoglobin α chain
Lysozyme				
lzm^A	1.00	0	0	Milk; 14,400 MW; 130 aa; starch gel electrophoresis, *p*H 5.3 (*88*)
lzm^B	0	1.00		
α_2-Macroglobulin; $N = 123$				
Xm^A	1.00	0	0	Plasma; four subunits, each 196,000 MW; acrylamide electrophoresis, *p*H 8.9; X-linked antigenic polymorphism observed in human populations (*89*) but not detectable by electrophoresis; human and chimpanzee proteins immunologically indistinguishable (*14*)
Xm^B	0	1.00		
Plasma cholinesterase (3.1.1.8); $N = 111$				
$E_1{}^u$	1.00	0	0	Plasma; four subunits, each ~ 87,000 MW; see esterase A_1 for procedures; chimpanzee protein has four components with faster mobilities than analogous human components (*15*)
$E_1{}^{Pan}$	0	1.00		
Transferrin; $N = 133$				
Homo: Tf^C	0.99	0	0	Plasma; 73,000 to 92,000 MW; ~ 650 aa; acrylamide electrophoresis, *p*H 8.9; tris-glycine, *p*H 8.4, cellulose acetate electrophoresis (*77, 90*)
Tf^{D_1}	0.01	0		
Pan: Tf^A	0	0.08		
Tf^B	0	0.06		
Tf^C	0	0.70		
Tf^D	0	0.15		
Tf^E	0	0.02		

* Allelic frequencies for human populations are calculated from data summarized by Nei and Roychoudhury (*28*). Sample sizes generally greater than 1000. Only alleles with frequency > 0.01 are listed. The relative sizes of racial groups were estimated to be Caucasian, 45 percent; Black African, 10 percent; and Mongoloid-Amerind (combined), 45 percent. † See Eq. 2 in text. ‡Given in this column are: the tissue used, polypeptide chain length, electrophoretic conditions, and references to previous studies on people and chimpanzees. Genetic, population, and physiological studies of most human red cell and plasma proteins are summarized by Giblett (*56*) or Harris (*91*); studies of plasma proteins are summarized by Schultze and Heremans (*92*). References are for additional studies of chimpanzee or human proteins. § Not included in identity calculations. || Notation for the chimpanzee alleles at the PGM_1 locus differs in publishhed surveys. Ours is as follows: $PGM_1{}^{Pan}$ (which is chimpanzee $PGM_1{}^1$, of Goodman and co-workers and $PGM_1{}^{Pan}$ of Schmitt and co-workers) is the allele with slowest electrophoretic mobility; $PGM_1{}^1$ (which is human $PGM_1{}^1$, the chimpanzee $PGM_1{}^1$ of Schmitt, and the chimpanzee $PGM_1{}^2$ of Goodman) is intermediate; and $PGM_1{}^2$ (found only in human populations) has the fastest mobility.

systems used in this study is about 0.27 (*18*).

If we assume that, at a particular amino acid site on a given protein, amino acid substitutions have occurred (i) independently and (ii) at random with respect to species since the evolutionary divergence of humans and chimpanzees, then the number of proteins that have accumulated r amino acid substitutions since this divergence approximates a Poisson variate (*19*). That is, the probability that r substitutions have accumulated in a particular polypeptide is

$$P_r = \frac{(mc)^r\, e^{-mc}}{r!} \qquad (4)$$

where m is the expected number of amino acid substitutions per polypeptide (the mean of the Poisson distribution), and c is the proportion of those substitutions that are electrophoretically detectable. The probability that the polypeptides are electrophoretically identical (that is, that no electrophoretically detectable substitutions have occurred) is 0.52. Therefore,

$$P_0 = 0.52 = \frac{(mc)^0\, e^{-mc}}{0!} = e^{-mc} \qquad (5)$$

Thus $mc = 0.65$ and the expected number of amino acid differences per polypeptide is

$$m = 0.65/0.27 = 2.41 \qquad (6)$$

For comparative purposes, this value can also be expressed in terms of the expected number of amino acid differences per 1000 amino acids. The average number of amino acids per polypeptide for all the proteins analyzed electrophoretically is 293 ± 27 (standard error). Therefore the expected degree of amino acid difference between human and chimpanzee is

$$\frac{2.41 \times 1000}{293} = 8.2 \qquad (7)$$

substitutions per 1000 sites, with a range (within one standard error) of 7.5 to 9.1 differences per 1000 amino acids. The estimate based on amino acid sequencing and immunological comparisons (Eq. 1) agrees well with this estimate. Both estimates indicate that the average human protein is more than 99 percent identical in amino acid sequence to its chimpanzee homolog (*20*).

Comparison of nucleic acids. Another method of comparing genomes is nucleic acid hybridization. Several workers have compared the thermostability

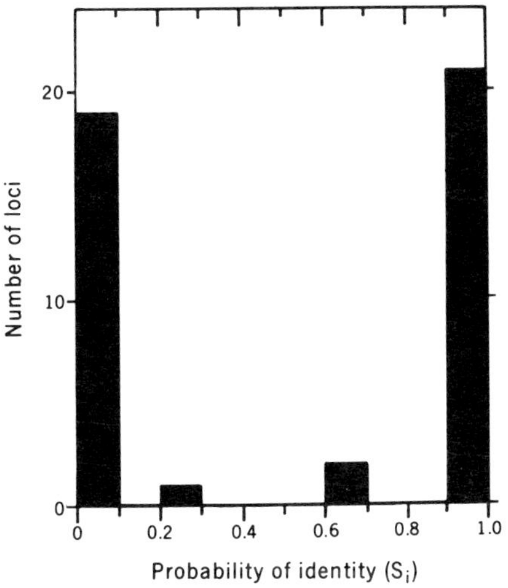

Fig. 2. Electrophoretic comparison of 43 proteins from humans and chimpanzees. The probability of identity (S_i) represents the likelihood that at locus i, human and chimpanzee alleles will appear electrophoretically identical.

of human-chimpanzee hybrid DNA formed in vitro with the thermostability of DNA from each species separately. By this criterion, human and chimpanzee mitochondrial DNA's appear identical (*21*). Working with "nonrepeated" DNA sequences, Kohne has estimated that human-chimpanzee hybrid DNA dissociates at a temperature (ΔT) 1.5°C lower than the dissociation temperature of reannealed human DNA (*22*). Hoyer *et al.*, on the other hand, have estimated that ΔT equals 0.7°C for human-chimpanzee hybrid DNA (*23*). If ΔT is the difference in dissociation temperature of reannealed human DNA and human-chimpanzee hybrid DNA prepared in vitro, then the percentage of nucleic acid sequence difference is $k \times \Delta T$ where the calibration factor k has been variously estimated as 1.5, 1.0, 0.9, or 0.45 (*22, 24*). Based on k being 1.0 and ΔT being 1.1°C, the nucleic acid sequence difference of human and chimpanzee DNA is about 1.1 percent. In a length of DNA 3000 bases long (representing 1000 amino acids), there will be about 0.011 × 3000, or 33 nucleotide sequence differences between the two species.

The evidence from the DNA annealing experiments indicates that there may be more difference at the nucleic acid level than at the protein level in human and chimpanzee genomes. For every amino acid sequence difference observed, about four base differences are observed in the DNA. Li *et al.* (*25*) found the same distinction between amino acid and nucleic acid differences in the tryptophan synthetase of several bacterial species: the nucleic acid sequences were about three times as different as the amino acid sequences. A similar result has been observed in three related RNA bacteriophages, as well as in studies of the relative rates of DNA and protein evolution in cow, pig, and sheep (*26*).

There are a number of probable reasons for this discrepancy (*25, 26*). First, more changes may appear in DNA than in proteins because of the redundancy of the code and consequently the existence of third-position nucleotide changes which do not lead to amino acid substitutions. The nature of the code indicates that if first-, second-, and third-position substitutions were equally likely to persist, then about 30 to 40 percent of potential base replacements in a cistron would not be reflected in the coded protein; that is, 1.4 to 1.7 base substitutions would occur for each amino acid substitution (*27*). However, it is likely that a larger proportion of the actual base substitutions in a cistron are third-position changes, since base substitutions that do not affect amino acid sequence are more likely to spread through a population. In addition, many of the nucleic acid substitutions may have occurred in regions of the DNA that are not transcribed and are therefore not conserved during evolution. Proteins analyzed by electrophoresis, sequencing, or microcomplement fixation techniques, on the other hand, all have definite cellular functions and may therefore have been conserved to a greater extent during evolution.

Genetic Distance and the Evolution of Organisms

The resemblance between human and chimpanzee macromolecules has been measured by protein sequencing, immunology, electrophoresis, and nucleic acid hybridization. From each of these results we can obtain an estimate of the genetic distance between humans and chimpanzees. Some of the same approaches have been used to estimate the genetic distance between other taxa, so that these estimates may be compared to the human-chimpanzee genetic distance.

First, we consider genetic distance estimated from electrophoretic data, using the standard estimate of net codon differences per locus developed by Nei and Roychoudhury (*28*). Other indices have been suggested for handling electrophoretic data (*29*) and give the same qualitative results, though somewhat different underlying assumptions are required. Nei and Roychoudhury's standard estimate of genetic distance between humans and chimpanzees can be written:

$$D = D_{HC} - \frac{D_C + D_H}{2} \qquad (8)$$

where

$$D_{HC} = -\log_e \bar{S}$$

$$D_H = -\log_e \left(\frac{1}{L} \sum_{i=1}^{L} \sum_{j=1}^{A_i} x_{ij}^2 \right)$$

$$D_C = -\log_e \left(\frac{1}{L} \sum_{i=1}^{L} \sum_{j=1}^{A_i} y_{ij}^2 \right)$$

according to the notation of Table 2 and Eqs. 2 and 3. Therefore, D is an estimate of the variability between human and chimpanzee populations (D_{HC}), corrected for the variability within human populations (D_H) and within chimpanzee populations (D_C). D_C and D_H are also measurements of the degree of heterozygosity in human and chimpanzee populations (*30*). Based on the data of Table 2, D_{HC} is 0.65, D_C is 0.02, and D_H is 0.05, so that:

$$D = 0.62 \qquad (9)$$

In other words, there is an average of 0.62 electrophoretically detectable codon differences per locus between homologous human and chimpanzee proteins.

This distance is 25 to 60 times greater than the genetic distance between human races (*28, 31*). In fact, the genetic distance between Caucasian, Black African, and Japanese populations is less than or equal to that between morphologically and behaviorally identical populations of other species. In addition, these three human populations are equally distant from the chimpanzee lineage (Fig. 3).

However, with respect to genetic distances between species, the human-chimpanzee D value is extraordinarily small, corresponding to the genetic distance between sibling species of *Drosophila* or mammals (Fig. 4). Nonsibling species within a genus (referred to in the figure as congeneric species) generally differ more from each other, by electrophoretic criteria, than humans and chimpanzees. The genetic distances

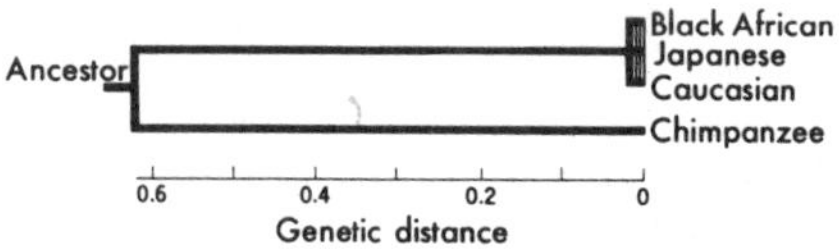

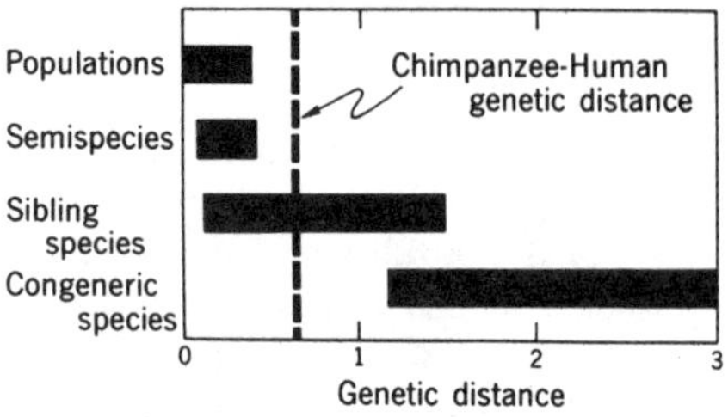

Fig. 3 (left). Phylogenetic relationship between human populations and chimpanzees. The genetic distances are based on electrophoretic comparison of proteins. The genetic distances among the three major human populations ($D = 0.01$ to 0.02) that have been tested are extremely small compared to those between humans and chimpanzees ($D = 0.62$). No human population is significantly closer than another to the chimpanzee lineage. The vertically hatched area between the three human lineages indicates that the populations are not really separate, owing to gene flow. Fig. 4 (right). The genetic distance, D, between humans and chimpanzees (dashed line) compared to the genetic distances between other taxa. Taxa compared include several species of *Drosophila* [*D. willistoni* (*94*), *D. paulistorum* (*95*), and *D. pseudoobscura* (*96*)], the horseshoe crab *Limulus polyphemus* (*97*), salamanders from the genus *Taricha* (*98*), lizards from the genus *Anolis* (*99*), the teleost fish *Astyanax mexicanus* (*100*), bats from the genus *Lasiurus* (*101*), and several genera of rodents [*Mus, Sigmodon, Dipodomys, Peromyscus*, and *Thomomys* (*99*), *Geomys* (*101*), and *Apodemus* (*102*)]. Selander and Johnson (*99*) summarize most of the data used in this figure. The great majority of proteins in these studies are intracellular.

among species from different genera are considerably larger than the human-chimpanzee genetic distance.

The genetic distance between two species measured by DNA hybridization also indicates that human beings and chimpanzees are as similar as sibling species of other organisms. The difference in dissociation temperature, ΔT, between reannealed human DNA and human-chimpanzee hybrid DNA is about 1.1°C. However, for sibling species of *Drosophila*, ΔT is 3°C; for congeneric species of *Drosophila*, ΔT is 19°C; and for congeneric species of mice (*Mus*), ΔT is 5°C (*32*).

Immunological and amino acid sequence comparisons of proteins lead to the same conclusion. Antigenic differences among the serum proteins of congeneric squirrel species are several times greater than those between humans and chimpanzees (*33*). Moreover, antigenic differences among the albumins of congeneric frog species (*Rana* and *Hyla*) are 20 to 30 times greater than those between the two hominoids (*34, 35*). In addition, the genetic distances among *Hyla* species, estimated electrophoretically, are far larger than the chimpanzee-human genetic distance (*36*). Finally, the human and chimpanzee β chains of hemoglobin appear to have identical sequences (Table 1), while the β chains of two *Rana* species differ by at least 29 amino acid substitutions (*37*). In summary, the genetic distance between humans and chimpanzees is well within the range found for sibling species of other organisms.

The molecular similarity between chimpanzees and humans is extraordinary because they differ far more than sibling species in anatomy and way of life. Although humans and chimpanzees are rather similar in the structure of the thorax and arms, they differ substantially not only in brain size but also in the anatomy of the pelvis, foot, and jaws, as well as in relative lengths of limbs and digits (*38*). Humans and chimpanzees also differ significantly in many other anatomical respects, to the extent that nearly every bone in the body of a chimpanzee is readily distinguishable in shape or size from its human counterpart (*38*). Associated with these anatomical differences there are, of course, major differences in posture (see cover picture), mode of locomotion, methods of procuring food, and means of communication. Because of these major differences in anatomy and way of life, biologists place the two species not just in separate genera but in separate families (*39*). So it appears that molecular and organismal methods of evaluating the chimpanzee-human difference yield quite different conclusions (*40*).

An evolutionary perspective further illustrates the contrast between the results of the molecular and organismal approaches. Since the time that the ancestor of these two species lived, the chimpanzee lineage has evolved slowly relative to the human lineage, in terms of anatomy and adaptive strategy. According to Simpson (*41*):

> *Pan* is the terminus of a conservative lineage, retaining in a general way an anatomical and adaptive facies common to all recent hominoids *except Homo*.

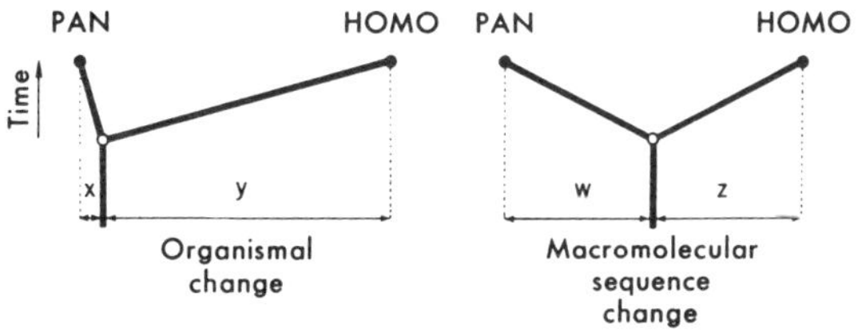

Fig. 5. The contrast between biological evolution and molecular evolution since the divergence of the human and chimpanzee lineages from a common ancestor. As shown on the left, zoological evidence indicates that far more biological change has taken place in the human lineage (y) than in the chimpanzee lineage ($y \gg x$); this illustration is adapted from that of Simpson (*41*). As shown on the right, both protein and nucleic acid evidence indicate that as much change has occurred in chimpanzee genes (w) as in human genes (z).

> *Homo* is both anatomically and adaptively the most radically distinctive of all hominoids, divergent to a degree considered familial by all primatologists.

This concept is illustrated in the left-hand portion of Fig. 5. However, at the macromolecular level, chimpanzees and humans seem to have evolved at similar rates (Fig. 5, right). For example, human and chimpanzee albumins are equally distinct immunologically from the albumins of other hominoids (gorilla, orangutan, and gibbon) (*10, 42, 43*), and human and chimpanzee DNA's differ to the same degree from DNA's of other hominoids (*21, 22*). Construction of a phylogenetic tree for primate myoglobins shows that the single amino acid difference between the sequences of human and chimpanzee myoglobin occurred in the chimpanzee lineage (*7*). Analogous reasoning indicates that the single amino acid difference between the sequences of human and chimpanzee hemoglobin δ chains arose in the human lineage (*8*). It appears that molecular change has accumulated in the two lineages at approximately equal rates, despite a striking difference in rates of organismal evolution. Thus, the major adaptive shift which took place in the human lineage was probably not accompanied by accelerated protein or DNA evolution.

Such an observation is by no means peculiar to the case of hominid evolution. It appears to be a general rule that anatomically conservative lineages, such as frogs, have experienced as much sequence evolution as have lineages that have undergone rapid evolutionary changes in anatomy and way of life (*34, 35, 44*).

Molecular Basis for the Evolution of Organisms

The contrasts between organismal and molecular evolution indicate that the two processes are to a large extent independent of one another. Is it possible, therefore, that species diversity results from molecular changes other than sequence differences in proteins? It has been suggested by Ohno (*45*) and others (*46*) that major anatomical changes usually result from mutations affecting the expression of genes. According to this hypothesis, small differences in the time of activation or in the level of activity of a single gene could in principle influence considerably the systems controlling embryonic development. The organismal differences between chimpanzees and humans would then result chiefly from genetic changes in a few regulatory systems, while amino acid substitutions in general would rarely be a key factor in major adaptive shifts.

Regulatory mutations may be of at least two types. First, point mutations could affect regulatory genes. Nucleotide substitutions in a promoter or operator gene would affect the production, but not the amino acid sequence, of proteins in that operon. Nucleotide substitutions in a structural gene coding for a regulatory protein such as a repressor, hormone, or receptor protein, could bring about amino acid substitutions, altering the regulatory properties of the protein. However, we suspect that only a minor fraction of the substitutions which accumulate in regulatory proteins would be likely to alter their regulatory properties.

Second, the order of genes on a chromosome may change owing to inversion, translocation, addition or deletion of genes, as well as fusion or fission of chromosomes. These gene rearrangements may have important effects on gene expression (*47*), though the biochemical mechanisms involved are obscure. Evolutionary changes in gene order occur frequently. Microscopic studies of *Drosophila* salivary chromosomes show, as a general rule, that no two species have the same gene order and that inversions are the commonest type of gene rearrangement (*48*). Furthermore, there is a parallel between rate of gene rearrangement and rate of anatomical evolution in the three major groups of vertebrates that have been studied in this respect, namely birds, mammals, and frogs (*46*). Hence gene rearrangements may be more important than point mutations as sources for evolutionary changes in gene regulation.

Although humans and chimpanzees have rather similar chromosome numbers, 46 and 48, respectively, the arrangement of genes on chimpanzee chromosomes differs from that on human chromosomes. Only a small proportion of the chromosomes have identical banding patterns in the two species. The banding studies indicate that at least 10 large inversions and translocations and one chromosomal fusion have occurred since the two lineages diverged (*49*). Further evidence for the possibility that chimpanzees and humans differ considerably in gene arrangement is provided by annealing studies with a purified DNA fraction. An RNA which is complementary in sequence to this DNA apparently anneals predominantly at a cluster of sites on a single human chromosome, but at widely dispersed sites on several chimpanzee chromosomes (*50*). The arrangement of chromosomal sites at which ribosomal RNA anneals may also differ between the two species (*50*).

Biologists are still a long way from understanding gene regulation in mammals (*51*), and only a few cases of regulatory mutations are now known (*52*). New techniques for detecting regulatory differences at the molecular level are required in order to test the hypothesis that organismal differences between individuals, populations, or species result mainly from regulatory differences. When the regulation of gene expression during embryonic development is more fully understood, molecular biology will contribute more significantly to our understanding of the evolution of whole organisms. Most important for the future study of human evolution would be the demonstration of differences between apes and humans in the timing of gene expression during development, particularly during the development of adaptively crucial organ systems such as the brain.

Summary and Conclusions

The comparison of human and chimpanzee macromolecules leads to several inferences:

1) Amino acid sequencing, immunological, and electrophoretic methods of protein comparison yield concordant estimates of genetic resemblance. These approaches all indicate that the average human polypeptide is more than 99 percent identical to its chimpanzee counterpart.

2) Nonrepeated DNA sequences differ more than amino acid sequences. A large proportion of the nucleotide differences between the two species may be ascribed to redundancies in the genetic code or to differences in nontranscribed regions.

3) The genetic distance between humans and chimpanzees, based on electrophoretic comparison of proteins encoded by 44 loci is very small, corresponding to the genetic distance between sibling species of fruit flies or mammals. Results obtained with other biochemical methods are consistent with this conclusion. However, the substantial anatomical and behavioral differences between humans and chimpanzees have led to their classification in separate families. This indicates that macromolecules and anatomical or behavioral features of organisms can evolve at independent rates.

4) A relatively small number of genetic changes in systems controlling the expression of genes may account for the major organismal differences between humans and chimpanzees. Some of these changes may result from the rearrangement of genes on chromosomes rather than from point mutations (*53*).

References and Notes

1. S. L. Washburn, Ed., *Classification and Human Evolution* (Aldine, Chicago, 1963). That there were striking similarities in blood proteins between apes and humans was known in 1904 [G. H. F. Nuttall, *Blood Immunity and Blood Relationships* (Cambridge Univ. Press, London, 1904)].
2. Sibling species are virtually identical morphologically.
3. R. F. Doolittle, G. L. Wooding, Y. Lin, M. Riley, *J. Mol. Evol.* **1**, 74 (1971).
4. M. O. Dayhoff, Ed., *Atlas of Protein Sequence and Structure* (National Biomedical Research Foundation, Georgetown Univ. Medical Center, Washington, D.C., 1972), vol. 5.
5. S. H. Boyer, E. F. Crosby, A. N. Noyes, G. F. Fuller, S. E. Leslie, L. J. Donaldson, G. R. Vrablik, E. W. Schaefer, T. F. Thurmon, *Biochem. Genet.* **5**, 405 (1971).
6. W. W. W. DeJong, *Biochim. Biophys. Acta* **251**, 217 (1971).
7. A. E. Romero Herrera and H. Lehmann, *ibid.* **278**, 62 (1972).
8. W. W. W. DeJong, *Nat. New Biol.* **234**, 176 (1971).
9. E. M. Prager and A. C. Wilson, *J. Biol. Chem.* **246**, 5978 and 7010 (1971).
10. V. M. Sarich and A. C. Wilson, *Science* **158**, 1200 (1967).
11. A. C. Wilson and V. M. Sarich, *Proc. Natl. Acad. Sci. U.S.A.* **63**, 1088 (1969); J. E. Cronin and V. M. Sarich, personal communication; R. Palmour, personal communication.
12. L. Nonno, H. Herschman, L. Levine, *Arch. Biochem. Biophys.* **136**, 361 (1970).
13. N. Hanke, E. M. Prager, A. C. Wilson, *J. Biol. Chem.* **248**, 2824 (1973).
14. A variety of immunological techniques have been used to compare chimpanzee proteins with their human counterparts [N. Mohagheghpour and C. A. Leone, *Comp. Biochem. Physiol.* **31**, 437 (1969); M. Goodman and G. W. Moore, *Syst. Zool.* **20**, 19 (1971); K. Bauer, *Humangenetik* **17**, 253 (1973)]. The immunodiffusion techniques employed in these studies are less sensitive to small differences in amino acid sequence than is microcomplement fixation [E. M. Prager and A. C. Wilson, *J. Biol. Chem.* **246**, 5978 (1971)]. Nevertheless, their results are generally consistent with those in Table 1. The few cases of large antigenic differences between human and chimpanzee proteins are probably not indicative of large sequence differences. For example, the haptoglobin difference reported by Mohagheghpour and Leone is due mainly to the fact that the haptoglobin 2 polypeptide is nearly twice the length of the haptoglobin 1 polypeptide [J. A. Black and G. H. Dixon, *Nature (Lond.)* **218**, 736 (1968)]. Human haptoglobin 1 is immunologically very similar to chimpanzee haptoglobin [J. Javid and H. H. Fuhrmann, *Am. J. Hum. Genet.* **207**, 496 (1971)]. The immunoglobulin differences reported by Bauer may be due to comparison of peptide chains that are not strictly homologous. In addition, Bauer's kappa chain results conflict with quantitative studies which detected no immunological difference [A. C. Wang, J. Shuster, A. Epstein, H. H. Fudenberg, *Biochem. Genet.* **1**, 347 (1968)]. Finally, the large Xh factor difference that Bauer reported might result from the fact that the chimpanzees in his studies were not pregnant and thus lacked Xh factor.
15. M.-C. King, thesis, University of California, Berkeley (1973).
16. D. Stollar and L. Levine, in *Methods in Enzymology*, S. P. Colowick and N. O. Kaplan, Eds. (Academic Press, New York, 1967), vol. 6, p. 928.
17. F. J. Ayala, M. L. Tracey, L. G. Barr, J. F. McDonald, S. Perez-Salas, *Genetics* **77**, 343 (1974).
18. See figure 9.3 in Dayhoff (*4*). We determined the proportion of amino acid substitutions causing a charge change during vertebrate evolution for several additional proteins: cytochrome c, lysozyme, myoglobin, α and β hemoglobin chains, triosephosphate dehydrogenase, α-lactalbumin, growth hormone, trypsin, and insulin. The average for these proteins is also 0.27. Our estimate of 0.16 for hemoglobin alone is very similar to that of Boyer *et al.* [S. H. Boyer, A. N. Noyes, C. F. Timmons, R. A. Young, *J. Hum. Evol.* **1**, 515 (1972)], who calculated that the ratio between electrophoretically silent and electrophoretically detectable hemoglobin alleles in primates is about 5.5; that is, about 15 percent of amino acid substitutions in primate hemoglobin evolution would be electrophoretically detectable.
A change in charge at a single amino acid site may have little or no effect on the net charge of a protein unless the substituted amino acid is on the exposed surface of the protein. Lee and Richards [B. Lee and F. M. Richards, *J. Mol. Biol.* **55**, 379 (1971)] determined the degree of exposure of each of the amino acid residues of lysozyme, ribonuclease, and myoglobin, based on the three-dimensional structure of these molecules. Their data indicate that 100 percent of the lysine residues, 100 percent of the arginine residues, 95 percent of the aspartic acid residues, 100 percent of the glutamic acid residues, and 70 percent of the histidine residues are on exposed surfaces of the proteins. Thus more than 90 percent of the substitutions involving charged amino acids would have affected the net charge of the protein and would, therefore, be detectable by electrophoresis.
19. A negative binomial variable may better describe the distribution of amino acid substitutions along lineages, since substitutions occur in proteins which are subject to varying selective pressures. That is, since different proteins evolve at different rates, the probability of a particular protein accepting a mutation varies from protein to protein [T. Uzzell and K. W. Corbin, *Science* **172**, 1089 (1971)]. However, for small values of m the negative binomial distribution agrees substantially with the Poisson distribution [C. I. Bliss and R. A. Fisher, *Biometrics* **9**, 176 (1953)]. Thus, for this calculation, the Poisson distribution should provide a very good estimate of the true m.
20. Further evidence regarding the correlation between electrophoretic and immunological measures of genetic distance has been obtained in studies involving many taxa of mammals, reptiles, and amphibians, by S. M. Case, H. C. Dessauer, G. C. Gorman, P. Haneline, K. H. Keeler, L. R. Maxson, V. M. Sarich, D. Shochat, A. C. Wilson, and J. S. Wyles.
21. W. M. Brown and R. L. Hallberg, *Fed. Proc.* **31** (2), Abstr. 1173 (1972).
22. D. E. Kohne, *Q. Rev. Biophys.* **3**, 327 (1970); ———, J. A. Chiscon, B. H. Hoyer, *J. Hum. Evol.* **1**, 627 (1972).
23. B. H. Hoyer, N. W. van de Velde, M. Goodman, R. B. Roberts, *J. Hum. Evol.* **1**, 645 (1972).
24. N. R. Kallenbach and S. D. Drost, *Biopolymers* **11**, 1613 (1972); J. R. Hutton and J. G. Wetmur, *Biochemistry* **12**, 558 (1973); J. S. Ullman and B. J. McCarthy, *Biochim. Biophys. Acta* **294**, 416 (1973).
25. S. L. Li, R. M. Denney, C. Yanofsky, *Proc. Natl. Acad. Sci. U.S.A.* **70**, 1112 (1973).
26. H. D. Robertson and P. G. N. Jeppesen, *J. Mol. Biol.* **68**, 417 (1972); C. Laird, B. L. McConaughy, B. J. McCarthy, *Nature (Lond.)* **224**, 149 (1969).
27. *Cold Spring Harbor Symp. Quant. Biol.* **31**, 1 (1966).
28. M. Nei and A. K. Roychoudhury, *Am. J. Hum. Genet.* **26**, 431 (1974).
29. L. L. Cavalli-Sforza and A. W. F. Edwards, *ibid.* **19**, 233 (1967); N. E. Morton, *Annu. Rev. Genet.* **3**, 53 (1969); V. Balakrishnan and L. D. Sangvhi, *Biometrics* **24**, 859 (1968); T. W. Kurczynski, *ibid.* **26**, 525 (1970); P. W. Hedrick, *Evolution* **25**, 276 (1971); R. R. Sokal and P. H. A. Sneath, *Principles of Numerical Taxonomy* (Freeman, San Francisco, 1973); J. S. Rogers, "Studies in Genetics," *Univ. Texas Publ. No. 7213* (1972), vol. 7, p. 145.
30. The average heterozygosity estimates for the loci in this study are quite low, especially for chimpanzee populations. To obtain comparable heterozygosity estimates for humans and chimpanzees, we included only loci that have been surveyed for both species and only polymorphisms detectable by techniques used for surveying both species. Thus some confirmed polymorphisms in human populations were excluded. There are at least three reasons for the difference between the heterozygosity estimates for human and chimpanzee populations. First, many more humans than chimpanzees have been surveyed at each locus, so that the variability estimate for humans is biased insofar as it is based on alleles present at low frequency in human populations. Second, there are many more humans than chimpanzees alive today, living in a greater variety of environments and with a larger number of gene pools. As a result, more mutants reach appreciable frequencies in human populations. Third, and probably most important, the chimpanzees in colonies available for study are based on even fewer gene pools and are highly inbred in many cases. The discrepancies in real population size and sampling technique between human and chimpanzee populations probably account for the greater number of polymorphic loci, the larger number of alleles at polymorphic loci, and the higher average heterozygosity

estimates in human populations.
31. M. Nei and A. K. Roychoudhury, *Science* **177**, 434 (1972).
32. N. R. Rice, *Brookhaven Symp. Biol.* **23**, 44 (1972); C. D. Laird, *Annu. Rev. Genet.* **3**, 177 (1973).
33. M. E. Hight, M. Goodman, W. Prychodko, *Syst. Zool.* **23**, 12 (1974).
34. D. G. Wallace, L. R. Maxson, A. C. Wilson, *Proc. Natl. Acad. Sci. U.S.A.* **68**, 3127 (1971); D. G. Wallace, M.-C. King, A. C. Wilson, *Syst. Zool.* **22**, 1 (1973).
35. L. R. Maxson, and A. C. Wilson, *Syst. Zool.*, in press.
36. R. K. Selander, personal communication; L. R. Maxson and A. C. Wilson, *Science* **185**, 66 (1974).
37. T. O. Baldwin and A. Riggs, *J. Biol. Chem.* **249**, 6110 (1974).
38. G. H. Bourne, Ed., *The Chimpanzee* (Karger, New York, 1970).
39. Th. Dobzhansky, in *Classification and Human Evolution*, S. L. Washburn, Ed. (Aldine, Chicago, 1963), p. 347; E. Mayr, in *ibid.*, p. 332; E. L. Simons, *Primate Evolution* (Macmillan, New York, 1972); G. G. Simpson, *Principles of Animal Taxonomy* (Columbia Univ. Press, New York, 1961). L. Van Valen [*Am. J. Phys. Anthropol.* **30**, 295 (1969)] has suggested that, based on differences in their adaptive zones, humans and chimpanzees be placed in separate suborders.
40. On the basis of some protein evidence available in 1970, Goodman and Moore proposed that humans and African apes be placed in the same subfamily [M. Goodman and G. W. Moore, *Syst. Zool.* **20**, 19 (1971)]. By analogy, the protein evidence now available would lead to placement of chimpanzees and humans in the same genus. However, as protein evolution and organismal evolution apparently can proceed independently, it is questionable whether organismal classifications should be revised on the basis of protein evidence alone.
41. G. G. Simpson, in *Classification and Human Evolution*, S. L. Washburn, Ed. (Aldine, Chicago, 1963).
42. V. M. Sarich, in *Old World Monkeys*, J. R. Napier and P. H. Napier, Eds. (Academic Press, New York, 1970), p. 175.
43. ——— and A. C. Wilson, *Proc. Natl. Acad. Sci. U.S.A.* **58**, 142 (1967).
44. V. M. Sarich, *Syst. Zool.* **18**, 286 and 416 (1969); *Nature* (*Lond.*) **245**, 218 (1973).
45. S. Ohno, *J. Hum. Evol.* **1**, 651 (1972).
46. A. C. Wilson, L. R. Maxson, V. M. Sarich, *Proc. Natl. Acad. Sci. U.S.A.* **71**, 2843 (1974); A. C. Wilson, V. M. Sarich, L. R. Maxson, *ibid.*, p. 2028; E. M. Prager and A. C. Wilson, *ibid.* **72**, 200 (1975).
47. E. Bahn, *Hereditas* **67**, 79 (1971); B. Wallace and T. L. Kass, *Genetics* **77**, 541 (1974).
48. M. J. D. White, *Annu. Rev. Genet.* **3**, 75 (1969).
49. J. de Grouchy, C. Turleau, M. Roubin, F. C. Colin, *Nobel Symp.* **23**, 124 (1973); B. Dutrillaux, M.-O. Rethoré, M. Prieur, J. Lejeune, *Humangenetik* **20**, 343 (1973); D. Warburton, I. L. Firschein, D. A. Miller, F. E. Warburton, *Cytogenet. Cell Genet.* **12**, 453 (1973); C. C. Lin, B. Chiarelli, L. E. M. de Boer, M. M. Cohen, *J. Hum. Evol.* **2**, 311 (1973); J. Ecozcue, M. R. Caballin, C. Goday, *Humangenetik* **18**, 77 (1973); M. Bobrow and K. Madan, *Cytogenet. Cell Genet.* **12**, 107 (1973).
50. In situ annealing studies have been performed with RNA complementary to purified human satellite DNA [K. W. Jones, J. Prosser, G. Carneo, E. Ginelli, M. Bobrow, *Symp. Med. Hoechst* **6**, 45 (1973)] and with human ribosomal RNA [A. Henderson, D. Warburton, K. C. Atwood, *Chromosoma* **46**, 435 (1974)].
51. J. E. Darnell, W. R. Jelinek, G. R. Molloy, *Science* **181**, 1214 (1973); E. H. Davidson and R. J. Britten, *Q. Rev. Biol.* **48**, 565 (1973); C. A. Thomas, Jr., in *Regulation of Transcription and Translation in Eukaryotes*, E. K. F. Bautz, Ed. (Springer-Verlag, Berlin, 1973).
52. D. J. Weatherall and J. B. Clegg, *The Thalassaemia Syndromes* (Blackwell, Oxford, ed. 2, 1972).
53. Additional inferences can be drawn from the comparison of human and chimpanzee macromolecules; some of these will be discussed elsewhere.
54. I. N. H. White and P. J. Butterworth, *Biochim. Biophys. Acta* **229**, 193 (1971).
55. J. Schmitt, K. H. Lichte, W. Fuhrmann, *Humangenetik* **10**, 138 (1970); G. Tariverdian, H. Ritter, J. Schmitt, *ibid.* **11**, 323 (1971).
56. E. R. Giblett, *Genetic Markers in Human Blood* (Blackwell, Oxford, 1969).
57. J. Schmitt, G. Tariverdian, H. Ritter, *Humangenetik* **11**, 100 (1971).
58. R. E. Tashian, *Am. J. Hum. Genet.* **17**, 257 (1965).
59. H. Matsubara and E. L. Smith, *J. Biol. Chem.* **237**, 3575 (1962).
60. R. Schiff and C. Stormont, *Biochem. Genet.* **4**, 11 (1970).
61. A. Yoshida, *Proc. Natl. Acad. Sci. U.S.A.* **57**, 835 (1967); *Biochem. Genet.* **1**, 81 (1967); J. Kömpf, H. Ritter, J. Schmitt, *Humangenetik* **11**, 342 (1971); M. Goodman and M. D. Poulik, "Genetic variations and phylogenetic properties of protein macromolecules of chimpanzees" (6571st Aeromedical Research Laboratory, Rep. ARL-TR-68-3, Holloman Air Force Base, New Mexico, 1968).
62. J. L. Brewbaker, M. D. Upadhya, Y. Mäkinen, T. Macdonald, *Physiol. Plant* **21**, 930 (1968); C. R. Shaw and R. Prasad, *Biochem. Genet.* **4**, 297 (1970).
63. J. Kömpf, H. Ritter, J. Schmitt, *Humangenetik* **13**, 72 (1971).
64. J. C. Kaplan, *Nature (Lond.)* **217**, 256 (1968); W. K. Long, *Science* **155**, 712 (1967).
65. D. Rifkin and W. Konigsberg, *Biochim. Biophys. Acta* **104**, 457 (1965).
66. H. Harris, *J. Med. Genet.* **8**, 444 (1971).
67. R. G. Davidson, J. A. Cortner, M. C. Rattazzi, F. H. Ruddle, H. A. Lubs, *Science* **196**, 391 (1970).
68. G. Beckman, E. Lundgren, A. Tärnvik, *Hum. Hered.* **23**, 338 (1973); B. B. Keele, Jr., J. M. McCord, I. Fridovich, *J. Biol. Chem.* **246**, 2875 (1971); H. M. Steinman and R. L. Hill, *Proc. Natl. Acad. Sci. U.S.A.* **70**, 3725 (1973).
69. G. S. Bailey and A. C. Wilson, *J. Biol. Chem.* **243**, 5843 (1968).
70. A. L. Koen and M. Goodman, *Biochem. Genet.* **3**, 457 (1969).
71. R. L. Kirk, E. M. McDermid, N. M. Blake, R. L. Wight, E. H. Yap, M. J. Simons, *Humangenetik* **17**, 345 (1973); G. Tariverdian, H. Ritter, J. Schmitt, *ibid.* **11**, 339 (1971).
72. G. Tariverdian, H. Ritter, G. G. Wendt, *ibid.*, p. 75.
73. J. Schmitt, G. Tariverdian, H. Ritter, *ibid.*, p. 95.
74. P. C. Hudgins, C. M. Whorton, T. Tomoyoshi, A. J. Riopelle, *Nature (Lond.)* **212**, 693 (1966).
75. W. H. P. Lewis, *Ann. Hum. Genet.* **36**, 267 (1973).
76. S. Povey, G. Corney, W. H. P. Lewis, E. B. Robson, J. M. Parrington, H. Harris, *ibid.* **35**, 455 (1972).
77. M. Goodman and R. E. Tashian, *Hum. Biol.* **41**, 237 (1969).
78. G. Tariverdian, H. Ritter, J. Schmitt, *Humangenetik* **12**, 105 (1971).
79. H. Rubinson, M. C. Meienhofer, J. C. Dreyfus, *J. Mol. Evol.* **2**, 243 (1973); P. H. Corran and S. G. Waley, *Biochem. J.* **139**, 1 (1974).
80. W. E. Marshall, *J. Biol. Chem.* **241**, 4731 (1966).
81. Y. T. Li and S. C. Li, *ibid.* **245**, 825 (1970).
82. A. L. Tarnoky, B. Dowding, A. L. Lakin, *Nature (Lond.)* **225**, 742 (1970).
83. G. Kellermann and H. Walter, *Humangenetik* **10**, 145 (1970); G. Kellermann and H. Walter, *ibid.*, p. 191.
84. M. H. K. Shokeir and D. C. Shreffler, *Biochem. Genet.* **4**, 517 (1970).
85. C. A. Alper and F. S. Rosen, *Immunology* **14**, 251 (1971); E. A. Azen, O. Smithies, O. Hiller, *Biochem. Genet.* **3**, 214 (1969).
86. H. Cleve, *Hum. Hered.* **20**, 438 (1970); F. D. Kitchin and A. G. Bearn, *Am. J. Hum. Genet.* **17**, 42 (1965).
87. B. S. Blumberg, *Proc. Soc. Exp. Biol. Med.* **104**, 25 (1960); S. H. Boyer and W. J. Young, *Nature (Lond.)* **187**, 1035 (1960); M. Cresta, *Riv. Antrop. Roma* **47**, 225 (1961); V. Lange and J. Schmitt, *Folia Primatol.* **1**, 208 (1963); O. Mäkelä, O. V. Rekonen, E. Salonen, *Nature (Lond.)* **185**, 852 (1960); W. C. Parker and A. G. Bearn, *Ann. Hum. Genet.* **25**, 227 (1961); J. Javid and M. H. Fuhrman, *Am. J. Hum. Genet.* **23**, 496 (1971); B. S. Shim and A. G. Bearn, *ibid.* **16**, 477 (1964); J. Planas, *Folia Primatol.* **13**, 177 (1970).
88. Although human and chimpanzee lysozymes have been reported to be electrophoretically identical [N. Hanke, E. M. Prager, A. C. Wilson, *J. Biol. Chem.* **248**, 2824 (1973)], more refined techniques indicate that their mobilities in fact differ (E. Prager, personal communication).
89. K. Berg and A. G. Bearn, *Annu. Rev. Genet.* **2**, 341 (1968).
90. M. Goodman, R. McBride, E. Poulik, E. Reklys, *Nature (Lond.)* **197**, 259 (1963); M. Goodman and A. J. Riopelle, *ibid.*, p. 261; M. Goodman, W. G. Wisecup, H. H. Reynolds, C. H. Kratochvil, *Science* **150**, 98 (1967).
91. H. Harris, *The Principles of Human Biochemical Genetics* (Elsevier, New York, 1970).
92. H. E. Schultze and J. F. Heremans, *Molecular Biology of Human Proteins* (Elsevier, New York, 1966), vol. 1.
93. M.-C. King and A. C. Wilson, in preparation.
94. F. J. Ayala, J. R. Powell, M. L. Tracey, C. A. Mourão, S. Pérez-Salas, *Genetics* **70**, 113 (1972).
95. R. C. Richmond, *ibid.*, p. 87.
96. S. Prakash, R. C. Lewontin, J. L. Hubby, *ibid.* **61**, 841 (1969).
97. R. K. Selander, S. Y. Yang, R. C. Lewontin, W. E. Johnson, *Evolution* **24**, 402 (1970).
98. D. Hedgecock and F. J. Ayala, *Copeia* (1974), p. 738.
99. R. K. Selander and W. E. Johnson, *Annu. Rev. Ecol. Syst.* **4**, 75 (1973).
100. C. R. Shaw, *Biochem. Genet.* **4**, 275 (1970).
101. R. K. Selander, D. W. Kaufman, R. J. Baker, S. L. Williams, *Evolution* **28**, 557 (1974).
102. W. Engel, W. Vogel, I. Voiculescu, H. Ropers, M. T. Zenges, K. Bender, *Comp. Biochem. Physiol.* **44B**, 1165 (1973).
103. Samples of chimpanzee blood for this study were obtained from the Laboratory for Experimental Medicine and Surgery in Primates, New York University Medical Center, P.O. Box 575, Tuxedo, N.Y. 10987; M. Goodman, Wayne State University School of Medicine, Detroit, Mich. 48201; and H. Hoffman, National Institutes of Health, Bethesda, Md. 20014. This work was supported by grant GM-18578 from NIH. Many colleagues helped us with this project. We thank S. S. Carlson, R. K. Colwell, L. R. Maxson, J. Maynard Smith, E. M. Prager, V. M. Sarich, and G. S. Sensabaugh for advice and ideas; M. Nei for unpublished data; and E. Bradley, D. Healy, P. Lozar, K. Pippen, and R. Wayner for expert technical assistance.

Reprinted from *Science* **189**: 760–763.

The Metabolic Code

Biological symbolism and the origin of intercellular communication is discussed.

Gordon M. Tomkins

This article presents a model for the evolution of biological regulation and the origin of hormone-mediated intercellular communication. Because of certain similarities with the processes of genetic coding, the present hypothesis is termed the "metabolic code." The formulations are almost entirely speculative and should therefore be regarded, at least for the moment, primarily as a pedagogical device for organizing a number of facts about cellular control. However, since analogies are drawn between regulation in unicellular prokaryotes and in multicellular eukaryotes, the ideas put forward might also prove useful for suggesting new experimental approaches to understanding sophisticated control mechanisms in complicated higher organisms.

Because of the ubiquity of cyclic adenosine monophosphate (cyclic AMP) in biological regulation, considerable attention is directed toward its function and possible evolution. Nevertheless, as treated in the present context it may represent only a model for other as yet undiscovered intracellular effectors which also operate according to the principles outlined below.

Cyclic AMP, originally discovered during studies on the mechanism of epinephrine action (*1*), has subsequently been shown to mediate the intracellular actions of almost all those hormones that interact with the cell membrane (*2*). Cyclic AMP also controls the "catabolite repression" mechanism in bacteria (*3*) and other microorganisms (*4*), and more recently—together with the related nucleotide cyclic GMP (cyclic guanosine monophosphate) has been implicated in the modulation of growth and development of a large number of cell types in both prokaryotic and eukaryotic organisms. For example, it has been suggested that these molecules play important roles in the immune response (*5*), the nervous system (*6*), and the process of malignant transformation (*7*).

The author was professor of biochemistry and biophysics, at the University of California, San Francisco 94143. This article was completed shortly before his death on 22 July 1975.

Such ubiquity implies one of two explanations. The cyclic nucleotides, by virtue of some intrinsic chemical or physical properties, could be absolute requirements for the living state. This seems unlikely, however, in view of the viability of mutant organisms lacking adenylate cyclase, the enzyme that catalyzes the formation of cyclic AMP (*8, 9*). Alternatively, once the cyclic nucleotides had formed (as the result of a biosynthetic accident), their universality derived from the adaptive advantages conferred upon descendants of the organisms in which they first appeared.

In this article I adopt the latter point of view which, together with other assumptions about the evolutionary origins of biological control, suggests a model for the organization and strategy of regulatory mechanisms in modern unicellular organisms as well as indicating the principles which underlie intercellular communication in metazoa.

Simple and Complex Regulation and the Metabolic Code

Even the most ancient molecular assemblies possessing recognizable cellular properties must have been capable of self-duplication, implying the prior existence of DNA and the machinery necessary for its replication. Obviously, these organisms also contained mechanisms for the expression of genetic information. Since both nucleic acid and protein synthesis are endergonic reactions, primordial cells were almost certainly endowed with the capacity to capture the necessary energy from the environment and to transform it into usable form, presumably ATP (adenosine triphosphate).

The biosynthetic capabilities of primitive cells were, however, probably quite limited. Changes in the environment which diminished the supply either of the monomeric units required for polymer synthesis, or compromised the formation of ATP might have easily proven lethal. Survival would therefore have required the evolution of regulatory mechanisms that could maintain a relatively constant intracellular environment in the face of changes in external conditions.

In this discussion, I shall define two modes of regulation, "simple" and "complex," both present in modern organisms, which differ from each other in their relative sophistication as well as, most likely, the order in which they evolved.

The essential feature of simple regulation is a direct chemical relationship between the regulatory effector molecules and their effects. Thus, substrates or end products affect their own metabolism, independent of the biochemical mechanism employed. Simple regulation may be positive, as in enzyme induction, or negative, as in feedback inhibition of enzyme activity and repression of enzyme biosynthesis.

If regulation were limited only to simple mechanisms, survival might be tenuous, since the regulatory effector molecules are themselves important metabolic intermediates. Dramatic changes in the intracellular environment could therefore follow rapid depletion or replenishment of essential nutrients. However, present-day organisms also display more sophisticated regulatory behavior, presumably of later evolutionary origin than the simple mechanisms, which confer greater stability on the internal environment. I define these as "complex" control mechanisms.

Complex regulation is characterized by two entities not operating in simple mechanisms: metabolic "symbols" and their "domains." The term "symbol" refers to a specific intracellular effector molecule which accumulates when a cell is exposed to a particular environment (*10*). For example, cyclic AMP in most microorganisms acts as a symbol for carbon-source starvation, and ppGpp (guanosine 5′-diphosphate 3′-diphosphate) (*11*), a symbol for nitrogen or amino acid deficiency.

Metabolic symbols need bear no structural relationship to the molecules which promote their accumulation in a nutritional or metabolic crisis (that is, cyclic

AMP is not a chemical analog of glucose). Another important propety of intracellular symbols is metabolic lability, which allows their concentrations to fluctuate quickly in response to environmental change. For instance, cyclic AMP and ppGpp are both rapidly formed and inactivated by specific enzymic reactions (*12, 13*).

Since a particular environmental condition is correlated with a corresponding intracellular symbol, the relationship between the extra- and intracellular events may be considered as a "metabolic code" in which a specific symbol represents a unique state of the environment.

A second essential concept in complex regulation is that of the "domain" of a symbol, defined as all the metabolic processes controlled by the symbol. For instance, the responses of a glucose-starved *Escherichia coli* constitute the domain of cyclic AMP in this organism, and the reactions of the "stringent response" (*14*) of amino acid–deprived bacteria form the domain of ppGpp. A comparison of these responses illustrates that the biochemical reactions included in the domain of a symbol are related by their biological effects rather than their chemical mechanisms. Thus, ppGpp may interact with a variety of cellular macromolecules to coordinate the reactions involved in the stringent response; these reactions include gene transcription (*15*), membrane transport (*16*), and a variety of enzyme-catalyzed metabolic interconversions (*17*). Cyclic AMP, on the other hand, may interact with only a single bacterial receptor protein (CRP) which, in turn, associates with specific DNA sequences regulating the transcription of genes under catabolite repression control (*18*).

In *E. coli*, flagellin synthesis requires cyclic AMP (*19*). As a result, bacteria become motile, presumably as an adaptation to nutritional stress. The catabolite repression domain therefore contains elements that affect gross aspects of bacterial behavior as well as metabolism. Furthermore, a given process may be included in a particular domain under only special circumstances. For instance, the transcription of an inducible, catabolite-repressible operon will take place only in the presence both of its specific inducer as well as of cyclic AMP. Moreover, a given process might be part of several different domains. These considerations indicate that the symbol-domain relationship endows a cell with considerable regulatory sophistication, allowing a relatively simple environmental change to bring about a complex coordinated cellular response.

Evolution and Universality of the Metabolic Code

To explore these ideas further, and in particular to extend them to intercellular communication, it is useful to speculate about the evolutionary basis of complex regulation. Despite the virtual impossibility of obtaining relevant experimental data about historical origins, an examination of present-day biochemical reactions suggests reasonable possibilities. For example, ppGpp is made from GTP (guanosine triphosphate) during the "idling" of protein synthesis (*20*), while the ribosome-messenger complex is temporarily deprived of the amino acid specified by a particular codon (*21*). This mechanism also suggests the evolutionary origin of ppGpp. Since GTP is used extensively in protein synthesis, inhibition of this process by amino acid starvation of a primitive organism could have led to the formation of ppGpp, which for accidental (but genetically determined) reasons had favorable regulatory consequences for the organism in question and its progeny.

Similarly, cyclic AMP might have become associated with carbon-source starvation as a result of the "idling" of a primordial kinase which normally catalyzed the phosphorylation of glucose with ATP. In the absence of glucose, ATP might have been converted to the cyclic phosphate, again with favorable evolutionary consequences for the descendants of the organism in which it originally occurred.

The origins of the domains are equally obscure. However, one might assume that when cyclic AMP or ppGpp first appeared in evolution, nucleotide binding sites on proteins (*22*) already existed. As analogs of the nucleoside triphosphates, the regulatory molecules would thus have had a number of potential enzyme binding sites, interaction with which might have simultaneously influenced a number of biochemical reactions. The aggregate of those modulations that proved adaptively useful would ultimately have evolved into the complex regulatory domains.

By whatever means, complex domains probably evolved initially by the gradual accretion of new elements. Having once attained a certain level of complexity, however, they may have undergone significant changes in the rate and manner of their evolution. As illustrated by the universality of the genetic code (*23*), biological networks that interconnect a number of important cellular processes tend to attain evolutionary stability (*24*). This arises because mutations damaging one element of a system have pleiotropic effects that might imperil the entire organism. If this reasoning is applied to the symbol-domain relationship of the metabolic code, then once a domain has reached a certain degree of complexity at least some of its regulatory interactions would become constant with respect to evolutionary change.

This stability might be one of the explanations not only for the ubiquity of the cyclic nucleotides but also for an apparent generality of the metabolic code. For instance, cyclic AMP symbolizes carbon-source starvation in *E. coli*, whereas glucagon and epinephrine, hormones which stimulate cyclic AMP production in vertebrates, mobilize metabolic stores such as glycogen (*2*) and triglycerides (*25*) as if these organisms were also subjected to acute starvation. These aspects of the mammalian and bacterial responses to cyclic AMP seem quite similar. This and other apparent similarities between prokaryotic and eukaryotic regulation (*26*) indeed suggest a sort of universality in metabolic coding.

Intercellular Transfer of Metabolic Information: Origin of Hormones

According to the ideas already presented, the overall functional state of any cell is determined by the activities of the reactions in the various domains. In a colony of unicellular organisms, each individual cell responds independently to the environment by generating appropriate intracellular metabolic symbols. In most multicellular organisms, however, only certain cells are stimulated directly by the environment. These in turn secrete specific effector molecules, the hormones, which signal other cells (perhaps insulated from the environment) to respond metabolically to the initial stimulus. In higher organisms, such a chain of cellular communication may involve many intermediate steps. For example, in vertebrates, many environmental conditions are processed as nerve impulses impinging on the hypothalamus, from which hormones travel to the pituitary. From there, other hormones are transmitted to a variety of different cells, many of which manufacture specific products in response to the initial stimulus originating in the nervous system. Here I propose a mechanism by which metabolic coding in unicellular organisms might have evolved into the endocrine system of the metazoa.

Dictyostelium discoidium, a cellular slime mold, serves as an excellent model for how the transition might have come about. Given sufficient nutrients, this organism exists as independent myxamoebas. Upon starvation, they generate cyclic AMP and release it into the surrounding medium (*27*). This substance serves as a chemical attractant that causes the aggregation of a large number of myxamoebas (*28*) to form a multicellular "slug." In this case, as in *E. coli*, cyclic AMP acts as an intracellular symbol of carbon-source starvation. In addition, however, the cyclic nucleotide is released from the *Dictyostelium* cells in which it is formed and diffuses to other nearby cells, promoting the aggregation response. Cyclic AMP thus acts in these organisms both as an intracellular symbol of starvation and as a hormone which carries this metabolic information from one cell to another.

These phenomena raise the question of why the cyclic nucleotides do not commonly play extracellular, hormonal roles in organisms more complex than the slime mold. Although it has been suggested (*29*) that these compounds mediate some short-range intercellular transactions, their locus of action is largely intracellular, while long-range chemical communication between cells is effected by other types of molecules. One reason for the predominantly intracellular action of the cyclic nucleotides may be their metabolic lability. It was pointed out above that the rapid turnover of intracellular symbols is advantageous for adaptation in unicellular organisms. Because of their sensitivity to hydrolysis the extracellular lifetime of the cyclic nucleotides is probably not long enough to allow them to travel the relatively long distances required for intercellular communication in large metazoa. It has been argued (*30*) that an increase in the size of organisms is favored by evolution. If this be the case, metabolic information must be transmissible over longer and longer distances (up to several meters in large vertebrates). Because of the unsuitability of intracellular symbols for this purpose, I propose that the hormones, which are more metabolically stable, took on this role.

These substances carry information from "sensor" cells in direct contact with environmental signals, to more sequestered responder cells. Specifically, the metabolic state of a sensor cell, represented by the levels of its intracellular symbols, is "encoded" by the synthesis and secretion of corresponding levels of hormones. When the hormones reach responder cells, the metabolic message is "decoded" into corresponding primary intracellular symbols. Thus, hormones apprise responder cells of the concentrations of intracellular symbols in the sensor cells, allowing relatively protected internal organs to respond coordinately to external perturbations.

These reactions are clearly illustrated in the vertebrate endocrine system. Pituitary cells generate intracellular cyclic AMP in response to the polypeptide "releasing factors" from the hypothalamus (*31*). The cyclic AMP, in turn, stimulates the pituitary cells to release specific trophic hormones such as ACTH (adrenocorticotrophic hormone) or TSH (thyroid-stimulating hormone) (*32*). These diffuse to their target cells (adrenal or thyroid, for instance) where, after interaction with specific membrane receptor proteins, they stimulate the production of cyclic AMP. Depending on the nature of the target cell, the stimulation may cause the release of yet other hormones (steroids or thyroxin). In this way, endocrine cells act as both sensors and responders, that is, intermediates in the transmission of metabolic information from primary sensor cells to the tissues in which the final chemical responses take place.

Neural Transmitters as Hormones

In many organisms the nervous and endocrine systems are intimately connected, and hormone release is often activated by neural stimulation. Moreover, intercellular communication within the nervous system is mediated by hormones, the neurotransmitters, that operate over very short distances. In view of the previous discussion, it is interesting to speculate on the possible metabolic origins and significance of the transmitters. Some of these substances, such as acetylcholine, occur in organisms under circumstances where they serve no apparent neural function (*33*). This suggests that the evolutionary appearance of the transmitters preceded that of the nervous system. Their hormonal function in modern organisms implies that they might have arisen as regulatory molecules—perhaps metabolic symbols, in the sense defined here. A possible clue to the biochemical origins of the neurotransmitters is provided by the fact that all the compounds currently accepted as transmitters are either amino acid metabolites—for example, the catecholamines, serotonin, γ-aminobutyric acid, acetylcholine—or are themselves amino acids (for example, glycine). Thus, perhaps the transmitters acted in primitive cells as intracellular symbols representing changes in environmental amino acid concentration. Eventually, these primordial nerve cells might have utilized the symbols in short-range intercellular (hormonal) roles, originally concerned with transducing information related to amino acid accumulation, and gradually with many other aspects of the environment.

Conclusions

Quite obviously, the formulations presented in this article are largely speculative. They represent an attempt to understand regulation in complicated multicellular organisms in terms of the evolution and function of seemingly comparable processes which occur in much simpler systems. Clearly, a number of gaps remain in the scheme. Nowhere have I dealt, for example, with the origins and significance of the steroids or thyroxin.

These molecules act as hormones in the sense that they transmit information about cyclic AMP levels from the sensor cells where they are produced to specific responder cells. Nevertheless, the effectors themselves appear to function primarily as direct intracellular modulators of gene activity in responder cells (*34, 35*). Thus, like cyclic AMP in *D. discoidium*, steroids and thyroxin act as both hormone and intracellular symbols. Their effects tend to be more protracted than those of the membrane-bound hormones which regulate cyclic AMP concentrations, perhaps because of relatively slow breakdown in the responder cells. Clearly, further work is required to begin to understand the evolutionary and metabolic significance of the thyroid hormones and the steroids.

A further difficulty relates to the fact that, although I have frequently referred to "symbols" in the plural, the only substances that might legitimately be so termed in eukaryotic cells at the present are the cyclic nucleotides. Nevertheless, there are increasing indications that other substances including certain ions (Ca^{2+}, Na^{+}, and K^{+}) might also function in this way (*36*).

Despite these evident deficiencies, it seems to me that there are significant advantages in presenting a general hypothesis at this time. One is that a great deal of heretofore unrelated information is unified for pedagogical reasons. Another is based on the likelihood that both the molecular mechanisms and overall strategy of regulation will be understood first in simple organisms. If this is the case, then the general scheme outlined here might suggest new experimental approaches to the study of intercellular communication in more complex organisms.

References and Notes

1. E. W. Sutherland, *Science* **177**, 401 (1972).
2. G. A. Robison, R. W. Butcher, E. W. Sutherland, *Cyclic AMP* (Academic Press, New York, 1971).
3. R. L. Perlman, B. de Crombrugghe, I. Pastan, *Nature (Lond.)* **223**, 810 (1969); B. de Crombrugghe, R. L. Perlman, H. E. Varmus, I. Pastan, *J. Biol. Chem.* **244**, 5828 (1969).
4. R. Van Wijk and T. Konijn, *FEBS (Fed. Eur. Biochem. Soc.) Lett.* **13**, 184 (1971); J. Sy and D. Richter, *Biochemistry* **11**, 2788 (1972); G. Schlanderer and H. Dellweg, *Eur. J. Biochem.* **49**, 305 (1974).
5. W. Braun, L. M. Lichtenstein, C. W. Parker, Eds., *Cyclic AMP, Cell Growth, and the Immune Response* (Springer-Verlag, New York, 1974).
6. A. G. Gilman and M. Nirenberg, *Nature (Lond.)* **234**, 356 (1971); P. Furmanski, D. J. Silverman, M. Lubin, *ibid.* **233**, 413 (1971); K. N. Prasad and A. Vernadakis, *Exp. Cell Res.* **70**, 27 (1972).
7. J. Otten, J. Bader, G. Johnson, I. Pastan, *J. Biol. Chem.* **247**, 1632 (1972); J. R. Sheppard, *Nat. New Biol.* **236**, 14 (1972).
8. R. Perlman and I. Pastan, *Biochem. Biophys. Res. Commun.* **37**, 151 (1969).
9. H. R. Bourne, P. Coffino, G. M. Tomkins, *Science* **187**, 750 (1975).
10. Many of the ideas on the role of these indicator molecules in bacterial cells originated from discussions with B. N. Ames. He calls these molecules "alarmones" (J. C. Stephens, S. W. Artz, B. N. Ames, *Proc. Natl. Acad. Sci. U.S.A.*, in press) and considers them in detail in B. N. Ames and S. W. Artz, in preparation.
11. M. Cashel and T. J. Gallant, *Nature (Lond.)* **221**, 838 (1969); M. Cashel, *J. Biol. Chem.* **244**, 3133 (1969); ——— and B. Kalbacher, *ibid.* **245**, 2309 (1970).
12. E. W. Sutherland, T. W. Rall, T. Menon, *J. Biol. Chem.* **237**, 1220 (1962); E. W. Sutherland and T. W. Rall, *ibid.* **232**, 1077 (1958).
13. W. A. Haseltine, R. Block, W. Gilbert, K. Weber, *Nature (Lond.)* **238**, 381 (1972); J. Sy, Y. Ogawa, F. Lipmann, *Proc. Natl. Acad. Sci. U.S.A.* **70**, 2145 (1973); T. Laffler and J. Gallant, *Cell* **1**, 27 (1974); G. Stamminger and R. A. Lazzarini, *ibid.*, p. 85.
14. G. S. Stent and S. Brenner, *Proc. Natl. Acad. Sci. U.S.A.* **47**, 2005 (1961); A. M. Ryan and E. Borek, *Prog. Nucleic Acid Res. Mol. Biol.* **11**, 193 (1971).
15. R. A. Lazzarini and A. E. Dahlberg, *J. Biol. Chem.* **246**, 420 (1971); J. Gallant and G. Margason, *ibid.* **247**, 2289 (1972).
16. G. Edlin and J. Neuhard, *J. Mol. Biol.* **24**, 225 (1967); D. P. Nierlich, *Proc. Natl. Acad. Sci. U.S.A.* **60** 1345 (1968); Y. Sokawa and Y. Kaziro, *Biochem. Biophys. Res. Commun.* **34**, 99 (1969).
17. Y. Sokawa, E. Nakao, Y. Kaziro, *Biochem. Biophys. Res. Commun.* **33**, 108 (1968); J. Irr and J. Gallant, *J. Biol. Chem.* **244**, 2233 (1969); Y. Sokawa, J. Sokawa, Y. Kaziro, *Nat. New Biol.* **234**, 7 (1971).
18. M. Emmer, B. de Crombrugghe, I. Pastan, R. Perlmann, *Proc. Natl. Acad. Sci. U.S.A.* **66**, 480 (1970); G. Zubay, D. Schwartz, J. Beckwith, *ibid.* p. 104.
19. T. Yokota and J. S. Gots, *J. Bacteriol.* **103**, 513 (1970).
20. M. Cashel and J. Gallant, *J. Mol. Biol.* **34**, 317 (1968).
21. W. A. Haseltine and R. Block, *Proc. Natl. Acad. Sci. U.S.A.* **70**, 1564 (1973); F. S. Pedersen, E. Lund, N. O. Kjeldgaard, *Nat. New Biol.* **243**, 13 (1973).
22. M. G. Rossmann, D. Moras, K. W. Olsen, *Nature (Lond.)* **250**, 194 (1974).
23. C. R. Woese, *The Genetic Code* (Harper & Row, New York, 1967).
24. C. H. Waddington, *Organisers and Genes* (Cambridge Univ. Press, London, 1940); *The Strategy of the Genes* (Allen & Unwin, London, 1957).
25. M. Vaughan and D. Steinberg, *J. Lipid Res.* **4**, 193 (1963).
26. A. Hershko, P. Mamont, R. Shields, G. M. Tomkins, *Nat. New Biol.* **232**, 206 (1971).
27. D. S. Barkley, *Science* **165**, 1133 (1969); T. M. Konijn, *J. Bacteriol.* **99**, 503 (1969).
28. T. M. Konijn, J. G. C. van de Meene, J. T. Bonner, D. S. Barkley, *Proc. Natl. Acad. Sci. U.S.A.* **58**, 1152 (1967); J. T. Bonner, *Annu. Rev. Microbiol.* **25**, 75 (1971).
29. D. B. P. Goodman, F. E. Bloom, E. R. Battenberg, H. Rasmussen, W. L. Davis, *Science* **188**, 1023 (1975).
30. J. T. Bonner, *On Development* (Harvard Univ. Press, Cambridge, Mass., 1974).
31. F. Labrie *et al.*, *Adv. Cyclic Nucleotide Res.* **5**, 787 (1975).
32. J. F. Wilber, G. T. Peake, R. D. Utiger, *Endocrinology* **84**, 758 (1969); N. Fleischer, R. A. Donald, R. W. Butcher, *Am. J. Physiol.* **217**, 1287 (1969).
33. G. A. Buznikov, I. V. Chudakova, L. V. Berdysheva, N. M. Vyazmina, *J. Embryol. Exp. Morphol.* **20**, 119 (1968); T. Gustafson, in *Cellular Recognition*, R. T. Smith and R. A. Good, Eds. (Appleton-Century-Crofts, New York, 1969), pp. 47–60; G. A. Buznikov, B. N. Manukhin, A. V. Sakharova, L. N. Markova, *Sov. J. Dev. Biol. (Engl. Transl. Ontogenez)* **3**, 257 (1972).
34. J. H. Oppenheimer, D. Koerner, H. L. Schwartz, M. I. Surks, *J. Clin. Endocrinol. Metab.* **35**, 330 (1972); H. H. Samuels and J. S. Tsai, *Proc. Natl. Acad. Sci. U.S.A.* **70**, 3488 (1973).
35. R. J. B. King and W. I. P. Mainwaring, *Steroid-Cell Interactions* (University Park Press, Baltimore, 1974).
36. H. Rasmussen, *Science* **170**, 404 (1970); D. McMahon, *ibid.* **185**, 1012 (1974).

 Reprinted from *J. Mol. Evol.* **16**: 73-94.

Evolution of Patterns of Gene Expression in Hawaiian Picture-winged *Drosophila*

W. J. Dickinson

Genetics Department, University of Hawaii and Biology Department, University of Utah, Salt Lake City, Utah 84112, USA*

Summary. The tissue and stage specificity of expression of five enzymes was examined by electrophoretic analysis of relative enzyme levels in extracts of 13 larval and adult tissues in 27 species of Hawaiian picture-winged *Drosophila*. The developmentally regulated patterns of enzyme expression thus characterized were compared to a modal standard phenotype. About 30% of the pattern features analyzed differed significantly from the standard in one or more species. Many of these regulatory differences are essentially qualitative, with tissue specific differences in enzyme activity in excess of 100 fold for some species pairs. The adaptive significance of these pattern differences is unknown, but the results provide strong direct evidence for rapid evolution of new patterns of gene regulation in this group of organisms.

Key words: Enzymes – Evolution – Gene regulation – Hawaiian *Drosophila*

Introduction

There has been substantial interest in recent years in the role of regulatory genes in evolution. It has been suggested that regulatory changes are more important in the evolution of new adaptations than are changes in structural (protein coding) genes (Zuckerkandl 1963; Zuckerkandl and Pauling 1965; King and Wilson 1975; Markert et al. 1975; Wilson 1976; Kolata 1975). That is, changes in the location, timing and amount of synthesis of a protein may be more significant than changes in its amino acid sequence. However, in contrast to the extensive documentation of evolutionary changes in amino acid sequences and the population genetics of structural gene polymorphisms (Dayhoff 1972; Lewontin 1974; Ayala 1976; Nevo 1978; Wilson et al. 1977), the evidence for changes in gene regulation during evolution is scant and largely indirect. The basic argument advanced by King and Wilson (1975) is that there is little or no correlation between the rate of amino acid substitutions in proteins, studied either by direct sequence comparison or by immunological methods, and the rate of evolution of morphological and ecological differences. The latter differences are therefore presumed to be based on some other kind of genetic change, probably in regulatory genes. This essentially negative argument is not compelling. For example, if a majority of structural gene substitutions are adaptively neutral, this majority could mask any meaningful correlation even if most new morphological and ecological adaptations were based on a minority of significant structural mutations.

Attempts to obtain more direct evidence can present problems as well. Gould (1977) has suggested that changes in relative timing of developmental events, leading ultimately to altered adult morphology, may be examples of evolution by regulatory modification. However, it is easy to propose reasonable models in which such timing changes depend on mutations in structural genes. Similar objections would hold for almost any approach that starts from differences in gross morphological phenotypes. At the other extreme, comparisons of the rate of evolutionary divergence of nucleotide sequences in protein coding and non-coding regions of the genome (Rosbash et

*Permanent address

al. 1975) are interesting but hard to interpret until the functions of noncoding regions (which need not be restricted to regulation) are better understood. Even direct examination of known regulatory regions, if that were possible, would not be conclusive at present since we do not know what kinds of sequence changes are functionally significant.

Direct examination of the patterns of expression of specific genes with known products avoids these problems and seems to be the most certain way of detecting evolutionary changes in gene regulation at present. This is roughly analogous to detecting structural gene diversity and change by electrophoretic methods (Lewontin and Hubby 1966). A number of workers interested in mechanisms of gene regulation in eukaryotes have studied natural genetic variants that alter the developmental pattern of expression of specific enzymes (reviewed by Paigen 1971, 1979). Such variants have been found by comparing the patterns of expression of the selected enzyme(s) (i.e., relative amounts of the enzyme present in various differentiated tissues and/or at different developmental stages) in a series of inbred lines of a given species. Substantial strain to strain variability in these patterns of expression has been found in surveys of quite modest size, at least in maize (Schwartz 1962, 1976); mice (Paigen 1961; Boubelik et al. 1975; Paigen et al. 1975, 1976; Lusis and Paigen 1975) and *Drosophila* (Dickinson 1975, 1978; Chovnick et al. 1976; Thompson et al. 1977; Abraham and Doane 1978). Recently, Powell and Lichtenfels (1979) and Powell (1979) have confirmed the existence of substantial regulatory polymorphism affecting amylase expression in *D. pseudoobscura* in a more systematic survey of natural populations, and Snyder (1978) has done so for variants affecting hemoglobin expression in mice. Such polymorphisms could be the raw material on which natural selection acts to generate the new patterns of regulation hypothesized to be important in adaptive evolution. It therefore seems reasonable to ask how frequently new patterns of expression of specific enzymes become established during the evolution of a group of organisms.

A number of examples of evolutionary change in regulatory patterns of specific genes are known (Wilson et al. 1977) but I am aware of only a few systematic surveys comparing such patterns in a number of related species (e.g. Shows et al. 1969; Shaklee et al. 1973; Markert et al. 1975; Fisher and Whitt 1978; Ferris and Whitt 1979). The most extensive of these studies have been concerned primarily with divergence of pattern of expression of paralogous loci following gene duplication, situations that might be expected to facilitate more rapid regulatory change than could occur in the absence of a "spare" copy. These studies also concern taxa that have diverged over considerable evolutionary periods and most of the reported differences in patterns of gene expression were found at the level of families or at least genera.

I report here a preliminary effort to apply methods like those used to identify regulatory variants within a species to a systematic survey of regulatory differences between species of the picture-winged group of Hawaiian *Drosophila*. This extensively studied group (Carson and Kaneshiro 1976) is extremely well suited to this project. A large number of morphologically and ecologically diverse species have arisen in a relatively short time. Their phylogenetic relationships have been documented in detail. Their large size (typically 10–15 times the weight of *D. melanogaster*) facilitates the comparisons of regulatory patterns. The results reported here show that a surprisingly large number of very dramatic regulatory changes affecting the selected enzymes have occurred during the evolution of this group, even between very closely related species, and suggest that the general approach illustrated here will be a useful one.

Materials and Methods

The species surveyed are listed in Table 1 along with information about the specific stocks used. Figure 1 depicts the currently accepted relationships between these species, which are distributed throughout the picture-winged phylogeny. Except where noted, all flies were laboratory reared by a standard culture procedure (Wheeler and Clayton 1965) with constant temperature and humidity. Tissues were obtained for electrophoretic analysis by dissecting staged individuals in ice-cold *Drosophila* Ringers.

Table 1. Species, stock numbers[a] and collection sites

Species	Stock No.	Collection site
1. *D. adiastola*	M55G17	Kaulalewelewe, Maui
D. adiastola	T79B7	Kaulalewelewe, Maui
D. adiastola	U44J7	Kaulalewelewe, Maui
D. adiastola	U53B7	Kaulalewelewe, Maui
D. adiastola	U53B8	Kaulalewelewe, Maui
D. adiastola	U53B10	Kaulalewelewe, Maui
D. adiastola	U54B3+5*	Waikamoi, Maui
2. *D. affinidisjuncta*	S36G1	Kaulalewelewe, Maui
3. *D. bostrycha*	S30B5	Apee, Molokai
4. *D. clavisetae*	U84Y26*	Waikamoi, Maui
D. clavisetae	U84Y27	Waikamoi, Maui
D. clavisetae	U84Y30	Waikamoi, Maui
5. *D. crucigera*	U70 (mass)	Mt. Haupu, Kauai
D. crucigera	U72Y3*	Mt. Tantalus, Oahu
D. crucigera	U72Y5	Mt. Tantalus, Oahu
D. crucigera	U72Y8	Mt. Tantalus, Oahu
6. *D. cyrtoloma*	U78Y (wild)	Waikamoi, Maui
7. *D. differens*	U43V1	Hanalilolilo, Molokai
8. *D. discreta*	U84Y41	Waikamoi, Maui
9. *D. disjuncta*	S56H12	Kipahulu Valley, Maui
10. *D. engyochracea*	U77G (mass)	Bird Park, Hawaii
11. *D. formella*	M87G1	Puuwaawaa. Hawaii
12. *D. grimshawi*	G1*	Auwahi, Maui
D. grimshawi	S30G10	Apee, Molokai
13. *D. heteroneura*	Q71G12	Olaa Forest Reserve, Hawaii
D. heteroneura	T94B18	Olaa Forest Reserve, Hawaii
D. heteroneura	U26B52	Kahuku Ranch, Hawaii
D. heteroneura	U35B38*	Hualalai Ranch, Hawaii
D. heteroneura	U52B10	Waihaka Gulch, Hawaii
14. *D. neopicta*	U80Y (wild)	Hanalilolilo, Molokai
15. *D. ornata*	U73Y1	Mt. Kahili, Kauai
16. *D. orthofascia*	U55G11	Olaa Forest Reserve, Hawaii
D. orthofascia	U58G1*	Auwahi, Maui
D. orthofascia	U58B3+G6	Auwahi, Maui
17. *D. paucipuncta*	U64Y (mass)	Olaa Forest Reserve, Hawaii
18. *D. picticornis*	U70 (mass)	Mt. Haupu, Kauai
19. *D. planitibia*	U84Y14	Waikamoi, Maui
D. planitibia	U84Y17	Waikamoi, Maui
D. planitibia	U84Y18	Waikamoi, Maui
D. planitibia	U84Y (mass)*	Waikamoi, Maui
20. *D. primaeva*	U73Y2	Mt. Kahili, Kauai
21. *D. prostopalpis*	S15B33	Kaulalewelewe, Maui
22. *D. punalua*	U72Y13*	Mt. Tantalus, Oahu
D. punalua	U72Y15	Mt. Tantalus, Oahu
23. *D. setosimentum*	U69G12	Waihaka Gulch, Hawaii
24. *D. silvestris*	T94B7	Olaa Forest Reserve, Hawaii
D. silvestris	U26B9	Kahuku Ranch, Hawaii
D. silvestris	U28T2*	Kilauea Forest Reserve, Hawaii
D. silvestris	U34B4	Kehena Ditch, Hawaii
25. *D. silvarentis*	U83G2	Pohakuloa State Park, Hawaii
26. *D. sproati*	U64G6	Olaa Forest Reserve, Hawaii
D. sproati	U64G8	Olaa Forest Reserve, Hawaii
D. sproati	U64G10*	Olaa Forest Reserve, Hawaii
27. *D. villosipedis*	U70 (wild)	Mt. Haupu, Kauai
D. villosipedis	U74Y (wild)	Kokee State Park, Kauai

[a] Unless otherwise designated, stocks are derived from single wild-caught females. "Mass" indicates a stock derived from several females and "wild" indicates that no laboratory stock was successfully established but wild caught adults were analyzed. In the case of species for which more than one stock was used, an asterisk marks the stock first analyzed and used in constructing the modal standard phenotype given in Table 3

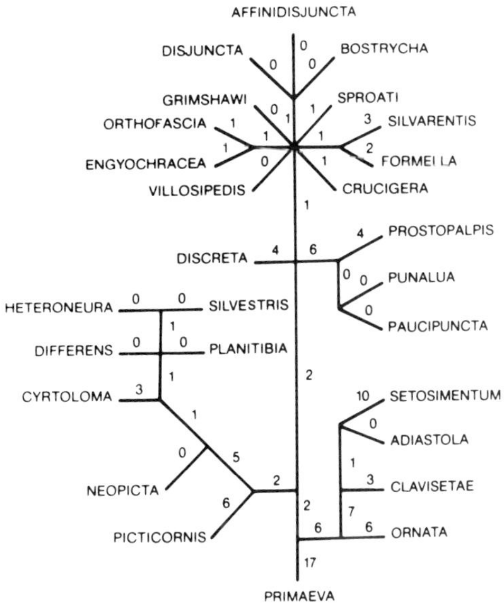

Fig. 1. Phylogenetic relationships of the picture-winged *Drosophila* surveyed for enzyme regulatory patterns. This dendrogram is greatly simplified from the one given by Carson and Kaneshiro (1976) which shows relationships between about 100 picture-winged species. It is based primarily on chromosome inversions but is supported by a variety of morphological, ecological and behavioral studies to which the above reference provides access. The number adjacent to each line segment represents the number of fixed inversion differences between branch points or between a branch point and the designated species. The major subgroups represented are the planitibia group (lower left), the adiastola group (lower right), the punalua group (center right) and the grimshawi group (top)

The stages and tissues used are detailed in Table 2 along with standard abbreviations. Extracts were prepared by grinding each organ in a small volume of distilled water (10 μl for larval carcass and adult head and 6 μl for all other tissues). Three μl aliquots of these extracts were electrophoresed on agarose gels (Dickinson and Carson 1979). These gels yield adequate resolution of the enzymes selected for study, are much easier and faster to prepare and run than either starch or acrylamide gels, provide greater staining sensitivity with very small samples and are easily preserved by drying (with no

Table 2. Standard set of tissues

Stage and tissue	Abbreviation
A. Feeding third instar larvae[a]	
1. carcass (hypodermis and muscle)	LCA
2. midgut	LMG
3. hindgut	LHG
4. Malpighian tubules	LMT
5. fat body	LFB
B. Late larvae[b]	
1. salivary glands	LSG
C. Mature adult females[c]	
1. head	AHD
2. midgut	AMG
3. hindgut	AHG
4. Malpighian tubules	AMT
5. fat body	AFB
6. ovary	AOV
D. Mature adult males[d]	
1. reproductive system (testes, ejaculatory bulb, and paragonia)	AMR

[a] These larvae were taken from the food in cultures in which some individuals had begun to pupate. They had a midgut distended with food and salivary glands of moderate size

[b] These larvae had left the food to pupate and had a much smaller midgut and enlarged salivary glands

[c] Females were aged 4–6 weeks from eclosion and had large ovaries with mature oocytes

[d] Males were aged 4–6 weeks from eclosion

special equipment). Preliminary studies with acrylamide gels at several concentrations revealed no additional isozymes not resolved in agarose. Two gels were run with each set of samples. One was stained for aldehyde oxidases (AO) as described (Dickinson 1970) except with 0.1 μl/ml benzaldehyde as substrate, and the other for alcohol dehydrogenase (ADH), octanol dehydrogenase (ODH) and xanthine dehydrogenase (XDH) as described for ADH and ODH (Dickinson and Carson 1979) but with the addition of 1 mg/ml of hypoxanthine, dissolved by boiling, to the buffer. Staining of the AO gel was for 30 min and of the dehydrogenase gel for 60 min, both at 37°C. The gels were then fixed and washed in water: ethanol: acetic acid (5:5:1) and air dried on a glass plate to make a permanent record of the pattern of expression of these enzymes in the standard set of tissues.

Semiquantitative estimates of the amount of each enzyme in each tissue were obtained by visual comparison of the survey gels to a set of standards prepared by making serial two fold dilutions of extracts of appropriate tissues and electrophoresing and staining each dilution series exactly as was done with the survey samples. The most dilute step in each standard series giving a visually detectable stain was assigned a value of one. The successive (two-fold) more concentrated steps in the series were designated by successive higher integers. The staining intensity of each sample for each enzyme in the survey was given the numerical value of the step in the appropriate dilution series that it most closely resembled. Since successive integers represent two-fold differences in enzyme activity, the numerical scores are a logarithmic (base 2) function of relative enzyme activity. This procedure is simpler than that of Klebe (1975) in that a single standard dilution series is prepared for each enzyme and the survey samples are compared to that series rather than preparing a dilution series from each sample to be evaluated. It is undoubtedly the case that the range of enzyme concentrations covered exceeds the range over which staining intensity is linear with concentration. However, this should have no serious effect on the scoring since standards and samples were prepared and stained under identicial conditions and for the same time and both should be similarly affected by any non-linearity. There is no difficulty in visually distinguishing adjacent members of each standard series, even at the upper end of the concentration range.

Results

Enzyme Homology

The basic form of the data obtained for each species is illustrated in Fig. 2. It is a set of gels displaying the relative activities of several enzymes in each of thirteen larval and adult tissues. To properly evaluate differences in the patterns obtained with different species, we obviously must recognize products of orthologous structural genes in these patterns. The cytogenetic similarity of all the picture-winged *Drosophila* (Carson and Kaneshiro 1976) would strongly suggest that they carry equivalent sets of structural genes. We used three criteria to confirm this expectation and to recognize the various enyzmes: (1) substrate specificity; (2) relative electrophoretic mobility; (3) basic features of the regulatory pattern conserved in all species examined. Based on these criteria, we recognized six enzymes stained in our procedures of whose species to species correspondence we are confident.

1. Alcohol dehydrogenase (ADH) is active with ethanol and isopropanol but not with octanol or in the absence of NAD. It stays near the origin or migrates slightly toward the cathode in our electrophoretic procedure, is consistently present in larval and adult fat body (though at variable levels) and is never present in ovary.
2. Octanol dehydrogenase-1 (ODH-1) is active with octanol but not with ethanol or isopropanol or in the absence of NAD. Its electrophoretic mobility is intermediate between ADH and XDH. It is always prominent in ovary.
3. Octanol dehydrogenase-2 (ODH-2) is detectable only in some species. It has substrate requirements like ODH-1 but moves further toward the anode during electrophoresis.

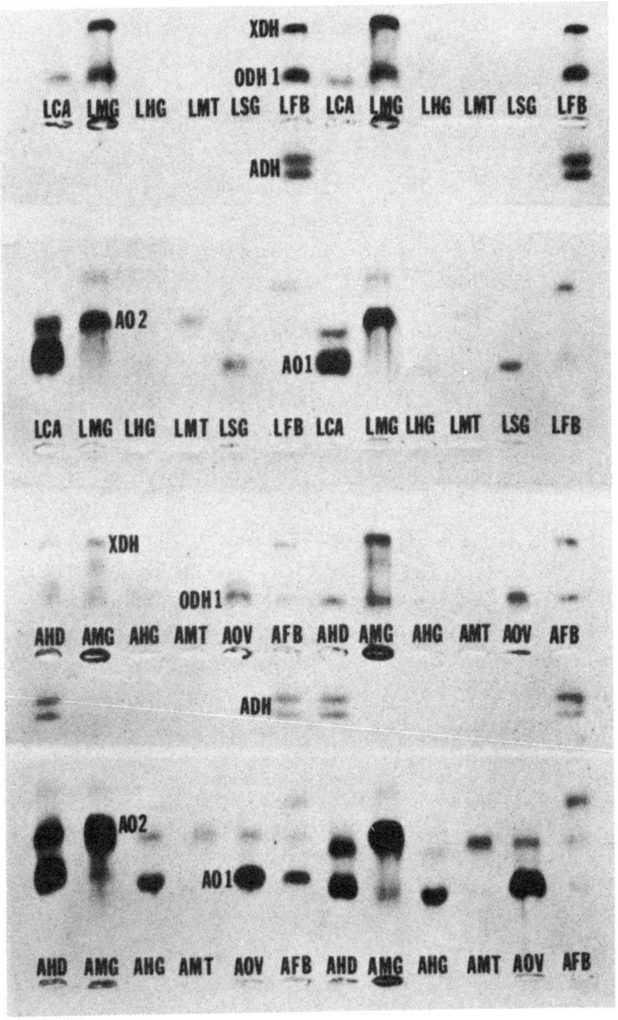

Fig. 2. Pattern of enzyme expression in *D. formella*. The four gels shown here are, from top to bottom: (1) dehydrogenases in larvae, (2) aldehyde oxidases in larvae, (3) dehydrogenases in adult females, and (4) aldehyde oxidases in adult females. Anodal migration is toward the top of the figure in each case. The positions of the enzyme of interest are marked on each gel. The multiple bands in the ADH region are products of a single gene (Dickinson and Carson 1979). The tissue analyzed in each track is labeled just above the sample wells (origin) using the abbreviations given in Table 2. The left and right halves of each gel provide duplicate samples derived from different individuals, demonstrating the reproducibility of the overall pattern

4. Xanthine dehydrogenase (XDH) is active with hypoxanthine in the presence of NAD and with aldehydes in the absence of NAD. It forms the most anodal band in both dehydrogenase and AO gels. It is consistently present in larval midgut and fat body.
5. Aldehyde oxidase-1 (AO-1) is active with a variety of aldehydes and does not require NAD. It is the major band closest to the origin in AO gels. It is always prominent in adult head and ovary.
6. Aldehyde oxidase-2 (AO-2) has substrate requirements similar to AO-1. It is intermediate in mobility between AO-1 and XDH (AO-3). It is always prominent in larval and adult midgut and absent or very weak in ovary.

More detailed evidence for homology has been reported for ADH, AO-1 and ODH-1 (Dickinson and Carson 1979; Dickinson 1980a) and substantially similar evidence is available for the other enzymes. In addition, electrophoretic variants were used to establish that at least five of these enzymes are products of distinct structural genes and that corresponding bands found with extracts of different tissues from an individual fly are indeed products of the same structural locus. One enzyme, ODH-2, is not identifiable in all species and is less clearly a distinct gene product (Dickinson 1980a). It is not included in the numerical tabulations in this paper but is listed above because it shows some very interesting regulatory features discussed later. Two or three bands can sometimes be seen in gels stained for AO that are not included in the set of enzymes considered. AO-3 is identical to XDH (present in the same tissues and always co-migrates) and is scored only as XDH. AO-4 is a product of the same gene as AO-1 (affected coordinately by electrophoretic variants) and is not scored separately. An additional AO is sometimes seen in midgut and may be a modification of AO-2 but it appears too sporadically to test. It has been ignored when present.

Standard Phenotype

To facilitate comparisons between what proved to be rather diverse patterns, a standard phenotype was constructed as follows. When gels had been run on two to four individuals of one stock of each species, they were scored to generate, for each species, a five (enzymes) by thirteen (tissues) matrix of numerical values representing relative staining intensities. In general, the scores of different individuals from a given stock were nearly identical (see Fig. 2). Where significant differences were found among the first set of individuals analyzed (compare, for example, AO-1 in adult fat body of the two individuals in Fig. 2), the relevant tissues and enzymes were examined in at least six additional individuals of the same stock and the modal value was used in the matrix for that species. The standard phenotype was then derived by taking the modal value (across all species) at each position in a matrix of the same format. This standard phenotype is given in the top line of each of the five rows in Table 3.

Table 3. Standard regulatory phenotype for five enzymes and deviations therefrom[a]

	LCA	LMG	LHG	LMT	LFB	LSG	AHD	AMG	AHG	AMT	AFB	AOV	AMR
	5	4	0	2	7	0	4	0	0	0	4	0	0
AHD	0-8	0-8	0	0-3	4-9	0-1	0-7	0-2	0	0-1	1-7	0	0
	R	R	N	R	C	N	R	R	N	N	R	N	N
	2	3	0	0	3	0	2	2	0	0	1	4	0
ODH-1	2-4	2-4	0	0-1	2-5	0-2	1-3	0-3	0-2	0-1	0-4	3-5	0-1
	C	N	N	N	C	R	N	U	U	N	U	N	N
	0	3	0	0	2	0	1	3	0	1	2	0	0
XDH	0-1	2-5	0	0-3	2-5	0-2	0-3	2-4	0	0-2	2-4	0-2	0
	N	C	N	R	C	U	C	N	N	N	C	R	N
	3	0	0	0	0	0	6	0	3	0	0	8	3
AO-1	2-8	0	0-4	0	0-7	0-6	5-8	0	2-5	0-3	0-8	7-9	2-4
	R	N	R	N	R	R	C	N	C	R	R	N	N
	0	4	0	0	2	0	0	6	0	2	2	0	0
AO-2	0-3	3-9	0	0-3	0-5	0-1	0-3	4-9	0-1	0-4	1-4	0-1	0
	N	C	N	R	R	N	R	C	N	R	C	N	N

[a]Tissues (abbreviated as in Table 2) are listed across the top and enzymes down the left side. There are three lines for each enzyme. The top line lists the standard phenotype (see text), the second line shows the range of values observed in the survey of all species, and the bottom line is an evaluation of the variability employing the following notation

- N – No significant variation
- C – Qualitatively constant. Differences greater than two fold from the mode are observed but the affected tissue(s) fall in the usual rank order (relative to other tissues in the same species)
- R – Clear regulatory differences that are significant, reproducible and tissue or stage specific in effect
- U – Unconfirmed. Not consistently reproducible or additional individuals unavailable for confirmation

Deviations from the Standard

Once the standard phenotype was generated, the phenotype for each species was compared to it. For our purposes, we considered as significant only deviations greater than one two-fold step above or below the standard value. Thus, given a modal value of three, two or four would be counted as "normal" but one or five would be noted as significantly different. This is a conservative procedure that may well have resulted in some real differences being ignored but it seemed justified in view of the semiquantitative scoring in two-fold steps and the variability between species in individual weight, which covers about a two fold range.

Features judged significantly different from the standard were rechecked in two steps:

1. The relevant enzymes and tissues were analyzed in at least six additional individuals from the same stock.
2. When additional stocks of the same species were available, the same feature(s) were checked in several individuals from each stock.

Table 3 is a summary of the extent of variability in each of the 65 characters scored in this study. Nineteen of the 65 (29%) are significantly variable between species. In a number of cases, the differences are apparently qualitiative and dramatic. For example, ADH is undetectable in larval caracass and midgut of some species and ranges up to a level about 128 fold higher than the minimum readily detectable level in other species. AO-1 in adult fat body shows a similar range of values. Several less extreme but still dramatic cases can be found by inspecting Table 3. In both of the foregoing cases (and most others) the species with low or undetectable activity in the affected tissue(s) have activities of the same enzyme not significantly lower than the standard in one or more other tissues. Thus there is no defect in the structural gene resulting in drastically lowered activity in general but rather differences that are tissue or stage specific.

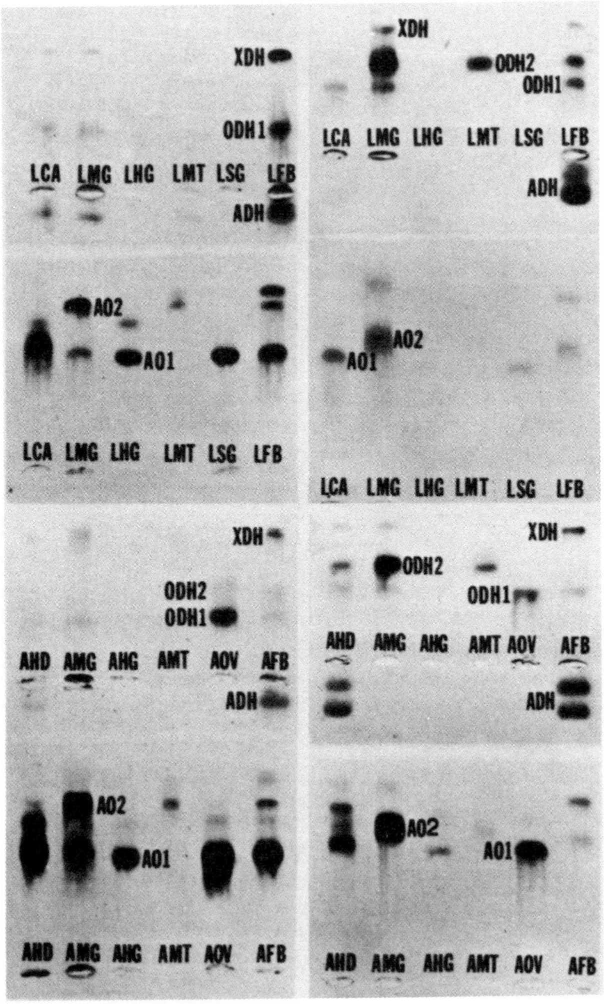

Fig. 3. Regulatory patterns in *D. silvestris* and *D. orthofascia*. Gels are arranged and labeled as in Fig. 2 except *D. silvestris* patterns on the left are compared to *D. orthofascia* patterns on the right

This point is reinforced in Fig. 3, which provides a side by side comparison of two species (*D. silvestris* and *D. orthofascia*) showing several regulatory differences. For example, ADH is undetectable in the larval midgut of *D. orthofascia* but readily detectable in that tissue of *D. silvestris* even though the level in fat body is higher in the former species. AO-1 is undetectable in adult fat body but present at a high level in head and ovary tissues of *D. orthofascia* while it is present at about equal levels in all three tissues of *D. silvestris*. This enzyme is likewise present at relatively lower levels

Table 4. Summary of variant phenotypes

Phen. No.	Affected feature Enzyme	Tissue	Activity level variant	Standard	Species in which found	Comments
1	ADH	LCA	0-2	5	8,10,11,16, 17,20,25,26	Co-ordinate with phenotypes 2 and 3 in species 8,11,16,20,25. Cis-acting control in 12/16 hybrids (Dickinson and Carson 1979).
2	ADH	LMG	0-1	4	8,11,16,20, 23,25,26	See above
3	ADH	LMT	0	2	8,11,15,16, 20,25,26	See above
4	ADH	AHD	0	4	6,23	ADH activity lower than average in other tissues but readily detectable.
5	ADH	AMG	2	0	2,7	Cis-acting control in 13/7 hybrids (Dickinson 1980b).
6	ADH	AFB	1-2	4	1,6,15, 18,23	In species 1,6,15 and 23 associated with low activity in other adult tissues but species 18 has above average ADH in AHD.
7	ODH-1	LSG	2	0	21	Only one stock available
8	XDH	LMT	3	0	21	Only one stock available
9	XDH	AOV	2	0	1,15	Both members of adiastola subgroup
10	AO-1	LCA	5-8	3	5,8,10,11, 13,24,25,26	Species 8,10,11,25 and 26 have high activity in this tissue associated with low activity in LFB. Cis-acting control in 12/11 hybrids (Dickinson 1980a).
11	AO-1	LHG	2-4	0	13,19,24	All members of planitibia subgroup.
12	AO-1	LFB	2-5	0	5,12,13, 21,24	Cis-acting control in 13x7 (Dickinson 1979b) and trans-acting control in 12x11 (Dickinson 1980a). Variable within some species.
13	AO-1	LSG	2-6	0	5,7,10,11, 13,16,19, 24,25	Trans-acting control in 12/11 hybrids (Dickinson 1980a).
14	AO-1	AMT	2-3	0	1,9	Polymorphic in both species. Cis-acting control in 1 (unpublished).
15	AO-1	AFB	3-8	0	2,3,5,9,11, 12,13,14,15, 19,20,21,24, 26	Scattered through the phylogeny. Variable within species.
16	AO-2	LMT	2-3	0	7,11,19,25	All but 11 are in planitibia subgroup.
17A	AO-2	LFB	0	2	2,10,18,21 22,26	
17B	AO-2	LFB	5	2	7	Compared to other planitibia subgroup, AO-2 is high and AO-1 is low in this tissue.
18	AO-2	AHD	2-3	0	5,13,19	
19	AO-2	AMT	0	2	8,14,17	

in all larval tissues of *D. orthofascia*. This species also shows a strong ODH-2 band in Malipighian tubules while *D. silvestris* has no ODH in this tissue. This enzyme is not listed in Table 3 for reasons given above but this is nevertheless a clear regulatory difference. The reader can undoubtedly locate additional differences in the patterns of enzyme expression of these two species. Other specific examples have been documented elsewhere (Dickinson and Carson 1979; Dickinson 1980a, b). Table 4 provides a description of the deviant phenotypes that have been confirmed in this survey of 27 species and lists the species (from Table 1) in which they have been found. Complete listings of the phenotypes of all species appear in an appendix.

Discussion

It is clear that the method used here readily detects differences in patterns of enzyme expression between species. The number of quite dramatic differences detected is surprising. The estimate that about 30% of the regulatory features scored differed significantly from the standard in at least one species is very conservative. Beyond the fact that only differences greater than 2 fold were counted, many of the features scored as "qualitatively constant" (C) in Table 3 may actually have significant differences with a regulatory basis. It has been shown, for example, that the overall higher level of ADH and lower level of AO-1 in *D. differens* compared to *D. heteroneura* are probably due to differences in the number of enzyme molecules present, not to differences in catalytic properties (Dickinson 1980b). In several cases, intraspecific variants that produce an overall change in enzyme level not known to be tissue or stage specific have been interpreted as being due to regulatory sites (e.g. Chovnick et al. 1976; Thompson et al. 1977). These would not be counted as confirmed regulatory variants by the criteria used here. Thus, the number of traits displaying regulatory variability may well be almost 50% (31 of 65).

In addition it seems likely that some of the variant patterns of expression have arisen independently more than once. Consider, for example, the absence of ADH from larval carcass and midgut. The species showing this combination of traits (Table 4) are scattered in the phylogeny with intervening species that have the more common phenotype. Likewise, drastic differences in levels of AO-1 in fat body are found within several of the subgroups. Given the small sample sizes for some species, it remains possible that some of the variant patterns are more widespread, as intraspecific polymorphisms, than is evident from the data available. Nevertheless, these and similar cases reinforce the conclusion that changes in regulatory pattern, or at least dramatic changes in the frequencies of alternative alleles affecting regulatory patterns are common in this group of species for the set of enzymes examined.

Several questions raised by the current work merit further investigation. Perhaps foremost is whether these pattern differences reflect changes in distinct regulatory loci. It is conceivable that mutations in the structural gene coding for an enzyme can somehow influence its tissue or stage specific expression. From a mechanistic point of view this is clearly an interesting question the answer to which would shed light both on how patterns are normally generated and on how new patterns evolve. From an evolutionary point of view, there seems to be no doubt that the *consequences* of the variants we have detected are "regulatory" whether or not the *mechanisms* conform to some formal definition of that term. If it turns out that structural mutations can result in dramatically altered patterns of expression, this will add a new dimension to the debate concerning the significance of structural gene polymorphism, which has generally been couched in terms of the functional properties (kinetic characteristics, heat stability, temperature or pH response etc.) of the alternate forms of the enzyme.

Some genetic analysis of several of the regulatory differences reported here has already been completed (Dickinson and Carson 1979; Dickinson 1980a, b and unpublished). Of 17 traits analyzed thus far in three hybridizable species pairs and two intraspecific variants, 11 are definitely under control of cis-acting genes. That is, in heterozygotes, each structural allele is expressed independently according to the pattern characteristic of the parent from which it is derived regardless of the allele present on the homologous chromosome. Three others show intermediate expression in heterozygotes, an observation that is consistent with cis-acting control, but that could result from a more complex situation. Indeed, even the unambiguous cases of cis-acting control should not be interpreted as meaning that other kinds of elements are unimportant in producing the normal patterns of expression. It simply means that the major *differences* in the patterns observed in the species compared are due to differences in cis-acting genetic elements. In the final three cases, the pattern characteristic of one parent dominates in the hybrid. This implies control by trans-acting (diffusible) factors but provides no information on the number and location of relevant genetic loci. If we eliminate duplications of the same trait analyzed in different species pairs and combine cases where several features of expression of the same enzyme differ between species and show a similar genetic basis (i.e., assume a multi-functional regulator), we are left with six cases of cis-acting control, some quite complex in effect (Dickinson 1980b) and still three of trans-acting control. In all sufficiently analyzed cases of cis-acting control (Dickinson 1980b), the mechanism apparently involves differential accumulation of enzyme molecules, not catalytic differences. All of these results are quite consistent with what is known about intraspecific regulatory variants (Paigen 1971, 1979). Efforts to define, at a biochemical level, the mechanisms involved in producing these pattern differences are under way.

A second major question is whether the observed pattern differences are physiologically and ecologically significant. Although intuitively it seems likely that drastic differences in relative amounts of an enzyme present in different cell types would "matter" to an organism more than subtle physical/chemical differences in the properties of the enzyme, we have no data that rigorously excludes extension of a neutralist interpretation to the present regulatory differences. The conservation of certain regulatory features throughout the phylogeny (ADH always present in fat body, ODH and AO-1 always present in ovary, AO-2 always the major AO in gut etc.) suggests that these enzymes do have significant functions and their presence in at least some tissues has adaptive significance. Indeed we have some evidence (unpublished) that certain major regulatory features are conserved between picture-winged *Drosophila* and *D. melanogaster*. However, it is possible that enzyme levels in other tissues are not functionally significant and that interspecific differences do not reflect adaptive responses. We have not found obvious correlations between the observed pattern differences and ecological parameters. However, it may be possible to get experimental evidence for a functional significance for some of these pattern differences. There is now good evidence that overall ADH levels in other species of *Drosophila* influence tolerance for ethanol in the environment (David 1977) and that low XDH activity results in sensitivity to purines (Finnerty et al. 1970). Some of the other enzymes considered here may have similar functions with respect to other potentially toxic compounds. We might therefore ask whether changes in tissue distribution of these enzymes also can affect sensitivity to potentially toxic substrates. If differences are observed, correlations with differences in the food sources utilized by various species could be examined.

A third question concerns the generality of the findings reported here. In this initial study, no serious effort was made to include a range of enzymes fulfilling a variety of metabolic roles. Indeed, null mutations are known in *D. melanogaster* for all of these enzymes except ODH and homozygotes survive perfectly well, at least under laboratory conditions (Dickinson and Sullivan 1975). One might therefore worry that expression of these enzymes is under, at best, weak selection pressure and hence many variant patterns have appeared. (However, see Zuckerkandl 1978, for a discussion of the distinction between lack of an essential function and selective neutrality). Undoubtedly, additional enzymes should be examined. Nevertheless, the present evidence supports the hypothesis that novel regulatory patterns arise with sufficient frequency to be significant in the evolution of new adaptations if they do have selective significance. Note in particular that the variants include addition of new characteristics, not just loss of functions. For example, *D. orthofascia* has a high level of ODH-2 in Malpighian tubules (Fig. 3) while no other species, including close relatives, has significant ODH in this tissue. Thus, if presence of ODH in this tissue opens up new adaptive possibilities (e.g. use of a previously toxic food source), that opportunity can be exploited without sacrificing the functions previously performed by the same enzyme in other tissues. It is relevant to note that even with the generally modest sample size examined so far for each species, we have already detected at least three cases of presumptive regulatory polymorphism within species (AO-1 presence or absence in Malpighian tubules of *D. adiastola* and *D. bostrycha* and AO-1 level in fat body of *D. heteroneura*). The phenotypic effects in these cases are qualitatively quite similar to observed interspecific differences and clearly are due to simple genetic polymorphisms at cis-acting loci (unpublished results). This confirms the expectation, as discussed in the introduction, that intraspecific regulatory variability that could serve as the basis for the evolution of new species specific patterns is present in natural populations.

A fourth question is whether overall regulatory similarity or difference between species correlates with phylogenetic distance. A detailed treatment of this question, possible by methods of numerical taxonomy, is beyond the scope of this paper. In general, it is the case that closely related species have very similar overall patterns. Thus, *D. disjuncta, D. affinidisjuncta, D. bostrycha* and *D. grimshawi* are all very similar, *D. sylvestris* and *D. heteroneura* are essentially indistinguishable etc. On the other hand, *D. orthofascia* seems to have several regulatory features not present in near relatives. Looked at from another point of view, some variant patterns have been found only in members of a single subgroup but others are rather widespread (see Table 4).

Finally, we can return to the proposition that prompted this study. If we attribute much adaptive evolution to regulatory change because of the poor correlation between morphological and ecological differences and changes in structural loci, it is relevant to ask whether there is a better correlation with regulatory changes. A couple of contrasts between the present results and those obtained in somewhat comparable studies on patterns of enzyme expression in a range of fish species are worth noting. It appears that fish species in the same genus or even in closely related genera have qualitatively very similar patterns of enzyme expression (Phillip et al. 1979) with major pattern differences usually noted only at the level of families or orders (Markert et al. 1975; Fisher and Whitt 1978; Ferris and Whitt 1979). In the present study, quite marked pattern differences are found even within closely related subgroups of the picture-winged *Drosophila*, some members of which have almost certainly diverged within the last few million years or less (Carson and Kaneshiro 1976). It also appears that in some pairs of fish species, quite similar patterns of regulation may depend on somewhat different regulatory mechanisms, with the result that expression is abnormal in hybrids (Whitt et al. 1977). In contrast in some crosses between species pairs of picture-winged *Drosophila*, the parental patterns are quite distinct but each allele appears to be expressed in a perfectly normal way in the hybrids (Dickinson and Carson 1979; Dickinson 1980b). In fact, alleles of ADH and AO-1 derived from *D. heteroneura* continue to be expressed according to their normal distinct patterns after backcrossing hybrids to *D. differens* for two generations (unpublished results). Thus it appears possible either to conserve a pattern while mechanisms diverge or to conserve *most* elements of the regulatory machinery while patterns diverge. More extensive studies employing similar methods with perhaps a constant set of enzymes and taxonomic groups less divergent than fish and *Drosophila* (e.g. a more conservative group of *Drosophila* compared to the picture-winged group) should provide much needed direct evidence on the relative importance of regulatory change in adaptive evolution.

Acknowledgments. This research was supported by NSF grant DEB 78-10679. Much of the work was done while the author was a guest in the laboratory of H.L. Carson, Genetics Department, University of Hawaii and employed stocks maintained in that laboratory or collected and identified with the assistance of H.L. Carson and his colleagues.

References

Abraham I, Doane WW (1978) Proc Nat Acad Sci USA 75:4446–4450
Ayala FJ (1976) Molecular evolution. Sinauer, Sunderland, Mass
Boubelik M, Lengerova A, Bailey DW, Matousek V (1975) Dev Biol 47:206–214

Carson HL, Kaneshiro KY (1976) Annu Rev Ecol Syst 7:311–345
Chovnick A, Gelbart W, McCarron M, Osmond B, Candido EPM, Baillie DL (1976) Genetics 84:232–255
David J (1977) Annee Biol 16:451–472
Dayhoff MO (1972) Atlas of protein sequence and structure. Vol 5, National Biomed Research Foundation, Silver Spring, Maryland.
Dickinson WJ (1970) Genetics 66:487–496
Dickinson WJ (1975) Dev Biol 42:131–140
Dickinson WJ (1978) J Exp Zool 206:333–342
Dickinson WJ (1980a) Science 207:995–997
Dickinson WJ (1980b) Devel Gen 1:229–240
Dickinson WJ, Carson HL (1979) Proc Nat Acad Sci USA 76:4559–4562
Dickinson WJ, Sullivan DT (1975) Gene-enzyme systems in *Drosophila*. Springer, Berlin, Heidelberg, New York
Ferris SD, Whitt GS (1979) J Mol Evol 12:267–317
Finnerty V, Duck P, Chovnick A (1970) Genet Res 15:351–355
Fisher SE, Whitt GS (1978) J Mol Evol 12:25–55
Gould SJ (1977) Ontogeny and phylogeny. Harvard Univ Press, Cambridge, Mass, p 405
King MC, Wilson AC (1975) Science 188:107–116
Klebe RJ (1975) Biochem Genet 13:805–812
Kolata GB (1975) Science 189:446–447
Lewontin RC (1974) The genetic basis of evolutionary change. Columbia University Press, New York
Lewontin RC, Hubby JL (1966) Genetics 54:595–609
Lusis AJ, Paigen K (1975) Cell 6:371–378
Markert CL, Shaklee JB, Whitt GS (1975) Science 189:102–114
Nevo E (1978) Theoret Popul Biol 13:121–177
Paigen K (1961) Proc Nat Acad Sci USA 47:1641–1649
Paigen, K (1971) The genetics of enzyme realization. In: Rechcigl M (ed), Enzyme synthesis and degradation in mammalian systems. University Park Press, Baltimore, p 1
Paigen K (1979) Genetic factors in developmental regulation. In: Scandalios JG (ed) Physiological genetics. Academic Press, New York, p 1
Paigen K, Meisler M, Felton J, Chapman V (1976) Cell 9:533–539
Paigen K, Swank RT, Tomino S, Ganschow RE (1975) J Cell Physiol 85:379–392
Phillip DP, Childers WF, Whitt GS (1979) J Exp Zool 210:473–488
Powell JR (1979) Genetics 92:613–622
Powell JR, Lichtenfels TM (1979) Genetics 92:603–612
Rosbash M, Campos MS, Gummerson KS (1975) Nature 258:682–686
Schwartz D (1962) Genetics 47:1609–1615
Schwartz D (1976) Proc Nat Acad Sci USA 73:582–584
Shaklee JB, Kepes KL, Whitt GS (1973) J Exp Zool 185:217–240
Shows TB, Massara EJ, Ruddle FH (1969) Biochem Genet 3:525–536
Snyder LRG (1978) Genetics 89:531–550
Thompson JN, Ashburner M, Woodruff RC (1977) Nature 270:363
Wheeler MR, Clayton FE (1965) Dros Inf Serv 40:98
Whitt GS, Phillip DP, Childers WF (1977) Differentiation 9:97–109
Wilson AC (1976) Gene regulation in evolution. In: Ayala FJ (ed), Molecular evolution. Sinauer, Sunderland, Mass, p 225
Wilson AC, Carlson SS, White TJ (1977) Annu Rev Biochem 46:573–639
Zuckerkandl E (1963) Perspectives in molecular anthropology. In: Washburn S (ed), Classification and human evolution. Aldine Publishing, Chicago, p 243
Zuckerkandl E (1978) J Mol Evol 12:57–89
Zuckerkandl E, Pauling L (1965) In: Bryson V, Vogel HJ (eds), Evolving genes and proteins. Academic Press, New York, p 97

Received February 19, 1980; Revised May 21, 1980

PROBLEMS

1. Define the following: heterochrony; neoteny; homoeotic mutation; selector gene; genetic distance; allometry; cis-acting control; trans-acting control; sibling species; congeneric species; electromorph; stringent response; catabolite repression; microcomplement fixation; immunoelectrophoresis; immunodiffusion; chromosome banding; gel electrophoresis; activity stain.

2. Why are genetic distances, estimated as described in King and Wilson (1975, this chapter), underestimates of DNA divergence?

3. Table 2 of King and Wilson (1975) lists the various electrophoretic enzyme variants for a large number of enzymes examined in humans and chimpanzees.
(a) Which species appears to be more polymorphic?
(b) How does your conclusion compare to that presented by Ferris *et al.* (1981a, Chapter 2) on the basis of mitochondrial DNA comparisons?
(c) What are the strengths and weaknesses of estimating degrees of polymorphism by the two methods?

4. A fundamental distinction among regulatory genetic elements is between trans-acting regulator genes, which specify diffusible controlling molecules, thereby switching on or off the contiguous structural genes (Jacob and Monod, 1961). Imagine that you are given two strains of Hawaiian picture-winged *Drosophila* whose octanol dehydrogenases differ with respect to both electrophoretic mobility and activity. The two strains are found to be capable of interbreeding. How would you go about distinguishing whether the regulatory difference(s) involved cis- or trans-acting control elements?

5. What sort of evidence would you look for in the kind of survey conducted by Dickinson (1980, this chapter) that would bear on the question of whether a specific enzyme that appears at different developmental stages is the product of the same or of different structural genes?

6. If the proposal of Tomkins (1975, this chapter) is correct
(a) What general prediction does it make about the requirement for adenylate cyclase (the enzyme that synthesizes cyclic AMP) as one moves up the evolutionary scale from prokaryotes to complex multicellular eukaryotes and about the extent of pleiotropy that adenyl cyclase mutants would produce in this evolutionary progression? Do higher plants use cyclic AMP?
(b) What might you expect about the utilization of cyclic AMP, and perhaps ppGpp, in prokaryotes simpler than *E. coli*--possibly mycoplasmas or some of the archaebacteria (see Chapter 8)?

7. Figure 6-7 shows the relative karyotypic stability of two groups of vertebrates (frogs and mammals) as a function of time of divergence measured with the albumin "molecular clock". More than 300 interspecific comparisons were made within each group with respect to immunological difference of the albumins (generally considered to be a reasonable reflection of sequence divergence) and chromosome numbers. Plotted are the percentage of species pairs with a given albumin immunological difference that have the same number of chromosomes.

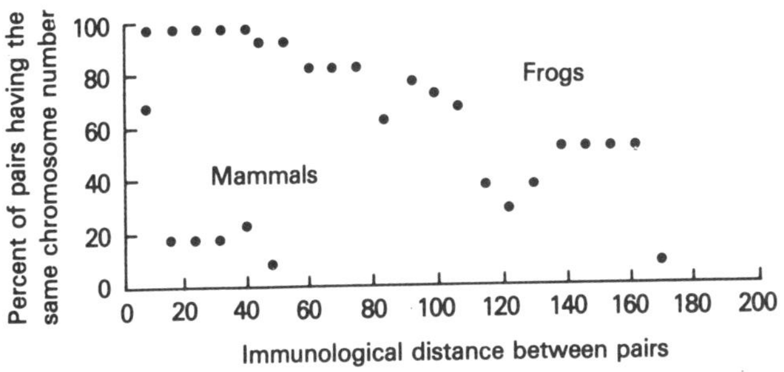

Figure 6-7

(a) Summarize the difference, in your own words, between these two groups.
(b) Comment on the potential significance of this difference in terms of the role of regulatory gene change in evolution and the possible connection between rearrangement or change in number of the chromosomes and genetic regulatory changes. (Relate the karyotypic changes to what is known about genome organization in eukaryotes.)

(c) Would you expect there to be a fixed relation between regulatory gene changes and divergence of the structural genes that affect immunological distance?

8. A recent comparison of high-resolution chromosome bands of human and chimpanzee chromosomes (Yunis *et al.*, 1980) shows that the banding patterns of the two species are nearly identical; the only differences are one chromosome fusion, differences in the amount of (presumably nonexpression) constitutive heterochromatin, and nine pericentric inversions (inversions encompassing the centromere). Virtually all the nonheterochromatic sub-bands were found to be homologous.
(a) Does this finding necessarily invalidate the conclusion that you drew in answering Problem 7?
(b) If karyotypic changes have played a major role in primate divergence, what does this imply about the number of significant genetic changes responsible for the differences between man and chimpanzee?

9. The possibility that middle repetitive sequences function directly in the regulation of gene expression through the activation of batteries of genes contiguous to these elements (Britten and Davidson, 1969) has been discussed in earlier chapters. Young (1979) has found by means of in situ hybridization of cloned middle repetitive DNA that the locations of members of different middle repetitive families can be strikingly different between strains of *Drosophila*. How does this observation reflect on the hypothesized role of these elements in the regulation of gene expression?

10. "The data of heterochrony are data about regulation; they define changes in the timing of development for features held in common by ancestors and descendants." (Gould, 1977). Relate this statement about "regulation" to the concept of genetic regulation as used by molecular biologists (see, for instance, Britten and Davidson, 1971). (Suggestion: it is advisable in thinking about developmental rate changes to make the distinction between such rate changes within localized embryonic fields and systemic growth-rate changes that often create changes between organ systems. Thinking about the genetic regulatory basis of different kinds of heterochrony is a challenging problem; you are grappling with what may well be the major issue in developmental biology and in evolution over the next decade or so.)

7 EUKARYOTIC ORIGINS AND PHYLOGENY

Organized beings represent a tree, irregularly branched; some branches far more branched—hence genera. As many terminal buds dying, as new ones generated.
Charles Darwin, Notebooks (1837, p. 43)

The gulf between prokaryotic and eukaryotic cellular forms of organization is recognized as one of the most fundmental distinctions in terrestrial biology. The existence of this discontinuity raises a fundamental problem in evolution. It is undoubtedly correct that the first true cells had a prokaryotic type of organization, yet the manner by which eukaryotic cells evolved from these cells is by no means clear. Two general classes of hypotheses ahve been proposed: autogenous ("self - generated") hypotheses, which propose that nuclei, mitochondria and chloroplasts arose gradually with membranes enclosing parts of the protoplasm, and endosymbiotic theories, which propose that at least some of these organelles (and possibly also flagella) arose from a symbiotic association between two or more prokaryotic organisms.

Until 1968 most of the discussion centered around a possible autogenous transition from the prokaryotic blue-green algae (*Cyanobacteria*) to the eukaryotic red algae, because of the remarkable similarity of photosynthesis in blue-green and red algae. Red algae were regarded as primitive eukaryotes because they lack flagella and because it was believed that they lack mitochondria. However, electron microscopic studies have subsequently shown that red algae have normal mitochondria, as well as advanced mitotic and meiotic cell divisions (Kubai, 1975). Although some sequence information about red algae is available (Andrew *et al.*, 1981), it is desirable to have more data in view of the central importance of the red algae in distinguishing between autogenous and symbiotic theories.

The development of the alternative theory stemmed from the similarity between bacteria and mitochondria, a resemblance that had been pointed out last century (see Margulis, 1970). However, the endosymbiotic theory was not taken seriously until the early 1960s when mitochondria and chloroplasts were shown to contain both DNA and RNA. This discovery led to the serial symbiosis hypothesis of Margulis (1968, Chapter 1), which proposed that the eukaryotic cell arose by a series of symbioses. In particular, she suggested that mitochondria, chloroplasts, and flagellae were derived from previously free-living prokaryotes. This paper, together with a more detailed description of the hypothesis (Margulis, 1970; 1981), described the possibility of producing a phylogeny for both prokaryotes and eukaryotes that satisfactorily accounted for the main features of both groups.

In principle, many possible theories for the origin of the chloroplast, mitochondrion, flagellum, and nucleus are possible. Several combinations of number and sequence of symbiotic events have been suggested including: (1) no symbioses, fully autogenous, e.g., Cavalier-Smith (1975): (2) chloroplast symbiotic, others autogenous, e.g., Taylor (1976); (3) only chloroplast and mitochondria symbiotic (many authors); (4) chloroplast, mitochondrion and flagellum symbiotic (Margulis, 1968, chapter 1). In addition, others have suggested that both the chloroplasts and mitogchondria have arisen more than once (Raven, 1970; Kuntzel and Kochel, 1981).

Margulis (1968, Chapter 1) has pointed out that taxonomic schemes should lead to specific predictions and ask new questions. To what extent has subsequent evidence (particularly protein and nucleic acid sequences) been in agreement with the symbiotic hypotheses? From what group of prokaryotes are the symbionts derived? Schwartz and Dayhoff (1978, this chapter) present the most complete study available of these questions and use sequences from several proteins and nucleic acids in the analysis. This paper contains a wealth of information and ideas about both prokaryotic and eukaryotic evolution, and it provides additional evidence from several proteins that blue-greenlike algae were the ancestors of the chloroplast. Other sequences such as of tRNA, 5S rRNA (Sankoff *et al.*, 1982), and oligonucleotides from 16S rRNA (Bonen and Doolittle, 1976) also confirm the relationship between chloroplasts and blue-green algae. Since in autogenous theories the entire eukaryotic cell is derived from a blue-green algal cell, the similarity of the blue-green algae and chloroplasts will not discriminate between the autogenetic and symbiotic hypotheses.

The cytochrome c sequences give more convincing support for the symbiotic hypothesis because they are consistent with a specific suggestion of John and Whatley (1975). From a detailed study of electron transport and oxidative phosphorylation they proposed that the mitochondrion is derived from a bacterium like *Paracoccus* or from the related photosynthetic nonsulphur purple bacterial group (*Rhodospirillaceae*). Indeed, Woese (1977) has suggested that the symbiont may have originally been photosynthetic but later became specialized for oxidative respiration after the evolution of the more efficient oxygen-releasing photosynthesis of the blue-green algae. Because cytochrome c, a mitochondrial protein, is encoded in nuclear DNA in contemporary organisms, a corollary of symbiotic theories is that some DNA from the early mitochondrion has been transferred into the nucleus (Weeden, 1981).

Schwartz and Dayhoff (1978, this chapter) also present an analysis of 5S rRNA sequences that supports the idea that the cytoplasm and nucleus had a different origin from that of the chloroplast and from the mitochondrion. This idea will be considered in more detail in the next chapter. In addition, a discussion of the origin of photosynthesis and respiration will be presented, pointing out that the evidence generally supports the hypothesis that anaerobic photosynthesis arose early in the history of living organisms and that aerobic respiration was derived from that process.

Recent evidence supporting the serial endosymbiosis hypothesis for mitochondria and chloroplasts is the finding that the DNA in these organelles encode their own tRNA molecules (Barrell *et al.*, 1980; Bonitz *et al.*,

1980). The existence of a complete set of tRNA and rRNA molecules is strong evidence against the autogenous hypothesis of Raff and Mahler (1972) that the mitochondrial DNA was derived from a plasmid carrying some of the genes necessary for respiration. In plants, the 5S rRNA and all tRNA molecules are distinct in the chloroplast, mitochondrion and cytoplasm (Kuntzel *et al.*, 1981). At this time the symbiotic origin of chloroplasts and mitochondria is well supported (Gray and Doolittle, 1982), but the position is uncertain for flagella, though a good case has been made by Margulis (1981). Experimental support for a symbiotic origin for the nucleus may be very difficult to obtain. The question of eukaryotic origins has become a popular subject and several recent symposia are available (Proc. Royal Soc. London Vol. 204; BioSystems Vol. 10; Frederick, 1981), as well as a book (Margulis, 1981). These references contain information on the large number of naturally occurring symbiotic associations (both intra- and extracellular).

Molecular information has also been used in many studies on the phylogeny of higher eukaryotes. De Jong, Gleaves and Boulter (1977, this chapter) analyzed sequences of the eye lens protein (alpha-crystallin A) from a broad range of mammals. This paper was chosen because of the selection of animals and the discussion of the results in relation to the previous understanding of mammalian phylogeny. The sequence information contained therein supports the division of mammals into marsupials and placentals and, in addition, gives more details on the orders of placental mammals--for example, evidence is provided that elephants and the hyrax (paenungulates) form an early branch of the placentals. A later paper (de Jong *et al.*, 1981) places the manatee and the aardvark in this same group and also provides evidence that the *Edentata* (sloths and South American anteaters) are another early placental branch. Dene *et al.* (1980) combined sequence data from beta-globin, myoglobin and the fibrinopeptides and provided results that support the separation of the paenungulate line from the ungulate line; however, the data did not support an early separation of the paenungulates from the rest of the placental mammals, as initially suggested. The approach of combining sequence data does limit the number of taxa that can be included in a study but should increase the reliability of the results. More careful selection of material for sequencing is needed, so that maximum use can be made of the information by comparing sequences from more than one molecule from the same collection of organisms (Sankoff *et al.*, 1982).

With vertebrates, trees constructed from molecular sequences agree quite well with those constructed from paleontology and comparative anatomy. In contrast, there has been little agreement from morphological studies on the interrelationships of flowering plants and it was hoped that sequence data would help resolve this difficulty. Boulter, Ramshaw, Richardson, and Brown (1972, this chapter) report on cytochrome c sequences from higher plants and give a good description of their methods; the results are discussed in the light of previous knowledge. One of the unexpected findings was that a morphologically diverse and distinct group, the *entrospermae* (spinach, rhubarb, cacti (Mabry and Behnke, 1976)) appear to have diverged from the other flowering plants even before the separation of monocotyledons and the remaining dicotyledons. This suggestion has been almost universally rejected by botanists (Cronquist, 1980); Baba *et al.* (1981) explain these distinct *Centrospermae* sequences by postulating a duplication of the cytochrome c gene followed by inactivation of one of the duplicates, which can then undergo relatively rapid divergence. The later reexpression of this gene as cytochrome c in the ancestor of the *Centrospermae* would allow a later branch point for this group. However, by postulating cycles of gene duplication, suppression, and reexpression, almost any tree topology could be generated, thereby making it difficult to falsify phylogenetic trees. Again, it would be useful to have additional sequences from the same taxa to see whether they give phylogenetic trees similar to that of cytochrome c. Studies of 5S rRNA and from ferredoxin (Boulter, 1980) support the suggestion of Boulter *et al.* (1972, this chapter) that the *Centrospermae* are a very early group of flowering plants. Recent work with plants has concentrated on the evolution of photosynthetic proteins such as plastocyanin. This protein evolves at a faster rate than cytochrome c and consequently has been useful in examining detailed questions about the evolution of major plant families (Boulter *et al.*, 1979; Grund *et al.*, 1981).

Molecular data are particularly suitable for

investigating questions of phylogeny between large groups of organisms that diverged early--such as the relationship between animals, plants, and fungi. If each of these groups is monophyletic, then protein and nucleic acids common to all three groups can, in principle, be used to distinguish between the three main possibilities for their evolutionary relationships (Figure 7-1).

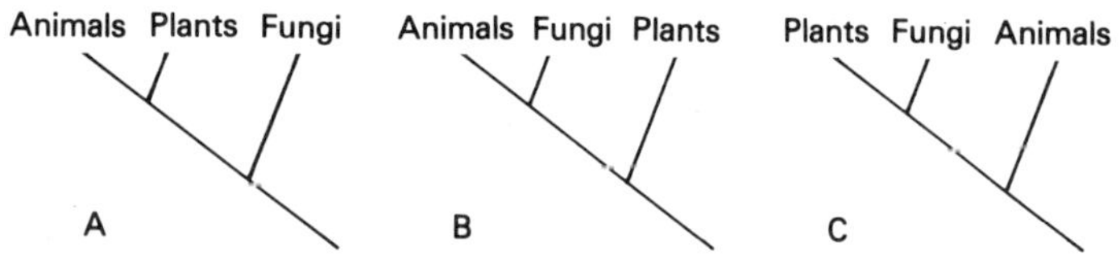

Figure 7-1

Notice in Schwartz and Dayhoff (1978, this chapter) that the cytochrome tree is unable to distinguish between possibilities A, B, and C but that the 5S-RNA tree favours alternative A; several other authors have come to the same conclusion (Cedergren *et al.*, 1980; Hori and Osawa, 1979). Kuntzel *et al.* (1981) favored C, though protozoa were included with the metazoa; in contrast, most phylogenies based on molecular data have suggested that protozoa are a diverse group, no more closely related to higher animals than to higher plants or fungi (Schwartz and Dayhoff, 1978, this chapter; Baba *et al.*, 1981). We must be careful on this point because it is important to distinguish between gene phylogeny and organism phylogeny, as discussed in Chapter 4. For example, the latter conclusion about protozoa assumes that with cytochrome c (a mitochondrial function), sequences are orthologous--that is, all contemporary eukaryotic cytochrome c genes are related by direct descent to the corresponding gene in a single ancestral symbiont. In such a case, the evolutionary history of cytochrome c genes is congruent with the evolutionary history of eukaryotes. However, if mitochondria have arisen independently in several different lines of eukaryotic descent (that is, if they are polyphyletic), then by using cytochrome c sequences alone, only gene phylogeny would be examined, and such a treatment would yield nothing about the relatedness between groups of organisms that were derived from different endosymbiotic events. Nevertheless, the nuclear-encoded 5S rRNA sequences support the conclusion that protozoa (and protists generally) comprise an independent but very diverse group that is not specifically allied to any of the major groups of animals, fungi, or higher plants.

In the papers discussed so far, relationships between major groups of organisms have been considered. At the other end of the taxonomic scale are studies of the classification of single species. For example, an early study of this type used immunological comparisons of both albumins and transferrins and indicated that the giant panda is more closely related to bears than to raccoons (Sarich, 1973). The possibility now exists for making immunological comparisons of proteins extracted from the preserved remains of recently extinct species such as the hairy mammoth and the marsupial Tasmanian wolf, in order to determine more precisely their taxonomic status (Lowenstein *et al.*, 1981).

Many important crop plants have arisen through hybridization and often the parent species have been identified by chromosomal analysis. Because the large subunit of ribulose bis-phosphate carboxylase is encoded in the chloroplast DNA, it shows maternal inheritance and, consequently, in several cases it has been possible to identify the species that is the maternal parent of the original hybrid (Gray, 1980).

The last question to be considered is the fascinating problem of the relationship between humans and the apes (chimpanzee, gorilla, orangutang, gibbon, and siamang). Darwin had claimed that chimps and gorillas are the nearest relatives of humankind and that therefore our ancestors should be looked for in Africa; indeed, some immunological comparisons from early in this century supported a close relationship between humans, chimps, and gorillas. Modern work began with that of Goodman from 1961-1963 (see Goodman *et al.*, 1971; Doolittle *et al.*, 1971) and results from both immunological methods and protein sequencing came to the same conclusion; it appeared that the last common ancestor of humans, chimps, and gorillas must have existed about five million years ago. Goodman has consistently argued that chimps and gorillas should be placed in the same taxonomic family as humans, though there is still little agreement on the validity of using molecular similarity as a basis for taxonomic categories (King and Wilson, 1975, Chapter 6). Attempts are being made to isolate collagen from early human fossils; this material could then be used for examining evolutionary relationships with other higher primates (Lowenstein, 1981).

Paleontologists had concluded that the last common ancestor of humans and the great apes had lived about thirty million years ago, and in order to reconcile this long time period with the closeness of the genetic relationship, Goodman *et al.* (1971) proposed that the molecular clock had slowed down in the higher primates. This apparent decrease could have been a result of selection, caused by possible effects of immune reactions in the placenta or from a longer generation time. The alternative possibility suggested by Wilson and Sarich (1969) is that the molecular evidence is correct (divergence being about five million years ago), and that paleontologists had misinterpeted their data (or rather lack of data). The argument is given in more detail in Wilson *et al.* (1977), who point out that the fossil record does not give information about the ancestry of chimps and gorillas. The paleontological interpretation rests on the assumption that changes to teeth and jaw features are irreversible. It is known that teeth are particularly adapted to diet and if there had been any change in the diets of chimp and gorilla, then jaw and teeth structure might have been altered during evolution. Furthermore, changes in the rate of development (neoteny in humans and/or anaboly in the great apes; see Gould, 1977) would make comparisons unreliable if they were based solely on adult bone structure.

All recent work has reinforced the conclusion that chimps and gorillas are our closest living relatives and that the time of divergence is much closer to five million years than to the earlier paleontological estimate. The currently favored tree of descent has the first branch leading to siamang and gibbons and the next to the orangutang; the order of branching among humans, gorillas, and chimps remains uncertain. The most detailed evidence comes from restriction enzyme analysis (Ferris *et al.*, 1981b) and sequence studies of mitochondrial DNA (Brown *et al.*, 1982), which indicate that for these three species the earliest branch point gives a human line and a chimp-gorilla line. However, this first branch and the succeeding branch point leading to gorillas and chimps are separated by a comparatively short interval. All current evidence supports the suggestion of Goodman that one taxonomic family could accomodate these three species.

 Reprinted from *Proc. Roy. Soc. Lond.* B **181**: 441–455.

A phylogeny of higher plants based on the amino acid sequences of cytochrome *c* and its biological implications

By D. Boulter, J. A. M. Ramshaw, E. W. Thompson, M. Richardson and R. H. Brown
Department of Botany, University of Durham, Durham

(*Communicated by D. Lewis, F.R.S. – Received 2 February* 1972)

Higher plant phylogenetic trees were constructed from the amino acid sequences of cytochrome c from fifteen plants using the 'ancestral sequence' and 'flexible numerical' methods. The validity of these methods is discussed and the results obtained are compared with existing phylogenies based mainly on morphological characters.

Introduction

Recently, amino acid sequences of animal proteins have been used to construct phylogenies, which support and complement those established by so-called classical means (Dayhoff 1969). In these instances, the groups considered have good fossil records, and the phylogenetic tree established using molecular data could be checked against them. In the case of plants, an adequate fossil record does not exist, and it is of utmost importance therefore, when using molecular data, to state the assumptions involved in constructing the tree and give an estimate of its accuracy. If the findings of the molecular tree for a group of taxa conflict with other biological evidence, a re-investigation of the relationships of the group would appear profitable.

Methods

Tree construction

The 'ancestral sequence method' used to construct the phylogenetic tree (figure 1), was based on the scheme of Dayhoff & Eck (1966).

The tree consists of branches, the junctions of which are called nodes. Each node has three branches which lead either to an adjacent node or to a determined sequence. In addition, by using the rules described below, it was possible to reconstruct the most probable ancestral sequences of each node, so that the tree contains not only the determined sequences but also the computed ancestral sequences. Since the sequences themselves do not indicate the point of earliest time within the tree, this point has been established from biological considerations.

The computing strategy employed was based on that described by Dayhoff (1969), and the program used in calculating the tree consisted of three main procedures:

Procedure 1

The construction starts initially with any three sequences; only one tree exists relating these. The next sequence is added to the existing tree in all possible positions, and each of these trees is evaluated by using Procedure 2. From these alternatives the best tree is selected (see Procedure 2) and used for further construction. Additional sequences are added successively to the 'best' tree obtained at the previous step, until all have been added and a final tree is obtained.

Procedure 2

This procedure is used to evaluate trees. It starts by inferring the 'ancestral sequence' at each of the nodes which relate the sequences in the tree. Each position along the sequences is considered in turn. For each node three lists are made which consist of the amino acids found in this position along the three branches which attach to the node. Thus, at each nodal position all the sequences in the tree are considered. This is repeated in turn for all the positions in the sequence. If for a given node and at a given position in the sequence, only one amino acid is found which occurs on more branch-lists than any other, then it is selected as the nodal amino acid; otherwise, at this stage the nodal position is left blank. When all the nodes have been assigned an amino acid or a blank for every position in the sequence, the situation at each blank position is reassessed. If a blank position has at least two of its three adjacent neighbours (either node or sequence) the same, then this amino acid is assigned to the position. The process is repeated until no more additions occur at any of the blank positions. Finally, the nodal sequences are checked so that if the amino acid at the node is not the same as at least two of its adjacent neighbours, it is changed to a blank. This process will give a definite assignment of the ancestral amino acid whenever one choice is clearly preferable, but will leave blanks when reasonable doubt exists.

When the ancestral sequences at all the nodes have been determined, the tree is evaluated as follows: the numbers of amino acid changes along every branch of the tree are counted by comparing each sequence with each adjacent sequence, position by position. All the branch changes are then totalled to give the evaluation for the whole tree. However, this calculation is complicated by the existence of any blanks in the ancestral sequences. The blank means that two or more equally probable amino acids exist for that position. In such cases it does not matter for the overall evaluation which amino acid is chosen to fill the blank, as the number of total changes on the tree will be the same if any of the alternatives is used unless two parallel mutations occur at adjacent nodes. Because we assume a minimum route for evolution, in this case the same amino acid is chosen for both blank nodal positions. When the branch lengths were evaluated for the final tree (figure 1), the minimum number of mutations counted around a blank or series of blanks was divided equally among all the independent branches.

Procedure 3

Since a sequence, once fixed to the tree, does not change its relative position, the fusion strategy of Procedure 1 effectively limits the amount of evaluation required. Because no account is made of the remaining sequences still to be added when each new sequence is fixed, this means that, in retrospect, a wrong decision may be made at any step and the final tree may only be a close approximation to the 'best' tree. For this reason a third procedure was used which systematically relocates the branches of the final tree into all alternative positions. Each new tree is then evaluated using the second procedure to see if a better tree exists.

Sequence data

The methods used to purify and to determine the sequence of cytochrome *c* are given in Richardson *et al.* (1970), Richardson *et al.* (1971 *a, b*), Thompson, Laycock, Ramshaw & Boulter (1970) and Ramshaw, Thompson & Boulter (1971).

Results

Figure 1 is a representation of the phylogeny of some higher plants derived from cytochrome *c* sequences using the 'ancestral sequence' method (Dayhoff & Eck

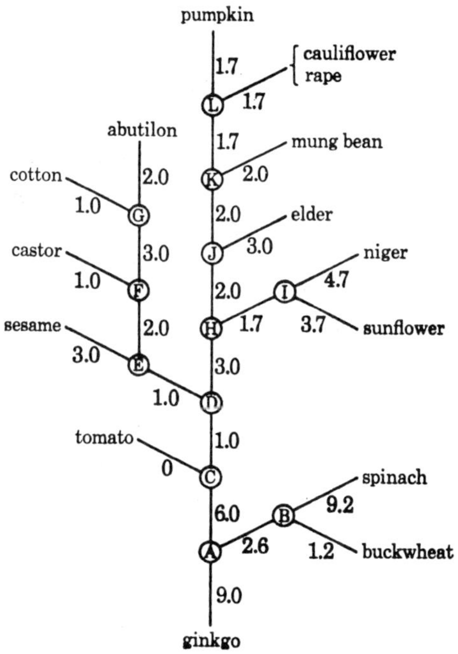

FIGURE 1. A phylogenetic tree relating fifteen plant species constructed using the 'ancestral sequence' method. The sequences used in constructing the tree are:

Phaseolus aureus L. = *Vigna radiata*	mung bean	Thompson, Laycock, Ramshaw & Boulter (1970)
Helianthus annuus L.	sunflower	Ramshaw, Thompson & Boulter (1970)
Ricinis communis L.	castor	Thompson, Richardson & Boulter (1970)
Sesamum indicum L.	sesame,	
Cucurbita maxima L.	pumpkin	Thompson, Richardson & Boulter (1971*a*)
Fagopyrum esculentum Moench.	buckwheat	Thompson, Richardson & Boulter (1971*b*)
Brassica oleracea L.	cauliflower	
Abutilon theophrasti Medic.	abutilon	Thompson, Notton, Richardson & Boulter (1971)
Gossypium barbadense L.	cotton	
Ginkgo biloba L.	ginkgo	Ramshaw, Richardson & Boulter (1971)
Brassica napus L.	rape	Richardson, Ramshaw & Boulter (1971*a*)
Guizotia abyssinica Cass.	niger	J. A. M. Ramshaw, unpublished
Lycopersicum esculentum L.	tomato	R. Scogin, unpublished
Spinacea oleracea L.	spinach	R. Brown, unpublished
Sambucus nigra L.	elder	

1966; Dayhoff 1969). Figure 2 is a construction for the same data using the 'flexible numerical matrix' method of Lance & Williams (1967).

DISCUSSION

Similarities between sequences may arise by chance, convergence or homology, but only homologous sequences can be used to construct species phylogenies (Nolan & Margoliash 1968; Dayhoff 1969). Fitch & Margoliash (1967*a*) and Fitch & Markowitz (1970) have devised a semi-rigorous method which indicates that the cytochromes of animals and fungi, are homologous. Similar calculations, including the plant sequences, indicate that the mitochondrial cytochrome *c* of animals, fungi and plants are homologous (J. A. M. Ramshaw & R. H. Brown, unpublished).

Dickerson *et al.* (1971), using X-ray diffraction data, have established a three-dimensional structure of horse-heart cytochrome *c*, and shown it to be essentially the same as that of *Bonito*, a fish. Using their projections and the plant sequences, Boulter & Ramshaw (1972) have indicated that the three-dimensional structure of the plant cytochrome *c* approximates to that of the animal, although the method used did not have the rigour of the full X-ray diffraction method. Since there are

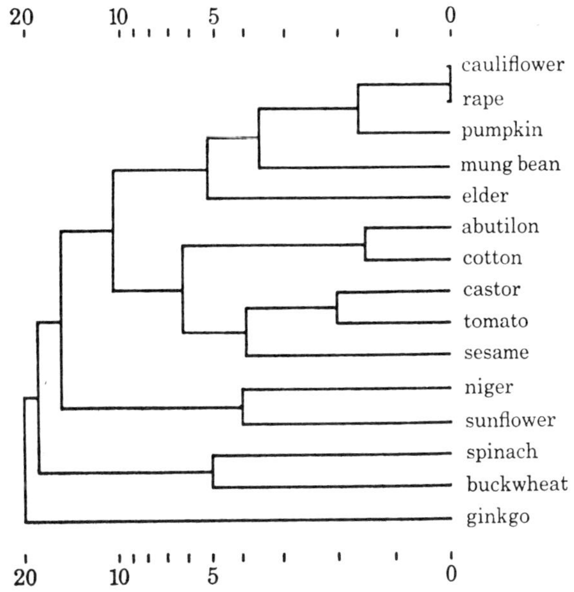

FIGURE 2. A phylogenetic tree relating fifteen plant species constructed using the 'flexible numerical' method. The sequences used are given in figure 1. The tree was constructed using amino acid differences to relate the species. The value chosen for the variable parameter β was -0.1. This value appears to give better results, e.g. ginkgo as a single line descent, when used with protein sequence comparisons, that the value of -0.25 suggested by Lance & Williams (1967) for use with the method (J. A. M. Ramshaw & D. Boulter, unpublished).

about thirty residues out of a total of over one hundred, which are invariant (Fitch & Markowitz 1970), it is unlikely that the coincidence of three-dimensional structure between plant and animal cytochrome *c* would have occurred by convergence rather than by common ancestry.

The 'ancestral sequence' method involves two assumptions: first, that evolution has taken place by the minimum number of amino acid substitutions, and secondly, that the final tree accepted is that which has the minimum number of amino acid substitutions of all possible trees. With regard to the first of those assumptions, it is quite clear that back and parallel mutations have occurred in the evolution of cytochrome *c*. The results given in tables 1 and 2 estimate that about 27 % parallel mutations and 6 % back mutations were detected; these changes have been taken into consideration in constructing the tree. The problem, however, is that there is no certain way of knowing the extent of undetected back and parallel mutations, especially as the number of sequences considered is extremely small when compared with the number of flowering plant taxa. However, the likelihood of relating two sequences which are similar, as a consequence of parallel and/or back mutations rather than because of common ancestry, is limited, since in each comparison all positions of the sequence were considered (see tables 1 and 2). Examples of positions where parallel changes occur are position 52 Ala → Pro; position 13 Ala → Ser; position 58 Ala → Thr. Examples of positions where back mutations have occurred are position 63 Lys → Met → Lys; position 109 Ala → Ser → Ala.

Fitch & Margoliash (1969) have calculated that 20 % parallel and 1 % back mutations have occurred in cytochrome *c* during the evolution of animals (Fitch & Margoliash 1967*b*). Although their tree represents a slightly larger data set, the findings with plant cytochromes are of the same order.

With reference to the second assumption, there are 8×10^{12} possible trees in considering fifteen taxa and relatively few (about 10^3) have been examined here. Clearly one cannot be certain that the tree selected has the fewest possible amino

TABLE 1. DIFFERENCES BETWEEN PLANT CYTOCHROMES *c*

Positions where differences are found between the sequences and the nodal sequences. The main nodal line of descent is constant and shown below each position. The numbers given refer to the positions in the complete sequences.

	8	12	13	15	23	41	51	52	53	58	59	71	104	108	111	112	113	
pumpkin								Pro										p
cauliflower, rape																		{c, r}
mung bean			Ser							Thr								mb
elder					Asn													e
sunflower	Ala	Thr	Thr	Ala										Thr				s
niger	Ala											Ser		Ala				n
abutilon								Pro										a
cotton																		c
castor														Gln				c
sesame			Ser					Pro										s
tomato																		t
spinach		Asp	Val	Ala			Ala		Ser					Asp	Gln			s
buckwheat			Ser											Ser	Gln			b
ginkgo						His				Thr	Gly		Ser	Gln	Ser	Gln	Glu	g
	Pro	Lys	Ala	Glu	Ala	Asn	Thr	Ala	Gly	Ala	Ala	Thr	Ala	Glu	Ala	—	—	

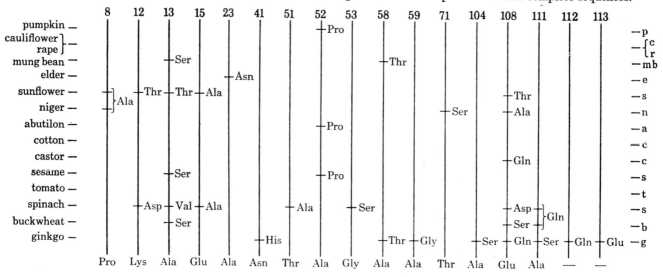

TABLE 2. DIFFERENCES BETWEEN PLANT CYTOCHROMES *c*

Positions where differences in the sequences occur, which lead to differences along the nodal line of descent. The nodal residue for each group of nodes is shown. In cases where a sequence differs from the adjacent node the amino acid is indicated in brackets. The lettering and topology of the nodes is as given in figure 1, and the numbers refer to the positions in the complete sequence.

Node	Species	2	4	10	11	29	63	66	68	70	74	98	109	
node L	pumpkin						Arg						Ala	p
node L	cauliflower, rape		Asp	Asn	Ser	Asp	Lys	(Glu)		Lys		Asp		{c, r}
node K	mung bean													mb
node J	elder							(Asn)	Glu					e
node I	niger		Ala	Asp	(Ala)		(Lys)	(Ala)					Ser	n
node I	sunflower													s
node H		Ser			Pro	Glu	Met	Ile				Glu		
node G	abutilon		Gln					Gln						a
node G	cotton				Ala	(Asp)		(Asn)		Asn	Asp	Asp	Ser	c
node F	castor													c
node E	sesame		Asp		Val	(Asp)								s
node D				Asn				Ile	Gly					
node C	tomato				Pro			Asn				Glu	Ala	t
node B	spinach				(Lys)	(Asp)		(Ile)	(Ser)			(Asp)	(Ser)	s
node B	buckwheat	Thr	Ser		(Ile)		Lys	(Thr)		Asp	Glu		(Glu)	b
node A	ginkgo			Asp						Gln				g

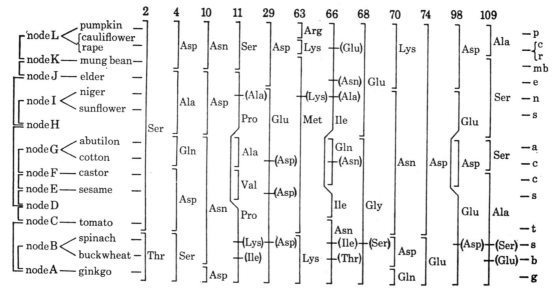

acid substitutions. Computer Procedure 3, however, is specifically designed to reduce the number of comparisons which have to be made in order to obtain with reasonable certainty, the absolute minimum tree.

Using the same data the matrix method gives a tree (figure 2) which is very similar to the 'ancestral sequence' method tree (figure 1). It differs in that, in the matrix tree sunflower and niger come before the abutilon/castor/etc. group, whereas they come after this group in the 'ancestral sequence' method tree. Also, in the matrix tree, tomato joins the castor/sesame group. Since the rules for constructing the two types of tree are very different, their similarity supports the suggestion that computer Procedure 3 was effective and that the tree given in figure 1 is an absolute minimum tree and not just a 'local' minimum. However, the number of amino acid substitutions involved in the best matrix topology is greater than that for the ancestral sequence tree.

Of the two methods of tree construction, the 'ancestral sequence' method is to be preferred, since in addition to the two assumptions already discussed, the matrix method also assumes that the rate of evolution has remained constant (Jardine, van Rijsbergen & Jardine 1969). It is reasonable to use this assumption with proteins such as cytochrome *c* in calculating the times of divergence of major groups, for example, mammals and insects which involve time periods in excess of 200 million years. This is certainly not the case when considering the shorter time periods covered by the evolution of the flowering plants.

The tree depicted in figure 1 is the one which has the smallest number of amino acid substitutions (see later, figure 3). This same tree is generated whatever the order of addition of species to the growing topology but, again, it has not been possible to try more than a very limited number of the total possible combinations.

We have compared this phylogeny with that of Takhtajan (1969) by calculating the number of amino acid substitutions that would be required if the species were arranged according to his scheme. This number is 77 as opposed to the 68 required in the tree given in figure 1. Estimates indicate that at least one-third of the possible trees with these 15 taxa have less amino acid substitutions than that of Takhtajan.

The tree represented in figure 1 consists of a main axis from which taxa diverge at successive points in time. Since the method gives a closed topology, the point of earliest time cannot be determined from the sequences themselves. It was taken to be the point of divergence of ginkgo since this is the only gymnosperm investigated, and this group is known from fossil evidence to have evolved earlier than the angiosperms (Walton 1953). According to these results, the Compositae diverged from the common ancestor of sunflower/niger and pumpkin/cauliflower/mung bean and elder, more recently than the abutilon/sesame group diverged from the common ancestor of all these taxa. Looked at in another way, the period of common ancestry of the abutilon/sesame group with all the above taxa, was less than the period of common ancestry of sunflower/niger/pumpkin/cauliflower/mung bean and elder. Using the 'ancestral sequence' method ensures that this is correct so far as the sequence evidence goes, irrespective of the rates of evolution in the separate lines.

Phylogenetic schemes for higher plants are based, in the main, on morphological characters, although the most recent ones of Takhtajan (1969), Cronquist (1968) and Thorne (1968), take into account other types of information, e.g. chemical data, where these are available.

The main weaknesses of trying to establish a phylogeny using morphological characters in the absence of a continuous fossil record are first, that similarities, due to convergence, are not easily detected; secondly, that morphological characters have evolved in the same and in different groups at different rates.

In comparison, the 'ancestral sequence' method detects some of the convergence, although at present that which is not detected cannot be quantified. Also, the problem of different rates of evolution in different parts of the tree is largely overcome. Even if rates are different, two taxa will still be most closely related to their own common ancestor. Along a branch where a rather faster than average rate of evolution has occurred, there will be more changes between the present-day sequence and the ancestral sequence, with the result that fewer positions in the sequence may be significant in linking the species to the tree. However, unless by chance the majority of the changes relate to another sequence, the species will still link to its own ancestor. In practice certain species tend to 'wander' to some extent, but as more sequences are added they become 'fixed'. Thus, all the species in the figure 1 tree, except for buckwheat and spinach, are now 'fixed'. The species have remained in the same relationship to each other since an eleven-species tree was generated. Their relationships have not been changed by the addition to the eleven-species tree of spinach, tomato, rape and elder, to give the latest fifteen-species tree. Figure 1 is not a unique best tree and the two additional possibilities, due to movements by buckwheat and spinach, are summarized in figure 3. These alternatives are due to the fact that the branch lengths associated with these species, are long as a result of many changes having occurred, and the ancestral sequences are consequently not well-defined, i.e. they contain several blank positions. The topology in figure 1 was chosen from the three possible best trees using biological considerations. We see no reason why such biological considerations

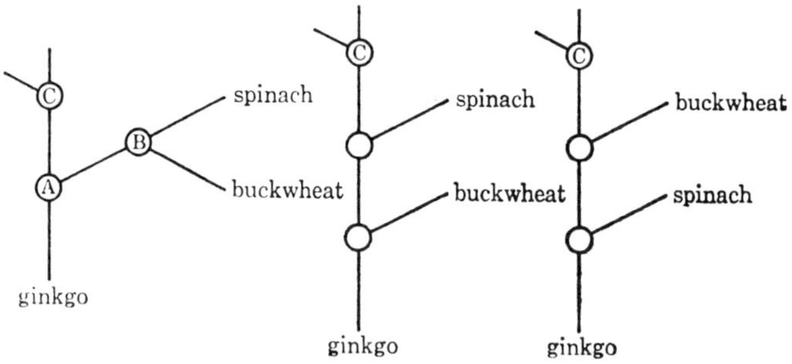

FIGURE 3. A summary of the three minimum amino acid substitution phylogenetic trees relating fifteen plant species constructed using the 'ancestral sequence' method. Node C and the remaining unshown topology is common to all three trees, and as shown in figure 1.

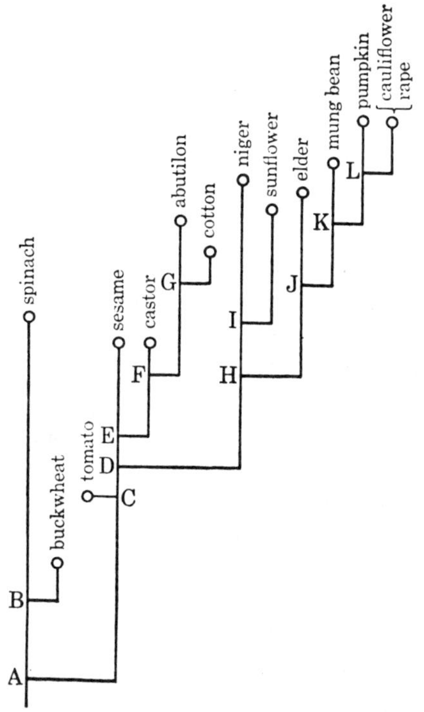

FIGURE 4. A phylogenetic tree relating fifteen plant species constructed using the 'ancestral sequence' method, drawn to indicate relative branch lengths.
The phylogenetic tree in figure 1 has been drawn with all branch lengths drawn to scale in a vertical direction. The node lettering is as for figure 1.

should not be used in conjunction with chemical data. A well established tenet of taxonomy is that one should use all available kinds of data to the best advantage and not restrict oneself exclusively to one type. The history of taxonomy is replete with examples where the ignoring of this principle has led to misleading results.

Figure 4 is derived by plotting the branch lengths of figure 1 linearly. If the rate of evolution had remained constant and if all parallel and back mutations had been detected, the taxa at the ends of the branches, should all lie in a single plane. This is not so however. In certain cases the main reason for the divergence is that not all parallel and back mutations have been detected. Thus, for example, spinach and buckwheat diverged early and as there are no other members on this line of descent between the point of divergence and the present-day sequences, it is only possible to calculate a single ancestral sequence. This means that very few parallel or back mutations have been detected along this line of descent. Since, in the plant tree, 27 % parallel and 6 % back mutations have been detected overall, it is probable that more changes have occurred on these branches than have been detected. More species belonging to this line of descent would be required in order to discover any additional changes. The difference in height between spinach and buckwheat reflects the large number of blank positions in the common ancestor which by chance favours similarity to buckwheat. On the other hand, the variations in height of members at the top of the tree reflect the random differences expected over the short periods of time which are being considered.

If one examines the tree and determines not the number of amino acid substitutions but, instead, the number of base changes as predicted from the accepted amino acid codons (Dayhoff 1969), one arrives at a total of 80 changes. The additional number of base changes compared with the 68 amino acid changes result from certain of the amino acid changes requiring a minimum of two base changes. In these cases, additional sequence determinations should lead to the discovery of the intermediates, since present evidence using haemoglobin, tobacco mosaic virus coat-protein and tryptophan synthetase mutants (Crick 1963), indicates that the majority of mutations are single base changes. The absence of triple base changes and the relatively few double changes, shows that our results are in accordance with these findings.

The sequence positions, where double base changes may have taken place, are 4, 11, 12, 23, 66, 70, 108 and 111. The number of possible double base changes could be less than this, since in using a closed tree the amino acid sequence of the primitive nodes contain several blanks and do not take into account the sequences ancestral to the closed tree. If these were available, alternative amino acids to those present might reduce the number of double base changes. For example, with regard to position 11, on the existing tree, nodes A and C are Pro and the branches contain Lys for spinach and Ile for buckwheat. By the ancestral sequence method node B has either Lys or Ile and this requires a double base change between nodes A and B, and a single base change along one of the species branches, making a total of three in all. If, however, undetermined ancestral sequences were postulated to give Thr for node B, then the total number of base changes required would be three as before, but in this case a double base change would not be involved. A similar uncertainty could be postulated for positions 70 and 108.

In one instance, it is possible to rearrange the tree and eliminate a double base change at position 70 while not increasing the overall number of amino acid substitutions and base changes. This is achieved by placing ginkgo close to the elder/mung bean/pumpkin/cauliflower line. Other rearrangements of the tree, for example, moving buckwheat/spinach to a sub-branch of sunflower/niger or vice versa, whilst removing a double base change at 108, would increase the number of changes at positions 2, 68, 74 and 109, giving a greater overall number of changes in the tree and create an additional double change at position 109. So far as positions 12 and 23 are concerned, it is not possible to rearrange the tree so as to reduce the

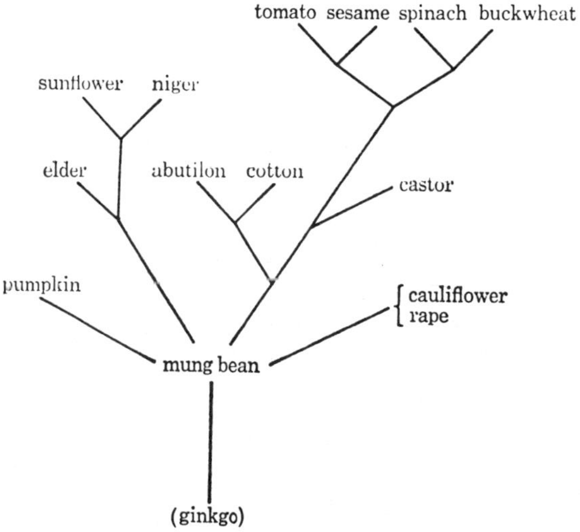

FIGURE 5. The phylogenetic scheme for the species examined according to Bessey (1915).

double base changes. In all these instances, therefore, the most likely possibility is to postulate an intermediate, the amino acid of which can be deduced if a minimum route is to be followed. For example, if one considers position 23, all sequences, except elder, have Ala in this position; elder has Asn. The postulated intermediate, therefore, would have to be either Thr or Asp; thus, the route would be Ala to either Thr or Asp, thence to Asn, all involving single base changes. The determination of further sequences may or may not reveal these postulated intermediates.

It is generally accepted, as suggested by Bessey (1915) (see figure 5), that there has been a trend during evolution from flowers with a large indefinite number of parts to those with a small definite number. This trend has been accompanied by fusion of parts. Sporne (1971) has listed many of the characters which he considers primitive, and has shown that a high proportion of these characters occurs in the lists of fossil angiosperms which have been published (Sporne 1971). He suggests that the Magnoliales complex was of the earliest origin. The species investigated in this paper, which most resemble the hypothetical primitive angiosperms therefore, are mung bean, cauliflower and pumpkin. The 'ancestral sequence' method places these at the top of the tree (figure 1).

There are two ways of considering what constitutes 'primitiveness'. Engler (1926; see Cronquist 1968), for example, regarded the Amentiferae as of ancient origin, since the reduced, simple flowers must have been the result of a long and separate evolutionary history. Cronquist (1968), on the other hand, states that: 'An essential requirement for any phylogenetic system is that one starts with the groups which are least modified from the ancestral prototype, rather than with those which have undergone the most change. All groups are of equal age, if one takes in all the ancestors as well as the members of the group. It is only if one bases concepts of age on the members that would actually be referred to a particular group that groups differ in age and a phylogenetic system becomes possible.' This is the crux of the matter. The Englerian position is that the possession of advanced characters by a group probably means that it diverged early, and therefore, should be considered primitive, and Cronquist's viewpoint is that groups which resemble the

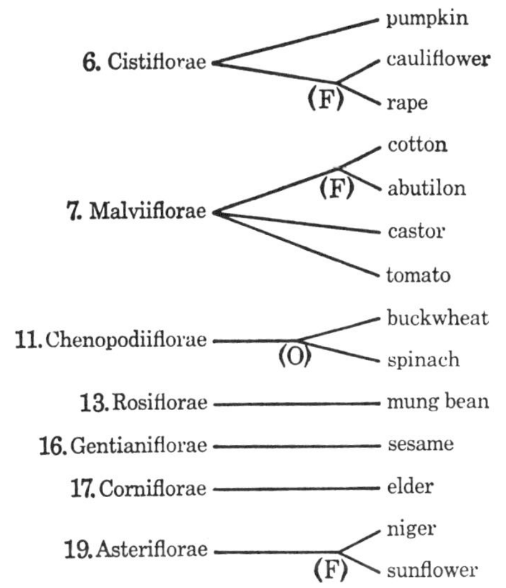

FIGURE 6. The phylogenetic scheme for the species examined according to Thorne (1968). Species placed within the same Order (O) or same Family (F) within Superorders are indicated. Numbers show the order of the nineteen dicotyledonous Superorders given in Thorne (1968).

ancestral prototype should be considered primitive, even though the possibility exists that they diverged late. This dilemma of interpretation is very appropriate to the results given in figure 1, since the 'ancestral sequence' method defines the times of divergence of groups relative to each other. The simplest interpretation in our view, is that there was a basic flowering plant stock, which has remained rather constant during evolution and from which the various groups have diverged at successive intervals of time. Subsequent to their divergence, groups have evolved at varying rates to become, to a greater or lesser extent, advanced.

Thorne (1968) (figure 6) differs from Cronquist (1968) and Takhtajan (1969) (figure 7), in his treatment of the relationships of the flowering plants by the removal of the Solanales from the Sympetalae to his Malviflorae. This group includes Malvales, Euphorbiales and Solanales; our results are in agreement with this alinement. We do not agree, however, with Thorne, Cronquist and Takhtajan, by including sesame (Pedaliaceae) as a member of this group. It will be of considerable interest therefore, to examine a member of the Plantaginaceae, since this family has affinities both with the Pedaliaceae and also the Polemoniales (Cronquist 1968).

We also disagree with Cronquist (1968) and Takhtajan (1969) in their suggestion that the Asteridae were derived from the Rosidae, our results indicating that the Compositae, in fact, diverged prior to the Rosidae, although this is based solely on the results with mung bean (Rosidae) and sunflower and niger (Asteridae). However, as pointed out by Cronquist (1968), the morphological evidence for the derivation of the Asteridae from the Rosidae is not strong; in fact, the Dilleniidae are probably equally good candidates on morphological grounds. The chemical evidence, while favouring a Rosidian origin, is far from conclusive or consistent (Gibbs 1962; Hegnauer 1964). There is also some serological evidence that connects *Cornus* with the family Caprifoliaceae (Hillebrand & Fairbrothers 1965; Hillebrand 1966).

We would also like to comment on the early divergence of buckwheat (Polygonaceae) and spinach (Chenopodiaceae). While all taxonomists consider these

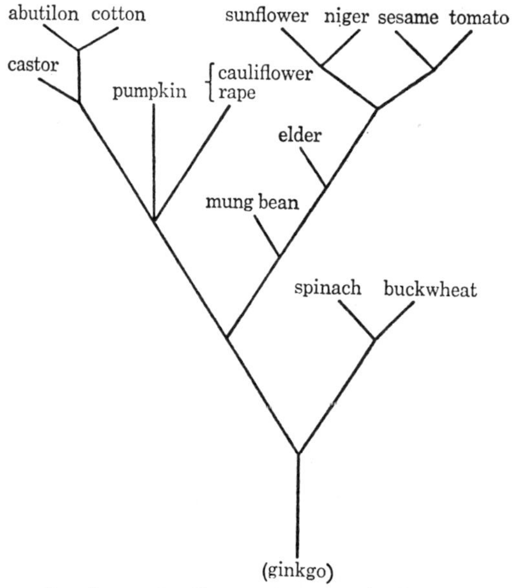

FIGURE 7. The phylogenetic scheme for the species examined according to Takhtajan (1969). Cronquist (1968) is very similar, except castor is placed between mung bean and elder.

two families as closely related, they have not generally been considered to have evolved as early as suggested by the results in figure 1 (Lawrence 1951).

The sequences of cytochrome *c* from five monocotyledonous species have also been determined (Stevens, Glazer & Smith 1967; R. H. Brown & D. Richardson, unpublished), and present indications are that this group diverged from the main dicotyledonous stock at about the same time as the Compositae. Consideration of these monocotyledon sequences with those given in the paper, does not affect the present findings of relationships of dicotyledonous families. However, sufficient sequences have not been determined to establish clearly the familial relationships of the monocotyledons.

The 'ancestral sequence' method gives results within taxa which are consistent with classical methods. Thus, the following: sunflower, niger; spinach, buckwheat; abutilon, cotton; cauliflower, rape; are more closely related by both methods to one another, than to any other of the taxa investigated. Whilst, as a result of the present work, new phylogenetic insights are limited, and we do not consider phylogenetic speculation to be profitable at this time, we are hopeful that work on further sequences now in progress will confirm and develop the ideas put forward here.

We wish to thank Mr A. A. Young, Computing Laboratory, University of Durham, for writing the program used in constructing phylogenetic trees by the 'ancestral sequence' method, and Professor D. Lewis, F.R.S., for reading the manuscript and for suggestions made. We also wish to thank the Nuffield Foundation, the Science Research Council, the Agricultural Research Council and the Royal Society, for financial support.

REFERENCES

Bessey, C. A. 1915 The phylogenetic taxonomy of flowering plants. *Ann. Mo. bot. Gdn.* **2**, 109–164.

Boulter, D. & Ramshaw, J. A. M. 1972 Structure-function relationships in plant cytochrome *c*. *Phytochem.* **11**, 553–561.

Crick, F. H. C. 1963 The recent excitement in the coding problem. *Progr. nucl. acid Res.* **1**, 163–217.

Cronquist, A. 1968 *The evolution and classification of flowering plants*. London and Edinburgh: Nelson.

Dayhoff, M. O. 1969 *Atlas of protein sequence and structure*, vol. 4. Maryland: National Biomedical Research Foundation.

Dayhoff, M. O. and Eck, R. V. 1966 *Atlas of protein sequence and structure*, vol. 2. Maryland: National Biomedical Research Foundation.

Dickerson, R. E., Takano, T., Eisenberg, D., Kallai, O. B., Samson, L., Cooper, A. & Margoliash, E. 1971 Ferricytochrome c. I. General features of the horse and bonito proteins at 2.8 Å resolution. *J. biol. Chem.* **246**, 1511–1535.

Engler, A. 1926 *Die naturlichen Pflanzenfamilien* (2nd ed.) **14a**, 136–137. Leipzig: Englemann.

Fitch, W. M. & Margoliash, E. 1967*a* A method for estimating the number of invariant amino acid coding positions in a gene using cytochrome c as a model case. *Biochem. Genet.* **1**, 65–71.

Fitch, W. M. & Margoliash, E. 1967*b* Construction of phylogenetic trees. *Science, N.Y.* **155**, 279–284.

Fitch, W. M. & Margoliash, E. 1969 The construction of phylogenetic trees. II. How well do they reflect past history? *Brookhaven Symp. Biol.* **21**, 217–241.

Fitch, W. M. & Markowitz, E. 1970 An improved method for determining codon variability in a gene and its application to the rate of fixation of mutations in evolution. *Biochem. Genet.* **4**, 579–593.

Gibbs, R. D. 1962 Comparative chemistry of plants as applied to a problem of systematics: the Tuberiflorae. *Trans. R. Soc. Can.* (Sect, III, Sci. Pt. 2 Biol. Sci.) **56**, 143–159.

Hegnauer, R. 1964 *Chemotaxonomie der Pflanzen*. III. Basel and Stuttgart: Birkhauser Verlag.

Hillebrand, G. R. 1966 Phytoserological systematic investigation of the genus *Viburnum*. Ph.D. Thesis, Rutgers University.

Hillebrand, G. R. & Fairbrothers, D. 1965 Phytoserological correspondence among selected genera of the Cornales, Garryales, Rosales, Rubiales and Umbellales as an indication of the taxonomic position of the genus *Viburnum*. *Am. J. Bot.* **52**, 648.

Jardine, N., van Rijsbergen, C. J. & Jardine, C. J. 1969 Evolutionary rates and the inference of evolutionary tree forms. *Nature, Lond.* **224**, 185.

Lance, G. N. & Williams, W. T. 1967 A general theory of classificatory sorting strategies. I. Hierarchical systems. *Comp. J.* **9**, 373–380.

Lawrence, G. H. M. 1951 *Taxonomy of vascular plants*. New York: Macmillan.

Nolan, C. & Margoliash, E. 1968 Comparative aspects of primary structures of proteins. *A. Rev. Biochem.* **37**, 727–790.

Ramshaw, J. A. M., Thompson, E. W. & Boulter, D. 1970 The amino acid sequence of *Helianthus annuus* L. (sunflower) cytochrome c deduced from chymotryptic peptides. *Biochem. J.* **119**, 535–539.

Ramshaw, J. A. M., Richardson, M. & Boulter, D. 1971 The amino acid sequence of the cytochrome c of *Ginkgo biloba* L. *Europ. J. Biochem.* **23**, 475–483.

Richardson, M., Laycock, M. V., Ramshaw, J. A. M., Thompson, E. W. & Boulter, D. 1970 Isolation and purification of cytochrome c from some species of higher plants. *Phytochem.* **9**, 2271–2280.

Richardson, M., Ramshaw, J. A. M. & Boulter, D. 1971*a* The amino acid sequence of rape (*Brassica napus* L.) cytochrome c. *Biochim. biophys. Acta* **251**, 331–333.

Richardson, M., Richardson, D., Ramshaw, J. A. M., Thompson, E. W. & Boulter, D. 1971*b* An improved method for the purification of cytochrome c from higher plants. *J. Biochem.* **69**, 811–813.

Sporne, K. R. 1971 *The mysterious origin of flowering plants*. Oxford: University Press.

Stevens, F. C., Glazer, A. N. & Smith, E. L. 1967 The amino acid sequence of wheat germ cytochrome c. *J. biol. Chem.* **242**, 2764–2779.

Takhtajan, A. 1969 *Flowering plants – origin and dispersal*. Edinburgh: Oliver and Boyd.

Thompson, E. W., Laycock, M. V., Ramshaw, J. A. M. & Boulter, D. 1970 The amino acid sequence of *Phaseolus aureus* L. (mung-bean) cytochrome c. *Biochem. J.* **117**, 183–192.

Thompson, E. W., Notton, B. A., Richardson, M. & Boulter, D. 1971 The amino acid sequence of cytochrome c from *Abutilon theophrasti* Medic. and *Gossypium barbadense* L. (cotton). *Biochem. J.* **124**, 787–791.

Thompson, E. W., Richardson, M. & Boulter, D. 1970 The amino acid sequence of sesame (*Sesamum indicum* L.) and castor (*Ricinus communis* L.) cytochrome c. *Biochem. J.* **121**, 439–446.

Thompson, E. W., Richardson, M. & Boulter, D. 1971*a* The amino acid sequence of cytochrome c from *Cucurbita maxima* L. (pumpkin). *Biochem. J.* **124**, 779–781.

Thompson, E. W., Richardson, M. & Boulter, D. 1971*b* The amino acid sequence of cytochrome c of *Fagopyrum esculentum* Moench (buckwheat) and *Brassica oleracea* L. (cauliflower). *Biochem. J.* **124**, 783–785.

Thorne, R. F. 1968 Synopsis of a putatively phylogenetic classification of the flowering plants. *Aliso* **6**, 57–66.

Walton, J. 1953 *An introduction to the study of fossil plants* (2nd ed). London: A. and C. Black.

Reprinted from *J. Mol. Evol.* **10**: 123–135.

Evolutionary Changes of α-Crystallin and the Phylogeny of Mammalian Orders

Wilfried W. de Jong[1] *, J. Timothy Gleaves[2], and Donald Boulter[2]

[1] Department of Biochemistry, University of Nijmegen, Nijmegen, The Netherlands
[2] Department of Botany, University of Durham, Durham DH1 3LE, United Kingdom

Summary. The sequences of the A chains of the eye lens protein α-crystallin from seventeen mammalian species were compared. They showed a generally slow rate of evolution, but with marked variations in different lineages. Most substitutions have occurred in the C-terminal part of the chain, which probably forms part of the surface of the α-crystallin aggregate. The ancestral sequence method of Dayhoff revealed interesting indications about the phylogenetic relationships between the eleven mammalian orders that were represented by the investigated species. Most evident was the divergence of marsupial and placental orders. A notable resemblance between the hyrax and elephant sequences was observed, setting them apart from the ungulates, including whale. Primates, rodents, lagomorphs, insectivores and tupaiids seem to derive from a common stem group. These phylogenetic inferences are discussed in relation to current palaeontological and taxonomical opinions, and compared to evidence from other protein sequence data.

Key words: Ancestral sequence – Eye lens protein – Evolution – Phylogenetic tree.

Comparative studies of amino acid sequences of homologous proteins have revealed in the past few years some of the principles of protein evolution (Dickerson, 1971; Fitch, 1973; Kimura and Ohta, 1974) and have been used to assess phylogenetic relationships between species (Dayhoff, 1972; Williams, 1974; Peacock and Boulter, 1975). It is important for the further understanding of protein evolution to study proteins of widely differing functions and properties. It is for this reason that we present in this paper a reconstruction of the evolutionary history of the A chain of α-crystallin, a major protein of the vertebrate eye lens. α-Crystallin is distinguished from previously investigated proteins since it is a structural protein occuring in large aggregates (average molecular weight 800,000), composed of two types of chains, A and B (for review see Harding and Dilley, 1976; Bloemendal, 1977). The results are discussed in relationship to some of the existing ambiguities concerning the phylogeny of mammalian taxa and included therefore in our investigation are not only some commonly studied mammals, but also representatives of groups for which little or no protein sequence data exist.

Substitutions in Mammalian A Chains

α-Crystallin occurs in the eye lenses of vertebrates from all classes as aggregates of high molecular weight and low isoelectric point (around pH 5.0) (De Jong et al., 1976). The two types of chains, A and B, of which it is built up, occur in the ox in a ratio of approximately 2:1, and show 57% sequence homology (Van der Ouderaa et al., 1974). Recently, the amino acid sequences of the α-crystallin A chains of seventeen mammalian species, representing eleven orders, have been established (Van der Ouderaa et al., 1973; De Jong et al., 1975a; 1975b; 1977; De Jong and Terwindt, 1976). The results of these

* To whom offprint requests should be sent

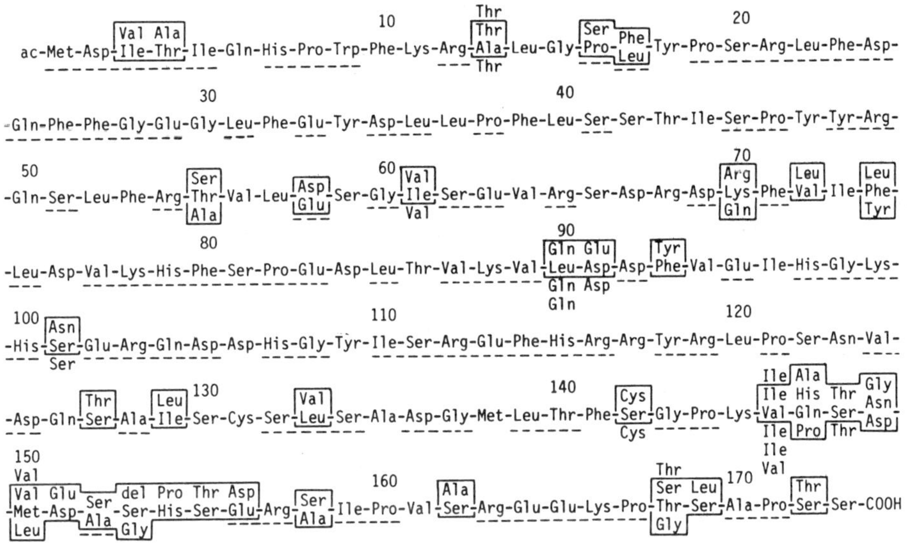

Fig. 1. Composite sequence of the α-crystallin A chains of seventeen mammalian species. The variable positions are taken from Table 1. Residues in line with the main sequence represent the most likely A chain sequence at the time of placental-marsupial divergence. Where no decision could be made about the identity of the ancestral residue in a given position, the two alternatives are shown off-line. In position 147 His, Gln, Pro and Met are possible ancestral residues. In position 158 also proline can be the ancestral residue but is not shown because it has not been observed in any present-day mammal. The actually observed variability at each mammalian position is represented by the residues enclosed in solid lines. The residues outside the solid lines represent probable convergent and back substitutions, as inferred from the branching pattern in Figure 2a. The dotted line below the sequence indicates those residues which are identical in the bovine A and B chains (Van der Ouderaa et al., 1974). It shows that substitutions in the mammalian A chain mostly occur in regions where the bovine A and B chains are different. In position 55 of the hyrax A chain equal amounts of alanine and threonine are present (De Jong et al., 1977), where alanine probably represents a genetic allele which has not yet reached fixation in the population. The deletion in position 153 of primate A chain is indicated by "del"

sequence determinations are combined in Table 1 and Figure 1. Table 1 summarizes which residues are found in the variable positions in all seventeen A chains. Figure 1 is a composite sequence of these mammalian A chains and contains all observed amino acid replacements.

Figure 1 shows that the C-terminal end of the chain, especially from position 146–156, is the most variable region, whereas other parts, e.g. position 18 through 54, have not accepted a single substitution. The presence of variable and conservative residues and chain regions is well known from other proteins. It is generally accepted that residues which are conserved during evolution are important for the maintenance of conformational and functional integrity of the protein (Dickerson, 1971). Changes at the variable positions are probably due to a considerable extent to selectively neutral, or nearly neutral, substitutions (Kimura and Ohta, 1974). In some cases, however, arguments have been given for the adaptive character of amino acid substitutions, as in haemoglobin (Goodman et al., 1975), insulin (Blundell and Wood, 1975) and cytochrome *c* (Margoliash et al., 1976).

Because neither the 3-dimensional configuration, nor the functional significance of particular residues of α-crystallin is known, it cannot be decided to what extent the observed substitutions are selectively neutral or reflect Darwinian adaptations to changing requirements of, for instance, visual acuity, nocturnal life or lens accomodation. It is, however, conceivable that the most variable regions of the A chain, especially from position 146–156, form part of the surface of the α-crystallin aggegate, as it is observed that surface areas in other proteins, e.g. haemoglobin (Goodman et al., 1975), pancreatic ribonuclease (Barnard et al., 1972), alcohol dehydrogenase (Eklund et al., 1976) and insulin (Blundell and Wood, 1975), contain the most highly variable areas. The possibility that parts of the C-terminal ends of the A chains are exposed

Table 1. Variable positions in the α-crystallin A chains of seventeen mammalian species

	Positions 3 4 13 16 17 18 20 26 31 32 33 37 39 40 52 55 56 58 61 70 72 74 86 89 90 91 93 101 122 124 125 126 127 128 129 130 133 135 138 142 146 147 148 149 150 151 152 153 154 155 156 158 162 168 169 172
HORSE	IAAPFYSFLFELFLLTVDIKVFTVQEFNSVDQTALSVAMSIPSGMDAGHSEASGSS
RHINOCEROS	IATPFYSFLFELFLLSVDVKVFTVQEFNSVDQTALSLAMSIPSGMDAGHSEASSSS
PIG	IAAPFYSFLFELFLLTVDVKVFTVQEFNSVDQSALSLAMSVPSGVDAGHSEASSST
OX	IATPFYSFLFELFLLTVDIKVFTVQEFNSVDQSALSLAMSIPSGVDAGHSEASSSS
WHALE	IAAPFYSFLFELFLLTVDIKVFTVQEFNSVDQSALSLAMSVPSGMDAGHSEASSSS
DOG, CAT	IAAPFYSFLFELFLLTVDIKVFTVLEFNSVDQSALSLAMSVPSGVDAGHSEASSSS
HEDGEHOG	VTAPFYS.......LTVDIK..TVLEFS........LAMSVASGLDAGPSEASSSS
RAT, TUPAIA	VTAPFYSFLFELFLLTVDIKVFTVLEFNSVDQSALSLAMSVQSGLDAGHSEASSSS
RABBIT	VTTPFYSFLFELFLLTVDIKVFTVQEFNSVDQSALSLAMSVQSGLDAGHSEASSSS
HUMAN	VTTPFYSFLFELFLLTVDIKVFTVQDFNSVDQSALSLAMCIQTGLDA-HTEASTSS
MONKEY	VTTPFYSFLFELFLLTVDIKVFTVQDFNSVDQSALSLAMSIQTGLDA-HTEAASSS
ELEPHANT	VTAPFYSFLFELFLLTVDIQVLTVQDFNSVDQSALSLAMCIQSGMDASHSFASSSS
HYRAX	VTAPFYSFLFELFLLTVDIQLLTVLDFNSVDQSALSLAMCVQSGMDASHSEASSSS
KANGAROO	ITASLYSFLFELFLLTVEIKVFTVLDFSSVDQSAISLAMSIHSDMDASHSDSSTLS
OPOSSUM	ITASLYSFLFELFLLTVEIRVYTVLDYSSVDQSAISLAMSIHTNMESSHSDSSTLS
FROG	ITAPFYNVMFDFFLFGFDIR..TILDFSSLNESSISLAISLMSSLDSSHGEPSTSS
CHICKEN	ITAPLISFLLELLFLSVEIK..SIIDFSAVDQSAITLSMSVPSNMDPSHSEPSTSS
	***** * *** * *** * * * ****** ** *** **

The A chains are from horse (*Equus caballus,* Order *Perissodactyla*); white rhinoceros (*Ceratotherium simum, Perissodactyla*); pig (*Sus scrofa, Artiodactyla*); ox (*Bos taurus, Artiodactyla*); minke whale of lesser rorqual (*Balaenoptera acutorostrata, Cetacea*); dog (*Canis familiaris, Carnivora*); cat (*Felis catus, Carnivora*); European hedgehog (*Erinaceus europaeus, Insectivora*)[a]; rat (*Rattus norvegicus, Rodentia*); treeshrew or tupaia (*Tupaia belangeri, Primates* or *Scandentia*)[a]; rabbit (*Oryctolagus cuniculus, Lagomorpha*); human (*Homo sapiens, Primates*)[b], rhesus monkey (*Macaca mulatta, Primates*); African elephant (*Loxodonta africana, Proboscidea*); Cape hyrax or rock dassie (*Procavia capensis, Hyracoidea*); red kangaroo (*Macropus rufus, Marsupialia*); North American opossum (*Didelphis marsupialis, Marsupialia*) (De Jong et al., 1975a, 1975b, 1977; De Jong and Terwindt, 1976; Van der Ouderaa et al., 1973). Chicken (*Gallus gallus*)[a] and frog (*Rana esculenta*)[a] A chain positions are included for comparison in order to establish the ancestral residues at the time of marsupial-placental divergence (Fig. 1) and are used in the construction of phylogenetic trees (Fig. 2 and 3). Only those positions marked by * are used in computation for phylogenetic tree-construction; the other positions are trivial, i.e. they do not change the topology of a tree. The one-letter notation for amino-acids has been used: A, alanine; C, cysteine; D, aspartic acid; E, glumatic acid; F, phenylalanine; G, glycine; H, histidine; I, isoleucine; K, lysine; L, leucine; M, methionine; N, asparagine; P, proline; Q, glutamine; R, arginine; S, serine; T, threonine; V, valine; Y, tyrosine; (–) means deletion and (.) not determined.

a The A chain sequences of hedgehog, tupaia, chicken and frog are unpublished data from W.W. de J., E.C.Nuij-Terwindt and A. Zweers. Tupaia lenses were provided by the Battelle-Institut, Frankfurt am Main. Positions 22–49, 71–78 and 120–131 of hedgehog A chain have not yet been determined; the usual placental mammalian residues have been assumed to be present in these positions for purposes of calculation. Positions 71–76 in the frog and 72–74 in the chicken A chains are also unknown; trivial residues have been assumed to be present there

b The published sequence of human A chain (De Jon et al., 1975b) has been corrected to contain 136-Asp and 142-Cys (Kramps, 1977)

at the surface of the α-crystallin aggregate is supported by the observation that *in vivo* breakage of A chains occurs between residues 151 and 152 (Van Kleef et al., 1975) and that limited tryptic hydrolysis of the native α-crystallin aggregate specifically cleaves the bond between residues 157 and 158 (Siezen and Hoenders, 1977). On the other hand it is likely that invariable regions of the A chain, like position 18–54, occur in the interior of the molecule or are involved in contacts between neighbouring chains in the aggregate.

Rate of Evolution

The average rate of evolution of the A chain has been estimated from the observed numbers of substitutions between the different A chains and the proposed times of divergence of the concerned species (De Jong et al., 1975a; 1977; De Jong and Terwindt, 1976). The rate of 2.9 amino acid substitutions per 100 residues in 100 Myr places the α-crystallin A chain among the slowly evolving proteins like cytochrome *c*

(3% substitution in 100 Myr) and insulin (4% in 100 Myr) (Dayhoff, 1972). Comparison of bovine and human α-crystallin B chains indicates that the B chain may evolve even more slowly (Kramps et al., 1977). In cytochrome *c* and insulin the slow rate of evolutionary change has been attributed to constraints resulting from their biological functions, which require the conservation of the chemical properties of most side chains. In α-crystallin no such explanation is evident, and it seems reasonable to attribute the slow rate of evolution at least partially to the requirements of many side-chain interactions in the closely-packed interior of the large α-crystallin aggregate. In this connection it is interesting to point out that the interior positions of haemoglobin, which only serve to stabilize the tertiary structure, evolved at a rate af about 9% in 100 Myr (Goodman et al., 1975).

The chemically conservative nature of most of the observed replacements in the α-crystallin A chain is in agreement with the rigid conservation of structure of this protein throughout mammalian evolution. Of the at least 60 substitutions in the phylogenetic tree of the mammalian A chains, since the marsupial-placental divergence (cf. Fig. 2), only two give rise to a change in electrophoretic mobility at pH 8.5: 70-Gln in the elephant-hyrax line and 149-Asp in the ancestry of kangaroo. This is far below the approximately 25% of electrophoretically detectable substitutions estimated to occur in protein evolution (King and Wilson, 1975; Nei, 1975) and demonstrates the limitations of electrophoretic data in evolutionary and population genetical studies.

The trees to be described in Figure 2 suggest considerable differences in rates of substitution in the A chains in different evolutionary lineages. Since the divergence of the ancestors of man and rat, some 75 Myr ago (Young, 1962; Romer, 1966; McKenna, 1969), the first appears to have accumulated nine substitutions in its A chain and the latter none. Among the marsupials the kangaroo A chain accepted at most two substitutions, against five to six in the opossum, since their common ancestry, possibly 70 Myr ago (Romer, 1966). Although overall rates of evolution for each protein are roughly constant over extended geological periods (Dickerson, 1971; Kimura and Ohta, 1974), considerable variation within evolutionary lines is well demonstrated in many proteins, like haemoglobin (Holmquist et al., 1976; Langley and Fitch, 1974), cytochrome *c* (Margoliash et al., 1976; Moore et al., 1976; Langley and Fitch, 1974), myoglobin (Holmquist et al., 1976; Romero-Herrera et al., 1973), insulin (Blundell and Wood, 1975) and pancreatic ribonuclease (Van den Berg and Beintema, 1975). Accelerated rates of evolution have been attributed to the occurrence of adaptive substitutions in a given line (Goodman et al., 1975; Blundell and Wood, 1975; Ohta and Kimura, 1971), or to increased rates of fixation of nearly neutral substitutions occurring in small populations at the time of speciation (Ohta, 1974; Bush, 1975).

It appears that fluctuations in the evolutionary rates of different proteins do not occur synchronously in the same lines. Examples of asynchronous protein evolution are the exceptionally slow rate of change in globin chains in the higher primates (Holmquist et al., 1976; Langley and Fitch, 1974), where the α-crystallin A chain shows a relatively high number of substitutions, and the slow rate of change of myoglobin in the opossum line (Romero-Herrera and Lehmann, 1975), where the haemoglobin α chain (Stenzel, 1974) and α-crystallin A chain evolve more rapidly.

The conservation during vertebrate evolution of large regions of the α-crystallin molecule is also evident from the fact that 27 of the 35 variable positions observed in mammalian A chains, accounting for at least 51 substitutions, occur among the 74 positions where bovine A and B chain differ, whereas only eight of the variable positions, accounting for eight substitutions, appear among the 99 positions which are identical in bovine A and B chain (see Fig. 1).

Phylogenetic Trees Derived from the A Chain Sequences

From the differences between the A chain sequences, as given in Table 1, phylogenetic trees have been constructed by using the ancestral sequence method of Dayhoff and

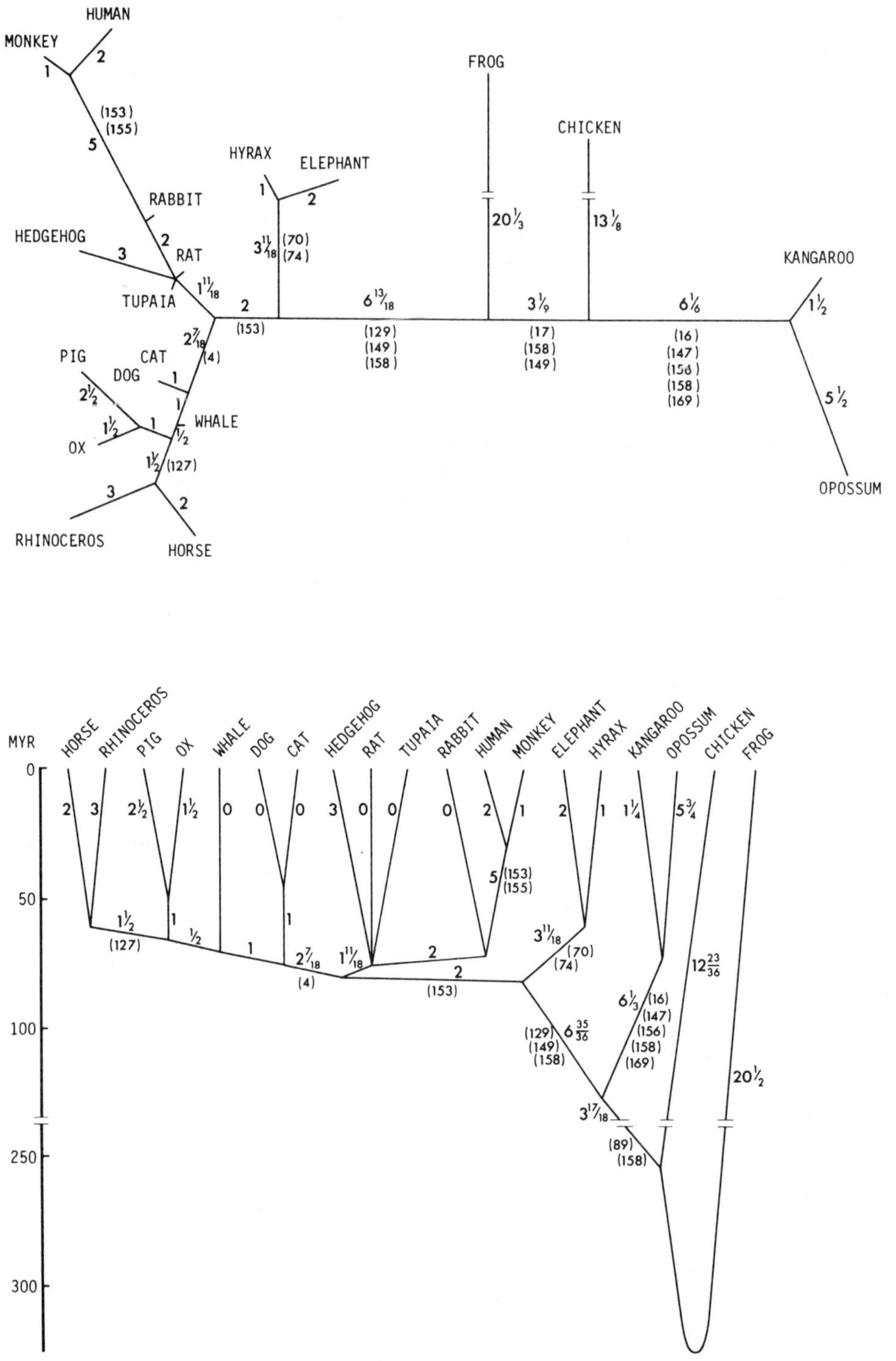

Fig. 2a–b. Phylogenetic trees of α-crystallin A chains under the criterion of parsimony. The number of substitutions necessary for their evolution was minimised by using the ancestral sequence method of Dayhoff and Eck (1966), with modification by D.P. Peacock and A.A. Young (unpublished). Branch lengths are expressed in substitutions. No value is given for zero distances and fractional values arise where there is doubt as to the ancestral sequence at a node. Positions are shown (in brackets) of substitutions other than back, parallel or trivial (giving no cladistic information).
a The most parsimonious phylogenetic tree, involving a total of 96 substiutions. Several other equally parsimonious minor variants of this tree exist and are considered in the text.
b A taxonomically more acceptable phylogenetic tree with chicken and frog on a separate branch. This tree requires 97 substitutions. The branching pattern of the mammalian species is as in Figure 2a. A geological time scale is shown on the ordinate to give an impression of the different rates of substitutions in various lineages. Estimated times of divergence are taken from palaeontological literature. (Romer, 1966; Romero-Herrera et al., 1973; L. van Valen in Langley and Fitch, 1974; Young, 1962)

Eck (1966) since this procedure gives an accurate result when the sequence data contain a small number of amino acid differences (Peacock and Boulter, 1975). The most parsimonious phylogenetic tree, involving 96 substitutions, is depicted in Figure 2a. There were several other equally parsimonious minor variants of this tree which are taken into account in the text. Furthermore, only one additional substitution gives the biologically more acceptable tree of Figure 2b, with chicken and frog on a separate branch. An ancestral sequence of the A chain at the time of placental-marsupial divergence is given in Figure 1. The data were also subjected to the method of Estabrook and Landrum (1975). This procedure searches for sets of characters (residue positions), which do not contain convergent substitutions when a tree is specified. The resulting phylogenetic trees were very similar to Figure 2a, but they suggest a closer link between the primates and elephant and hyrax than does Figure 2a.

Phylogenetic Inferences

Of the several equally parsimonious trees (e.g. Fig. 2a) certain features are constant and are unlikely to change when additional A chain sequences become available. A fundamental dichotomy of the mammals, separating the marsupials opossum and kangaroo from all placental species occurs in all trees. The ungulates, including whale, always come together as a group, with the carnivores as an earlier offshoot of this line. Horse and rhinoceros share the unique substitution 127 Ser-Thr. Man and monkey firmly go together, having a unique substitution and a deletion. The rabbit joins the primate line at an earlier stage. A special affinity of elephant and hyrax is reflected in the fact that their pairing is a constant feature of all trees. The connection of this pair to the tree is variable, but never with the ungulates. The positions of hedgehog, tupaia and rat are more variable, but they usually group together and join the primate-lagomorph branch.

It is evidently of interest to compare the topology of the α-crystallin A chain tree with current opinions about the phylogeny of mammalian orders. The eutherian *Placentalia* and the metatherian *Marsupialia* apparently are monophyletic groups (sensu Hennig, 1966) which arose independently from common pantotherian ancestors in the early Cretaceous (McKenna, 1969; Lillegraven, 1975; Hoffstetter, 1973). The sequence data support the monophyletic character of the investigated placentals on the one hand and marsupials on the other.

The radiation of the eutherian orders in the late Cretaceous and the Paleocene occurred rapidly and fossil evidence is scarcely available to reveal this process. Late Cretaceous eutherian mammals have been divided, on the basis of dentition and ear bones (McKenna, 1969; Lillegraven, 1969; MacIntuire, 1972) into two groups, which might be ancestral to ungulates and carnivores on the one hand, and primates, rodents and the true insectivores on the other. Our sequence data are compatible with such a division, although there is some ambiguity about the position of rat, tupaia and hedgehog.

A different dichotomy of the placental orders has been proposed on the basis of the number of bony elements in the basicranial axis (Hoffstetter, 1973). One group includes the ungulates and *Proboscidea*, and the other, having a fourth basicranial element as supposed shared derived (synapomorphous (Hennig, 1966)) character, the *Hyracoidea, Primates, Rodentia, Lagomorpha* and *Carnivora.* The A chain sequences would rather favour the exclusion of the carnivores from the latter group, and strongly argue against the separation of *Proboscidea* and *Hyracoidea.*

There is general agreement about the common condylarth origin of *Artiodactyla, Perissodactyla, Hyracoidea* and *Proboscidea. Proboscidea* and *Hyracoidea* are mostly considered to be more closely related to each other than to other orders, on the basis of similar morphological structures (Thenius, 1969; Romer, 1966; Colbert, 1969), and immunological cross-reactivity (Weitz, 1953). However, some very recent classifications of mammalian taxa, based on fossil evidence, group the *Hyracoidea, Proboscidea* and *Perissodactyla* in the same superorder (Szalay, 1977), or place the *Hyracoidea* closer to the *Perissodactyla* than to the *Proboscidea* (McKenna, 1975). The A chain sequences leave little doubt that *Hyracoidea* and *Proboscidea* form a monophyletic group and exclude a special relationship of the *Hyracoidea* to the *Perissodactyla*. The sequence data strongly argue against the usually assumed relationship of elephant and hyrax to the ungulates.

The *Cetacea* were previously thought to have a separate origin (Simpson, 1945), or suggested to be possible branched from primitive carnivorous stock (Romer, 1966), but they are now considered to share the condylarth origin of the ungulates (Romero-Herrera et al., 1973; Van Valen, 1971; McKenna, 1975; Szalay, 1977) or even to be especially related to the *Artiodactyla* on the basis of immunological and karyological findings (Thenius, 1969). Our sequence data lend strong support to the relationship of the *Cetacea* to the ungulates.

The *Carnivora* are thought to derive by a distinct line from ancestral insectivore stock (Romer, 1966; Colbert, 1969). The sequence data support Simpson (1945), who included them with the ungulates in the cohort *Ferungulata*. The once-favoured grouping together of the *Lagomorpha* with the *Rodentia* Simpson, 1945) is now generally abandoned. The morphological resemblances between rabbits and rodents are thought to be due to convergent development, and the fossil record indicates a geographically widely separated origin (McKenna, 1969). This leaves both lagomorphs and rodents as isolated groups of largely obscure origin (Romer, 1966; Colbert, 1969). Lagomorphs show some odontologic relation to *Condylarthra* or leptictid *Insectivora* (Thenius,, 1969; Colbert, 1969), whereas serological and ethological relations to the *Artiodactyla* have been mentioned (Thenius, 1969). McKenna (1975) considers the lagomorphs as a separate, early offshoot of the main placental line. The *Rodentia* may be related to the stemgroup of the *Primates* (Thenius, 1969), although the most recent classifications (McKenna, 1975; Szalay, 1977) are unable to assign them a position in relation to the other mammalian orders. The sequence data tend to associate the lagomorphs with the primate line, but are more equivocal as to the position of the rodents.

Although a close affinity of present-day primitive primates and insectivores seems obvious (Thenius, 1969), the palaeontological record of the early stages of primate origin is very limited (Simons, 1969). It seems, however, that primates might derive in the latest Cretaceous from erinaceoid insectivore lineage (Clemens, 1974). Tupaiids are sometimes included in the order *Primates* (Simpson, 1945; Clark, 1971) but several characters argue for classification as a separate order *Scandentia* (McKenna, 1975; Szalay, 1977; Butler, 1972) which shows some relation to leptictid insectivores. On balance the sequence data support a common stemline for primates, lagomorphs, rodents, tupaiids and insectivores. The identical sequences of rat and tupaia A chains do not necessarily imply a special relation of the latter to the rodents. The phylogenetic tree (Fig. 2) rather indicates that rat and tupaia A chains have remained unchanged since they originated from their common stemgroup, i.e. they form a plesiomorphous (Hennig, 1966) character.

The current opinions about the phylogeny of mammalian orders are reflected in the tree shown in Figure 3a. The bifurcating trees shown in Figure 3b are based on the topology of Figure 3a and both require 105 substitutions, which is nine more than the most parsimonious tree of Figure 2. This shows the "cost", in terms of additionally required substitutions, if one wants to adhere to the prevailing ideas about mammalian relationships. These additional substitutions are especially due to the moving of elephant and hyrax to the ungulate branch, and the removing of the rabbit from the primate branch.

Comparison with other Protein Sequence Data

Comparative sequence data of the same mammalian orders as examined here are available from a number of other proteins, namely cytochrome *c*, myoglobin, haemoglobin α and β chain, pancreatic ribonuclease and fibrinopeptides, and have been used for the construction of phylogenetic trees (Beintema and Fitch, 1977; Dayhoff, 1972; Goodman, 1975 and 1976; Goodman et al., 1975; Holmquist et al., 1976; Moore et al., 1976; O'Neil and Doolittle, 1974; Penny, 1974; Romero-Herrera et al., 1973). The significance of these trees is difficult to evaluate, because different tree-construction programmes, using the same data, can yield different topologies. Quite different topologies are often equally, or almost equally, parsimonious, and the addition of new sequence data can significantly alter a previously proposed pattern of branching.

Despite these problems some features of the protein-based mammalian phylogenies are quite general. The kangaroo proteins always separate this marsupial from the placental orders, except sometimes in the case of cytochrome *c*. *Artiodactyla* and *Perissodactyla* are closely related on the basis of sequence data from cytochrome *c*, myoglobin, α and β haemoglobin and fibrinopeptides. *Cetacea* are found to join this ungulate branch in the case of cytochrome *c*, myoglobin and ribonuclease. Cytochrome *c*, α and β haemoglobin and fibrinopeptides from primates and lagomorphs show them to be more closely related to each other than to the ungulate branch. The rodents are in the case of fibrinopeptides and α haemoglobin always connected to the primate-lagomorph branch, but not so with the β haemoglobin data set. Tupaia tends to be excluded from the *Primates* by α haemoglobin and myoglobin sequence data (Romero-Herrera and Lehmann, 1974). The carnivores either diverge from the branch leading to ungulates, including *Cetacea*, or from the branch leading to the rodent-lagomorph-primate group, or as an earlier separate eutherian branch.

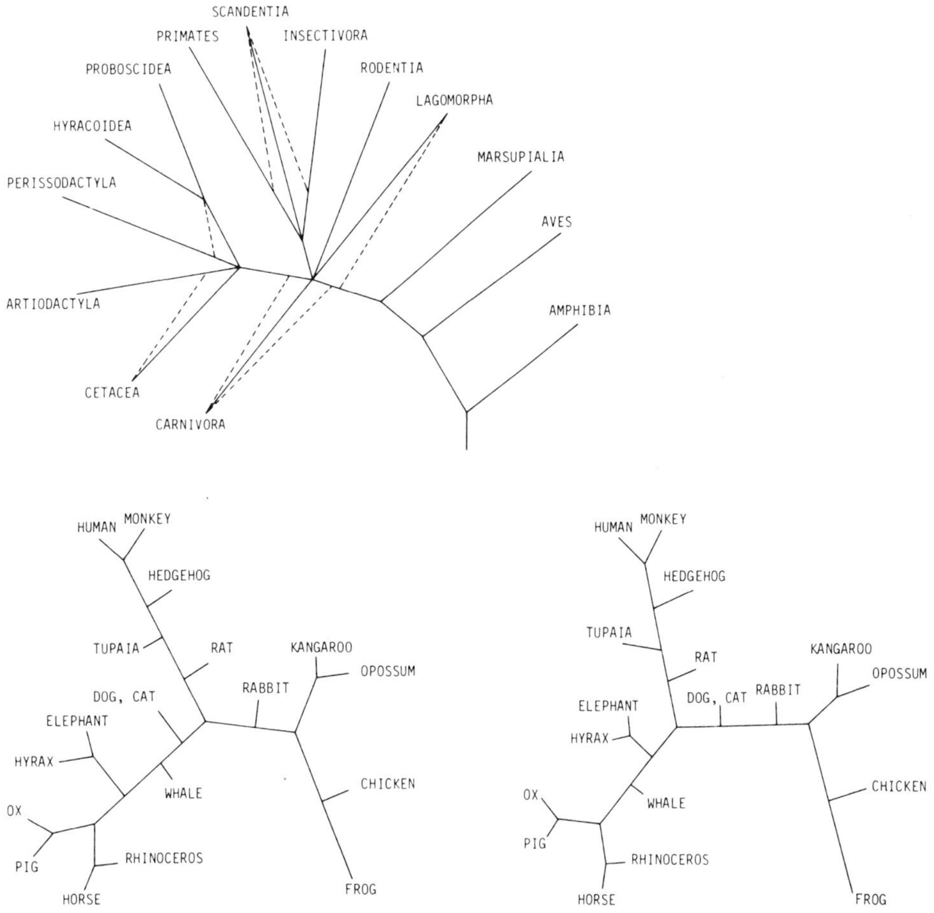

Fig. 3. a Schematic representation of phylogenetic relationships between mammalian orders based on a consensus of palaeontological evidence. Relations suggested by some authors (– – – – – – –)
b Two examples of bifurcating trees comprising the species given in Table 1 and based on the topology of Figure 3a. According to the ancestral sequence method these trees both require 105 substitutions

For some species only a single protein has as yet been included in phylogenetic trees (Homquist et al., 1976): elephant β haemoglobin is placed as a separate branch of the eutherians, after the splitting off of carnivores and rodents; hedgehog myoglobin (Romero-Herrera et al., 1975) is connected to the primate line, and opossum α haemoglobin (Stenzel, 1974) joins the kangaroo. The sequences of the fibrinopeptides from rhinoceros and elephant are known and apparently connect them to the other *Perissodactyla* and as an earlier branch of the carnivore-ungulate group, respectively (O'Neil and Doolittle, 1974).

It appears from the combined protein sequence data that indeed valuable information about phylogenetic affinities of the mammalian orders has been obtained. Unfortunately, no protein data are yet available from a number of small but interesting mammalian orders of which the evolutionary position is till disputed. The α-crystallin A chain promises to be a suitable tool to study the phylogenetic relationship of such orders as the *Edentata, Pholidota, Tubulidentata, Chiroptera* and *Dermoptera.*

In considering the present α-crystallin data set uncertainty stems firstly from the fact that not all trees have been generated and that no method exists to decide which to choose between the alternative trees which have been generated. Also, there is no way of knowing if a phylogenetic tree is a true cladogram. However, there is theoretical and practical evidence which indicates that amino acid sequence data are useful in phylogenetic studies, but that at present it is not possible to be sure which information is the most useful.

In our view the data should be accepted at face-value with caution and be used by taxonomists interested in other types of character to institute reinvestigation where the sequence data appear firm and go against present ideas.

Acknowledgements. Lens-material was kindly provided by Drs. A. Schwaier (Battelle-Institut, Frankfurt a.M.), P. Stenzel and R.T. Jones (Portland, Or.), Å. Jonsgård and I. Christensen (Norway), U. de V. Pienaar, V. de Vos, P. van Wyk and M.E. Keep (South-Africa). W.W. de J. thanks the participants in the 1976 NATO Advanced Study Institute on Major Patterns in Vertebrate Evolution for valuable discussions. This work was supported in part by the Netherlands Foundation for Chemical Research (S.O.N.) and with financial aid from the Netherlands Organization for Pure Research (Z.W.O.).

References

Barnard, E.A., Cohen, M.S., Gold, M.H., Kim, J.K. (1972). Nature **240**, 395–398

Beintema, J.J., Fitch, W.M. (1977). J. Mol. Evol. (in press)

Bloemendal, H. (1977). Science **197**, 127–138

Blundell, T.L., Wood, S.P. (1975). Nature **257**, 197–203

Bush, G.L. (1975). Ann. Rev. Ecol. Syst. **6**, 339–364

Butler, P.M. (1972). In: Studies in vertebrate evolution, K.A. Joysey, Kemp, T.S., eds., pp. 253–265. New York: Winchester Press

Clark, W.E. Le Gros (1971). The antecedents of man, 3rd ed. Chicago: Quadrangle

Clemens, W.A. (1974). Science **184**, 903–905

Colbert, E.H. (1969). Evolution of the vertebrates, 2nd ed.. New York: Wiley

Dayhoff, M.O. (1972). Atlas of protein sequence and structure, **5**. National Biomedical Research Foundation, Silver Spring, Maryland

Dayhoff, M.O., Eck, R.V. (1966). Atlas of protein sequence and structure, **2**, National Biomedical Research Foundation, Silver Spring, Maryland

De Jong, W.W., Van der Ouderaa, F.J., Versteeg, M., Groenewoud, G., Van Amelsvoort, J.M., Bloemendal, H. (1975a). Eur. J. Biochem. **53**, 237–242

De Jong, W.W., Terwindt, E.C., Bloemendal, H. (1975b). FEBS Lett. **58**, 310–313

De Jong, W.W., Terwindt, E.C. (1976). Eur. J. Biochem. **67**, 503–510

De Jong, W.W. Terwindt, E.C., Groenewoud, G. (1976). Comp. Biochem. Physiol. **55B**, 49–56

De Jong, W.W., Nuij-Terwindt, E.C., Versteeg, M. (1977). Biochim. biophys. Acta **491**, 573–580

Dickerson, R.E. (1971). J. Mol. Evol. **1**, 26–45
Eklund, H., Brändén, C.J., Jörnvall, H. (1976). J. Mol. Biol. **102**, 61–73
Estabrook, G.F., Landrum, L. (1975). Taxon, **24**, 53–57
Fitch, W.M. (1973). Ann. Rev. Genet. **7**, 343–380
Goodman, M. (1975). In: Phylogeny of the primates, W.P. Luckett, F.S. Szalay, eds., pp. 219–248. New York-London: Plenum Press
Goodman, M. (1976). In: Molecular evolution. F.J. Ayala, ed., pp. 141–159. Sunderland: Sinauer
Goodman, M., Moore, G.W., Matsuda, H. (1975). Nature **253**, 603–608
Harding, J.J., Dilley, K.J. (1976). Exp. Eye Res. **22**, 1–73
Hennig, W. (1966). Phylogenetic systematics. Urbana: University of Illinois Press
Hoffstetter, R. (1973). Ann. Paléontol. (Paris) **59**, 137–169
Holmquist, R., Jukes, T.H., Moise, H., Goodman, M., Moore, G.W. (1976). J. Mol. Biol. **105**, 39–74
Kimura, M., Ohta, T. (1974). Proc. Natl. Acad. Sci. **71**, 2848–2852
King, M.C., Wilson, A.C. (1975). Science **188**, 107–115
Kramps, J.A. (1977). Ph.D. Thesis, University of Nijmegen
Kramps, J.A., De Man, B.M., De Jong, W.W. (1977). FEBS Lett. **74**, 82–84
Langley, C.H. Fitch, W.M. (1974). J. Mol. Evol. **3**, 161–177
Lillegraven, J.A. (1969). Paleont. Contrib. Univ. Kansas **50**, 1–122
Lillegraven, J.A. (1975). Evolution **29**, 707–722
MacIntuire, G.T. (1972). In: Evolutionary biology, T. Dobzhansky, M.K. Hecht, Steere, W.C., eds., Vol. 6, pp. 275–303, New York: Appleton-Century-Crofts
Margoliash, E., Ferguson-Miller, S., Kang, C.H., Brautigan, D.L. (1976). Fed. Proc. **35**, 2124–2130
McKenna, M.G. (1969). Ann. N.Y. Acad. Sci. **167**, 217–240
McKenna, M.G. (1975). In: Phylogeny of the primates, W.P. Luckett, F.S. Szalay, eds., pp. 21–46. New York-London: Plenum Press
Moore, G.W., Goodman, M., Callahan, C., Holmquist, R., Moise, H. (1976). J. Mol. Biol. **105**, 15–37
Nei, M. (1975). Molecular population genetics and evolution. In: Frontiers of biology, Vol. 40, p. 26. Amsterdam: North-Holland
Ohta, T. (1974). Nature **252**, 351–354
Ohta, T., Kimura, M. (1971). J. Mol. Evol. **1**, 18–25
O'Neil, P.B., Doolittle, R.F. (1974). Syst. Zool. **22**, 590–595
Peacock, D., Boulter, D. (1975). J. Mol. Biol. **95**, 513–527
Penny, D. (1974). J. Mol. Evol. **3**, 179–188
Romer, A.S. (1966). Vertebrate paleontology, 3rd ed.. University of Chicago Press
Romero-Herrera, A.E., Lehmann, H. (1974). Biochim. biophys. Acta **359**, 236–241
Romero-Herrera, A.E., Lehmann, H. (1975). Biochim. biophys. Acta **400**, 387–398
Romero-Herrera, A.E., Lehmann, H. Joysey, K.A., Friday, A.E. (1973). Nature **246**, 389–395
Romero-Herrera, A.E., Lehmann, H., Fakes, W. (1975). Biochim. biophys. Acta. **379**, 13–21
Siezen, R.J., Hoenders, H.J. (1977). FEBS Lett. (in press)
Simons, E.L. (1969). Ann. N.Y. Acad. Sci. **167**, 319–331
Simpson, G.G. (1945). Bull. Am. Mus. Nat. Hist. **85**, 1–350
Stenzel, P. (1974). Nature **252**, 62–63
Szalay, F.S. (1977). In: Major Patterns in Vertebrate Evolution, M.K. Hecht, P.C. Goody, B. Hecht, eds., NATO ASI Series. New York: Plenum
Thenius, E. (1969). Phylogenie der Mammalia. Berlin: Walter de Gruyter
Van den Berg, A., Beintema, J.J. (1975). Nature **253**, 207–210
Van der Ouderaa, F.J., De Jong, W.W., Bloemendal, H. (1973). Eur. J. Biochem. **39**, 207–222
Van der Ouderaa, F.J., De Jong, W.W., Hilderink, A., Bloemendal, H. (1974). Eur. J. Biochem. **49**, 157–168

Van Kleef, F.S.M., De Jong, W.W., Hoenders, H.J. (1975). Nature **258**, 264–266
Van Valen, L. (1971). Evolution **25**, 420–428
Weitz, B. (1953). Nature **171**, 261
Williams, J. (1974). In: Chemistry of macromolecules, biochemistry, H. Gutfreund, ed., Ser. 1, Vol. 1, pp. 1–56. MTP International Reviews of Science, London: Butterworth; Baltimore: University Park Press
Young, J.Z. (1962). The life of vertebrates, 2nd ed. Oxford University Press

Received March, 16, 1977; Revised Version May 19, 1977

Origins of Prokaryotes, Eukaryotes, Mitochondria, and Chloroplasts

A perspective is derived from protein and nucleic acid sequence data.

Robert M. Schwartz and Margaret O. Dayhoff

Many proteins and nucleic acids are "living fossils" in the sense that their structures have been dynamically conserved by the evolutionary process over billions of years (*1*). Their amino acid and nucleotide sequences occur today as recognizably related forms in eukaryotes and prokaryotes, having evolved from common ancestral sequences by a great number of small changes (*2*). These sequences may still carry sufficient information for us to unravel the early evolution of extant biological species and their biochemical processes. There are two principal computer methods that can be used to treat sequence data in order to elucidate evolutionary history. These were first described more than 10 years ago (*3, 4*) and have been used to construct a vertebrate phylogeny from each of a number of proteins (*5–7*). This phylogeny is generally consistent with both the fossil record and morphological data. Only recently has enough sequence information become available from diverse types of bacteria and blue-green algae, and from the cytoplasm and organelles of eukaryotes, for us to attempt the construction of a biologically comprehensive evolutionary tree. These sequences include ferredoxins, 5*S* ribosomal RNA's, and c-type cytochromes. We describe here an evolutionary tree derived from sequence data that extends back close to the time of the earliest divergences of the present-day bacterial groups.

Knowledge of the evolutionary relationships between all species would have great predictive advantage in many areas of biology, because most systems within the organisms would show a high degree of correlation with the phylogeny. Knowing the relative order of the divergences of prokaryote types and their protein constituents is important to understanding the evolution of metabolic pathways. With such information the long-standing question of how eukaryote organelles originated might be resolved.

Before we consider the phylogenies based on sequence data, we will briefly review the fossil record and describe the time scale during which the various prokaryote groups diverged (*8, 9*). The early fossil record is sparse and subject to some uncertainty of interpretation. The oldest known bacterium-like structures that could possibly be biogenic are preserved in the Swaziland sediments and are more than 3.1 billion years old (*8, 10*). Stromatolitic structures have been dated at nearly 3.0 billion years old (*11*). These structures are widely assumed to be evidence for the antiquity of blue-green algae but might equally well be interpreted as products of communities of photosynthetic bacteria, gliding flexibacteria, or non-oxygen-releasing ancestors of the blue-green algae (*8*). Both coccoid and filamentous microfossils from 2.3 billion years ago have been identified. The fossil record of microorganisms from about 2 billion years ago clearly shows a great abundance and diversity of morphological types, many resembling present-day bacteria and blue-green algae. Geological evidence suggests that free oxygen began to accumulate in the atmosphere about 2.0 billion years ago (*12*). Eukaryote cells may have originated as early as 1.4 billion years ago, because there is an abrupt increase in cell size and diversity in microfossils of this age (*13*). The multicellular eukaryote kingdoms, plants, animals, and fungi, are thought to have diverged between 1 billion and 700 million years ago. Fossils of metazoans nearly 700 million years old have been found.

There is good evidence that, in many bacterial sequences of basic metabolic importance, the rate of accumulation of point mutations accepted in the wild-type population is even slower than it is in higher plants and animals (*14*). Thus, even though the time span for the evolution of prokaryotes is more than two times as long, we hope to infer correct evolutionary trees for them as well as we do for eukaryotes. Because the morphological evidence of biologists has proved to be inadequate to the task of organizing the major groups of bacteria and because the fossil record is difficult to interpret, sequence data may prove to be essential.

In this article, we assume that the major types of bacteria have conserved the integrity of the groups of sequences performing basic metabolic functions; we also assume that the substitution of a new sequence for one already functioning in a group through genetic transfer is sufficiently rare to be discounted. Frequent transfer between closely related species should not impair our ability to deduce the course of evolution of the major bacterial types. Only sequences that were transferred will lead to conflicting evolutionary histories for the species involved; sequences from any of the close species would be equally useful in deducing the evolutionary position of the bacterial type.

Reconstructing Evolution on the Basis of Sequence Data

Evolutionary history can be conveniently represented by a tree on which each point corresponds to a time, a macromolecular sequence, and a species within which the sequence occurred. Although we may not yet be able to infer its exact location, there is one point that corresponds to the earliest time and the original ancestral sequence and species.

Dr. Schwartz is a senior research scientist at the National Biomedical Research Foundation. Dr. Dayhoff is an associate professor in the Department of Physiology and Biophysics at Georgetown University Medical School and Associate Director of Research at the National Biomedical Research Foundation at Georgetown University Medical Center, Washington, D.C. 20007.

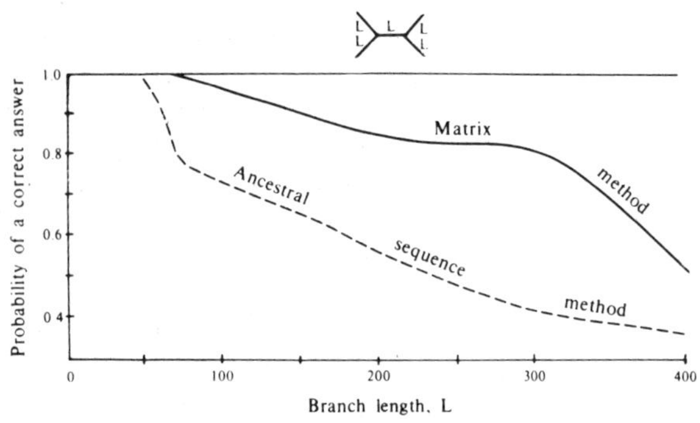

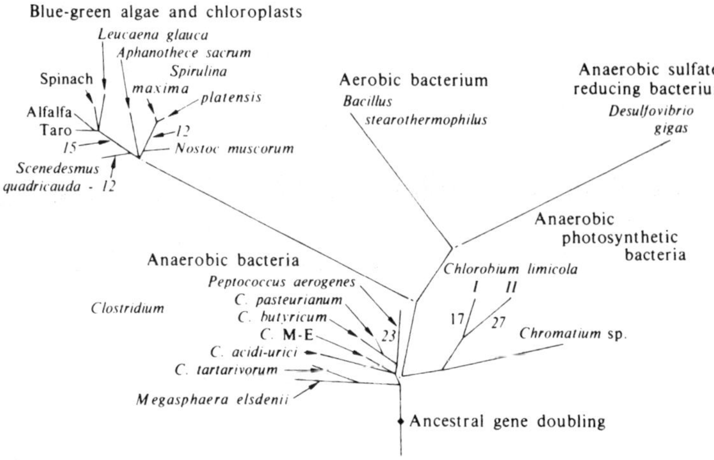

Fig. 1 (left). Probability of inferring the correct topology from sequences of simulated evolutionary connection. A history of five equal intervals of mutational distance was used, as shown at the top of the figure. Ten sets of sequences of 100 residues were generated for each mutational distance (L = 25, 50, 75, 100, 125, 150, 175, 200, 225, 250, 300, 400). Random events were assigned according to the average mutability and mutation pattern of each amino acid (*40*). In totaling the number of correct topologies inferred, a single correct answer counted 1, a wrong answer 0, a two-way tie 0.5, and a three-way tie 0.33. Smoothed curves through the data points are shown. The matrix method is clearly superior. This figure is adapted from (*2*). Fig. 2 (right). Evolutionary tree derived from ferredoxin sequences. Two subtrees were constructed separately on the basis of matrices of percentage differences between sequences from bacteria and from plants and the blue-green algae. A matrix based on an alignment omitting multiple-residue insertions and deletions was used to estimate the evolutionary distance and the connection between the subtrees. Mutiple-residue insertions in the sequences were omitted in constructing the bacterial subtree. The numbers on the tree are evolutionary distances in units of accepted point mutations per 100 residues. The order of divergence of the *Bacillus* and *Desulfovibrio* lines is unclear, as indicated by the dashed connection of the *Desulfovibrio* branch. An insertion of three residues in only these two sequences is consistent with the topology shown, but the topology with the *Desulfovibrio* branch coming directly off the ancestral line to the anaerobic bacteria does have nearly as short an overall branch length. For this alternative topology, the calculated branch lengths in the neighborhood of the *Desulfovibrio* connection would be slightly different.

Time advances on all branches of the tree emanating from this point. During evolution, sequences in different species have gradually and independently accumulated changes, yielding the sequences found today in the extant species represented at the ends of the branches. The absolute chronology, of course, cannot be inferred from sequences; the topology of the branches gives the relative order of events. On the evolutionary trees in this article the lengths of the branches are proportional to the inferred amount of change in the sequences. It is not usually possible to infer the position of the point of earliest time on a tree from the sequence data. An exception occurs when a gene doubling produces a reiterated sequence that is ancestral to all the sequences on the tree and when this duplication has been well preserved. For some biological groups, such as the vertebrates, the fossil record can be used to fix the location of the point of earliest time as well as its approximate chronological date.

In constructing an evolutionary tree on the basis of sequence data, we treat each amino acid or nucleotide residue in a sequence as an inherited biological trait. This assumption implies our ability to align sequences so that changes reflect substitutions of one residue for another during the course of evolution. Insertions and deletions of genetic material affect our ability to align sequences. Superimposed and parallel point mutations limit the accuracy with which we can infer the amount of evolutionary change. (Superimposed mutations are multiple changes at the same site in a sequence; parallel mutations are independent changes in two or more sequences resulting in the same amino acid or nucleotide at corresponding positions.) Experimental and methodological errors in sequencing present further difficulties. These problems affect the overall perspective less as more data become available.

One of the computer methods used for constructing phylogenetic trees proceeds by generating ancestral sequences (*15*); the other produces a least-squares fit to a matrix of evolutionary distances between the sequences (*2, 4*). The ancestral sequence method is a problem in double-minimization of inferred changes. For each possible configuration of the evolutionary tree, a set of ancestral sequences is determined that minimizes the number of inferred changes. Of these configurations, the one that minimizes the total number of changes between ancestral and known sequences is selected as the best representation of the evolutionary tree. This method also yields a set of ancestral sequences corresponding to the branch points of the tree.

In the matrix method, used to construct the trees described here, we begin by calculating a matrix of percentage differences between sequences in an overall alignment. Large unmatched regions, either internal or at the ends of sequences, do not correspond to point mutations and therefore are omitted from our calculations. The matrix elements are corrected for inferred parallel and superimposed mutations according to a scale based on the average way amino acids change during evolution (*16*); this gives evolutionary distances in accepted point mutations per 100 residues. For a given matrix, the determination of the best tree is a problem in double-minimization. For each possible configuration, a set of branch lengths is determined that provides a weighted least-squares fit between the distances given by the reconstructed matrix and those of the original matrix. The configuration that has the minimal total branch length is selected as the best solution.

To test the accuracy of both of these methods, we produced a number of families of sequences that were related through simulated evolutionary change that included amino acid replacements but no insertions or deletions (*2*). The results (Fig. 1) show that for sparse trees of distantly related sequences the matrix method is clearly superior to the ancestral sequence method. Because this is precisely the type of tree with which we are concerned, all of the individual trees shown here were constructed by the matrix method.

The matrix method is much more accurate for distantly related sequences because the information utilized is degraded more slowly by superimposed and parallel mutations than that utilized in the ancestral sequence method. If genetic material has been deleted or inserted, there is an additional factor favoring the use of the matrix method: the results obtained are less affected by variations in gap placement. The number of matching residues is not critically dependent on the exact alignment of two sequences. Typically, several alignments varying in the placement of gaps give the same number, and there are usually many more alignments corresponding to slightly smaller numbers of matches.

Our alignments are based on a computer program that determines the best alignment of two sequences (*17, 18*). The residue-by-residue comparisons match amino acids according to a model of the point mutation process that takes into account amino acid mutabilities and replacement probabilities (*16*). These pairwise sequence comparisons are adjusted to produce a comprehensive alignment. In the alignments used here, the relative magnitudes of terms in the matrices of percentage differences closely reflect the order of similarity of the pairs of sequences. Other criteria can be used in determining gap placement. For example, because overall conformation appears to be well conserved by evolution, Dickerson *et al.* (*19*) propose an alignment of c-type cytochromes that matches residues according to their positions in the three-dimensional structure of the molecules. Alignments based on mutations are more appropriate here because the programs for reconstructing phylogeny seek a minimum number of genetic events.

The topological configuration obtained is not very sensitive to the correction method used for the matrix elements. In the simulated problems (Fig. 1), the curve of correctly inferred topologies obtained directly with the matrix of percentage differences is almost identical with that obtained with the values corrected for presumed superimposed and parallel mutations. The reconstructed branch lengths approached an asymptote when the matrix of percentage differences was used, whereas they were correct on the average (but only very approximately in any particular case) when the corrected matrix was used.

Four protein superfamilies include sequences from several prokaryotes and eukaryotes: ferredoxin, 5*S* ribosomal RNA, c-type cytochromes, and azurin-plastocyanin. Two other superfamilies, flavodoxin and rubredoxin, have sequences that are known from at least four types of bacteria in common with the first superfamilies. Each of these groups of sequences can be used to construct an evolutionary tree; generally, the information that an individual tree provides corresponds closely to the evolution of the biochemical system within which the molecule functions. Plastocyanin, for example, functions in oxygen-releasing photosynthesis, and these sequences provide information about the evolution of photosynthesis in the blue-green algae and in the chloroplasts of higher plants. Together with azurin sequences they also depict the divergence of the blue-green algae from the bacteria.

None of these individual trees, however, gives an overall picture of the course of evolution and the development of new biochemical adaptations from the appearance of the earliest living forms to the divergence of the eukaryote kingdoms. For example, no one tree contains sequences from both cytoplasm and chloroplasts of higher plants; thus, individual trees leave unresolved the question of whether or not the eukaryote organelles had symbiotic origins. Understanding the development of new biochemical pathways requires information that cannot come from a tree derived from data on a single type of molecule. Fortunately, the groups of organisms and eukaryote organelles from which sequences are available overlap in such a way that the trees can be correlated and a composite tree can be constructed. This composite tree depicts more fully the relationships between the major developments in early biological evolution.

Ferredoxins

The ferredoxins are small, iron-containing proteins that are found in a broad spectrum of organisms and that participate in such fundamental biochemical processes as photosynthesis, oxidation-reduction respiratory reactions, nitrogen fixation, and sulfate reduction. The amino acid sequences of these proteins have been elucidated by a number of workers, including particularly K. T. Yasunobu, H. Matsubara, and their coworkers [see (*20, 21*)]. The tree (Fig. 2) derived from these sequences provides a framework for the events outlined in the other evolutionary trees presented here; moreover, a gene-doubling shared by all the ferredoxin sequences makes it possible to deduce the point of earliest time in these trees. The clostridial-type ferredoxins, in particular, are still very similar in sequence to the extremely ancient protein that duplicated. Most of these ferredoxins are composed of fewer than the 20 coded amino acids, and they lack those amino acids that are thermodynamically least stable, such as tryptophan and histidine (*22*).

The clostridial-type ferredoxins show the strongest evidence of gene-doubling. Using our computer program RELATE (*17*), we compared all pairs of different segments that were 15 residues long within each of these proteins and calculated a probability of less than 10^{-9} that the repetitive character of the two halves occurred by chance. From an alignment of the first and second halves of these sequences, we inferred an ancestral half-chain sequence. This sequence, doubled, was included in the computations of the evolutionary tree. Because all of the ferredoxin sequences show some evidence of gene-doubling, this event must have occurred prior to the species divergences shown here. The doubled sequence is therefore located at the base of the tree. All organisms near the base of the tree (*Clostridium*, *Megasphaera*, and *Peptococcus*) are anaerobic, heterotrophic bacteria (*23*). Most species of these groups lack heme-containing proteins, such as the cytochromes and catalase. It has long been thought that, of the extant bacteria, these species most closely reflect the metabolic capacities of the earliest species (*24*). In Fig. 2, *Chloro-*

bium and *Chromatium* are pictured as having diverged very early from the ancestral heterotrophic bacteria, although the exact point of this divergence is not clearly resolved. *Chromatium* and *Chlorobium* are anaerobic bacteria capable of photosynthesis using H_2S as an exogenous electron donor. Of the anaerobic bacteria shown, only *Chlorobium limicola* cannot live fermentatively, and it is reasonable to suppose that this ability is primitive and was lost by this bacterium. The two ferredoxins in *Chlorobium* are the result of a gene duplication within this line.

In Fig. 2, the *Bacillus* and *Desulfovibrio* lines diverge next from the line leading to the blue-green algae. Members of the genus *Bacillus* are either strictly aerobic or facultatively aerobic, capable of respiring aerobically but living fermentatively under anaerobic conditions. *Desulfovibrio* is a sulfate-reducing bacterium; it respires anaerobically using sulfate as the terminal electron acceptor. The use of sulfate by *Desulfovibrio* contrasts sharply with the use of oxygen as the terminal electron acceptor of respiration by *Bacillus*. This difference suggests that the divergence of these bacteria occurred after some components in the respiratory chain had developed. The topology pictured here indicates that the final components in the chain evolved separately in the *Bacillus*, *Desulfovibrio*, and blue-green algal lines.

The plant-type ferredoxins are all very closely related. Those from the green alga *Scenedesmus* and the higher plants are found in the chloroplasts of these organisms. *Spirulina* and *Nostoc* belong to one major division of the blue-green algae, the filamentous type; *Aphanothece* represents the other major division, the coccoid type (*25*).

The tree in Fig. 2 indicates that the structure of ferredoxin has changed much more in some lines than in others; this, at least in part, reflects changes in ferredoxin function. The clostridial sequences are little changed over the entire time represented, possibly more than 3 billion years. The adjustment to bacterial photosynthesis required somewhat more change, the adjustment to an aerobic metabolism required still more, and the most change occurred in the adaptation to oxygen-releasing photosynthesis. Unlike the situation in eukaryotes, the rate of acceptance of point mutations in prokaryote sequences is very uneven and cannot provide a useful evolutionary clock.

Origins of Eukaryote Organelles

There are two schools of thought concerning the origin of the eukaryote mitochondria and chloroplasts: one is that they arose by the compartmentalization of the DNA within the cytoplasm of an evolving protoeukaryote (*26*); the other is that they arose from free-living forms that established symbiotic relationships with host cells (*27*). According to the first theory, all genes arose within a single ancestral line, and homologs found in both the nucleus and the organelles arose by gene duplication. Thus, in the evolutionary tree for ferredoxin, the animals and fungi would appear together with the higher plants in the upper portion of the tree, after the divergence of the blue-green algae.

According to the symbiotic theory, the chloroplasts descended from free-living blue-green algae; other symbionts would include the mitochondrion, which was originally a free-living aerobic bacterium, and the flagellum and mitotic apparatus, which may have descended from spirochetes. It is proposed that these prokaryotes separately invaded protoeukaryote host cells with which they became symbiotic and continued to evolve to their current status as organelles. If the symbiotic theory is accurate, mitochondrial and chloroplast genes should show evidence of recent common ancestry with the separate types of contemporary free-living prokaryote forms. The host and organelles should occur on different branches that also contain free-living forms. The tree of plant-type ferredoxins would thus depict the radiation of blue-green algae followed by the development of symbiosis between one of these organisms and an ancestor of *Scenedesmus* and the higher plants. Although the appearance of the ferredoxin tree, particularly its branch lengths, is more consistent with this theory than with the first explanation, the tree by itself does not enable us to distinguish between the two theories because no ferredoxin sequences are available from the eukaryote cytoplasm or mitochondria.

5*S* Ribosomal RNA

The 5*S* ribosomal RNA molecule has a low molecular weight and is about 120 nucleotides in length. It is associated with the larger ribosomal subunit and is thought to function in the nonspecific binding of transfer RNA to the ribosome during protein synthesis (*28*). Because this function is independent of the kind of amino acid, this type of molecule could be extremely ancient, predating the contemporary form of the genetic code. Sequences of 5*S* ribosomal RNA have been determined by a number of workers, including B. J. Forget, S. M. Weissman, and C. R. Woese [see (*29, 30*)]; they have been taken from a wide variety of sources, including aerobic and anaerobic bacteria, blue-green algae, and the cytoplasm of several eukaryotes. The cytoplasmic sequences, in particular, present the possibility that an evolutionary tree based on this molecule will provide further insight into the origin of the eukaryotes.

Aligning nucleotide sequences in a way that reflects their evolution is more difficult than aligning amino acid sequences because there are only four kinds of bases. However, knowledge of the secondary structure alleviates the alignment problem somewhat because we can assume that positions involved in the base-paired regions of the molecule were highly conserved during evolution. We have aligned the known sequences to reflect their natural division into groups from prokaryotes and eukaryotes and to match a model of their secondary structure (*29*) adapted from Nishikawa and Takemura (*31*).

We derived an evolutionary tree (Fig. 3) on the basis of this alignment and placed its base on the branch to the anaerobic bacterium *Clostridium* in conformance with the ferredoxin tree. All the eukaryote 5*S* ribosomal RNA's were isolated from cytoplasmic ribosomes, one of the three ribosomal systems found in eukaryotes.

The branch leading to these cytoplasmic sequences diverges from the prokaryotes at a point that is close to the origin of the *Bacillus* branch. Like some members of the genus *Bacillus*, *Escherichia* is also facultatively aerobic and has both fermentative and respiratory metabolisms. Unless aerobic respiration arose separately in the *Bacillus* and *Escherichia* lines, their most recent common ancestor had this bimodal metabolic capacity also.

The two lines leading to organisms

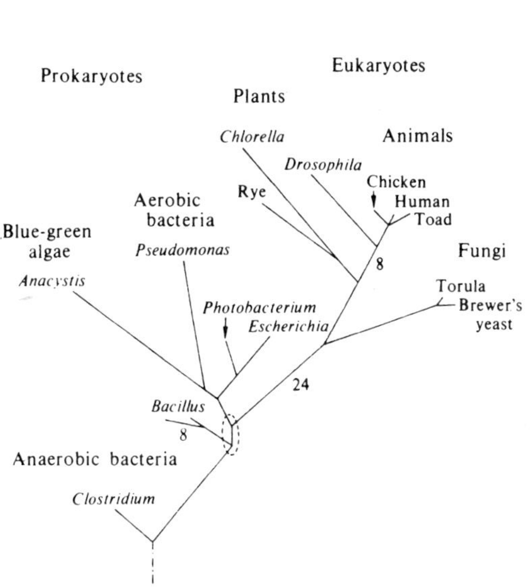

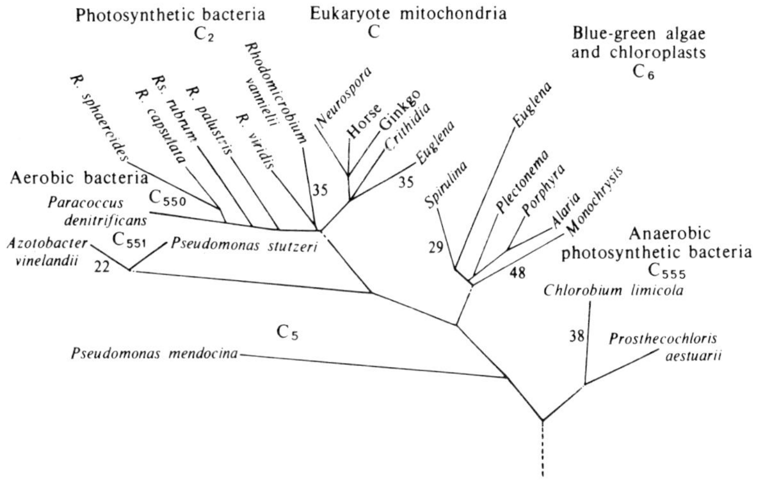

Fig. 3 (left). Evolutionary tree derived from 5*S* ribosomal RNA. This tree was derived by the matrix method. Branches are drawn proportional to the amount of evolutionary change they represent; selected branch lengths are indicated in numbers of accepted point mutations per 100 residues. The order of divergence for the branches leading to the eukaryotes and to *Bacillus* is not clearly resolved; the tree whose topology reverses the order of these branches has nearly as short an overall length. The prokaryote species shown are *Clostridium pasteurianum*, *Bacillus megaterium*, *B. licheniformis*, *Escherichia coli*, *Pseudomonas fluorescens*, and *Anacystis nidulans*. The *Chlorella* species is *C. pyrenoidosa*. Fig. 4 (right). Evolutionary tree derived from c-type cytochromes. The subtrees pictured here, such as the cytochrome c_6 tree, were derived separately from matrices of percentage differences between the complete sequences. Branches connecting the subtrees were estimated from a matrix calculated from an alignment that omitted multiple-residue insertions and deletions. Branch lengths are given in accepted point mutations per 100 residues. The points of earliest divergence in the individual subtrees cannot be precisely determined and, therefore, the bases of the subtrees are represented by dashed lines. Cytochrome c_{550} and all of the c_2's differ from the other c-type cytochromes in having a deletion close to the heme-binding cysteine in their sequences and on this basis were placed on a branch separate from the cytochrome c sequences. *Euglena* and *Crithidia* have been placed on a single branch together, a configuration slightly less than optimal, because they share a unique mutation of the active cysteine. The connections of the c_{551} and c_{555} sequences have been centered. The genera *Rhodospirillum* and *Rhodopseudomonas* are abbreviated *Rs.* and *R.*, respectively. Cytochrome c_6 sequences were taken from the following species: *Spirulina maxima*, *Monochrysis lutheri*, *Porphyra tenera*, *Euglena gracilis*, *Alaria esculenta*, and *Plectonema boryanum*; cytochrome c sequences were taken from protists *Euglena gracilis* and *Crithidia oncopelti*. This tree is adapted from (*2*).

that are capable of oxygen-releasing photosynthesis appear on opposite sides of this tree. One leads to rye and *Chlorella*, a eukaryote green alga, the other to *Anacystis*, a blue-green alga. *Anacystis* is grouped in the same family with *Aphanothece* (*25*), which appears on the ferredoxin tree. These coccoid blue-green algae are certainly more closely related than the blue-green algal orders represented by *Aphanothece* and *Spirulina*. Thus, we predict that *Aphanothece* would be found to diverge near the end of the *Anacystis* branch, preceded slightly by the divergence of the chloroplast branches. The very separate history of the cytoplasmic sequences points to a symbiotic origin of the chloroplasts, with the cytoplasmic sequences representing the evolution of the organism that was invaded by a blue-green alga.

C-Type Cytochrome

The evolutionary tree based on c-type cytochromes (Fig. 4) is important to an understanding of the origin and evolution of the mitochondrion. R. Ambler, E. Margoliash, D. Boulter, M. Kamen, E. Smith, and G. Pettigrew, in particular, are responsible for sequencing many of these proteins [see (*32*)]. Cytochrome c is coded in the nucleus but functions in the mitochondrion. This is usually explained in the symbiotic theory by transfer of genetic information, including the gene for cytochrome c, from the invading aerobic bacterium to the protoeukaryote host during the development of their current relationship. In the nonsymbiotic theory, genetic rearrangement is also an essential feature.

The eukaryote mitochondrion is placed in the portion of the tree (Fig. 4) that includes the aerobic bacteria after their divergence from the blue-green algae and chloroplasts. The cytochrome c sequences are on a branch that most recently diverged from cytochrome c_2 of the nonsulfur, purple, photosynthetic bacteria; together these diverged from the branch leading to cytochrome c_{551} from strict aerobes such as *Pseudomonas*. This contrasts with the evolution of the eukaryote cytoplasmic constituents depicted by the tree derived from 5*S* ribosomal RNA (Fig. 3). There the eukaryotes diverged with the facultatively aerobic bacteria, such as *Bacillus*, from the line leading toward the blue-green algae. *Pseudomonas* diverged from this line somewhat later. This contrasting picture reinforces arguments favoring a symbiotic origin for the mitochondrion. The cytoplasmic 5*S* ribosomal RNA sequences appear to describe the evolution of the protoeukaryote host. The subtree for animals that was derived from cytochrome c (*5*) is consistent with that derived from 5*S* ribosomal RNA sequences. According to the symbiotic theory, this would be because these divergences were subsequent to the mitochondrial invasion of host cells.

Cytochrome c_2 is found in bacteria such as *Rhodomicrobium* and *Rhodopseudomonas*. These bacteria photosynthesize anaerobically but respire aerobically. *Paracoccus denitrificans* possesses cytochrome c_{550}, which is very

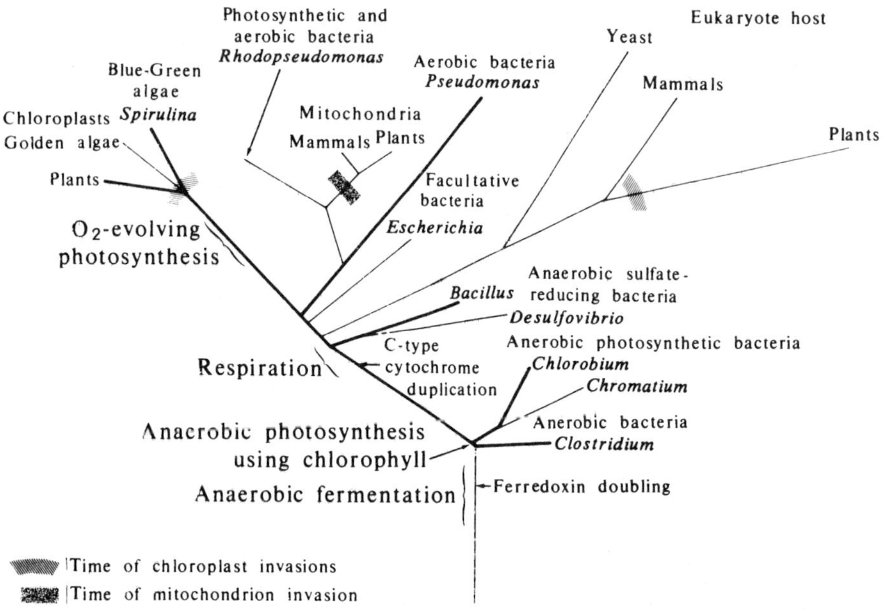

Fig. 5. Composite evolutionary tree. This tree presents an overview of early evolution based on ferredoxin, c-type cytochromes, and 5*S* ribosomal RNA sequences. The heavy lines represent a tree calculated from a matrix of evolutionary distances combining two or more of the individual trees. The lighter lines represent branches scaled from a single tree and added to the combined tree. The point in evolution at which the mitochondrial symbiosis occurred is stippled; the chloroplast symbiosis is shaded.

similar to cytochrome c_2 along its entire length. The position of *Paracoccus* in the tree suggests that it arose from a nonsulfur photosynthetic bacterium by the loss of its photosynthetic ability (*19*).

Two different c-type cytochromes, c_5 and c_{551}, are found in *Pseudomonas*, a nonphotosynthetic bacterium. These may be the result of a gene duplication early in the tree with subsequent loss of the c_5 gene in some lines. For species in which the gene was not lost, we would expect that the topology of the main tree subsequent to the duplication would be reiterated on the branch that included the c_5 sequence. A transfer of genetic material from a bacterium that contained the c_5 gene to *Pseudomonas* could also explain this branch.

The presumed duplication of the c-type cytochrome gene in *Pseudomonas* brings up a problem in interpreting evolutionary trees that is especially acute here. If we are treating two products of an unsuspected gene duplication, we will assume that the two lines of protein evolution correspond to a single pattern of species evolution. This can lead us to believe that two closely related organisms are quite distant from one another. For example, if cytochrome c_{551} had gone undetected in species of *Pseudomonas*, the tree in Fig. 4 would suggest that *Azotobacter* and *Pseudomonas* were very distantly related. At present, metabolic function guides our selection of homologous proteins and helps avoid this difficulty. In the tree of c-type cytochromes, protein function has changed, and this presents an added difficulty in interpretation. As a more complete picture of the protein complement of each species becomes known, this problem will become unimportant.

Cytochrome c_6 is found in the photosynthetic lamellae of blue-green algae and in the chloroplasts of eukaryotes where it functions in the electron transport chain between photosystems I and II. As in the ferredoxin tree, there is a close similarity between the sequences from the blue-green alga *Spirulina* and the various eukaryote algal chloroplasts; as in the 5*S* ribosomal RNA tree, the blue-green algae are most closely related to strictly aerobic bacteria, such as *Pseudomonas*. It is not possible to locate precisely the point at which the main tree connects to the subtree of cytochrome c_6. However, the topology and branch lengths reflect a symbiotic origin for photosynthesis in eukaryotes. There are separate branches leading to the two filamentous blue-green algae, *Spirulina* and *Plectonema*, intermixed with the eukaryote algae branches. The most direct explanation for this is that the cytochrome c_6 subtree reflects, at least in part, the evolutionary relationships among the blue-green algae that became symbionts rather than the speciation of eukaryotes. Some of the eukaryote algal chloroplasts appear to be derived from different symbiotic associations, as Raven has suggested (*33*).

The point of earliest time on the tree for c-type cytochromes was placed near the divergence of *Chlorobium* and *Prosthecochloris*; both of these are anaerobic, obligate, photosynthetic bacteria. This is consistent with the position of the sequence from *Chlorobium* in the ferredoxin tree. As in the tree of 5*S* ribosomal RNA's, prokaryote branch lengths are strikingly unequal, probably reflecting changes in protein function, and cannot be used to estimate time reliably.

Composite Evolutionary Tree

Each of the individual trees we have presented contains information about the early course of biological evolution. We have used the topologies and evolutionary distances derived for these trees to construct a composite tree (Fig. 5) which, although it is based on sparse data from a few species, begins to provide a coherent evolutionary framework that can be expanded as new sequence data become available.

We constructed the composite evolutionary tree from a composite matrix according to the same methods that we used to construct the individual trees. The matrix included the six species that appear on at least two of the three individual trees (their branches are represented by the heavy lines in Fig. 5). First, the trees were scaled so that distances were comparable. To do this, we compared the overall length of the distances each tree had in common with each other tree. These ratios were adjusted slightly to give a consistent set of scale factors. A combined matrix of distances between the six species was calculated by averaging the scaled contributions from each of the individual trees. As previously, for each possible configuration of the combined trees, a set of branch lengths was determined that provided a weighted least-squares fit between the matrix of distances between species and a matrix reconstructed from the tree. The configuration with the shortest overall branch length was chosen. This configuration is the one that is also consistent with all of the individual trees. Finally, branches found in only one tree were scaled in length and added to the composite tree, maintaining the relative internodal distances (these are represented by light lines in Fig. 5).

The genetic doubling of the clostridial-type ferredoxins allows us to locate the base on this tree. Moreover, because the species whose sequences have changed least since this doubling event were all anaerobic, heterotrophic bacteria, it is likely that the ability to live fermentatively is primitive.

The composite tree describes the evolution of photosynthesis using chlorophyll, starting with the development of the ability to synthesize this class of compounds. Three families of photosynthetic bacteria are represented: Chromatiaceae, Chlorobiaceae, and Rhodospirillaceae. The divergence of the Chromatiaceae and the Chlorobiaceae from the other anaerobic bacteria was quite early, and it is clear that this type of photosynthesis arose at a very early stage in evolution and has not changed much. The Rhodospirillaceae provide an example of the confusion that morphological criteria can cause. As Stanier *et al.* (*23*) point out, these bacteria are indistinguishable in structure and pigments from members of the Chromatiaceae under anaerobic conditions; however, when grown under strictly aerobic conditions, they appear to be the same as nonphotosynthetic bacteria of similar form, such as the pseudomonads. The c-type cytochrome sequences clearly place them in a portion of the tree surrounded by strictly aerobic forms on a branch leading to *Pseudomonas*.

If we assume that it is very hard to achieve a photosynthetic metabolism with its many coordinated macromolecules, but relatively easy to lose one through any one of many genetic changes, then it is reasonable to suppose that the main trunk of this tree represents a continuum of photosynthetic forms. We would not be surprised to find photosynthetic forms branching off at any point. Except for the early anaerobes, all nonphotosynthetic bacteria would be descended from a few ancestral forms that have independently lost their photosynthetic ability. This is very different from biological classifications where all photosynthetic forms are grouped together, separate from the nonphotosynthetic forms.

The next major event shown on the main trunk of the tree is the development of aerobic respiration. As we noted in discussing the ferredoxin evolutionary tree, the divergence of *Bacillus* and *Desulfovibrio* probably marks the appearance of some components of this adaptation. The final elements in this adaptation were evolved separately because these groups differ in their terminal electron acceptor in respiration. Additionally, the divergence of cytochrome c_5 occurred just prior to this time. In the tree of the c-type cytochrome, it is unclear whether this divergence represents a gene duplication or a recent genetic transfer; in the context of the evolution of a respiratory metabolism, a duplication of the ancestral gene could have provided the genetic material and relaxed evolutionary constraints necessary for the development of aerobic respiration.

The clearest and most direct interpretation of the sequence data is provided by the symbiotic theory for the origin of the eukaryotes. The branch we identify as the eukaryote host diverged at about the same time as *Bacillus* and *Escherichia*. Both of these bacteria are facultative aerobes. This suggests that the ancestral eukaryote host was also facultatively aerobic at the time of its divergence.

The bacterium that became the mitochondrion was most closely related to the third family of photosynthetic bacteria, the Rhodospirillaceae. The topology of the tree invites the speculation that this ancestral bacterium was photosynthetic until shortly before or just after it invaded the host. Because a single cytochrome c is found in most eukaryotes, it seems reasonable to suppose that the aerobic respiratory metabolism of this invading protomitochondrion was more effective than that of the host and that the host lost any primitive system that it might have had.

The final biochemical adaptation depicted here is the development of photosystem II. The blue-green algae and chloroplasts of the eukaryotes are capable of oxygen-releasing photosynthesis, and the available sequence data point to this capacity having evolved only once. It appears to have combined the new biochemical adaptation, photosystem II, with proteins modified from two earlier adaptations, bacterial photosynthesis and respiration. The chloroplasts, like the mitochondria, are grouped together with free-living prokaryotes, a result consistent with the symbiotic theory.

It is frequently suggested that aerobic respiration developed after oxygen-releasing photosynthesis as a protective mechanism in response to atmospheric oxygen produced by blue-green algae (*24, 34*). The composite tree clearly suggests that many components of the respiratory chain predate oxygen-releasing photosynthesis; all of the organisms on the tree above *Desulfovibrio* are aerobic, whereas the earlier branches all lead to anaerobic forms. Although possible, it seems unlikely that the use of oxygen as the terminal electron acceptor in respiration evolved separately in the lines leading to the blue-green algae, the facultative aerobic bacteria, and the strictly aerobic bacteria after the development of photosystem II in the blue-green algal line. As Schopf (*8*) has pointed out, it is difficult to imagine the development of oxygen-releasing photosynthesis prior to the development of a rudimentary mechanism for coping with oxygen. Oxygen is produced intracellularly in photosynthesis, and it is intracellular components that must be protected from oxidation. A rudimentary form of aerobic respiration most probably arose at a time near the divergence of the *Bacillus* line from that leading to the blue-green algae.

The composite tree makes it particularly clear that the three branches that contribute to the eukaryote host and organelles are distinctly separate; each is closely related to free-living prokaryotes. The chloroplasts share a recent ancestry with the blue-green algae; the mitochondrion shares a recent ancestry with certain respiring and photosynthetic bacteria, the Rhodospirillaceae; whereas the eukaryote host diverged from the other groups at a considerably earlier time along with *Bacillus* and *Escherichia*.

Corroborative Sequence Data

A limited amount of sequence data from four or more species is available from other proteins. Azurin and plastocyanin are blue, copper-containing proteins whose sequences show statistical evidence of their common evolutionary origin (*20*). Moreover, they appear to be parts of homologous electron transport systems because each of them exchanges electrons with c-type cytochromes. Azurin is thought to exchange an electron with cytochrome c_{551} in bacterial respiration (*35*); plastocyanin exchanges an electron with cytochrome c_6 in the electron transport chain between photosystems I and II.

Plastocyanin sequences from both eukaryote chloroplasts and a blue-green alga have been determined by a number of workers (*20, 36*) including D. Boulter and A. Aitken, in particular; azurin sequences from aerobic bacteria have been determined by R. Ambler (*20*). We constructed an evolutionary tree (Fig. 6) based on these sequences. The relationships depicted are consistent with those presented in the other trees. The aerobic bacteria, on the right side of the tree, are very closely related, and the extents of their divergences are comparable to those of the higher plants, shown on the left side of the tree. *Anabaena*, a filamentous blue-green alga, is closely related to chloroplasts of the eukaryote green alga *Chlorella* and of the higher plants. The topology of this tree again is consistent with the symbiotic origins for chloroplasts. The divergences of the higher plant chloroplasts are close enough to classical phylogenies to be explained by a single symbiotic association that predated plant divergences. Lewin (*37*) has recently proposed a new division, the Prochlorophyta, for a group of bright green, generally spherical, prokaryote algae. These had previously been classified with the blue-green algae. However, like the higher plant chloroplasts and unlike the blue-green algae they contain both chlorophylls a and b and no detectable bilin pigments. Sequences from these organisms might be especially helpful in extending our understanding of the evolution of higher plant photosynthesis.

Flavodoxins are low-molecular weight proteins that have been isolated so far from only bacteria and algae. They have one flavin mononucleotide prosthetic group per molecule and substitute for ferredoxins in a variety of reactions, including the phosphoroclastic splitting of pyruvate and the photosynthetic reduction of nicotinamide-adenine dinucleotide phosphate. Flavodoxin sequences are known from three anaerobic bacteria, *Clostridium*, *Megasphaera*, and *Desulfovibrio*, and an aerobic bacterium, *Azotobacter*; most of these sequences were determined by K. T. Yasunobu, J. L. Fox, and their co-workers [see (*38*)]. The matrix of the percentage differences between these flavodoxin sequences (Table 1) reveals the close relationship of the highly conserved *Megasphaera* and *Clostridium* sequences and of the more highly evolved *Desulfovibrio* and *Azotobacter* sequences, supporting the phylogeny we have drawn.

Rubredoxin, another protein that participates in electron transport, has also been used as a basis for constructing an evolutionary tree (*20, 39*). This tree includes sequences from *Megasphaera elsdenii, Peptococcus aerogenes, Clostridium pasteurianum, Pseudomonas oleovorans, Desulfovibrio gigas*, and *D. vulgaris*, and is also consistent with the ones we have constructed except that the closely related *Desulfovibrio* sequences diverge close to the earliest point on the tree and are very highly conserved. That the placement of the *Desulfovibrio* branch is inconsistent with that depicted in the trees for ferredoxin and flavodoxin suggests that there might be an unsuspected gene duplication in the rubredoxins, a rare occurrence of an accepted gene transfer, or a misleading concatenation of evolutionary events in these short sequences.

Summary

If current estimates of the antiquity of life on earth are correct, bacteria very much like *Clostridium* lived more than 3.1 billion years ago. Bacterial photosynthesis evolved nearly that long ago, and it seems reasonable, in view of our composite tree, to attribute the most ancient stromatolites, formed nearly 3.0 billion years ago, to early photosynthetic bacteria. Blue-green algae appear to have evolved later. The tree shows that by the

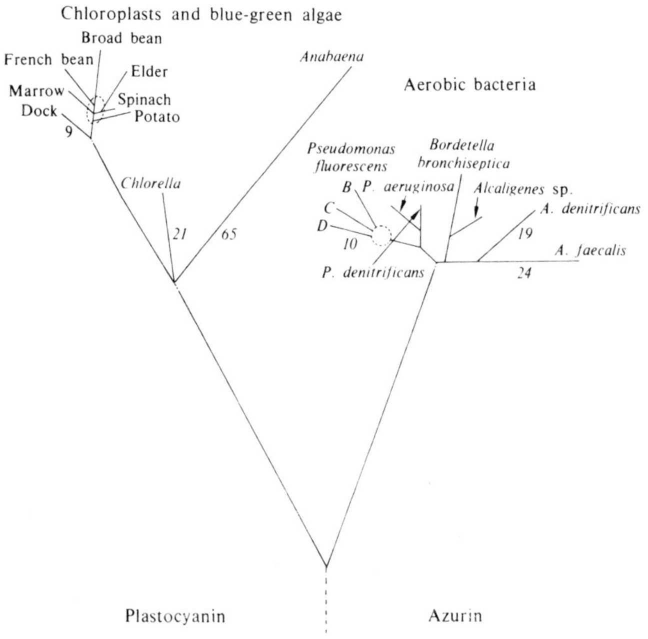

Table 1. Flavodoxins

	Number of differences			
Megasphaera elsdenii		76	107	155
Clostridium MP	55		104	152
Desulfovibrio vulgaris	71	69		151
Azotobacter vinelandii	85	83	82	
	Percentage difference			

Fig. 6. Evolutionary tree derived from azurin and plastocyanin. Subtrees for the two proteins were derived separately from matrices calculated from complete sequences. In order to estimate the evolutionary distance between the two families, we used a matrix calculated from an alignment that omitted multiple-residue insertions and deletions between the families. Branch lengths are drawn proportional to amounts of evolutionary change in the units of accepted point mutations per 100 residues. The positions of the connecting branch within the subtrees cannot be precisely determined from the data, and therefore the connections are represented by dashed lines. The order of divergence of the three biotypes of *Pseudomonas fluorescens* could not be resolved. The order of divergence of the chloroplasts of higher plants is also unclear; the topology with the minimal overall length is shown. The algal species shown are *Chlorella fusca* and *Anabaena variabilis*.

time oxygen-releasing photosynthesis originated in the blue-green algal line, there must have been a great diversity of morphological types, including bacteria that are ancestral to most of the major groups pictured on the composite tree. This time probably corresponded to the great increase in complexity of the fossil record about 2 billion years ago. Our composite tree suggests that aerobic respiration preceded oxygen-releasing photosynthesis. This may mean that the formation of oxygen from water in the upper atmosphere was important to evolving prokaryotes prior to 2 billion years ago. Oxygen-releasing photosynthesis arose later and was, in large measure, responsible for the final transition to the present-day oxygen level in the atmosphere. Judging from the relative branch lengths on the tree, the mitochondrial invasion occurred during this transition. Finally, perhaps 1.1 billion years ago, several independent symbioses between protoeukaryotes and various blue-green algae gave rise to photosynthetic eukaryotes; some of these developed into modern eukaryote algae, whereas a single line, possibly from an ancestral green alga, appears to have evolved into the higher plants.

By combining the information from evolutionary trees based on several types of sequences, we have developed a broad outline of early events in the emergence of life that can be refined as new sequence information becomes available. The schema presents a working hypothesis for relating the many observations from metabolic and morphological studies of bacteria and from paleogeology. Eventually all of the biochemical components of intermediary metabolism may be correlated with the development of the prokaryote types and their metabolic capacities.

References and Notes

1. V. Bryson and H. J. Vogel, Eds., *Evolving Genes and Proteins* (Academic Press, New York, 1965); M. O. Dayhoff and R. V. Eck, Eds., *Atlas of Protein Sequence and Structure* (National Biomedical Research Foundation, Washington, D.C., 1968), vol. 3.
2. M. O. Dayhoff, *Fed. Proc. Fed. Am. Soc. Exp. Biol.* **35**, 2132 (1976).
3. R. V. Eck and M. O. Dayhoff, Eds., *Atlas of Protein Sequence and Structure* (National Biomedical Research Foundation, Washington, D.C., 1966), vol. 2.
4. W. M. Fitch and E. Margoliash, *Science* **155**, 279 (1967).
5. P. J. McLaughlin and M. O. Dayhoff, *J. Mol. Evol.* **2**, 99 (1973).
6. W. M. Fitch, *ibid.* **8**, 13 (1976); M. Goodman, G. W. Moore, J. Barnabas, G. Matsuda, *ibid.* **3**, 1 (1974).
7. M. O. Dayhoff, L. T. Hunt, P. J. McLaughlin, D. D. Jones, in *Atlas of Protein Sequence and Structure*, M. O. Dayhoff, Ed. (National Biomedical Research Foundation, Washington, D.C., 1972), vol. 5, p. 17.
8. J. W. Schopf, *Annu. Rev. Earth Planet. Sci.* **3**, 213 (1975).
9. P. Cloud, *Paleobiology* **2**, 351 (1976).
10. J. W. Schopf, *Origins Life* **7**, 19 (1976).
11. T. R. Mason and V. Von Brunn, *Nature (London)* **266**, 47 (1977).
12. P. E. Cloud, Jr., *Science* **160**, 729 (1968).
13. J. W. Schopf and D. Z. Oehler, *ibid.* **193**, 47 (1976).
14. M. O. Dayhoff, W. C. Barker, P. J. McLaughlin, *Origins Life* **5**, 311 (1974).
15. M. O. Dayhoff, C. M. Park, P. J. McLaughlin, in *Atlas of Protein Sequence and Structure*, M. O. Dayhoff, Ed. (National Biomedical Research Foundation, Washington, D.C., 1972), vol. 5, p. 7; W. M. Fitch and J. S. Farris, *J. Mol. Evol.* **3**, 263 (1974).
16. M. O. Dayhoff, Ed., *Atlas of Protein Sequence and Structure* (National Biomedical Research Foundation, Washington, D.C., 1976), vol. 5, suppl. 2, p. 311.
17. M. O. Dayhoff, in *ibid.*, p. 1.
18. S. B. Needleman and C. D. Wunsch, *J. Mol. Biol.* **48**, 443 (1970).
19. R. E. Dickerson, R. Timkovich, R. J. Almassy, *ibid.* **100**, 473 (1975).
20. W. C. Barker, R. M. Schwartz, M. O. Dayhoff, in *Atlas of Protein Sequence and Structure*, M. O. Dayhoff, Ed. (National Biomedical Research Foundation, Washington, D.C., 1972), vol. 5, p. 51. Respiratory protein sequences and our alignments of them are collected here as are tabulations of computer alignment scores for the ferredoxins.
21. T. Hase, K. Wada, H. Matsubara, *J. Biochem. (Tokyo)* **79**, 329 (1976); M. Tanaka, M. Haniu, K. T. Yasunobu, K. K. Rao, D. O. Hall, *Biochemistry* **13**, 5284 (1974); ———; *ibid.* **14**, 5535 (1975); K. Wada, T. Hase, H. Tokunaga, H. Matsubara, *FEBS Lett.* **55**, 102 (1975); M. Tanaka, M. Haniu, K. T. Yasunobu, M. C. W. Evans, K. K. Rao, *Biochemistry* **14**, 1938 (1975); T. Hase, N. Ohmiya, H. Matsubara, R. N. Mullinger, K. K. Rao, D. O. Hall, *Biochem. J.* **159**, 55 (1976); T. Hase, K. Wada, M. Ohmiya, H. Matsubara, *J. Biochem. (Tokyo)* **80**, 993 (1976).
22. R. V. Eck and M. O. Dayhoff, *Science* **152**, 363 (1966).
23. R. E. Buchanan and N. E. Gibbons, Eds., *Bergey's Manual of Determinative Bacteriology* (Williams & Wilkins, Baltimore, Md., ed. 8, 1974); R. Y. Stanier, M. Doudoroff, E. A. Adelberg, *The Microbial World* (Prentice-Hall, Englewood Cliffs, N.J., ed. 3, 1970). Nutritional and ecological descriptions of the bacterial genera used in the text were taken from these references.
24. E. Broda, *The Evolution of the Bioenergetic Process* (Pergamon, Oxford, 1975).
25. T. V. Desikachary, in *The Biology of Blue-Green Algae*, N. G. Carr and B. A. Whitton, Eds. (Univ. of California Press, Berkeley, 1973), p. 473.
26. R. A. Raff and H. R. Mahler, *Science* **177**, 575 (1972); T. Uzzell and C. Spolsky, *Am. Sci.* **62**, 334 (1974).
27. L. Margulis, *Origin of Eukaryotic Cells* (Yale Univ. Press, New Haven, Conn., 1970); ———, in *Handbook of Genetics*, R. C. King, Ed. (Plenum, New York, in press), vol. 1.
28. R. Monier, in *Ribosomes*, M. Normura, A. Tissieres, P. Lengyel, Eds. (Cold Spring Harbor Laboratory, Cold Spring Harbor, N.Y., 1974), p. 141.
29. R. M. Schwartz and M. O. Dayhoff, in (*16*), p. 293. Sequences of 5*S* ribosomal RNA's, a model of their secondary structure, and our alignment of them are collected here, as are tabulations of computer alignment scores.
30. C. R. Woese, C. D. Pribula, G. E. Fox, L. B. Zablen, *J. Mol. Evol.* **5**, 35 (1975); C. D. Pribula, G. E. Fox, C. R. Woese, *FEBS Lett.* **64**, 350 (1976); H. A. Raue, T. J. Stoff, R. J. Panta, *Eur. J. Biochem.* **59**, 35 (1975); J. Benhamou and B. R. Jordan, *FEBS Lett.* **62**, 146 (1976); P. I. Payne and T. A. Dyer, *Eur. J. Biochem.* **71**, 33 (1976).
31. K. Nishikawa and S. Takemura, *J. Biochem. (Tokyo)* **76**, 935 (1974).
32. M. O. Dayhoff and W. C. Barker, in (*16*), p. 25. Sequences of c-type cytochromes and our alignment of them are collected here, as are tabulations of computer alignment scores; R. P. Ambler, T. E. Meyer, M. D. Kamen, *Proc. Natl. Acad. Sci. U.S.A.* **73**, 472 (1976); A. Aitken, *Nature (London)* **263**, 793 (1976).
33. P. H. Raven, *Science* **169**, 641 (1970).
34. L. Margulis, J. C. G. Walker, M. Rambler, *Nature (London)* **264**, 620 (1976).
35. M. Brunori, C. Greenwood, M. T. Wilson, *Biochem. J.* **137**, 113 (1974).
36. B. Haslett, C. R. Bailey, J. A. M. Ramshaw, M. D. Scawen, D. Boulter, *Biochem. Soc. Trans.* **2**, 1329 (1974); J. A. M. Ramshaw, M. D. Scawen, D. Boulter, *Biochem. J.* **141**, 835 (1974); M. D. Scawen and D. Boulter, *ibid.* **143**, 237 (1974); M. D. Scawen, J. A. M. Ramshaw, D. Boulter, *ibid.* **147**, 343 (1975); A. Aitken, *ibid.* **149**, 675 (1975).
37. R. A. Lewin, *Nature (London)* **261**, 697 (1976).
38. W. C. Barker and M. O. Dayhoff, in (*16*), p. 67. Flavodoxin sequences are collected here; M. Tanaka, M. Haniu, K. T. Yasunobu, D. C. Yoch, *Biochem. Biophys. Res. Commun.* **66**, 639 (1975).
39. H. Vogel, M. Bruschi, J. Le Gall, *J. Mol. Evol.* **9**, 111 (1977).
40. M. O. Dayhoff, R. V. Eck, C. M. Park, in *Atlas of Protein Sequence and Structure*, M. O. Dayhoff, Ed. (National Biomedical Research Foundation, Washington, D.C., 1972), vol. 5, p. 89.

41. Supported by NASA contract NASW 3019 and NIH grants GM-08710 and RR-05681. We thank W. C. Barker and B. C. Orcutt for their helpful discussions and criticism and M. J. Gantt for technical assistance.

PROBLEMS

1. Define the following terms: eukaryote, prokaryote, serial symbiotic hypothesis, autogenous theories, ferredoxin, polyphyletic origin, gene phylogeny, organism phylogeny.

2. By reference to a standard text, list contemporary photosynthetic organisms (prokaryotic and eukaryotic) and their main photosynthetic pigments.

3. In Chapter 4 the point is made that molecular evolutionary theory makes the testable prediction that similar phylogenetic trees should be found when considering different protein and nucleic acid sequences from the same collection of organisms. If you had to defend the scientific status of evolutionary theory, how would you cope with the nonconcordance of the ferredoxin and 5S rRNA trees reported in Schwartz and Dayhoff (1978, this chapter), but without falling victim to explaining too much, as discussed in the Introduction.

4. Raven (1970) has suggested that a prokaryotic blue-green alga once existed that only had chlorophyll a as its photosynthetic pigments. This hypothetical organism gave rise to at least three biochemically distinct relatives: (1) the living blue-green algae with phycobilins; (2) the "green prokaryotes" with chlorophyll b; (3) the "yellow prokaryotes" in which chlorophyll c evolved. All of these groups entered into symbiotic associations with primitive eukaryotic cells to yield chloroplasts. It is thought that groups 2 and 3 no longer exist as free living organisms. In light of this reconstruction, how would you interpret the discovery of the green prokaryotic photosynthetic organism *Prochloron*? (See Lewin, 1976; 1981.) What predictions would you make about their protein and nucleic acid sequences, compared to higher plants and to blue-green algae? Which sequences (nuclear, chloroplast, or mitochondrial) do you think would give the most definite answers? See also the papers in Frederick (1981).

5. Chloroplasts of green algae and plants are surrounded by double membranes. On the symbiotic theory the inner chloroplast membrane is interpreted as the outer membrane of a prokaryotic ancestor and the outer chloroplast membrane is considered to be derived from the host cell. However, the eukaryote *Euglena* has three membranes around its chloroplasts. Gibbs (1978) suggests that *Euglena* chloroplasts may be derived from a symbiotic green alga (eukaryotic) and that the third (middle) membrane is derived from the outer membrane of this symbiotic alga. Assume green algae (for example, *Chlorella*) are ancestral to higher green plants and that this divergence occurred before the supposed symbiosis to give *Euglena*. Construct the expected phylogenetic trees for blue-green and green algae, *Euglena*, and higher plants based on (1) nuclear derived rRNA and (2) chloroplast rRNA. What other evidence could help to test the hypothesis? (See papers in Frederick, 1981.) Would you expect mitochondrial rRNA to give a pattern like nuclear rRNA or like chloroplast rRNA?

6. Pickett-Heaps (1974) has discussed two possible origins for the eukaryotic nucleus, stating, "perhaps ... it also arose as the result of a symbiosis, i.e., that an ingested organism somehow became the nucleus perhaps following the loss of much of its own cytoplasm ... even now, some dinoflagellates have two nuclei, one characteristic of dinoflagellates and the other a more typical eukaryotic nucleus. Another possibility is that perhaps the mesosome of a primordial bacteria-like cell came to envelope the genome ... and then this enclosed structure became separate from the cell membrane." Consider carefully what predictions you could make from these two hypotheses. How could you decide between them?

7. In mitochondrial DNA (mtDNA) the triplet code for amino acids has some differences from all other systems (Heckman *et al.*, 1980; Barrell *et al.*, 1980; Bonitz *et al.*, 1980). These authors worked with three different species and there are some differences in codon usage between their mitochondria, though in each case the codon UGA codes for tryptophan rather than for chain termination. However, the

mitochondrial DNA codes for only a few proteins such as cytochrome b (Anderson *et al.*, 1981). At least three hypotheses have been proposed to explain these differences: (1) mitochondria are direct descendants from some early form of life that existed before the code was finalized, (2) the main changes occurred in the mitochondrial ancestor group, and (3) the main changes occurred after the mitochondria were formed. Evaluate these hypotheses after considering evidence from sequences, using the examples of cytochromes (Schwartz and Dayhoff, 1978, this chapter), tRNA, and 5S rRNA (Kuntzel *et al.*, 1981), and considering the argument of Eck and Dayhoff (1966a, Chapter 1) that basic properties cannot change if they are expressed in many places in metabolism. Do the alternative assumptions of the neutralist-panselectionist debate influence your answer, or does the observation that only a few proteins are encoded make a difference? What other information would you need for a more definite answer?

8. Phylogenetic trees have been reconstructed from the RNA of the small ribosomal subunits of mitochondria, prokaryotes, and eukaryotes (Kuntzel and Kochel, 1981). The position of fungal mitochondria supports the symbiotic hypothesis, but the suggestion that animal mitochondria have a separate origin from fungal mitochondria has been made. The cytochrome c results (Schwartz and Dayhoff, 1978, this chapter; Almassy and Dickerson, 1978, Chapter 4) do not support the notion of different origins. Taking into account the observation that mitochondrial DNA has a faster rate of evolution (Brown *et al.*, 1979), design a set of analyses of experiments that may resolve the apparent anomaly.

8 PROKARYOTIC ORIGINS AND PHYLOGENY

Therefore, on the principle of natural selection with divergence of character, it does not seem incredible that, from some such low and intermediate form, both animals and plants may have been developed; and, if we admit this, we must likewise admit that all the organic beings which have ever lived on this earth may be descended from some one primordial form.
Charles Darwin, The Origin of Species (1859)

Until recently, the reconstruction of prokaryotic evolutionary trees has presented the student of evolution with some particularly difficult problems. Prokaryotes show little of the morphological differentiation that has been so helpful in assessing phylogenetic relationships among multicellular organisms, and apart from the *Cyanobacteria* (blue-green algae), few have left an interpretable fossil record. Possible phylogenetic relationships between prokaryotes can be constructed on the basis of assorted phenotypic characteristics (Sneath and Sokal, 1973) such as serotypes, ecological niches, or metabolic requirements, but such phenotypic estimates of genetic relationships have not led to general agreement in the phylogenetic schemes based upon them.

The importance of prokaryotic phylogeny is almost proportional to its obscurity. Not only are prokaryotes important in their own right, but they are associated with the origin of the first eukaryotic cells (Margulis, 1970; 1981). Furthermore, they are believed to be early life forms and thus may provide insight into the origins of life itself. The successful reconstruction of prokaryotic phylogenetic trees cannot help but be informative about these two fundamental issues. The recent application of the methods of comparative biochemistry and molecular biology has not yet provided definitive answers but has reduced confusion and led to the formulation of some testable hypotheses about prokaryotic origins and relationships.

The first molecular methods applied to the analysis of prokaryotic evolution were those of comparative biochemistry. These studies centered on similarities among metabolic pathways in different organisms and on the enzymes and pigments of these pathways. The comparison of the photosynthetic pigments and associated compounds was an outstanding example of this research. The discovery of phycobilins in cyanobacteria and in certain algal chloroplasts (see also Chapter 7) was an early indication of a possible relationship between cyanobacteria and chloroplasts, while the detailed study of the chlorophylls has led to a much better understanding of the origins of specific groups of algae, and even to the prediction (Raven, 1970) of the existence of a class of prokaryotes, the *Prochlorophyta*, which was subsequently discovered (Lewin, 1976).

The analysis of proteins has also contributed to the assessment of prokaryotic phylogenetic relationships. Amino acid sequence comparisons, such as those of the cytochrome c family (Schwartz and Dayhoff, 1978, Chapter 7) are the most informative of the various kinds of analyses, but are also the most labor-intensive. However, true relationships may be obscured if plasmid-mediated transfer of genes between distantly related groups is common (Reanney, 1978). Such an event is often called "horizontal" or "lateral" gene transfer, because it involves a lateral movement of genes between even distant branches of a phylogenetic tree. The significance of this type of gene transfer as a complication in the deduction of prokaryotic phylogenetic relationships is yet to be established (Wilson *et al.*, 1977). Use of several different protein or nucleic acid sequences from different positions in the genome is perhaps the best precaution against being misled by possible horizontal gene transfer.

Molecular methods in which nucleic acids are analyzed--for example, nucleic acid hybridization--are of particular value. The use of DNA hybridization analyses has clarified the relation between members of the *Agrobacterium-Rhizobium* group of bacteria (De Ley, 1974, this chapter; Crow *et al.*, 1981) and of the *Enterobacteriaceae* (Brenner, 1973). These and other analyses illustrate the directness and usefulness of quantitative measures of overall genetic similarity or difference. The simplicity and utility of this approach stands in striking contrast to the various phenotypic measures of similarity that have been used as an indirect estimate of genetic relatedness and that have dominated the field until recently. De Ley (1974, this chapter) compares the patterns of relationships deduced from a variety of approaches.

Applications of RNA-DNA hybridization to questions of bacterial phylogeny often make use of highly conserved rRNA molecules; by concentrating on such conserved genes the relationships between somewhat more distantly related species can be discerned (Dubnau *et al.*, 1965; Pace and Campbell, 1971). Hybridization methods are less useful for determining distant relationships, even when conserved ribosomal cistrons are examined. In part, the problem is intrinsic to the hybridization reaction because the rate of reannealing falls off rapidly as the proportion of residue differences between the reannealing strands increase (Bonner *et al.*, 1973). Another difficulty is that similar degrees of hybridization between different pairs of species can be produced by very different patterns of sequence divergence. Thus, measures of overall similarity may overestimate phylogenetic relationships. relationships.

Sequencing of homologous nucleic acids is free of the problems just described and is probably the most informative of contemporary nucleic acid-based techniques. The usefulness of nucleotide sequencing techniques in addressing questions of eukaryotic gene evolution has been indicated in earlier chapters--for example, the work on pseudogenes (Lacy and Maniatis, 1980, Chapter 2), the analysis of rates of nucleotide sequence change (van Ooyen *et al.*, 1979, Chapter 4), and the reconstruction of the patterns of mutational change (Sankoff *et al.*, 1973, Chapter 2). In prokaryotes, sequence analysis has been applied to the less rapidly diverging RNA molecules, such as tRNA and rRNA. The smaller molecules--the tRNAs (about 85 nucleotides) and the 5S rRNA (116-120 nucleotides)--can be sequenced and used for the construction of phylogenetic trees.

Comparison of tRNA sequences reveals numerous differences, as well as many instances of identical independent substitutions on different lines of descent; however, with careful selection of sequences phylogenetic information can be extracted (Cedergren *et al.*, 1980; 1981). The results confirm an early separation of prokaryotes and eukaryotes and a relation between chloroplasts and cyanobacteria. The 5S rRNA sequences are also informative about relations between prokaryotes and suggest an early divergence leading to the precursor of the Gram-positive bacteria and a common precursor of the cyanobacteria and Gram-negative bacteria (Kuntzel *et al.*, 1981). Sankoff *et al.* (1982) have compared the schemes of Schwartz and Dayhoff (1978, Chapter 7), Hori and Osawa (1979), Fox *et al.* (1980) and Kuntzel *et al.* (1981). Although considerable agreement about the composition of the main groups of bacteria is found between these workers, the order of branching of these major groups is more controversial.

The 16S and 23S rRNA molecules are so large that these molecules have been sequenced only recently. Instead oligonucleotides derived from T1 RNase digests of the 16S rRNA molecule have been sequenced and listed to yield an oligonucleotide catalog (Woese *et al.*, 1975; 1976). The 16S rRNA molecule is large enough to provide reliable comparative information over a wide phylogenetic range, yet small enough to yield a manageable oligonucleotide catalog for each species. A table is then drawn up that records the presence or absence of specific oligonucleotides with six or more bases, similarity coefficients are calculated for each pair of species and, from these values, phylogenetic trees are constructed. One of the first uses of the technique was to examine chloroplast origins; the analysis permitted a tracing of red algal chloroplasts to particular cyanobacterial groups (Bonen and Doolittle, 1976; Gray and Doolittle, 1982).

In the realm of prokaryotic phylogeny, the most important advance resulting from the use of this technique was the identification of a large group of anomalous bacteria, the *Archaebacteria*. This group comprises the methanogens and certain acidophiles and thermophiles, and appears to be as distantly related to the remaining groups of bacteria as the latter are to the eukaryotes (Fox, Magrum, Balch, Wolfe and Woese, 1977, this chapter; Fox *et al.*, 1980; Van Valen and Maiorana, 1980, this chapter). This finding has led to the proposal that the prokaryotes represent at least two groups of organisms whose divergence from one another is as ancient as the divergence of the line leading to eukaryotes. Thus, prokaryotes may be a group united by similarity of cellular organization rather than by common phylogeny. By this view, all contemporary organisms are considered to have descended from three original "kingdoms": (1) the group that gave rise to the true bacteria (eubacteria), (2) the group from which the archaebacteria have been derived, and (3) the group that was the source of the original eukaryotic cell (urkaryotes); these are described in Woese and Fox (1977a, this chapter). The complete classification of extant prokaryotes by the oligonucleotide-cataloging procedure and the general congruence of the results with those produced by other molecular phylogenetic methods have been discussed and reviewed by Fox *et al.* (1980).

The three-kingdom hypothesis has not been universally accepted. In the last paper of this chapter, Van Valen and Maiorana (1980, this chapter) have argued on several grounds that the progenitor of the eukaryotic cell is likely to have come from the archaebacteria. This suggestion receives its principal support from the discovery of archaebacteria with some distinctively eukaryotic properties: histonelike proteins, nucleosomelike particles in their chromosomes, and contractile proteins (Searcy *et al.*, 1978). More recently, archaebacteria have been shown to have introns within genes (Luehrsen *et al.*, 1981), some repetititve DNA (Sapienza and Doolitle, 1982), and a eukaryote-like RNA polymerase (Zillig *et al.*, 1982).

A further possible objection to the hypothesis of Woese and Fox is based on the discreteness of the proposed three kingdoms. The very low similarity coefficients (on the order of 0.10 to 0.20) that separate the most distantly related members of the proposed archaebacterial and eubacterial kingdoms (Fox *et al.*, 1980) approach the values that separate the three kingdoms (Woese and Fox 1977a, this chapter). At issue here is the relation between divergence time and similarity coefficients for progressively more ancient separations. If different rates of evolution have occurred, then methods using similarity coefficients can produce errors in divergence times. Whether the origins of eukaryotes are to be sought in a group that has now vanished as an independent entity (the urkaryotes) or among the archaebacteria (as Van Valen and Maiorana propose) must still be considered an open question (Woese and Gupta, 1981).

The three-kingdom hypothesis of Woese and Fox has encouraged speculation about the kinds of organisms or protoorganisms that must have preceded cells of present day prokaryotic complexity, the "one primordial form", as Darwin said, from which all more complex forms have descended. Such hypothetical organisms have been termed progenotes (Woese and Fox, 1977a, this chapter) and must have had a less accurate translation apparatus than any contemporary cell, each of which possesses highly evolved genetic systems. The possibility of translational (or transcriptional) sloppiness in the earliest protocells may appear unattractive. However, it should be noted that contemporary cells can survive high levels of experimentally generated translational ambiguity (Gallant and Palmer, 1979) and the mitochondrion may prove to be a naturally sloppy system (Borst and Grivell, 1981). The possible nature of some of the earliest postulated progenote-type organizations and hypotheses as to their origins are the subject of the next chapter.

Reprinted from *Taxon* **23**: 291-300.

PHYLOGENY OF PROCARYOTES

J. De Ley
Laboratory of Microbial Genetics,
Faculty of Science, State University,
Ledeganckstraat 35, 9000 Gent, Belgium

Summary

We first list the main experimental methods used to get an insight into the phylogenetic relationships of the presently existing bacterial genera. Their scope and relative merits are briefly discussed.

As a model system we briefly summarize the application of most of those methods to elucidate the phylogenetic relationships within the genus Agrobacterium. The evolutionary development of this genus is described.

Similar experiments with Zymomonas are summarized.

INTRODUCTION

The phylogeny of procaryotes, in particular of bacteria, remained an enigma until about ten years ago. The phylogeny of the blue-green algae is still an enigma.

In this paper we shall be concerned solely with bacteria. The relative lack of knowledge on the phylogeny of these organisms is largely due to their very small size. Phylogenetic relationships of higher organisms have been unraveled because of comparative morphology, anatomy and embryology, of paleontology, stratigraphy and several other disciplines. In the bacterial world, however, diversity in morphology and anatomy is very limited and the other disciplines, mentioned above, are of no or very limited applicability. Therefore the insight in the phylogeny of the higher organisms is fair to excellent, but for bacteria it remained a near void.

Several schemes of bacterial phylogeny and evolution have been put forward in the past. We quote only the most important ones. Stanier and van Niel (1941) imagined that a coccus was the simplest bacterium from which the others evolved in separate lines. According to Bisset (1962), a spirillum was the primitive starting point for bacteria evolution. Knight (1938) and Lwoff (1943) proposed a picture of bacterial evolution based on a nutritional and physiological background. None of these hypotheses are satisfactory because of the lack of data. From modern molecular-biological experiments on DNA:RNA hybridization (De Ley, De Smedt and others, to be published) it appears that the hypotheses of Stanier and van Niel may contain a nucleus of truth. Recently Margulis (1970, 1972) proposed a phylogenetic tree of bacteria. It contains a number of correlations between bacterial genera, which are not supported by extant physiological, biochemical and molecular-biological data.

DETECTION OF BACTERIAL PHYLOGENY

Experimental data from different approaches are now accumulating, permitting one to see phylogenetic relationships within and between several groups of genera, such as *Escherichia with Shigella*, *Salmonella*, *Enterobacter* and other *Enterobacteriaceae*; *Agrobacterium* with *Rhizobium*; *Pseudomonas* with *Xanthomonas*, etc. However, the phylogenetic relationships between most other genera are unknown.

Several methods allow one to obtain an insight in phylogenetic relationships amongst bacteria.

1. *The classical or orthodox taxonomy* uses generally at most 25 morphological, physiological, biochemical, serological and ecological criteria for determining a strain. A number of strains have been grouped together in taxa in an almost true picture of their real phylogenetic relationships. In many cases, however, the small number of criteria used (probably only a few percent of the genetic potential) led to many mistakes and to confusion. Examples of the latter are, e.g. the chaotic classification of the so-called genera *Achromobacter*, *Alcaligenes*, *Flavobacterium*, etc. There are many examples of mistaken identities, misidentified strains, wrongly labeled taxa, etc. For example, strains included in *Agrobacterium* as the species *A. gyposphilate* and *A. pseudotsugae*, are in reality *Erwinia herbicola* and a coryneform, respectively.

2. *Adamsonian or numerical analysis*. This method uses more strains and more features, involving probably up to one third of the genome. The large amount of data is handled by computers and grouped by mathematical clustering methods. The probability is thus much greater that the strains are grouped together according to their real relationship. The groups obtained in this fashion are taxonomically and phylogenetically more significant. This approach is most powerful for closely related organisms. Similarity of phenotypic features does not

always mean identity of the genes concerned. For example. organisms which ferment glucose may do this by at least five different enzymic pathways. Enzymes may have the same function, but be structurally quite different. Therefore numerical analysis probably exaggerates the similarity and decreases the phylogenetic distance of remotely related genera.

3. Comparative metabolism. Knowledge of the different enzymic pathways of intermediary metabolism quite frequently allows a broad insight in intergeneric relations. This field was first explored by De Ley (1962) where several examples can be found. More recently, an interesting comparison of energy metabolism in the bacterial world was presented by Decker et al. (1970). Very many data on a great variety of enzymes and enzymic reactions in many bacteria are dispersed all over the literature. Computer-assisted data storage and comparison would yield much information.

4. Electrophorograms of soluble proteins (protein band profiles). In this laboratory we worked out a procedure for the comparison of electrophoretic protein band profiles of large numbers of bacteria (Kersters and De Ley, to be published). It consists essentially of a very reproducible technique for making the profiles and a computer-assisted clustering method. We examine thus a kind of protein copy of each genome. Our experience shows that our method has very much resolution. It detects small individual differences between strains and it is able to resolve a species or a small genus into its genetic races. It can detect intergeneric relationships but not quantitatively.

5. Average DNA base ratios (average % GC). This subject has been so extensively dealt with in the literature that a few words on it may suffice. The average DNA base ratio is very important as it is one of the very few constants in biology. In itself it does not measure the degree of relationship. When two organisms have an identical or similar % GC, they may be, but they are not necessarily related. Other criteria are then needed. However, when two organisms are 25 or more % GC apart, their genomes are mutationally widely different or even totally unrelated. Several examples have been given by De Ley (1969).

6. DNA:DNA hybridization. Similarity in genome DNA base sequences are measured by DNA:DNA or DNA:cRNA hybridizations. The interpretation of the different methods is apparently very simple: the fraction of two genomes which hybridizes is considered to be ancestral and a direct measure of relatedness. In reality, the situation is much more complex. Except in very closely related strains (e.g. two strains of *E. coli*)

mutations affected the genomes unevenly and often quite drastically in different sites over their length. Mismatchings in heterologous DNA duplexes are inversely proportional to the amount of DNA duplexed. The percent of mismatchings determines the thermal stability of the hybrids. Unfortunately, there is a strong tendency among some workers to carry out hybridizations at temperatures above Tm - 25°C. In this fashion one artificially selects for a stable fraction of the hybrid, which may be unrepresentative of the degree of relationship; the less-stable DNA hybrids are not detected, but they may be phenotypically at least as important. Quantitative conclusions on homology are only valuable for a group of organisms with a high degree of relatedness. Several bacterial taxa have been clarified in this fashion. When organisms have a homology of less than 70% at Tm - 25°C, the experimental degrees of relatedness are too low. This applies to many interspecies, and intergeneric DNA hybridizations which have been reported with the stringent method. For the moment the true genetic similarity between most bacterial genera cannot yet be determined with certainty by DNA:DNA hybridization. Molecular biological aspects of bacterial phylogeny have been discussed by De Ley (1968).

7. Sequencing of small proteins and RNA molecules. A number of bacterial cytochromes c (Ambler, personal communication), ferridoxins, and other small molecules have been sequenced. This field is very promising to provide insight into intergeneric evolution. Fragments of the relationships which have been resolved, have been summarized by Dayhof (1972).

PHYLOGENETIC RELATIONSHIPS WITHIN THE GENUS *AGROBACTERIUM*

We decided to determine experimentally the phylogenetic relationships of strains within one group of bacteria as a model system. We needed a rather small genus for this study. *Agrobacterium* was selected for several reasons. Firstly, the number of strains known, about 300, is not too large and experimentally manageable. We have most of them in our collection. Secondly, they have been isolated all over the world. Thirdly, this genus seems to be identified phenotypically rather easily. Fourthly, the number of species is rather small: it contains the crown gall bacteria *A. tumefaciens*, the cane gall bacteria *A. rubi*, the hair root disease bacteria *A. rhizogenes*, and nonpathogenic strains called *A. radiobacter*. About eleven other species such as *A. gypsophilae*, *A. stellulatum*, *A. pseudotsugae*, etc. are taxonomic mistakes, belonging in other genera, and will not be considered

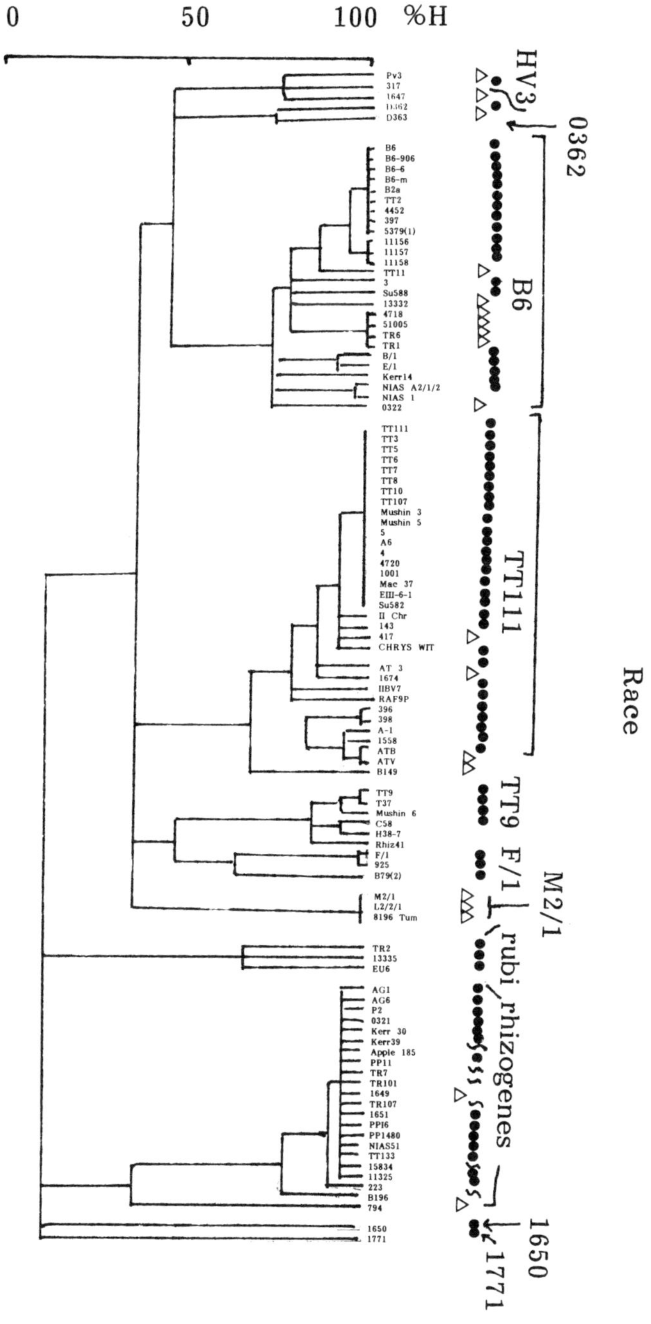

Fig. 1. Relatedness in the genus *Agrobacterium*, determined with DNA: DNA hybridization.

This tree was determined with the initial renaturation rate method (De Ley et al., 1970). Many data were confirmed with the DNA membrane filter method. The existence of the eleven genetic races is obvious. Symbols: • tumorigenic strain, ▽ doubtfully tumorigenic strain, ʃ excessive root-forming strain. The figure is extended and modified from De Ley (1972).

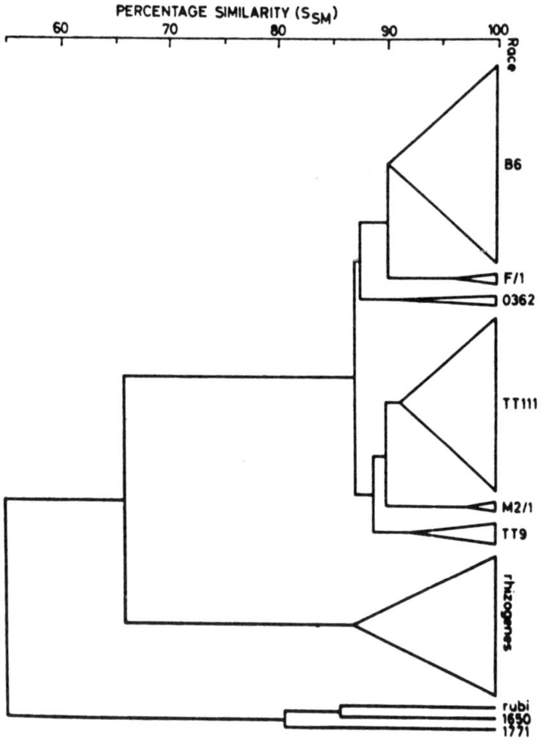

Fig. 2. Numerical analysis of 120 phenotypic features in 70 *Agrobacterium* strains. Unweighted pair group method. The full data are given by Kersters, De Ley, Sneath and Sackin in J. gen. Microbiol. 1973.

differs by some 20 features from the tumefaciens-radiobacter cluster. The three other races at 15% homology differ more phenotypically. One clearly sees, that even in a well-defined and homogeneous genus, DNA hybridization in stringent conditions diminishes phylogenetic relationships. The % similarity levels, calculated by numerical analysis are >93, >85 and >65%, the homology levels are >80, 50 and 15%. This quantitative paradox can be explained if mutations are located mainly outside the active center of the enzymes.

3. Electrophoretic protein profiles (Kersters and De Ley, to be published). If the above hypothesis is correct, one should expect great differences in the electrophoretic protein profiles between the different genetic races. The experiments confirm this hypothesis completely. We find that strains of each genetic race display similar protein patterns and that each race has its own typical pattern. One example has been given by De Ley (1972). The phenotypic explanation has thus been given. The electrophorograms can replace DNA hybridization techniques in this genus; they are extremely well suited as a simple and quick identification method.

here. Fifthly, many strains cause the formation of plant tumors. The real molecular cause of this disease is not yet known. A better knowledge of these bacteria is required for the elucidation of the pathogenic mechanism. We used a variety of methods.

1. DNA:DNA hybridization. It was our purpose to determine firstly quantitative similarities between genomes. Preliminary research (Heberlein, De Ley, Tytgat, 1967) showed that the genus is genetically heterogeneous. To establish the degree of DNA homology amongst, for example, 200 strains, theoretically 20,100 hybridizations are required. Fortunately it turned out we could do with much less, but still a few thousands, including the controls, were required. We first developed a new method, based on initial renaturation rate measurements, which required no radioactive DNA and took only 0.5 hr instead of the usual 16 hr (De Ley, Cattoir and Reynaerts, 1970). If classical taxonomy is correct, one would expect four groups of hybrids, each corresponding to a species. Reality is quite different (Fig. 1). The genus consists of 11 genetic races. Three are large and contain each about 70 strains. They are called here the B6, the TT111 and the rhizogenes race. The remaining eight races contain between 1 and 10 strains. Within each race the strains are at least 80% DNA-homologous. The B6 race, the TT111 race and five small races hybridize at ca. 50% homology. We call it the tumefaciens-radiobacter cluster. This large aggregate, the rhizogenes race, the rubi race and two individual strains 1650 and 1771 hybridize with about 15% DNA homology. Hybridizations were carried out at Tm - 25°C. Other temperatures are being explored. It is quite clear that many mutations happened within this genus. We shall see below that these genetic races are real, but that the phylogenetic distances are exagerated. One expects that strains, which are 50% or 20% DNA-homologous should show considerable phenotypic differences.

2. Numerical analysis of phenotypic features (Kersters, De Ley, Sneath and Sacking, J. gen. Microbiol. 1973). We used a random selection of 70 strains. From the Agrobacterium literature we chose over a hundred different phenotypic features, such as growth on amino acids, organic acids, aromatic compounds, growth and acid production on a series of carbohydrates, 3-ketolactose test, litmus milk test, nitrate and nitrite reduction, phytopathogenicity, etc. The presence or absence of each feature was determined in each strain. The results are represented in Fig. 2. Organisms within each genetic race show no significant phenotypic differences. Genetic races at the 50% homology level differ by at most 5 features. The rhizogenes race

4. Racial differences between cytochrome c (Van Beeumen, Van den Brande and De Ley, to be published). The previous conclusions call for a direct experimental confirmation. The obvious choice is the comparison of amino acid sequences of cytochrome c from each of the main races. We prepared cytochrome c 556 from the races B6, TT111 and rhizogenes in pure state. Table 1 summarizes some of their features. The differences in isoelectric point and in electrophoretic mobility strongly suggests that there must be differences in amino acid composition. These experiments are now being extended.

5. DNA base composition and genome size. All strains have a % GC within the range 58 to 62. There are few noticeable

Table 1. Some properties of pure cytochrome c from strains of the three main genetic races in *Agrobacterium* (Van Beeumen, Van den Brande en De Ley, to be published).

Genetic race	B6		TT111	rhizogenes
strain	B2a	B6	II Chrys	Apple 185
$\lambda\alpha$	556	556	556	556
Mol. weight in SDS	11.000	11.000	11.000	11.000
Isoelectric point	5.67	5.67	6.23	5.27
el. phor. mobility	0.57	0.57	0.40	0.73

differences between the genetic races. The molecular weight of the genome DNA of many strains was determined with a new method developed in our laboratory and based on initial renaturation rates (Gillis, De Ley and De Cleene, 1970). The genome DNA of strains from each of the three main races is the same within the experimental error. The average molecular weight is 3.4 x 10^9, corresponding to about 5.6 million base pairs. The Agrobacterium genome is about 50% longer than the E. coli genome. Genomes of some of the smaller races have sometimes a slightly (up to 15%) different molecular weight. In summary one can say that the size and base composition of the Agrobacterium genome has not changed noticeably during evolution.

6. Thermal stability of DNA:DNA hybrids (De Ley, Tytgat, De Smedt and Michiels, J. gen. Microbiol. 1973). Up to 80% of the genome length has been sufficiently changed by mutation to prevent hybridization. One can wonder whether the common parts which constitute the hybrids, have been affected by mutation, and if so, how much. This can be determined from the thermal

stability curve of the hybrid (Fig. 3). Depending on the temperature used, two types of hybrids are formed (Fig. 3). One is a labile one of unknown nature which will not be considered here. The other type is more or less stable, depending on its

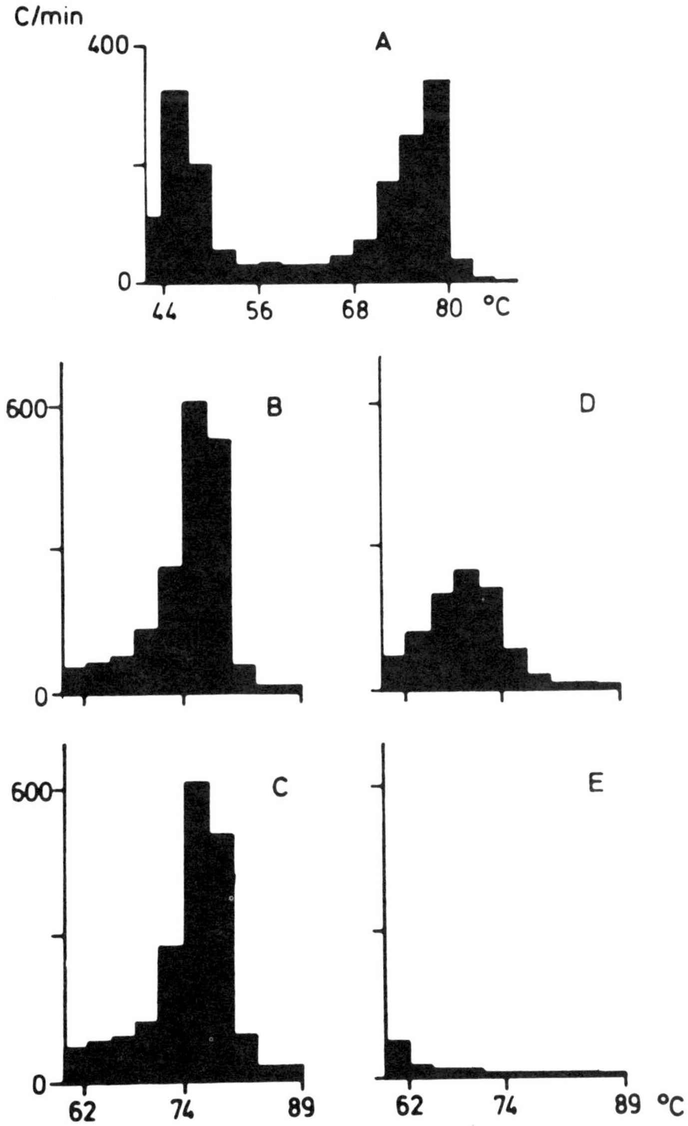

Fig. 3. Thermal stability of DNA: DNA hybrids in *Agrobacterium*.

Fig. A. The presence of a labile and a stable hybrid fraction in a heterologous hybridization at T_m–42° C.

Fig. B.-E. different types of stable hybrids formed at T_m–25° C. Fig. B. a control homologous renaturation for comparison. Fig. C. the stability of the hybrid formed between two DNA's of strains within the same race. Fig. D. idem between two DNA's of strains from races with 50% homology. Fig. E. idem between two DNA's of strains from races with ca 15% homology. The full data by De Ley et al. are given in J. gen. Microbiol. 1973.

origin. Withn each of the main Agrobacterium races thermal stability of the DNA hybrid was about the same, within 2°C, as for the homoduplex (Fig. 3B and C). Between races of 15% homology the stability was at least 13°C lower (Fig. 3E). Obviously, the less two races are evolutionary related, the more mutations occurred within the common part. DNA relatedness between the B6, TT111 and some of the small races show that single-stranded stretches, totalling about 2.8 x 10^6 bases are still able to hybridize. About 10^5 of these bases would be mismatched. The duplexing stretches between the rhizogenes and the other races total about 8 x 10^5 bases. The mismatching would be over 50,000 bases.

7. Intergeneric relationships and the evolutionary origin of Agrobacterium. Previous results (Heberlein et al, 1967) showed that there is a definite DNA relationship between Agrobacterium and Rhizobium. The relationship with other genera is more remote and hard to detect by DNA:DNA hybridization. In the Enterobacteriaceae and a few other organisms the base sequences in the rRNA cistrons are conserved, i.e. they have mutated less than the rest of the genome (e.g. Moore and McCarthy, 1967). It seemed to us that this is an approach to establish remote relationships between bacterial genera. A great number of intergeneric DNA:rRNA hybridizations substantiated this view (De Smedt and De Ley, to be published). We report here only on hybridizations with ^{14}C-rRNA from A. tumefaciens TT111 as an example (Fig. 4). One can distinguish, but less clearly, the same subdivision in Agrobacterium as with the previous methods. The rRNA cistrons from the peritrichously flagellated fast growing Rhizobium strains are almost identical. The rRNA cistrons of the subpolarly flagellated, slow growing rhizobia, the acetic acid bacteria and Zymomonas are already clearly different. Other genera such as Pseudomonas, Bacillus, the Enterobacteriaceae, many gram positives and others mentioned in Fig. 4, are only remotely related.

8. The evolution of Agrobacterium. We have now enough data to visualize the origin of Agrobacterium and the evolution within this genus. After the rise of the dicotyledonous plants some 200 million years ago, the ancestors of the present rhizobia and agrobacteria probably started to develop in two directions.

One group of bacteria acquired the property to grow in many plants and to produce tumors by an as yet unknown mechanism. They were the ancestors of the presently living Agrobacterium.

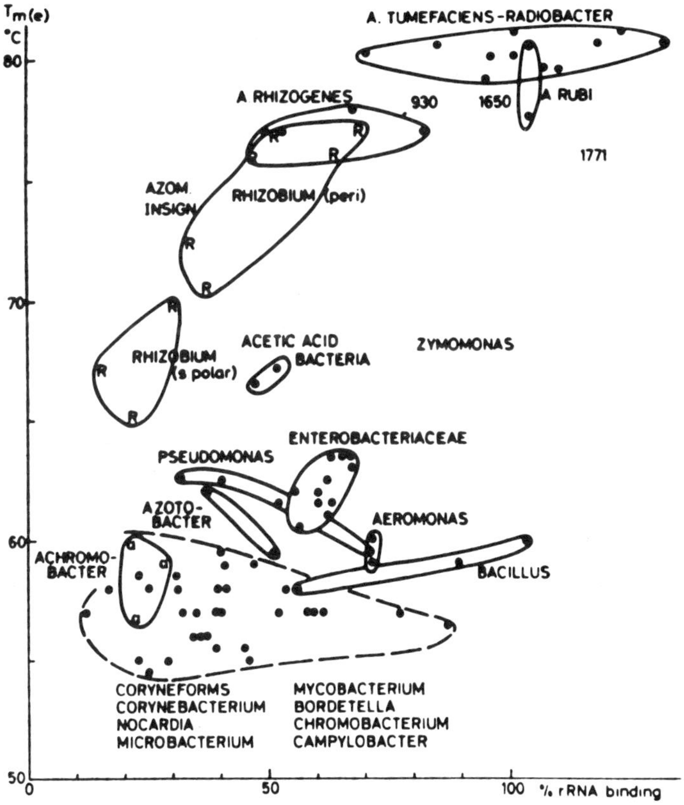

Fig. 4. The amount and the thermal stability of heterologous hybrids between ^{14}C-rRNA from *Agrobacterium tumefaciens* TT111 with DNA from many other bacteria.

The second group started to live symbiontically with the leguminous plants and were probably responsible for their present great importance. These bacteria formed at least two branches, the ancestors of the presently living subpolarly and peritrichously flagellated rhizobia. Alternatively it is possible that the present agrobacteria arose from existing ancestral rhizobia or vice versa. But it is certain that they are all twigs of the same branch.

Let us consider here only the ancestral agrobacteria. The first steps in the evolution within this group probably was nearly the same as we see now within each genetic race: a limited number of mutations, mainly outside the active center of enzymes, perhaps mainly in the third base of the triplets or permitting substitution of similar amino acids in a few proteins. Their phenotype was almost identical. For some unknown reason (geographical, ecological?) this population was dispersed into at least five groups, the ancestors of the present tumefaciens-radiobacter cluster, the rhizogenes, rubi, 1650 and 1771 races. Under different conditions they evolved away from each other, probably very similar to the differences between the present B6 and TT111 races: more mutations modifying the structural features of many proteins, mutations in active centers of a few enzymes. Their phenotype was first slightly different. More and more mutations affected active centers until the groups reached their present diversity: ca. 15% DNA homology, considerable mutations in the common DNA parts and at least 20% difference in their phenotypic features. The races rubi 1650 and 1771 remained extremely small in number. The rhizogenes race spread out over the entire world, mainly in warm or arid climates; some strains developed an unknown system to induce excessive root growth. This race is either genetically very stable or its geographical spreading occurred rather recently, because almost all strains are genotypically and phenotypically very similar. The tumefaciens-radiobacter cluster started a further evolutionary diversification all on its own. It evolved into seven races, all with 50% DNA homology, moderate mutations in the common parts, considerable mutations outside the active center changing the properties of proteins, and less that 5% differences in their phenotype. Two races spread out. The B6 race apparently prevails around the Pacific Ocean, as it was isolated so far only in the U.S.A., Japan, New Zealand and Australia. The USA, Europe, South Africa. The five other races remained small in number of strains. Most of them started to diversify already slightly, as testified by their 80-100% DNA homoogy.

Fig. 1 also shows that there is no correlation between the species and the genetic races. Strains named A. tumefaciens occur in all races, strains named A. rubi form one race with strains named A. tumefaciens, A. radiobacter occurs in seven races, strains named A. rhizogenes and A. tumefaciens occur at very high DNA homology in the same race. It is obvious that the present species concept and species differentation in Agrobacterium is not justified. A nomenclatorial revision will have to be carried out. We can now also assess the value of the four current Latin species names: they refer only to the state of phytopathogenicity, to the presence or absence of a disease-provoking property; they have only the value of formae speciales, pathotypes; they do not represent a true biological unit.

PHYLOGENETIC RELATIONSHIPS WITHIN THE GENUS ZYMOMONAS

Not all bacterial taxa diversified in the same fashion as Agrobacterium. Some populations diversified phenotypically much more profoundly, as in the family of the Enterobacteriaceae. Others remained more homogeneous as we shall see in the next example of Zymomonas.

These gram-negative anaerobic rods ferment glucose or fructose through a modification of the Entner-Doudoroff mechanism to ethanol, CO_2 and a little lactic acid. They have been isolated from fermenting plant material and in breweries. We have many of the existing strains in our collection: they are from fermenting agave juice (pulque, Mexico), palm wines (Zaire) and breweries (England). We studied the same parameters as for Agrobacterium (Swings and De Ley, to be published). We shall report our detailed results elsewhere. Zymomonas is a very tight group with over 80% DNA homology (except one strain), almost identical phenotypic features and protein profiles, % GC and genome size. These bacteria, because of their strict requirements (presence of glucose or fructose, growth factors, amino acids and anaerobiosis) occur only in approximately the same selective conditions. they hardly diversified except in minor features, whether they occur in Zaire, in Mexico or in European breweries. So far we have not yet found generic relatives. Apparently this genus evolved and remained all by itself in the large group of gram-negative bacteria.

Acknowledgement

We are indebted to the Fonds voor Kollektief Fundamenteel Onderzoek (Belgium) for supporting our research with several grants.

References

BISSET, K.A. 1962 - The phylogenetic concept in bacterial taxonomy, 12th Symp. Soc. Gen Microb. "Bacterial Classification" 361-373. Cambridge University Press.

DAYHOFF, M. O. 1972 - Atlas of protein sequence and structure. NBFR, Washington.

DECKER, K., JUNGERMANN, K. & R. K. THAUER 1970 - Wege der Energiegewinnung in Anaerobiern. Angew. Chem. 82:1553-173.

DE LEY, J. 1962 - Comparative biochemistry and enzymology in bacterial classification. 12th Symp. Soc. Gen. Microb. "Bacterial Classification" 164-195. Cambridge University Press.

DE LEY, J. 1968 - Molecular Biology and Bacterial Phylogeny. Evolutionary Biology. Vol 2. Ed. Th. Dobzhansky, M. K. Hecht and Wm. C. Steere, 103-156. Appleton-Century-Crofts, New York.

DE LEY, J. 1969 - Compositional nucleotide distribution and the theoretical prediction of homology in bacterial DNA. J. Theor. Biol. 22; 89-116.

DE LEY, J. 1972 - Agrobacterium: intrageneric relationships and evolution. Proc. 3d Int. Conf. Plant Pathog. Bacteria. April 1971. Wageningen, the Netherlands, Pudoc.

DE LEY, J., CATTOIR, H. & A., REYNAERTS, 1970 - The quantitative measurement of DNA hybridization from renaturation rates. Eur. J. Biochem. 12: 133-142.

DE LEY, J., TYTGAT, R., DE SMEDT, J. & M. MICHIELS 1973 - Thermal stability of DNA:DNA hybrids within the genus Agrobacterium. J. Gen. Microbiol. 78:241-252.

GILLIS, M., DE LEY, J. & M. DE CLEENE 1970 - The determination of molecular weight of bacterial genome DNA from renaturation rates. Eur. J. Biochem. 12: 143-153.

HEBERLEIN, G., DE LEY, J. & R. TYTGAT 1967 - Deoxyribonucleic acid homology and taxonomy of Agrobacterium, Rhizobium and Chromobacterium. J. Bacteriol. 94: 116-124.

KERSTERS, K., DE LEY, J., SNEATH, P. H. A. & M. SACKIN 1973 - Numerical Taxonomic Analysis of Agrobacterium. J.Gen. Microbiol. 78: 227-239.

KNIGHT, B. C. 1938 - Bacterial Nutrition. H. M. Stationery Office, London.

LWOFF, A. 1943 - L'evolution physiologique, Hermann et Cie, Paris.

MARGULIS, L., 1970 - Origin of Eukaryotic Cells. Yale University Press, New Haven.

MARGULIS, L. 1972 - Early cellular evolution. Exobiology, Ed. C. Ponnamperuma. North Holland Public. Co. 342-368. Amsterdam.

MOORE, R. L. & B. McCARTHY 1967 - Comparative study of ribosomal ribonucleic acid cistrons in Enterobacteria and Myxobacteria. J. Bacteriol. 94: 1066-1074.

STANIER, R. Y. & C. B. VAN NIEL 1941 - The main outlines of bacterial classification. J. Bacteriol. 42:437-466.

Classification of methanogenic bacteria by 16S ribosomal RNA characterization

(comparative oligonucleotide cataloging/phylogeny/molecular evolution)

GEORGE E. FOX* †, LINDA J. MAGRUM*, WILLIAM E. BALCH‡, RALPH S. WOLFE‡, AND CARL R. WOESE*‡

Departments of *Genetics and Development and ‡ Microbiology, University of Illinois, Urbana, Illinois 61801

Communicated by H. A. Barker, August 10, 1977

ABSTRACT The 16S ribosomal RNAs from 10 species of methanogenic bacteria have been characterized in terms of the oligonucleotides produced by T_1 RNase digestion. Comparative analysis of these data reveals the methanogens to constitute a distinct phylogenetic group containing two major divisions. These organisms appear to be only distantly related to typical bacteria.

The methane-producing bacteria are a poorly studied collection of morphologically diverse organisms that share the common metabolic capacity to grow anaerobically by oxidizing hydrogen and reducing carbon dioxide to methane (1–3). Their relationships to one another and to other microbes remain virtually unknown. Protein and nucleic acid primary structures are perhaps the most reliable indicators of phylogenetic relationships (4–6). By using a molecule, such as the 16S ribosomal RNA, that is readily isolated, ubiquitous, and highly constrained in sequence (7), it is possible to relate even the most distant of microbial species. To date, approximately 60 bacterial species have been characterized in terms of their 16S ribosomal RNA primary structures (refs. 6–9, unpublished data). We present here results of a comparative study of the methanogens by this method, which shows their relationships to one another and to typical bacteria.

METHODS

Methanobacterium ruminantium strain *PS*, *Methanobacterium* strain *M.o.H.*, *Methanobacterium formicicum*, and *Methanosarcina barkeri* were provided by M. P. Bryant. *Methanobacterium arbophilicum* (10) was obtained from J. G. Zeikus. Two new marine isolates, Cariaco isolate JR-1 and Black Sea isolate JR-1, were provided by J. A. Romesser. *Methanospirillum hungatii* (11) and the above methanogens were cultivated in the following low-phosphate medium (values in g/liter): $(NH_4)_2SO_4$, 0.22; NaCl, 0.45; $MgSO_4 \cdot 7H_2O$, 0.09; $CaCl_2 \cdot H_2O$, 0.06; $FeSO_4 \cdot 7H_2O$, 0.002; resazurin, 0.001; sodium formate, 3.0; sodium acetate, 2.5; $NaHCO_3$, 6.0; trace mineral solution and vitamin solution (12), 10 ml each; and dephosphorylated yeast extract (Difco) and Trypticase (BBL), 2.0 each. For growth of marine isolates, NaCl was added to a final concentration of 15 g/liter. Procedures for preparation of media, growth of organisms, ^{32}P labeling, extraction of labeled 16S ribosomal RNA, and analysis of T_1 RNase digests of this RNA have been published (13–17).

The resulting oligonucleotide catalogs were examined with standard clustering techniques (18). An association coefficient for each binary couple is defined as follows: $S_{AB} = 2N_{AB}/(N_A + N_B)$, in which N_A, N_B, and N_{AB} are the total number of residues represented by hexamers and larger in catalog *A* and in catalog *B* and their overlap of common sequences, respectively. The association coefficient, S_{AB}, so defined provides what is generally an underestimate of the true degree of homology between two catalogs because related but nonidentical oligomers are not considered. The matrix of S_{AB} values for each binary comparison among the members of a given set of organisms is used to generate a dendrogram by average linkage (between the merged groups) clustering. The resulting dendrogram is, strictly speaking, phyletic because no "ancestral catalog" has been postulated. However, it is clear from the molecular nature of the data that the topology of this dendrogram would closely resemble, if not be identical to, that of a phylogenetic tree based upon such ancestral catalogs.

RESULTS

The 10 organisms whose 16S ribosomal RNA oligonucleotide catalogs are listed in Tables 1 and 2 cover all of the major types of methanogens now in pure culture except for 2; we have been unable to obtain a culture of *Methanococcus vannielii* (19), and *Methanobacterium mobile* (20) has proven difficult to grow and label. The sequences in Table 1 bear little resemblance to those for typical bacteria (refs. 6–9; unpublished data). Fig. 1 is a dendrogram derived from the S_{AB} values in Table 3. It can be seen that the methanogens comprise two major divisions. The first contains the *Methanobacterium* species; the second contains *Methanosarcina*, *Methanospirillum*, and the two marine isolates. Each division has two subgroups: group IA comprises coccobacillus-like Gram-positive rods, IB comprises long Gram-positive rods, and IIA comprises various Gram-negative forms; group IIB contains one member, a Gram-positive sarcina. Table 2 lists the post-transcriptionally modified sequences found in these RNAs. Most of the modifications are unique to the methanogens, and variations in their pattern correlate strongly with the grouping shown in Fig. 1, providing independent evidence for this grouping.

DISCUSSION

Because of their diverse morphologies and different Gram reactions, some microbiologists have considered the methanogens to be a heterogeneous group of organisms. Their scattered classification in the seventh edition of *Bergey's Manual* reflected this attitude. On this view, the commonality of their biochemistry, if it required explanation, could be rationalized in terms of a reticulate evolution, involving an appropriate plasmid. However, the above evidence indicates that this type of relationship among the methanogens is certainly not the case. The basis for classification used herein—i.e., ribosomal RNA—is

The costs of publication of this article were defrayed in part by the payment of page charges. This article must therefore be hereby marked "*advertisement*" in accordance with 18 U. S. C. §1734 solely to indicate this fact.

† Present address: Department of Biophysical Sciences, University of Houston, Houston, TX 77004.

Table 1. Oligonucleotide catalogs for 16S rRNA of 10 methanogens

Oligonucleotide sequence	Present in organism number
5-mers	
CCCCG	1–10;1,5,8
CCCAG	6
CCACG	10
ACCCG	10
CCAAG	9
CACAG	9
CAACG	1–10;8–9
ACACG	7–9
ACCAG	7
AACCG	1–10;10
ACAAG	1–6;1,5
AAACG	7–9
AAAAG	1,6,9–10
CUCCG	4,7
CCCUG	9
UCCAG	6–8,10
CUCAG	1–10
CCAUG	1–10
UCACG	1–2,4–5
UACCG	1–6,8
ACCUG	4–5;5
ACUCG	6
AUCCG	9
UAACG	4–9
CAAUG	1–6;4
ACUAG	2–3,8–9
ACAUG	10
AUACG	7
AAUCG	10
UAAAG	2
AUAAG	3–10,3,6–9;7
AAAUG	4
UUCCG	1–6,8;4
CUUCG	5–6,8
UCCUG	1–6;4
CCUUG	1
CUCUG	6,8
UCUAG	7
UUCAG	5,7–9;9
CUAUG	5
UACUG	7–10;8–10
UAUCG	7–8
ACUUG	1–6,10
AUCUG	3–5,7–8
AUUCG	2–3,10
UUAAG	1–10;1–2,4,6,8,10
UAAUG	1–2,5,10;2
AUAUG	3–4,9
AAUUG	1–10;1–2,4–6,9
AUUAG	1–10;1–7,9;7
UUUCG	4,7,9
UUCUG	3
UCUUG	8–9
CUUUG	1–3,5,10
UUUAG	2,7
UUAUG	4,9;9
UUUUG	2,9
6-mers	
CCCCAG	4,6
CCCAAG	6,10
CAACCG	8–9
ACCACG	7–9
ACACCG	6–10
AAACCG	8–10
CCCUCG	5,8,10
CCUCAG	5
CUCCAG	5
UCCCAG	2,7–10
CCACUG	4–5,9
ACCUCG	9
CCUAAG	1–3,5,10
CUCAAG	4–6

Oligonucleotide sequence	Present in organism number
CCAUAG	4
CAUACG	1
ACACUG	4–5,7–9
AACCUG	1–6,10;1
AAUCCG	7,9–10
CUAAAG	7–9
UAAACG	1–6,8–10
ACUAAG	9
ACAAUG	1–10
AUAACG	10
AAUACG	1–6,10
AACAUG	10;10
AAACUG	1–10;8–9
AAAUCG	1–3,7
AAUAAG	1–2,4–6
CCCUUG	6,10
CCUCUG	7–9
UCCCUG	1–10
CCUUAG	4,7–8
CUCUAG	1–3
CUUCAG	9
UCCUAG	1–2
UUCCAG	1–6
CCUAUG	3
CUACUG	1–3,6
UCACUG	3,7–9
CUAUCG	7–10
UCAUCG	7,9
CAUCUG	7
ACUCUG	7–8
ACCUUG	4–6
AUCCUG	1–10
UCUAAG	7–8
UUACAG	8
UAUCAG	9
UAUACG	7
UAAUCG	1–10
AUACUG	3,7–8,10
ACAUUG	1
AACUUG	3
AAUCUG	5–9
UAAAUG	4
AUUAAG	1–8
AAUAUG	9
CCUUUG	1–2,5
CUUUCG	10
UCUCUG	1–2,4–6
UUCCUG	5
UCUUAG	5
CUAUUG	1–4,6
UUACUG	10
UAUUCG	3
AUUCUG	2,8–10
ACUUUG	2
UAUAUG	8
CUUUUG	1–5;1
UCUUUG	1,4
UUUUUG	7
7-mers	
ACCCACG	1–9
ACCACCG	7
AACCCCG	7
CCAACAG	7–8
CAACACG	1–2,5–6
CAAACCG	8–9
CCCUACG	1–10
CCCACUG	10
UCCACCG	4–6
CCACCUG	10
CCCUAAG	7–8
UCACACG	3
CUACACG	4,7–10
UAACCCG	5–6

Oligonucleotide sequence	Present in organism number
AUACCCG	1–10
AACCUCG	8
CCUAAAG	1–6
UAACACG	1–10
AUAACCG	7
AAUCCAG	8–10
AACAUCG	10
AAAUCCG	7–9
UAAAAAG	1,3–6
CCCUUAG	1,3–6
CAUCCUG	7–10
UACUCCG	7
AUCUCCG	8
ACCUUCG	9
UCCUAAG	7
UUACCAG	1–2,4–6,10
CUAACUG	3–4
UAACUCG	1–4,7–8,10
AUUCCAG	7
AUCAUCG	6
AAUCUCG	3
AACCUUG	6
UCUAAAG	10
CUUAAAG	7–9
CAAUAUG	10
AUACUAG	1
AAUCUAG	1–2,4,7–8
AAAUCUG	10
UAAAAUG	10
CUCCUUG	1–3,5–10
UCCCUUG	9
UUCUCCG	7
CUCUUAG	2
UACUUCG	8
UACUCUG	10
UCAUAUG	10
UAAUCUG	4
AAUUUAG	3
UUCUUCG	10
UCUCUUG	7–8
CUUUAUG	10
UUUAUCG	1
UAUUUCG	1
AUUAUUG	10
UUCUUUG	4–6
UAUUUUG	3
UUUUUUG	1–3
8-mers	
CCACAACG	1–3,5–6,9–10
ACCCCAAG	1,5
AAACCCCG	9
UCCACCAG	9
CCCACAUG	7–8
CUCAACCG	8
ACCCUCAG	7
ACCACCUG	1,3–6,8,10
UAACACCG	1–6,10
AUCCCAAG	2–3
AAAUCCCG	1
CCCUCAUG	1,3–4
UACUCCCG	4
AU(CCUC)CG	5
CCUAUCAG	10
CCUAACUG	5
CUUAACCG	4,7,9
UAAUCCCG	9
CUACAAUG	1–10
UACUACAG	10
UAAUACCG	7–9
AUUACCAG	3
AUAACCUG	6–8,10

Table 1 continues on following page.

Table 1. (continued)

Oligonucleotide sequence	Present in organism number
ACAAUCUG	9
AAAUCCUG	1–2,6–9
AUAAACUG	3–6
AUAAAUAG	2
(CU,CCUU)CG	4
AUCCUUCG	4
UCUAACUG	1
CUUAACUG	2–3,5–6
UAAUCCUG	1–3,6
UCUAAAUG	1
UUAAAUCG	10
CAUAUAUG	10
AAAUCUUG	10
AAAUUCUG	2–3
AUAAAUUG	1
CUUUUCAG	6
UUCUCAUG	2
UUUAAUCG	9
UAUCAUUG	9
UUUAAAUG	2–3
UUUAAUUG	1–8,10
UUUUUUCG	2–3
UUUUAUUG	1
9-mers	
CCCACCAAG	4–5
CACACACCG	1–10
(CCA,CAA)CAG	8
CCCAACAAG	7–9
AACCCCAAG	6
AAACCCAAG	4
CCUCACCAG	8
CCUACCAAG	6
CCUACAACG	10
AUAACCCCG	6,8,10
AAACCUCCG	1–6
CACACUAAG	1–6
AUAAACCCG	6
UACUCCCAG	1–3,5–6
UAAUCCCCG	7
AAUCCCCUG	1,3–6
CUUACCAAG	1–3
(UC)ACACAUG	3
(UC)ACAAUCG	2–3
UCAUAACCG	4
CUAAUACCG	3
ACCCUUAAG	7
AUAAUCCCG	9
AUAACCCUG	1–5
AUAAUACCG	4–5
AUAUACAAG	9
UCUUACCAG	10
UCACUAUCG	6
UAAUCCCUG	10
UAAUCCUCG	8
AAUUUCCCG	10
AAUCCUCUG	2
UCAUAAUCG	1,5
CUAAUACUG	1
CAUCAUAUG	10
AUAAUUCCG	10
AAUUAUCCG	7–9
UUUAAAACG	7
UAAACUAUG	7
AUAAUACUG	2
CUAUUACUG	9
UUAAAUUCG	1
UUUAAUAAG	2
UUAUAUUCG	2
UAUUUCUAG	9
UUUAUUAAG	1
CUUUUAUUG	6
UUUUUAUUG	2,4
UUUUUUUCG	1
10-mers	
AAUAACCCCG	7
ACCACCUAUG	9
AAUCUCACCG	8
AAAUCUCACG	4
UAACUCAAAG	8
AAACUUAAAG	1–10
ACCUUACCUG	10
UUACCAUCAG	3
UACCUACUAG	10
AAUCACUUCG	5
AACCCUUAUG	6
UAAAUAACUG	9
UUCUUCACCG	6
ACUCUACUUG	9
CUUAACUAUG	1
AUACUAUUAG	2,4–5
UUCCCUAUUG	4
UCUUCUUAAG	4
AUUUUUUUCG	1
UUUUCUUUUG	5
11-mers	
ACAACUCACCG	10
AAAUCCCACAG	6
CAUCUCACCAG	7,9
UAACUCACCCG	9
AAAUCUCACCG	7,9
AAACACCUUCG	6
AAAUCCCAUAG	5
UCCCUCCCCUG	10
CAUAUCCUCCG	10
AAAUCCUAUAG	3
UUUCAACAUAG	7,9
A(UA,UCA,CUA)UG	6
UUUCAAUAUAG	10
CUUUUCUUAAG	1,3
CUUUUCAUUAG	2
UUCUUUAAUCG	7
UUUUUUUUCCUG	1
UUUUUUUUUAAG	2
12-mers	
CCACCCAAAAAG	1–2,4,6
UCAAACCACCCG	8–10
UCAAACCAUCCG	7
ACAUCUCACCAG	1–6
CCACUCUUAACG	4–6
CCAUUCUUAACG	1–3
CUCAACUAUUAG	10
CCACUAUUAUUG	7
CAAUUAUUCCUG	2
CCACUUUUAUUG	8
CCAUUUUUAUUG	5
(CUA,CUUUUA)UUG	3
13-mers	
UAAACUACACCUG	10
(CAA,CCA)CAUUCUG	6
UAAUACUCCAUAG	9
UUUCAAAAUAACG	8
AUAAUUUUUCCUG	3
(UUU,CUU,CU)AAAUG	5
14-mers	
AAAACUUUACCAUG	9
AAAACUUUACAAUG	7–8,10
AUUUUU(CCU,CU)UUG	2
15-mers	
UCUAAAACACACCUG	8
AUAACCUACCCUUAG	1–3
AUAACCUAACCUUAG	4
AAUAAUACCCUAUAG	8
AAUAAUACUCCAUAG	7
AUAAUCUACCCUUAG	5
16-mers	
UAAUCCCCUAAACCAG	6
AAAUCCUAUAAUCCCG	4
AAUCUCCUAAACAUAG	5
CAAUCUCUUAAACCUG	7
UAAUCUCCUAAACCUG	4
AAAUCCUAUAAUCCUG	5
17-mers	
CAAUCUUUUAAACCUAG	3
UAAU(CCU,CU)AAACUUAG	1–2
AUAAU(CCU,CU)AAACCUG	9
18-mer	
AACAAUCUCCUAAACCUG	8
24-mer	
(AAACA,UAAUCUCA)— CCCAUCCUUAG	10
termini	
5′ end	
pAG	4,6
pAAUCCG	5
pAAUCUG	1,3
pAUUCUG	2,7–10
3′ end	
AUCACCUCCU$_{OH}$	1–10

First column is oligonucleotide sequence; second column shows organisms in which that sequence is found. Organisms are designated by number (see Fig. 1) as follows: 1, *M. arbophilicum;* 2, *M. ruminantium* strain *PS;* 3, *M. ruminantium* strain *M-1;* 4, *M. formicicum;* 5, *M.* sp. strain *M.o.H.;* 6, *M. thermoautotrophicum;* 7, Cariaco isolate JR-1; 8, Black Sea isolate JR-1; 9, *Methanospirillum hungatii;* 10, *Methanosarcina barkeri.* Multiple occurrences of a sequence in a given organism are denoted by repeating the organism's number in column 2: e.g., *1–4,6–8;3,7,;3* signifies a double occurrence in organism 7 and a triple occurrence in organism 3.

independent of particular biochemistries and, as representative of the cellular information processing systems, should be considered idionomonic of the organism. By means of this approach we have shown not only that methanogens are a coherent phylogenetic grouping but also that they are quite distinct from other bacteria as well. Just how distinct they may be is indicated in Fig. 1; even enterics and blue-green algae appear closely related by comparison.

Table 2. Post-transcriptionally modified sequences and likely counterparts

Sequence	Occurrence in methanogens IA	IB	IIA	IIB	Occurrence in typical bacteria
1. ȦȦCCUG	+	+	–	–	30%
ȦȦUCUG	–	–	+	+	None
ȦȦG	–	–	–	–	55%
2. UȦȦCAAG	+	+	–	–	None
UAȦCAAG	–	–	+	+	None
U̇AACAAG	–	–	–	–	>95%
3. AUṄCAACG	+	+	–	–	None
ACṄCAACG	–	–	+	+	None
AXĠĊAACG	–	–	–	–	>90%
4. ṄCCG	+	+	–	–	None
C(Ċ,C)G	–	–	+	+	Nonc
Ṅ′CCG	–	–	–	–	>95%
5. CCĊCG	–	–	–	+	>95%

Post-transciptionally modified sequences in methanogens and their likely counterparts in the bacteria that have been examined. In group 1, Ȧ is *N*-6-diMe (21), identified by electrophoretic mobilities of Ȧ and ȦȦ and by total resistance to U_2 nuclease. In group 2, U̇ is partially resistant to pancreatic nuclease, the first Ȧ when modified is still U_2 nuclease sensitive; the second Ȧ is *N*-6-diMe. Ṅ in group 3 is resistant to pancreatic nuclease but is electrophoretically U-like. X stands for U or A. In group 4, Ṅ and Ṅ′ are not cleaved by endonucleases; ṄC and Ṅ′C are electrophoretically distinguishable; Ċ is cleaved by pancreatic nuclease and has C-like electrophoretic properties. In group 5, Ċ (21, 22) is not cleaved by pancreatic nuclease and is readily deaminated by NH_4OH.

A phylogenetic distinction of this apparent magnitude is suspect unless substantiated by other evidence. In fact, a distinction of this magnitude reasonably demands that there be many and striking differences in corresponding phenotypes. Consider the following points.

(*i*) Methane production involves a highly unique biochemistry. In probing its details, the biochemist is beginning to uncover an unusual spectrum of coenzymes. For example, coenzyme M, involved in methyl transfer in methane formation, is the smallest of all known coenzymes; it is unique in its sulfur content and acidity (24). One of us (W.E.B.) has examined a wide variety of tissues and organisms fcr the presence of this cofactor and found it to be confined to the methanogens. Similarly, coenzyme F_{420}, which handles low-potential electrons, is present in all methanogens but so far is not found elsewhere (25).

(*ii*) We have been unable to detect cytochromes in these organisms, and R. Thauer obtained no evidence for the presence of quinones in *M. thermoautotrophicum* (personal communication). The extent to which their overall biochemistry is unique remains to be determined.

(*iii*) All other bacterial cell walls so far examined, with the single exception of the extreme halophiles, contain peptidoglycan (26, 27). However, cell walls of the methanogens (eight examples) do not contain this compound (ref. 28; O. Kandler, personal communication).

(*iv*) Table 2 shows that the pattern of base modification in 16S ribosomal RNA in methanogens is, for the most part, different from that in typical bacteria. This holds for the 23S rRNA as well (D. Stahl, personal communication). Moreover, methanogens are the first major group of organisms characterized (prokaryote or eukaryote) whose tRNAs lack the so-called "common sequence," TΨCG. Division I methanogens contain a ΨΨĊĠ sequence, whereas in division II it becomes U̇ΨĊG (the dot above a base signifies an unidentified modification; U̇ ≠ T) (L. Magrum and D. Stahl, unpublished data).

It should be noted that three of these four points appear to be completely unrelated to the production of methane or to the requirement of a strictly anaerobic niche. These differences become the more impressive when it is realized that methanogens have been characterized but little in terms of their general biochemistry and molecular biology, and not at all genetically. It would appear that methanogens ultimately may have to be classified as a systematic group distinct from other bacteria (inclusive of the blue-green algae).

Although it cannot be unequivocally concluded that methanogens represent the most ancient divergence yet encountered in the bacterial line of descent, the possibility is certainly likely. How ancient, then, could the methanogenic phenotype be? It may well be older than the blue-green algal one, which fossil evidence suggests to be close to 3 billion years (29). On the assumption that equivalent S_{AB} values measure the same physical time, the most ancient divergence within the methanogens proper ($S_{AB} \sim 0.25$) is comparable to that which separates blue-green algae from most of the other bacteria (Fig. 1). Methanogens might then have existed at a time when an anaerobic atmosphere, rich in carbon dioxide and hydrogen, enveloped the planet and, if so, could have played a pivotal role in this planet's physical evolution.

Note Added in Proof: Preliminary characterization of *Methanobacterium mobile*, a motile, Gram-negative, short rod, places this organism in group IIA. *Methanobacterium* sp. strain AZ (30) has been shown to be a strain of *M. arbophilicum*; $S_{AB} = 0.87$ for the pair.

The work reported herein was performed under National Aeronautics and Space Administration Grant NSG-7044 and National

Table 3. S_{AB} values for each indicated binary comparison

Organism	1	2	3	4	5	6	7	8	9	10	11	12	13
1. *M. arbophilicum*	—												
2. *M. ruminantium PS*	.66	—											
3. *M. ruminantium M-1*	.60	.60	—										
4. *M. formicicum*	.50	.48	.49	—									
5. *M.* sp. *M.o.H.*	.53	.49	.51	.60	—								
6. *M. thermoautotrophicum*	.52	.49	.51	.54	.60	—							
7. Cariaco isolate JR-1	.25	.27	.25	.26	.23	.25	—						
8. Black Sea isolate JR-1	.26	.28	.26	.28	.27	.29	.59	—					
9. *Methanospirillum hungatii*	.20	.24	.21	.23	.23	.22	.51	.52	—				
10. *Methanosarcina barkeri*	.29	.26	.24	.24	.26	.25	.33	.41	.34	—			
11. Enteric-vibrio sp.	.08	.08	.11	.09	.09	.10	.05	.06	.07	.10	—		
12. *Bacillus* sp.	.10	.10	.14	.11	.11	.12	.08	.10	.10	.08	.27	—	
13. Blue-green sp.	.10	.10	.10	.10	.10	.11	.08	.09	.08	.11	.24	.26	—

The values given for enteric-vibrio sp., *Bacillus* sp., and blue-green sp. represent averages obtained from 11 (9), 7 (6), and 4 (23) individual species, respectively.

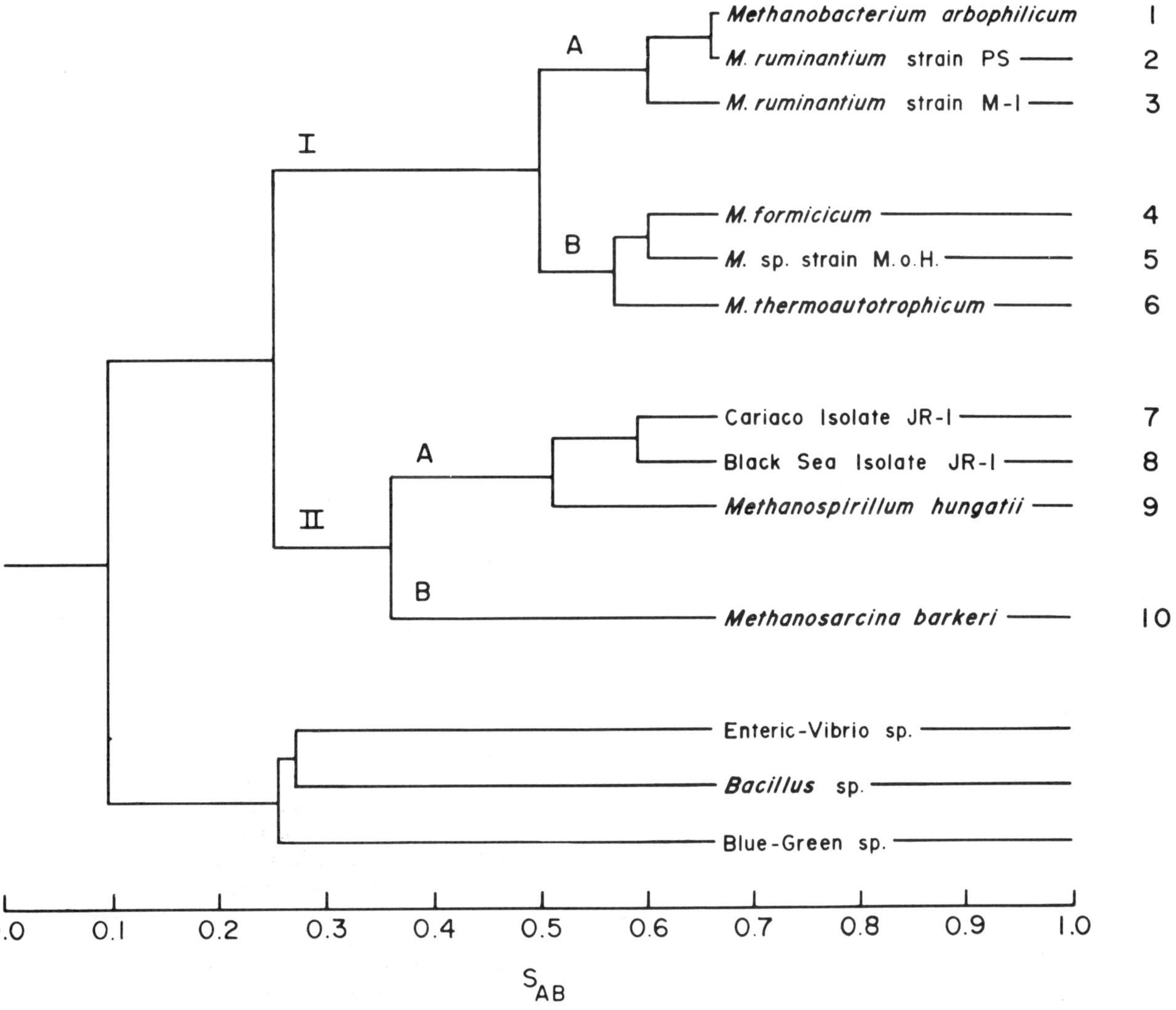

FIG. 1. Dendrogram of relationships of methanogens and typical bacteria. The figure was constructed by average linkage clustering (between the merged groups) from the S_{AB} values given in Table 3.

Science Foundation Grant PCM 74-15227 to C.R.W. and U.S. Public Health Service Grant AI-12277 and National Science Foundation Grant PCM 76-02652 to R.S.W.

1. Wolfe, R. S. (1972) *Adv. Microbiol. Physiol.* **6**, 107–146.
2. Zeikus, J. G. (1977) *Bacteriol. Rev.* **41**, 514–541.
3. Zeikus, J. G. & Bowen, V. G. (1975) *Can. J. Microbiol.* **21**, 121–129.
4. Zuckerkandl, E. & Pauling, L. (1965) *J. Theor. Biol.* **8**, 357–366.
5. Fitch, W. M. (1976) *J. Mol. Evol.* **8**, 13–40.
6. Fox, G. E., Pechman, K. R. & Woese, C. R. (1977) *Int. J. Syst. Bacteriol.* **27**, 44–57.
7. Woese, C. R., Fox, G. E., Zablen, L., Uchida, T., Bonen, L., Pechman, K., Lewis, B. J. & Stahl, D. (1975) *Nature* **254**, 83–86.
8. Zablen, L. & Woese, C. R. (1975) *J. Mol. Evol.* **5**, 25–34.
9. Zablen, L. (1975) Ph.D. Thesis, University of Illinois, Urbana, IL.
10. Zeikus, J. G. & Henning, D. L. (1975) *Antonie van Leeuwenhoek* **41**, 543–552.
11. Ferry, J. G., Smith, P. H. & Wolfe, R. S. (1974) *Int. J. Syst. Bacteriol.* **24**, 465–469.
12. Wolin, E. A., Wolin, M. J. & Wolfe, R. S. (1963) *J. Biol. Chem.* **238**, 2882–2886.
13. Balch, W. E. & Wolfe, R. S. (1976) *Appl. Environ. Microbiol.* **32**, 781–791.
14. Balch, W. E., Magrum, L. J., Fox, G. E., Wolfe, R. S. & Woese, C. R. (1977) *J. Mol. Evol.* **9**, 305–311.
15. Uchida, T., Bonen, L., Schaup, H. W., Lewis, B. J., Zablen, L. & Woese, C. (1974) *J. Mol. Evol.* **3**, 63–77.
16. Sanger, F., Brownlee, G. G. & Barrell, B. G. (1965) *J. Mol. Biol.* **13**, 373–398.
17. Woese, C., Sogin, M., Stahl, D., Lewis, B. J. & Bonen, L. (1976) *J. Mol. Evol.* **7**, 197–213.
18. Anderberg, M. R. (1973) *Cluster Analysis for Applications* (Academic Press, New York).
19. Bryant, M. P. (1974) in *Bergey's Manual of Determinative Bacteriology*, eds. Buchanan, R. E. & Gibbons, N. E. (Williams and Wilkins, Baltimore, MD), pp. 472–477.
20. Paynter, M. J. B. & Hungate, R. E. (1968) *J. Bacteriol.* **95**, 1943–1951.
21. Fellner, P. (1969) *Eur. J. Biochem.* **11**, 2–27.
22. Sogin, M. L., Pechmann, K. J., Zablen, L., Lewis, B. J. & Woese, C. R. (1972) *J. Bacteriol.* **112**, 13–16.
23. Bonen, L. & Doolittle, W. F. (1976) *Nature* **261**, 669–673.
24. Taylor, C. D. & Wolfe, R. S. (1974) *J. Biol. Chem.* **249**, 4879–4885.
25. Cheeseman, P., Toms-Wood, A. & Wolfe, R. S. (1972) *J. Bacteriol.* **112**, 527–531.
26. Brown, A. D. & Cho, K. Y. (1970) *J. Gen. Microbiol.* **62**, 267–270.
27. Reistad, R. (1972) *Arch. Mikrobiol.* **82**, 24–30.
28. Kandler, O. & Hippe, H. (1977) *Arch. Microbiol.* **113**, 57–60.
29. Schopf, J. W. (1972) in *Exobiology—Frontiers of Biology—Vol. 23* (North Holland, Amsterdam), pp. 16–61.
30. Zehnder, A. J. B. & Wuhrmann (1977) *Arch. Microbiol.* **111**, 199–205.

Reprinted from *Proc. Natl. Acad. Sci. USA* **74**: 5088–5090.

Phylogenetic structure of the prokaryotic domain: The primary kingdoms

(archaebacteria/eubacteria/urkaryote/16S ribosomal RNA/molecular phylogeny)

CARL R. WOESE AND GEORGE E. FOX*

Department of Genetics and Development, University of Illinois, Urbana, Illinois 61801

Communicated by T. M. Sonneborn, August 18, 1977

ABSTRACT A phylogenetic analysis based upon ribosomal RNA sequence characterization reveals that living systems represent one of three aboriginal lines of descent: (*i*) the eubacteria, comprising all typical bacteria; (*ii*) the archaebacteria, containing methanogenic bacteria; and (*iii*) the urkaryotes, now represented in the cytoplasmic component of eukaryotic cells.

The biologist has customarily structured his world in terms of certain basic dichotomies. Classically, what was not plant was animal. The discovery that bacteria, which initially had been considered plants, resembled both plants and animals less than plants and animals resembled one another led to a reformulation of the issue in terms of a yet more basic dichotomy, that of eukaryote versus prokaryote. The striking differences between eukaryotic and prokaryotic cells have now been documented in endless molecular detail. As a result, it is generally taken for granted that all extant life must be of these two basic types.

Thus, it appears that the biologist has solved the problem of the primary phylogenetic groupings. However, this is not the case. Dividing the living world into *Prokaryotae* and *Eukaryotae* has served, if anything, to obscure the problem of what extant groupings represent the various primeval branches from the common line of descent. The reason is that eukaryote/prokaryote is not primarily a phylogenetic distinction, although it is generally treated so. The eukaryotic cell is organized in a different and more complex way than is the prokaryote; this probably reflects the former's composite origin as a symbiotic collection of various simpler organisms (1–5). However striking, these organizational dissimilarities do not guarantee that eukaryote and prokaryote represent phylogenetic extremes.

The eukaryotic cell *per se* cannot be directly compared to the prokaryote. The composite nature of the eukaryotic cell makes it necessary that it first be conceptually reduced to its phylogenetically separate components, which arose from ancestors that were noncomposite and so individually are comparable to prokaryotes. In other words, the question of the primary phylogenetic groupings must be formulated solely in terms of relationships among "prokaryotes"—i.e., noncomposite entities. (Note that in this context there is no suggestion *a priori* that the living world is structured in a dichotomous way.)

The organizational differences between prokaryote and eukaryote and the composite nature of the latter indicate an important property of the evolutionary process: Evolution seems to progress in a "quantized" fashion. One level or domain of organization gives rise ultimately to a higher (more complex) one. What "prokaryote" and "eukaryote" actually represent are two such domains. Thus, although it is useful to define phylogenetic patterns within each domain, it is not meaningful to construct phylogenetic classifications between domains: Prokaryotic kingdoms are not comparable to eukaryotic ones. This should be recognized by an appropriate terminology. The highest phylogenetic unit in the prokaryotic domain we think should be called an "urkingdom"—or perhaps "primary kingdom." This would recognize the qualitative distinction between prokaryotic and eukaryotic kingdoms and emphasize that the former have primary evolutionary status.

The passage from one domain to a higher one then becomes a central problem. Initially one would like to know whether this is a frequent or a rare (unique) evolutionary event. It is traditionally assumed—without evidence—that the eukaryotic domain has arisen but once; all extant eukaryotes stem from a common ancestor, itself eukaryotic (2). A similar prejudice holds for the prokaryotic domain (2). [We elsewhere argue (6) that a hypothetical domain of lower complexity, that of "progenotes," may have preceded and given rise to the prokaryotes.] The present communication is a discussion of recent findings that relate to the urkingdom structure of the prokaryotic domain and the question of its unique as opposed to multiple origin.

Phylogenetic relationships cannot be reliably established in terms of noncomparable properties (7). A comparative approach that can measure degree of difference in comparable structures is required. An organism's genome seems to be the ultimate record of its evolutionary history (8). Thus, comparative analysis of molecular sequences has become a powerful approach to determining evolutionary relationships (9, 10).

To determine relationships covering the entire spectrum of extant living systems, one optimally needs a molecule of appropriately broad distribution. None of the readily characterized proteins fits this requirement. However, ribosomal RNA does. It is a component of all self-replicating systems; it is readily isolated; and its sequence changes but slowly with time—permitting the detection of relatedness among very distant species (11–13). To date, the primary structure of the 16S (18S) ribosomal RNA has been characterized in a moderately large and varied collection of organisms and organelles, and the general phylogenetic structure of the prokaryotic domain is beginning to emerge.

A comparative analysis of these data, summarized in Table 1, shows that the organisms clearly cluster into several primary kingdoms. The first of these contains all of the typical bacteria so far characterized, including the genera *Acetobacterium, Acinetobacter, Acholeplasma, Aeromonas, Alcaligenes, Anacystis, Aphanocapsa, Bacillus, Bdellovibrio, Chlorobium, Chromatium, Clostridium, Corynebacterium, Escherichia, Eubacterium, Lactobacillus, Leptospira, Micrococcus, Mycoplasma, Paracoccus, Photobacterium, Propionibacterium,*

The costs of publication of this article were defrayed in part by the payment of page charges. This article must therefore be hereby marked "*advertisement*" in accordance with 18 U. S. C. §1734 solely to indicate this fact.

* Present address: Department of Biophysical Sciences, University of Houston, Houston, TX 77004.

Table 1. Association coefficients (S_{AB}) between representative members of the three primary kingdoms

	1	2	3	4	5	6	7	8	9	10	11	12	13
1. *Saccharomyces cerevisiae*, 18S	—	0.29	0.33	0.05	0.06	0.08	0.09	0.11	0.08	0.11	0.11	0.08	0.08
2. *Lemna minor*, 18S	0.29	—	0.36	0.10	0.05	0.06	0.10	0.09	0.11	0.10	0.10	0.13	0.07
3. L cell, 18S	0.33	0.36	—	0.06	0.06	0.07	0.07	0.09	0.06	0.10	0.10	0.09	0.07
4. *Escherichia coli*	0.05	0.10	0.06	—	0.24	0.25	0.28	0.26	0.21	0.11	0.12	0.07	0.12
5. *Chlorobium vibrioforme*	0.06	0.05	0.06	0.24	—	0.22	0.22	0.20	0.19	0.06	0.07	0.06	0.09
6. *Bacillus firmus*	0.08	0.06	0.07	0.25	0.22	—	0.34	0.26	0.20	0.11	0.13	0.06	0.12
7. *Corynebacterium diphtheriae*	0.09	0.10	0.07	0.28	0.22	0.34	—	0.23	0.21	0.12	0.12	0.09	0.10
8. *Aphanocapsa* 6714	0.11	0.09	0.09	0.26	0.20	0.26	0.23	—	0.31	0.11	0.11	0.10	0.10
9. Chloroplast (*Lemna*)	0.08	0.11	0.06	0.21	0.19	0.20	0.21	0.31	—	0.14	0.12	0.10	0.12
10. *Methanobacterium thermoautotrophicum*	0.11	0.10	0.10	0.11	0.06	0.11	0.12	0.11	0.14	—	0.51	0.25	0.30
11. *M. ruminantium* strain M-1	0.11	0.10	0.10	0.12	0.07	0.13	0.12	0.11	0.12	0.51	—	0.25	0.24
12. *Methanobacterium* sp., Cariaco isolate JR-1	0.08	0.13	0.09	0.07	0.06	0.06	0.09	0.10	0.10	0.25	0.25	—	0.32
13. *Methanosarcina barkeri*	0.08	0.07	0.07	0.12	0.09	0.12	0.10	0.10	0.12	0.30	0.24	0.32	—

The 16S (18S) ribosomal RNA from the organisms (organelles) listed were digested with T1 RNase and the resulting digests were subjected to two-dimensional electrophoretic separation to produce an oligonucleotide fingerprint. The individual oligonucleotides on each fingerprint were then sequenced by established procedures (13, 14) to produce an oligonucleotide catalog characteristic of the given organism (3, 4, 13–17, 22, 23; unpublished data). Comparisons of all possible pairs of such catalogs defines a set of association coefficients (S_{AB}) given by: $S_{AB} = 2N_{AB}/(N_A + N_B)$, in which N_A, N_B, and N_{AB} are the total numbers of nucleotides in sequences of hexamers or larger in the catalog for organism A, in that for organism B, and in the intersection of the two catalogs, respectively (13, 23).

Pseudomonas, *Rhodopseudomonas*, *Rhodospirillum*, *Spirochaeta*, *Spiroplasma*, *Streptococcus*, and *Vibrio* (refs. 13–17; unpublished data). The group has three major subdivisions, the blue-green bacteria and chloroplasts, the "Gram-positive" bacteria, and a broad "Gram-negative" subdivision (refs. 3, 4, 13–17; unpublished data). It is appropriate to call this urkingdom the *eubacteria*.

A second group is defined by the 18S rRNAs of the eukaryotic cytoplasm—animal, plant, fungal, and slime mold (unpublished data). It is uncertain what ancestral organism in the symbiosis that produced the eukaryotic cell this RNA represents. If there had been an "engulfing species" (1) in relation to which all the other organisms were endosymbionts, then it seems likely that 18S rRNA represents that species. This hypothetical group of organisms, in one sense the major ancestors of eukaryotic cells, might appropriately be called *urkaryotes*. Detailed study of anaerobic amoebae and the like (18), which seem not to contain mitochondria and in general are cytologically simpler than customary examples of eukaryotes, might help to resolve this question.

Eubacteria and urkaryotes correspond approximately to the conventional categories "prokaryote" and "eukaryote" when they are used in a phylogenetic sense. However, they do not constitute a dichotomy; they do not collectively exhaust the class of living systems. There exists a third kingdom which, to date, is represented solely by the methanogenic bacteria, a relatively unknown class of anaerobes that possess a unique metabolism based on the reduction of carbon dioxide to methane (19–21). *These "bacteria" appear to be no more related to typical bacteria than they are to eukaryotic cytoplasms.* Although the two divisions of this kingdom appear as remote from one another as blue-green algae are from other eubacteria, they nevertheless correspond to the *same* biochemical phenotype. The apparent antiquity of the methanogenic phenotype plus the fact that it seems well suited to the type of environment presumed to exist on earth 3–4 billion years ago lead us tentatively to name this urkingdom the *archaebacteria*. Whether or not other biochemically distinct phenotypes exist in this kingdom is clearly an important question upon which may turn our concept of the nature and ancestry of the first prokaryotes.

Table 1 shows the three urkingdoms to be equidistant from one another. Because the distances measured are actually proportional to numbers of mutations and not necessarily to time, it cannot be proven that the three lines of descent branched from the common ancestral line at about the same time. One of the three may represent a far earlier bifurcation than the other two, making there in effect only two urkingdoms. Of the three possible unequal branching patterns the case for which the initial bifurcation defines urkaryotes vs. all bacteria requires further comment because, as we have seen, there is a predilection to accept such a dichotomy.

The phenotype of the methanogens, although ostensibly "bacterial," on close scrutiny gives no indication of a specific phylogenetic resemblance to the eubacteria. For example, methanogens do have cell walls, but these do not contain peptidoglycan (24). The biochemistry of methane formation appears to involve totally unique coenzymes (23, 25, 26). The methanogen rRNAs are comparable in size to their eubacterial counterparts, but resemble the latter specifically in neither sequence (Table 1) nor in their pattern of base modification (23). The tRNAs from eubacteria and eukaryotes are characterized by a common modified sequence, TΨCG; methanogens modify this tRNA sequence in a quite different and unique way (23). It must be recognized that very little is known of the general biochemistry of the methanogens—and almost nothing is known regarding their molecular biology. Hence, although the above points are few in number, they represent most of what is now known. There is no reason at present to consider methanogens as any closer to eubacteria than to the "cytoplasmic component" of the eukaryote. Both in terms of rRNA sequence measurement and in terms of general phenotypic differences, then, the three groupings appear to be distinct urkingdoms.

If a third urkingdom exists, does this suggest that many more such will be found among yet to be characterized organisms? We think not, although the matter clearly requires an exhaustive search. As seen above, the number of species that can be classified as eubacteria is moderately large. To this list can be added *Spirillum* and *Desulfovibrio*, whose rRNAs appear typically eubacterial by nucleic acid hybridization measurements (27). Because the list is also phenotypically diverse, it seems unlikely that many, if any, of the yet uncharacterized prokaryotic groups will be shown to have coequal status with the present three. Conceivably the halophiles whose cell walls contain no peptidoglycan, are candidates for this distinction (28, 29).

Eukaryotic organelles, however, could be a different matter. There can be no doubt that the chloroplast is of specific eubacterial origin (3, 4). A question arises with the remaining organelles and structures. Mitochondria, for example, do not conform well to a "typically prokaryotic" phenotype, which has led some to conclude that they could not have arisen as endosymbionts (30). By using "prokaryote" in a phylogenetic sense, this formulation of the issue does not recognize a third alternative—that the organelle in question arose endosymbiotically from a separate line of descent whose phenotype is not "typically prokaryotic" (i.e., eubacterial). It is thus conceivable that some endosymbiotically formed structures represent still other major phylogenetic groups; some could even be the only extant representation thereof.

The question that remains to be answered is whether the common ancestor of all three major lines of descent was itself a prokaryote. If not, each urkingdom represents an independent evolution of the prokaryotic level of organization. Obviously, much more needs to be known about the general properties of all the urkingdoms before this matter can be definitely settled. At present we can point to two arguments suggesting that each urkingdom does represent a separate evolution of the prokaryotic level of organization.

The first argument concerns the stability of the general phenotypes. The general eubacterial phenotype has been stable for at least 3 billion years—i.e., the apparent age of blue-green algae (31). The methanogenic phenotype seems to be at least this old in that branchings within the two urkingdoms are comparably deep (see Table 1). The time available to form each phenotype (from their common ancestor) is then short by comparison, which seems paradoxical in that the two phenotypes are so fundamentally different. We think that this ostensible paradox implies that the common ancestor in this case was not a prokaryote. It was a far simpler entity; it probably did not evolve at the "slow" rate characteristic of prokaryotes; it did not possess many of the features possessed by prokaryotes, and so these evolved independently and differently in separate lines of descent.

The second argument concerns the quality of the differences in the three general phenotypes. It seems highly unlikely, for example, that differences in general patterns of base modification in rRNAs and tRNAs are related to the niches that organisms occupy. Rather, differences of this nature imply independent evolution of the properties in question. It has been argued elsewhere that features such as RNA base modification generally represent the final stage in the evolution of translation (32). If these features have evolved separately in two lines of descent, their common ancestor, lacking them, had a more rudimentary version of the translation mechanism and consequently, could not have been as complex as a prokaryote (6).

With the identification and characterization of the urkingdoms we are for the first time beginning to see the overall phylogenetic structure of the living world. It is not structured in a bipartite way along the lines of the organizationally dissimilar prokaryote and eukaryote. Rather, it is (at least) tripartite, comprising (*i*) the typical bacteria, (*ii*) the line of descent manifested in eukaryotic cytoplasms, and (*iii*) a little explored grouping, represented so far only by methanogenic bacteria.

The ideas expressed herein stem from research supported by the National Aeronautics and Space Administration and the National Science Foundation. We are grateful to a number of colleagues who have helped to generate the yet unpublished data that make these speculations possible: William Balch, Richard Blakemore, Linda Bonen, Tristan Dyer, Jane Gibson, Ramesh Gupta, Robert Hespell, Bobby Joe Lewis, Kenneth Luehrsen, Linda Magrum, Jack Maniloff, Norman Pace, Mitchel Sogin, Stephan Sogin, David Stahl, Ralph Tanner, Thomas Walker, Ralph Wolfe, and Lawrence Zablen. We thank Linda Magrum and David Nanney for suggesting the name "archaebacteria."

1. Stanier, R. Y. (1970) *Symp. Soc. Gen. Microbiol.* **20**, 1–38.
2. Margulis, L. (1970) *Origin of Eucaryotic Cells* (Yale University Press, New Haven).
3. Zablen, L. B., Kissel, M. S., Woese, C. R. & Buetow, D. E. (1975) *Proc. Natl. Acad. Sci. USA* **72**, 2418–2422.
4. Bonen, L. & Doolittle, W. F. (1975) *Proc. Natl. Acad. Sci. USA* **72**, 2310–2314.
5. Bonen, L., Cunningham, R. S., Gray, M. W. & Doolittle, W. F. (1977) *Nucleic Acid Res.* **4**, 663–671.
6. Woese, C. R. & Fox, G. E. (1977) *J. Mol. Evol.*, in press.
7. Sneath, P. H. A. & Sokal, R. R. (1973) *Numerical Taxonomy* (W. H. Freeman, San Francisco).
8. Zuckerkandl, E. & Pauling, L. (1965) *J. Theor. Biol.* **8**, 357–366.
9. Fitch, W. M. & Margoliash, E. (1967) *Science* **155**, 279–284.
10. Fitch, W. M. (1976) *J. Mol. Evol.* **8**, 13–40.
11. Sogin, S. J., Sogin, M. L. & Woese, C. R. (1972) *J. Mol. Evol.* **1**, 173–184.
12. Woese, C. R., Fox, G. E., Zablen, L., Uchida, T., Bonen, L., Pechman, K., Lewis, B. J. & Stahl, D. (1975) *Nature* **254**, 83–86.
13. Fox, G. E., Pechman, K. R. & Woese, C. R. (1977) *Int. J. Syst. Bacteriol.* **27**, 44–57.
14. Uchida, T., Bonen, L., Schaup, H. W., Lewis, B. J., Zablen, L. B. & Woese, C. R. (1974) *J. Mol. Evol.* **3**, 63–77.
15. Zablen, L. B. & Woese, C. R. (1975) *J. Mol. Evol.* **5**, 25–34.
16. Doolittle, W. F., Woese, C. R., Sogin, M. L., Bonen, L. & Stahl, D. (1975) *J. Mol. Evol.* **4**, 307–315.
17. Pechman, K. J., Lewis, B. J. & Woese, C. R. (1976) *Int. J. Syst. Bacteriol.* **26**, 305–310.
18. Bovee, E. C. & Jahn, T. L. (1973) in *The Biology of Amoeba*, ed. Jeon, K. W. (Academic Press, New York), p. 38.
19. Wolfe, R. S. (1972) *Adv. Microbiol. Phys.* **6**, 107–146.
20. Zeikus, J. G. (1977) *Bacteriol. Rev.* **41**, 514–541.
21. Zeikus, J. G. & Bowen, V. G. (1975) *Can. J. Microbiol.* **21**, 121–129.
22. Balch, W. E., Magrum, L. J., Fox, G. E., Wolfe, R. S. & Woese, C. R. (1977) *J. Mol. Evol.*, in press.
23. Fox, G. E., Magrum, L. J., Balch, W. E., Wolfe, R. S. & Woese, C. R. (1977) *Proc. Natl. Acad. Sci. USA*, **74**, 4537–4541.
24. Kandler, O. & Hippe, H. (1977) *Arch. Microbiol.* **113**, 57–60.
25. Taylor, C. D. & Wolfe, R. S. (1974) *J. Biol. Chem.* **249**, 4879–4885.
26. Cheeseman, P., Toms-Wood, A. & Wolfe, R. S. (1972) *J. Bacteriol.* **112**, 527–531.
27. Pace, B. & Campbell, L. L. (1971) *J. Bacteriol.* **107**, 543–547.
28. Brown, A. D. & Cho, K. Y. (1970) *J. Gen. Microbiol.* **62**, 267–270.
29. Reistad, R. (1972) *Arch. Mikrobiol.* **82**, 24–30.
30. Raff, R. A. & Mahler, H. R. (1973) *Science* **180**, 517–521.
31. Shopf, J. W. (1972) *Exobiology—Frontiers of Biology* (North Holland, Amsterdam), Vol. 23, pp. 16–61.
32. Woese, C. R. (1970) *Symp. Soc. Gen. Microbiol.* **20**, 39–54.

Reprinted from *Nature* **287**: 248–250.

The Archaebacteria and eukaryotic origins

Leigh M. Van Valen & Virginia C. Maiorana

Biology Dept (Whitman), University of Chicago, 915 E. 57th Street, Chicago, Illinois 60637

Critical analysis of the phylogeny of prokaryotes is in its infancy. Woese and others[1–5] have made the startling proposal that methane-producing bacteria and a few others form a phyletically unified group, the Archaebacteria, as old and as diverse (although not now as numerous) as all other bacteria. The only critique[6] of this proposal is inadequate[7]. Here we present an alternative view, that the Archaebacteria were derived from other bacteria and contain the ancestor of a cell which engulfed others, eventually to become the first eukaryote.

The main evidence used for the relative time of divergence of the Archaebacteria is that their 16S ribosomal RNA is very different from that of other bacteria. However, as Woese recognizes[3], this can be the result of rapid early evolution. The origin of new adaptations often occurs rapidly, as exemplified by the evolution of bats[8], whales[9] and modern placental mammals as a whole[10]. Evolutionary rates vary widely[11], and an attempt[12] to extrapolate the average rate of evolution of cytochrome *c* from vertebrates to angiosperms gave results which contradicted the fossil record[13]. (We do not mean to dispute the usual approximate constancy of protein evolution.) Proteins themselves often evolve unusually rapidly when they undergo a major change in function[14,15]. Even the supposedly nearly invariant histone H4 is very different between ciliates and other eukaryotes[16]. The suggestion[5] that the diversity of Archaebacteria argues against rapid evolution near their origin is a non sequitur; placental mammals are a familiar counter-example in which both phenomena occur. The only studied genomes of archaebacteria are small[17,18], although far from the smallest known for prokaryotes[19].

Our proposal (Fig. 1), based in part on work by Margulis[20], Hall[21] and others, takes the evolution of mechanisms for acquisition of free energy to be central. We discuss the general topic elsewhere[22]. Any phylogeny must be based on a clear recognition of primitive states and derived states, for shared primitive states give no phyletic information[23,24]. A set of differences is not in itself an adequate basis for a phylogeny, in part because the base of the phylogeny can then be anywhere. We take glycolysis alone (as in some lactic-acid bacteria), especially its later stages, to be primitive among surviving organisms because it is almost universally distributed and provides materials for other, presumably derived, processes such as the completed tricarboxylic–acid cycle.

The Archaebacteria are moderately derived prokaryotes in this view, although most surviving prokaryotes are even more derived. Why then should they be so divergent from the others? The answer may be lost in the Archaean, but we can offer an initial suggestion. One of the free-living Archaebacteria, *Thermoplasma*, lacks a cell wall, while the other known Archaebacteria have a wide diversity of cell walls[25] all very different in composition from the walls of all other walled cells. Thus it seems as though the ancestral lineage of Archaebacteria lost its cell wall, possibly for flexibility in locomotion or for energy conservation. Such a change could be related to other changes in adaptation.

Moreover, *Thermoplasma* itself shows several specific and derived similarities with eukaryotes[26]. It has a histone-like protein which produces nucleosome-like condensations of DNA. It has actin- and myosin-like proteins and microbody-like respiration. It can invaginate its plasma membrane, although apparently not with fully developed endocytosis. Such an ability is necessary for a cell which engulfs others.

However, like other surviving Archaebacteria[27], *Thermoplasma* has lipids with branched chains and with ether linkages instead of the esterifications of other prokaryotes and eukaryotes[28]. Unless eukaryote lipids came from another member of the presumptive symbiosis, this suggests that the surviving Archaebacteria evolved their unique lipids and other traits after the ancestral eukaryote diverged.

Halobacterium, a surviving archaebacterium[29] with independently derived light-driven ATP synthesis[18,30], has other specific and derived similarities with eukaryotes. Its light-transducing pigment is almost identical to rhodopsin[18]. Its initiator tRNA resembles that of eukaryotes in some ways[18] and produces a non-formylated methionine[18], as in eukaryotes. The only ribosomal protein studied has a partial amino acid sequence nearly intermediate between those of the bacterium *Escherichia* and the yeast *Saccharomyces*[18]. *Halobacterium* has a ferredoxin rather similar to that of plants[18], and its cell wall is composed of a glycoprotein resembling those of eukaryotic cell walls[18]. It even seems to have more than one non-plasmid chromosome, perhaps several[18]. Another archaebacterium, *Methanobacterium*[17], but not *Halobacterium*[18], may have some repeated DNA.

Perhaps those resemblances to eukaryotes (except probably the cell wall) and those of *Thermoplasma* were present together in an ancestral form (*Thermoplasma* has not been studied for all of these characters) and some have been modified in each lineage. All enzymes of the tricarboxylic–acid cycle occur in archaebacteria, but not all together in the same species[5]. Parasitic mycoplasms also show some specific resemblances to *Thermoplasma*[25,31]; their relation to Archaebacteria needs clarification.

We therefore provisionally regard the Archaebacteria as one of several phyla of prokaryotes, most as yet undefined. The Proalgae (new name) are another, comprising the Cyanophyta and Prochlorophyta[32]—the oxygen-releasing prokaryotic algae. (Eukaryotic algae, not a natural group, can be called eualgae, a term which makes adjectives easily. Proalgae are true algae adaptively; to call them bacteria[33] is like calling mammals and birds reptiles[34] because both groups come from the evolutionary radiation of reptiles.)

How can we test such hypotheses? One way is to investigate further the extent to which the proposed phylogeny, and any alternatives, make adaptive sense, using as many characters as possible. A second, more decisive when it can be applied, is to accumulate characters from which the topology of evolutionary relationships can be inferred, whenever possible with information on what state is primitive and with recognition of the possibility of convergence, especially by loss. Sequences of proteins and nucleic acids give the greatest density of such characters, but the data on rRNA (not sequences because only isolated oligonucleotides have been studied) suggest that so great changes may have occurred in the origins of both archaebacteria and eukaryotes that even such slowly evolving molecules show only random similarity to those of their ancestors. Primitive eukaryotes such as *Eimeria*[35] and *Pelomyxa*[36] also deserve intensive scrutiny.

The translation apparatus of Archaebacteria differs from that of other organisms[4]. Perhaps there is an unbridgeable adaptive valley in this difference[3]. However, we think that this matter cannot be discussed adequately until much more is known about

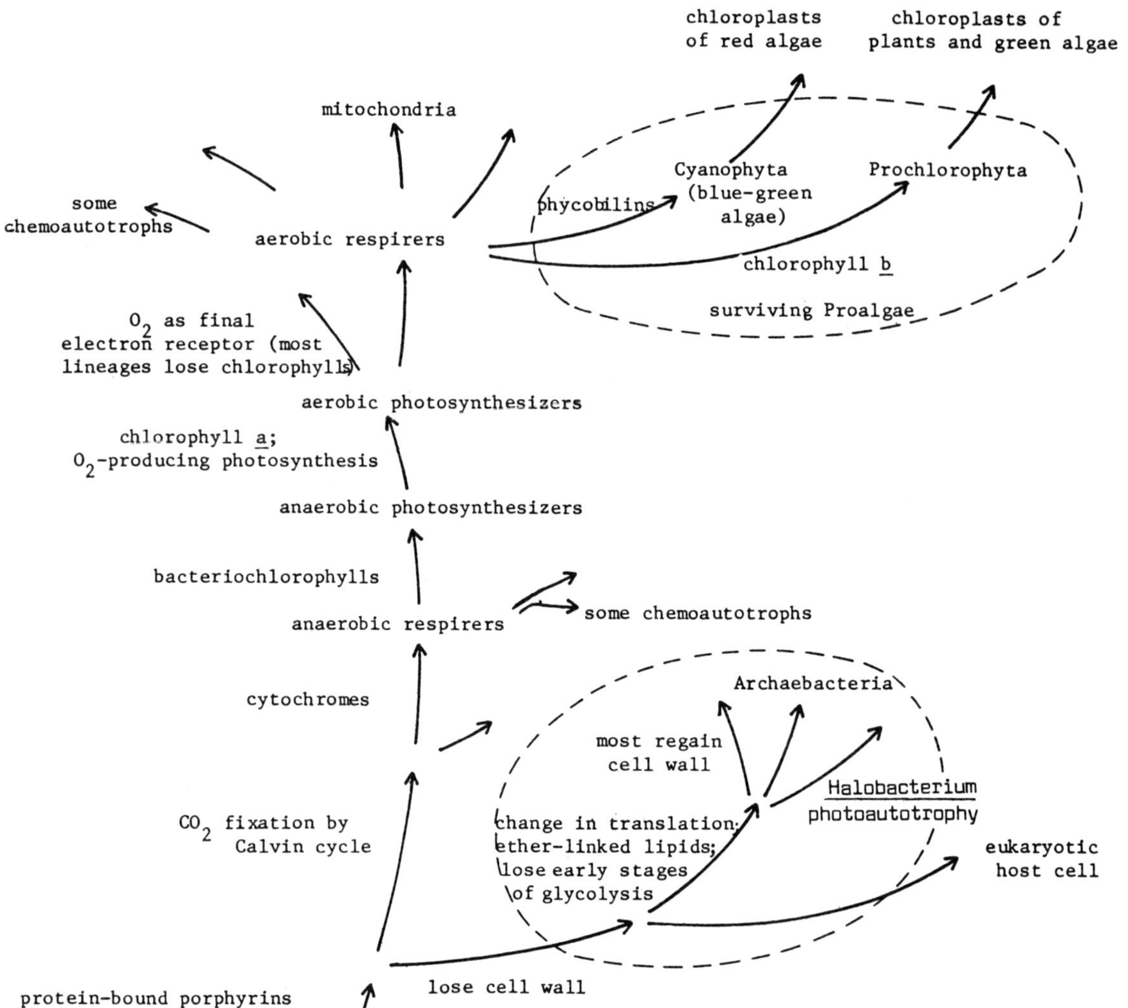

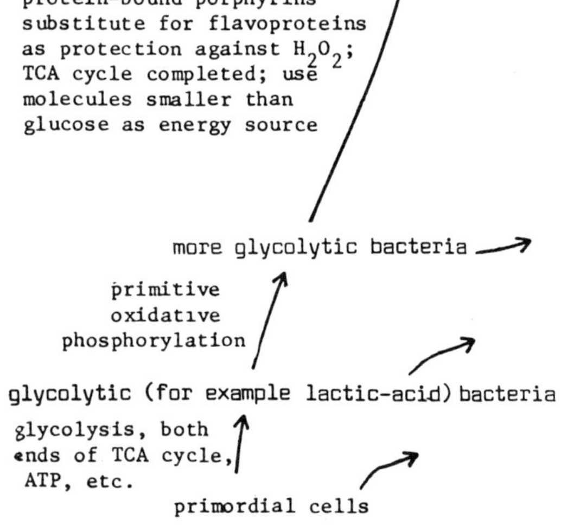

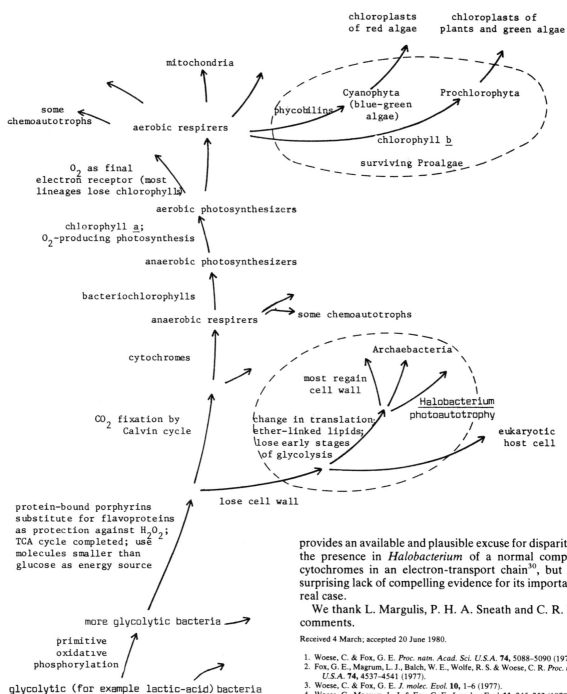

Fig. 1 Provisional phylogeny of prokaryotes.

the functioning of existing translation apparatuses, and of potential intermediates, in realistic conditions. Until then we prefer to emphasize other evidence; it is dangerous to claim something is impossible in biology. The same function can be performed very differently in different selective conditions[37].

Nevertheless, Fig. 1 is consistent with existing evidence from macromolecular sequences[38] if primitive aerobic photosynthesizers are now extinct. Reticulate evolution by gene transfer provides an available and plausible excuse for disparities such as the presence in *Halobacterium* of a normal complement of cytochromes in an electron-transport chain[30], but there is a surprising lack of compelling evidence for its importance in any real case.

We thank L. Margulis, P. H. A. Sneath and C. R. Woese for comments.

Received 4 March; accepted 20 June 1980.

1. Woese, C. & Fox, G. E. *Proc. natn. Acad. Sci. U.S.A.* **74,** 5088–5090 (1977).
2. Fox, G. E., Magrum, L. J., Balch, W. E., Wolfe, R. S. & Woese, C. R. *Proc. natn. Acad. Sci. U.S.A.* **74,** 4537–4541 (1977).
3. Woese, C. & Fox, G. E. *J. molec. Evol.* **10,** 1–6 (1977).
4. Woese, C., Magrum, L. J. & Fox, G. E. *J. molec. Evol.* **11,** 245–252 (1978).
5. Balch, W. E., Fox, G. E., Magrum, L. J., Woese, C. R. & Wolfe, R. S. *Microbiol. Rev.* **43,** 260–296 (1979).
6. Steitz, J. A. *Nature* **273,** 10 (1978).
7. Woese, C. R. & Fox, G. E. *Nature* **273,** 101 (1978).
8. Van Valen, L. M. *Evol. Theory* **4,** 103–121 (1979).
9. Van Valen, L. M. *Evolution* **22,** 37–41 (1968); *Bull. Am. Mus. nat. Hist.* **132,** 1–126 (1966).
10. Van Valen, L. M. *Evol. Theory* **4,** 45–80 (1978).
11. Simpson, G. G. *The Major Features of Evolution* (Columbia University Press, 1953).
12. Ramshaw, J. A. M. *et al. New Phytol.* **71,** 773–779 (1972).
13. Van Valen, L. M. *J. molec. Evol.* **3,** 89–101 (1974).
14. Goodman, M. in *Molecular Evolution* (ed. Ayala, F. J.) 141–159 (Sinauer, Sunderland, 1976).
15. Kimura, M. & Ohta, T. *Proc. natn. Acad. Sci. U.S.A.* **71,** 2848–2852 (1974).
16. Glover, C. V. C. & Gorovsky, M. A. *Proc. natn. Acad. Sci. U.S.A.* **76,** 585–589 (1979).
17. Mitchell, R. M., Loeblich, L. A., Klotz, L. C. & Loeblich, A. R. III *Science* **204,** 1082–1084 (1979).
18. Bayley, S. T. & Morton, R. A. *CRC Crit. Rev. Microbiol.* **6,** 151–205 (1978).
19. Schwemmler, W. *Cytobiologie* **3,** 427–429 (1971).
20. Margulis, L. *Origin of Eukaryotic Cells* (Yale University Press, 1970).
21. Hall, J. B. *J. theor. Biol.* **30,** 429–454 (1971).
22. Maiorana, V. C. & Van Valen, L. *Cells: Their Lives and Evolution* (Harper and Row, New York, in the press).

23. Hennig, W. *Phylogenetic Systematics* (University of Illinois Press, 1966).
24. Van Valen, L. M. *Evol. Theory* **3,** 285–299 (1978).
25. Kandler, O. & König, H. *Archs Microbiol.* **118,** 141–152 (1978).
26. Searcy, D. G., Stein, D. B. & Green, G. R. *BioSystems* **10,** 19–28 (1978).
27. Tornabene, T. G. & Langworthy, T. A. *Science* **203,** 51–53 (1979).
28. Tornabene, T. G., Langworthy, T. A., Holzer, G. & Oro, J. *J. molec. Evol.* **13,** 73–83 (1979).
29. Magrum, L. J., Luehrsen, K. R. & Woese, C. R. *J. molec. Evol.* **11,** 1–8 (1978).
30. Lanyi, J. K. *Microbiol. Rev.* **42,** 682–706 (1978).
31. Barile, M. F. & Razin, S. (eds) *The Mycoplasmas* Vol. 1 (Academic, New York, 1979).
32. Lewin, R. A. *Nature* **261,** 697–698 (1976).
33. Stanier, R. Y. & Cohen-Bazire, G. *A. Rev. Microbiol.* **31,** 225–274 (1977).
34. von Huene, F. R. in *Robert Broom Commemorative Volume* (ed. Du Toit, A. L.) 65–106 (Royal Society of South Africa, Cape Town, 1948).
35. Wang, C. C. *Comp. Biochem. Physiol.* B**61,** 571–579 (1978).
36. Bovee, E. C. & Jahn, T. L. in *The Biology of Amoeba* (ed. Jeon, K. W.) 37–82 (Academic, New York, 1973).
37. Ghilarov, M. S. *Zakonomernosti Prisposoblenii Chleneistonogikh k Zhizni na Sushe* (Izdatel'stvo "Nauka", Moscow, 1970).
38. Schwartz, R. M. & Dayhoff, M. O. *Science* **199,** 395–403 (1978).

PROBLEMS

1. Define the following terms: methanogen, numerical taxonomy, thermophile, hybridization competition, *Archaebacteria*, heterologous RNA:DNA hybrids, *Cyanobacteria*, oligonucleotide catalog, tRNA "common sequence", similarity coefficient, peptidoglycan, progenote, Gram-positive, urkaryote, Gram-negative.

2. Table 8-2, abstracted from Pace and Campbell (1971), lists the competitive abilities of several purified bulk RNA samples (largely consisting of ribosomal RNA) in hybridization experiments of *E. coli* rRNA with *E. coli* DNA. In each experiment, 8 micrograms of ^{3}H-labeled *E. coli* DNA and 3 micrograms of P-32-labeled *E. coli* rRNA were mixed with up to 108 micrograms of cold competitor RNA. The extent of reduction in hybridization of the *E. coli* nucleic acid was measured and taken as an index of homology between *E. coli* and the other species.

TABLE 8.2

Organism	*% homology*
Aerobacter aerogenes	77
Vibrio marinus	52
Alcaligenes faecalis	30
Spirillum itersonii	23
Bacillus stearothermophilus	17
Thermophile 194	12
Tetrahymena pyriformis	9

Can these data be used to construct a phylogenetic tree? What relations can be deduced?

3. The 16S rRNA molecule contains approximately 1600 nucleotides, and the enzyme T1 RNase always cuts after a G residue. Explain why only oligomers consisting of six or more nucleotides were selected for the construction of phylogenetically meaningful similarity coefficients.

4. Draw the shape of the plot of ribonucleotide catalog similarity coefficients versus time of divergence if (1) every nucleotide is equally replaceable and (2) if there is a conserved set of sequences, as argued by Woese *et al.* (1975). Do you have to make any additional assumptions? Explain.

5. The filamentous cyanobacteria (blue-green algae) are commonly divided into two orders, *Nostocales* and *Stigonematales*, whereas the unicellular forms are traditionally grouped in a single order, the *Chroococcales*. The dendrogram in Figure 8-5 was constructed by Bonen and Doolittle (1978) from an analysis of 16S rRNA oligoribonucleotide catalogs of two filamentous forms (*Nostoc* and *Fischerella*, members of the *Nostocales* and *Stigonematales*, respectively) and several unicellular cyanobacterial strains. Similarity coefficients were calculated as described in Fox *et al.* (1977, this chapter). What information does this dendrogram contain about the traditional classification of the blue-green algae on the basis of morphology? What precautions are suggested in interpreting the fossil record for this group of prokaryotes?

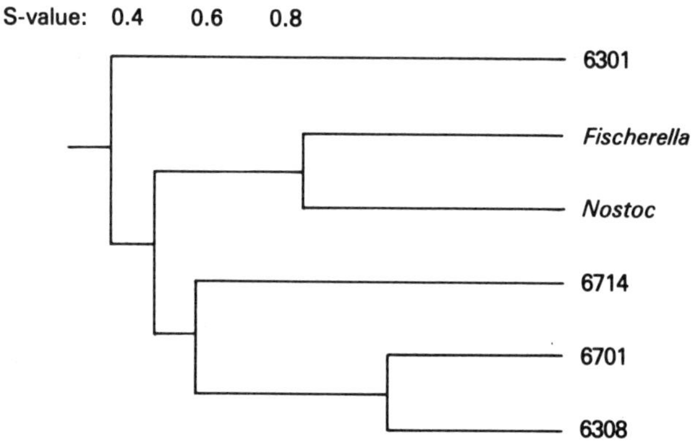

Figure 8-5

6. If "lateral" or horizontal" (plasmid-mediated) transfer of prokaryotic 16S rRNA genes between distantly related groups were discovered, how would this observation influence the phylogenetic scheme elaborated by Woese and Fox (1977a, this chapter)? Is this problem likely to affect other sequences?

7. It is usually assumed that the prokaryotic genome, typified by that of *E. coli*, is primitive and that of eukaryotes is "advanced". Doolittle (1978) has questioned this assumption and has suggested that the *E. coli* genome is probably an evolved, streamlined genome, whose efficiency derives in part from the elimination of introns from the genome of an ancestral form. Some support does come

from the idea of Carlile (1982) that prokaryotes are specialized for "r selection" (short cell cycles in which possession of presumably "unnecessary" DNA is disadvantageous), whereas eukaryotes are more specialized for "K selection". Might the investigation of gene structure in presumptive archaebacterial strains clarify this matter? If introns are found in archaebacterial genes, how would it affect the debate over the three-kingdom hypothesis? Would such a discovery prove decisive?

8. You have been given a thermophilic bacterium isolated from one of the hot springs of Yellowstone National Park. You have all the requisite tools for the analysis of macromolecules at hand (and the expertise) and a set of cultures of different reference prokaryotic strains covering a wide phylogenetic spectrum, but no microscope, and no Bergey's Manual. What strategy would you employ for classifying the unknown strain?

9. The bacterial genus *Pseudomonas*, which contains hundreds of strains, has been subjected to intensive phenotypic and genotypic characterization. From various phenotypic characteristics the species *Pseudomonas fluorescens* is conventionally divided into strain clusters, or biotypes. The trees depicted in Figure 8-9 have been constructed for a group of seven strains in biotype C by Champion *et al.* (1980) from DNA homologies, cytochrome c sequences, and relatedness of the electron-transport protein azurin, as measured by the microcomplement fixation procedure.

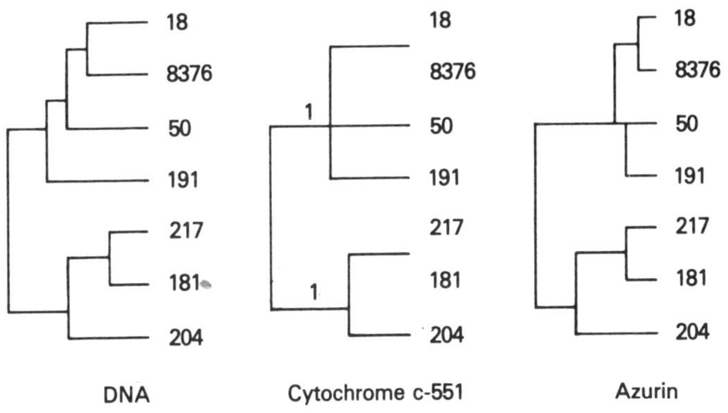

Figure 8-9

(a) Which tree contains the most information and which contains the least? Can you suggest possible reasons?
(b) What do the three trees reveal about the overall rates of gene change throughout the genome, and the role, if any, of horizontal transfer of genes between the members of this biotype?

10. Until recently it had been assumed that the genetic code is universal among all contemporary life forms. The discovery of coding differences between mitochondria and the standard "universal" pattern has led to a reevaluation of this point (see previous chapter). In the light of the material presented in this chapter, can you suggest how possible extensions of studies on the code to the *Archaebacteria* might be fruitful in evaluating the respective strengths of the three-kingdom hypothesis and the proposal of Van Valen and Maiorana?

9 ORIGIN OF LIFE

In our own minds we have followed a long path from the incandescent atoms of carbon of the earliest nebula to the living things of our times. We have seen how it is possible to explain the origin of life while basing our ideas all the time on scientifically established facts.
A.I. Oparin (1924)

To explain the origin of life is perhaps the most difficult challenge to the imagination of the evolutionary biologist or biochemist, because of the apparent historical (archival) and experimental inaccessibility of these early events. Jacob (1977) suggested that perhaps the best we might hope for is to perceive the historical constraints--largely from the laws of physics and chemistry--and to identify a range of possible steps towards current life forms. In this view, the actual series of steps taken would correspond to the "frozen accidents" described by Crick (1968, Chapter 1) and there would be little chance of reproducing early events in the history of life. However, recent work gives confidence that eventually the origin of life can be understood as a predictable outcome of a specifiable sequence of events.

The study of the origin of life has been approached in several ways. One theory, a statistical theory, assumes that life arose more or less all at once by the chance combination of atoms into a self-reproducing system (see Hoyle and Wickramasinghe, 1981). This idea is rejected by most modern workers because it predicts an impossibly low probability for the origin of life and because it contradicts our understanding of complex events occurring through a progressive series of contingent events. The other approach assumes many intermediate forms, each of which follows the earlier stages according to known physical and chemical principles. The distinction between the two approaches is important and an analogy may be helpful. If a supersaturated copper sulfate solution is left standing, crystals will appear. You could calculate the probability of the crystals forming all at once by all of the atoms just happening to be simultaneously in the right place. The calculation would show that crystals would not form in this manner. Instead, a small nucleus forms, then more atoms add to it, and then more again; innumerable intermediate steps occur, each consistent with thermodynamic principles. The rest of this chapter is devoted to a discussion of the possible intermediates in the origin of life.

The search for intermediate states can proceed by examining conditions required to generate larger molecules from smaller ones (a chemical approach) or by simplifying the contemporary systems in an effort to define the requirements and properties of minimal living systems (a biological approach). Both approaches will be examined, beginning with a review of attempts to mimic experimentally prebiotic synthesis of biologically important monomers and polymers.

The first scientific discussions of early life centered on an analysis of prebiotic chemical conditions and an attempt to form chemical systems exhibiting some features of life. In 1922 the Russian biochemist A. I. Oparin gave a lecture (revised and expanded in Oparin (1938)), to the Botanical Society in Moscow in which he developed a testable theory involving a gradually increasing complexity of organic molecules and their aggregates to generate simple living forms (Keosian, 1965). Haldane (1929) developed similar ideas and stressed the importance of an atmosphere lacking significant amounts of free oxygen. Oparin (1938) was able to form simple vesicles consisting of various polymers (these vesicles were called coacervates). However, the generation of amino acids and nucleotides, the building blocks of proteins and nucleic acids, was a failure, primarily because the experimental systems lacked sufficient nitrogen, a major component of both amino acids and nucleotides.

The first successful synthesis of prebiotic molecules were the experiments of Miller (1953)--working with Urey--who showed that important biological monomers could be generated in great profusion and variety under conditions generally thought to be possible on the prebiological earth. Since these experiments were done, other workers have examined many aspects of the formation of both amino acids and nucleotides and the formation of chemically activated forms of these monomers. This work has been reviewed extensively by Miller and Orgel (1974) and briefly by Dickerson (1978) and Schuster (1981). The mechanisms of the chemical reactions are now understood--for example, adenine is synthesized by condensing five molecules of hydrogen cyanide (Miller and Orgel, 1974). The significance of these experiments has been reinforced by the finding that many of these simple organic molecules occur in space (Turner, 1980) and also that they are stable (apart from ultraviolet photodestruction) once they are formed. For example, it is estimated that the half-life for decarboxylation of alanine (in the absence of oxygen and enzymes) is about three billion years at 25°C (Miller and Orgel, 1974). The conclusion of most of the chemical experiments is that under the right circumstances, the building blocks for proteins and nucleic acids will form strictly in accordance with chemical and physical principles. However, what is lacking is certainty about the conditions that prevailed early in the history of the earth (4.6 to 3.7 billion years ago); in fact, current ideas suggest that the atmosphere was not as reducing as previously thought, though free oxygen was at a very low concentration (Levine, 1982).

The polymerization of amino acids and nucleotides to give polypeptides and nucleic acids respectively has also been studied. Several approaches have been taken to resolve the kinetic and thermodynamic difficulties inherent in polymerization reactions in an aqueous medium. Living organisms use high-energy phosphorylated intermediates to ensure that the reaction goes towards synthesis; thus, possible prebiotic high-energy derivatives of amino acids and nucleotides have been studied (Lohrmann and Orgel, 1973). Polymerization has been demonstrated to result from heating dry mixtures of amino acids at 150-200°C (Fox and Dose, 1972); similar results were obtained at 65<C if the reaction mixture was left for three

months (Rohlfing, 1976).

A new approach examines a fluctuating clay environment--"drying cycles" in which clay and monomers undergo a regular cycle of wetting, drying, and heating (as may occur on a beach). It is assumed that as the water evaporates, the monomers absorb onto the clay surface and their polymerization is catalyzed (Lahav *et al.*, 1978). Such catalysis on a structurally regular surface provides a mechanism for the selection of stereoisomers, and this could lead to the well-known and poorly understood chirality of amino acids in proteins. Recent papers on enzyme-free "instructed" polymerization include work by van Roode and Orgel (1980) on oligonucleotides and by White and Erickson (1980) on oligopeptides.

The potential enzymatic activity of oligonucleotides (Brewin, 1972) and of the beta-peptide conformations, which have been proposed as a very early coded product (Brack and Orgel, 1975), has encouraged speculation about the nature of the earliest precellular predecessors of living systems. Indeed, enzymatic activity has been found associated with products of the thermal polymerization of amino acids (see Fox and Dose, 1972) and a catalysis of the formation of peptide bonds by histidinylhistidine has been demonstrated (White and Erickson, 1980). A program that allows the rapid screening for enzymic activity in small peptides is outlined in Ninio (1982). A later stage in this precellular evolution is the aggregation (or phase separation) of polymers, some with at least weak enzymatic activity, into "protocells" or coacervates that show the rudiments of metabolic activity (Fox, 1980). These proposed steps occur according to known physical and chemical principles and so meet the criteria for an acceptable scheme of early evolution.

A complementary approach to reconstruction of possible early events is to start with a contemporary duplicating system and extrapolate it back to its minimum elements in order to learn about the properties of a primitive duplicating system. It is not yet certain whether the earliest living system was protein, nucleic acid, or both, though it seems more likely that very early life was based on RNA and protein, without DNA (Reanney, 1979; Eigen *et al.*, 1981). The basic argument is that RNA is intimately involved in all principal aspects of biological information handling--priming of DNA synthesis, storage and transfer of information (mRNA), and translation (tRNA and rRNA)--whereas DNA is utilized only for information storage. Normally, RNA is copied by a single multisubunit polymerase enzyme, whereas the double-stranded DNA helix requires a much larger complex of enzymes to separate the strands, copy the strands, rejoin the fragments, and even to synthesize an RNA primer. Indeed, nonenzymatic template-directed RNA synthesis has been demonstrated (Schuster, 1981). Finally, the deoxyribonucleotides in contemporary organisms are all synthesized from ribose intermediates, an important indication of a later evolution of the deoxy form.

"Evolution" of RNA has been shown in the work of Mills *et al.* (1967), who found that the RNA polymerase of *E. coli* Q-beta phage could catalyze *in vitro* the replication of a substantially reduced Q-beta RNA. Orgel (1979, this chapter) has described the ways in which this system exhibits the primary elements of Darwinian evolution: reproduction, and mutation coupled to selection. This polymerase can even catalyse the *de novo* synthesis of polyribonucleotides from mixtures of mononucleoside triphosphates (Sumper and Luce, 1975; Biebricher *et al.*, 1981). Under specific experimental conditions identical reaction mixtures eventually generate stable populations of molecules that are similar to one another in length but possess small variations in base sequence. The nature of the final product varies with environmental conditions. The significance of the experiments lies in the demonstration that variation can arise in such a simple system and that the variants contribute differentially to the progeny molecule pool. These experiments also show that a population of nucleic acid molecules reproducing under error-prone conditions might be more usefully regarded as a quasi-species cluster rather than as a collection of absolutely identical sequences. In such a population, while replication errors lead to degeneration in the homogeneity of the sequences, this is countered by their differential reproduction, such that the most common sequence, the master sequence, characterizes a population of progeny molecules. If the frequency of appearance of "errors" exceeds the homogenizing capacity of selection, an error catastrophe occurs and the population loses its identity, unless an improved sequence arises as a new quasispecies.

Perhaps the most difficult questions concern the origin of a complete genetic system able to specify its own replication; such a system would require its own primitive code, but the origin of the code raises a special problem. Is the code an accidental one or an optimal one? Crick (1968, Chapter 1) and Orgel (1968) were among the first to combine the best guesses about the primitive earth environment with knowledge of contemporary cellular mechanisms to present plausible speculations about early steps in the development of the universal code. Since that time, there has been no shortage of attempts at reconstructing possible steps in the development of the current coding scheme. Examples of considerations that have been regarded as important are a possible stereochemical fit between copolymerizing amino acids and nucleotides (Nelsestuen, 1978) or discerned regularities in the code table (reflecting relative frequencies of amino acids and nucleotides), together with a requirement for diversity of peptide products (Ishigami and Nagano, 1975). The discovery of departures from the universal code, particularly in mitochondria, provided the evidence to allow Jukes (1981; 1983, this chapter) to specify a sequence of steps leading from a primitive archetypal scheme encoding 15 amino acids with 15 anticodon sequences to our contemporary scheme encompassing 20 amino acids and at least 46 anticodons.

Speculations on the origin of the code must be coupled to a model for primitive ribosomeless RNA-directed peptide synthesis. Crick *et al.* (1976) suggested that the first codons were of the form 5'RRY3' (R = purine, Y = pyrimidine) and that the corresponding anticodon sequence was YYR bracketed by 3'UG....UU5'. Such a model defines a reading frame (thus providing a comma-free code), allows a transient five-base interaction between tRNA and mRNA, as required in the absence of the stabilizing effect of a ribosome, and could explain the regular appearance of related anticodon bracketing sequences in modern tRNAs. These general features have been incorporated into a more specific model of the primitive RNA (both messenger and adaptor) presented by Eigen and Schuster (1978b, this chapter) in which the repeating RRY motif is replaced by RNY (N = purine or pyrimidine). This model has the additional advantages of allowing stabilization by the formation of double-stranded regions, as well as providing anticodon sequences.

A complete sequence of demonstrable steps that led from polymerized material to a small collection of molecules with the properties of replication, mutation, and susceptibility to natural selection, is not yet available. At present, the most complete study is a series of three theoretical papers by Eigen and Schuster (1977; 1978a; 1978b, this chapter), in which the concept of a hypercycle is developed, whereby replication of two or more RNA molecules are coupled through their primitively coded peptide products. It is hypothesized that these RNA molecules are at the same time both messenger and adaptor. Perhaps the greatest virtue of this model is that by appropriate mutational variation the number of hypercyclically coupled molecules can be increased, and hence the complexity of the system can also increase without exceeding the error catastrophe threshold imposed by the fidelity of the very primitive replicating system. The first two papers of the series are a mathematical examination of the size and complexity limits on a self-replicating system, in terms of maximum error rate and selection intensity parameters. The last paper of the series, which is reprinted here, presents a model based on the quantitative parameters identified in the previous two papers; the model reflects both the conditions believed to have existed in the primitive environment and features of contemporary cellular systems that are thought to be relics of the primitive condition. In contrast to the models in which microspheres are suggested to be the earliest form of organization (Fox, 1980), in the hypercycle model, compartmentalization is believed to come later (Schuster, 1981), though this order is not demanded by either logic or experimental data. An important point in the paper is the parallelism between a stable multi-member hypercycle, in which a faulty element may be rejected without destroying the system, and the biological structural principle of using a large number of small identical subunits rather than one big polypeptide chain (Caspar and Klug, 1962). Maynard-Smith (1979), Schuster (1981), and Eigen *et al.* (1981) have provided summaries of the hypercycle model set in a prebiotic evolutionary context. King (1977), White (1980) and White and Raab (1982) have also proposed semiquantitative pathway, for traversing the difficult terrain between nonlife and life, which incorporate certain elements of the hypercycle model.

All of these models consider polymers that are much shorter than nearly all existing genes and proteins--for example, peptides no more than thirty amino acids long (Eigen and WWinkler-Oswatitsch, 1981b) and oligomers containing no more than ten amino acid or nucleotide residues (White, 1980). This assumption is not an unreasonable one because in eukaryotes, domains of proteins are encoded in relatively short exons (see Chapters 4 and 5). Even if small peptides were initially adequate catalysts, a large protein is more efficient because of increased specificity, flexibility of conformation, exclusion of water from the reaction site, and the possibility of regulatory sites on individual protein chains or on multisubunit enzymes. Hence, strong selection toward larger molecules is to be expected (Koshland, 1976).

In previous chapters, we have noted the usefulness of nucleic acids and proteins as documents of gene history, but are there any such records of the earliest genes? The virtually ubiquitous ferredoxin molecule studied by Eck and Dayhoff (1966a, Chapter 1) may be one such example; it reveals an internally repetitive structure that includes the amino acids believed to have been most abundant in prebiological times. The contemporary genetic code table reveals that the two amino acids that are formed most abundantly in a Miller experiment (glycine and alanine) utilize codons rich in guanine and cytosine, bases that, for reasons of stability of the nucleotide pair, are presumed to have predominated in primitive oligonucleotides (Eigen and Schuster, 1978b, this chapter).

The tRNAs are a highly conserved class of molecules (Holmquist *et al.*, 1973) and have therefore been extensively examined for evidence of primitive structure (Cedergren *et al.*, 1981). Comparison of the sequences of the same tRNA from different species indicates that the molecules are related to each other by the expected phylogenetic tree. However, comparison of all tRNA molecules from a single species suggest that they are best described as descendants from a single "quasi-species" family whose master sequences perhaps represented a primitive RNA molecule (Eigen and Winkler-Oswatitsch, 1981a). Of specific interest is the sequence of the methionyl tRNA of the archaebacterium *Thermoplasma* (Searcy, 1982); this is remarkably similar to the sequence that Eigen and Winkler-Oswatitsch (1981a) suggested as the original master sequence for tRNA. Finally, a striking three-base periodicity has been discerned in a wide variety of both prokaryotic and eukaryotic coding sequences (Brown, 1980; Shepherd, 1981); this periodicity has been interpreted as the vestige of the primitive comma-free code in which G and C occupied the first and last positions of the codon respectively (Eigen and Winkler-Oswatitsch, 1981b). However, an alternative explanation, couched in terms of contemporary functional requirements, may be found.

In this discussion and in the selected reprinted papers, we have focused on that earliest stage in the origin of life in which order with evolutionary potential first arose. We have assumed an adequate supply of chemical intermediates for these early steps, and have ignored the subject of the later development of metabolic strategies needed to transform the energy required to increase and maintain local order. The subject is of some interest since it is likely that the utilization of new energy sources provides an important pressure for sequence alterations. With growing numbers of protein sequences known and increasing sophistication of techniques for sequence comparison, it may soon be possible to reconstruct a metabolic history inferred from the protein sequences, with hints from contemporary metabolic pathways. Similarly, the possibility that nucleotide cofactors (White, 1976) or physiological signals (Tomkins, 1975, Chapter 6) may be vestiges of primitive metabolic systems may contribute to that reconstruction. For further information, the reader is referred to Broda (1975) for a discussion of energy metabolism and evolution, and to Hartman (1975), Clarke and Elsden (1980), Baldwin and Krebs (1981), and Krebs (1981) for discussions of the evolution of metabolic pathways.

In retrospect, the statement of Oparin that opened this chapter may have seemed unduly optimistic when viewed in the context of the state of knowledge at that time. However, a wealth of specific experimental data now available make questions concerning the origin of life appear accessible.

Reprinted from *Naturwissenschaften* **65**: 341–369.

The Hypercycle

A Principle of Natural Self-Organization

Part C: The Realistic Hypercycle

Manfred Eigen
Max-Planck-Institut für biophysikalische Chemie, D-3400 Göttingen

Peter Schuster
Institut für theoretische Chemie und Strahlenchemie der Universität, A-1090 Wien

The proposed model for a 'realistic hypercycle' is closely associated with the molecular organization of a primitive replication and translation apparatus. Hypercyclic organization offers selective stabilization and evolutive adaptation for all geno- and phenotypic constituents of the functionally linked ensemble. It originates in a molecular quasi-species and evolves by way of mutation and gene duplication to greater complexity. Its early structure appears to be reflected in: the assignment of codons to amino acids, in sequence homologies of tRNAs, in dual enzymic functions of replication and translation, and in the structural and functional organization of the genome of the prokaryotic cell.

XI. How to Start Translation?

"The origin of protein synthesis is a notoriously difficult problem. We do not mean by this the formation of random polypeptides but the origin of the synthesis of polypeptides directed, however crudely, by a nucleic acid template and of such a nature that it could evolve by steps into the present genetic code, the expression of which now requires the elaborate machinery of activating enzymes, transfer RNAs, ribosomes, factors, etc."

Our subject could not be characterized more aptly than by these introductory phrases, quoted from a recent paper by F.H.C. Crick, S. Brenner, A. Klug and G. Pieczenik [3].

Let us for the time being assume that a crude replication and translation machinery, functioning with adequate precision, and adapted to a sufficiently rich alphabet of molecular symbols, has come into existence by some process not further specified, e.g., by self-organization or creation, in Nature or in the laboratory. Let us further suppose an environment which supplies all the activated, energy-rich material required for the synthesis of macromolecules such as nucleic acids and proteins, allowing both reproduction and translation to be spontaneous processes, i.e., driven by positive affinities. Would such an ensemble, however it came into existence, continue to evolve as a Darwinian system? In other words, would the system preserve indefinitely the information which it was given initially and improve it further until it reaches maximal functional efficiency?

In order to apply this question to a more concrete situation let us consider the model depicted in Figure 45. The plus strands of a given set of RNA molecules contain the information for a corresponding number of protein molecules. The products of translation can fulfill at least the following functions: (1) One protein acts as an RNA-polymerase similar to the specific replicases associated with various RNA phages. Its recognition site is adapted to a specific sequence or structure occurring in all plus and minus strands of the RNAs; in other words, it reproduces efficiently only those RNA molecules which identify themselves as members of the particular ensemble. (2) The other translation products function as activating enzymes, which assign and link various amino acids uniquely to their respective RNA adaptors, each of which carries a defined anticodon. The number of different

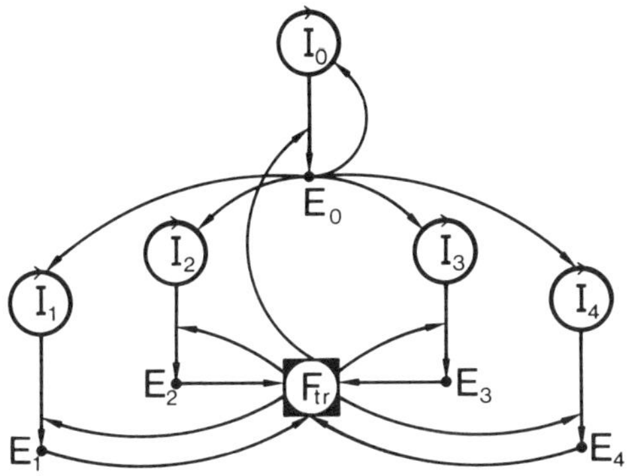

Fig. 45. A minimum model of primitive translation involves a messenger I_0 encoding a replicase E_0, which is adapted to recognize specifically the sequences I_0 to I_4. The plus strands of I_1 to I_4 encode four synthetase functions E_1 to E_4, while the minusstrands may represent the adapters (tRNAs) for four amino acids. Such a system, although it includes all functions required for translation and self-reproduction, is unstable due to internal competition. Coherent evolution is not possible, unless I_0 to I_4 are stabilized by a hypercyclic link

amino acids and hence of adaptors is adjusted to match the variety of codons appearing in the messenger sequences, i.e., the plus strands of the RNAs, so as to yield a 'closed' translation system with a defined code. It does not necessarily comprise the complete genetic code, as it is known today, but rather may be confined to a – functionally sufficient – smaller number of amino acids (e.g., four), utilizing certain constraints on the codon structure in order to guarantee an unambiguous read-off. The adaptors may be represented by the minus strands of the RNA constituents, or, if this should be too restrictive a condition, they could be provided along with further machinery, such as ribosomes, in the form of constant environmental factors similar to the host factors assisting phage replication and translation inside the bacterial cell.

At first glance, we might find comfort in the thought that the system depicted in Figure 45 appears to be highly functionally interwoven; all I_i are supported catalytically by the replicase E_0, which in turn owes its existence to the joint function F_t of the translation enzymes E_1 to E_4 without which it could not be translated from I_0. The enzymes E_1 to E_4, of course, utilize this translation function for their own production too, but being the translation products of I_1 to I_4, they are finally dependent also upon E_0 or I_0 respectively. However, a detailed analysis shows that the couplings present are not sufficient to guarantee a mutual stabilization of the different genotypic constituents I_i. The general replicase function exerted by E_0 and the general translation function F_{tr} are represented in all differential equations by the same term. The equations then reduce to those for uncoupled competitors, multiplied by a common time function $f(t)$. The system, which initially functions quite well, is predestined to deteriorate, owing to internal competition. A typical set of solution curves, obtained by numerical integration of the rate equations, is shown in Figure 46.

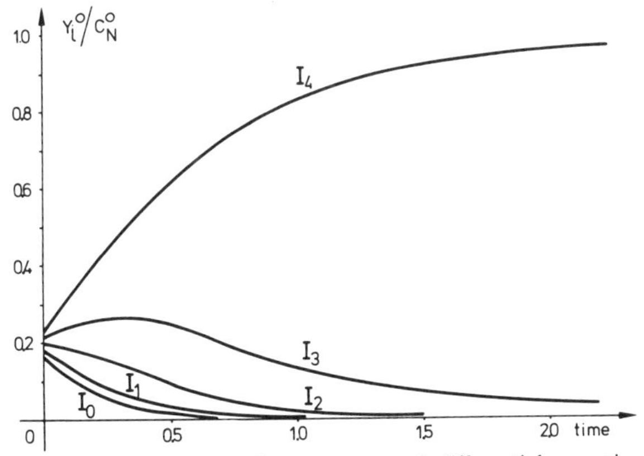

Fig. 46. Solution curves for a system of differential equations simulating the model represented in Figure 45. In this particular example, it is assumed that initial concentrations and autocatalytic-reproduction-rate constants increase linearly from I_0 to I_4, while the other parameters – such as translation-rate constants ($I_i \xrightarrow{k_i} E_i$), amino acid assignments (contribution of E_1, E_2, E_3 E_4 to F_{tr}) or enzyme-substrate-complex stabilities ($I_i + E_0 \overset{K_i}{\rightleftharpoons} I_i \cdot E_0$), etc. – are identical for all reaction partners. The time course of the relative population numbers (y_i^0/c_N^0) reflects the competitive behavior. The most efficiently growing template (I_4) will supersede all others and finally dominate ($y_4^0/c_N^0 \rightarrow 1$). However, since both replication (represented by E_0) and translation function (contributions of E_1, E_2 and E_3 to F_{tr}) disappear, I_4 will also die out. The total population is bound to deteriorate ($c_N^0 \rightarrow 0$)

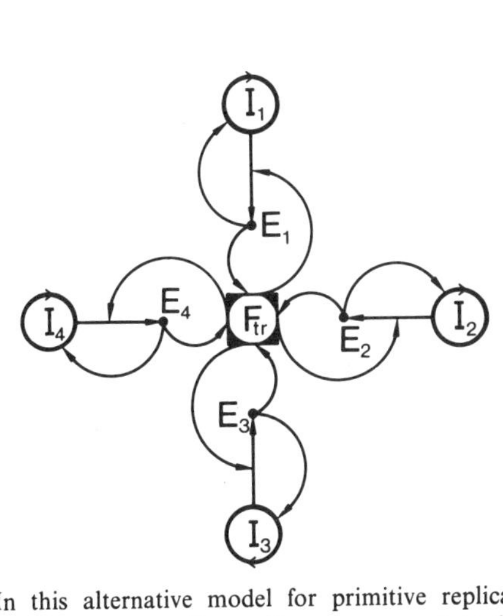

Fig. 47. In this alternative model for primitive replication and translation, the enzymes E_1 to E_4 are assumed to have dual functions, i.e., as specific replicases of their own messengers and as synthetases for four amino acid assignments. The fate of the system is the same as that of the system depicted in Figure 45, since the messengers are highly competitive

Another example of this kind is represented in Figure 47. Here all messengers produce their own specific replicases E_1 to E_4, which also provide synthetase functions (F_{tr}). Again, this coupling by means of a concomitant translation function does not suffice to stabilize the ensemble. The answer to our question, whether the mere presence of a system of messengers for replicase and translation functions and of translation products is sufficient for its continuous existence and evolution, is that unless a particular kind of coupling among the different replicative constituents I_i is introduced, such systems are not stable, despite the fact that they contain all required properties for replication and translation. Even if all partners were selectively equivalent (or nearly equivalent) and hence were to coexist for some time (depending on their population size), they could not evolve in a mutually controlled fashion and hence would never be able to optimize their functional interaction. Their final fate would always be deterioration, since an occasional selective equivalence cannot be coherently maintained over longer periods of evolution unless it is reinforced by particular couplings.

Knowing the results of part B, we are, of course, not surprised by this answer. A closer inspection of the particular linkages provided by the functions of replication and translation enzymes does not reveal any hypercyclic nature. Therefore these links cannot establish the mutual control of population numbers that is required for the interrelated evolution of members of an organized system. The couplings present in the two systems studied can be reduced to two common functions, which, like environmental factors, influence all partners in exactly the same way and hence do not offer any possibility of mutual control.

The above examples are typical of what we intend to demonstrate in this article, namely, that

1. In the early phases of evolution, characterized by low fidelities of replication and translation as well as by the initially low abundance of efficiently replicating units, hypercyclic organization offers large relative advantages over any other kind of (structural) organization (Sect. XV), and

2. That hypercyclic models *can* indeed be built to provide realistic precursors of the reproduction and translation apparatus found in present prokaryotic cells (Sect. XVI).

How could we envisage an origin of translation, given the possible existence of reproducible RNA molecules as large as tRNA and the prerequisites for the synthesis of proteins in a primitive form, utilizing a limited number of (sufficiently commonly occurring) amino acids?

XII. The Logic of Primordial Coding

XII. 1. The RRY Code

A most appealing speculative model for the origin of template-directed protein synthesis, recently proposed [3], is based on a number of logical inferences that are related to the problem of comma-free and coherent read-off. A primordial code must have a certain frame structure, otherwise a message cannot be read off consistently. Occasional phase slips would produce a frame-shifted translation of parts of the message and thereby destroy its meaning. The authors therefore propose a particular base sequence to which all codons have to adhere. Or, in other words, only those sequences of nucleotides that resemble the particular pattern could become eligible for messenger function. Uniformity of pattern could arise through instruction conferred by the exposed anticodon loop of tRNAs as well as by internal self-copying. Among the possible patterns that guarantee nonoverlapping read-off, the authors chose the base sequence purine-purine-pyrimidine, or, in the usual notation, RRY, to be common to all codons specifying a message. The particular choice was biased by a sequence regularity found in the anticodon loop of present tRNAs, which reads 3′NR$\alpha\beta\gamma$UY, $\alpha\beta\gamma$ being the anticodon, N any of the four nucleotides, and R and Y a purine and a pyrimidine, respectively. Another prerequisite of ribosome-free translation is the stability of the complex formed by the messenger and the growing polypeptide chain. A peptidyl-t-RNA must not fall off before the transfer to the subsequent aminoacyl-tRNA has been accomplished, that is, until the complete message is translated. Otherwise, only functionally inefficient protein fragments would be obtained. It is obvious from known base-pair stabilities that a simple codon-anticodon interaction does not guarantee the required stability of the messenger-tRNA complex. Therefore the model was based essentially on three auxiliary assumptions.

1. The structure of the anticodon loop of the adaptor (tRNA precursor) is such, that – given the particular and common codon pattern – an RNA can always form five base pairs with the messenger. The primitive tRNA is then assigned the general anticodon-loop sequence

3′ ∿∿∿∿∿ U-G-[Y]-[Y]-[R]-U-U ∿∿∿∿∿ 5′

where YYR is the anticodon.

2. The anticodon loop of each primitive tRNA can assume two different conformations, which are detailed in Figure 48. Both configurations had been described in an earlier paper by C. Woese [60] who

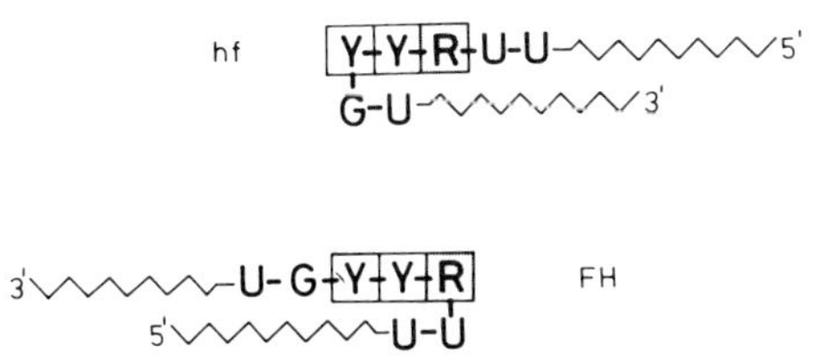

Fig. 48. Two possible configurations of the anticodon loop of tRNAs: FH according to Fuller and Hodgson [61] and hf according to Woese [60]. The anticodon pattern (framed) refers to the model of Crick et al. [3]

named them FH and hf. (FH refers to Fuller and Hodgson [61] who originally proposed that five bases at the 3′-end of the unpaired seven-base sequence in the anticodon loop are stacked on top of each other, while hf, according to Woese, designates a complementary configuration keeping the five bases at the 5′-end of the loop in a stacked position.) Woese assumed that the transition between both configurations plays an important role in ribosomal protein synthesis, but also referred to its possible significance in past, more primitive mechanisms.

3. Which of the two configurations is actually present depends on whether the tRNA is attached to an amino acid or to a peptide chain. By transferring an amino acid to the peptide chain, the tRNA flips from the hf to the FH configuration (cf. Fig. 49).

An additional, fourth postulate, not absolutely necessary as a prerequisite of the model, invokes an interaction between two adjacent tRNAs at the messenger, which assures that the required configration will be energetically the most favorable and contributes further to the stability of the polypeptide-messenger complex.

Figure 49 shows in more detail how polypeptide synthesis may be facilitated on the basis of the arguments given. The growing polypeptide chain is transported along the messenger, utilizing as 'fuel' the free energy of the transfer reaction. This reaction may be aided by a general nonspecific catalyst, but the mechanism does not require any sophisticated machinery, such as is nowadays provided by the ribosomes. Although interaction between codon and anticodon is stabilized by five base pairs, it is essential that the actual code make use of base triplets. It had been emphasized previously [62] that a primitive code utilizing anything but three bases is useless in explaining the present code.

The model explains well how in the absence of a sophisticated translation apparatus a message can be read off

spontaneously,
sequentially,
completely, i.e., unfragmented, and

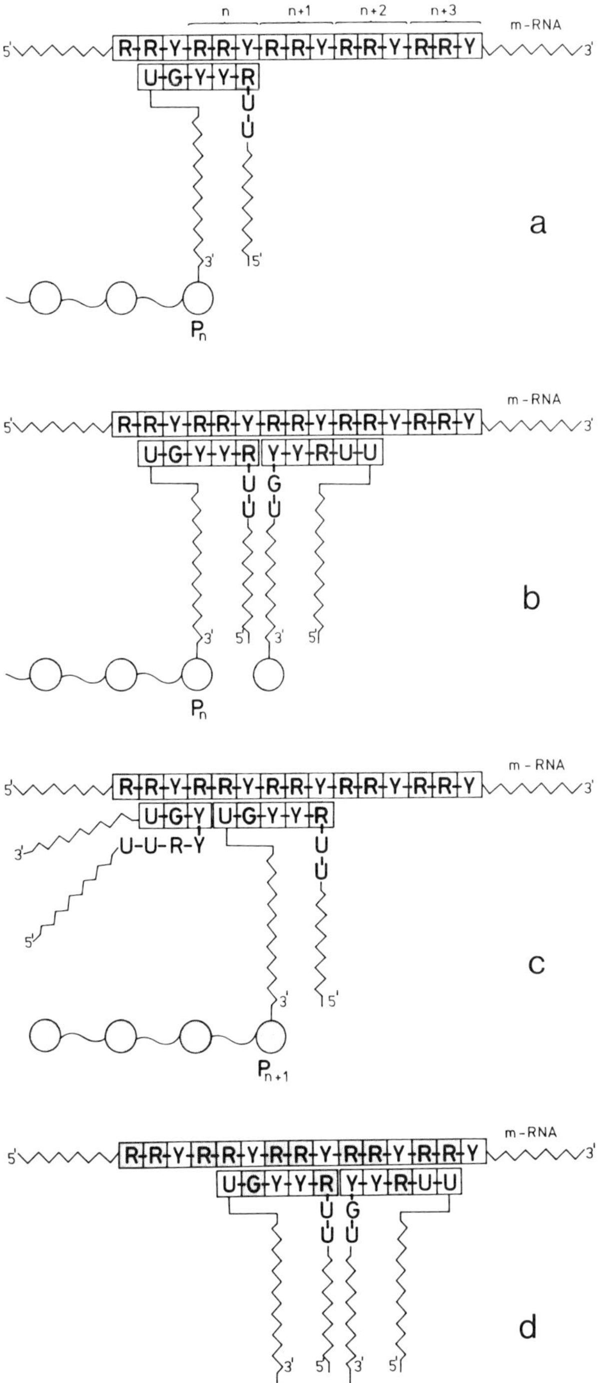

Fig. 49. The primitive translation mechanism requires 'sticky' interactions between the messenger and the peptidyl-tRNA. It thereby allows the growing peptide chain to remain in contact with the message until translation is completed. According to Crick et al. [3] the transport is effected by a flip mechanism involving conformational changes of the tRNA (FH⇌hf). The nascent peptide chain is always connected with the messengers via five base pairs with some additional stabilization by the adjacent aminoacyl-tRNA. The partial overlap of base pairing guarantees a consistent reading of a message encoded in base triplets

reproducibly, i.e., strictly maintaining a given codon frame.

The code is inherently related to certain structural features of the anticodon loop of present tRNAs, suggesting that these molecules are the descendents of the first functionally organized entities. The four amino acids assigned by this model are:

GG^{U}_{C} ;	GA^{U}_{C} ;	AG^{U}_{C} ;	AA^{U}_{C} ;
glycine	aspartic acid	serine	asparagine

Some of which were very abundant in the primitive soup [63].

On the other hand, the model also introduces some difficulties. A uniform RRY sequence has a large excess of purine over pyrimidine and therefore does not easily lend itself to stable internal folding.
As a consequence, such sequences

are quite labile toward hydrolysis
(if present as single strands),
have a greater tendency to form duplices
(which do not easily replicate by primitive mechanisms),
lack internal symmetry, and
produce minus strands of a different general code pattern (i.e., 5′RYY).

XII.2. The RNY Code

Before going into a more detailed discussion of the points listed above, let us now offer an alternative model that is free of these particular difficulties. The suggestion of using the general codon pattern RNY, where N stands for any of the four nucleotides A, U, G, C, is also to be credited to Crick et al. [1]. However, it was disfavored by its authors on the grounds of a disadvantage: If N represents a pyrimidine, then the anticodon loop, having the general sequence 3′ⓊGYNRUⓊ, can in some cases use only its five central nucleotides to form stable base pairs with the messenger. This argument may, however, be counteracted by the observation that an RNY code can assign eight amino acids, so that one may exclude certain combinations that do not fulfil the stability requirements for the messenger-peptidyl-tRNA complex (cf. below).

What are the advantages of a general code pattern RNY? First of all, the RNY code, like its RRY analog, is comma-free. Moreover, it is symmetric with respect to plus and minus strands. If read from 5′ to 3′, the general frame structure of both the plus and the minus strand is $R^{N'}_{N}Y$ where N and N′ are complementary, situated at mirror-image positions in the sequences of both strands. Similar symmetries can also develop internally within a single strand, allowing the formation of symmetric secondary folding structures. Typical examples of (almost) symmetric foldings are the present-day tRNAs. Single-stranded phage genomes and their derivatives (such as the midi-variant of Qβ-RNA) are also distinguished by such elements of symmetry. Here the selective advantage of a symmetric structure is obvious. If the molecules are to be reproduced by a polymerase, which recognizes specifically some structural feature, only a symmetric structure would allow the plus and the minus strands to be equally efficient templates. Such an equivalence of efficiency is required for selection. Thus the symmetry of tRNA may well be a relic from a time when these molecules still had to reproduce autonomously.

Internal folding also enhances the molcule's resistance to hydrolysis and offers an easy way of instructing the correct read-off of the message. In an open structure in which many nucleotides remain unpaired (e.g., for RRY sequences), replication and translation could start at any unpaired position of the sequence, leading to fragmentary products. In a completely paired structure, unmatched sticky ends are predestined to be the starting-points of replication and translation. In this way the complete message can be read off, requiring only transient partial unfoldings of the template structure, which may be enforced by interactions with the growing chain. Symmetry, although not absolutely required, would in this case again be of advantage (cf. Fig. 50).

From a logical point of view the RNY code seems to be more attractive than the RRY code, on the basis of three arguments.

1. Selective enhancement of RNA molecules must be effective for the plus as well as the minus strand. Symmetric RNY patterns can fulfil this requirement more easily than RRY sequences, which differ from their minus strands (RYY) and hence cannot both become equivalent targets for specific recognition by enzymes.

2. In view of the high complexity, there is little chance finding the very few sequences that offer useful properties for replication and translation. If these sequences, being symmetric structures, fulfil the requirements listed under (1), both the plus and the minus strands may be candidates for representing such functions.

3. Evolution of the translation apparatus with its various tRNAs and messengers requires a mutual stabilization of all replicative molecules. As will be shown below, hypercycles may emerge more easily from a quasi-species if this, on account of its symmetry, offers two complementary functions.

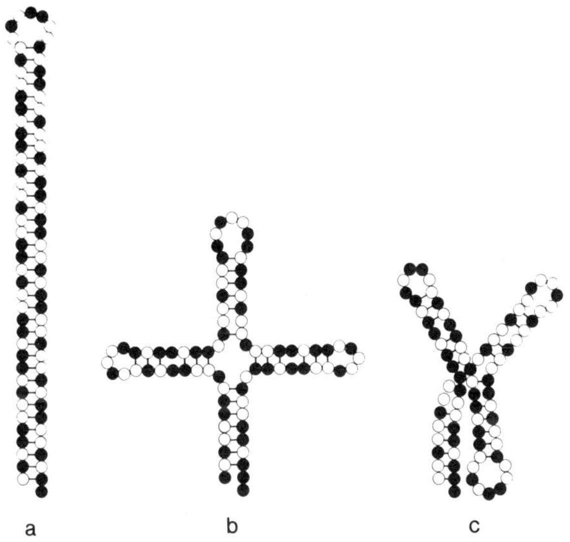

Fig. 50. Symmetric secondary structures of RNA require a corresponding internal complementarity (cf. Fig. 14 in Part A). Plus and minus strands will then exhibit similar foldings. RNY-codon patterns are able to produce such structures. A game has been devised, the rules of which take into account the physical interactions observed with oligonucleotides and tRNAs [17, 18]. It shows which of the structures is most likely to occur. Hairpins require two complementary halves of the molecule, e.g., a 5'-RRY sequence linked up to its minus strand involving a 5'RYY pattern

A system can accumulate information and eventually evolve to higher complexity only if it adopts the 'Darwinian logic' of selective self-organization. However, this logic must find its basis and its expression in material properties. All that the constituents can recognize at the beginning is natural abundance and strength of interaction. These are the properties with which we shall have to deal in order to understand the start of translation.

XIII. Physics of Primordial Coding

XIII.1. The Starting Conditions

Self-organization as a multimolecular process requires monomers as well as polymers to be present in sufficiently high concentrations. Its onset therefore must have been preceded by an extended phase of prebiotic synthesis, during which all the material necessary for yielding a 'highly enriched soup' was accumulated. We do not intend to dwell on these processes of primordial chemistry, nor do we quarrel about details of historical boundary conditions. Questions as to whether the 'soup' originated in the oceans, in a pond, or even in small puddles, or whether interfaces, coarse-grained, or porous surfaces were involved, may be important if absolute rates of the *historical* processes are to be estimated.

Here we simply start from the assumption that when self-organization began all kinds of energy-rich material were ubiquitous, including in particular:
amino acids in varying degrees of abundance,
nucleotides involving the four bases A, U, G, C,
polymers of both preceding classes, i.e., proteinoids as well as tRNA-like substances, having more or less random sequences.

'Less random' in this context means the existence of nearest-neighbor and more complex folding interactions, leading to a preference of certain structures, while 'more random' refers to their primary unrelatedness to any functional destination, which – if initially present – could only be coincidental.

On the other hand, we definitely do not suppose the presence of any adapted protein machinery, such as specific polymerases,
adapted synthetases, or any of the ribosomal functions.

This does not exclude an involvement of non-instructed, poorly adapted protein catalysts, in facilitating the start of replication and translation. However, not being able to reproduce and improve selectively, those proteins must be subsumed together with other catalytic surfaces under 'constant environmental factors'.

XIII.2. Abundance of Nucleotides

Since nucleic-acid-like structures are the only ones that offer inherent self-reproductive properties, it is important to analyze first their abundances as well as their mutual interactions in more detail.
The monomeric nucleotides, especially their energy-rich oligophosphate forms, are more difficult to obtain (using possible prebiotic mechanisms) than are amino acids. Quantitative statements of relative abundance are therefore scarce. S. Miller and L. Orgel ([63], p. 104) emphasize the central role of adenine nucleotides both in genetic processes as well as in energy transfer and correlate it with the relative ease with which this substance is formed. J. Oró and his co-workers [64] found that adenine can be obtained in yields of 0.5% in concentrated aqueous solutions of ammonium cyanide, while Miller and Orgel ([63], p. 105) showed that even hydrogen cyanide alone, in a reaction catalyzed by sunlight, yields the important intermediate

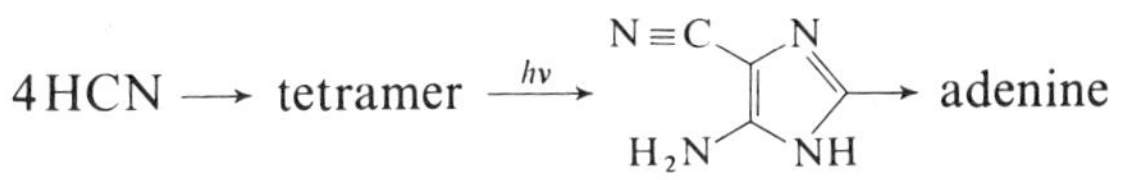

The same intermediate can also react with cyanate, urea, or cyanogen to give guanine. Less well understood are the mechanisms of pyrimidine synthesis. A pathway could be demonstrated for the synthesis of cytosine, using cyanoacetylene – an electric-discharge product of mixtures of methane and nitrogen – in combination with cyanate. Uracil, on the other hand, appears to be a hydrolysis product of cytosine and may possible owe its primordial existence to this source.

Only little can be said about the primordial natural abundance of the purines and pyrimidines. The rate of template-instructed polymerization is proportional to the concentration of the monomer to be included. For complementary instruction, at least two nucleotides are required and they will have to be equivalently represented in informational sequences. Therefore the inclusion of the less abundant nucleotide will always be the rate-limiting step, at least for chain elongation. A possible large excess of A over U in the primordial monomer distribution would then have been of little help in favoring AU over GC copolymers, except in cases where the nucleation of oligo-A primers is the rate-limiting step. The capacity for replicative growth is limited to the template function of the less abundant member of the complementary nucleotide pair. If under primordial conditions the abundance of G and C was intermediate to that of A and U, then GC- and AU-rich copolymers may well have formed with comparable rates.

We therefore cannot maintain the earlier, speculative view that the first codons were recruited exclusively from a binary alphabet, made up of AU copolymers alone.

XIII.3. Stability of Complementary Structures

The stability of base pairing may provide clues that are more illuminating with respect to the question of the first codons. Stabilities and rates of base pairing have been studied using various nucleotide combinations. These results have been discussed in detail in earlier reviews [44, 4]. They reveal quantitatively the generally accepted view that GC pairs within a cooperative stack provide considerably more stability than AU pairs.

The stability constant of a continuous and homogeneous oligomeric sequence of n nucleotide pairs can be represented by the relation

$$K_n = \beta s^n \tag{91}$$

which refers to a linear Ising model. The factor β is a cooperativity parameter, which for both the AU and GC pairs amounts to an order of magnitude of 10^{-3}*, while s is the stability constant of a single pair within the cooperative stack. For homogeneous sequences of AU pairs this parameter s is about one order of magnitude smaller than for homogeneous sequences of GC pairs, or in a rough quantitative representation:

$$s_{AU} \approx 10 \quad \text{while} \quad s_{GC} \approx 100.$$

Higher absolute stabilities than predicted by relation (91) are found if one of the complementary strands can assume the particular stacking configuration present in the anticodon loop of tRNA. Presumably the cooperativity parameter β is changed in this case. Yet O.C. Uhlenbeck, J. Batter and P. Doty [65, 66] found that tri- and tetranucleotides, complementary to the anticodon region of a tRNA and differing in one AU pair, exhibit a difference of one order of magnitude in their stability constants, quite in agreement with the figure quoted above. It is also reasonable that the largest absolute values of stability constants found are those for the interaction of two tRNAs that are complementary in their anticodons [67].

The data obtained with defined short sequences may serve at least for a comparison between various replication and translation models and for conclusions about their relative significance. It is obvious that single isolated AU or GC pairs are unstable under any realistic conditions of concentration. The start of replication, therefore, requires special help in the form of primer formation, and it is particularly this step that calls first for enzymic support. Present-day phage RNA replicases are also specifically adapted to a primary sequence pattern of the phage genome. For chain elongation, the incoming nucleotide is bound cooperatively on top of a stack of base pairs of the growing chain. Here the data suggest that the GC pair is about ten times more stable than the AU pair, resulting in a relatively higher fidelity q for G and C than for A and U. If the rate of replication is limited by the formation of the covalent link in the polynucleotide backbone (rather than by base-pair formation) the fidelity can be correlated with the monomer concentrations m_R and m_Y and the pair-stability constants K_{RY}, K_{RR}, and K_{YY}. The reproduction fidelity for any given nucleotide then may be obtained from the geometric mean of the fidelities for both complementary processes,

$$R \rightarrow Y; \quad Y \rightarrow R:$$

$$q_{RY} = \frac{m_Y K_{RY}}{\sum_N m_N K_{RN}} \quad \text{and} \quad q_{YR} = \frac{m_R K_{YR}}{\sum_N m_N K_{YN}} \tag{92}$$

* Such a relation is formally valid for both the internal base pairing within a given sequence and the binary association of two complementary sequences, where β has the dimension M^{-1}.

where $K_{RY} \approx K_{YR}$ and summation is extended over all N=A, U, G, C. Those q values are identical for A and U or G and C, respectively, since the error can appear either in the plus or in the minus strand. If the monomeric concentrations are of equal magnitudes, the stability constants determine what fidelities are obtainable. Then it follows that G and C reproduce considerably more accurately than A and U. The ratio of the error rates for GC and AU reproduction in mixed systems, however, does not exactly resemble the (inverse) ratio of the corresponding stability constants, owing mainly to the presence of GU wobble interactions, which are the main source of reproduction errors – even in present-day RNA phage replication [34].

We have made a guess of q values based on various sets of data for enzyme-free nucleotide interactions. They are summarized in Table 15. The first three sets refer to equal monomeric concentrations of A, U, G, and C. This assumption may be unrealistic and is therefore modified in the last three examples. One may object to the application of stability data that were obtained from studies with oligonucleotides. However, the inclusion of a single nucleotide in the replication process involves cooperative base-pair interactions and hence should resemble the relative orders found with oligonucleotides. All that is required for calculating the q values are *relative* rather than *absolute* stabilities.

The conclusion from the different estimates presented in Table 15 is: G and C reproduce with an appreciably higher fidelity than A and U. Depending on the superiority of the selected sequences (σ, cf. Eq. (28), Part A), the reproducible information content of GC-rich sequences in early replicative processes is limited to about twenty to one hundred nucleotides, i.e., to tRNA-like molecules, while that of AU-rich sequences can hardly exceed ten to twenty nucleotide residues per replicative unit. It should be emphasized at this point, that longer sequences of *any* composition may well have been present. However, they were *not* reproducible and therefore could not evolve according to any functional requirement.

From an analysis of experimental data for phage replication we concluded in Part A that even well-adapted RNA replicases would not allow the *reproducible* accumulation of more than 1000 to 10000 nucleotides per strand. This is equivalent to the actual gene content of the RNA phages.

We may now complete our statement regarding primordial replication mechanisms: A size as large as tRNA is *reproducibly* accessible only for GC-rich structures. Hence:

GC-rich sequences qualify as candidates for early tRNA adapters and for reproducible messengers, at least as long as replication is not aided by moderately adapted enzymes.

A similar conclusion can be drawn with respect to the start of *translation*. As was emphasized by Crick et al. [3], stability of the peptidyl-tRNA-messenger complex is critical for any primitive translation mo-

Table 15. Estimates of fidelities and error rates for G and C vs. A and U reproduction

Monomer concentrations	Stability constants of base pairs	Fidelity q		Error rate 1-q	
		GC	AU	GC	AU
$m_A = m_G = m_C = m_U$	$K_{RR} = K_{YY} = 1$ $K_{AC} = 1$; $K_{GU} = 10$ $K_{AU} = 10$; $K_{GC} = 100$	0.93	0.59	0.07	0.41
$m_A = m_G = m_C = m_U$	$K_{RR} \approx K_{YY} \ll 1$ $K_{AC} = 1$; $K_{GU} = 10$ $K_{AU} = 10$; $K_{GC} = 100$	0.95	0.67	0.05	0.33
$m_A = m_G = m_C = m_U$	$K_{RR} = K_{YY} \ll 1$ $K_{AC} = 1$; $K_{GU} = 5$ $K_{AU} = 10$; $K_{GC} = 100$	0.97	0.78	0.03	0.22
$m_A = 10 m_G$ $m_G = m_C$ $m_C = 10 m_U$	$K_{RR} \approx K_{YY} \ll 1$ $K_{AC} = 1$; $K_{GU} = 5$ $K_{AU} = 10$; $K_{GC} = 100$	0.93	0.81	0.07	0.19
$m_A = 10\ m_G$ $m_G = m_C$ $m_C = 10\ m_U$	$K_{RR} \approx K_{YY} \ll 1$ $K_{AC} = 1$; $K_{GU} = 5$ $K_{AU} = 10$; $K_{GC} = 100$	0.95	0.69	0.05	0.31
$m_A = 10\ m_G$ $m_G = m_C$ $m_C = 10\ m_U$	$K_{RR} = K_{YY} = 1$ $K_{AC} = 2$; $K_{GU} = 10$ $K_{AU} = 10$; $K_{GC} = 100$	0.86	0.25	0.14	0.75

del. Applying the data quoted above, the stability constant of a complex including five GC pairs amounts to

$$K_{5GC} \approx 10^{7}\ \mathrm{M}^{-1}$$

while that for five AU pairs is five orders of magnitude lower:

$$K_{5AU} \approx 10^{2}\ \mathrm{M}^{-1}$$

Again these values must be seen as relative; they might actually be somewhat larger if the stacked-loop region or tRNA (as we know it today) were involved, which, however, would not invalidate the argument based on relative magnitudes.

We might also evaluate the models on the basis of lifetime data. The recombination-rate constants, as measured for complementary chains of oligonucleotides, were found consistently to lie in the order of magnitude of

$$k_R \approx 10^{6}\ \mathrm{M}^{-1}\ \mathrm{s}^{-1}.$$

Given the stability constants quoted above, the lifetimes of the respective complexes would amount to

$$\tau_{5GC} \approx 10\ \mathrm{s} \quad \text{and} \quad \tau_{5AU} \approx 10^{-4}\ \mathrm{s}.$$

Again, these numbers might shift to larger values if stabilities turned out to be higher, and if two adjacent tRNAs are able to stabilize each other when attached to the messenger chain. Then lifetimes might just suffice for GC-rich sequences to start primitive translation. The lifetimes are certainly much too short if AU pairs are in excess. We see now that the slight disadvantage of the RNY relative to the RRY code resulting from stabilities can be balanced by utilizing primarily G and C at least for part of the R and Y positions. A four-membered GC structure is definitely more stable than any five-membered structure, including more than two AU pairs.

The conclusion is:

The start of translation is highly favored by GC-rich structures both of the tRNA precursors and of the messengers.

XIV. The GC-Frame Code

XIV.1. The First Two Codons

If we combine the conclusions drawn from stability data with the arguments produced by Crick et al., we can predict which codon assignments were probably the first ones.

The only sufficiently long sequences that are able to reproduce themselves faithfully must have been those in which G and C residues predominated. The first codons were then exclusively combinations of these two residues. The requirement for a comma-free read-off excludes the symmetric combinations GGG/CCC and GCG/CGC. This may be easily verified by writing down such sequences. Adaptors with the correct anticodon combinations can bind in various overlapping positions. This will have even more deleterious consequences, if further codon combinations, derived from symmetric precursors, are introduced. We are thus left with two complementary pairs of combinations, namely, GGC/GCC and CCG/CGG (all patterns to be read from 5′ to 3′). From the point of view of symmetry both appear completely equivalent. There is, however, a slight asymmetry based on wobble in the third position. Let us compare messenger sequences consisting exclusively of either CNG or GNC codons. In the first case the wobble base G is always situated in the third codon position, while in the second case it is restricted to the first position, in both the plus and the minus strands (if consistently read from 5′ to 3′). For replication the different codon positions are not distinguishable. Hence, wherever a wobble base occurs, an ambiguity may be introduced in the complementary strand, which, when it comes to translation, in one case affects the first, in the other case the third codon position:

$$5'\text{----CNG CNG CNG----}3'$$
$$\downarrow$$
$$5'\text{----}{\genfrac{}{}{0pt}{}{\mathrm{C}}{\mathrm{U}}}\mathrm{N'G}{\genfrac{}{}{0pt}{}{\mathrm{C}}{\mathrm{U}}}\mathrm{N'G}{\genfrac{}{}{0pt}{}{\mathrm{C}}{\mathrm{U}}}\mathrm{N'G}\text{----}3'$$

and

$$5'\text{----GNC GNC GNC----}3'$$
$$\downarrow$$
$$5'\text{----}\mathrm{GN}{\genfrac{}{}{0pt}{}{\mathrm{C}}{\mathrm{U}}}\mathrm{GN}{\genfrac{}{}{0pt}{}{\mathrm{C}}{\mathrm{U}}}\mathrm{GN}{\genfrac{}{}{0pt}{}{\mathrm{C}}{\mathrm{U}}}\ \text{----}3'$$

Only in the second case are the reproduced sequences correctly translated, i.e., if wobble interactions with the adapter occur preferentially in the third rather than in the first codon position.

In other words, an adapter with the anticodon 3′CNG can read both 5′GN′C and 5′GN′U, whereas an adapter with the anticodon 3′GNC will only read 5′CN′G, but reject 5′UN′G. The argument might be weak if five base pairs are required to keep the adapter bound to the messenger, since then wobble positions may not be as clearly distinct. Nevertheless, this asymmetric relationship between the first and third codon positions does exist and is obvious in the present genetic code.* Relatively small selective advantages are usually sufficient to bias the course of evolution. Crick et al.

---

* Our argument is aided by the fact that in the stationary distribution G is more persistent than C.

obviously preferred the RRY (or RNY) model on the basis of such arguments, too.
We are now able to make a unique assignment for the first two codons, namely

5′GGC and 5′GCC

which are complementary if aligned in an antiparallel fashion. This choice was dictated by four arguments, viz.,

stability of adapter-messenger interaction and
fidelity of replication, both favoring GC combinations to start with,
comma-free read-off in translation requiring an unsymmetric GC pattern, and
consistency of translation restricting *wobble* ambiguities to the third codon position.

We should like to emphasize that these arguments are based exclusively on the properties of nucleic acids. It is satisfying to notice that the two codons GGC and GCC in the present genetic code refer to the two simplest amino acids, glycine and alanine, which in experiments simulating primordial conditions indeed appear with by far the greatest abundance.
One may object that translation products consisting of these two residues only, will hardly represent catalysts of any sophistication. We shall return to this question in Section XVI. At the moment it suffices to note that translation at this stage is not yet a property required for the conservation of the underlying messengers. The first GC-rich strands are selected solely on the basis of structural stability and their ability to replicate faithfully. Many different GC sequences would serve this purpose equally well and hence may have become jointly selected as (more or less degenerate) partners of one quasi-species. Symmetric structures are greatly favored here, because they can fulfil the criteria of stability for the plus and for the minus strand simultaneously.
Among stable members, perhaps induced by template function of anticodon loops and subsequent pattern duplications, comma-free code sequences may have occurred and then started translation. If their translation products add any advantage to the stability or to the reproduction rates of their messengers, they will evolve further by a Darwinian mechanism and thereby change continuously the quasi-species distribution. Before we come back to such a stabilization by means of translation products, let us enlarge somewhat more on the question of stability of structure versus efficiency of replication, because it seems that both required properties are based on conflicting prerequisites.

XIV.2. The 'Aperiodic Linear GC Lattice'

The tRNA-like molecule with its internal folded structure strengthened by hydrogen bonds may be considered a microcrystallite. If it involves longer stretches of complementary GC pattern, the resulting internal structure may be quite inert. From melting curves of tRNA loops or corresponding oligonucleotides we know that an uninterrupted sequence of only four GC pairs is already quite stable. S. Coutts [68] studied an oligonucleotide corresponding to the extra loop of $tRNA^{Ser}_{1.2}$ from yeast, which contains 4 GC pairs (and was obtained by partial digestion of the tRNA molecule). He found a melting temperature of $84 \pm 1°$ C and a ΔH of 44 ± 4 kcal/mol. This is equivalent to a stability constant of about 2×10^5 at 25° C, in good agreement with the figures mentioned above.
What is rated in the *selection* of a quasi-species is not merely the structural stability of the strands, but rather a favorable combination of structural stability with reproductive efficiency. Efficient template function requires a quick partial unwinding of a loop, a procedure for which excessively long stretches of GC pairs are prohibitive. However:

Natural sequences are not perfect anyway.

Given a high abundance of A-monomers and the limited fidelity of base pairing, the GC microcrystallites will always be highly doped with A-residues, acting like imperfections in the linear GC lattice. A priori, there may be any kind of sequence from high to low A, U, G, or C content. What is to be selected and then reproducibly mulitplied, will be a sequence rich in GC, but not perfect. If, for instance, every fifth position in such a sequence is substituted by an A or U residue, then base-paired regions, depending on internal complementarity, will involve on the average no more than four GC pairs (cf. present tRNA). Those structures can melt locallly with ease, especially if the replication process is aided by a protein, which then represents the most primitive form of a replicase.

Note: *A-U imperfections in the aperiodic GC lattice are selectively advantageous.*

As Thomas Mann* said: 'Life shrinks back from absolute perfection.'

XIV.3. From GNC to RNY

Given a certain abundance of A (and complementary U) imperfections in the GC-rich strands of the

* Th. Mann: Der Zauberberg (The Magic Mountain)

selected quasi-species, the next step in the evolution of a code seems to be preprogrammed. Mutations might occur in any of the three codon positions, but their consequences are quite different. Substitution of the middle base of a codon would enforce a complementary substitution of the middle base of the corresponding anticodon occurring in the minus strand and hence immediately introduce two new codons, GAC and GUC. Changes in the first or third position, on the other hand, would be complemented by changes in the third (or first) position, respectively, of the minus strand and – by wobble arguments – finally lead to only one further assignment. Moreover, the GC frame for comma-free reading would be perturbed.

Stability requirements do not initially allow for a substitution of more than one AU pair in the five-base-pair region of the messenger-tRNA complex. Hence the most likely codons to occur next are 5′GAC and 5′GUC. Being mutants of the pre-existing pair 5′GGC/5′GCC, they may be abundantly present as members of the selected GC-rich quasi-species.

However, if these mutants are assigned a function in translation, they have to become truly equivalent to the dominant 5′GGC/5′GCC species. It is at this stage that hypercyclic stabilization of the four codon adapters (and the messengers which encode the coupling factors) becomes an absolute requirement. Without such a link the different partners of the primary translation system may coexist for some time, but they would never be able to evolve or to optimize their cooperation in any coherent fashion.

The four codons allow four different amino-acid assignments, which can now offer a rich palette of functions. The resulting proteins therefore could become efficient coupling factors. Messengers and tRNAs, as members of the same quasi-species, might have emerged from complementary strands of the same RNA species, thus sharing both functions.

On the other hand, this may be too restrictive a constraint for their further evolution. We then have to assume that they were derived from a common precursor, but later on diverged into different seqences owing to their quite different structural and functional requirements.

The assignments for GAC and GUC, according to the present table of the genetic code, are aspartic acid and valine. Before we discuss the amino-acid aspect in more detail we may look briefly at some further steps in the evolution toward a more general code.

High stability of codon-anticodon interaction is required less and less as the translation products become better adapted. Wobble interactions are finally admitted and the GC frame code can evolve to the more general RY frame code. All together this brings four more amino acids onto the scene. The first substitution still occurs under the stability constraint, which forces the AU content to be as low as possible. Hence it introduces the two codons 5′AGC (=serine) and 5′ACC (=threonine). Their complementary sequences affect the third codon position, yielding 5′GCU and 5′GGU, which reproduce the assignments for alanine (GCC) and glycine (GGC). The degeneracy, accounting for the wobble interactions in the reproduction of these latter codons, may have been the primary cause of the appearance of AGC and ACC codons and their assignments.

If with the evolution of an enzymic machinery more than one AU pair is allowed in the codon region we arrive at two more new assignments, namely, AA^{U}_{C}(asparagine) and AU^{U}_{C} (isoleucine). This completes all possible assignments for an RNY code. The further evolution of the genetic code requires a relaxation of the constraint of nonoverlapping frames. Adaptation of ribosomal precursors is therefore now imperative.

XIV.4. The Primary Alphabet of Amino Acids

Quite reliable estimates can be made for the *primordial abundance* of various amino acids. Structure and composition already provide the main clues for a guess about the likelihood of synthesis under primordial conditions. In Figure 51 the family tree of the first dozen nonpolar aliphatic amino acids as well as a few branches demonstrating the kinship relations for the simpler polar side chains are shown. Interesting questions concerning Nature's choice of the protein alphabet arise from this diagram.

The two simplest amino acids, glycine and alanine, are 'natural' representatives. It was apparently easier to fulfil requirements for hydrophobic interaction by adding some of the higher homolog, such as valine, leucine, and iso leucine. This specific choice may have been subject to chance; perhaps it was biased also by discriminative interactions with the adapters available. Among the polar side chains we find some obvious aliphatic carboxylic acids (aspartic and glutamic acid) as well as alcohols (serine and threonine), but not the corresponding amines (α, β-diaminopropionic acid and α,γ-diaminobutyric acid). Only the second next homolog (lysine) appears among the twenty 'natural' amino acids, while the intermediate (ornithine) still shows its traces. The reason may be that upon activation of the second amino group lactame formation or elimination occurs, which terminates the polymerization. Moreover, the second amino group may lead to a branching of the polypeptide chain (although

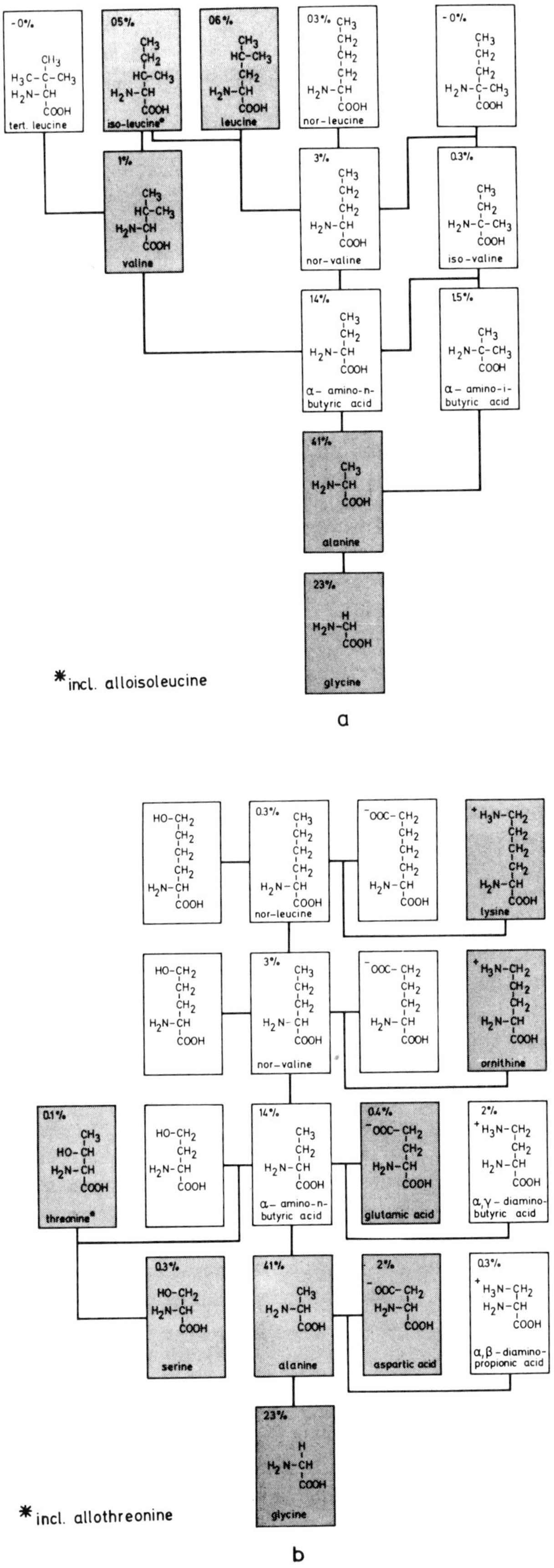

a similar argument may be raised for the carboxylic groups). Positively charged side chains may well have been dispensible in the first functional polypeptides. Even in present sea water the concentration of Mg^{2+} is high enough (~ 50 mM) to cause an appreciable complexation with carboxylic groups. Under reducing conditions even more divalent ions (such as Fe^{2+}) may have been dissolved in the oceans. Those metal ions, attached to carboxylic groups and still having free coordination sites, are especially important for close interactions between early proteins and (negatively charged) polynucleotides. From this point of view, side chains containing negatively charged ligands seem to be less dispensible than those containing positive charges.

The 'natural' amino acids not appearing in Figure 51 bear considerably more complex sidechains and were therefore present in the primordial soup at comparatively low concentrations.

These guesses from structure and composition are excellently confirmed by experiments simulating the prebiotic synthesis of amino acids, carried out by S. Miller and others (reviewed in [63]). The yields obtained for the natural amino acids (but also for other branches of the family tree) correspond grossly to the chemist's expectations (cf. numbers in Fig. 51), although many interesting items of detailed information are added by these experiments. The results are, furthermore, in good agreement with data obtained from meteorite analysis [69, 70] reflecting the occurrence of amino acids in interstellar space. Table 16 contains a compilation of data (taken from [63]), which are relevant for our discussion.

There is no doubt that the primordial soup was very rich in glycine and alanine. In Miller's experiments these amino acids appear to be about twenty times more frequent than any of the other 'natural' representatives. The next two positions in the abundance scale of natural amino acids are held by aspartic acid and valine with a clear gap between these and leucine, glutamic acid, serine, isoleucine, threonine, and proline.

There is every reason to assume that assignments of codons to amino acids actually followed the abundance scale. If glycine and alanine are by far the most abundant amino acids, why should they not have been assigned first, as soon as chemical mecha-

Fig. 51. The family tree of the first aliphatic amino acids and some branches for the simplest polar side chains. The number in the left upper corner of each plate refers to Miller's data of relative yields under primordial conditions [63] (i.e., molar yield of the particular amino acid divided by the sum of yields of all amino acids listed in their Table 7–2 on p. 87). The plates for the natural amino acids are shaded

Table 16. Abundance of natural amino acids in simulated prebiotic synthesis and in the Murchison meteorite. The first column refers to those amino acids which appear in the proteins. The data in the second column are typical results from S. Miller's experiments (as reviewed in [63]). They were obtained by sparking 336 mMol of methane in the presence of nitrogen and water. The total yield of amino acids (including those which do not appear in the proteins) based on carbon was 1.9%, the corresponding yields of glycine and alanine 0.26% and 0.71%, resp. The yields in the table refer to molar abundance. Similar data are obtained under different conditions, gly, ala, asp, val usually occurring as the most abundant natural amino acids. The meterorite data were reported by J. Oró et al. [69] and by K.A. Kvenvolden et al. [70], D-isomers appeared in all cases to be close to 50% Further literature can be found in [63]

Compound	Yield [μM]	μg/g meteorite
Glycine	440	6
Alanine	790	3
Aspartic acid	34	2
Valine	19.5	2
Leucine	11.3	
Glutamic acid	7.7	3
Serine	5.0	
Isoleucine	4.8	
Threonine*	1.6	
Proline	1.5	1

* Including allothreonine

nisms of activation became available? The first primordial polypeptides – the instructed as well as the noninstructed – must then have been largely gly-ala copolymers with only occasional substitutions by other amino acids, probably including also those which finally did not become assigned.

The agreement between the abundances of natural amino acids and the order of the first four codon assignments is striking. It should be emphasized that our choice of codons is based exclusively on arguments that are related to the structure of nucleic acids. Not only do the first four GC frame codons coincide exactly with the order of abundance, but furthermore the four additional RNY codons are assigned to amino acids which – with the exception of asparagine – are well represented with appreciable yields in Miller's table. One may well ask whether the assignment of AA_C^U to asparagine is a primary one or – as suggested by the intimate neighborhood to the lysine codon – was originally related to any of the lower diamino acid homologs, which appear in Miller's table with a reversed order of yields as compared with aspartic acid (α,γ-diaminobutyric acid) and glutamic acid (α,β-diaminopropionic acid). Without further evidence, however, this might be a too far-reaching speculation.

Also of importance in this respect are some recent results from the protein-sequence analysis of nucleotide-binding enzymes, which are believed to have existed more than $3 \cdot 10^9$ years ago under precellular conditions [71, 72]. The data suggest the existence of a precursor sequence of the nucleotide-binding surface, which include the amino acids valine, aspartic (and glutamic) acid, alanine and glycine besides isoleucine, lysine and threonine (although these data actually refer to a later stage of precellular evolution than is discussed in this paper).

XV. Hypercyclic Organization of the Early Translation Apparatus

Any model for the evolution of an early code and translation apparatus will have to provide conditions that allow tRNA-like adapters as well as gene precursors (or messengers) for various enzymic factors not only to coexist, but also to grow coherently and to evolve to optimal function. In Parts A and B it was shown that such a self-organization requires cyclically closed reactive links among all individual partners, unless they all can be integrated structurally into one replicative unit. In this section we have to show how realistic code models are correlated with an hypercyclic organization, and how such systems can evolve. An apparent difficulty of the hypercycle is how it is to originate. In plain language, the abundant presence of all members of the hypercycle seems to be a prerequisite for its origin. In more scientific terms, the hypercycle, being a higher-order-reaction network, has to 'nucleate' by some higher-order mechanism in order to come into existence.

Consider, for comparison, a simple replicative unit that grows according to a first-order autocatalytic rate law. In a solution buffered with energy-rich building material one copy is sufficient to start the multiplication process. Those experiments have been carried out with phage RNA or its noninfectious variants [7, 8, 32, 34, 73]. One template strand is sufficient to produce – within a few minutes – a large population of identical copies (cf. Part A).

A hypercycle never could start in this way. A single template copy would not multiply unless a sufficiently large number of its specific catalytic correspondents were present. These in turn are encoded by templates which themselves could not have multiplied without the help of their translation products. The growth of all templates in the system is dependent upon catalytic support, but the catalysts cannot grow unless the templates multiply. How large is the probability that nucleation will occur through some accidental fluctuation? Let us assume a test tube with 1 ml of sample solution. Diffusion-controlled reactions of macromolecules may have rate constants of the order of magnitude of 10^8 M^{-1} s^{-1}. Hence at least 10^8 identical copies of a given catalytic reaction partner have to be present in order to start template multiplication with a half-time of about one day. There is no chance that correlated functions among several such partners could result from coincidences of such giant fluctuations. It may, of course, be possible that the various templates multiply according to mixed-order terms; in other words: that first-order (enzyme-free) autocatalytic terms (representing template multiplication without catalytic help by other members) are superimposed upon the second-order catalytic replication terms. The hypercyclic link would then become effective only after concentrations have risen to a sufficiently high level. However, the system cannot know in advance which of the many alternative sequences multiplying according to a first-order autocatalysis are the ones which provide the useful information for the catalysts required at the later stages of organization.

There is only one solution to this problem:

The hypercycle must have a precursor, present in high natural abundance, from which it originates gradually by a mechanism of mutation and selection.

Such a precursor, indeed, can be the quasi-species consisting of a distribution of GC-rich sequences. All members of a stable quasi-species will grow until they are present in high concentrations. As was shown in Section XIV, some GC-rich sequences may be able to start a translation by assigning amino acids to defined anticodons. At this stage the translation products are really not yet necessary for conserving the system, so translation can still be considered a game of trial and error. If, however, it happens that one of the translation products offers advantages for the reproduction of its own messenger, this messenger may become the dominant representative of the quasi-species distribution.

A single RNA species could at best assign a two-amino-acid alphabet, if both the plus and the minus strands act as adapters for two complementary codons (e.g., GGC and GCC). If adapter sequences are sufficiently abundant, there is also a finite chance that coexisting mutants assign the two or even four codons (including GAC and GUC for aspartic acid and valine), again possibly utilizing both plus and minus strands. All this may still happen during the quasi-species phase.

Such a system, however, can evolve only if the different RNA species stabilize each other with the help of their translation products. We defer a discussion of the details of assignments – e.g., as to whether plus and minus strands of a given RNA species can evolve concomitantly and thereby become two adapters for complementary codons, or whether the plus strand as messenger encodes for the coupling factor, while only the minus strand acts as an adapter – to Section XVI. Here we study the problem of how hypercyclic organization can gradually evolve out of a quasi-species distribution.

Figure 52 shows how such a process can be envisaged. Assume two abundant mutants of the quasi-species, whose plus and minus strands are able to act as adapters of (at most) two amino acid pairs (e.g., gly/ala and asp/val), and which at the same time may be translated into a protein made up of (at most) four classes of amino acids. If the translation products offer any catalytic function in favor of the reproduction of their messengers, one would probably encounter one of the situations represented in Figures 52 or 53.

Both messengers, being closely related mutants, encode for two proteins with closely related functions. If one is a specific replicase, the other will be too, both functions being self- as well as mutually enhanc-

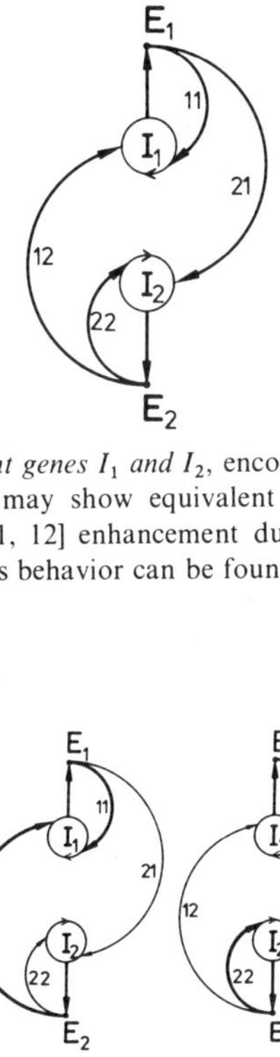

Fig. 52. *Two mutant genes I_1 and I_2*, encoding for their own replicases E_1 and E_2, may show equivalent couplings for self- [11, 22] and mutual [21, 12] enhancement due to their close kinship relation. Analogous behavior can be found in present RNA-phage replicases

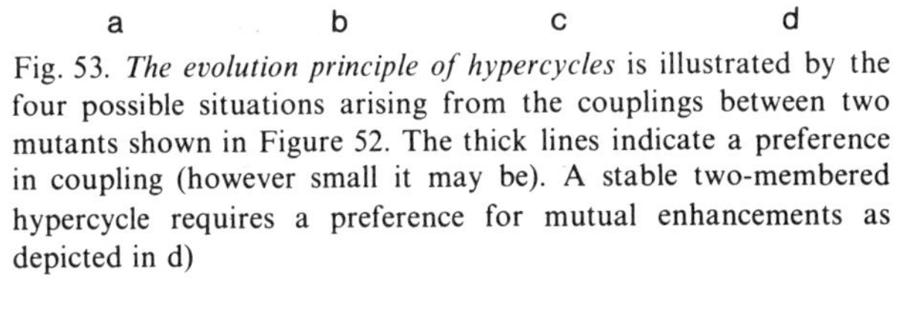

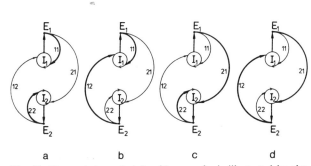

Fig. 53. *The evolution principle of hypercycles* is illustrated by the four possible situations arising from the couplings between two mutants shown in Figure 52. The thick lines indicate a preference in coupling (however small it may be). A stable two-membered hypercycle requires a preference for mutual enhancements as depicted in d)

Table 17. Fixed-point analysis of the two-member hypercycle

represented in Figure 52 has been carried out using the simplified rate equations

$$\dot{x}_i = \sum_{k=1,2} k_{ik} x_i x_k - \frac{x_i}{c} \sum_{l=1,2} \sum_{m=1,2} k_{lm} x_l x_m$$

$i = 1, 2; \quad x_1 + x_2 = c$

yielding the three fixed points and their eigenvalues:

$\bar{\mathbf{x}}_1 = (c, 0); \quad \omega^{(1)} = (k_{21} - k_{11})c$

$\bar{\mathbf{x}}_2 = (0, c); \quad \omega^{(2)} = (k_{12} - k_{22})c$

$$\bar{\mathbf{x}}_3 = (k_{22} - k_{12}, k_{11} - k_{21}) \frac{c}{k_{11} - k_{21} + k_{22} - k_{12}};$$

$$\omega^{(3)} = \frac{(k_{11} - k_{21})(k_{22} - k_{12})}{k_{11} - k_{21} + k_{22} - k_{12}} c.$$

Four cases may be distinguished

a) $k_{11} > k_{21}$; $k_{22} > k_{12}$ yielding competition between I_1 and I_2
b) $k_{11} > k_{21}$; $k_{22} < k_{12}$ yielding selection of I_1
c) $k_{11} < k_{21}$; $k_{22} > k_{12}$ yielding selection of I_2
d) $k_{11} < k_{21}$; $k_{22} < k_{12}$ yielding hypercyclic stabilization of I_1 and I_2

The fixed-point diagrams of these four cases are (cf. Part B)

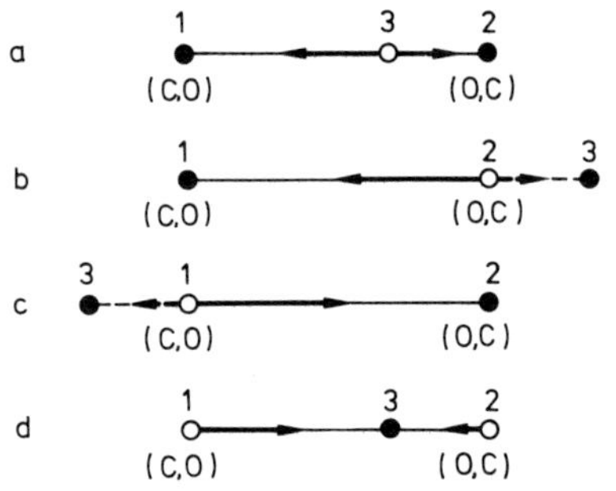

A unified representation can be achieved if the two coordinates: $\alpha = k_{12} + k_{21} - k_{11} - k_{22}$ and $\beta = k_{12} - k_{21} + k_{11} - k_{22}$ are introduced.

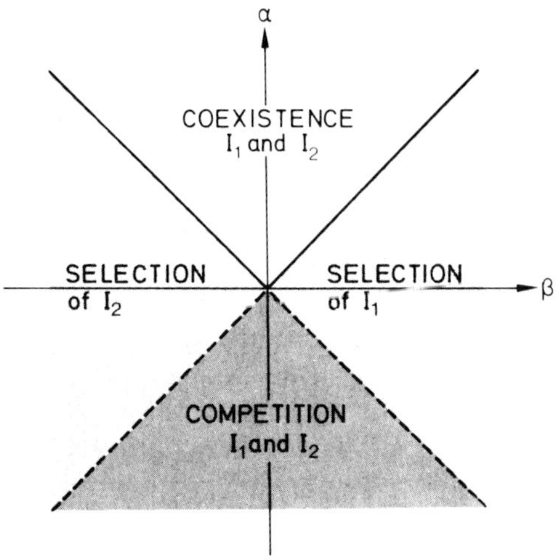

If in addition to the second-order term of the rate equations a linear autocatalytic term is introduced (yielding a growth function of the form $\Gamma_i = k_i x_i + \sum_{j=1,2} k_{ij} x_i x_j$), the region of stable hypercyclic coexistence of both species I_1 and I_2 is the space above the folded sheet in the three-dimensional parameter space with the coordinates:

$\alpha = k_{12} + k_{21} - k_{11} - k_{22}$

$\beta = k_{12} - k_{21} + k_{11} - k_{22}$

$\gamma = \frac{2}{c}(k_1 - k_2)$

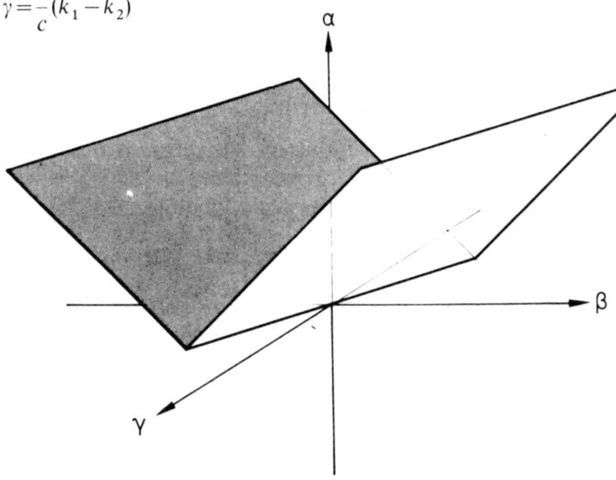

ing. There may, however, be specificity, too, because both proteins do not necessarily recognize both sequences equally well, nor do they recognize unrelated sequences at all. The sequences provide a specific binding site for initiating replication. The differences in binding strength for the four possible interactions of E_1 and E_2 with I_1 and I_2 may be slight. These differences, indicated by the line strengths, however small they are, will have drastic consequences, as follows from an inspection of the corresponding fixed-point diagrams (Table 17). We may distinguish four cases:

(1) E_1 favors I_1 over I_2, and E_2 favors I_2 over I_1 (Fig. 53a).
Consequence: I_1 and I_2 both are hypercyclically enforced by their respective enzymes, leading to strong competition. Only one of the competitors can survive, even if they are selectively equivalent.
(2) E_1 favors I_1 over I_2, and E_2 also favors I_1 over I_2 (Fig. 53b).
Consequence: I_1 will win the competition and I_2 will die out.
(3) E_2 favors I_2 over I_1, and so does E_1 (Fig. 53c).
Consequence: I_2 is now the winner, while I_1 dies out.

(4) E_1 favors I_2 over I_1, and E_2 favors I_1 over I_2 (Fig. 53d).
Consequence: Here we obtain a mutual, hypercyclic stabilization of I_1 and I_2.
It is important to note that small differences suffice to produce the behavior outlined above. In this respect it is of interest to see what happens if both E_1 and E_2 are exactly equivalent in their treatment of I_1 and I_2. Here we have complete impartiality, regardless of how much the population numbers x_i or y_i differ. I_1 or I_2 can die out in consequence of a fluctuation catastrophe, since there is no mutual stabilization present as in case 4. On the other hand, the fluctuations do not amplify themselves, and if the population numbers are large enough, a fluctuation catastrophe will practically never occur. Quite different in this respect is case (1), mentioned above. Only for exactly equal population numbers of I_1, I_2, E_1, and E_2 is here the system in a dynamically balanced state. A small fluctuation may disturb the balance and then, through self-amplification, inevitably will lead to selection of one of the two species. The same is true for any ensemble in which each messenger provides help for its own replicase only (cf. Fig. 47). The coupling resulting from a common translation function – all replicases (utilizing their RNA-recognition sites) may function simultaneously as activating enzymes – is not sufficient to enforce a coexistence. As in the system shown in Figure 45, there will be only one survivor, and translation function will subsequently break down.
The exact criteria for hypercycle formation are derived in Table 17. The figures give a clear representation of the stability ranges in terms of generalized coordinates referring to the rate parameters.
We have thus obtained an evolution principle for hypercycles. This kind of organization can emerge from a single quasi-species distribution, as soon as means of reaction coupling develop. The prerequisites for coexistence of precursors can be met generally only by closely related mutants. Thus the emergence of hypercycles requires the pre-existence of a molecular Darwinian system, but it will then lead to quite new consequences. The evolution principle is effective even with very small differences in rate parameters and hence responds sensitively to small changes brought about by mutations. Given a quasi-species distribution with developing interactions among the constituents, regardless, of how weak these interactions are, a hypercyclic organization will inevitably emerge whenever such interactions arise.
The hypercycle will also grow inevitably by way of mutations toward larger complexity (Fig. 54, 55). The evolution principle can be generalized by induction so as to apply to any n-membered hypercycle. A mutant I' then may either replace I, die out, or enlarge the hypercycle to a size comprising $n+1$ members (cf. XVI. 10.). More general evolution criteria can be derived as indicated in Table 18.

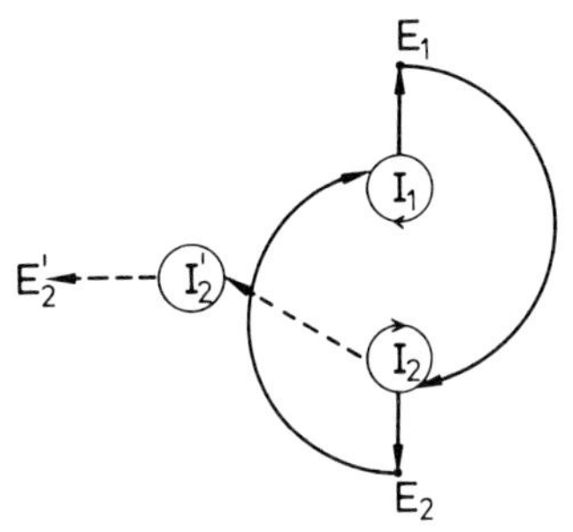

a

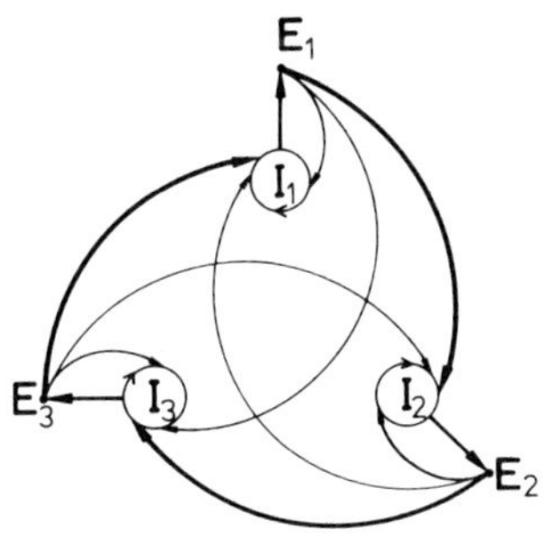

b

Fig. 54. *The generalization of the evolution principle* of hypercycles is illustrated in this diagram. The couplings have to fulfil the criteria derived in Tables 17 and 18, i.e., mutual enhancement has to prevail over self-enhancement (cf. thick lines). (a) A mutant of I_2 appears (I'_2). (b) The mutant (now I_3) is incorporated in the hypercycle

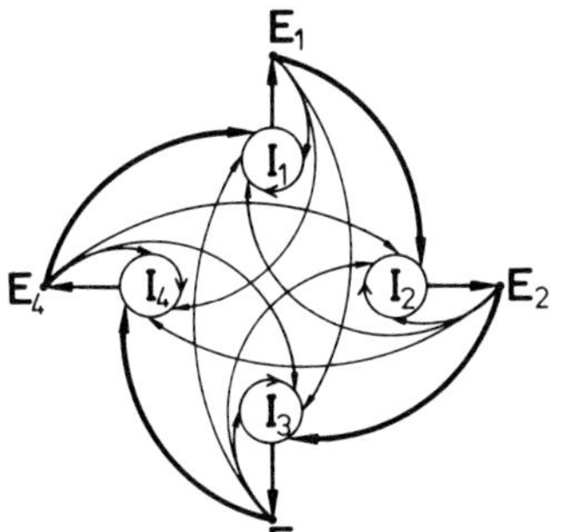

Fig. 55. *The 'realistic' four-membered hypercycle* assigns four messengers I_1 to I_4 (being mutants of a common precursor) to encode for four replicases E_1 to E_4 with common function, but slight preferences in specificity. The minus strands of I_1 to I_4 may concomitantly act as amino acid adapters

Table 18. Generalization of the evolution principle of hypercycles

is explained by the transition of a two-membered to a three-membered system as depicted in Figure 54.
The general rate equations are of the same form as in Table 17. Starting from a stable two-membered hypercycle, introduction of a third member I_3 (e.g., being a mutant of I_2) will frequently change the previously stable fixed point to a saddle point. This is most clearly seen if cyclic symmetry is assumed. Under these conditions, the notation can be simplified to:

$$k_{11}=k_{22}=k_{33}=k_D; \quad k_{13}=k_{21}=k_{32}=k_+; \quad k_{12}=k_{23}=k_{31}=k_-;$$

yielding the following matrix of rate coefficients.

$$\mathbf{K}=\begin{vmatrix} k_D & k_- & k_+ \\ k_+ & k_D & k_- \\ k_- & k_+ & k_D \end{vmatrix}$$

The fixed points and eigenvalues then are:

Corners: $\bar{\mathbf{x}}_1=(c,0,0)$; $\omega_1^{(1)}=(k_- - k_D)c$, $\omega_2^{(1)}=(k_+ - k_D)c$

$\bar{\mathbf{x}}_2, \bar{\mathbf{x}}_3$ analogous

Edges: $\bar{\mathbf{x}}_4=(0, k_D-k_-, k_D-k_+)\dfrac{c}{2k_D-k_+-k_-}$

$$\omega_1^{(4)}=\frac{k_-(k_D-k_-)+k_+(k_D-k_+)+k_+k_- -k_D^2}{2k_D-k_+-k_-}c$$

$$\omega_2^{(4)}=\frac{(k_D-k_+)(k_D-k_-)}{2k_D-k_+-k_-}c$$

$\bar{\mathbf{x}}_5, \bar{\mathbf{x}}_6$ analogous

Interior: $\bar{\mathbf{x}}_7=\left(\dfrac{c}{3},\dfrac{c}{3},\dfrac{c}{3}\right)$,

$$\omega_{1,2}^{(7)}=\{2k_D-k_+-k_- \pm i\sqrt{3}(k_+-k_-)\}\frac{c}{6}.$$

Again four cases are of special interest:

a) $k_D>k_+, k_-$
b) $k_+>k_->k_D$
c) $k_->k_+>k_D$
d) $k_+=k_->k_D$

yielding the following fixed-point diagrams:

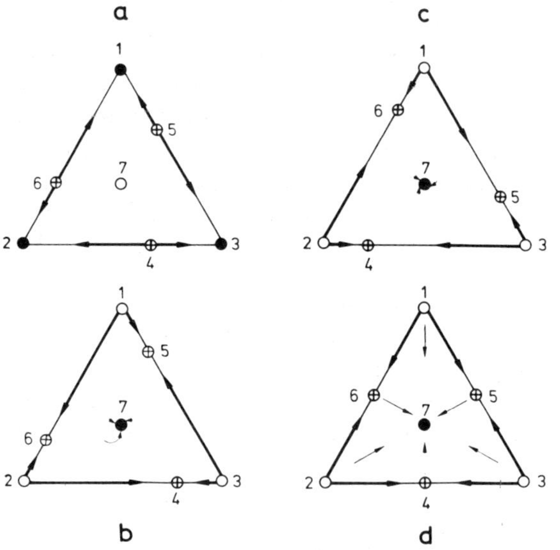

Large diagonal terms ($k_D \gg k_+, k_-$) lead to competition (diagram a). In the opposite situation, i.e., with large off-diagonal elements of **K**, the three species show cooperative behavior. The sense of rotation around the spiral sink in the center of the simplex is determined by the larger of the two constants k_+ and k_-. No rotational component is observed for equal constants $k_+=k_-$. The central fixed point is then a focus.
The example treated in this table provides a good illustration of the evolution to more complex hypercycles. In the absence of simplifying assumptions concerning the rate constants the analysis becomes quite involved. We refer to a more detailed representation [98], which includes a generalization to arbitrary dimensions.

XVI. Ten Questions

concerning our earliest molecular ancestors and the traces which they have left in the biosynthetic apparatus of present cells.

XVI.1. One RNA precursor?

This question is concerned with the complexity of the first molecules starting any reproducible function. A nucleotide chain of 100 residues corresponds to a complexity of about 10^{60} alternative sequences. If on grounds of stability we restrict ourselves to (AU-doped) GC copolymers only, we are still left with about 10^{30} possible arrangements. In order to achieve one or a few defined sequences, faithful self-reproduction is a necessary prerequisite. It will inevitably lead to Darwinian behavior, with selection of one defined quasi-species. The selected products are determined plainly by an optimal selective efficiency, but their structure depends on their historical route, which is strongly biased by self-copying of smaller oligonucleodtide patterns.

XVI.2. What Does Selective Advantage Mean to a Molecule?

Selective value is defined as an optimal combination of structural stability and efficiency of faithful replication. It can be expressed in quantitative terms related to the physical properties of a molecule in a given environment. Structural stability, resistance toward hydrolysis, and the development of cooperative properties call for elongation. Small oligonucleotides can-

not fold in any stable manner and may therefore easily be hydrolyzed. Furthermore, they do not offer sufficient adhesive strength for faithful copying or for translation. Length, on the other hand, is limited by replication rates and by copying fidelity. The properties of GC-rich sequences have been shown to be advantageous for forming stable copies with extended length. Whether these lengths resemble the sizes of present-day tRNA is uncertain. Sequence homologies have been found in tRNA [74], which indicate some self-copying of internal regions. This, however, may well have happened before codons became assigned. The onset of translation requires strong interactions between adapters and messengers, and these cannot be provided by molecules which are too small.
As soon as translation yields reproducible functions, selective value achieves a new dimension. It must, nevertheless, be expressed, for any given messenger, in terms of structural stability and efficiency of faithful reproduction. These properties now, however, also depend on the qualities (and concentrations) of the translation products. Specific coupling – as required for hypercyclic organization – is hence necessary for any system in which translation products are to be rated for selection and thereby become eligible for evolution. Such coupling is of a catalytic or protective nature.

XVI.3. Why Hypercyclic Organization of Single Mutant Genes Rather than One Steadily Growing Genome?

The answer to this question has been largely given in Part A. For a very primitive translation apparatus an amount of information would be required that corresponds to (or even exceeds) that of present RNA phages. The information of the phage genome can be preserved only with the help of a phage-specific enzyme complex, the availability of which is based on the efficiency of a complete translation machinery, provided by the host cell. If we accept the answers given to the first and second questions, the information needed to start translation must arise from cooperation among several mutants coexisting in the quasi-species distribution, rather than from a mere elongation of one sequence for which primarily no selection pressure would exist.
The hypercyclic stabilization of several coexisting mutants is equivalent to evolution by gene duplication. Originally, mutants appeared as single strands rather than as covalently linked duplicates. Fidelity restrictions would not allow for such an extension of length. Moreover, the probability of obtaining the required mutant combinations in one strand is very low. Sequences consisting of 100 G and C residues have

100 one-error mutants,
4950 two-error mutants
161 700 three-error mutants, etc., or
$N_k = \binom{100}{k}$ k-error mutants

The number of strands containing n mutant genes, each differing from the other in k specified positions (which may be necessary in order to qualify for a function) amounts to

$$\binom{N_k + n - 1}{n} \approx \frac{N_k^n}{n!} \quad (\text{for } n \ll N_k),$$

e.g., for $n=4$ and $k=3$ to 3×10^{19} alternative sequences. Given even these small deviations in the multiplied genes, the chance of finding a copy with a favorable combination within one giant strand is almost nil for any reasonably sized population. Each of the *isolated* mutant genes containing three substitutions, however, would be abundantly present in any macroscopic population.
Last but not least, the tRNAs being the adapters for translation must have been present anyway as separate strands. Evolution of a unified genome would have required complicated transcription control right at the start.
The isolated RNA strands, on the other hand, have a natural origin in the quasi-species distribution. All sequences were similar and so must have been their translation products. Whenever one translation product provides coupling functions, all of them will do so, owing to their similarities. Cyclic coupling – as required for hypercyclic organization – may then occur as well. We might even say that hypercyclic organization is most naturally associated with any realistic primitive translation model.
Does the present genome organization, established in prokaryotic cells, offer any clue as to its early structure? Present genes are certainly much larger than the early messengers. Gene elongation, as well as duplication, provided an advantage whenever the steadily increasing fidelity of the enzymic machinery allowed for it. The translation products could gain in sophistication, and more complex multienzyme mechanims could evolve, utilizing differentiated enzymes which had descended from a common precursor. Recombinant mechanisms as utilized by present-day cells will not have been available in primitive systems. The present structure of the prokaryotic genome therefore may have been achieved through elongation of isolated genes, their duplication and triplication to operons and their final mapping onto DNA, which can utilize more advanced means of reproduction so

as to allow for the formation of a unified genome. The present operon sizes correspond well to those which can be handled by a sophisticated RNA replicase (e.g., 1000 to 10000 nucleotides).

XVI.4. Are tRNAs Necessary to Start With?

This question may be alternatively posed as: Why not small oligonucleotide adapters?

Adapters without messengers do not make much sense. Short nucleotide sequences do not qualify as messengers. Decapeptides are already equivalent to almost half a tRNA molecule. Furthermore, short oligonucleotides may be unstable since they lack any tertiary structure. The simplest symmetric structure, i.e., a single loop stabilized by four or five base pairs, requires as many as fifteen nucleotides. Enzyme-free specific recognition of an amino acid involving simultaneously the anticodon loop and the 3′-end of the adapter is possible only with more extended structures. The same is true for interactions between two adjacent adapters, necessary for the stabilization of the messenger-peptidyl-tRNA complex, or for the conformational change (e.g., from HF to fh) which may facilitate the transport of the growing peptide chain along the messenger. G. Maass et al. [99] recently reported such a conformational change in the anticodon loop of tRNA, which they recorded by observing a fluorescence change of the Y base. The effect appeared to be absent in the anticodon-loop fragment (i.e., a decanucleotide having the sequence of the anticodon loop). All of this suggests that insufficiently long RNA sequences do not qualify for adapter function.

We may then ask: What distinguishes an adapter from a messenger? They both require comparable minimum lengths. They both have to be specifically folded structures, through which they may become reproducibly recognizable by coupling factors.

Since each tRNA *and* each messenger require a coupling factor, e.g., a replicase, favoring their selective stabilization, dual functions of the RNA sequences are indispensable. Plus and minus strands of a given RNA sequence may thus be utilized jointly as messenger and adapter.

XVI.5. Do Present-day tRNAs Provide Clues about their Origin?

Similarities in structure might either be the consequence of adaptation to a common goal or, alternatively, indicate a common ancestor. Present tRNAs show many points of correspondence [75] in their structures. Are we able to infer a common ancestor from these analogies? According to an analysis carried out by T.H. Jukes [76], this question may be answered with a cautious 'yes'. Why one must be cautious may be illustrated with an example. One of the common features exhibited by all prokaryotic and eukaryotic tRNAs studied so far is the sequence TΨCG in the so-called T-loop, a common recognition site for ribosomal control. Recent studies of methanogenic bacteria [77] revealed that these microorganisms, which are thought to be the 'most ancient divergences yet encountered in the bacterial line', lack this common feature of tRNA, but rather contain a sequence $\Psi\Psi$CG in one and UΨCG in another group. Although this finding does not call in question but rather underlines the close evolutionary relations of this class of microorganisms with other prokaryotes, it shows definitely that whole classes may concordantly adopt commont features. This is especially true for those molecules that are produced by a common machinery, such as the ribosome, which is the site of synthesis of all protein molecules.

Figure 56 shows an alignment of the sequences of four tRNAs from *E. coli*, which we think are the present representatives of early codon adpaters. Unfortunately, the sequence of the alanine-specific tRNA adapted to the codon GCC was not available. If we compare this species, which has the anticodon ^{5A}UGC, with its correspondent for valine, which has the anticodon ^{5A}UAC, we observe a better agreement

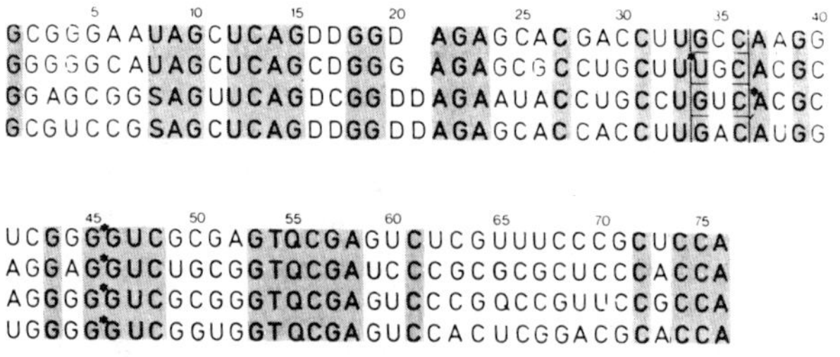

Fig. 56. *Alignment of the sequences of tRNAs* for the amino acids gly, ala, asp, and val. Unfortunately, the sequence referring to the codon GCC (for ala) is not yet available. Correspondences between gly- and ala-tRNAs are supposed to be closer for the correct sequence referring to the anticodon GCC (as suggested by the similarities between the two sequences for ala and val, referring to the anticodons *UGC and *UAC, resp.). The sequences show that base-paired regions consist predominantly of GC, and that close correspondences indicate the kinship between gly/ala and asp/val (cf. S in position 8 for asp and val instead of U for gly and ala, or the insertion of D between position 20 and 21 for asp and val). A = adenosine, *A = 2MA = N(2)-methyladenosine, C = cytidine, D = 5,6-dihydrouridine, G = guanosine, *G = 7MG = N(7)-methylguanosine, Q = Ψ = pseudouridine, S = thiouridine, T = ribosylthymine, U = uridine, *U = 5AU = 5-oxyacetyluridine

with the latter than with the one listed in the alignment (57 vis-à-vis 54 identical positions). Hence the correct alanine-tRNA with the anticodon GGC may have more coincidences with the gly-tRNA listed than for the 44 positions shown. Apart from this 'corporal defect' the data reveal

1. That all representatives agree in more than half of the positions (33 including the 'wrong' ala or 41 for gly, asp, and val),
2. That the subgroup gly/ala is distinguished from the subgroup asp/val by several features (thio-uridine 'S' instead of U in position 8, insertion of a 5,6-dihydro-U between position 20 and 21),
3. That all representatives have a pronounced excess of G and C over the A and U residues (or their derivatives), especially in the base-paired regions.

A comparison with other tRNA sequences, furthermore, indicates that these features—although certainly not uncommon among most tRNAs—are especially pronounced for this group. In particular, correspondences are as close as for different adapters of the same amino acid in the same organism.

One finding is particularly illuminating. If we compare the sequences of two adapters with complementary anticodons (e.g., asp and val) the coincidences between both plus strands of the tRNAs are much more pronounced than those between one plus strand (read from 3′ to 5′) and the other minus strand (read from 5′ to 3′). Actually, if we compare in this way the plus and minus strands of the same tRNA, the agreement is better. These coincidences are only the expression of the remarkable internal symmetry of tRNA, which places the anticodon almost exactly in the middle of the sequence and thereby allows for the formation of a symmetric two-dimensional pattern. We may rate this property as an indication of the early appearance of tRNA as an independent replicative unit. The requirement for the plus and minus strands to assume a similar pattern is important only if these represent independent replicative units rather than being structurally integrated into a large sequence of a genome, as they appear to be nowadays. We find a similar effect for phage RNAs or their variants, which have to multiply as single replicative units [78].

On the other hand, adaptation of tRNA to a common machinery must have brought about common deviations from symmetries originally required. The fact that the mirror images for plus and minus strands of the same tRNA show still more symmetric resemblance than those for plus and minus strands of tRNAs with complementary anticodons suggests that both tRNAs evolved as mutants of the same rather than of two complementary strands. We may then conclude that the present adapters for the codons GGC (gly), GCC (ala), GAC (asp), and GUC (val) derived from one quasi-species as single error mutants of a common ancestor. However, the original symmetry was not sufficient (why should it have been?) to allow adapter functions to derive from both the plus *and* the minus strand of a given RNA.

XVI.6. How Could Comma-free Messenger Patterns Arise?

The first messengers must have been identical with the first adapters (or their complementary strands). There is, indeed, a structural congruence behind adapter and messenger function. Whatever codon pattern occurs in the messenger sequence, it must have its complementary representation at the adapter. In primitive systems such a requirement could be most easily met by utilization of a common structural pattern for both types of molecules, such that the first adapters are the minus strands of the first messengers (if we define the plus strand as always being associated with a message) and that certain symmetries of structure allow both the plus and the minus strand to be recognized by the same replicase.

The first extended RNA molecules were rich in G and C, a consequence of selection based on criteria of structural stability and fidelity of copying. Molecules with a common codon pattern, such as GGC/GCC, require primer instruction (e.g., via catalysts or via exposed loops of RNAs present) with subsequent internal duplication. This will inevitably lead to structures that contain at least two codon patterns with internal complementarity, e.g., 5′GGC3′ and 3′CCG5′.

There is a good example for the efficiency of internal pattern duplication in the de-novo synthesis and amplification of RNA sequences by phage replicases. If Qβ replicase is severely deprived from any template, it starts to 'knit' its own primers, which it then duplicates and amplifies (selectively) until finally a uniform macroscopic population of RNA sequences—a few hundred nucleotides in length—appears. Under different environmental conditions, different (but uniform) sequence distributions are obtained [8]. S. Spiegelman, D. Mills and their co-workers have sequenced some of these 'midivariants,' all of which contain the specific recognition site for Qβ replicase [78]. Further experiments [73] have thrown light on the mechanism of this de-novo synthesis, showing that small pieces corresponding to sequences that are recognized by the enzyme are made as primers and then internally duplicated and selectively amplified. Earlier studies [22] have shown that, in particular, the sequences CCC(C) and UUCG can be recognized

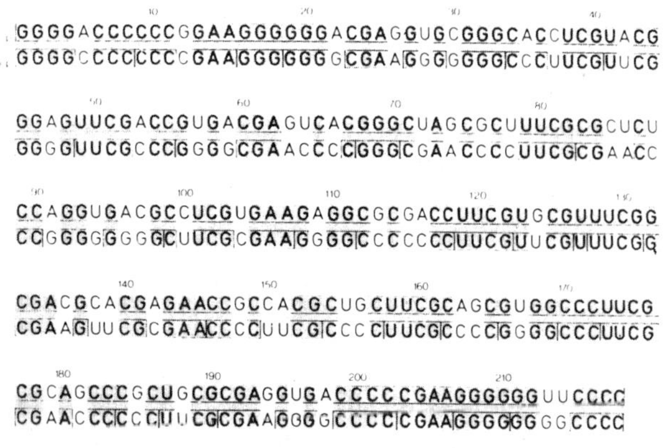

Fig. 57. *Alignment of the sequence of Qβ-midivariant* (determined by S. Spiegelman et al. [78]) with an artificial sequence composed of CCC(C)- and UUCG-blocks, as well as their complements [GGG(G) and CGAA]. Agreement at 169 of 218 positions suggests that midivariant is a de-novo product made by the enzyme Qβ-replicase, which possesses recognition sites for CCC(C) and UUCG (EF Tu). The kinetics of de-novo synthesis indicates a tetramer formation at the enzymic recognition sites, followed by some internal self-copying with occasional substitutions. The specific midivariant usually wins the competition among all appearing sequences and hence seems to be the most efficient template. The process demonstrates how uniform patterns can arise in primitive copying mechanisms

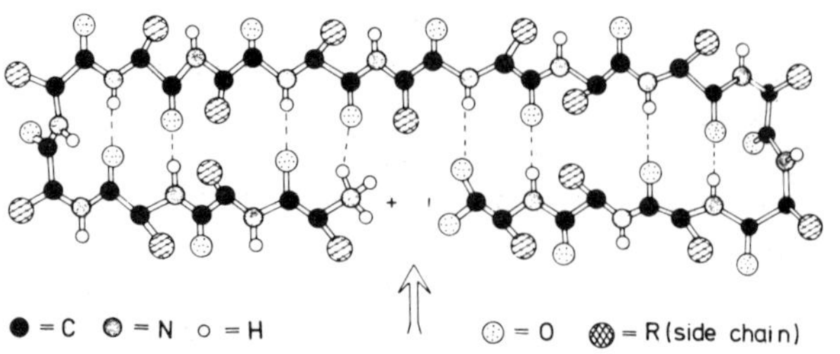

Fig. 58. *A simple enzyme precursor* is represented by a β-folded structure of some 15 to 25 amino acids (requiring messengers of 45 to 75 nucleotides). The active site includes a terminal amino group that is a very efficient proton donor (pK ~ 8), a terminal carboxylic group that acts as proton acceptor, and a catalytically active side chain (e.g., asp or ser). Many alternatives could be designed, only some of which have the correct pitch of the twisted chains to yield an efficient active site

by the enzyme. UUCG corresponds to the sequence TΨCG common to all tRNAs and known to interact specifically with the ribosomal elongation factor EF Tu, which acts as a subunit in the Qβ-replicase complex. An alignment of the midivariant with a sequence made up solely of the two oligonucleotides mentioned and their complementary segments – GGG(G) and CGAA – shows agreement in more than three-quarters of the positions, indicating the efficiency of internal copying of primer sequences (Fig. 57).

In a similar way we may think of the existence of primordial mechanisms of uniform pattern production. If among the many possible patterns 5′GGC/5′GCC and possibly also 5′GAC/5′GUC appeared, those messenger patterns could have started a reproducible translation according to the mechanism of Crick et al. [3] and have been capable of selective amplification with the help of their reproducible translation products.

XVI.7. What Did the First Functionally Active Proteins Look Like?

The simplest protein could be a homogeneous polypeptide, e.g., polyglycine. Does it offer any possible catalytic activity? This is a question that can and should be answered with experiments. With mixed sequences, including a sufficiently large number of residues, say about fifteen to thirty, β-sheet structures may form with an active center, in which the terminal carboxylic group is placed in a defined position near the terminal amino group (Fig. 58). The proximal distance varies with the chain length, since the β-structure involves a twist among both antiparallel chains [79]. The pK of the terminal amino group is around eight, hence the catalytic site contains at least an efficient proton donor-acceptor system. Alternating gly-ala residues are very favorable for the formation of β-structures. However, there are serious solubility problems for chains consisting exclusively of gly and ala, which would restrict them to interfaces only.

The folding of β-sheets has been studied by P.Y. Chou and G.D. Fasman [80], who analyzed X-ray data for 29 proteins in order to elucidate 459 β-turns in regions of chain reversal. The three residues with the highest β-turn potential in all four positions of the bend include gly and asp, while in regions following the β-turn, hydrophobic residues are predominant.

An important prerequisite of catalytic efficiency is the defined spatial arrangement of the terminal groups. The utilization of two or more classes of amino acids may be necessary for stabilizing a reproducible folding. β-Sheets have long been known to be important building elements of protein structure. According to M. Levitt [81], they may be utilized in a very general manner to stabilize active conformations of proteins.

The large abundance of glycine and alanine might have determined in essence the appearance of the first proteins, but polar side chains are indispensible for the solubility of longer sequences. Four amino acid classes would of course offer much more flexibility. If aspartic acid and valine were the next two candidates, globular structures might have formed, stabilized by hydrophobic interactions of the side chains of valine and alanine and solubilized by the carboxylic side chains of aspartic acid. This residue further offers many possibilities for forming specific catalytic sites with the participation of divalent metal ions.

Our imagination is taxed to estimate the vast number of possibilities. Experiments that are supposed to test various structures with respect to their efficiency in discriminating between RNA sequences and their structural features are under way. Results obtained with ribonucleases [82] encourage one to seek a 'minimum structure', able to recognize RNA sequences specifically.

XVI.8. Are Synthetases Necessary to Start With?

In the three-dimensional structure of present-day tRNAs (cf. Part A, Fig. 14) the anticodon loop is fixed at a considerable distance from the amino acyl site. Such a structure is adapted to the functional needs of present tRNA molecules, imposed by the ribosomal mechanism and by the structure of synthetases. On the other hand, it is known that tRNA can undergo conformational changes that drastically alter its shape and dimensions. R. Rigler and his coworkers [83] studied conformational lifetimes as well as rotational relaxation times by fluorescence methods and concluded the existence of at least three different rapidly interconverting conformational states. Analogous results were obtained by T. Olson et al. [84], who used laser light-scattering techniques. The population of the different conformational states depends strongly on magnesium-ion concentration. It is important, again, to note that under conditions that correspond to those present in sea water (Mg^{2+}; ~ 50 mM), a conformer is present that differs in shape from the L-form found by crystallographic studies, being considerably more cylindric.

This point is stressed because it is most relevant to the question raised. Early enzymes were made of only a very limited number of amino acid residues and therefore cannot have been very bulky globular structures. In order to guarantee a unique assignment of an amino acid to an anticodon, either enzymes as sophisticated as present-day aminoacyl synthetases had to be available, or else the tRNA structure had to allow a much closer contact between the aminoacyl and anticodon sites than the L-form does, in order to admit a simultaneous checking of both sites. The high mutation rate at early stages would otherwise very soon have destroyed any unique coincidental correspondence between these two sites. On the other hand, the conformational transition is still required since the mechanism of peptide-bond formation (cf. Fig. 48) calls for a well-defined separation of the messenger and the growing peptide chain. The data quoted invite reflection about such possibilities. If, on the other hand, a structure similar to the pattern c) shown in Figure 49 is likely to arise, the first amino-acid assignments might even have been made without enzymic help. The tRNA structure as such certainly offers sufficient subtlety for specific recognition. It has been noted [85] that the fourth base from the 3′-end (i.e., the one following 3′ACC) is somehow related to the anticodon. The primary expectations regarding a unique correlation for all tRNAs finally did not materialize. However, such a correlation may have played a key role in the early specific recognition of amino acids by tRNAs. Referring to data from *E. coli* and T_4 phage, the nucleotides in the position following 3′ACC are: U for gly, A for ala, G for asp, and A for val. It was certainly important for early adapters to ensure unique assignment by sufficiently discriminative sites. This property might have been partially lost during later phases of evolution. This is admittedly a speculation and calls for experimental confirmation.

To conclude: Synthetases may have been dispensible at the very early stages, but tRNAs finally turned out to be an unsatisfactory attempt by Nature to make enzymes from nucleic acids. More efficient recognition may have evolved from the coupling factors, which were predestined to recognize tRNA-like structures specifically.

XVI.9. Which Were the First Enzymes?

If synthetases are not really necessary for a *start* of translation (and this is a big 'if!'), we are left with the coupling factors, probably replicases, as the only absolute primary requirements for a coherent evolution of translation. Via such a function, a selective advantage occurring in a translation product can be most efficiently fed back onto the messenger. Hence specific replicases (all belonging to one class of similar protein molecules) not only provide the prerequisites for hypercyclic coupling, but also turn out to be most important for the further evolution of proteins, since only they can tell the messenger what is phenotypically advantageous and how to select for it at the genotypic level, i.e., by enhanced synthesis of the particular messenger. As will be seen in the next paragraph, such a selective coupling between geno- and phenotypic levels works best in combination with spatial separation or compartmentation.

Next, of course, we have to look for catalytic support for the various translation functions. If replicases have established a defined relationship with tRNA-like messengers (including both the plus *and* the minus strands), their recognition properties may well be utilized for synthetase and translatase (i.e., preribosomal) functions. In other words, a gene duplicate of a replicase may well be the precursor of a synthetase messenger as well as of a translation factor such as EF Tu, the more so since the chemistry of replicase and transfer function is very similar and in present systems appears to be effected by similar residues.

Dual functions with gradual divergence may have been a very early requisite of replication and translation mechanisms, just as gene duplication was one of the main vehicles of evolution at later stages.

Those dual functions have clearly left their traces in present cell organelles, and viruses have utilized them as well for their postbiotic evolution in the host cell. The genome of the phage $Q\beta$ encodes for only one subunit of its replicase, but utilizes three more factors of the host cell, which have been identified as the ribosomal protein S_1 and the elongation factors EF Tu and EF Ts [87, 88].

Ch. Biebricher [89] has studied the properties of these factors and found that they are involved simultaneously in several functions of ribosomal control, utilizing their acquired property, namely, to recognize tRNA molecules. He argues that also the β-factor of $Q\beta$ replicase, which is encoded by the phage genome, has its precursor in the *E. coli* cell, and this seems, indeed, to be the case. Using immunologic techniques, he was able to identify a protein containing EF Tu and EF Ts that behaves like a precursor of the $Q\beta$ replicase in uinfected *E. coli* and that appears to be involved in an – as yet unspecified – RNA-synthesis function of *E. coli* [87]. Further, similar correspondences may yet be found with synthetases. It seems that once a certain function has been developed – such as the ability to recognize certain types of RNA – then Nature utilizes this function wherever else it is needed (e.g., specific replication, ribosomal transport and control, amino acid activation).

In some respects the formation of RNA phages may well have mimicked the evolution of early RNA messengers. Phages utilize as many host cell functions as possible except one, namely, specific recognition of their own genome (i.e., coupling via specific replication). Different phages (e.g., $Q\beta$, Ms2, R17) inherit different recognition factors [9], although they all derive from a common ancestor in the host cell. In Part A it was also shown that the primary phase of RNA-phage infection is equvalent to a simple hypercyclic amplification process.

XVI.10. Why Finally Cells with Unified Genomes?

Hypercycles offer advantages for enlarging the information content by functional integration of a messenger system, in which the single replicative unit is limited in length due to a finite fidelity of copying. The increase in information content allows the build-up of a reproducible replication and translation apparatus, by which the translation products can evolve to higher efficiency. This will allow better fidelities, which in turn will increase the information content of each single replicative unit and thereby, again, enhance the quality of the enzymes. Simultaneously, as shown in Section XV, the hypercycle itself will evolve to higher complexity by integrating more differentiated mutant genes.

Increase of information content will not only produce better enzymes; it may also allow each replicative unit to inherit information for more than one enzyme. Dual functions can thereby be removed from the list of earlier evolutionary constraints, i.e., duplicated messengers may develop independently, according to the particular functional needs of their translation products. This may have been the origin of operon structures with control mechanisms for simultaneous replication of several structural genes. Replicases may thus have evolved to common polymerases associated with specific control factors for induction or repression.

After having realized the advantages of functional coupling, which seems to be a requirement for any start of translation, we should ask why functional coupling has finally been replaced by complete structural integration of all genes, the genomes of even the most primitive known cells being structural units. So where are the limitations of hypercyclic organization and what improvements can be made in it?

In a system controlled by functional links we have to distinguish two kinds of mutations. One class will primarily change the phenotypic properties of the messenger itself and thereby alter its target function with respect to a specific replicase or control factor. These mutations are especially important in the early phases of evolution owing to the important role of phenotypic properties of RNA structures. Those target mutations will immediately become selectively effective, advantageous mutations will be fixed, and disadvantageous ones will be dismissed.

The second kind of mutation – which may or may not be neutral with respect to the target function – refers to phenotypic changes in the translation products. The more precisely specified the messengers are, the more specifically a mutation may alter the function of the translation product.

Whether or not a mutant is specifically favored by

selection depends only on the target function, regardless of whether the translation product is altered in a favorable or an unfavorable sense or whether it remains neutral. For the later stages of precellular evolution the most common consequence of a mutation will be a phenotypic change in the translation product coupled with an unaffected target function. The mutant may then proliferate further, but it is not specifically selected against its former wild type, nor would the system select against the mutant, if its translation product proves to be unfavorable. What should really be achieved is a rating of the system as a whole. This may be accomplished by spatial separation of the messenger systems, by niches or – even more efficiently – by compartmentation. A messenger in a given compartment can enrich its environment with its own translation products and compete with other compartments using its efficiency of proliferation. To a limited extent this is also possible simply by spatial separation. However, a compartment without hypercyclic organization does not work at all. The enhanced competition among all messengers in the limited living space of the compartment would destroy any cooperative function.

A compartment could proliferate more efficiently by correlating its own reproduction with the re-duplication of its total gene content. This, of course, requires a fairly involved control mechanism, which could be facilitated by the integration of all genes into one giant replicative unit. Such an individualization of the total compartment requires high fidelity in the replication machinery. In Part A we compared the information content of various stages of life with their corresponding (and observed) replication fidelities (cf. Table 4).

The individualization of compartments is probably connected with the transition from RNA genes or operons to DNA genomes, since only the mechanism of DNA replication could guarantee sufficiently high fidelity. The new individualized unit was the integrated Proto-cell. The previous functional organization of genes and gene products has been superseded and amended by a coupled structural and functional organization. A closer study of the cyclic arrangement of genetic maps may still reveal some remnants of the origins of structural organization, although recombinative epigenetic effects may have covered many of the traces.

As a consequence of unification and individualization, the net growth of (asexual) multiplication of cells obeys a first-order autocatalytic law (in the absence of inhibitory effects). The Darwinian properties of such systems allow for selective evolution as well as for coexistence of a large variety of differentiated species. The integrated unit of the cell turns out to be superior to the more conservative form of hypercyclic organization.

On the other hand, the subsequent evolution of multicellular [90] organisms may again have utilized analogous or alternative forms of hypercyclic organization (nonlinear networks) applied to cells as the new subunits, and thereby have resembled in some respect the process of molecular self-organization.

XVII. Realistic Boundary Conditions

A discussion of the 'realistic hypercycle' would be incomplete without a digression on realistic boundary conditions. We shall be brief, not because we disregard their importance in the historical process of evolution – the occurrence of life on our planet is after all a historical event – but because we are aware of how little we really can say. While the *early stages of life,* owing to evolutionary coherence, have left at least some traces in present organisms, there are no corresponding remnants of the *early environment.* In our discussion so far we have done perhaps some injustice to experiments simulating primordial, template-free protein synthesis, which were carried out by S.W. Fox [91] and others (cf. the review by K. Dose and H. Rauchfuss [92]). It was the goal of our studies to understand the early forms of organization that allowed self-reproduction, selection, and evolutionary adaptation of the biosynthetic machinery, such as we encounter today in living cells. Proteins do not inherit the basic physical prerequisites for such an adaptive self-organization, at least not in any obvious manner as nucleic acids do. On the other hand, they do inherit a tremendous functional capacity, in which they are by far superior to the nucleic acids. Since proteins can form much more easily under primordial conditions, the presence of a large amount of various catalytic materials must have been an essential environmental quality. Research in this field has clearly demonstrated that quite efficient protein catalysis can be present under primordial conditions.

Interfaces deserve special recognition in this respect. If covered with catalytically active material they may have served as the most favorable sites of primordial synthesis. The restriction of molecular motion to the dimensions of a plane increases enormously the efficiency of encounters, especially if sequences of high-order reactions are involved.

L. Onsager [93] has emphasized that under primordial conditions the oceans must have been extensively covered with layers of deposited hydrophobic and hydrophilic material cf. also [94]). Those multilayers must have offered favorable conditions for a primordial preparative chemistry. In view of the obvious

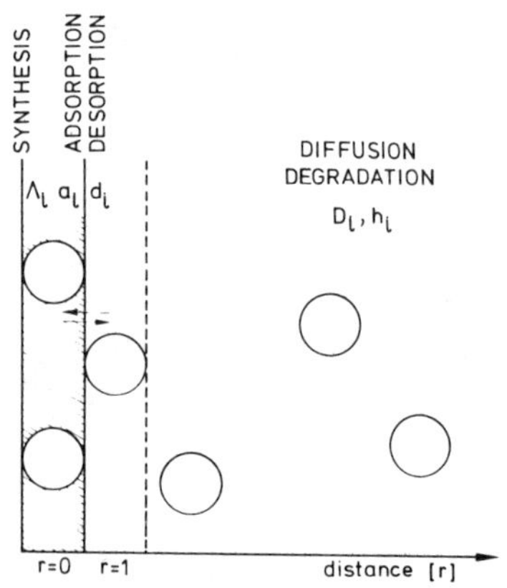

Fig. 59. Schematic representation of a heterogeneous reaction model including hypercyclic coupling. Three spatial regions are distinguished: $r=0$ bound to interface, $r=1$ transition layer at interface, $r>1$ bulk of solution phase. Diffusion to and from interface is superimposed on chemical reactions proceeding according to a hypercyclic scheme

advantages offered by interfaces we have examined the properties of hypercycles under corresponding environmental boundary conditions.

As a simple model we consider a system such as that depicted schematically in Figure 59. Polymer synthesis is restricted to a surface layer only ($r=0$), which has a finite binding capacity for templates and enzymes. The kinetic equations are similar to those applying to homogeneous solutions except that we have to account explicitly for diffusion. We distinguish a growth function that refers to the surface concentrations of replicative molecules and enzymes. Diffusion within the surface is assumed to be fast and not rate-determining. Adsorption and desorption of macromolecules is treated as an exchange reaction between the surface layer ($r=0$) and a solution layer next to the surface ($0<r\leqq 1$). Decomposition may occur at the interface and/or (only) in the bulk of the solution. Finally, transport to and from the interface is represented by a diffusion term.

Depending on the mechanism of synthesis assumed, it may be necessary to consider independent binding sites for both templates and enzymes. We used this model to obtain some clues about the behavior of hypercycles with translation (cf. Sect. IX in Part B). Numerical integration for several sets of rate parameters was performed according to a method described in the literature [95]. Three characteristic results – two of which are in complete analogy to the behavior of hypercycles in homogeneous solutions – can be distinguished:

(A) At very low concentrations of polynucleotides and polypeptides or large values of K_i [see Eqs. (73), (75), and (79) in Part B], the surface densities of polymers do not approach a steady-state value but decrease either monotonically or in damped oscillations. Consequently, the macromolecules die out after some time (Fig. 60).

(B) Above a certain threshold value of total concentration, we find limit cycle behavior in systems with $n<4$. The situation is analogous to the low-concentration limit in a homogeneous solution (Fig. 61).

(C) At sufficiently high concentrations we finally obtain a stationary state:

$$\lim_{t\to\infty} \frac{\partial x_i}{\partial t}=0, \qquad \lim_{t\to\infty} \frac{\partial y_i}{\partial t}=0$$

and $\bar{x}_i>0$, $\bar{y}_i>0$, $i=1,2 \ldots n$ (Fig. 62), x_i and y_i being the concentrations of enzymes and messengers, respectively, $\bar{x}_i$ and $\bar{y}_i$ their final stationary values, and t the time.

In systems of lower dimensions ($n\leqq 4$) behavior of types (A) and (C) only was observed.

These model calculations were supplemented by several studies of closely related problems using stochastic computer-simulation techniques. The results again showed the close analogy of behavior of hypercycles at interfaces and in homogeneous solution (as described in detail in Part B).

Consideration of realistic boundary conditions is a point particularly stressed in papers by H. Kuhn [96]. We do not disagree with the assumption of a 'structured environment', nor do we know whether we can agree with the postulation of a very particular environment, unless experimental evidence can be presented that shows at least the usefulness of such postulates.

Our models are by no means confined to spatial uniformity (cf. the above calculations). In fact, the logical inferences behind the various models – namely, the existence of a vast number of structural alternatives requiring natural selection, the limitation of the information content of single replicative units due to restricted fidelities, or the need for functional coupling in order to allow the coherent evolution of a complete ensemble – apply to *any* realistic environment. Kuhn's conclusion that the kind of organization proposed is 'restricted to the particular case of spatial uniformity' is beside the point. Who would claim today, that life could *only* originate in porous material, or at interfaces, or within multilayers at the surface of oceans, or in the bulk of sea water? The models show that it may originate – with greater or lesser likelihood – under any of those boundary conditions, if – and only if – certain criteria are fulfilled. These criteria refer to the problem of generation and accumulation of information and do not differ qualitatively when different boundary conditions are applied.

Much the same can be said with respect to temporal uniformity or nonuniformity. It has been shown in Part B that selection criteria may assume an especially simple form if they apply to steady-state conditions. Since they refer to relative rather than to absolute reaction rates, they are qualitatively the same, regardless of whether the system is growing, oscillating, or in a stationary state.

It is true that annealing is a useful procedure for many problems related to phase separations. Whether, however, thermal fluctuations serve equally well for selection of longer polynucleotides, remains to be shown by experiments.

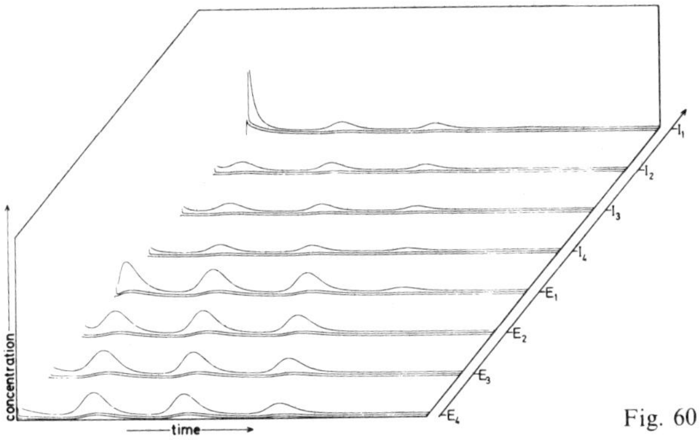

Fig. 60

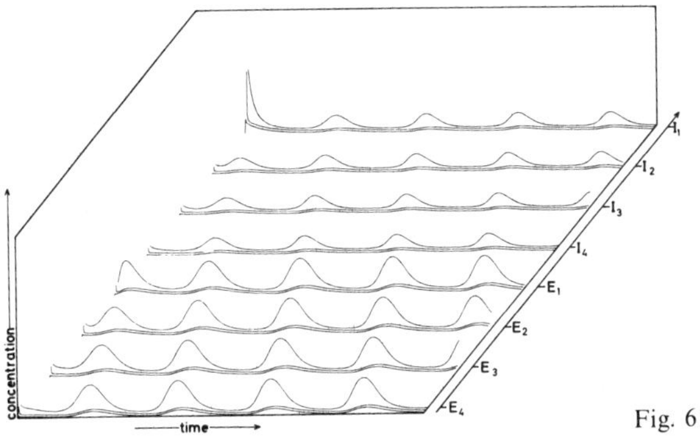

Fig. 61

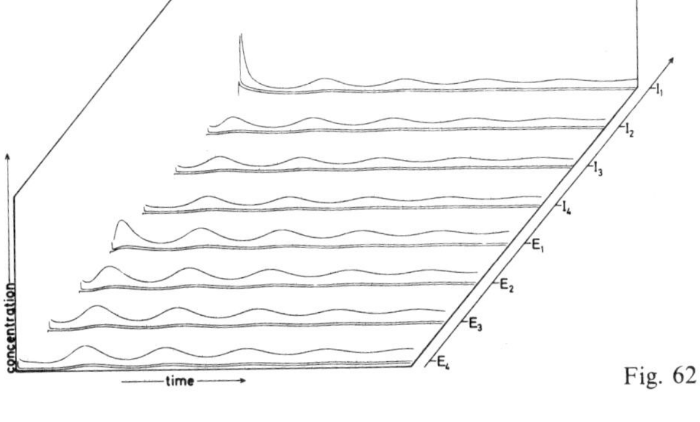

Fig. 62

In order to decide whether fluctuations of temperature improve the selection of strands with higher information content, one must analyze carefully the relative temperature coefficients of all processes involved. The tempererature coefficient of hydrolysis is likely to be the largest of all. *Instructed* replication is by no means generally enhanced at high temperatures. The incoming nucleotide has to bind cooperatively to its complementary base at the template, at the same time utilizing the stacking interaction with the top bases in the growing chain. This is not possible above the melting point of the templates. These considerations apply to any kind of environment, be it an aqueous bulk phase, a surface layer, or a compartment in a coarse-grained or porous material.

The important point for raising the information content is the relative strength of complementary vis-à-vis noncomplementary interactions. Discrimination generally works better at lower than at higher temperatures. S. Miller and L. Orgel ([63], p. 126) conclude from their experimental data:

"We do not know what the temperature was in the primitive ocean, but we can say that the instability of various organic compounds and polymers makes a compelling argument that life could not have arisen in the ocean unless the temperature was below 25° C. A temperature of 0° C would have helped greatly, and −21° C would have been even better. At such low temperatures, most of the water on the primitive earth would have been in the form of ice, with liquid sea water confined to the equatorial oceans.

There is another reason for believing that life evolved

Figs. 60 to 62. Solution curves, obtained by numerical integration, for a system of partial differential equations corresponding to the model depicted in Figure 59. The rate equations account for a growth function A_i as introduced in Part B, which refers to a four-membered hypercycle with translation (B. IX) and which has nonzero catalytic-rate terms only at the interface ($r=0$). The equations furthermore take care of adsorption and desorption (a_i, d_i describing the transition of particles between $r=0$ and $r=1$), hydrolysis (effective at $r \geqq 1$), and diffusion in the bulk of solution ($r>1$, i.e., to and from the transition layer at $r=1$). Each set of three curves refers to the three spatial positions $r=0$ (upper), $r=1$ (medium) and $r=2$ (lower). Figures 60 to 62 differ only in the assumption of different values for the stability constants of the catalytically active complexes $I_i \times E_{i-1}$, which are highest (0.16) in Figure 60, intermediate (0.06) in Figure 61, and lowest (0.04) in Figure 62. The balance between production and removal is sufficient to make the assumption of a dilution flux dispensable. As a consequence of the values chosen (relative to the uniform parameters f_i, k_i — according to B,IX — a_i, d_i and D) autocatalytic production cannot compete with removal by transport and decomposition in Figure 60, where all partners I_i and E_i die out. In both other cases a stable hypercyclic organization is established at the interface, where population numbers are either oscillatory (Fig. 61) or stationary (Fig. 62)

at low temperatures, whether in the oceans or lakes. All of the template-directed reactions that must have led to the emergence of biological organization take place only below the melting temperature of the appropriate organized polynucleotide structure. These temperatures range from 0° C, or lower, to perhaps 35° C, in the case of polynucleotide-mononucleotide helices.
The environment in which life arose is frequently referred to as a warm, dilute soup of organic compounds. We believe that a cold, concentrated soup would have provided a better environment for the origins of life."

XVIII. Continuity of Evolution

It has been the object of this final part of the trilogy to demonstrate that hypercycles may indeed represent realistic systems of matter rather than merely imaginary products of our mind.
Evolution is conservative and therefore appears to be an almost continuous process—apart from occasional drastic changes. Selection is in fact based on instabilities brought about by the appearance of advantageous mutants that cause formerly stable distributions to break down. The descendents, however, are usually so closely related to their immediate ancestors that changes emerge very gradually. Prebiotic evolution presents no exception to the rule.
Let us summarize briefly what we think are the essential stages in the transition from the nonliving to the living (cf. Fig. 63).
1. The first appearance of macromolecules is dictated by their structural stability as well as by the chemical abundances of their constituents. In the early phase, there must have been many undetermined protein-like substances and much fewer RNA-like polymers. The RNA-like polymers, however, inherit physically the property of reproducing themselves, and this is a necessary prerequisite for systematic evolution.
2. The composition of the first polynucleotides is also dictated by chemical abundance. Early nucleic acids are anything but a homogeneous class of macromolecules, including L- and D-compounds as well as various ester linkages, predominantly 2′–5′ besides 3′–5′. Reproducibility of sequences depends on faithfulness of copying. GC-rich compounds can form the longest reproducible sequences. On the other hand, AU substitutions are also necessary. They cause a certain structural flexibility that favors fast reproduction. Reproducible sequences form a quasi-species distribution, which exhibits Darwinian behavior.
3. Comma-free patterns in the quasi-species distribution qualify as messengers, while strands with exposed

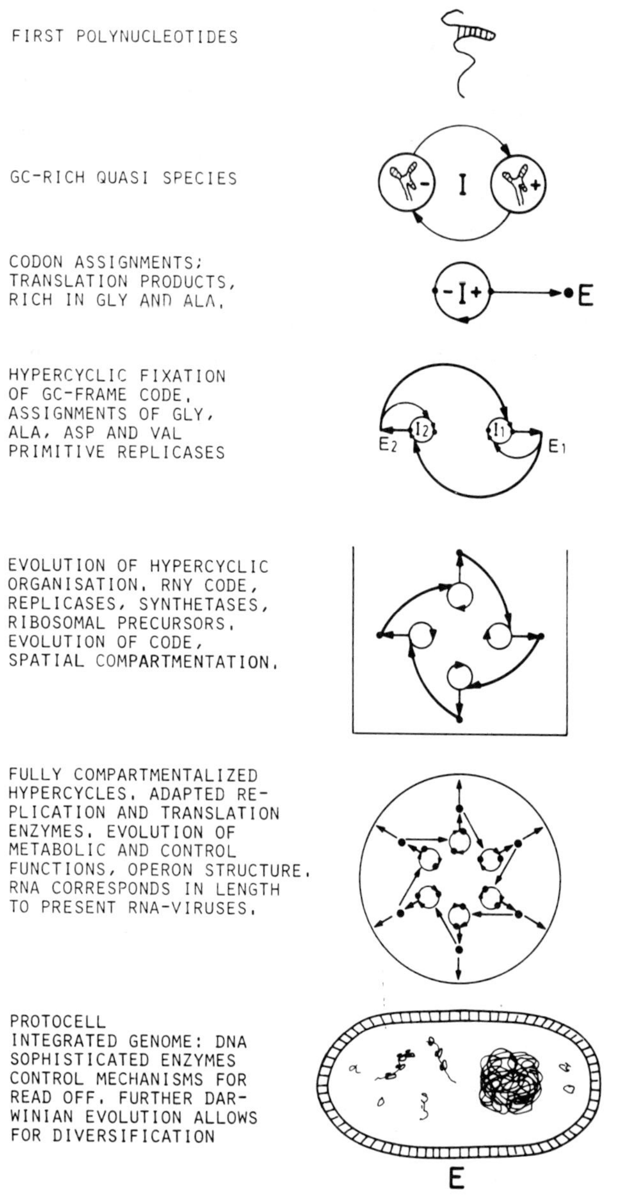

Fig. 63. Hypothetical scheme of evolution from single macromolecules to integrated cell structures

complementary patterns (possibly being the minus strands of messengers) represent suitable adapters. The first amino acids are assigned to adapters according to their availabilities. Translation products look monotonous, since they consist mainly of glycine and alanine residues. The same must be true for the bulk of noninstructed proteins.
4. If any of the possible translation products offers catalytic support for the replication of its own messenger, then this very messenger may become dominant in the quasi-species distribution and, together with

its closely related mutants, will be present in great abundance. The process may be triggered by some of the *noninstructed* environmental proteins, which in their composition reflect the relative abundance of amino acids and hence may mimic primitive *instructed* proteins in their properties.

5. The mutants of the dominant messenger – according to the criteria for hypercyclic evolution – may become integrated into the reproduction cycle, whenever they offer further advantages. Thus hypercyclic organization with several codon assignments can build up. Such a hypercyclic organization is a prerequisite for the coherent evolution of a translation apparatus. More and more mutants become integrated, and the steadily increasing fidelities will allow a prolongation of the sequences. Different enzymic functions (replicases, synthetases, ribosomal factors) may emerge from joint precursors by way of gene duplication and subsequently diverge. Units, including several structural genes, i.e., which are jointly controlled by one coupling factor.

6. The complex hypercyclic organization can only evolve further if it efficiently utilizes favorable phenotypic changes. In order to favor selectively the corresponding genotypes, spatial separation (either by compartmentation or by complex formation) becomes necessary and allows selection among alternative mutant combinations. Remnants of complex formation may be seen in the ribosomes.

We do not know at which stage such a system was able to integrate its information content completely into one giant genome molecule. For this a highly sophisticated enzymic machinery was required, and the role of information storage had to be gradually transferred to DNA (which might have happened at quite early stages).

These glimpses into the historical process of precellular evolution may suffice to show in which direction a development, triggered by hypercyclic integration of self-replicative molecular units, may lead, and how the developing system may finally converge to give an organization as complex as the prokaryotic cell. We want to stress the speculative character of part C. The early phase of self-organization left traces, but no witnesses, so that many important steps still remain in the dark.

It was not even our intention to uncover historical truth. For a process so largely dependent upon chance – where indeterminate microscopic events, such as mutations, amplify and finally determine the course of macroscopic development – a complete reconstruction of history is not possible at all. Even in biology there is a 'poverty of historicism'. On the other hand, the principles governing the historical process of evolution – even in their finer details – may well be susceptible to our understanding. The traces of history in present systems may provide enough clues to allow one day the construction of 'those n equations for the n unknowns'.

All we wanted to show in this part is that the unique class of reaction networks, which we have termed hypercycles, is indeed the simplest realistic molecular organization that ensures the coexistence of functionally related self-replicative units. Self-replication is required for the conservation of information. Hence the hypercycle is the simplest system that can allow the evolution of reproducible functional links. It can originate from one self-replicating unit and its mutants, i.e., from a single (molecular) quasi-species. Its emergence was inevitable, whenever the conditions laid down by Nature allowed it. And yet:

"If anyone can name a more beautiful triangle underlying the composition of bodies, we will greet him not as an opponent but as a friend in the right."

(Plato, Timaios) [97]

This work was greatly stimulated by discussions with Francis Crick, Stanley Miller, and Leslie Orgel; which for us meant some 'selection pressure' to look for more continuity in molecular evolution. Especially helpful were suggestions and comments by Ch. Biebricher, I. Epstein, B. Gutte, D. Pörschke, K., Sigmund, P. Woolley, and R. Wolff.

The work at Vienna was supported by the Austrian 'Fonds zur Förderung der wissenschaftlichen Forschung' (Project Nr. 3502).

Ruthild Winkler-Oswatitsch designed most of the illustrations and was always a patient and critical discussant.

Thanks to all for their help.

60. Woese, C.R.: Nature *226*, 817 (1970)
61. Fuller, W., Hodgson, A.: ibid. *215*, 817 (1967)
62. Crick, F.H.C.: J. Mol. Biol. *38*, 367 (1968)
63. Miller, S.L., Orgel, L.E.: The Origins of Life on Earth. Englewood Cliffs, N.J.: Prentice Hall 1973
64. Oró, J., Kimball, A.P.: Biochem. Biophys. Res. Commun. *2*, 407 (1960); Oró, J.: Nature *191*, 1193 (1961)
65. Lewis, J.B., Doty, P.: ibid. *225*, 510 (1970)
66. Uhlenbeck, O.C., Baller, J., Doty, P.: ibid. *225*, 508 (1970)
67. Grosjean, H., Söll, D.G., Crothers, D.M.: J. Mol. Biol. *103*, 499 (1976)
68. Coutts, S.M.: Biochim. Biophys. Acta *232*, 94 (1971)
69. Kvenvolden, K.A., et al.: Nature *228*, 923 (1970)
70. Oró, J., et al.: ibid. *230*, 105 (1971); Cromin, J.R., Moore, C.B.: Science *172*, 1327 (1971)
71. Rossman, M.G., Moras, D., Olsen, K.W.: Nature *250*, 194 (1974)
72. Walker, G.W.R.: BioSystems *9*, 139 (1977)
73. Biebricher, Ch., Eigen, M., Luce, R.: in preparation
74. Erhan, S., Greller, L.D., Rasco, B.: Z. Naturforsch. *32c*, 413 (1977)
75. Dayhoff, M.O.: Atlas of Protein Sequence and Structure, Vol. 5, Suppl. 2, p. 271 (1976)
76. Jukes, T.H.: Nature *246*, 22 (1973); Holmquist, R., Jukes, T.H., Pangburn, S.: J. Mol. Biol. *78*, 91 (1973)
77. Fox, G.E., et al.: Proc. Nat. Acad. Sci. USA *74*, 4537 (1977)

78. Uramer, F.R., et al.: J. Mol. Biol. *89*, 719 (1974)
79. Chothia, C.: ibid *75*, 295 (1973)
80. Chou, P.Y., Fasman, G.D.: ibid. *115*, 135 (1977)
81. Levitt, M., Chothia, C.: Nature *261*, 552 (1976)
82. Gutte, B.: J. Biol. Chem. *252*, 663 (1977)
83. Rigler, R.: personal communication; Rigler, R., Ehrenberg, M., Wintermeyer, W., in: Molecular Biology, Biochemistry and Biophysics Vol. 24, p. 219 (ed. I. Pecht, R. Rigler). Berlin-Heidelberg-New York: Springer 1977
84. Olson, T., et al.: J. Mol. Biol. *102*, 193 (1976)
85. Crothers, D.M., Seno, T., Söll, D.G.: Proc. Nat. Acad. Sci. USA *69*, 3063 (1972)
86. Biebricher, Ch., Druminski, M.: ibid (submitted)
87. Wahba, A.J., et al.: J. Biol. Chem. *249*, 3314 (1974)
88. Blumenthal, T., Landers, T.A., Weber, K.: Proc. Nat. Acad. Sci. USA *69*, 1313 (1972)
89. Biebricher, Ch.: in preparation
90. Meinhardt, H., in: Synergetics, p. 214 (ed. H. Haken), Berlin-Heidelberg-New York: Springer 1977
91. Fox, S.W., Dose, K.: Molecular Evolution and the Origin of Life. San Francisco: Freeman 1972
92. Dose, K., Rauchfuss, H.: Chemische Evolution und der Ursprung lebender Systeme. Stuttgart: Wissensch. Verlagsges. 1975
93. Onsager, L., in: Quantum Statistical Mechanics in Natural Sciences, p. 1 (ed. B. Kursunoglu). New York-London: Plenum Press 1973
94. Lasaga, A.C., Holland, H.D., Dwyer, M.O.: Science *174*, 53 (1971)
95. Schmidt, E.: Z. Ges. Eis- u. Kälteindustrie *44*, 163 (1937)
96. Kuhn, H., in: Synergetics, p. 200 (ed. H. Haken). Berlin-Heidelberg-New York: Springer 1977
97. Plato: Timaios
98. Schuster, P., Sigmund, K.: in preparation
99. Urbanke, W., Maass, G.: to be published

Received March 28, 1978

Reprinted from *Proc. Roy. Soc. Lond.* B **205**: 435–442.

Selection *in vitro*

By L. E. Orgel, F.R.S.

The Salk Institute for Biological Studies, P.O. Box 1809,
San Diego, California 92112, *U.S.A.*

The Qβ-polymerase–Qβ-RNA system is reviewed and the evolution of resistance to inhibitors is discussed. It is suggested that this system provides a useful model of the evolution of haploid genomes under natural selection. Consequences for theories of the origins of life are discussed.

1. Introduction

Many writers have suggested that the non-enzymic replication of an RNA- or DNA-like molecule must have been an early and essential step in the origins of life on the Earth. Attempts to demonstrate such a reaction experimentally have not progressed very far (Orgel & Lohrmann 1974), so it would not be useful to review work in this area at a meeting mainly concerned with the evolution of higher organisms. However, there is an enzymic system much less complex than a living organism whose behaviour is likely to throw light on prebiotic and biotic evolution.

The RNA polymerase of the RNA-containing bacteriophage Qβ was first isolated by Haruna & Spiegelman (1965). It is a specific polymerase in the sense that it will bring about the replication of Qβ-RNA but, in general, will not act on unrelated RNA molecules. The enzyme is robust: given a single strand of a suitable RNA substrate, it will bring about the in vitro synthesis of an indefinite number of copies. RNA synthesis is exponential so long as nucleoside triphosphates are available and the enzyme is in excess.

Spiegelman *et al.* (1975) established that, under appropriate conditions, the product made when Qβ-polymerase uses viral RNA as template is active biologically: it can direct the production of normal viruses. On the other hand, the polymerase does occasionally make mistakes, both *in vivo* and *in vitro*. The error rate is probably about 1 in 10^4 under physiological conditions.

These simple properties of Qβ-polymerase and its RNA substrates open up a new domain for the study of evolution under the influence of mutation and natural selection. One no longer needs a living organism to provide the link between the nucleic acid and its expression: one can study natural selection in a test-tube. In particular, protein synthesis, which is involved in one way or another in determining the selective value of most mutations in living organisms, is completely bypassed. Of course, selection now acts at a completely different level: normally, selection 'tests' the appropriateness of a nucleic acid sequence only after the sequence has been 'interpreted' by the protein synthetic apparatus, etc. In our in-vitro system, selection 'tests' directly the physico-chemical properties of an RNA molecule or the interaction between the RNA molecule and its polymerase. The in-vitro system provides a much less rich domain for investigation, but one which is much more easily subjected to analysis at the molecular level.

I should like, first, to make these abstract ideas clear by concentrating on a single example, the development of resistance to the drug ethidium bromide in an in-vitro system. I shall describe how the experiments were done, and then turn to the interpretation of the observed selective values in terms of molecular properties. A great deal of background material must be omitted; it is covered in an extensive review by Spiegelman *et al.* (1975).

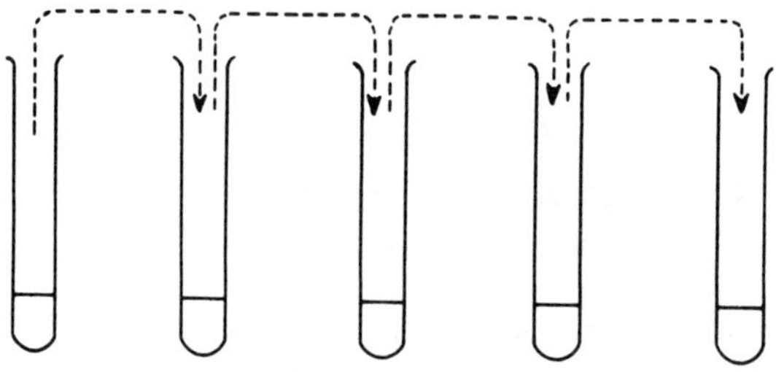

FIGURE 1. Life-cycles (4) of RNA substrates.

2. RESISTANCE TO ETHIDIUM BROMIDE

In early experiments, Spiegelman and his colleagues established that Qβ-RNA and replicating variants derived from it 'adapt' rather rapidly to unfavourable incubation conditions. They showed, for example, that RNA grown repeatedly in the presence of inhibiting concentrations of the ATP analogue tuberdicin became resistant to the drug, presumably by 'mutating away' those sites where substitution of adenosine by tubercidin is particularly damaging. Similarly, mutants that would grow at very low levels of nucleoside triphosphates were obtained by repeated growth in incubation mixtures containing abnormally low levels of the nucleoside triphosphates.

In our laboratory, the first experiments on the selection of RNAs resistant to ethidium bromide were carried out with a highly purified RNA variant called V2 (Saffhill *et al.* 1970). This RNA is replicated much faster than Qβ RNA by the polymerase: it has only about 550 bases in its sequence compared with the 3300 present in Qβ-RNA. The design of the experiment is illustrated in figure 1.

Each of the first few tubes contains a standard amount of enzyme and incubation mixture and an inhibiting concentration of ethidium bromide (2 μg/ml: 40% inhibition). An aliquot of V2 RNA is introduced into the first tube and allowed to replicate at 38 °C for 30 min. Then 9% (one-eleventh) of the sample from the first tube is added to the second tube and that tube incubated for 30 min. Transfers of this type are repeated until the amount of RNA synthesized per incubation approaches the uninhibited value. At this point a higher concentration of ethidium bromide (4 μg/ml) is added to the incubation mixture in subsequent tubes. The result is shown in figure 2. Synthesis of RNA drops at first but soon recovers (11 transfers). The concentration of drug is then doubled again. By repeating this procedure we obtained, after about 100 transfers, an RNA (V40) which would grow in the presence of 40 μg/ml of drug.

Standard inhibition studies showed that the RNA becomes progressively more resistant to ethidium bromide as the experiment progresses (figure 3). The resistance is not lost rapidly on transfer in the absence of ethidium bromide. Most interestingly, fluorescence studies show that the adapted V40 RNA binds less ethidium bromide than does the starting V2 RNA (figure 4). In the absence of ethidium bromide, V40 grows at only about 40% of the rate achieved by V2 under the same conditions.

These experiments suggest that V2 RNA molecules bind ethidium bromide strongly at a few special positions in the molecule in such a way as to slow down their replication. Mutant RNAs arise from time to time in which these special sites are modified so that they no longer bind the drug. Selection in the presence of the drug then leads to the outgrowth of resistant mutants. However, the mutants are less well adapted than the wild type to growth in the absence of the drug. These speculations were confirmed in an elegant series of experiments carried

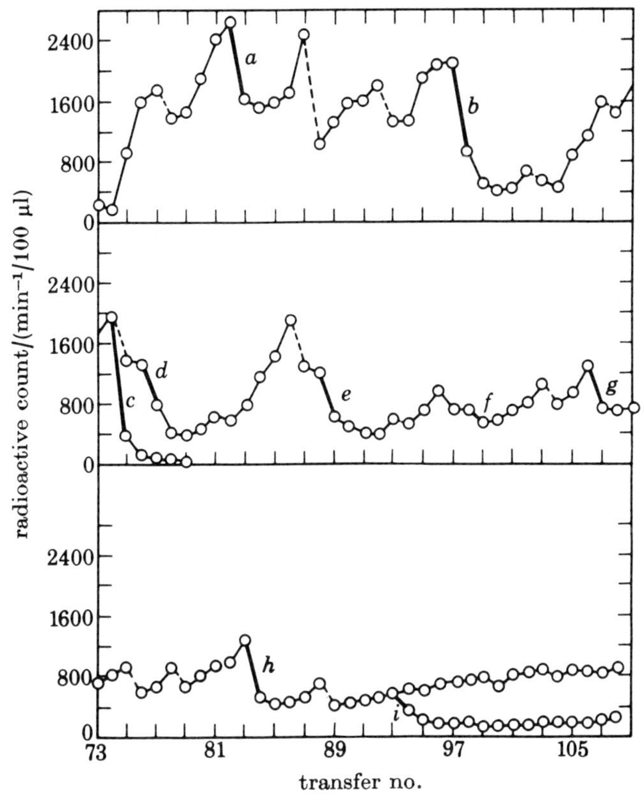

FIGURE 2. Radioactivity incorporated in precipitated material after successive transfers of Qβ variant RNA in the presence of increasing concentrations of ethidium bromide. Broken lines indicate overnight storage at − 15 °C between transfers. Thick lines indicate an increase in ethidium bromide concentration: (*a*) 2 to 4 μg/ml; (*b*) to 8 μg/ml; (*c*) to 16 μg/ml; (*d*) to 12 μg/ml; (*e*) to 16 μg/ml; (*f*) to 20 μg/ml; (*g*) to 30 μg/ml; (*h*) to 40 μg/ml; (*i*) to 50 μg/ml. (Reproduced from Saffhill *et al.* (1970).)

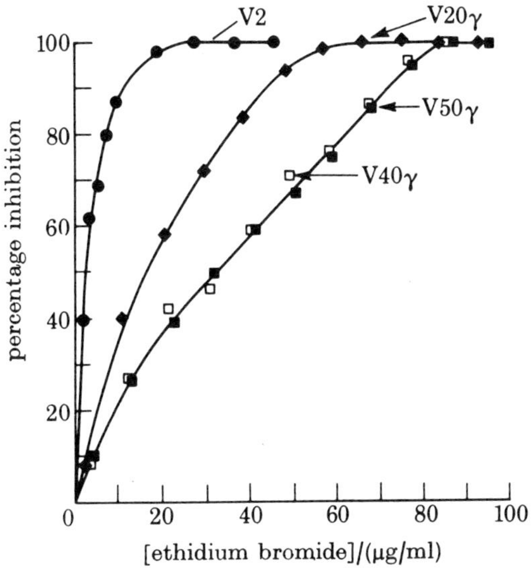

FIGURE 3. Inhibition of replication by ethidium bromide. V2 is the input sequence. V20γ, V40γ and V50γ are variants resistant to 20, 40 and 50 μg/ml ethidium bromide, respectively. (Reproduced from Saffhill *et al.* (1970).)

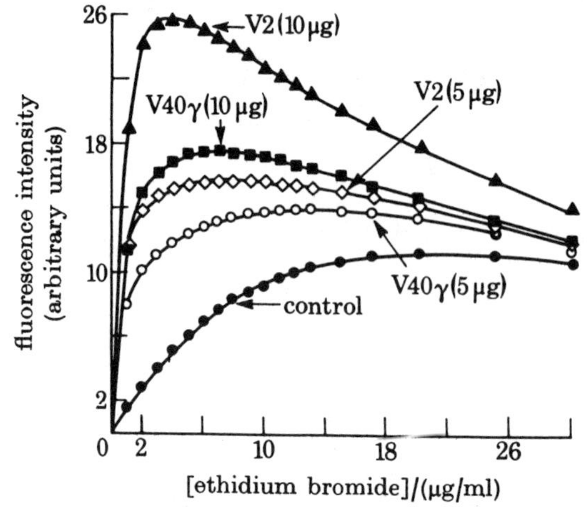

FIGURE 4. Fluorescence of ethidium bromide in the presence of V2 and V40γ in 0.1 M Tris buffer (pH 7.4). The control has no added RNA. (Reproduced from Saffhill *et al.* (1970).)

out in Spiegelman's laboratory (Kramer *et al.* 1974).

This set of experiments used another RNA variant, MDV-1, which contains only 218 nucleotides in its sequence. The complete sequence of the RNA was determined by Mills *et al.* (figure 5). A series of 25 sequential transfers was carried out in 15 μM ethidium bromide. In these experiments only 10^{-5} of the material in each tube was transferred to initiate the next reaction, so as to maximize the period of exponential growth.

At the end of this series of transfers an RNA population was obtained that was more resistant than the wild-type to ethidium bromide (figure 6). The sequence of this RNA was determined and found to differ from that of MDV-1 at just three

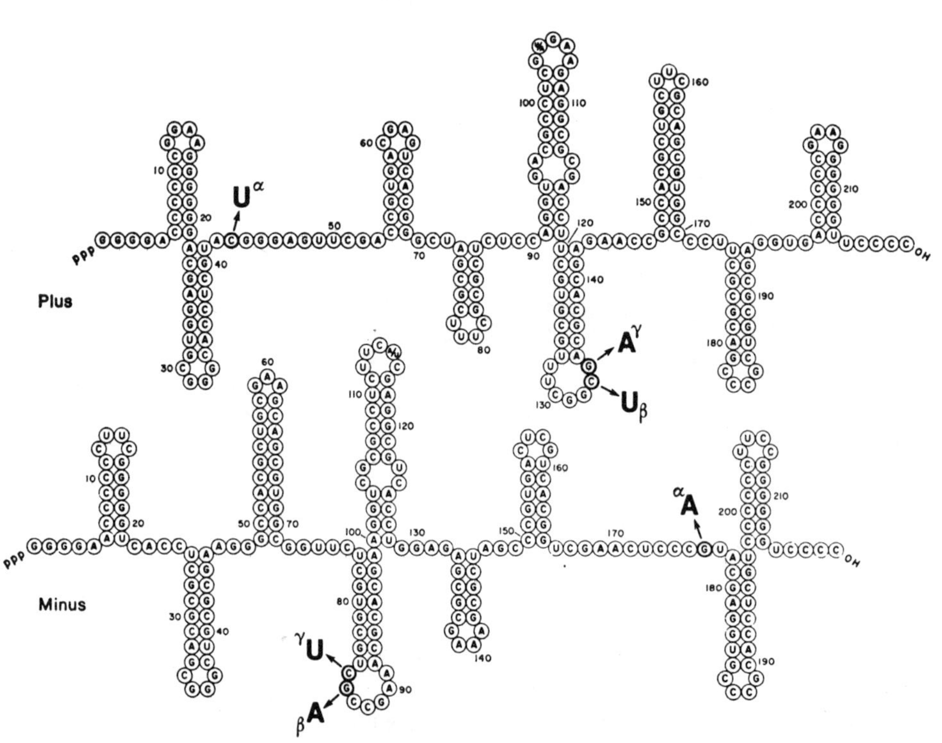

FIGURE 5. Nucleotide sequence of mutant RNA. Both strands of MDV-1 RNA are illustrated. The bold letters and arrows indicate the identity and location of each nucleotide substitution that occurred. The three mutations are arbitrarily labelled α, β and γ. (Reproduced from Kramer *et al.* (1974).)

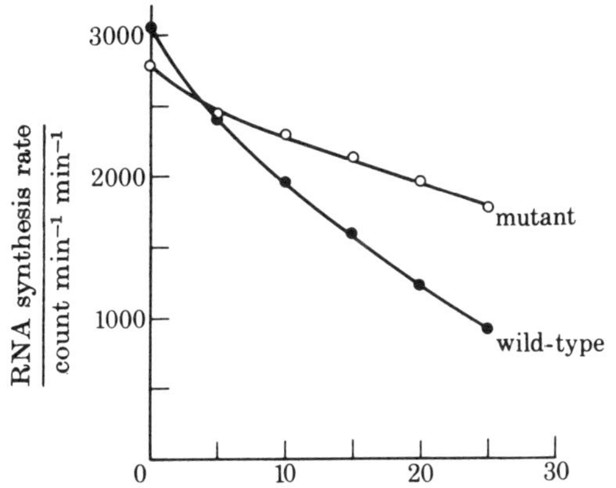

FIGURE 6

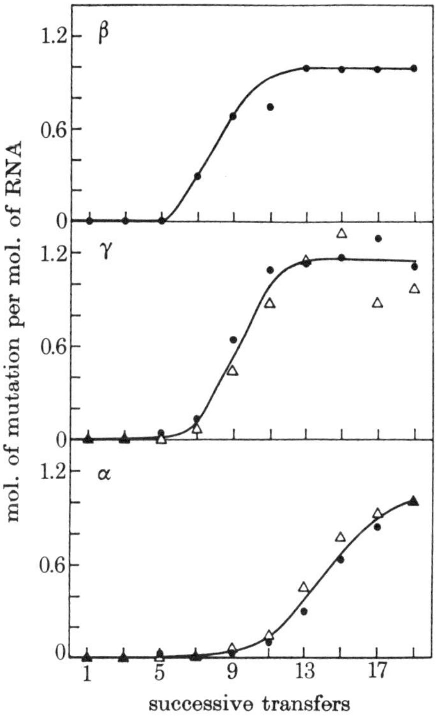

FIGURE 7

FIGURE 6. Effect of ethidium bromide on mutant and wild-type RNA synthesis. (Reproduced from Kramer *et al.* (1974).)

FIGURE 7. The appearance of mutations α, β and γ as a function of the number of transfers. (Reproduced from Kramer *et al.* (1974).)

positions (figure 5). The amount of drug bound to MDV-1 and to the adapted variant was determined by equilibrium dialysis. The three mutations had reduced drug-binding by about 35 %.

A further series of experiments was carried out to decide whether the final adapted RNA was derived from a single variant strand in the original MDV-1 preparation, or whether it arose *de novo* during adaptation. In the former case, the three mutations α, β and γ should have been present in equal amounts at each stage of adaptation; in the latter, the mutations would have appeared at different stages in the transfer series. The results shown in figure 7 indicate clearly that the mutation α appeared after β and γ showing that the resistant variant arose *de novo* A more detailed analysis showed that β appeared before γ.

3. Evolution of RNAs

The ease with which we can select RNA molecules that have lost some of their binding sites for ethidium bromide is highly suggestive. Many important control mechanisms involve the interaction of nucleic acids with proteins, for example the interaction of DNA with RNA polymerase and the interaction of RNA with ribosomal proteins. The alteration of the sequence of an RNA so as to change its interaction with a protein should be achieved at least as easily as the alterations which change the RNA–ethidium bromide interaction. As a result of our work with ethidium bromide, we concluded that many aspects of cellular control are likely to evolve through the modification of the sequences of promotors, operators, etc., more readily than by the modification of the proteins with which they interact.

The direct interaction of polynucleotides with small molecules is likely to have been particularly important at the earliest stages in the evolution of life. The origin of the genetic code, for example, may have depended on the stereospecific interaction of amino acids with tRNA-like molecules. Unfortunately, virtually nothing is known about the strength and specificity of these interactions, and the subject is not easily attacked by conventional methods. The selective experiments with ethidium bromide, although only indirectly relevant, strongly suggest that specific positive interactions could evolve relatively quickly, at least in the case of amino acids such as tryptophan and tyrosine which have large, planar side chains. I suspect that a fair level of discrimination could be achieved between other pairs of amino acids.

Finally, I should like to emphasize the wide range of experiments in natural selection that could be done with the use of the Qβ system. It should be trivial for example to select for molecules that (1) stick (or fail to stick) to a chosen surface; (2) pass through (or fail to pass through) a porous barrier; (3) are resistant to destruction by an enzyme or other physical or chemical agent.

With slightly greater ingenuity it should be possible to select for molecules that 'recognize' small molecules in their environment and utilize such molecules to increase their own lifetime or rate of replication or even to interfere with the replication of their competitors. Here perhaps lies an approach to the origin of the 'selfish gene'. I find it surprising that so little has been done along these lines; perhaps it is because experiments in this field are difficult to fund.

The Qβ polymerase system has been studied more extensively than any other in the context of natural selection *in vitro*. However, it is not unique. We found that the DNA-dependent polymerase of *Escherichia coli*, under appropriate conditions, will select from a totally random mixture of polynucleotides one or more strands which it is capable of replicating indefinitely. This suggests (Biebricher & Orgel 1973) that any RNA polymerase will find some substrates on which it acts as a replicase, even if the enzyme does not have a replicase function in nature.

4. Population structure

The theory of natural selection is concerned with the development of populations in time. At first sight, this does not seem to be so for the Qβ system *in vivo* or *in vitro*: the wild-type sequence appears on the printed page as a unique sequence of bases. However, a more detailed consideration of the experimental situation shows that the idea of a unique, wild-type sequence is an oversimplification. A typical analysis of an RNA would surely pick up ambiguity at a position in the sequence if the two alternative bases at that position were present in comparable amounts, but would probably miss ambiguity below the 10% level and would certainly miss it at the 1% level. A great deal of variation could, therefore, be present without any being detected by the normal procedures of sequence analysis. Eigen has emphasized that a wild-type population might contain only a very small fraction of wild-type sequences (Eigen & Schuster 1977).

The mutation rate for an RNA polymerase is typically 10^{-4} per base per replication. Suppose that the selective disadvantage of a typical mutation is 10%. Then, at equilibrium, one must expect a mutant frequency of about 10^{-3} at a typical site. The majority of Qβ phages in a wild type population would have at least 1 mutation among their *ca.* 3000 bases. Under in-vitro conditions mutation rates are likely to be higher and selection pressure may well be much weaker. Thus the amount of variation could be much larger. Population structure of this kind will certainly prove important for the analysis of the rate of evolution of RNA populations.

REFERENCES (Orgel)

Biebricher, C. K. & Orgel, L. E. 1973 *Proc. natn. Acad. Sci. U.S.A.* **70**, 934.
Eigen, M. & Schuster, P. 1977 *Naturwissenschaften* **64**, 541.
Haruna, I. & Spiegelman, S. 1965 *Proc. natn. Acad. Sci. U.S.A.* **54**, 579.
Kramer, F. R., Mills, D. R., Cole, P. E., Nishihara, T. & Spiegelman, S. 1974 *J. molec. Biol.* **89**, 719.
Orgel, L. E. & Lohrmann, R. 1974 *Acct. chem. Res.* **7**, 368.
Saffhill, R., Schneider-Bernloehr, H., Orgel, L. E. & Spiegelman, S. 1970 *J. molec. Biol.* **51**, 531.
Spiegelman, S., Mills, D. R. & Kramer, F. R. 1975 In *Stability and origins of biological information* (ed. I. R. Miller), p. 183. New York: Halsted Press.

Reprinted from *J. Mol. Evol.* **19**: 219–225.

Evolution of the Amino Acid Code: Inferences from Mitochondrial Codes

Thomas H. Jukes

Space Sciences Laboratory, University of California, Berkeley, California 94720, USA

Summary. The amino acid code is usually presented as a table of 64 codons. Actually the code results from the action of tRNA molecules that carry amino acids to codons in mRNA by means of codon-anticodon pairing. The tRNA molecules are transcribed from genes that undergo evolution and the number of anticodons can therefore increase during evolution, but the number of codons is fixed at 64. Mammalian mitochondrial codes contain only 22 anticodons for 20 amino acids as compared with 54 anticodons for 20 amino acids in the universal code. It is proposed that an archetypal code containing 16 anticodons for 15 amino acids evolved into the universal code by gene duplication, followed by mutations that modified the anticodons and amino acid acceptor sites. In substantiation of this proposal, it is noted that the mammalian mitochondrial code is simplified by comparison with the universal code. For example, single anticodons are used for each of eight amino acids in the mammalian mitochondrial code. This simplification may represent an evolutionary retrogression towards the proposed archetypal code.

Key words: Amino acid code – Molecular evolution – Mitochondrial code – Anticodons – Archetypal code

The Universal Amino Acid Code

The universal code is commonly presented as showing the assignment of codons to amino acids (Table 1). These assignments were established by various procedures, especially by the alignment of DNA and RNA sequences with polypeptide sequences of proteins (Jukes 1977). Translation of mRNA into proteins is by means of codon-anticodon pairing in which third bases of codons pair with the first bases of anticodons in tRNA molecules according to wobble rules of pairing (Crick 1966). The first two codon bases of codons pair conventionally with the second and third anticodon bases. Thus codons UUU and UUC both pair with phenylalanine anticodon GAA in opposite (antiparallel) directions.

The prototype of Table 1 was first published by Eck (1963). The early studies on synthetic polynucleotides and the amino acid code had shown a tendency for a single amino acid to have two or more codons with a common pair of bases, a "shared doublet." For example, threonine was incorporated into polypeptides by using polynucleotides containing A and C, with or without U. Eck took the bold step of suggesting that the entire code was patterned on the basis of "shared doublets," even though no information was available on polynucleotides that contained G.

Wobble rules (Table 2) enable a table of predicted anticodons to be made (Table 3). Most of these have been verified by their presence in the anticodon loops of tRNA molecules from various prokaryotic and eukaryotic organisms. The missing anticodons (in parentheses) will presumably by discovered in other tRNAs, either in completed molecules or in their genes. The second and third bases of anticodons have been identified in a tRNA of each amino acid. The first, or wobble-pairing bases, are the distinctive feature upon which the arrangement of the code into 16 quartets (Table 1) depends. In many cases, these bases undergo post-transcriptional modification (Jukes 1977).

U, G and C are all used in tRNAs in first anticodon positions, but A is deaminated to H (hypoxanthine) which is the base of the nucleoside I (inosine), pairing as shown in Table 2. Because I pairs with U, C and A, anticodons starting with I can be used only in "family boxes," which are the boxes in which one amino acid has four codons (three in the case of box 3). For exam-

Table 1. Codons in the amino acid code arranged in numbered "boxes". Boxes 2, 4, 5, 6, 7, 8, 14 and 16 are "family boxes"

Box	Codon	Amino acid	Box	Codon	Amino acid
1.	UUU	phenylalanine	9.	UAU	tyrosine
	UUC	phenylalanine		UAC	tyrosine
	UUA	leucine		UAA	terminator
	UUG	leucine		UAG	terminator
2.	CUU	leucine	10.	CAU	histidine
	CUC	leucine		CAC	histidine
	CUA	leucine		CAA	glutamine
	CUG	leucine		CAG	glutamine
3.	AUU	isoleucine	11.	AAU	asparagine
	AUC	isoleucine		AAC	asparagine
	AUA	isoleucine		AAA	lysine
	AUG	methionine		AAG	lysine
4.	GUU	valine	12.	GAU	aspartic acid
	GUC	valine		GAC	aspartic acid
	GUA	valine		GAA	glutamic acid
	GUG	valine		GAG	glutamic acid
5.	UCU	serine	13.	UGU	cysteine
	UCC	serine		UGC	cysteine
	UCA	serine		UGA	terminator
	UCG	serine		UGG	tryptophan
6.	CCU	proline	14.	CGU	arginine
	CCC	proline		CGC	arginine
	CCA	proline		CGA	arginine
	CCG	proline		CGG	arginine
7.	ACU	threonine	15.	AGU	serine
	ACC	threonine		AGC	serine
	ACA	threonine		AGA	arginine
	ACG	threonine		AGG	arginine
8.	GCU	alanine	16.	GGU	glycine
	GCC	alanine		GGC	glycine
	GCA	alanine		GGA	glycine
	GCG	alanine		GGG	glycine

Table 2. Wobble pairing in the universal code (Crick 1966)

First anticodon base	Pairs with third codon bases
U	A, G
G	U, C
H (I)	U, C, A
C	G

Crick (1966) added the following footnote: "It seems likely that inosine will be formed enzymically from an adenine in the nascent sRNA. This may mean that A in this position will be rare or absent, depending upon the exact specificity of the enzyme(s) involved"

ple, anticodon IAA is excluded because it would pair with two phenylalanine and one leucine codons, which would give rise to ambiguity in protein synthesis.

Some explanation of the pairing between codons in Table 1 and anticodons in Table 3 is in order. As an example, in the first box (Tables 1 and 3), UUU and UUC pair with GAA, UUA and UUG pair with UAA, and UUG pairs with CAA. Box 4 has a set of four anticodons. According to wobble rules, IAC pairs with GUU, GUC and GUA, GAC with GUU and GUC; UAC with GUA and GUG; and CAC with GUG. However, the U in anticodons UNN is often modified to pair only with A. In box 9, UAU and UAC pair with GUA, while UAA and UAG have no anticodons with which to pair, and they are chain-terminating ("stop") codons. In box 13, UGU and UGC pair with GCA, UGA has no anticodon, and is a stop codon, and UGG pairs with CCA. In box 3, AUA pairs with CAU, in which the first C is modified, and AUG pairs only with CAU. The other boxes follow pairing procedures described for boxes 1 and 4. In many cases, the first anticodon base U is modified to pair only with A. Other modifications of first anticodon bases were listed by Jukes (1977).

Mitochondrial Codes

Complete nucleotide sequences have been determined for mitochondria of three mammals; human, bovine and mouse (Anderson et al. 1981; Anderson et al. 1982; Bibb et al. 1981) so that all of their anticodons have been identified. In addition, sequences of a number of mitochondrial tRNAs have been described for yeast, *Neurospora* and *Aspergillus* (Bonitz et al. 1980; Heckman et al. 1980; Köchel et al. 1981). The anticodons in the mammalian mitochondrial code are in Table 4. A, I, and C are absent from first anticodon positions except for CAU, methionine, box 3. Wobble pairing for G and U follows the rules in Table 2 in boxes 1, 3, 9, 10, 11, 12, 13 and 15, but in the eight family boxes,

Table 3. Anticodons in the universal amino acid code. Anticodons in parentheses have not yet been discovered in tRNAs that have been isolated and sequenced

		IGA	Ser				
GAA	Phe	(GGA)	Ser	GUA	Tyr	GCA	Cys
UAA	Leu	UGA	Ser				
CAA	Leu	CGA	Ser			CCA	Trp
IAG	Leu	IGG	Pro			ICG	Arg
GAG	Leu	(GGG)	Pro	GUG	His	GCG	Arg
UAG	Leu	UGG	Pro	UUG	Gln	UCG	Arg
CAG	Leu	(CGG)	Pro	CUG	Gln	CCG	Arg
IAU	Ile	IGU	Thr				
GAU	Ile	GGU	Thr	GUU	Asn	GCU	Ser
*CAU	Ile	UGU	Thr	UUU	Lys	UCU	Arg
CAU	Met	(CGU)	Thr	CUU	Lys	(CCU)	Arg
IAC	Val	IGC	Ala			(ICC)	Gly
GAC	Val	(GGC)	Ala	GUC	Asp	GCC	Gly
UAC	Val	UGC	Ala	UUC	Glu	UCC	Gly
CAC	Val	CGC	Ala	CUC	Glu	CCC	Gly

*C is modified to pair only with A

Table 4. Mammalian mitochondrial code in the form of anticodons

GAA	Phe			GUA	Tyr	CGA	Cys
UAA	Leu	UGA	Ser			UCA	Trp
				GUG	His		
UAG	Leu	UGG	Pro	UUG	Gln	UCG	Arg
GAU	Ile			GUU	Asn	GCU	Ser
CAU	Met	UGU	Thr	UUU	Lys		
				GUC	Asp		
UAC	Val	UGC	Ala	UUC	Glu	UCC	Gly

2, 4, 5, 6, 7, 8, 14 and 16, the pairing system differs from the universal code because there is only one anticodon per box, starting with U, which pairs with U, C, A and G in the first position of codons. The mammalian mitochondrial code (MMC) may have been derived from the universal code by a retrogressive simplification in which many anticodons have been deleted (Jukes 1981), so that the MMC contains the bare minimum of anticodons needed for the 20 amino acids used in protein synthesis, and for all possible codons except for four stop codons. This minimum is reached by not using anticodons containing C or A (I) in first positions (except for CAU), and by also eliminating G from anticodons in family boxes, which contain only anticodons starting with U.

Polypeptide Chain Initiation and Termination

Mammalian mitochondria have less DNA and fewer tRNA genes than have yeast or *Aspergillus* (Bonitz 1980; Heckman 1980; Köchel 1981). This provides evidence that evolutionary simplification has been in progress since the separation of lines leading to Mammalia and Ascomycetes, and thus supports the proposal that simplification of the code has taken place in mitochondria. Mammalian mitochondria have only one gene for methionine tRNA, which resembles initiator tRNA (Anderson et al. 1981) and has anticodon CAU which normally pairs with AUG. Anderson and coworkers (1981) note that, in human mitochondria, AUA and probably AUU may be initiator codons in addition to AUG. Bibb et al. (1981) conclude that all four codons ATN (AUN) may initiate translation in mouse mitochondria and "there may be only one methionine tRNA gene from which the two tRNA species [initiator methionine tRNA and internal methionine tRNA] are produced by differential modification of a primary transcript."

Methionine tRNA and formylmethionine (initiator) tRNA, both with anticodon CAU, are present in yeast (*S. cerevisiae*) mitochondria, in which AUG is a methionine codon (Bonitz et al. 1980). *Aspergillus nidulans* has three methionine and two glycine tRNAs (Köchel 1981). The first and second methionine tRNAs "seem to be functionally analogous to *N. crassa* mitochondrial initiator and non-initiator" by sequence homologies (Köchel 1981). The isoleucine tRNA that pairs with codon AUA in the prokaryotic universal code is absent from mammalian mitochondria, which use both codons AUA and AUG for methionine. This isoleucine tRNA in prokaryotes has anticodon *CAU, in which the *C has been modified so that the anticodon pairs with AUA but not with AUG (Kuchino et al. 1980; Fukada and Abelson 1980).

The coding capacity of mammalian mitochondrial DNA is small, and much of it is known to be used for purposes other than that of coding for ribosomal proteins. Although the codon-anticodon pairing system in the mitochondrial code is simplified, ribosomes in mitochondria are large and complex. Mitochondrial ribosomes contain many proteins nearly all of which must originate in the cytoplasm and migrate into the mitochondria. In view of the large size and complexity of mammalian mitochondrial ribosomes, initiation and termination may be just as complex in mitochondria as in cytoplasm, in which termination involves release factor proteins, peptidyl transferase and binding factors. Nevertheless, there are clues in the mitochondrial code to the possible origin of terminator codons. AGR codons have apparently evolved from coding for arginine in ascomycete mitochondria to chain termination in human and bovine mitochondria, which lack the anticodon UCU for AGR codons. This indicates that a codon may acquire chain terminating property by loss of its anticodon. A second fact supports the conclusion: in mitochondrial codes, UGA and UGG are codons for tryptophan and both pair with anticodon UCA, while in the universal code, UGG is the sole codon for tryptophan, and pairs with CCA, which does not pair with stop codon UGA. Therefore, mutation of anticodon UCA to CCA would leave UGA without an anticodon, and it would become a stop codon. This may have taken place in evolution of the universal code from an earlier code that resembled the present mitochondrial code in using UGA and UGG as tryptophan codons.

The other stop codons, UAA and UAG, also lack the cognate anticodon UUA, in both the universal and mitochondrial codes. Balasubramanian (1982) has proposed that UAA and UAG were primitive stop codons because of conclusions derived from model-building.

The Archetypal Code

I proposed (Jukes 1966) that an archetypal code for 15 amino acids once existed in which a single anticodon was used in each of the 16 family boxes. It is now evident that such anticodons would have U in the first position. The reason for U, rather than C, A or G, is

Table 5. Anticodons of archetypal code, each pairing with four codons for a single amino acid, except for GUA, pairing only with AUU and AUC

UAA	Phe or Leu	UGA	Ser	GUA	Tyr	UCA	Cys or Trp
UAG	Leu	UGG	Pro	UUG	His or Gln	UCG	Arg
UAU	Ile or Met	UGU	Thr	UUU	Asn or Lys	UCU	Arg or Ser
UAC	Val	UGC	Ala	UUC	Asp or Glu	UCC	Gly

explained by Grosjean et al. (1978) and Barrell et al. (1980) as being that U is "the only base that can possibly form a stable base pair with either A, G, C or U." The discovery of the MMC shows that U would be the first anticodon base in such an archetypal system. It can also be deduced that fourfold pairing with U, as in the MMC, is indeed a retrogression to a primitive and simple coding system. This simplification evidently took place while the ancestor of mammalian mitochondria underwent drastic reduction of its DNA content. Woese (1981) has proposed that this ancestor was probably a purple photosynthetic bacterium. Presumably it used the universal code. Woese's proposal is based upon homologies found by him between 16-S ribosomal RNA sequences of various organisms. During evolution, deletion of tRNA genes from the mitochondrial genomes took place so that all anticodons starting with A or C were "pruned out," and so were all anticodons starting with G in family boxes. Only 22 anticodons (and tRNA genes) remained. Under wobble rules in Table 4, these 22 were sufficient to pair with 60 codons, leaving codons UAA, UAG, AGA and AGG unpaired as stop codons. It is noteworthy in Table 4 that U in the first anticodon position pairs with U, C, A and G in the family boxes, but only with A and G in other boxes. Heckman et al. (1980) found that a modified U, pairing only with A and G rather than with U, C, A and G, was present in the first position of the anticodons of *Neurospora crassa* mitochondrial tRNAs for glutamine, leucine (UUR codons) and tryptophan. Anderson et al. (1981) propose that modification of U in the first position of anticodons in the non-family boxes of the MMC has also taken place.

A feature of molecular evolution is the occurrence of gene duplication, followed either by differentiation of one of the duplicate genes to perform a new function (Jukes 1966; Li 1982), or by retaining duplicate genes for purposes of redundancy, or by the duplicates becoming non-functional pseudogenes. Duplication provides for increasing complexity of organisms with the progress of time. Li tabulated the number of tRNA genes in several haploid genomes, and notes that *E. coli* contained about 100. Rough estimates for *Saccharomyces cerevisiae* were 320 to 400, *Drosophila melanogaster* 750, *Xenopus laevis* 7,800, and human 1,300 (Li 1982).

The ancestry of living organisms is explored by studying the fossil record and by comparing existing organisms with each other to construct phylogenies. In molecular studies, comparisons are made of the sequences of homologous nucleic acids and proteins as a means of deducing the sequence of a common ancestor.

The mammalian mitochondrial code, by having retrogressed to a simplified and primitive form, provides a unique opportunity for detecting steps in the early evolution of the code. There are eight family boxes in the MMC (Table 4) in which a single anticodon pairs with four codons (boxes 2, 4, 5, 6, 7, 8, 14 and 16). Boxes 1, 3, 10, 11, 12, and 13 can be converted to a similar archetypal form by reducing the number of anticodons in each box from two to one. In this way the composition of an archetypal code can be postulated as having a single anticodon, starting with U and pairing with four codons, in each of 15 boxes (Table 5). Box 9 contains GUA rather than UUA, thus providing an anticodon for UAU and UAC, but none for UAA and UAG, the stop codons.

This postulation leads to an archetypal code containing 14 or 15 amino acids.

Expanding the Archetypal Code to the Universal Code

Having gone backward to this point, we can now go forward from it. The universal code could evolve from the archetypal code by gene duplication followed by mutational changes that would convert codon-anticodon pairing from the fourfold system (U with U, C, A and G) to the twofold system (G with U and C; U with A and G). The generalized procedure is as follows, where UNN is an anticodon in an archetypal tRNA molecule:

$$\text{UNN} \rightarrow 2\ \text{UNN} \rightarrow \text{UNN} + \text{GNN}.$$

Following the second of these steps, another mutation took place in seven of the boxes. This mutation changed the aminoacylation site in tRNA with either UNN or GNN so that a "new" amino acid was accepted. The other eight boxes continued to contain the same single amino acid, but with two anticodons instead of one. The proposal accounts for the fact that no box in the universal code contains more than two amino acids.

As an example, the gene for tRNA containing anticodon UUG in the archetypal code duplicated. Let us assume that this was $tRNA_{His}$. The anticodon in one of the duplicates mutated to GUG. The other duplicate tRNA retained anticodon UUG and underwent a mutation in the aminoacylation site so that it became charged with glutamine. In this way, the box originally containing only histidine now contained histidine and glutamine. Alternatively, UUG may originally have been the glutamine anticodon, in which case similar duplication and mutation procedures could have resulted in the addition of histidine to the code.

The change in aminoacylation postulated above has a known counterpart in evolution of yeast mitochondrial code. Yeast mitochondria use anticodon UAG for threonine (Bonitz et al. 1980), but the related ascomycetes *Neurospora* and *Aspergillus* use UAG for leucine, as do all other known codes. Therefore, following the divergence of the ancestor of yeast from the common ancestor of *Ascomycetes*, a mutation took place in the aminoacylation site of $tRNA_{Leu}$ so that it accepted threonine instead of leucine.

It is a matter of speculation, in the absence of other information, as to which was the original or "primitive" amino acid in each of the boxes in the universal code containing two amino acids. In the case of phenylalanine and leucine (box 1), archetypal anticodon UAA might have been for phenylalanine, and leucine might have "acquired" the UUA and UUG codons by the procedures described above. Conversely, leucine might have originally had both anticodons UAA and UAG in the archetypal code, with eight codons, and have "lost" codons UUU and UUC to a new amino acid, phenylalanine. Some "educated guesses" can be made as to which of a pair of amino acids is the more primitive, by noting which amino acids are formed more readily in abiotic syntheses that supposedly simulate conditions on the early earth. On this basis, isoleucine is probably more primitive than the more complex methionine, and isoleucine could have been used archetypally for initiation of translation. Asparagine is readily formed by amidation of aspartic acid, and may be more primitive than lysine. Aspartic acid is produced in abiotic synthesis in larger quantities than glutamic acid. Cysteine is likely more primitive than the complex and unstable tryptophan. The decision between serine and arginine for anticodon UCU in the archetypal code is an interesting one; each of these amino acids has another box to itself. If similarity of anticodons (UCU and UCG are more similar to each other than UCU and UGA) is a clue, then arginine would be the original occupant of the box containing anticodon UCU.

A second factor in deciding the entrance of new amino acids into the archetypal code must have been the relative ease with which the aminoacylation site could be modified by mutation to accept the new amino acid. This clue comes from the finding that threonine displaced leucine in aminoacylation of yeast mitochondrial tRNA with anticodon UAG (Bonitz et al. 1980). Perhaps a single nucleotide substitution was suffioient to cause the change, for it is difficult to think of it as resulting from a series of intermediate steps; aminoacylation is an all-or-none reaction. Such considerations may have

Table 6. Proposed "early" universal code in the form of anticodons. Each anticodon pairs with two codons by wobble pairing in the first anticodon position; G with U and C, U with A and G

GAA	Phe	GGA	Ser	GUA	Tyr	GCA	Cys
UAA	Leu	UGA	Ser			UCA	Trp
GAG	Leu	GGG	Pro	GUG	His	GCG	Arg
UAG	Leu	UGG	Pro	UUG	Gln	UCG	Arg
GAU	Ile	GGU	Thr	GUU	Asn	GCU	Ser
UAU	Met	UGU	Thr	UUU	Lys	UCU	Arg
GAC	Val	GGC	Ala	GUC	Asp	GCC	Gly
UAC	Val	UGC	Ala	UUC	Glu	UCC	Gly

been paramount as the code proceeded to evolve from Table 5 to Table 6. Lagerkvist (1981) has proposed that affinity of codon-anticodon pairing played a part, because codons beginning with CC, CG, GC and GG are all present in family boxes, while codons beginning with UU, UA, AU and AA are all in non-family boxes. This, of course, would correspond to anticodons ending with these two base pairs in Table 6. It is also possible that this distribution may be only a coincidence. The primary factor in deciding which of the family boxes in the archetypal code became a non-family box in the early universal code (Table 6) may have been the effect of mutations in the aminoacylation site. Another factor could have been the extent to which each amino acid was used in proteins. For example, cysteine is used sparingly and does not "need" four codons (Jukes et al. 1975), while alanine, glycine and valine are used abundantly enough that each requires four codons, which they now possess.

The question of which of the pairs of amino acids were more primitive should encourage the reader to further speculation.

Freezing the Code

Addition of new amino acids to the code could have proceeded until a total of 28 amino acids was reached. That this did not happen was presumably caused by a freezing of the code when the total was 20. The reason for freezing has been discussed by Crick (1968) as being the reaching of a point at which further changes in the code could not be tolerated.

There are two different ways of considering the origin of the universal amino acid code. The first approach says that the code is a "frozen accident," a term used by Crick (1968). The essence of this is that the present universal code cannot be changed without intolerable consequences to living organisms, because any change would be dispersed throughout all proteins in the cell. This would alter their properties to an extent that would be fatal. To quote Crick, "At the present time, any change would be lethal, or at least very strongly selected against ... because the code determines the amino acid sequences of so many highly evolved protein molecules .... " The idea of the "frozen accident" is that all living organisms have evolved from an ancient single ancestor, or a small group of ancestors, that displaced all contemporaries with other codes. After the evolutionary expansion of the descendants of this founder group had started, changes in the amino acid assignments of codons were not possible.

In support of the "frozen accident" model, we have pointed out that the code is frozen, rather than optimized, because it is not even the best code for presently existing organisms (Jukes et al. 1975). Arginine has six codons when it actually needs two or three in terms of the extent to which it is used, on the average, in proteins. Lysine has only two codons, but the extent of its use demands that it should really have four. Aspartic and glutamic acids should have more than two codons apiece. These conclusions come to light when the distribution of codons among amino acids is compared with the percentage occurrence of amino acids in proteins.

The second theory is that the code is a product of evolution and its composition was inevitable because it is the "best possible code." The theory often postulates a specific fit or affinity between each amino acid and its codon or anticodon. This is discussed as the "stereochemical theory" by Crick (1968). The number of codons for an amino acid, in this theory, depends upon the extent to which the amino acid is needed in proteins. For example, the simple amino acid, alanine, with four codons, is used four times as frequently as tryptophan, with one codon. It can be argued, in reverse, that the fact that alanine is used four times as often as tryptophan is because it has four times as many codons, and the composition of DNA governs the amino acid content of proteins.

Stereochemical affinity may have played a part in the early emergence of the code. However, in terms of the present code, the "stereochemical theory" is severely weakened by the finding that codons CUN can change from leucine to threonine in the yeast mitochondrial code and codon CGG can possibly change from arginine to tryptophan in maize mitochondrial code (Fox and Leaver 1981).

Discussion

We have reached a point in this account where an early universal code has appeared containing two anticodons per box, shown in Table 6. Further evolution took place, as shown in Table 3, so that more anticodons (tRNAs) were added by gene duplication without producing changes in the amino acids in the boxes listed in Table 1. The additions provided redundancy in codon-anticodon pairing, thus increasing protection against mutations in tRNA genes.

Two mitochondrial anticodons contain A in the first position. These are ACG, corresponding to anticodons CGN in yeast mitochondria (Bonitz et al. 1980), and ACC, pairing with GGU, in *Aspergillus* mitochondria (Köchel et al. 1981). Presumably this indicates absence of the enzyme anticodon deaminase (Kammen and Spengler 1970), which converts A to H in first anticodon positions in organisms that use the universal code (vide supra). It is perhaps of interest that CGN codes are used only sparingly in yeast mitochondrial genes (Coruzzi et al. 1981; Bonitz et al. 1980a).

The inclusion in this communication of arginine in the archetypal code is contrary to an earlier proposal that arginine was an "evolutionary intruder." This proposal that arginine is an "evolutionary intruder" into protein synthesis (Jukes 1973) was made because arginine is represented with such paucity in proteins by comparison with its large proportion of codons in the amino acid code. In contrast, lysine, with only two codons, occurs in proteins at an average of 6.6%, so that it really "needs" four codons. Rather than arginine being an "evolutionary intruder," it now seems more likely that lysine entered the code as a "new" amino acid by acquiring two codons through the tRNA with anticodon UUU, and its role in proteins subsequently expanded at the expense of that of arginine, which is the other strongly basic amino acid.

Anticodons starting with uridine-5-oxyacetic acid (oa^5U) are used in the universal code by *E. coli* for pairing with three bases – A, G and U in codons; thus oa^5UAC pairs with GUU, GUA, and GUG (Nishimura 1972). This may be a vestige of the primitive use of U in anticodons for fourfold pairing.

Conclusion

A proposal is made for steps in evolution of the amino acid code by duplication of tRNAs followed by substitutional changes in the first bases of anticodons, and also by changes in aminoacylation, so that a different amino acid becomes attached to a tRNA molecule without a change in the anticodon. Absence of anticodons may also occur because of their omission or deletion, or as a result of a mutation of the first base of the anticodon so that a different anticodon is formed.

The proposal is compatible with differences between the universal code and mitochondrial codes, and with the principle that gene duplication, followed by modifications, is largely responsible for molecular evolution. The list of 20 amino acids partaking in protein synthesis is "frozen," but evolution of the amino acid code has continued, and is characterized by substitutions and modifications of nucleotides that occur in the first position of anticodons.

The freezing of the code for 20 amino acids was a crucial event. These amino acids are responsible for the properties of the vast panoply of living organisms that inhabit the biosphere. A two-letter code could have specified only 15 amino acids, and might have given rise to a far more meager biota.

References

Anderson S, Bankier AT, Barrell BG, de Bruijn MHL, Coulson AR, Drouin J, Eperon IC, Nierlich DP, Roe BA, Sanger F, Schreier PH, Smith AJH, Staden R, Young IG (1981) Sequence and organization of the human mitochondrial genome. Nature 290:457–470

Anderson S, de Bruijn MHL, Coulson AR, Eperon IC, Sanger E, Young IG (1982) The complete sequence of bovine mitochondrial DNA: Conserved features of the mammalian mitochondrial genome. J Mol Biol 156:683–717

Balasubramanian R (1982) Origin of life: A hypothesis for the origin of adaptor-mediated ordered synthesis of proteins and explanation for the choice of terminating codons in the genetic code. Biosystems 15:99–104

Barrell BG, Anderson S, Bankier AT, de Bruijn MHL, Chen E, Coulson AR, Drouin J, Eperon IC, Nierlich DP, Roe BA, Sanger F, Schreier PH, Smith AJH, Staden R, Young IG (1980) Different pattern of codon recognition by mammalian mitochondrial tRNAs. Proc Natl Acad Sci USA 77:3164–3166

Bibb MJ, Van Etten RA, Wright CT, Walberg MW, Clayton DA (1981) Sequence and gene organization of mouse mitochondrial DNA. Cell 26:167–180

Bonitz SG, Berlani R, Coruzzi G, Li M, Macino G, Nobrega FB, Nobrega MP, Thalenfeld BE, Tzagoloff A (1980) Codon recognition rules in yeast mitochondria. Proc Natl Acad Sci USA 77:3167–3170

Bonitz SG, Coruzzi G, Thalenfeld BE, Tzagoloff A, Macino G (1980a) Assembly of the mitochondrial membrane system. J Biol Chem 255:11927–11941

Coruzzi G, Bonitz SG, Thalenfeld BE, Tzagoloff A (1981) Assembly of the mitochondrial membrane system: Analysis of the nucleotide sequence and transcripts in the *oxil* regions of yeast mitochondrial DNA. J Biol Chem 256:12780–12787

Crick FHC (1966) Codon-anticodon pairing: The wobble hypothesis. J Mol Biol 19:548–555

Crick FHC (1968) The origin of the genetic code. J Mol Biol 38:367–379

Eck RV (1963) Genetic code: Emergence of a symmetrical pattern. Science 140:477–482

Fox TD, Leaver CJ (1981) The Zea mays mitochondrial gene coding cytochrome oxidase subunit II has an intervening sequence and does not contain TGA codons. Cell 26:315–323

Fukada K, Abelson J (1980) DNA sequence of a T4 transfer RNA gene cluster. J Mol Biol 139:377–391

Groshean HJ, De Henan S, Crothers DM (1978) On the physical basis for ambiguity in genetic coding interactions. Proc Natl Acad Sci (USA) 75:610–614

Heckman JE, XX, XXX (1980) Novel features in the genetic code and codon reading patterns in *Neurospora crassa* mitochondria based on sequences of six mitochondrial tRNAs. Proc Natl Acad Sci USA 77:3159–3163

Jukes TH (1966) Molecules and Evolution. Columbia, New York

Jukes TH (1973) Arginine as a evolutionary intruder into protein synthesis. Biochem Biophys Res Commun 53:704–714

Jukes TH (1977) The amino acid code. In: Neuberger A (ed) Comprehensive biochemistry. Vol 24, Biological infomation transfer. Elsevier/North Holland, Amsterdam, p 235

Jukes TH (1981) Amino acid codes in mitochondria as possible clues to primitive codes. J Mol Evol 18:15–16

Jukes TH, Holmquist R, Moise H (1975) Amino acid composition of proteins: Selection against the genetic code. Science 189:50–51

Kammen HO, Spengler SJ (1970) The biosynthesis of inosinic acid in transfer RNA. Biochim Biophys Acta 213:352

Köchel HG, Küntzel H (1982) Mitochondrial L-rRNA from *Aspergillus nidulans*: Potential secondary structure and evolution. Nucl Acids Res 10:4795–4801

Köchel HG, Lazarus CM, Basak N, Küntzel H (1981) Mitochondrial tRNA gene clusters in Aspergillus nidulans: Organization and nucleotide sequence. Cell 23:625–633

Kuchino Y, Watanabe S, Harada F, Nishimura S (1980) Primary structure of AUA-specific isoleucine transfer ribonucleic acid from *Eschericia coli*. Biochemistry 19:2985–2098

Lagerkvist U (1981) Unorthodox codon reading and the evolution of the genetic code. Cell 23:305–306

Li W-H (1982) Ch. 3 in Evolution of Genes and Proteins (in press)

Nishimura S (1972) Minor components in transfer RNA: Their characterization, location and function. Prog Nucl Acid Res Mol Biol 12:49–85

Woese CR (1981) Archaebacteria. Sci Am 244:98–122

Received September 28, 1982/Revised December 26, 1982

PROBLEMS

1. Define the following terms: comma-free code, frozen accident, wobble, master sequences, quasi-species, Q-beta replicase, hypercycle, fluctuating clay environment.

2. What are the virtues of the RNY sequence as a repeating RNA unit in the primitive RNA molecule around which a "realistic hypercycle" could be constructed (Eigen and Schuster, 1978b, this chapter)?

3. In an *in vitro* RNA-synthesizing system using Q-beta replicase, what feature of the RNA molecule do you imagine would be selected for under conditions of exponential growth (enzyme excess) and under conditions of linear growth (template excess)? Following *de novo* syntheses of RNA by Q-beta replicase, Biebricher *et al.* (1981) observed that the first macroscopically visible population of molecules was heterogeneous with respect to electrophoretic mobility. If such a population is serially diluted into fresh medium, using short incubation times to maintain the exponential phase, population heterogeneity is maintained; however, if long incubation times are used (making sure that the total number of generations observed is approximately equivalent), just one species is ultimately dominant. Can you offer an explanation for this observation?

4. An important feature of a hypercyclically coupled system is its ability to expand and contract. The conditions for such events are investigated in this question. Figure 9-4 depicts a two-membered and a three-membered hypercycle; R_1, R_2, and R_3 designate replicating RNA molecules (+ and - strands); E_1, E_2, and E_3 designate primitively coded and translated peptide products of R_1, R_2, and R_3, repectively; and paths I, II, ... V indicate possible catalytic coupling of E to R.
(a) For the two-membered hypercycle A, what would be the consequences of the following sets of relative catalytic capacities:
   (1) II > IV, I > III
   (2) II < IV, I > III
   (3) II < IV, I < III

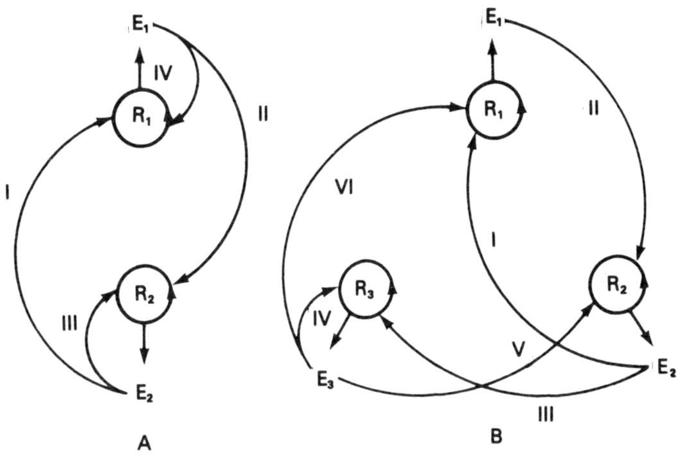

Figure 9-4

(b) Suppose a mutant of R_2 (designated R_3) should appear that potentially generates the three-member hypercycle B; what would be the consequences of the following sets of conditions:
   (1) I > III, IV > V > VI
   (2) I < III, IV < V > VI
   (3) I < III, IV < V > VI

5. In problem 4 the features of the system that constituted "catalytic coupling" are not specified. In the situation described for hypercycle B in problem 4b, in which R_3 is a mutant derivative of R_2, two rather different possibiities can be envisioned. (1) R_3 is unaltered with respect to R_2 in terms of its quality as a substrate for replication (and/or survival) but its product E_3 is functionally altered; or (2) R_3 is altered relative to R_2 with respect to substrate quality, but their products E_3 and E_2, respectively, are identical in their catalytic properties. (Eigen *et al.* 1981) and Schuster (1981) term these changes genotypic and phenotypic, respectively). Compare these two types of mutations with respect to their effect upon the survival of the hypercycle of which they are a part if (a) you consider the system in free solution and (b) you consider that the hypercycle has become compartmentalized into protocells.

6. Eigen *et al.* (1981) describe the following quantities relating to the duplication of a population of RNA molecules comprising sequences a, b, c, ... Q_i is the probability of generating a completely correct copy of sequence i; q_i is the mean replication

fidelity of each nucleotide position of molecule i (alternatively, 1 - q would be the mean per nucleotide error frequency); S_i is a measure of competitive advantage of sequence i (relative to other species in the competing pool) Q_iS_i is the error thresholdf or the species i. If species i is the master sequence for a stable quasi-species distribution, what range of values would you expect of the product Q_iS_i? If the value of this product should fall below 1, what would be the consequence for master sequence i and its quasi-species cluster? What features of a duplicating RNA molecule to you think would contribute to S?

The above defined quantities have been used to derive an expression (Schuster, 1981) for the maximum length of a molecule that could be expected to survive (i.e., keep below the error threshold):

$$L_i < L_{max} = -\frac{\ln S_i}{\ln \bar{q}_i} \left(= \frac{\ln S_i}{1 - \bar{q}_i} \text{ for } \bar{q}_i \text{ close to unity}\right)$$

(Schuster, 1981)

For each of the following values of q_i calculate the maximum length molecule that could be maintained as a master sequence with $S = 5$ and with $S = 50$:

$\bar{q}_i = 0.95$	$1 - \bar{q}_i = 5 \times 10^{-2}$	(enzyme-free RNA replication)
$\bar{q}_i = 0.9995$	$1 - \bar{q}_i = 5 \times 10^{-4}$	(RNA phage replicase)
$\bar{q}_i = 0.999999$	$1 - \bar{q}_i = 1 \times 10^{-6}$	(bacterial DNA polymerase plus proof-reading

How do your results compare with observed sizes of the Q-beta and *E. coli* genomes?

7. In the following three collections of internally related sequences, the topology of relatedness is distinctly different for each. Draw a diagram which best shows the relatedness of the sequences within each group.

1	GACGGUGCC	1	GAUGGUACC	1	AACGGCGUC
2	GAUGGUGCC	2	GACGGUACC	2	GACGGCGUC
3	GACGGCGCC	3	GAUAGUACC	3	AGCGGCGUC
4	GAUGAUGUC	4	GAUGAUAUC	4	AACGGCAUC
5	GAUGGUACC	5	AAUGGUACC	5	GACGGCAUC
6	AACGGCGUC	6	GAUGGCACC	6	AACGGCAUU
7	GACAGCGCU	7	GAUGGUGCU	7	GACGGCAUU
8	GGCGGUGCC	8	GGUGGUACC	8	AGCGGCGUU

These have been termed a (symmetric) tree, an ideal bundle, and a quasi-species, respectively (Eigen and Winckler-Oswatitsch, 1981a).

8. Eigen and Schuster (1978b, this chapter) made the suggestion that the primitive RNA molecule, which participated in a hypercyclic organization and may have served both as messenger and adaptor, may have had the repeating sequence RNY (where N = any nucleotide, R = purine and Y = pyrimidine). Shepherd (1981) made the remarkable discovery that in all coding sequences examined (prokaryotic and eukaryotic), written in terms of R and Y, this putative ancient repeating triplet appeared with a significantly higher frequency when the nucleotide sequence was read in the one correct frame, as opposed to either of the incorrect frames. From this observation he proposed a general method for establishing a reading frame in the absence of an amino acid sequence or of unambiguous punctuation, and suggested that this periodicity represents the remnants of the primitive coding sequence of Eigen and Schuster. With the following nucleotide sequence obtained, from ØX174, (Sanger *et al.*, 1978), determine the most likely reading frame, by the Shepherd criterion, and predict the amino acid sequence.

5'UGCGUUGAGGCUUGCGUUUAUGGUACGCUGGACUUUGUGGG
AUACCCUCGCUUUCCUGCUCCUGUUGAGUUUAUU
GCUGCCGUCAUUGCU 3'

Examine closely the codon assignments for glycine, alanine and valine. Do you see any asymmetry that may bear on the primitive status of the RNY family of triplets?

9. The work of Fresco *et al.* (1960) led Eigen (1971) to propose the following game relating to the secondary structure of RNA. Each player is given a random sequence of N digits of 4 classes--denoted by the letters A, U, G, C—and a tetrahedral die, each face of which represents one of the 4 letters. By throwing the die in turn and substituting a position in the sequence with the obtained digit, each player should try to approach a double-stranded structure with as many AU or GC pairs as possible. Each GC counts twice as many points as each AU pair. The constraint is that pairs must not be formed unless there is a succession of at least 2 GC, or 1 GC and 2 AU, or 4 AU pairs (cooperativity rule). Otherwise, one is free to form any kind of structure, e.g., hairpins, paperclips, clover leaves, etc., provided that one obeys the "cooperativity rule" for any started sequence. For each loop in the structure one has to leave at least 5 positions unpaired. The players throw in

turn. Each player can throw for any position they want, but must announce the position beforehand. The game ends whenever a player announces a complete structure. The winner is the one having the largest number of points

If one stipulates further that the value of N is 90, that the 5' and 3' ends must be aligned and that out-of-register loops are not permitted, what type of secondary structure is the most frequent winner?

10. What was the contribution of the determination of mitochondrial codon/anticodon sequences to the Jukes proposal for the evolution of the code (1983, this chapter)?

11. Compare the course of evolution of the tyrosine/stop, proline, and isoleucine/methionine clusters of anticodons according to the Jukes proposal (1983, this chapter). Include in your discussion any differences discerned between the prokaryotic and eukaryotic systems.

BIBLIOGRAPHY

Alberts, B., and R. Sternglanz (1977). "Recent excitement in the DNA replication problem." *Nature* 269: 655-661.

Almassy, R. J. and R. E. Dickerson (1978). "Pseudomonas cytochrome c551 at 2.0 A resolution: Enlargement of the cytochrome c family." *Proc. Natl. Acad. Sci. USA* 75: 2674-2678.

Anderson, S., A. T. Bankier, B. G. Barrell, M. H. L. de Bruijn, A. R. Coulson, J. Drouin, I. C. Eperon, D. P. Nierlich, B. A. Roe, F. Sanger, P. H. Schreier, A. J. H. Smith, R. Staden, and I. G. Young (1981). "Sequence and organization of the human mitochondrial genome." *Nature* 290: 457-465.

Andrew, P. W., L. J. Rogers, B. G. Haslett, and D. Boulter (1981). "Comparative properties of ferredoxins from a marine and freshwater species of Porphyridium." *Phytochemistry* 20: 1293-1298.

Anfinsen, C. B. (1959). *The Molecular Basis of Evolution*. Wiley.

Argos, P. and M. G. Rossmann (1979). "Structural comparisons of heme binding proteins." *Biochemistry* 18: 4951-4960.

Argos, P. and M. G. Rossman (1980). "The relationship between coding sequences and function in some heme-binding proteins." *J. Mol. Evol.* 16: 149-150.

Ayala, F. J. (1969). "Evolution of Fitness, V. Rate of evolution of irradiated populations of *Drosophila*." *Proc. Natl. Acad. Sci. USA* 63: 790-793.

Baba, M. L., L. L. Darga, M. Goodman, and J. Czelusniak (1981). "Evolution of cytochrome c investigated by the maximum parsimony method." *J. Mol. Evol.* 17: 197-213.

Baldwin, J. E. and H. Krebs (1981). "The evolution of metabolic cycles." *Nature* 291: 381-382.

Barker, W. C., L. K. Ketcham and M. O. Dayhoff (1978). "A comprehensive examination of protein sequences for evidence of internal gene duplication." *J. Mol. Evol.* 10: 265-281.

Barker, W. C. and M. O. Dayhoff (1979). "Evolution of homologous physiological mechanisms based on protein sequence data." *Comp. Biochem. Physiol.* 62B: 1-5.

Barrell, B. G., S. Anderson, A. T. Bankier, M. H. L. de Bruijn, E. Chen, A. R. Coulson, J. Drouin, I. C. Eperon, D. P. Nierlich, B. A. Roe, F. Sanger, P. H. Schreier, A. J. H. Smith, R. Staden, and I. G. Young (1980). "Different pattern of codon recognition by mammalian mitochondria tRNAs." *Proc. Nat. Acad. Sci. USA* 77: 3164-3166.

Barrie, P. A., A. J. Jeffreys, and A. F. Scott (1981). "Evolution of the beta-globin gene cluster in man and the primates." *J. Mol. Biol.* 149: 319-336.

Baumberg, S. (1981). "The evolution of metabolic regulation". In *Molecular and Cellular Aspects of Microbial Evolution*. 32nd Symp. Soc. Gen. Microbiol., Cambridge University Press.

Bedbrook, J., J. Jones and R. Flavell (1981). "Evidence for the involvement of recombination and amplification events in the evolution of Secale chromosomes." *Cold Spring Harbour Symp. Quant. Biol.* 45: 755-760.

Beintema, J. J., W. Gaastra, J. A. Lenstra, G. W. Welling, and W. M. Fitch. (1977). "The molecular evolution of pancreatic ribonuclease." *J. Mol. Evol.* 10: 49-71.

Benyajati, C., A. R. Place, D. A. Powers and W. Sofer (1981). "Alcohol dehydrogenase gene of Drosophila melanogaster: relationship of intervening sequences to functional domains in the protein." *Proc. Natl. Acad. Sci. USA* 78: 2717-2721.

Bernal, J. D. (1967). *The Origin of Life*. Weidenfeld and Nicolson.

Betz, J. L., P. R. Brown, M. J. Smyth and P. H. Clarke (1974). "Evolution in action." *Nature* 247: 261-264.

Biebricher, C. K., M. Eigen, and R. Luce (1981). "Product analysis of RNA generated de novo by Q-beta replicase." *J. Mol. Biol.* 148: 369-390.

Blake, C. C. F. (1979). "Exons encode protein functional units." *Nature* 277: 598.

Blundell, T. L. and R. E. Humbel (1980). "Hormone families: pancreatic hormones and homologous growth factors." *Nature* 287: 781-787.

Bonen, L. and W. F. Doolittle (1976). "Partial sequences of 16S rRNA and the phylogeny of blue-green algae and chloroplasts." *Nature* 261: 669-673.

Bonen, L. and W. F. Doolittle (1978). "Ribosomal RNA homologies and the evolution of the filamentous blue-green algae." *J. Mol. Evol.* 10: 283-291.

Bonitz, S. G., R. Berlani, G. Coruzzi, M. Li, G. Macino, F. G. Nobrega, M. P. Nobrega, B. E. Thalenfeld, and A. Tzagoloff (1980). "Codon recognition rules in yeast mitochondria." *Proc. Natl. Acad. Sci. USA* 77: 3167-3170.

Bonner, T. I., D. J. Brenner, B. R. Neufeld and R. J. Britten (1973). "Reduction in the rate of DNA reassociation by sequence divergence." *J. Mol. Biol.* 81: 123-135.

Borst, P. and L. A. Grivell (1981). "Small is beautiful--portrait of a mitchondrial genome."

Nature 290: 443-444.

Bouchard, R. A. (1982). "Moderately repetitive DNA in evolution." *Intern. Rev. Cytol.* 76: 113-193.

Boulter, D. (1980). "The evaluation of present results and future possibilities of the use of amino acid sequence data in phylogenetic studies with specific reference to plant proteins." In *Chemosystematics: Principles and Practice*. F. A. Bisby, J. G. Vaughan and C. A. Wright (eds.) Academic Press.

Boulter, D., D. Peacock, A. Guise, J. T. Gleaves, and G. Estabrook (1979). "Relationships between the partial amino acid sequences of plastocyanin from members of ten families of flowering plants." *Phytochemistry* 18: 603-608.

Boulter, D., J. A. M. Ramshaw, E. W. Thompson, M. Richardson, and R. H. Brown (1972). "A phylogeny of higher plants based on the amino acid sequences of cytochrome c and its biological implications." *Proc. Roy. Soc. London B* 181: 441-455.

Brack, A. and L.E. Orgel (1975). "Beta structures of alternating polypeptides and their possible prebiotic significance." *Nature* 256: 383-387.

Bradshaw, R. A. (1978). "Nerve growth factor." *Ann. Rev. Biochem.* 47: 191-216.

Brenner, D. J. (1973). "Deoxyribonucleic acid reassociation in taxonomy of enteric bacteria." *Int. J. Syst. Bact.* 23: 298-307.

Brewin, N. (1972). "Catalytic role for RNA in DNA replication." *Nature New Biol.* 236: 101.

Bridges, C. B. (1935). "Salivary chromosome maps." *J. Heredity* 26: 60-64.

Britten, R. J. and E. H. Davidson (1969). "Gene regulation for higher cells: A theory." *Science* 165: 349-357.

Britten, R. J. and E. H. Davidson (1971). "Repetitive and nonrepetitive DNA sequences and a speculation on the origins of evolutionary novelty." *Quart. Rev. Biol.* 46: 111-138.

Britten, R. J. and E. H. Davidson (1976). "DNA sequence arrangement and preliminary evidence on its evolution." *Fed. Proc.* 35: 2151-2157.

Britten, R.J. and D. E. Kohne (1968). "Repeated sequences in DNA." *Science* 161: 529-540.

Broda, E. (1975). *The Evolution of the Bioenergetic Process*. Pergamon Press.

Brown, A. P. (1980). "Evidence for remnants of an ordered codon sequence and a restricted codon composition in selected proteins." *J. Theor. Biol.* 83: 537-560.

Brown, D. D. and K. Sugimoto (1974). "The structure and evolution of ribosomal and 5S DNAs in Xenopus laevis and Xenopus mulleri." *Cold Spring Harbor Symp. Quant. Biol.* 38, 501-505.

Brown, W. M., M. George, Jr., and A. C. Wilson (1979). Rapid evolution of animal mitochondrial DNA. *Proc. Natl. Acad. Sci. USA* 76: 1967-1971.

Brown, W. M., E. M. Prager, A. Wang and A. C. Wilson (1982). "Mitochondrial DNA sequences of primates: tempo and mode of evolution." *J. Mol. Evol.* 18: 225-239.

Brutlag , D. L. (1980). "Molecular arrangement and evolution of heterochromatic DNA." *Ann. Rev. Genet.* 14: 121-144.

Calos, M. P. and J. H. Miller (1980). "Transposable elements." *Cell* 20: 579-595.

Carlile, M. J. (1982). "Prokaryotes and eukaryotes: strategies and successes." *Trends Biochem. Sci.* 7: 128-130.

Carlile, M. J., J. F. Collins and B. E. B. Mosely (eds) (1981). *Molecular and Cellular Aspects of Microbial Evolution*. 32nd Symp. Soc. Gen. Microb. Cambridge University Press.

Caspar, D. L. D. and A. Klug (1962). "Physical principles in the construction of regular viruses." *Cold Spring Harbor Symp. Quant. Biol.* 27: 1-24.

Cavalier-Smith, T. (1975). "The origin of

nuclei and of eukaryotic cells." *Nature* 256: 463-468.

Cavalier-Smith, T. (1978). "Nuclear volume control by nucleoskeletal DNA, selection for cell volume and cell growth rate, and the solution of the DNA C-value paradox." *J. Cell. Sci.* 34: 247-278.

Cedergren, R. J., B. LaRue, D. Sankoff, G. Lapalme, and H. Grosjean (1980). "Convergence and minimal mutation criteria for evaluating early events in tRNA evolution." *Proc. Natl. Acad. Sci. USA* 77: 2791-2795.

Cedergren, R. J., D. Sankoff, B. LaRue, and H. Grosjean. (1981). "The evolving tRNA molecule." *Critical Reviews in Biochemistry* 11: 35-104.

Champion, A. B., E. L. Barrett, N. J. Palleroni, K. L. Soderberg, R. Kunisawa, R. Contopoulou, A. C. Wilson, and M. Douderoff (1980). "Evolution in *Pseudomonas fluorescens*." *J. Gen. Microbiol.* 120: 485-511.

Childs, G., R. Maxson, R. H. Cohn and L. Kedes. (1981). "Orphons: dispersed genetic elements derived from tandem repetitive genes of eucaryotes." *Cell* 23: 651-663.

Clarke, P. H. (1974). "The evolution of enzymes for the utilization of novel substrates." In *Evolution in the Microbial World*. 24th Symp. Soc. Gen. Microb. Cambridge University Press.

Clarke, P. H. (1981). "Enzymes in bacterial populations." In *Biochemical Evolution*, H. Gutfreund (ed.), Cambridge University Press.

Clarke, P. H. and S. R. Elsden (1980). "The earliest catabolic pathways." *J. Mol. Evol.* 15: 333-338.

Costantini, F. D., R. H. Scheller, R. J. Britten, and E. H. Davidson (1978). "Repetitive sequence transcripts in the mature sea urchin oocyte." *Cell* 15: 173-187.

Cox, E. C. (1976). "Bacterial mutator genes and the control of spontaneous mutation." *Ann. Rev. Genet.* 10: 135-156.

Cox, E. C. and T. C. Gibson (1974). "Selection for high mutation rates in chemostats." *Genetics* 77: 169-184.

Crick, F. H. C. (1958). "On protein synthesis." In *The Biological Replication of Macromolecules*. 12th Symp. Soc. Expt. Biol. Cambridge University Press.

Crick, F. H. C. (1968). "The origin of the genetic code." *J. Mol. Biol.* 38: 367-379.

Crick, F. H. C., L. Barnett, S. Brenner and R. J. Watts-Tobin (1961). "General nature of the genetic code for proteins." *Nature* 192: 1227-1232.

Crick, F. H. C., S. Brenner, A. Klug, and G. Pieczenik (1976). "A speculation on the origin of protein synthesis." *Orig. of Life* 7: 389-397.

Cronquist, A. (1980). "Chemistry in plant taxonomy: an assessment of where we stand." In *Chemosystematics: Principles and Practice*. F. A. Bisby, J. G. Vaughan, and C. A. Wright (eds.). Academic Press.

Crow, V. L., B. D. W. Jarvis, and R. M. Greenwood (1981). "Deoxyribonucleic acid homologies among acid-producing strains of Rhizobium." *Internat. J. of Syst. Bact.* 31: 152-172.

Darwin, C. (1859). *Origin of Species*. 6th Edition. Reprinted by J. M. Dent and Sons. London.

Davidson, E. H. and R. J. Britten (1979). "Regulation of gene expression: possible role of repetitive sequences." *Science* 204: 1052-1059.

Davidson, E. H., G. A. Galau, R. C. Angerer and R. J. Britten (1975). "Comparative aspects of DNA organization in Metazoa." *Chromosoma* 51: 253-259.

Dayhoff, M. O. (1972). *Atlas of Protein Sequence and Structure*. National Biomedical Research Foundation, Silver Springs, Maryland.

Dayhoff, M. O. (1976). "The origin and evolution of protein superfamilies." *Fed. Proc.* 35: 2132-2138.

Dayhoff, M. O., P. J. McLaughlin, W. C. Barker and L. T. Hunt (1975). "Evolution of sequences within protein superfamilies." *Naturwissenschaften*. 62: 154-161.

de Beer, G. R. (1930). *Embryology and Evolution*. Clarendon Press.

de Jong, W. W., J. T. Gleaves, and D. Boulter (1977). "Evolutionary changes of alpha-crystallin and the phylogeny of mammalian orders." *J. Mol. Evol.* 10: 123-135.

de Jong, W. W. and L. Ryden (1981). "Causes of more frequent deletions than insertions in mutations and protein evolution." *Nature* 290: 157-159.

de Jong, W. W., A. Zweers, and M. Goodman (1981). "Relationship of aardvark to elephants, hyraxes and sea cows from alpha-crystallin sequences." *Nature* 292: 538-540.

De Ley, J. (1974). "Phylogeny of procaryotes." *Taxon*. 23: 291-300.

De Martelaere, D. A. and A. P. Van Gool (1981). "The density distribution of gene loci over the genetic map of Escherichia coli: its structural, functional and evolutionary implications." *J. Mol. Evol.* 17: 354-360.

Dene, H., M. Goodman, and A. E. Romero-Herrera (1980). "The amino acid sequence of elephant (Elephas maximus) myoglobin and the phylogeny of Proboscidea." *Proc. Roy. Soc. Lond B* 207: 111-127.

Dickerson, R. E. (1971). "The structure of cytochrome c and the rates of molecular evolution." *J. Mol. Evol.* 1: 26-45.

Dickerson, R. E. (1978). "Chemical evolution and the origin of life." *Sci. Amer.* 239(3): 62-78.

Dickinson, W. J. (1980). "Evolution of patterns of gene expression in Hawaiian picture-winged Drosophila." *J. Mol. Evol.* 16: 73-94.

Dobzhansky, T., F. J. Ayala, G. L. Stebbins, and J. W. Valentine (1977). *Evolution*. W. H. Freeman.

Doolittle, R. F. (1981). "Similar amino acid sequences: chance or common ancestry?" *Science* 214: 149-159.

Doolittle, R. F. and B. Blomback (1964). "Amino-acid sequence investigations of fibrinopeptides from various mammals: evolutionary implications." *Nature* 202: 147-152.

Doolittle, R. F., G. L. Wooding, Y. Lin, and M. Riley (1971). "Hominoid evolution as judged by fibrinopeptide structures". *J. Mol. Evol.* 1: 74-83.

Doolittle, W. F. (1978). "Genes in pieces: were they ever together?" *Nature* 272: 581-582.

Doolittle, W. F. and C. Sapienza (1980). "Selfish genes, the phenotypic paradigm and genome evolution." *Nature* 284: 601-603.

Dover, G. (1982). "Molecular drive: a cohesive model of species evolution." *Nature* 299: 111-117.

Dover, G. A. and R. B. Flavell (eds.) (1982). *Genome Evolution*. Academic Press.

Drake, J. W. (1974). "The role of mutation in microbial evolution." In *Evolution in the Microbial World*. 24th Symp. Soc. Gen. Microb. Cambridge University Press.

Dubnau, D., I. Smith, P. Morell, and J. Marmur (1965). "Gene conservation in Bacillus species, I. Conserved genetic and nucleic acid base sequence homologies." *Proc. Nat. Acad. Sci. USA* 54: 491-498.

Eaton, W. A. (1980). "The relationship between coding sequences and function in haemoglobin." *Nature* 284: 183-185.

Eck, R. V. and M. O. Dayhoff (1966a). "Evolution of the structure of ferredoxin based on living relics of primitive amino acid sequences." *Science* 152: 363-366.

Eck, R. V. and M. O. Dayhoff (1966b). "Inferences from protein sequence comparisons." In *Atlas of Protein Sequence and Structure*. 1966. National Biomedical Research Foundation.

Efstratiadis, A., J. W. Poskony, T. Maniatis,

R. M. Lawn, C. O'Connell, R. A. Spritz, J. K. DeRiel, B. G. Forget, S. M. Weissman, J. L. Slightom, A. E. Blechl, O. Smithies, F. E. Baralle, C. C. Shoulders, and N. J. Proudfoot (1980). "The structure and evolution of the human beta-globin gene family." *Cell* 21: 653-668.

Eigen, M. (1971). "Self-organization of matter and the evolution of biological macromolecules." *Naturwissenschaften* 58: 465-523.

Eigen, M., W. Gardiner, P. Schuster, and R. Winkler-Oswatitsch (1981). "The origin of genetic information." *Sci. Amer.* 244(4): 78-94.

Eigen, M. and P. Schuster (1977). "The hypercycle. A principle of natural self-organization. Part A: Emergence of the hypercycle." *Naturwissenschaften* 64: 541-565.

Eigen, M. and P. Schuster (1978a). "The hypercycle. A principle of natural self-organization. Part B: The abstract hypercycle." *Naturwissenschaften* 65: 7-41.

Eigen, M. and P. Schuster (1978b). "The hypercycle. A principle of natural self-organization. Part C: The realistic hypercycle." *Naturwissenschaften* 65: 341-369.

Eigen, M. and R. Winkler-Oswatitsch (1981a). "Transfer-RNA: The early adaptor." *Naturwissenschaften* 68: 217-228.

Eigen, M. and R. Winkler-Oswatitsch (1981b). "Transfer-RNA: an early gene?" *Naturwissenschaften* 68: 282-292.

Eldredge, N. and S. J. Gould (1972). "Punctuated equilibria: an alternative to phyletic gradualism." In *Models in Paleobiology*. T.J.M. Schopf (ed.) Freeman Cooper.

Engel, P. C. (1973). "Evolution of enzyme regulator sites. Evidence for partial gene duplication from amino-acid sequence of bovine glutamate dehydrogenase." *Nature* 241: 118-120.

Engels, W. R. (1981). "Hybrid dysgenesis in Drosophila and the stochastic loss hypothesis. *Cold Spring Harbor Symp. Quant. Biol.* 45: 561-565.

Feldman, M. and E. R. Sears (1981). "The wild gene resources of wheat." *Sci. Amer.* 241(1): 98-109.

Felsenstein, J. (1974). "The evolutionary advantage of recombination." *Genetics* 78: 737-756.

Felsenstein, J. (1982). "Numerical methods for inferring evolutionary trees." *Quart. Rev. Biol.* 57: 379-404.

Ferris, S. D., W. M. Brown, W. S. Davidson and A. C. Wilson (1981a). "Extensive polymorphism in the mitochondrial DNA of apes." *Proc. Natl. Acad. Sci. USA* 78: 6319-323.

Ferris, S. D., A. C. Wilson, and W. M. Brown (1981b). "Evolutionary tree for apes and humans based on cleavage maps of mitochondrial DNA." *Proc. Natl. Acad. Sci. USA* 78: 2432-2436.

Fisher, R. A. (1930). *The Genetical Theory of Natural Selection*. Clarendon Press.

Fitch, W. M. (1976). "The molecular evolution of cytochrome c in eukaryotes." *J. Mol. Evol.* 8: 13-40.

Fitch, W. M. and C. H. Langley. (1976). "Protein evolution and the molecular clock." *Fed. Proc.* 35: 2092-2097.

Fitch, W. M. and E. Margoliash (1967). "Construction of phylogenetic trees." *Science* 155: 279-284.

Fitch, W. M. and E. Markowitz (1970). "An improved method for determining codon variability in a gene and its application to the rate of fixation of mutations in evolution." *Biochemical Genetics* 4: 579-593.

Flavell, R. B., J. Rimpau and D. B. Smith (1977). "Repeated sequence DNA relationships in four cereal genomes." *Chromosoma* 63: 205-222.

Fox, G. E., L. J. Magrum, W. E. Balch, R. S. Wolfe, and C. R. Woese (1977). "Classification of methanogenic bacteria by 16S ribosomal RNA characterization." *Proc. Natl. Acad. Sci. USA* 74: 4537-4541.

Fox, G. E., E. Stackebrandt, R. B. Hespell, J. Gibson, J. Maniloff, T. A. Dyer, R. S. Wolfe, W. E. Balch, R. S. Tanner, L. J. Magrum, L. B. Zablen, R. Blakemore, R. Gupta, L. Bonen, B. J. Lewis, D. A. Stahl, K. R. Luehrsen, K. N. Chen, and C. R. Woese (1980). "The phylogeny of prokaryotes." *Science* 209: 457-463.

Fox, S. W. (1980). "Metabolic microspheres: Origins and evolution." *Naturwissenschaften* 67: 378-383.

Fox, S. W. and K. Dose (1972). *Molecular Evolution and the Origin of Life*. W. H. Freeman.

Frederick, J. F. (ed.) (1981). *Origins and Evolution of eukaryotic Intracellular Organelles*. *Ann. N.Y. Acad. Sci.* 361: 1-510.

Fresco, J. R., B. M. Alberts, and P. Doty (1960). "Some molecular details of the secondary structure of ribonucleic acid." *Nature* 188: 98-101.

Gallant, J. and L. Palmer (1979). "Error propagation in viable cells." *Mech. Age. Dev.* 10: 27-38.

Garcia-Bellido, A. (1977). "Homoeotic and atavic mutations in insects." *Am. Zool.* 17: 613-629.

Gething, M-J., J. Bye, J. Skehel, and M. Waterfield (1980). "Cloning and DNA sequence of double-stranded copies of haemagglutinin genes from H2 and H3 strains elucidates antigenic shift and drift in human influenza virus." *Nature* 287: 301-306.

Gibbs, S. P. (1978). "The chloroplasts of *Euglena* may have evolved from symbiotic green algae." *Can. J. Bot.* 56: 2883-2889.

Gilbert, W. (1978). "Why genes in pieces?" *Nature* 271: 501.

Gilbert, W. (1979). "Introns and exons: playgrounds of evolution." In *Eucaryotic Gene Regulation*, R. Axel, T. Maniatis, and C. F. Fox (eds.) Academic Press.

Gilbert, W. (1981). "DNA sequencing and gene structure." *Science* 214: 1305-1312.

Go, M. (1981). "Correlation of DNA exonic regions with protein structural units in haemoglobin." *Nature* 291: 90-92.

Goldschmidt, R. B. (1940). *The Material Basis of Evolution*. Yale University Press.

Goodman, M. (1981). "Globin evolution was apparently very rapid in early vertebrates: A reasonable case against the rate-constancy hypothesis." *J. Mol. Evol.* 17: 114-120.

Goodman, M., J. Barnabas, G. Matsuda, and G. W. Moore (1971). "Molecular evolution in the descent of man." *Nature* 233: 604-613.

Goodman, M., J. Czelusniak, G. W. Moore, A. E. Romero-Herrera, and G. Matsuda (1979). "Fitting the gene lineage into its species lineage, a parsimony strategy illustrated by cladograms constructed from globin sequences." *Syst. Zool.* 28: 132-163.

Goodman, M., G. W. Moore, J. Barnabas and G. Matsuda (1974). "The phylogeny of human globin genes investigated by the maximum parsimony method." *J. Mol. Evol.* 3: 1-48.

Gottlieb, L. D. (1974). "Gene duplication and fixed heterozygosity for alcohol dehydrogenase in the diploid plant *Clarkia franciscana*." *Proc. Nat. Acad. Sci. USA* 71: 1816-1818.

Gough, N. (1982). "Gene conversion and the generation of antibody diversity." *Trends Biochem. Sci.* 7: 307-308.

Gould, S. J. (1977). *Ontogeny and Phylogeny*. Harvard University Press.

Graham, D. E., B. R. Nuefeld, E. H. Davidson and R. J. Britten (1974). "Interspersion of repetitive and nonrepetitive DNA sequences in the sea urchin genome." *Cell* 1: 127-137.

Gray, J. C. (1980). "Fraction I protein and plant phylogeny." In *Chemosystematics: Principles and Practice*. F. A. Bisby, J. G. Vaughan, and C. A. Wright (eds.). Academic Press.

Gray, M. W. and W. F. Doolittle (1982). "Has the endosymbiont hypothesis been proven?" *Microbiol. Rev.* 46: 1-42.

Green, M. M. (1978). "Insertion mutants and the control of gene expression in *Drosophila melanogaster*." In *The Clonal Basis of Development*. S. Subtelny and I. M. Sussex (eds.). Academic Press.

Grund, C., J. Gilroy, T. Gleaves, U. Jensen, and D. Boulter (1981). "Systematic relationships of the Ranunculaceae based on amino acid sequence data." *Phytochemistry* 20: 1559-1565.

Grutter, M. G., L. H. Weaver, and B. W. Matthews (1983). "Goose lysozyme structure: an evolutionary link between hen and bacteriophage lysozyme ?" *Nature* 303: 828-831.

Haldane, J. B. S. (1929). "The origin of life." *The Rationalist Annual* 148: 3-10. Reprinted in J. D. Bernal (1967).

Hall, B. G. and N. D. Clarke (1977). "Regulation of newly evolved enzymes. III. Evolution of the ebg repressor during selection for enhanced lactase activity." *Genetics* 85: 193-201.

Hall, B. G. and T. Zuzel (1980). "Evolution of a new enzymatic function by recombination within a gene." *Proc. Natl. Acad. Sci. USA* 77: 3529-3533.

Harris, H. (1966). "Enzyme polymorphisms in man." *Proc. Roy. Soc. London B* 164: 298-310.

Harris, H. (1976). "Molecular evolution: the neutralist - selectionist controversy." *Fed. Proc.* 35: 2079-2082.

Hartigan, J. A. (1973). "Minimum mutation fits to a given tree." *Biometrics* 29: 53-65.

Hartley, B. S. (1966). "Enzymes are proteins." Advancement Science 22: 47-54.

Hartley, B. S. (1979). "Evolution of enzyme structure." *Proc. Roy. Soc. London B* 205: 443-452.

Hartley, B. S., I. Altosaar, J. M. Dothie and M. S. Neuberger (1976). "Experimental evolution of a xylitol dehydrogenase." In *Structure-Function Relationships of Proteins*. R. Markham and R. W. Horne (eds). North Holland.

Hartman, H. (1975). "Speculations on the origin and evolution of metabolism." *J. Mol. Evol.* 4: 359-370.

Heckman, J. E., J. Sarnoff, B. Alzner-DeWeerd, S. Yin and U. L. RajBhandary. (1980). "Novel features in the genetic code and codon reading patterns in Neurospora crassa mitochondria based on sequences of six mitochondrial tRNAs." *Proc. Natl. Acad. Sci. USA* 77: 3159-3163.

Hedrick, P. W. and J. F. McDonald (1980). "Regulatory gene adaptation: an evolutionary model." *Heredity* 45: 85-99.

Hendy, M. D., L. R. Foulds and D. Penny (1980). "Proving phylogenetic trees minimal with 1-clustering and set partitioning." *Math. Biosc.* 51: 71-88.

Hirsh, D., S. W. Emmons, J. G. Files and M. R. Klass (1979). "Stability of the C. elegans genome during development and evolution." In *Eukaryotic Gene Regulation*. R. Axel, T. Maniatis and C. F. Fox (eds.). Academic Press.

Hofmann, H., P. P. Fietzek and K. Kuhn (1980). "Comparative analysis of the sequences of the three collagen chains alpha 1(I), alpha2 and alpha1(III): Functional and genetic aspects." *J. Mol. Biol.* 141: 293-314.

Holmquist, R., T. H. Jukes, and S. Pangburn (1973). "Evolution of transfer RNA." *J. Mol. Biol.* 78: 91-116.

Honjo, T., S. Nakai, Y. Nishida, T. Kataoka, Y. Yamawaki-Kataoka, N. Takahashi, M. Obata, A. shimizu, Y. Yaoita, T. Nikaido and N. Ishida (1981). "Rearrangements of immunoglobulin genes during differentiation and evolution." *Immunol. Rev.* 59: 33-67.

Hood, L. E., J. H. Wilson and W. B. Wood (1975). *Molecular Biology of Eukaryotic Cells: A problems Approach*. W. A. Benjamin.

Hori, H. and S. Osawa (1979). "Evolutionary change in 5S RNA secondary structure and a phylogenic tree of 54 5S RNA species." *Proc. Natl. Acad. Sci. USA* 76: 381-385.

Horowitz, N. H. (1945). "On the evolution of biochemical syntheses." *Proc. Natl. Acad. Sci. USA* 31: 153-157.

Hoyer, B. H., B. J. McCarthy and E. T. Bolton (1964). "A molecular approach in the systematics of higher organisms." *Science* 144: 959-967.

Hoyle, F. and C. Wickramasinghe (1981). *Evolution from Space*. Dent and Sons.

Hseu, T. H., E. D. Jou, C. Wang and C. C. Yang (1977). "Molecular evolution of snake venoms." *J. Mol. Evol.* 10: 167-182.

Hubby, J. L. and R. C. Lewontin (1966). "A molecular approach to the study of genic heterozygosity in natural populations. I. The number of alleles at different loci in *Drosophila pseudoobscura*." *Genetics* 54: 577-594.

Hull, D. L., P. D. Tessner, and A. M. Diamond (1978). "Planck's principle." *Science* 202: 717-723.

Hunter, R. L. and C. L. Markert (1957). "Histochemical demonstration of enzymes separated by zone electrophoresis in starch gels." *Science* 125: 1294-1295.

Huxley, J. (1942). *Evolution: the Modern Synthesis*. George, Allen and Unwin.

Ingram, V. M. (1961). "Gene evolution and the haemoglobins." *Nature* 189: 704-708.

Ishigami, M. and K. Nagano (1975). "The origin of the genetic code." *Orig. of Life* 6: 551-560.

Jacob, F. (1977). "Evolution and tinkering." *Science* 196: 1161-1166.

Jacob, F. and J. Monod (1961). "Genetic regulatory mechanisms in the synthesis of proteins." *J. Mol. Biol.* 3: 318-356.

Jelinek, W. R. and C. W. Schmid (1982). "Repetitive sequences in eukaryotic DNA and their expression." *Ann. Rev. Biochem.* 51: 813-844.

Jelinek, W. R., T. P. Toomey, L. Leinwand, C. H. Duncan, P. A. Biro, P. V. Choudary, S. M. Weissman, C. M. Rubin, C. M. Houck, P. L. Deininger and C. W. Schmid (1980). "Ubiquitous, interspersed repeated sequences in mammalian genomes." *Proc. Natl. Acad. Sci. USA* 77: 1398-1402.

Jensen, E. O., K. Paludan, J. J. Hyldig-Nielsen, P. Jorgensen and K. A. Marcker (1981). "The structure of a chromosomal leghaemoglobin gene from soybean." *Nature* 291: 677-679.

John, P. and F. R. Whatley (1975). "*Paracoccus denitrificans* and the evolutionary origin of the mitochondrion." *Nature* 254: 495-498.

Jukes, T. H. (1966). *Molecules and Evolution*. Columbia University Press.

Jukes, T. H. (1981). "Amino acid codes in mitochondria as possible clues to primitive codes." *J. Mol. Evol.* 18: 15-17.

Kedes, L. H. (1979). "Histone genes and histone messengers." *Ann. Rev. Biochem.* 48: 837-870.

Keosian, J. (1965). *The Origin of Life*. Chapman and Hall.

Keyl, H. G. (1965). "A demonstrable local and geometric increase in the chromosomal DNA of *Chironomus*." *Experientia* 21: 191-193.

Kimura, M. (1968). "Evolutionary rate at the molecular level." *Nature* 217: 624-626.

Kimura, M. and T. Ohta (1974). "On some principles governing molecular evolution." *Proc. Natl. Acad. Sci. USA* 71: 2848-2852.

King, G. A. M. (1977). "Symbiosis and the origin of life. *Orig. of Life* 8: 39-53.

King, J. L. and T. H. Jukes (1969). "Non-Darwinian evolution." *Science* 164: 788-798.

King, M-C. and A. C. Wilson (1975). "Evolution at two levels in humans and chimpanzees." *Science* 188: 107-116.

Kohne, D. E., J. A. Chiscon and B. H. Hoyer (1972). "Evolution of primate DNA sequences." *J. Human Evol.* 1: 627-644.

Koshland, D. E. Jr. (1976). "The evolution of function in enzymes." *Fed. Proc.* 35: 2104-2111.

Krebs, H. (1981). "The evolution of metabolic pathways." In *Molecular and Cellular Aspects of Microbial Evolution.* (M. J. Carlile, J. F. Collins, and B. E. B. Mosely, eds.), 32nd Symp. Soc. Gen. Microbiol. Cambridge University Press.

Kubai, D. F. (1975). "The evolution of the mitotic spindle." *Intern. Rev. Cytology* 43: 167-227.

Kuntzel, H., M. Heidrich, and B. Piechulla (1981). "Phylogenetic tree derived from bacterial, cytosol and organelle 5S rRNA sequences." *Nucl. Acids Res.* 9: 1451-1461.

Kuntzel, H. and H. G. Kochel (1981). "Evolution of rRNA and origin of mitochondria." *Nature* 293: 751-755.

Lacy, E. and T. Maniatis (1980). "The nucleotide sequence of a rabbit beta-globin pseudogene." *Cell* 21: 545-553.

Lahav, N., D. White, and S. Chang (1978). "Peptide formation in the prebiotic era: Thermal condensation of glycine in fluctuating clay environments." *Science* 201: 67-69.

Lerner, S. A., T. T. Wu and E. C. C. Lin (1964). "Evolution of a catabolic pathway in bacteria." *Science* 146: 1313-1315.

LeRoith, D., J. Shiloach, J. Roth, and M. A. Lesniak (1980). "Evolutionary origins of vertebrate hormones: substances similar to mammalian insulins are native to unicellular eukaryotes." *Proc. Natl. Acad. Sci. USA* 77: 6184-6188.

Lesk, A. M. and C. Chothia (1980). "How different amino acid sequences determine similar protein structures: the structure and evolutionary dynamics of the globins." *J. Mol. Biol.* 136: 225-270.

Levine, J. S. (1982). "The photochemistry of the palaeoatmosphere." *J. Mol. Evol.* 18: 161-172.

Levine, M., H. Muirhead, D. K. Stammers, and D. I. Stuart (1978). "Structure of pyruvate kinase and similarities with other enzymes: possible implications for protein taxonomy and evolution." *Nature* 271: 626-630.

Lewin, R. A. (1976). "Prochlorophyta as a proposed new division of algae." *Nature* 261: 697-698.

Lewin, R. A. (1981). "Prochloron and the theory of symbiogenesis." *Ann. N. Y. Acad. Sci.* 361: 325-329.

Lewis, E. B. (1951). "Pseudoallelism and gene evolution." *Cold Spring Harbor Symp. Quant. Biol.* 16: 159-172.

Lewis, E. B. (1978). "A gene complex controlling segmentation in *Drosophila.*" *Nature* 276: 565-570.

Lewontin, R. C. (1974). *The Genetic Basis of Evolutionary Change.* Columbia University Press.

Lohrmann, R. and L. E. Orgel (1973). "Prebiotic activation processes." *Nature* 244: 418-420.

Long, E. O. and I. B. Dawid (1980). "Repeated genes in eukaryotes." *Ann. Rev. Biochem.* 49: 727-764.

Lowenstein, J. M. (1981). "Immunological reactions from fossil material." *Phil. Trans. Roy. Soc. London B* 292: 143-149.

Lowenstein, J. M., V. M. Sarich, and B. J. Richardson (1981). "Albumin systematics of the extinct mammoth and Tasmanian wolf." *Nature* 291: 409-411.

Luehrsen, K. R., D. E. Nicholson, D. C. Eubanks, and G. E. Fox (1981). "An archaebacterial 5S rRNA contains a long insertion sequence." *Nature* 293: 755-756.

Mabry, T. J. and H-D. Behnke (eds.) (1976). "Evolution of *Centrospermous* families." *Plant Syst. Evol.* 126: 101-6.

Malmberg, R. L. (1977). "The evolution of epistasis and the advantage of recombination in populations of bacteriophage T4." Genetics

86: 607-621.

Margulis, L. (1968). "Evolutionary criteria in *Thallophytes*: a radical alternative." *Science* 161: 1020-1022.

Margulis, L. (1970). *Origin of Eukaryotic Cells*. Yale University Press.

Margulis, L. (1981). *Symbiosis in Cell Evolution*. W. H. Freeman.

Margulis, L., L. P. To, and D. Chase (1981). "Microtubules, undulipodia, and Pillotina spirochaetes." *Ann. N. Y. Acad. Sci.* 361: 356-368.

Marsden, M. P. F. and U. K. Laemmli. (1979). "Metaphase chromosome structure: evidence for a radial loop model." *Cell* 17: 849-858.

Marx, J. L. (1981). "Genes that control development." *Science* 213: 1485-1488.

Maxam, A. and W. Gilbert (1977). "A new method for sequencing DNA." *Proc. Natl. Acad. Sci. USA* 74: 560-564.

Maynard Smith, J. (1978). *The Evolution of Sex*. Cambridge University Press.

Maynard Smith, J. (1979). "Hypercycles and the origin of life." *Nature* 280: 445-446.

Mayr, E. (1970). *Population, Species, and Evolution*. Harvard University Press.

Mayr, E. (1975). "Darwin and natural selection." *Am. Sci.* 65: 321-327.

Mayr, E. and W. B. Provine (eds) (1980). *The Evolutionary Synthesis: Perspectives in the Unification of Biology*. Harvard University Press.

McClintock, B. (1951). "Chromosome organization and genic expression." *Cold Spring Harb. Symp. Quant. Biol.* 16: 13-47.

Miller, S. L. (1953). "A production of amino acids under possible primitive earth conditions." *Science* 117: 528-529.

Miller, S. L. and L. E. Orgel (1974). *The Origins of Life on Earth*. Prentice-Hall.

Mills, D. R. R. L. Peterson and S. Spiegelman (1967). "An extracellular Darwinian experiment with a self-duplicating nucleic acid molecule." *Proc. Natl. Acad. Sci. USA* 58: 217-224.

Monod, J. (1971). *Chance and Necessity: An essay on the Natural Philosophy of Modern Biology*. Knopf.

Moore, G. P., R. H. Scheller, E. H. Davidson and R. J. Britten (1978). "Evolutionary change in the repetition frequency of sea urchin DNA sequences." *Cell* 15: 649-660.

Moore, G. P., F. D. Costantini, J. W. Posankony, E. H. Davidson and R. J. Britten (1980). "Evolutionary conservation of repetitive sequence expression in sea urchin egg RNA's." *Science* 208: 1046-1048.

Mortlock, R. P. (1976). "Catabolism of unnatural carbohydrates by micro-organisms." *Adv. Micro. Physiology* 13: 1-53.

Mortlock, R. P. (1982). "Regulatory mutations and the development of new metabolic pathways in bacteria." *Evol. Biol.* 14: 205-268.

Moyzis, R. K., J. Bonnet, D. W. Li and P. O. P. Ts'o. (1981). "An alternative view of mammalian DNA sequence organization I. Repetitive sequence interspersion in Syrian hamster DNA: a model system." *J. Mol. Biol.* 153: 841-864.

Muller, H. J. (1923). *Mutation. Eugenics, Genetics and the Family*. 1: 106-112.

Muller, H. J., (1932). "Some genetic aspects of sex." *Am. Naturalist* 66:118-138.

Nelsestuen, G. L. (1978). "Amino-acid directed nucleic acid synethesis: A possible mechanism in the origin of life." *J. Mol. Evol.* 11: 109-120.

Ninio, J. (1982). *Molecular Approaches to Evolution*. Pitman.

O'Donnell, J., L. Gerace, F. Leister, and W. Sofer (1975). "Chemical selection of mutants that affect alcohol dehydrogenase in *Drosophila*. II. Use of 1-Pentenyl-3-ol."

Genetics 79: 73-83.

Ohno, S. (1970). *Evolution by Gene Duplication*. Allen and Unwin.

Ohta, T. (1980). *Evolution and Variation in Multigene Families*. Springer-Verlag.

Oparin, A. I. (1924). *The Origins of Life*. Translated and reprinted in J. D. Bernal (1967).

Oparin, A. I. (1938) *Origin of Life*. Dover.

Orgel, L. E. (1968). "Evolution of the genetic apparatus." *J. Mol. Biol.* 38: 381-393.

Orgel, L. E. (1979). "Selection in vitro." *Proc. Roy. Soc. London B* 205: 435-442.

Orgel, L. E. and F. H. C. Crick (1980). "Selfish DNA: the ultimate parasite." *Nature* 284: 604-607.

Pace, B. and L. L. Campbell (1971). "Homology of ribosomal ribonucleic acid of diverse bacterial species with *Escherichia coli* and *Bacillus stearothermophilus*." J. Bact. 107: 543-547.

Palmer, J. D. (1982). "Physical and gene mapping of chloroplast DNA from *Atriplex triangularis* and *Cucumis sativa*." *Nucl. Acid Res.* 10: 1593-1605.

Pardoll, D. M., B. Vogelstein and D. S. Coffey (1980). "A fixed site of DNA replication in eucaryotic cells." *Cell* 19: 527-536.

Penny, D., L. R. Foulds and M. D. Hendy (1982). "Testing the theory of evolution by comparing phylogenetic trees constructed from five different protein sequences." *Nature* 297: 197-200.

Perutz, M. F. and K. Imai (1980). "Regulation of oxygen affinity of mammalian haemoglobins." *J. Mol. Biol.* 136: 183-191.

Pickett-Heaps, J. (1974). "The evolution of mitosis and the eukaryotic condition." *BioSystems* 6: 37-48.

Popper, K. R. (1976). *Unended Quest: An Intellectual Autobiography*. Fontana.

Prager, E. M. and A. C. Wilson (1971). "The dependence of immunological cross-reactivity upon sequence resemblance among lysozymes. I. Micro-complement fixation studies." *J. Biol. Chem.* 246: 5978-5989.

Provine, W. B. (1971). *The Origins of Theoretical Population Genetics*. University of Chicago Press.

Raff, R. A. and H. R. Mahler. (1972). "The nonsymbiotic origin of mitochondria." *Science* 177: 575-582.

Ramshaw, J. A. M., D. L. Richardson, B. T. Meatyard, R. H. Brown, M. Richardson, E. W. Thompson, and D. Boulter (1972). "The time of origin of the flowering plants determined by using amino acid sequence data of cytochrome c." *New Phytol.* 71: 773-779.

Raup, D. M. and A. Michelson (1965). "Theoretical morphology of the coiled shell." *Science* 147: 1294-1295.

Raven, P. H. (1970). "A multiple origin for plastids and mitochondria." *Science* 169: 641-646.

Reanney, D. C. (1978). "Coupled evolution: adaptive interactions among the genomes of plasmids, viruses and cells." *Intl. Rev. Cytol.* Suppl. 8: 1-68.

Reanney, D. C. (1979). "RNA splicing and polynucleotide evolution." *Nature* 277: 598-600.

Richardson, J. S. (1977). "Beta-sheet topology and the relatedness of proteins." *Nature* 268: 495-500.

Riley, M. and A. Anilionis (1978). "Evolution of the bacterial genome." *Ann. Rev. Microbiol.* 32: 519-560.

Rimpau, J., D. B. Smith, and R. B. Flavell (1980). "Sequence organization in barley and oats chromosomes revealed by interspecies DNA/DNA hybridization." *Heredity* 44: 131-149.

Rohlfing, D. L. (1976). "Thermal polyamino acids: Synthesis at less than 100°C." *Science* 193: 68-70.

Romero-Herrera, A. E., H. Lehmann, K. A. Joysey, and A. E. Friday (1978). "On the evolution of myoglobin." *Phil. Trans. Roy. Soc. London B* 283: 61-163

Rossman, M. G., D. Moras and K. W. Olsen (1974). "Chemical and biological evolution of a nucleotide-binding protein." *Nature* 250: 194-199.

Sakano, H., J. H. Rogers, K. Huppi, C. Brack, A. Traunecker, R. Maki, R. Wall and S. Tonegawa (1979). "Domains and the hinge region of an immunoglobulin heavy chain are coded in separate DNA segments." *Nature* 277: 627-633.

Sanger, F. (1952). "The arrangement of amino acids in proteins." *Adv. in Protein Chemistry* 7: 1-67.

Sanger, F. and A. R. Coulson (1975). "A rapid method for determining sequences in DNA by primed synthesis with DNA polymerase." *J. Mol. Biol.* 94: 441-448.

Sanger, F., A. R. Coulson, T. Friedmann, G. M. Air, B. G. Barrell, N. L. Brown, J. C. Fiddes, C. A. Hutchison III, P. M. Slocombe, and M. Smith. (1978). "The nucleotide sequence of bacteriophage OX174." *J. Mol. Biol.* 125: 225-246.

Sankoff, D., R. J. Cedergren and G. Lapalme (1976). "Frequency of insertion - deletion, transversion and transition in the evolution of 5S ribosomal RNA." *J. Mol. Evol.* 7: 133-149.

Sankoff, D., R. J. Cedergren, and W. McKay (1982). "A strategy for sequence phylogeny research." *Nucl. Acids Res.* 10: 421-431.

Sankoff, D., C. Morel and R. J. Cedergren (1973). "Evolution of 5S RNA and the non-randomness of base replacement." *Nature New Biology* 245: 232-234.

Sapienza, C. and W. F. Doolittle (1982). "Unusual physical organization of the Halobacterium genome." *Nature* 295: 384-389.

Sarich, V. M. (1973). "The giant panda is a bear." *Nature* 245: 218-220.

Schimke, R. T., F. W. Alt, R. E. Kellems, R. J. Kaufman and J. R. Bertino (1978). "Amplification of dihydrofolate reductase genes in methotrexate-resistant cultured mouse cells." *Cold Spring Harbor Symp. Quant. Biol.* 42: 649-657.

Schuster, P. (1981). "Prebiotic evolution." In *Biochemical Evolution*. H. Gutfreund (ed.). Cambridge University Press.

Schwartz, R. M. and M. O. Dayhoff (1978). "Origins of prokaryotes, eukaryotes, mitochondria and chloroplasts." *Science* 199: 395-403.

Searcy, D. G. (1982). "Thermoplasma: a primordial cell from a refuse pile." *Trends Biochem. Sci.* 7: 183-185.

Searcy, D. G., D. B. Stein, and G. R. Green (1978). "Phylogenetic affinities between eukaryotic cells and a thermophilic mycoplasma." *Biosystems* 10: 19-28.

Shepherd, J. C. W. (1981). "Method to determine the reading frame of a protein from the purine/pyrimidine genome sequence and its possible evolutionary justification." *Proc. Natl. Acad. Sci. USA* 78: 1596-1600.

Simpson, G. G. (1944). *Tempo and Mode in Evolution*. Columbia University Press.

Singer, M. F. (1982). "Highly repeated sequences in mammalian genomes." *Intern. Rev. Cytol.* 76: 67-112.

Sinsheimer, R. L. (1977). "Recombinant DNA." *Ann. Rev. Biochem.* 46: 415-438.

Smith, G. P. (1974). "Unequal crossover and the evolution of multigene families." *Cold Spring Harbor Symp. Quant. Biol.* 38: 507-513.

Smith, G. P. (1976). "Evolution of repeated DNA sequences by unequal crossover." *Science* 191: 528-535.

Sneath, P. H. A. (1967). "Trend surface analysis of transformation grids." *J. Zool.* 151: 65-122.

Sneath, P. H. A. and R. R. Sokal (1973). *Numerical Taxonomy: the Principles and Practice of Numerical Classification*. Freeman.

Southern, E. M. (1975). "Detection of specific sequences among DNA fragments separated by gel electrophoresis." *J. Mol. Biol.* 98: 503-517.

Sparrow, A. H. and A. F. Nauman (1976). "Evolution of genome size by DNA doublings." *Science* 192: 524-529

Spienza, C. an W. F. Doolittle (1982). "Unusual physical organization of the Halobacterium genome." *Nature* 295: 384-389.

Stanier, R. Y., M. Douderoff, and E. A. Adelberg (1963). *General Microbiology*. Macmillan.

Stanley, S. M. (1979). *Macroevolution: Pattern and Process*. W. H. Freeman.

Stebbins, G. L. (1966) "Chromosomal variation and evolution." *Science* 152: 1463-1469.

Stebbins, G. L. (1971). *Chromosomal Evolution in Higher Plants*. Arnold.

Stebbins, G. L. (1973). "Evolution of morphogenetic patterns." *Brookhaven Symp. Biol.* 25: 227-243.

Stebbins, G. L. (1974). "Adaptive shifts and evolutionary novelty: a compositionist approach." In *Studies in the Philosophy of Biology*. F. J. Ayala and T. Dobzhansky (eds.). Macmillan.

Sternberg, P. W. and H. R. Horvitz (1981). "Gonadal cell lineages of the nematode *Panagrellus redivivus* and implications for evolution by the modification of cell lineage." *Devel. Biol.* 88: 147-166.

Streisinger, G., Y. Okada, J. Emrich, J. Newton, A Tsugita, E. Terzaghi and M. Inouye (1966). "Frameshift mutations and the genetic code." *Cold Spring Harbor Symp. Quant. Biol.* 31: 77-84.

Strobel, E., P. Dunsmuir and G. M. Rubin (1979). "Polymorphisms in the chromosomal locations of elements of the 412, copia, and 297 dispersed repeated gene families in *Drosophila*." *Cell* 17: 429-439.

Sturtevant, A. H. (1937). "Essays on evolution. I. On the effects of selection on mutation rate." *Quart. Rev. Biol.* 12: 464-467.

Sumper, M. and R. Luce (1975). "Evidence for de novo production of self-replicating and environmentally adapted RNA structures by bacteriophage Q-beta replicase." *Proc. Natl. Acad. Sci. USA* 72: 162-166.

Tanaka, M., T. Nakashima, A. Benson, H. F. Mower, and K. T. Yasunobu (1964). "The amino acid sequcnce of *Clostridium pasteurianum* ferredoxin." *Biochem. Biophys. Res. Comm.* 16: 422-426.

Tang, J., M. N. G. James, I. N. Hsu, J. A. Jenkins and T. L. Blundell (1978). "Structural evidence for gene duplication in the evolution of the acid proteases." *Nature* 271: 618-621.

Tartof, K. D. (1979). "Evolution of transcribed and spacer sequences in the ribosomal RNA genes of *Drosophila*." *Cell* 17: 607-614.

Taylor, F. J. R. (1976). "Autogenous theories for the origin of eukaryotes." *Taxon*. 25: 377-390.

Thompson, D'A. W. (1917). *On Growth and Form*. Cambridge University Press.

Thompson, J. N. Jr. and R. C. Woodruff (1980). "Increased mutation in crosses between geographically separated strains of Drosophila melanogaster." *Proc. Natl. Acad. Sci. USA* 77: 1059-1062.

Tomkins, G. M. (1975). "The metabolic code." *Science* 189: 760-763.

Topal, M. D. and J. R. Fresco (1976). "Complementary base pairing and the origin of substitution mutations." *Nature* 263: 285-289.

Turner, B. E. (1980). "Interstellar molecules." *J. Mol. Evol.* 15: 79-101.

Valentine, J. W. and C. A. Campbell (1975). "Genetic regulation and the fossil record." *Amer. Sci.* 63: 673-680.

van Ooyen, A., J. van den Berg, N. Mantel and C. Weismann (1979). "Comparison of total sequence of a cloned rabbit beta-globin gene and its flanking regions with a homologous mouse sequence." *Science* 206: 337-344.

van Roode, J. H. G. and L. E. Orgel (1980). "Template-directed synthesis of oligoguanylates in the presence of metal ions." *J. Mol. Biol.* 144: 579-585.

Van Valen, L. M. (1974). "Molecular evolution as predicted by natural selection." *J. Mol. Evol.* 3: 89-101.

Van Valen, L. M. and V. C. Maiorana (1980). "The archaebacteria and eukaryotic origins." *Nature* 287: 248-250.

Vanin, E. F., G. I. Goldberg, P. W. Tucker and O. Smithies (1980). "A mouse alpha-globin-related pseudogene lacking intervening sequences." *Nature* 286:222-226.

Vogeli, G., H. Ohkubo, V. E. Avvedimento, M. Sullivan, Y. Yamada, M. Mudryj, I. Pastan, and B. de Crombrugghe (1981). "A repetitive structure in the chick alpha-2-collagen gene." *Cold Spring Harbor Symp. Quant. Biol.* 47: 777-783.

Waddington, C. H. (1957). *The Strategy of the Genes: a Discussion of Some Aspects of Theoretical Biology*. Allen and Unwin.

Wallace, B. (1975). "Gene control mechanisms and their possible bearing on the neutralist-selectionist controversy." *Evolution* 29: 193-202.

Wallace, D. C. (1982). "Structure and evolution of organelle genomes." *Microb. Rev.* 46: 208-240.

Wallace, D. G., L. R. Maxson and A. C. Wilson (1971). "Albumin evolution in frogs: a test of the evolutionary clock hypothesis." *Proc. Natl. Acad. Sci. USA* 68: 3127-3129.

Waterman, M. S. and T. F. Smith (1978). "On the similarity of dendrograms." *J. Theor. Biol.* 73: 789-800.

Watson, J. D. and F. H. C. Crick (1953a). "A structure for deoxyribose nucleic acid." *Nature* 171: 737-738.

Watson, J. D. and F. H. C. Crick (1953b). "Genetical implications of the structure of deoxyribonucleic acid." *Nature* 171: 964-967.

Weeden, N. F. (1981). "Genetic and biochemical implications of the endosymbiotic origin of the chloroplast." *J. Mol. Evol.* 17: 133-139.

White, D. H. (1980). "A theory for the origin of a self-replicating chemical system. I. Natural selection of the autogen from short, random oligomers." *J. Mol. Evol.* 16: 121-147.

White, D. H., and J. C. Erickson (1980). "Catalysis of peptide bond formation by histidyl-histidine in a fluctuating clay environment." *J. Mol. Evol.* 16: 279-290.

White, D. H. and M. S. Raab (1982). "A theory for the origin of a self-replicating chemical system. II. Computer simulation of the autogen." *J. Mol. Evol.* 18: 207-216.

White, H. B. III (1976). "Coenzymes as fossils of an earlier metabolic state." *J. Mol. Evol.* 7: 101-104.

White, M. J. D. (1978). *Modes of Speciation*. W. H. Freeman.

Wilde, C. D., C. E. Crowther, T. P. Cripe, M. G-S. Lee and N. J. Cowan (1982). "Evidence that a human beta-tubulin pseudo-gene is derived from its corresponding mRNA." *Nature* 297: 83-84.

Wiley, D. C., I. A. Wilson and J. J. Skehel (1981). "Structural identification of the antibody-binding sites of Hong Kong influenza haemaglutinin and their involvement in antigenic activity." *Nature* 289: 373-378.

Wills, C. and H. Jornvall (1979). "Amino acid substitutions in two functional mutants of yeast alcohol dehydrogenase." *Nature* 279: 734-736.

Wilson, A. C., S. S. Carlson and T. J. White (1977). "Biochemical evolution." *Ann. Rev. Biochem.* 46: 573-639.

Wilson, A. C., L. R. Maxson, and V. M. Sarich (1974a). "Two types of molecular evolution. Evidence from studies of interspecific hybridization." *Proc. Natl. Acad. Sci. USA* 71: 2843-2847.

Wilson, A. C. and V. M. Sarich (1969). "A molecular time scale for human evolution." *Proc. Natl. Acad. Sci. USA*. 63: 1088-1093.

Wilson, A. C., V. M. Sarich, and L. R. Maxson (1974b). "The importance of gene rearrangement in evolution: Evidence from studies on rates of chromosomal, protein and anatomical evolution." *Proc. Natl. Acad. Sci. USA*. 71: 3028-3030.

Woese, C. R. (1977). "Endosymbionts and mitochondrial origins." *J. Mol. Evol.* 10: 93-96.

Woese, C. R. and G. E. Fox (1977a). "Phylogenetic structure of the prokaryotic domain: the primary kingdoms." *Proc. Natl. Acad. Sci. USA* 74: 5088-5090.

Woese, C. R. and G. G. Fox (1977b). "The concept of cellular evolution." *J. Mol. Evol.* 10:1-6.

Woese, C. R., G. E. Fox, L. Zablen, T. Uchida, L. Bonen, K. Pechman, B. J. Lewis, and D. Stahl (1975). "Conservation of primary structure in 16S ribosomal RNA." *Nature* 254: 83-86.

Woese, C. R. and R. Gupta (1981). "Are archaebacteria merely derived prokaryotes?" *Nature* 289: 95-96.

Woese, C. R., M. Sogin, D. Stahl, B. J. Lewis, and L. Bonen (1976). "A comparison of the 16S ribosomal RNAs from mesophilic and thermophilic Bacilli: Some modifications in the Sanger methods for RNA sequencing." *J. Mol. Evol.* 7: 197-213.

Wolpert, L. (1969). "Positional information and the spatial pattern of cellular differentiation." *J. Theor. Biol.* 25: 1-47.

Wright, S. (1982). "The shifting balance theory and macroevolution." *Ann. Rev. Genetics* 16: 1-19.

Wu, R. (1978). "DNA sequence analysis." *Ann. Rev. Biochem.* 47: 607-634.

Wu, T. T., E. C. C. Lin and S. Tanaka (1968). "Mutants of *Aerobacter aerogenes* capable of utilizing xylitol as a novel carbon." *J. Bact.* 96: 447-456.

Yamada, Y., V.E. Avvedimento, M. Mudryj, H. Okhubo, G. Vogeli, M. Irani, I. Pastan and B. de Crombrugghe (1981). "The collagen gene: Evidence for its evolutionary assembly by amplification of a DNA segment containing an exon of 54bp." *Cell* 22: 887-892.

Ycas, M. (1972). "De novo origin of periodic proteins." *J. Mol. Evol.* 2: 17-27.

Ycas, M. (1973). "Self-complementarity of messenger RNAs of periodic proteins." *J. Mol. Evol.* 2: 329-338.

Young, M. W. (1979). "Middle-repetitive DNA: A fluid component of the *Drosophila* genome." *Proc. Natl. Acad. Sci. USA* 76: 6274-6278.

Yunis, J. J., J. R. Sawyer, and K. Dunham (1980). "The striking resemblance of high-resolution G-banded chromosomes of man and chimpanzee." *Science* 208: 1145-1148.

Zillig, W., R. Schnabel, J. Tu, and K. O. Stetter (1982). "The phylogeny of archaebacteria, including novel thermoacidophiles, in the light of RNA polymerase structure." *Naturwissenschaften* 69: 197-204.

Zimmer, E. A., S. L. Martin, S. M. Beverley, Y. W. Kan and A. C. Wilson (1980). "Rapid duplication and loss of genes coding for the alpha chains of hemoglobin." *Proc. Natl. Acad. Sci. USA* 77: 2158-2162.

Zuckerkandl, E. (1975). "The appearance of new structures and functions in proteins during evolution." *J. Mol. Evol.* 7: 1-57.

Zuckerkandl, E. (1976). "Evolutionary processes and evolutionary noise at the molecular level II. A selectionist model for random fixations in proteins." *J. Mol. Evol.* 7: 269-311.

Zuckerkandl, E. and L. Pauling (1962). "Molecular disease, evolution, and genic heterogeneity." In *Horizons of Biochemistry*. M. Kasha and B. Pullman (eds.) Academic Press.

Zuckerkandl, E. and L. Pauling (1965). "Molecules as documents of evolutionary history." *J. Theor. Biol.* 8: 357-366.

ANSWERS

CHAPTER 2

1. Definitions: *Goldberg-Hogness box*, also referred to as the TATA box is an AT-rich sequence that is located 5' to the initiation codons of structural eukaryotic genes, which may function as part of the promoter for transcription of the respective genes; the *Pribnow box* is the comparable promoter for prokaryotic genes. The *capping site* is the 5' end of a eukaryotic mRNA, which functions as the site of attachment of the special methylated guanine triphosphate attached to many eukaryotic mRNA molecules. The *Shine-Dalgarno sequence* is the part of the leader sequence of prokaryotic mRNA molecules that serves as the ribosome-binding sequence; a similar sequence is not present in eukaryotic mRNA molecules. An *intron-exon splicing junction* is the sequence at the junctions of introns and exons that apparently serves as the recognition and cutting site for the splicing system that removes introns from the transcripts. A *domain* is a functional region of a protein molecule that carries out a specific function for the protein, for example, for substrate- or ligand-binding. The *initiator codon* is the first triplet to be translated at the start of a structural gene-

-usually, AUG, which codes for methionine. A *terminator codon* is one of the three stop codons (UAG, UAA, UGA) that do not specify an amino acid and that signal the end of translation in an mRNA. A *coding sequence* is a sequence in a DNA or RNA molecule whose function is to specify a polypeptide. A *noncoding sequence* is any other nucleotide sequence, for example, one not involved in the specification of an amino acid sequence. A *flanking sequence* is any sequence on either side of a structural gene that is part of the control system for transcription or for processing information. A *pseudogene* is a structural gene related to and descended from a functional gene that has been rendered unexpressible by one or more mutations but that has been retained in the genome. A *transition* is a pyrimidine-pyrimidine or a purine-purine substitition. A *transversion* is a purine-pyrimidine or a pyrimidine-purine substitution. A *structural gene* is any gene that either codes for a polypeptide or specifies a tRNA or rRNA molecule. The *mutation rate* is the frequency with which new mutations occur; it is usually measured per cell or organismal generation. *Illegimate recombination* refers to recombination between nonhomologous DNA sequences. *Transposition* is the movement of a DNA sequence to a new location without loss of the sequence at the original location. A *plasmid* is any self-replicating extrachromosomal DNA molecule found in prokaryotic cells. A *Charon phage* is any one of a group of specifically engineered lambda strains designed to serve as vehicles for comparatively large DNA inserts. A *restriction enzyme* is a DNA endonuclease that makes a double-stranded cut, either staggered or blunt, at a particular short DNA sequence. A *ligase* is an enzyme that joins 5'P and 3'OH groups between adjacent nucleotides in a DNA chain. *Nick translation* is an effective method for making radioactive DNA molecules of high specific activity by polymerization from a single-strand break, using labeled nucleotides with concomitant displacement of the old strand. *Southern blotting* is a method of hybridizing electrophoresed restriction fragments to purified, labeled DNA fragments (probes). *In situ hybridization* is the annealing of either DNA or RNA to homologous sequences in fixed spread chromosomes in order to localize those sequences that share homology with the probe. *Polytene chromosomes* are multi-stranded, laterally expanded (and generally longitudinally extended) chromosomes; the ones most extensively studied are those of the salivary glands of *Drosophila* larvae.

2.

1-3	4-6	7-9	10-12	13-15	16-18	19-21	22-24	25-27
met	ser	ile	pro	glu	thr	gln	lys	ala
AUG	UCU	AUU	CCA	GAA	ACU	CAA	AAA	GCU

(a) From the universal code the amino acid sequence is Met-Ser-Ile-Pro-Glu-Thr-Gln-Lys-Ala. A transversion at -5 in the TATA box (from A to C or T) would decrease the frequency of initiation. A transition at 1 would change ATG to GTG (which codes for valine), changing the initiation codon, and should prevent initiation of translation of the mRNA. A single base addition at 21 might change the coding from Gln to His and would certainly change the reading frame at that point. A single base deletion at 19 would certainly change the Gln codon to Lys and would shift the reading frame, altering the rest of the protein. (b) The first three codons are AUG UCU AUU. Alteration in the first codon AUG would change the initiation codon and abolish translation. For any codon nine one-base-substitution changes are possible. For the second codon UCU (serine) three changes (all in the third codon) produce silent changes (that is, the coding specificity remains for serine), while the other six possible changes lead to substitutions for Thr, Ala, Pro, Tyr, Cys, and Phe. For the third codon AUU (Ile) there are two silent substitutions (to Ile) and seven missense substitutions, yielding Phe, Val, Leu, Asn, Ser, Thr, and Met. In the sequence given, for the first nine codons, two nonsense-producing transversions are possible: GAA (Glu) to UAA (ochre), and Lys (AAA to UAA (ochre). None of the missense mutations in the first three codons will yield variants that are electrophoreticallly detectable at pH 6-7.

3. (a) The globin genes are particualrly favorable material for genetic and molecular studies, for several reasons. First, many naturally occurring variants of the

globin genes are known, both in man and in other animals. Second, as one cell type, erythrocytes, specialize in the synthesis of globin on a large scale, both the protein and the mRNA are accessible to isolation and examination. The latter, when converted to complementary DNA (c-DNA) and cloned, can be used to detect genomic clones of globin sequences, by virtue of their homology. The ability to isolate genomic globin sequences in turn has facilitated the analysis of the flanking sequences that regulate the expression of these genes, and the isolation and analysis of globin pseudogenes. (b) The features that make the globin genes of special interest to the molecular evolutionist are their universal occurrence in chordates and their intermediate rate of change over long spans of evolutionary time. Although the globin genes do not change so slowly, like histone IV, that they are devoid of differences between many animal groups, nor so rapidly as the fibrinopeptides, they accumulate changes sufficiently often to permit the construction of both gene and organismsal phylogenetic trees. (c) The globins are especially suitable material for study by the developmental geneticist, because they constitute a gene family whose members are differentially regulated during embryonic and fetal development according to a fixed program. Analysis of the patterns of expression coupled to dissection of the flanking sequences should eventually produce valuable information about the mechanisms of control in these programs of expression.

4. (a) Each of the four bases can, in principle, undergo either of two transversions (for example, A to either C or T) but only one transition (A to G). Thus, if base substitution were a completely random event, there should be twice as many transversions as transitions. (b) Inspection of the data reveals that, for the 5S rRNA gene, more transitions than transversions have occurred, but not ten times as many. Clearly the data fit neither the strict stereochemical prediction not purely random occurrence. (c) One cannot say precisely why the observed pattern is the one seen, but the reasons presumably involve selective forces of some kind, the selective constraints permitting more transversions than expected on stereochemical grounds. An alternative explanation is that the reconstruction itself has been faulty, but a more extensive analysis by Sankoff *et al.* (1976) does support the reconstruction given. (d) For a coding sequence that is essential, rather different overall changes would be expected. Far fewer single base insertions and deletions should occur (since these change the reading frame) than transitions (since these would lead more often to synonomous coding changes than do transversion).

5. (a) For at least four of the five cloned fragments, there appears to be distinctly less hybridization to the DNA of the closely related strain Bergerac than to the parent strain Bristol, from which they were both derived. It appears that enough base substitutions have accumulated since the two stains diverged to have reduced the hybridizability of the Bristol fragments to Bergerac DNA. Furthermore, the sizes of several fragments have changed, indicating changes in restriction enzyme sites. The ability of the five DNA probes to hybridize to the DNA of the sister species *C. briggaea* is even further reduced, and three of the cloned sequences show no hybridization. (At the morphological level, these two species are similar.) It has been estimated that as few as 20 percent substitutions can completely abolish recognition in a Southern blotting experiment. (b) The simplest explanation for the existence of individual probes hybridizing to multiple bands is that these probes contain middle-repetitive sequences that, by definition, are present in multiple sites and that will turn up on multiple restriction-digest fragments in the Southern blot. (c) The pattern differences indicate that some sort of genome rearrangement must accompany development.

6. The fact that single or triple linked copies are occasionally found suggests that unequal crossing-over can occur between the duplicate copies. Randomly occurring point mutations can thus, in principle, become shuffled between the copies at the two positions, following crossing-over events, as discussed by Smith. Mating within comparatively small initial

populations would further increase the chance that particular populations could be identical, in both alpha-globin copies, for particular variant forms. Biased gene conversion associated with such unequal crossing-over events could result in the molecular drive described by Dover (1982).

7. The discovery of orphons suggests that the distinction between tandemly clustered and dispersed repeated gene families is at most a quantitative one of distribution rather than a quantitative distinction. It would appear that any DNA sequence can, at least on occasion, be transposed to a new location. The finding leaves open the question of how much or whether homology between the ends of such tranposed sequences and their insertion sites is required.

8. Possible selective disadvantages include the energy cost of splicing out the introns and the occurrence of splicing errors, possibly leading to dysfunctional proteins. Introns may serve to connect different protein domains, by rare recombinational events, but the selective advantage to the individual organism of introns can hardly lie in this prospect for future benefit to the species. Introns must either be selectively neutral or only slightly deleterious, or possess some as-yet-undocumented regulatory role in gene expression, which would give them a direct positive function selectable in itself. An interesting general fact about introns is that for all the genes that been examined so far, the positioning (though not the size) of the individual introns, is constant for a given gene, in widely differing organisms possessing that gene. Thus, the evidence is that introns must have been inserted early in evolution. Contemporary intron insertion must be a rare event.

9. (a) For the most part, mutator activity is higher in outcrosses involving natural populations than in intrapopulational crosses. (b) The one apparent exception is the cross between strains N-34 and M-4. (c) The direction of the cross does make a difference, which is particularly striking when the male parent is from a natural population and the female parent is from a laboratory strain. (d) It would seem that MR elements, one class of transposable element, generate little variability in crosses within populations, but may function as an important source of new variation in crosses between populations. The evolutionary significance, at least for *Drosophila* and any other organisms with similar genetic phenomena, is that matings between individuals of normally distinct populations may be a particularly potent source of stable de novo genetic variation. Thus, hybrid zones may be especially favorable regions for the generation of evolutionary novelty.

10. One approach to the molecular analysis of such variation is to collect mutant individuals from the population, then mate them to flies deficient for the gene in a set of crosses to generate flies homozygous for the selected mutations. The molecular analysis could be done by Southern blotting of whole mutant and wildtype genomic DNA (digested with a restriction enzyme), and probing with the cloned and labelled wildtype gene sequence. Large deletions or insertions would show up as a shift in the band pattern, while point mutations would generally give the wildtype pattern, except for the rare direct change in restriction enzyme site. By surveying even a moderately sized set of natural mutations, say, 10 to 15, one could obtain a fairly good idea of the frequency with which new mutations were associated with the insertion of transposable elements at or near the gene site.

CHAPTER 3

1. Definitions: A *first-order reaction* is a reaction whose rate depends only on the first power of the concentration of the reactant; a unimolecular reaction. A *second-order reaction* is a reaction whose rate depends either on the second power of the concentration of the reactant or on the concentration of two reactants; a bimolecular reaction. *Hypochromicity and hyperchromicity* are the increase of absorbance of DNA on melting and the decrease upon reannealing, respectively. A *nitrocellulose filter* is a paper to which dsDNA can be made to bind almost irreversibly under the appropriate

conditions of ionic strength, pH, and temperature. *Hydroxyapatite* is a chromatographic material to which dsDNA binds reversibly depending on the conditions of ionic strength, pH, and temperature. *S1 nuclease* is a DNase that acts specifically on single-stranded DNA. *High and low stringency* refer to conditions of DNA reannealing, generally regulated by the temperature; the lower the temperature, the less stringent are the homology requirements for the formation of a stable double-stranded molecule. A *cot curve* is a convenient display of the kinetics of reannealing, in which the proportion of dsDNA remaining is plotted against the product of the time (seconds) and the total concentration of DNA. The half value of Cot is the product of the concentration and the time required for a simple second-order reaction to go half way to completion. *Tm* is the temperature in a DNA melting curve at which half of the base pairs are broken.

2. (a) This instantaneously reannealing material which re-forms the double-stranded state with first-order kinetics, indicates palendromic or self-annealing "snap-back" sequences. (b) 25% highly repetitive, 35% mid-repetitive, and 40% low- or single-copy DNA. (c) The shift to the left of the $Cot_{1/2}$ value of the reisolated single-copy DNA is attributable to the fact that the C_0 relates exclusively to the single-copy class (rather than the entire genome) and hence the homologous sequences are at a relatively higher concentration. The total genome size is approximately 5 x 10^8, as deduced by the fact that the $Cot_{1/2}$ of the single-copy DNA in the complex genome curve is to the right, by a factor of 100, of the $Cot_{1/2}$ of the prokaryote C, which has a genome size of 5 x 10^6.

3. (a) While the reannealing of precisely repeated sequences should not be seriously affected by increasing stringency, those less precisely repeated sequences would experience increasing difficulty in annealing, behaving in the limit like single-copy DNA. Accordingly, portions of the curve would be displaced to the right (slower reannealing). (b) To the extent that the dispersed mid-repetitive sequences have diverged from one another, there would be a decrease in the number of reannealing mid-repetitive blocks, (hence an apparent decrease in amount of mid-repetitive sequence) and a modest increase in the apparent size of the interspersed single copy DNA. (c) Since the higher stringency permits the re-formation of only the better-matched sequences, a melting curve would show not only a sharper transition but also a higher Tm.

4. (a) The sharp transition for the satellite DNA is a reflection of a very homogeneous sequence and the lower Tm indicates a lower GC content than for the main peak DNA. The broader transition for the main peak DNA reflects sequence heterogeneity, with AT-rich regions melting at a lower temperature and GC-rich areas melting at a higher temperature. (b) The heavy solid line indicates that more than half of the DNA did not reanneal, while the portion that did reanneal has a Tm and sharpness of transition comparable to the native DNA, indicating an excellent sequence match in the reannealed portion. The dotted line represents divergent repetitive sequences manifesting a range of homology. (c) The series of species B to D represents a series of decreasing relatedness to A. The Tm values indicate that B nucleotide sequences differ from A by about 2.5%, C differs from A by about 5%, and D differs from A by about 7-1/2%.

5. That the long reannealed blocks more closely approximate the melting behaviour of the native DNA than do the short blocks indicates that the former have retained a high degree of sequence homology while the latter dispersed elements have, in general, diverged more widely.

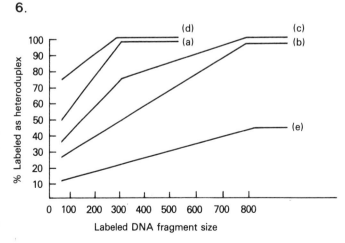

6.

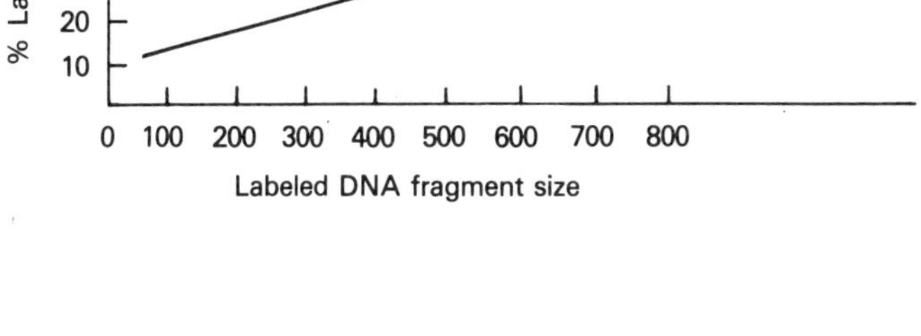

7. (a) The Alu family represents almost 3% of the genome and thus has an apparent repetition frequency of 2.4 x 10^5. If there has been sequence divergence in this family such that either homology has dropped below detection at the stringency level employed or the Alu site has been lost, then the Alu family of sequences could be even larger. (b) To determine the state of interspersion of the Alu family, a standard interspersion experiment could be performed by annealing excess purified Alu DNA (for example, from a cloned fragment) with labeled genomic DNA sheared to varying extents. Tandem repetition would be reflected by insensitivity to fragment size of the level of reannealing of the labeled probe. Interspersion with other sequences would be indicated by an increase of the level of reannealing of the label with increasing fragment size. If the latter was observed, the long labeled fragments annealing to the Alu sequence could be isolated, sheared and used as a probe in a standard Cot experiment. The $Cot_{1/2}$ value of annealing of the label is a measure of the repetitive frequency of the sequences adjacent to the Alu sequences.

8. The most straightforward interpretation of a regular ladder is that it results from a tandemly repeated sequence homologous to the probe, in which there has been random mutational alteration of the restriction site along the array. The result with a different restriction enzyme would suggest that in the evolutionary history of the tandem array there had been generated (by base mutation or insertion) a new restriction site in one of the repeats, followed by multiple tandem repeats. The higher-molecular-weight smear could correspond to mutational loss of the secondary restriction site.

9. (a) Group I = 20% (hybridization with oats). Group II = 23% (hybridization with wheat or rye minus Group I). Group IV = 28% (hybridization with itself minus Group I and Group II). About 10% of the DNA is in long runs of unique sequence DNA. (b) Group-VI sequences are found alone in about 10% of the genome (as Group I and II are found associated with 80% of the genome) and the repetitive blocks bracket single-copy blocks of about 700 bp. (c) 51% of the genome contains Group-I sequences, 37% interspersed with 700 bp blocks of non-Group-I sequences and 14% interspersed with 2500 bp non-Group-I sequences. To assess the proportion of nonrepeated DNA in this region, long (let us say 3500 bp), labeled barley DNA annealed with an excess of oats fragments to an intermediate Cot value may be isolated by hydroxyapatite chromatrography. This material may then be sheared, melted, reannealed with an excess of barley fragments to an intermediate Cot value and subject to hydroxyapatite chromatrography. The labeled single-stranded material passing through the hydroxyapatite would be a direct measure of the amount of nonrepeating sequence in the region of interest. The reader is referred to the paper of Rimpau *et al.* (1980) for an exhaustive explanation of what may be learned about genome organization by this general approach.

10. The utility of using a thoroughly studied cultivar like wheat is that a variety of useful constructed strains, like the wheat-rye addition series, are available. In this instance, the 2.2-kb complex was identified in each strain that carried one or another of the rye chromosomes but was totally absent from the wheat complement of chromosomes. An additional useful feature of a cultivar like wheat is that its recent (last 5,000 years) evolutionary history, and that of its nearest relatives, has been studied in detail. Thus some of the recent major genome events can be placed on a reasonable time scale.

CHAPTER 4

1. Definitions: *Orthologous sequences* are those whose descent is directly related to organismal descent; their time of divergence is the same as the organismal time of divergence, as in the case of the alpha globin gene of man and horse. *Paralogous sequences* can be related to each other through a gene duplicational event of uncertain antiquity, such as the alpha and beta globin genes of man. *Alpha globin chains* constitute two of the four components of the functional adult

hemoglobin; the other two are beta globin chains, which are structurally similar to, but distinct from, alpha chains. *Delta globin chains* are structurally and functionally very similar to, but distinct from, beta chains, and are found in their stead in a minority of adult hemoglobin molecules. *Gamma globin chains* are found in fetal hemoglobin molecules and are the functional equivalent of the beta or delta chains in adults. A *sigmoid oxygenation curve* reflects the cooperativity of oxygen uptake by hemoglobin; the binding of one oxygen molecule induces a subtle conformational alteration in the hemoglobin molecule such that the second molecule of oxygen is bound more tightly, and so on for each molecule of oxygen bound, up to the maximum of four observed. *Electrophoretic variants* are mutationally altered proteins that can be distinguished from one another by an electrophoretic technique; the differences of the variants are usually alterations in the net molecular charge of the protein molecule. *Polymorphism* refers to the stable coexistence of several alternative molecular variants in a population. *Synonymous nucleotide replacements* lead to no detectable amino acid replacements (typically codon third position replacements); *conservative replacements* lead to an amino acid replacement that is nondisruptive by virtue of similarity of functional groups. A *disruptive substitution* is one that leads to a substitution of an amino acid having a grossly different functional group. A *covarion* is a set of not necessarily contiguous amino acids in a protein chain that, at a particular time in the evolutionary history of the protein, are relatively free to change as the sequence and function of the molecule evolves. The *molecular clock* is the apparent long-term temporal regularity of amino acid substitutions in proteins whose function remains substantially unaltered during that period of time. *Neutralism and pan-selectionism* reflect the opposite ends of the scale of views of the driving force for progressive mutational alteration. Both views agree that deleterious alterations are eliminated, but in the neutralist view mutation and drift adequately account for most of the observed changes while the pan-selectionist insists that only changes leading to increased fitness can become fixed in a population. A *stochastic process* is a probabilistic (as opposed to a deterministic) event.

2. (a) An enzyme active site is generally an exquisitely "engineered" bit of molecular architecture in which the slightest alteration in terms of amino acid substitution is likely to lead to a decrease in functional efficiency. (b) An amino acid in an interdomain loop is unlikely to participate in the function of the protein molecule; however, it may be conserved because the corresponding nucleotide sequence is required for correct splicing at an exon-intron junction. (c) These regions can be presumed to have, like fibrinopeptides, no function other than perhaps as a "filler" in the amino acid or nucleotide sequence. (d) One would surmise that such a variable intron sequence has no function beyond being a filler between translated exons. (e) A conserved 5' upstream site is likely to be very interesting as a transcriptional, processing, or translational signal sequence. (f) This observation suggests that the entire mitochondrial replication apparatus is more error-prone than that of the nucleus, and may reflect the relatively sheltered existence of this cellular organelle.

3. This observation does not conflict with the covarion hypothesis. Beyond the fact that two different (though related) protein families are being considered (hemoglobin and cytochrome), the covarion constraint applies to a very restricted period of evolutionary time; comparison of the two most divergent members of a family encompass the aggregated changes accumulated over long evolutionary times and innumerable covarion shifts; more frequent parallel changes than might be expected in separate lines of descent are comprehensible within the context of chance maintenance of a particular covarion in those lines of descent. The hypothesis may be tested only with great difficulty. While the idea is conceptually very useful for thinking about the dynamics of protein evolution and by extension, organismal evolution, it has proven to be exceptionally refractive to critical test. For this reason the term has not gained wider recognition.

4 X-ray crystallography and the recognition of functional domains within molecules has provided valuable clues to appropriate alignments of segments of molecules. In the persistent absence of discernible sequence homology, the presence of introns in precisely homologous positions in genes may provide an alternative indication of common origin. Similarly, genes coding for common functions and located in identical positions in small genomes may be confidently assumed to have a common ancestor even in the absence of detectable sequence homology. In the absence of any of these clear indices of common origin, as in the case of the lysozymes, an independent indication of homology seems unlikely with current conceptual and technical tools.

5.

$u = 10^{-6}$	He	u(P)
$Ne = 10^{2}$	0.004	10^{-2}
$Ne = 10^{3}$	0.004	10^{-3}
$Ne = 10^{4}$	0.038	10^{-4}
$Ne = 10^{5}$	0.29	10^{-5}

As we are interested in evolution in nature, which is not experimentally controlled, the difficulty of assessing an "effective breeding population" (Ne) or the adaptive value (S) of an allele in the spectrum of microenvironments encountered by an organism seems insurmountable. For this reason, formal population genetics appears to be a contentious area. A level of heterozygosity lower than anticipated by the neutralist hypothesis would suggest at least temporary dominance of a superior allele.

6. An attractive hypothesis to explain the observation is that there are limited portions of the hemagglutinin molecule that are relatively free of functional constraints but nevertheless critical for recognition by the immune surveillance systems. These regions, under constant antibody challenge, may mutationally alter to generate a new variant that is modestly resistant until a modified immunity is established in the population. Such variants would generate local epidemics. The data suggest that occasionally a "hybrid" virus is generated, using a hemagglutinin from a wild reservoir of influenza-type virus. This might be expected to produce in a single step a virus that is totally foreign to the immune system of the human host population (world wide), and thus lead to a pandemic.

7. The selectionist explanation of the apparent high rates of change seen early in the several globin lineages is that in each case, a new "protein-environment space" is being explored. A number of amino acid substitutions are believed to provide functional improvements over the previous models, and hence are positively selectable. In particular, during the period of transition from the monomer to the allosteric tetramer, contact residues might well have been loci of active "evolutionary experimentation", especially if compared to those residues established much earlier for their optimal heme-binding capacity. The neutralist interpretation of these observations is that, following gene duplication, functional constraints are relaxed in one or the other of the duplicated pair of genes and accordingly a higher proportion of the randomly generated mutations would be permitted to drift to fixation. Dickerson (1971) was careful to point out that the concept of a molecular clock was generally valid for only a protein that had settled into a stable role. Inasmuch as the fossil record cannot shed light on the times of branching of paralogous lines of gene descent, one can entertain the possibility of much earlier gene-duplication events than show on the genealogical tree and commensurately lower rates of amino acid substitution that are more in line with the long-term average rate. To further confound the issue is the uncertainty of the dating of the fossil record and the possibility that pseudo-gene status was enjoyed, perhaps even only briefly, by one or another of the lineages early in their history.

8. In the absence of further information there are several possible responses to the discovery of the third intron in leghemoglobin. Perhaps the least interesting is that it is a recent addition brought about by the random insertion of a mobile element as discussed in both this chapter and in Chapter 2. More interesting is the suggestion of Go (1981) that the heme-binding domain may be comprised of two ancestral domains. Thus, the

organization of the leghemoglobin gene might reflect this earlier stop in evolution, the animal globins having lost the third intron. Indeed, Doolittle (1978) has argued that the presence of introns is a primitive condition, in the sense of reflecting an early stage in the evolution of a protein. Examination of the gene structure of some of the more advanced cytochromes (the presumed heme-binding ancestors of the globins) might shed some light on this question.

9. Noncongruence of trees is most readily explained by the hypothesis of lateral transfer of genetic information mediated, for example, by plasmids and/or transposons in the microbial world, or retroviruslike vectors in animals or Agrobacterium-like vectors in plants (see Reanney, 1978). Also, the symbiotic origin of cellular organelles (to be examined in more detail in Chapter 7) must give rise to noncongruence of specific trees.

CHAPTER 5

1. Definitions: *Historical reconstruction* is the process of estimating common ancestral (nodal) sequences on the basis of sequences of contemporary pairs of organisms. An *acid protease* is a member of a family of proteases characterized by low pH activity optima and two key aspartyl residues in the active site, formed by two structural domains that could have arisen from an ancestral duplication. Pepsin and chymotrypsin belong to this group. *Microevolutionary studies* are investigations in which population changes of presumed evolutionary significance can be directly observed experimentally, preferably under an identifiable selective regime. *Leghemoglobin* is the hemoglobin synthesized in the nitrogen-fixing nodules of leguminous plants, serving to protect nitrogenase from oxygen. It is encoded by the plant genome and has modest primary structural homology and strong tertiary structural homology to other globins. *Fixed heterozygosity* is a condition resulting from a duplication followed by recombination, for example,

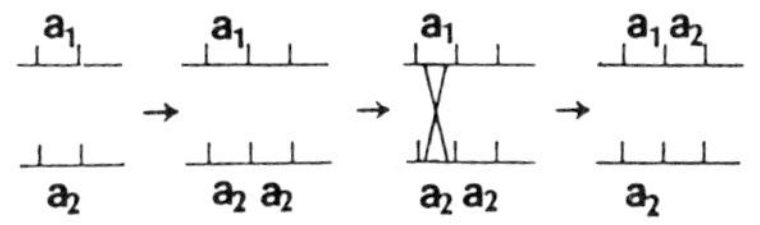

Evolved beta-galactosidase or Ebg protein is the product of the wildtype *ebgA* gene, that is induced by lactose even though the enzyme is unable to hydrolyse this sugar. Some mutations of the *ebgA* gene confer the ability to utilize various galactosides as the sole carbon and energy source. *Ferredoxin* is a ubiquitous nonheme Fe/S electron-transport protein with a reduction potential approaching that of molecular hydrogen. *Glutamate dehydrogenase* is the enzyme responsible for oxidative deamination of glutamic acid and accordingly occupies a key position in nitrogen metabolism. It is allosterically regulated by NADH, a cofactor of the reaction that it catalyzes. *Collagen* is a fibrous structural protein, found widely and in large quantities among higher animals, with the striking property that its primary structure is comprised almost exclusively of the repeating sequence gly-X-Y, in which X and Y are often proline and hydroxyproline. The *mononucleotide binding fold (mnbf)* is a conformationally identifiable domain responsible for binding mononucleotide cofactors; it is found as a structurally distinctive part of a variety of enzymes utilizing this class of cofactors. A *serine protease* is a member of a family of proteolytic enzymes, all of which have in common a very similar active site in which a serine residue becomes acylated and then deacylated during catalysis. Trypsin and elastase are representative of this group. *Flavodoxin*, like ferredoxin, is a low-molecular-weight protein that is an electron-transport cofactor; it has an exceptionally low reduction potential. Unlike ferredoxin, it has no iron atoms, but instead has a flavin prosthetic group.

2. G3PDH 198-221/GDH 112-135, 5 identities; G3PDH 198-221/GDH 267-290, 5 identities; GDH 112-135/GDH 267-290, 7 identities. Taking Engel's expression,

$$[n!19^{n-x}]/[x!(n-x)!20n]$$

in which n is the length of the sequence compared and x is the number of homologies, the probability of getting 5 and 7 observed identities between random sequences of 23 amino acids, respectively, is

2.2×10^{-4} and 8.4×10^{-5}

By the combination of information about the tertiary structure of other mononucleotide (NADH or NADPH) binding sites and a superior method of sequence alignment, Rossman *et al.* (1974) have found that the G3PDH NADH binding site shows the strongest homology with only one region of the GDH molecule, which corresponds to neither of the regions identified by Engle. This, the simplest interpretation of the homologies noted by Engle may require revision and suggests a more complex and perhaps more interesting evolutionary development of GDH than initially proposed. A three-dimensional structure for GDH should provide useful clues for resolving questions about the evolutionary history of this molecule.

3. The bovine cytochrome b5 must have arisen from an ancestral duplication of the cytochrome c gene. The duplicated gene could then have increased in size relative to either mitochondrial cytochrome c or *Pseudomonas aeruginosa* cytochrome c551, by either piecemeal terminal duplications, thereby expanding the heme-binding exon, or by the acquisition of one or two new exons by transpositions. Determination of the cytochrome b5 gene structure should help decide between these two possibilities in that the former would predict a single exon and the latter two or three exons.

4. The observation that homologous repeats (D) in different collagens are more similar than the different repeats of the same collagen suggest that a single primitive collagen evolved by an initial six-fold duplication of a sequence of nine base pairs (three amino acids) which became bracketed by introns. This block then appeared to undergo a thirteen-fold duplication to yield the basic protein repeat of D, which then underwent further tandem duplication. Then, this very long tandem array underwent another duplication and divergence to yield the contemporary variety of collagen molecules. An alternative scheme, which appears to be ruled out, is an earlier functional divergence commencing at the 9-bp, 54-bp, or 702-bp stage of evolution of this family of genes. (If you draw this out, you might find this conceptually easier to follow.)

5. In the paper of Levine *et al.* (1978), two relevant points were made. Firstly, mechanistic studies showed that although PK and TIM catalyze two very different types of reactions, the substrates pass through a very similar intermediate. Secondly, a picture of the evolution of modified or new protein functions that is emerging involves the reassortment of functional domains. Jeffcoat and Dagley (1973), in a survey of bacterial hydrolases and aldolases, reached conclusions similar to the first point above.

6. The alternative view is that the restricted number of stable beta-sheet conformations observed represent the descendants of one primitive gene product, which was initially judged more stringently by the criterion of stability rather than enzymatic efficiency. It is not difficult to imagine that a primitive stable peptide, or domain, would be fertile material for the generation of functional diversity. The evolutionary antiquity of these events makes the formulation of convincing conclusions very difficult. The lines of evidence that have been discussed are: (a) overall similarity of tertiary structure, (b) similarity of reaction intermediates, (c) membership in the same protein "super family," and (d) similarity of structure of the corresponding genes or portions thereof.

7. Constitutivity for the D-arabitol permease was an initial requirement in order to get xylitol into the cell where ribitol dehydrogenase, recruited from a different pathway, could act, albeit weakly, to produce xylulose (Wu *et al.* 1968). Ribitol-dehydrogenase superproducer strains arose by either gene duplication or presumed promoter-up mutations. Ribitol-dehydrogenase-specificity mutations arose with improved xylitol-dehydrogenase activity, with as few as single amino acid substitutions. D-arabitol permease mutants were isolated with improved xylitol-transport efficiency (Wu et al. 1968).

8. The first step was the selection of the constitutive $ebgR^{-}ebgA^{o}$ strain using phenyl-beta-galactoside, which is not an inducer of the *ebg* system, but can be sufficiently metabolized in the constitutive mutant to support growth. The next step, $ebgR^{-}ebgA^{+}$ was made by growing a *lacZ* (*ebgR*) deletion mutant on lactose-tetrazolium plates containing an IPTG (isopropylthiogalactoside) supplement. The lactose-tetrazolium medium permits the growth of lac^{-} cells to yield deep red colonies, while lactose-fermenting cells give white colonies. IPTG is a gratuitous inducer of the *lac* operon, and hence of Lac permease, which is required for transport of lactose (lactose is not the natural inducer for the *lac* operon). White sectors found in the occasional red colony were picked and repurified; a few of these yielded isolates that were able to grow (slowly) without a lag when switched to lactose as the sole carbon source in the presence of IPTG. Further selection using lactulose as the sole carbon and energy source yielded a second-site mutation, which alone was shown to be sufficient to permit the utilization of lactulose. The double mutant (lactose- and lactulose-fermenting) was able to ferment galactosylarabinoside, which neither single mutant was able to do.

9. Probably the most direct way of finding functionally diverged duplicated genes is to employ the current battery of recombinant DNA tools. Labeled DNA probes of known cloned genes of *E. coli* may be prepared and used for examining blotted restriction-enzyme patterns of whole genomic DNA. Assuming appropriate choice of enzyme for the preparation of the cloned probe and the genomic digest and appropriate stringency of hybridization, multiple spots on the autoradiogram could be taken as a strong initial indication of possibly cryptic gene duplication.

10. In order to be deemed mutationally altered, an enzyme should manifest a change in its catalytic activity with respect to its defined normal substrate or a change in the range of substrates with which it can interact. Since it is now clear that even single amino acid substitutions are sufficient to alter the activity of an enzyme, a silent pseudogene stage is not a necessary prerequisite for the development of a new function.

CHAPTER 6

1. Definitions: *Heterochrony* is the retardation or acceleration of growth rates of regions or of organ primordia in embryos, relative to ancestral development patterns. *Neoteny* is the retardation of somatic development relative to germline or gonadal maturation, leading to the retention of juvenile features in sexually mature organisms. A *homoeotic mutation* produces the substitution of one structure or region for another in development. A *selector gene* is a regulatory gene governing the activation of expression of genes characteristic of a given "compartment" in insect development. *Genetic distance* is a measure of the extent of gene sequence divergence between two species, based on protein electrophoretic differences and corrected for intraspecies diversity. *Allometry* is the measurement and comparison of altered morphology generated by altered differential growth rates. *Cis-acting control* is the regulation of gene expression specifically by neighboring contiguous genetic elements. *Trans-acting control* is the regulation of the expression of one or more genes by diffusible elements produced by other genes. *Sibling species* are closely related species within the same genus, often differing only slightly in morphological or other characters. *Congeneric species* are any species within the same genus. An *electromorph* is any protein variant differing in net electric charge, and hence in electrophoretic mobility, from the standard protein form. The *stringent response* is the regulatory response in bacteria, in which RNA synthesis is rapidly inhibited upon amino acid starvation. *Catabolite repression* is the inhibition of synthesis of many sugar-degrading enzymes during growth in glucose or other rich media, resulting from the inhibition of synthesis of cyclic AMP. *Microcomplement fixation* is a sensitive immunological method used for measuring extents of antigenic difference between homologous proteins from different sources, based on the amounts of complememt fixed in small

volumes during reaction with antibody to a standard form of the protein. *Immunoelectrophoresis* is the electrophoretic migration of antigen molecules within a supporting substratum to yield bands that are visualized by precipitation with cross-diffusing antibodies. *Immunodiffusion* is the cross migration by diffusion of antigen and antibody through agar or another gel, to yield precipitin lines; as with immunoelectrophoresis, it is a technique for the comparison of protein differences by antigenic means. *Chromosome banding* is a set of chromosome-staining techniques that produce characteristic band patterns along the length of individual chromosomes, the particular pattern being a function of the particular staining technique and the specific chromosome; it is an important cytological technique for the identification of individual chromosomes. *Gel electrophoresis* is a set of procedures for the separation of proteins or other macromolecules, on the basis of charge and/or size, in an electric field through some sort of gel matrix. An *activity stain* is a procedure for detecting separated electrophoresed proteins by means of their activity; the position of the enzymes is visualized following reaction with the substrate by means of a staining reaction for the product of the reaction.

2. All genetic distances determined from protein electrophoretic differences underestimate true genetic (i.e., DNA) change because electrophoretic separations cannot detect many charge-similar substitutions or codon-synonymous changes, the latter producing no amino acid substitutions at all.

3. (a) Humans appear to be more polymorphic for the enzymes examined. (b) Ferris *et al.* conclude that it is the other way around, chimpanzees being more polymorphic than humans. (c) The main strength of estimating extent of polymorphism from electrophoretic enzyme comparisons is that it is easy and inexpensive, making possible extensive data collection. Futhermore, since most enzymes are encoded by the nuclear genome, estimates of polymorphism obtained in this way provided data for estimation of overall genome polymorphism. The principle weakness of the method is that it samples only a portion of the underlying genetic diversity, for the reasons given in (2) above. An additional problem with the method is that it may spuriously overestimate the level of genetic difference, if post-translational modification of one or more of the enzyme genetic variants takes place. Ideally, one must also always check to make sure that the enzyme being examined is specified by one and only one Mendelian locus. In contrast, the estimation of underlying polymorphism by the mitochondrial DNA method is free of such possible complications, and permits a direct assessment of variation at the DNA level. Again, one is sampling only a portion of the total genetic variation (that which affects restriction-enzyme sites) but by using a variety of different enzymes, one can increase the portion of the genome that one is examining. The disadvantage of the method is its complexity and expense (compared to enzyme electrophoresis) and that it only provides a measure of nonnuclear genetic variation.

4. One would cross the two strains and examine the developmental profile of the two forms of the enzyme in the F_1, in comparison to that of the parental species. If both electromorphs follow their respective parental patterns, one would conclude that the regulatory difference for octonal dehydrogenase involves cis-acting control in both strains. Conversely, if either electromorph shows an altered pattern of increase in activity during development, then one would infer that there was an element of trans-acting control in either or both strains for this enzyme.

5. One would first determine whether the electrophoretic mobility of the enzyme was the same at all stages, or whether different bands of activity appeared. The latter result might also reflect differential post-translational modification. Secondly, one could do simple comparisons of the properties of the enzyme at different stages, such as test of thermal inactivation, comparisons of K_m, etc. Finally, one should isolate the enzyme from different stages, purify to homogeneity, and then do fingerprinting or direct

sequence analysis. Identical peptide maps and/or sequences would be good presumptive evidence of gene identity, while differences would establish that different structural genes are involved (at least in the sequence analysis). Small bound molecules might alter the peptide fingerprint; only sequencing of each oligopeptide would definitively establish the presence or absence of amino acid sequence differences.

6. (a) If the Tompkins hypothesis is correct, then cyclic AMP would be predicted to be found participating in an increasing number of cellular functions as one moves up the evolutionary scale. Correspondingly, adenyl cyclase deficiency should produce an increasingly pleiotropic response in this progression. Plants do not appear to have cyclic AMP, which raises the interesting question of what molecule(s) serve in its place. (b) One would expect that cyclic AMP, and ppGpp as well, would be utilized in even fewer regulatory reactions in organisms simpler than *E. coli*. Some of the archaebacteria, however, might not be simpler than standard eubacterial species, and such prokaryotes might use small regulatory molecules in as many, or more, reactions than the eubacteria.

7. (a) It would appear that for a given immunological distance (time of divergence) far fewer mammalian species pairs have maintained a constant chromosome number than have frogs. (b) No hard conclusions can be drawn, but to the extent that the more rapid morphological divergence of the mammals is attributable to "regulatory gene" change, the data are in accord with the possibility that rapid karyotypic change is associated with changes in regulatory networks affecting gene expression. This would be the case, for instance, if many chromosome rearrangements occur preferentially in regulatory DNA sequences, or if disrupting established linked-gene complexes tend to disrupt standard regulatory systems in some manner that promotes the selection of new gene expression networks. The first possibility, preferential breakage in regulatory DNA sequences, is not implausible if such breaks tend to occur within a near-repetitive DNA sequence, some of which may function directly in the regulation of gene expression. Dispersed repetitive sequences are a special feature of the eukaryotic genome. However, this explanation begs the question of why frogs and mammals, both of which possess grossly similar eukaryotic genomes, should differ in either their rates of karyotypic diversification or the evolutionary consequences of such diversification. (c) In the absence of better information about the nature of the genetic changes that lead to evolutionarily significant regulatory alternations, we certainly cannot expect a fixed relationship between the rate of change of structural genes and unknown regulatory elements.

8. (a) No, since the correlation between karyotypic change and organismal evolution is, at best a weak one. (b) If the karyotypic changes that have occurred since the divergence of chimpanzees and humans have been significant in the evolution of these branches, the small number of these changes would suggest that only a small number of regulatory changes have been responsible for the major differences between these species. These changes would presumably be ones of heterochrony, producing the relative neotenization of man.

9. The result would appear to rule out, at least for *Drosophila*, a fixed function for particular mid-repetitive elements in the activation of expression of specific structural genes. However, since the long blocks of mid-repetitive DNA in *Drosophila* appear to consist of tandem "scramble clusters" of shorter elements (Wensink, Shiro and Tabata, 1979), it could be the case that members of some different families may be able to substitute for one another partially or wholly, providing a certain informational redundancy for regulation. For those eukaryotes with the *Xenopus* pattern of genome organization, in which short (300-bp) mid-repetitive segments are interspersed with stretches of single-copy DNA, a comparable observation of mid-repetitive DNA fluidity would be harder to reconcile with simple regulatory schemes. However, recent data (Moyzis *et al.* 1981)

suggest that much of the sequence organization in genomes previously thought to contain the *Xenopus* pattern may, in fact, be in the *Drosophila* pattern. The observed movement of mid-repetitive sequences in the fruit fly genome, nevertheless, does raise questions about their role as specific gene-regulatory segments and may lend some support to the alternative notion that they function as structural elements in chromatin, perhaps for chromatin folding, rather than informational elements. (However, see Moore *et al.* 1980.)

10. The answer to this question is, as indicated in the question, open-ended at the present time.

CHAPTER 7

1. Definitions: A *eukaryote* is an organism with a membrane-bounded nucleus, whose DNA is organized in chromosomes, and with mitochondria and microtubules. A *prokaryote* is an organism lacking membrane-bound cellular organelles; its DNA is in the form of a single covalently closed circle. The *serial symbiotic hypothesis* is the theory that chloroplasts and/or mitochondria and/or flagella are derived from a series of symbioses with previously free-living organisms. *Autogenous theories* are theories that derive the entire eukaryotic cell from one prokaryotic cell. A *polyphyletic origin* refers to the arising of a group more than once, with the last common ancestor of the group not being a member of the group. A *gene phylogeny* is an evolutionary tree showing the history of genes derived from one original gene; the genes may or may not be in the same organism. An *organism phylogeny* is an evolutionary tree linking taxa back to a common ancestor.

2. PROKARYOTIC: (1) green bacteria (*Chlorobium*)--bacteriochlorophylls c, d, e and a, carotenoids; (2) purple sulphur bacteria (*Chromatium, Thiospirillum*)--bacteriochlorophylls a and b, carotenoids; (3) purple non-sulphur bacteria (*Rhodospirillum, Rhodopseudomonas*)--bacteriochlorophylls a and b, carotenoids; (4) blue-green bacteria--chlorophyll a, phycobilins, carotenoids. In addition, *Halobacterium* is able to synthesize ATP using light energy absorbed by bacteriorhodopsin.
EUKARYOTIC: (1) green algae and higher plants (*Euglena*)--chlorophyll a and b, carotenoids; (2) red algae--chlorophyll a, phycobilins, carotenoids; (3) brown algae, dinoflagellates--chlorophyll a and c, carotenoids.

3. There are several possible explanations for nonconcordance of trees from different sets of data, and it is important to find additional evidence that may support one of the explanations. In the present case, the difference between ferredoxin (chloroplast) and nuclear 5S rRNA was predicted from earlier evidence (Margulis, 1968; 1970), so there is no real problem of scientific status in this case. Several chloroplast sequences yield internally consistent results as do several non-chloroplast sequences, but like the ferredoxin and 5S rRNA sequences of this problem, the two sets do not agree with one another. In sum, there is rapidly accumulating a large amount of evidence supporting the hypothesis of separate origins of the nucleus and chloroplast. If only one sequence disagrees with others, one recalls that evolution is a stochastic process and that we do not expect identical trees from different sets of data, only similar.

4. *Prochloron* is definitely prokaryotic, but has many similarities with the chloroplasts of green algae and higher plants both in photosynthetic pigments and ultrastructure. This group is generally considered to provide strong support for the endosymbiotic hypothesis and has been separated from the blue-green algae into the *Prochlorophyta*, though *Chloroxybacteria* has been suggested to more clearly identify it as prokaryotic. What you might expect for a (say, 5S-rRNA) tree containing *Prochloron* and higher plants is:

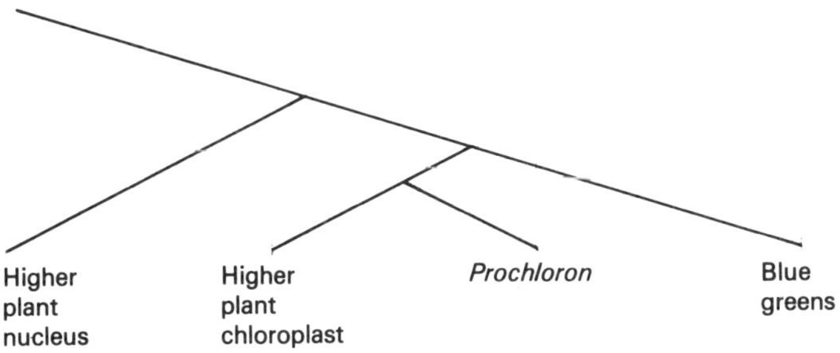

Note that you must be careful in identifying the root of the trees. A partial nucleotide catalogue for *Prochloron* has been reported (Seewaldt and Stackbrandt, 1981) and *Prochloron* sequences do come out close to the blue greens, but appear more different from higher plant chloroplasts than expected. However the tree-building method used will give misleading results with unequal rates of evolution, and it is to be expected that chloroplasts will show faster rates of evolution due to fewer constraints (Dickerson, 1971). Examination of the appropriate highly conserved amino acid sequences should lead to the same tree as predicted for the 5S rRNA, but, as in the case of nucleotide sequences, the antiquity of the events make interpretation difficult. While comparison of higher plant chloroplast sequences with *Prochloron* and blue-green algae sequences should yield the most direct information concerning chloroplast origins, examination of nuclear sequences can provide complementary information. Mitochondrial sequences would be less useful simply because they appear only in one member of the group under consideration.

5.

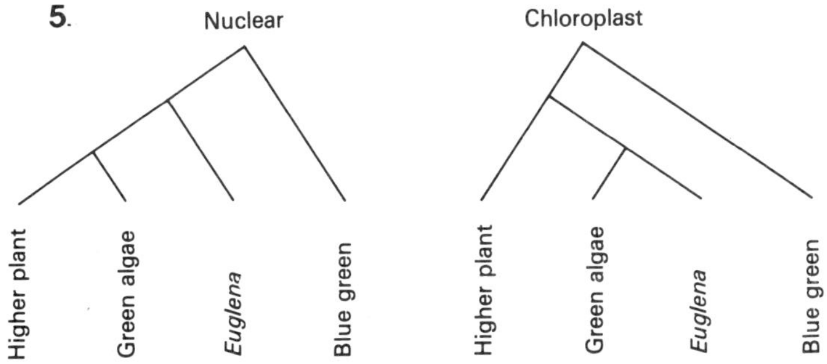

Other sequences and additional organisms would be useful, particularly if you could find a green algal group that had chloroplast rRNAs similar to *Euglena*. It is expected that mitochondrial rRNA would be similar to nuclear rRNA because the proposed green algal symbiont has lost its nucleus and cytoplasm. The 5S RNA sequence of *Euglena* nuclei supports the general hypothesis, because it is so different from green algae and higher plants (Delihas *et al*. 1981).

6. At present we do not see how to make testable predictions from these two hypotheses. With both mitochondria and chloroplasts it has been possible to identify sequences that appear to come from ancestors different from those of the nucleus. It is this observation, of course, which has supported the endosymbiont hypothesis. But if there are no sequences remaining from the cytoplasm host, then this approach is not useful. It is still possible that the nucleus does contain two original genomes, and that they have not yet been identified.

7. There is no definite answer yet, and there are several complicating factors. The cytochrome c sequences (Schwartz and Dayhoff, 1978) and its three-dimensional structure (Almassy and Dickerson, 1978) support an origin of mitochondria close to the purple and non-sulphur photosynthetic bacteria. This group can be examined for its codon usage. If they have the same usage as mitochondria, then this would support alternatives (1) and (2). If these bacteria have the same usage as other organisms, then (3) is supported. In most organisms, any change in codon usage would be expected to be lethal, because changing a few amino acids in every protein would almost certainly give several non-functional proteins. However, if only a few proteins are being coded for (as in mitochondria), and if many amino acid substitutions are neutral, then it is possible that a change in usage of a codon gives a fully viable individual; knowing the codon usage in more mitochondria could help test (3), and it should be possible to relate tRNAs in mitochondria to specific tRNAs in the bacteria. Notice that it has been assumed that cytochrome c, 5S rRNA, and the tRNAs all come from the same organism. A discussion of the problem may be found in Jukes (1981).

8. It does seem unlikely that two such similar metabolic units could have had completely different ancestors. The only anomaly seems to be in trees constructed by matrix methods from 5S rRNA sequences. These methods are known to give misleading results when comparing sequences that have evolved at different rates. Those that have evolved faster look to be less similar and so are not joined into the tree when they should be. To test this possibility, ancestral sequence methods could be tried on the same data.

9. Matrix Method, used in Schwartz and Dayhoff (1978) is from Fitch and Margoliash (1967). Sequences are first converted to a similarity matrix, a tree is constructed from this by comparing three subsets of taxa, the Average Percentage Standard Deviation (APSD) is calculated by comparing length on the tree with distances in the matrix, and finally branches are interchanged on the tree to see if a better tree can be found.

Ancestral Sequence used by Boulter et al. (1972) is derived from a method of Eck and Dayhoff (1966a). Three taxa are joined, then the next taxon is added at all positions and the best position selected. The process is repeated until all taxa are included. The algorithm of Fitch (1971) should be used to ensure that the minimum length is found for every tree. Finally, branches are interchanged in an attempt to improve the tree.

Both methods are heuristic so cannot guarantee to find the optimal tree for the criterion used. The matrix method discards a lot of the original information when forming the similarity matrix, but does seem to perform well with some sets of simulated data where rates of evolution are uniform (Schwartz and Dayhoff, 1978), but is likely to be misleading when they are unequal. Ancestral sequence methods are less sensitive to uneven rates, but one does not use the information when rates are even, It is not possible at present to say when one or the other is preferable. It is possible to use matrix methods with a minimal length criterion by going back to the original data, or using the methods of Farris (1972).

CHAPTER 8

1. Definitions: A *methanogen* is any one of the group of anaerobic prokaryotes that generate methane from carbon dioxide and hydrogen. *Numerical taxonomy* is a set of procedures for classifying organisms on the basis of the weighted or unweighted resemblances between them for a larger number of phenotypic characteristics. A *thermophile* is a bacterial species adapted to life in hot pools. *Hybridization competition* is the hybridization of a small amount of radioactive RNA or DNA with homologous DNA, in a series of reaction mixtures with increasing amounts of homologous or heterologous cold RNA or DNA, respectively. A method for assessing the proportion of common sequences among the homologous or heterologous renaturing species. *Archaebacteria* is the large group of prokaryotic species, adapted to peculiar ecological niches and characterized by unusual metabolisms, cell wall, etc., whose 16S rRNA show them to be only distantly related to the standard bacterial species in laboratory use. The latter (e.g., *Escherichia, Bacillus, Staphylococcus*) are now usually grouped, in contrast to the *Archaebacteria* in the Eubacteria. *Heterologous RNA:DNA hybrids* are renatured nucleic acid molecules consisting of base-paired RNA and DNA strands, in which the RNA is from a different species than that of the DNA source. *Cyanobacteria* is the photosynthesizing, oxygen-releasing prokaryotes of the Eubacteria; formerly designated as the "blue-green algae". An *oligonucleotide catalogue* is a complete set of oligonucleotides produced by RNase T1 digestion (all ending in G) of a defined RNA molecule. The catalogue for a given species provides a measure of the sequence information in the RNA for that species. The *tRNA "common sequence"* is the sequence T-pseudouridine-C-G, found in the tRNAs of all eubacteria and eukaryotes. The *association or similarity coefficient* is the degree of overlap (shared oligonucleotides) in the oligonucleotide catalogue for homologous RNAs from two different organisms. *Peptidoglycan* is a main component of the eubacterial cell wall, consisting of a polysaccharide chain of alternating N-acetyl-D-glucosamine and N-

acetylmuramic acid, linked to a tetrapeptide unit. A *progenote* is the hypothesized common ancestor of the three postulated central kingdoms (eubacteria, archaebacteria, and urkaryotes). *Gram-positive* bacteria are a large group of prokaryotic cells which stain dark blue when the Gram stain is applied; the color is caused by retention of the dye by their thick peptidoglycan walls. *Gram-negative* refers to the complementary staining reaction exhibited by non-Gram-positive prokaryotes, in which the stain is lost due to the high lipid content and thinner peptidoglycan walls. A *urkaryote* is any one of the group of organisms from which the eukaryotic cytoplasmic lineage can be traced, under the terms of the three-kingdom hypothesis.

2. One cannot construct a true phylogenetic tree from these data because the relationships of the listed organisms to one another is not specified, but only their degree of relatedness to *E. coli*. Thus, *A. faecalis* and *S. hersonii* might conceivably show up to 93% sequence relatedness in their rRNA genes (judging from their comparatively similar differences from *E. coli* and taking, for the moment, percent homology as a strict measure of sequence relatedness), or might be very different from one another if both had diverged from their common ancestor with *E. coli* fairly early, accumulating similar degrees of difference with respect to *E. coli* during this divergence. The main relationships that can be deduced concern relative time of divergence from the common ancestor with *E. coli*, assuming relatively constant rates of substitution in each line within the rRNA genes. Thus, the most distant divergence, for the organisms considered, would be that between the common ancestor for *E. coli* and *T. pyriformis*, while the most recent would be that between the lines for *A. aerogenes* and *E. coli*.

3. One wants oligonucleotides long enough that a shared sequence between two species is not likely to have been generated by chance alone. What is the probability that a pentamer shared by two species has been generated by chance and not by common descent? For a sequence five nucleotides long, there are $4^5 = 1024$ possibilities. The chance that any one sequence will be followed by a G is 1/4. The overall probability for finding a particular 6-mer sequence in a T1 digest is 1/4096. Because the number of sequences that are n nucleotides long in a molecule N nucleotides in length ($N > n$) is $N - n + 1$, the number of 6-mers that can be generated from a molecule of 1600 or so nucleotides is about 1600. Thus the joint occurrence of a particular 6-mer (or oligonucleotide of greater size) in two oligonucleotide catalogues from two different species is less likely to be produced by chance than by common descent. The repeated occurrence of that oligonucleotide in a number of different species is a strong indicator of shared genetic heritage.

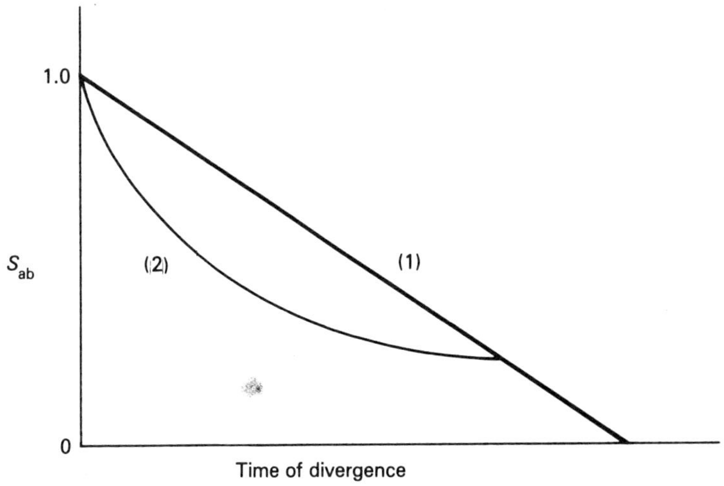

4. If there is a common core of conserved sequences due to the action of selection, then small differences in S_{ab} values at the low end of the scale become less meaningful as indicators of relative times of divergence. By excluding the conserved sequences (if they can be pinpointed unambiguously), one might be able to restore some linearity to this molecular clock.

5. The striking conclusion from the data is that *Fischerella* and *Nostoc* are more closely related to each other than either is to any of the unicellular cyanobacterial strains. Furthermore, the unicellular strains appear to be a much more diverse group than the filamentous strains, and if the latter two groups represent two separate

orders, the unicellular strains should accordingly be split up into several orders. The results suggest that caution is needed in interpreting the fossil morphologies in terms of phylogenetic relationships. Only where fossil groups can be unambiguously related to contemporary forms--an almost impossible criterion for prokaryotes--and the contemporary forms related to one another on the basis of their molecular phylogenies, will conclusions be warranted.

6. In principle, horizontal transfer of prokaryotic 16S rRNA genes between distantly related groups could greatly interfere with the construction of accurate phylogenetic trees; for instance, if a eubacterial species picked up and incorporated an archaebacterial 16s rRNA gene in place of its own, it could be falsely classified as being among the archaebacteria. Actually, such transfer is extremely unlikely, both because of the different habitats and ecological niches occupied by very different prokaryotic forms and by the cell surface differences that often distinguish such forms. That such transfer, even within the same genus, is probably of little importance in affecting phylogenies, either for the ribosomal RNA genes or other sequences, is indicated by other data (see Problem 8 below as an example).

7. The investigation of genome structure in *Archaebacteria* might prove very informative. If their genomes, or the genomes of some of these prokaryotes, should have eukaryotic-like properties, such as repetitive, noncoding DNA or introns, it would suggest, given the presumptive antiquity of these lines, that standard eubacterial species are indeed, streamlined, highly evolved organisms. If introns should be found in archaebacterial genes, it would not rule out the three-kingdom hypothesis, but it would be one more similarity between some archaebacterial members and eukaryotic cells. Should this and/or other such similarities be found, in addition to the ones noted in the introduction, the prior existence of a separate urkaryotic group would become a less tenable proposition. At that point, should it occur, attention would have to shift to questions about the origin of the nucleus and other signal eukaryotic features within a presumptive archaebacterial ancestor.

8. Clearly, you will have to employ the techniques of molecular phylogenetics to classify your strain. One simple initial approach would be to do a bulk DNA hybridization, using the DNA of your unknown strain (labeled in small amounts) and excess cold purified DNA of your reference strains, and determine the amount of hybridization under standard stringent conditions. If there was little or no hybridization with any of the strains, you might relax the hybridization criterion and repeat the experiment, looking for increased renaturation; those strains giving some hybridization are clearly more closely related to your unknown than the ones that do not. An alternative approach is to isolate the ribosomal RNA of your strain and hybridize to the DNA of the reference strains (as in Problem 2). Using this more conserved sequence, it might be possible to detect homologies more readily than with bulk DNA. Other specific macromolecular comparisons might also be made; for instance, if the strain produced a recognizable cytochrome c, you might do cytochrome c sequence comparisons, either by direct amino acid sequence analysis or by immunological comparisons. Finally and perhaps most definitively, you should do a 16S-rRNA oligonucleotide-catalogue comparison between your strain and the reference strains, calculate the values of S_{ab} and draw the resulting dendrogram. If reference dendrograms are available, it should be possible to classify your strain with high certainty in relation to the reference species.

9. (a) The DNA homology tree has the most information, the cytochrome c-551 the least. DNA homology tests, in principle, are subject to every base pair change during divergence of organisms (since each mismatched pair affects the re-annealing reaction); it is thus a very sensitive test of genetic change. The cytochrome c comparisons are intrinsically less sensitive because they cannot record third position or other synoymous codon changes. In addition, it is a fairly conserved protein,

because of selection pressures, and, for that reason also, would be less sensitive indicator of genetic change. (b) The trees are highly congruent. This suggests that the rates of change, though not identical for the sequences sampled, are at least highly correlated with one another. If there was much plasmid-mediated gene transfer occurring naturally between members of the group or from outside the group, one would not expect the trees to look nearly as similar as they do; the similarity suggests that this factor is not a major contributor to genome evolution in this bacterial group.

10. Comparisons of coding in members of the *Archaebacteria* with respect to the standard universal code should certainly be undertaken. If no differences are found between archaebacterial codes and the standard code, one may only conclude that all extant organisms do indeed derive from an ancestor that already has its code fixed in the standard pattern. However, if differences were to be found, it would give added weight to the three-kingdom hypothesis, strongly suggesting that the *Archaebacteria* are indeed a very separate group. It would also imply that there were either two different lines of progenotes or that the common ancestral progenote did not have its code fixed into one (standard) pattern.

CHAPTER 9

1. Definitions: A *comma-free code* is a codon dictionary from which polyuncleotide chains may be constructed whose correct reading frame is determined by the fact that all possible overlap triplets are nonsense (not to be found in the dictionary). The *frozen-accident* hypothesis is the general proposal that of a range of events possible at any given time in the course of evolution only one was chosen by chance; this chance event is frozen in the sense that it then specifies a new spectrum of possible further events in the progressive evolutionary process. *Wobble* is the steric freedom of the anticodon 5' position (corresponding to the codon 3' position) which permits all of the following incorrect base pairs between 5' anticodon and 3' codon positions, respectively: I = (U,C,A), A = (U,C) and U = (A,G). *Master sequence/quasi-species* refers to the condition when greater stability and/or superior substrate quality establish a given sequence as a master sequence; however, replication errors continually generate variants, which form a quasi-species cluster of sequences. *Q-beta replicase* is a tetrameric protein comprised of a single virus-encoded peptide and three host-encoded translation-related peptides (ribosomal protein S1 and elongation factors Tu and Ts). A *hypercycle* is that mutually interdependent relationship between nucleotide sequences and their primitively encoded products which leads to their co-selection. A *fluctuating clay environment* is an alternating wet and (hot) dry clay surface that has been proposed to be a particularly favorable location for the early abiotic polymerization of biological macromolecules. Such an environment provides a surface upon which to concentrate selected monomers from a dilute solution in a possibly stereospecific fashion, followed by dehydration, which kinetically favors polymerization by removal of the elements of water.

2. The superiority of an RNY repeating motif resides in the potential for internal complementarity which has the following virtues: (1) the resulting single-chain double-stranded molecule has greater stability, (2) complementary strands must have at least elements of identity that, in principle, provide identity of targets for the primitive replicase, and (c) the simultaneous utilization of a given sequence as both primitive mRNA and primitive tRNA is more easily imagined.

3. Under conditions of exponential growth (template-limiting), one might expect that sequences offering a superior substrate for elongation might predominate, while under conditions of linear growth (enzyme-limiting), superiority of enzyme-binding (initiation) would be favored. The reported results indicate that the binding/initiation property is a more stringent quality than is rate of chain elongation.

4. (a) (1) A stable hypercycle; (2) an unstable hypercycle in which R_2 would lose

out in the competition for monomers; (3) not a hypercycle at all, but rather two competing duplicating molecules, the outcome of which would be determined by the properties of R_1 and R_2 relative stability, rate of initiation and replication). (b) (1) Degen- erate to the original two-membered hyper- cycle consisting of R_1 and R_2; (2) stable three-membered hypercycle; (3) degenerate to a two-membered hypercycle consisting of R_2 and R_3.

5. In both free solution and in compartments, if R_3 is a superior substrate (termed a phenotypic mutant) with E_3 being unchanged in enzymatic properties, the R_1-R_2 hypercycle would be displaced by the R_2-R_3 hypercycle. However, should E_3 be ineffective as a catalyst, the hypercyclic coupling would break down in free solution whereas if the system was compartmentalized, those protocells populated by the original R_1-R_2 pair would survive. If the superiority of R_3 resides in an E_3 that is more active on R_2 than is E_1 (termed a genotypic mutant), only by compartmentalization could the R_3-R_2 hyper- cycle be selected over R_1-R_2. In free solution, the R_1-R_2 and R_3-R_2 hypercycles could coexist.

6. If we take the average value of S for the population of molecules comprising the quasi-species to be 1, then the Q_iS_i product must be greater than 1 to avoid extinction as a master species. If the value of Q_iS_i should fall below 1, than another member of the quasi-species would assume dominance and become the new master sequence for a new quasi-species distribution. Stability of the molecule, ability to bind the replicase (albeit primitive) and quality as a substrate for replication (chain elongation) may be expected to contribute to S.

	S = 5	S = 50
q = 0.95	$L < 1.6/5 \times 10^{-2} = 3.2 \times 10$	$L < 3.91/5 \times 10^{-2} = 7.8 \times 10$
q = 0.9995	$L < 1.6/5 \times 10^{-4} = 3.2 \times 10^{3}$	$L < 3.91/5 \times 10^{-4} = 7.8 \times 10^{3}$
q = 0.999999	$L < 1.6/1 \times 10^{-6} = 1.6 \times 10^{6}$	$L < 3.91/1 \times 10^{-6} = 3.9 \times 10^{6}$

We find that to a reasonably close approximation the genome size of Q-beta (3300 bases) and *E. coli* (4 x 10^6 base pairs) are at the size limit imposed by the fidelity of their respective polymerases.

7.

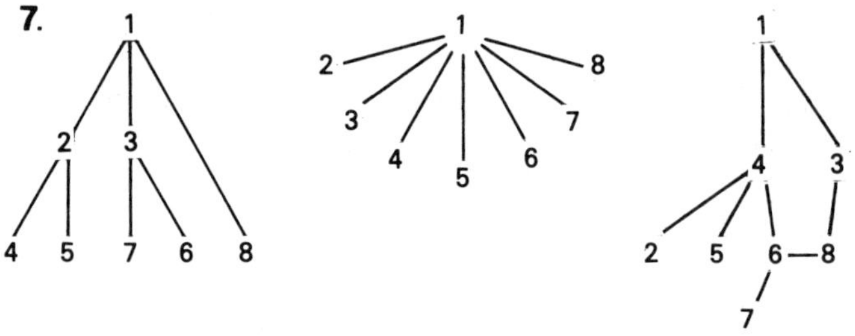

8. Reading from the first base, one finds 13 in register RNY triplets, from the second base 5 in register RNY triplets and from the third, 3 in register triplets. This suggests that the amino acid sequence encoded by this series of bases ought to be: N-cys-val-glu-ala-cys-val-tyr-gly-thr-leu-asp-phe-val-gly-tyr-pro-arg-phe-pro-ala-pro-val-phe-ile-ala-ala-val-ile-ala-C. Of the 12 gly, ala, and val codons, 10 are of the form RNY, while on a purely random codon assignment, only 50% ought to be of that form.

9. Try it, or even better, program your computer to do it. You will discover that the conventional tRNA cloverleaf is a winner!

10. In a very general sense, the mitochondrial coding scheme is a substantially reduced system which a priori might be expected to give some hints concerning how a system might function with much fewer anticodons at its disposal. In a more detailed sense, some of the specific economies effected in the mitochondrial system, and some of the alterations in the code correspond rather well with Jukes' proposals for specific steps in the evolution of the contemporary universal code. In particular, U in the anticodon wobble position can pair with all four bases in the codon 3' position; alteration of the meaning of an anticodon, presumably as a result of an alteration in the aminoacylation recognition site; deletion of an anticodon to generate two new chain termination signals; Anticodon alterations to generate suppressor tRNAs.

11. The archetypal coding scheme comprised of 16 anticodons, 15 of which begin with U (for universal wobble pairing), has the ad hoc feature of the 16th beginning with a G, to give GUA. The latter can pair only with the codons UAU and UAC, the proposed

archetypal and known universal contemporary codons for tyrosine. The missing UAG and UAA codons are the archetypal and contemporary terminator codons. Thus, according to the hypothesis under consideration the contemporary UAN codon cluster represents a direct vestige of the archetypal scheme. Similarly, the proline cluster of CCN codons observed in all contemporary systems is postulated to be a direct descendant of the archetypal UGG/CCN anticodon/codon sequences assigned (by chance?) to proline. The archetypal anticodon (tRNA) underwent at least one duplication and mutation to yield the contemporary eukaryotic AGG anticodon IGG following the action of anticodon deaminase. The AGG anticodon has not yet been identified in prokaryotes, which may mean that the duplication leading to AGG occurred relatively recently, early in the line leading to eukaryotes. An early duplication of the archetypal isoleucine UAU anticodon (tRNA) followed by mutation gave rise to the GAU anticodon, which recognizes the AUU and AUC codons. A further duplication and mutation (perhaps only in the eukaryotic line) gave rise to the AAU anticodon, which becomes IAU following deamination. The latter can pair with the codons AUU, AUC, and AUA. The original UAU anticodon (tRNA) underwent modification at the aminoacylation recognition site to become the methionine tRNA. The presence in at least some prokaryotic lines of a separate CAU anticodon (with a modified C) indicates that at some stage there was an additional duplication and modification to yield this particular species of tRNA. Assessment of the distribution of the modified and unmodified CAU anticodon should yield clues concerning the details of the evolution of the AUA and AUG codons.